HAMILTONIAN MECHANICS OF GAUGE SYSTEMS

The principles of gauge symmetry and quantization are fundamental to modern understanding of the laws of electromagnetism, weak and strong subatomic forces, and the theory of general relativity. Ideal for graduate students and researchers in theoretical and mathematical physics, this unique book provides a systematic introduction to Hamiltonian mechanics of systems with gauge symmetry.

The book reveals how gauge symmetry may lead to a non-trivial geometry of the physical phase space and studies its effect on quantum dynamics by path integral methods. It also covers aspects of Hamiltonian path integral formalism in detail, along with a number of related topics such as the theory of canonical transformations on phase space supermanifolds, non-commutativity of canonical quantization, and elimination of non-physical variables. The discussion is accompanied by numerous detailed examples of dynamical models with gauge symmetries, clearly illustrating the key concepts.

LEV V. PROKHOROV is a Leading Research Fellow at the V. A. Fock Institute of Physics at St-Petersburg State University and the acting head of the Laboratory of Quantum Networks. He is known for his work in the fields of effective Lagrangians, deep inelastic scattering at small transfer momenta, grand unification theory, path integrals, infrared, and collinear divergences. His current research focuses on the emergence of quantum mechanics and the nature of physical space. Among other honors, he received the V. A. Fock Prize from the Russian Academy of Sciences.

SERGEI V. SHABANOV is an Associate Professor of Mathematics and Affiliate Professor of Physics at the University of Florida, Gainesville. His research focuses on gauge theories and the path integral formalism, including the topological defects in lattice gauge theories and applications of knot solitons to effective infrared Yang–Mills theories. His recent research interests include nanophotonics and plasma physics. His achievements include receiving the V. A. Fock Prize from the Russian Academy of Sciences, and an Alexander von Humboldt fellowship.

CAMBRIDGE MONOGRAPHS ON MATHEMATICAL PHYSICS

General Editors: P. V. Landshoff, D. R. Nelson, S. Weinberg

S. J. Aarseth *Gravitational N-Body Simulations: Tools and Algorithms*
J. Ambjørn, B. Durhuus and T. Jonsson *Quantum Geometry: A Statistical Field Theory Approach*
A. M. Anile *Relativistic Fluids and Magneto-fluids: With Applications in Astrophysics and Plasma Physics*
J. A. de Azcárraga and J. M. Izquierdo *Lie Groups, Lie Algebras, Cohomology and Some Applications in Physics*†
O. Babelon, D. Bernard and M. Talon *Introduction to Classical Integrable Systems*†
F. Bastianelli and P. van Nieuwenhuizen *Path Integrals and Anomalies in Curved Space*
V. Belinski and E. Verdaguer *Gravitational Solitons*
J. Bernstein *Kinetic Theory in the Expanding Universe*
G. F. Bertsch and R. A. Broglia *Oscillations in Finite Quantum Systems*
N. D. Birrell and P. C. W. Davies *Quantum Fields in Curved Space*†
K. Bolejko, A. Krasiński, C. Hellaby and M-N. Célérier *Structures in the Universe by Exact Methods: Formation, Evolution, Interactions*
D. M. Brink *Semi-Classical Methods for Nucleus–Nucleus Scattering*†
M. Burgess *Classical Covariant Fields*
E. A. Calzetta and B.-L. B. Hu *Nonequilibrium Quantum Field Theory*
S. Carlip *Quantum Gravity in 2+1 Dimensions*†
P. Cartier and C. DeWitt-Morette *Functional Integration: Action and Symmetries*
J. C. Collins *Renormalization: An Introduction to Renormalization, the Renormalization Group and the Operator–Product Expansion*†
P. D. B. Collins *An Introduction to Regge Theory and High Energy Physics*†
M. Creutz *Quarks, Gluons and Lattices*†
P. D. D'Eath *Supersymmetric Quantum Cosmology*
F. de Felice and D. Bini *Classical Measurements in Curved Space-Times*
F. de Felice and C. J. S Clarke *Relativity on Curved Manifolds*
B. DeWitt *Supermanifolds*, 2nd edition†
P. G. O. Freund *Introduction to Supersymmetry*†
F. G. Friedlander *The Wave Equation on a Curved Space-Time*†
Y. Frishman and J. Sonnenschein *Non-Perturbative Field Theory: From Two-Dimensional Conformal Field Theory to QCD in Four Dimensions*
J. A. Fuchs *Affine Lie Algebras and Quantum Groups: An Introduction, with Applications in Conformal Field Theory*†
J. Fuchs and C. Schweigert *Symmetries, Lie Algebras and Representations: A Graduate Course for Physicists*†
Y. Fujii and K. Maeda *The Scalar–Tensor Theory of Gravitation*
J. A. H. Futterman, F. A. Handler, R. A. Matzner *Scattering from Black Holes*†
A. S. Galperin, E. A. Ivanov, V. I. Orievetsky and E. S. Sokatchev *Harmonic Superspace*
R. Gambini and J. Pullin *Loops, Knots, Gauge Theories and Quantum Gravity*†
T. Gannon *Moonshine beyond the Monster: The Bridge Connecting Algebra, Modular Forms and Physics*
M. Göckeler and T. Schücker *Differential Geometry, Gauge Theories and Gravity*†
C. Gómez, M. Ruiz-Altaba and G. Sierra *Quantum Groups in Two-Dimensional Physics*
M. B. Green, J. H. Schwarz and E. Witten *Superstring Theory Volume 1: Introduction*†
M. B. Green, J. H. Schwarz and E. Witten *Superstring Theory Volume 2: Loop Amplitudes, Anomalies and Phenomenology*†
V. N. Gribov *The Theory of Complex Angular Momenta: Gribov Lectures on Theoretical Physics*
J. B. Griffiths and J. Podolský *Exact Space-Times in Einstein's General Relativity*
S. W. Hawking and G. F. R. Ellis *The Large Scale Structure of Space-Time*†
F. Iachello and A. Arima *The Interacting Boson Model*
F. Iachello and P. van Isacker *The Interacting Boson-Fermion Model*
C. Itzykson and J. M. Drouffe *Statistical Field Theory Volume 1: From Brownian Motion to Renormalization and Lattice Gauge Theory*†
C. Itzykson and J. M. Drouffe *Statistical Field Theory Volume 2: Strong Coupling, Monte Carlo Methods, Conformal Field Theory and Random Systems*†
C. V. Johnson *D-Branes*†
P. S. Joshi *Gravitational Collapse and Spacetime Singularities*
J. I. Kapusta and C. Gale *Finite-Temperature Field Theory: Principles and Applications*, 2nd edition
V. E. Korepin, N. M. Bogoliubov and A. G. Izergin *Quantum Inverse Scattering Method and Correlation Functions*†
M. Le Bellac *Thermal Field Theory*†
Y. Makeenko *Methods of Contemporary Gauge Theory*
N. Manton and P. Sutcliffe *Topological Solitons*

N. H. March *Liquid Metals: Concepts and Theory*
I. Montvay and G. Münster *Quantum Fields on a Lattice*†
L. O'Raifeartaigh *Group Structure of Gauge Theories*†
T. Ortín *Gravity and Strings*†
A. M. Ozorio de Almeida *Hamiltonian Systems: Chaos and Quantization*†
L. Parker and D. J. Toms *Quantum Field Theory in Curved Spacetime: Quantized Fields and Gravity*
R. Penrose and W. Rindler *Spinors and Space-Time Volume 1: Two-Spinor Calculus and Relativistic Fields*†
R. Penrose and W. Rindler *Spinors and Space-Time Volume 2: Spinor and Twistor Methods in Space-Time Geometry*†
S. Pokorski *Gauge Field Theories*, 2nd edition†
J. Polchinski *String Theory Volume 1: An Introduction to the Bosonic String*
J. Polchinski *String Theory Volume 2: Superstring Theory and Beyond*
J. C. Polkinghorne *Models of High Energy Processes*†
V. N. Popov *Functional Integrals and Collective Excitations*†
L. V. Prokhorov and S. V. Shabanov *Hamiltonian Mechanics of Gauge Systems*
R. J. Rivers *Path Integral Methods in Quantum Field Theory*†
R. G. Roberts *The Structure of the Proton: Deep Inelastic Scattering*†
C. Rovelli *Quantum Gravity*†
W. C. Saslaw *Gravitational Physics of Stellar and Galactic Systems*†
R. N. Sen *Causality, Measurement Theory and the Differentiable Structure of Space-Time*
M. Shifman and A. Yung *Supersymmetric Solitons*
H. Stephani, D. Kramer, M. MacCallum, C. Hoenselaers and E. Herlt *Exact Solutions of Einstein's Field Equations*, 2nd edition
J. Stewart *Advanced General Relativity*†
T. Thiemann *Modern Canonical Quantum General Relativity*
D. J. Toms *The Schwinger Action Principle and Effective Action*
A. Vilenkin and E. P. S. Shellard *Cosmic Strings and Other Topological Defects*†
R. S. Ward and R. O. Wells, Jr *Twistor Geometry and Field Theory*†
J. R. Wilson and G. J. Mathews *Relativistic Numerical Hydrodynamics*

†Issued as a paperback

Hamiltonian Mechanics of Gauge Systems

LEV V. PROKHOROV
Saint-Petersburg State University

SERGEI V. SHABANOV
University of Florida

CAMBRIDGE
UNIVERSITY PRESS

CAMBRIDGE
UNIVERSITY PRESS

University Printing House, Cambridge CB2 8BS, United Kingdom

Cambridge University Press is part of the University of Cambridge.

It furthers the University's mission by disseminating knowledge in the pursuit of education, learning and research at the highest international levels of excellence.

www.cambridge.org
Information on this title: www.cambridge.org/9780521895125

First published 2011

A catalogue record for this publication is available from the British Library

ISBN 978-0-521-89512-5 Hardback

Contents

Preface

In the past four decades, gauge theories have become dominant in high energy physics. All known fundamental interactions – strong, weak, electromagnetic, and gravitational, are described by gauge-invariant Lagrangians (actions). In the framework of classical physics, gauge invariance does not cause any significant theoretical problems. In the case of electrodynamics, which has always served as the benchmark for all other field theories, gauge arbitrariness can be eliminated by adding a suitable supplementary (gauge) condition on the vector potential, e.g. the Lorenz condition, to the equations of motion. In gravity, the De Donder–Fock gauge is often used. But initial studies on quantization of electromagnetic fields showed that gauge theories required a special approach. Canonical quantization demands the existence of the Hamiltonian formalism, whose construction turned out to be not an easy task. It appeared that in electrodynamics the equations that relate generalized velocities and canonical momenta cannot be solved for the former, i.e. the velocities cannot be expressed as functions of the generalized coordinates and momenta (a consequence of the Lagrangian being singular (or degenerate)). As a result, conditions on canonical variables (constraints) occur. It was required, first, to formulate the theory of dynamical systems with constraints and, second, to find a consistent procedure for their quantization.

For electrodynamics these problems had already been solved by W. Heisenberg and W. Pauli in 1930. In a further development of modern quantum electrodynamics (QED), it appeared to be possible to avoid the essential problem of quantization in the presence of constraints. It was established that in QED one can develop perturbation theory insensitive to the peculiarities of Hamiltonian dynamics associated with the degeneracy of the Lagrangian. The only difference with a non-gauge theory was the existence of the Ward identities, while the only additional concern was the necessity to watch for the gauge invariance of the results. However, attempts to describe more complicated constrained systems (such as, e.g. gravity) demanded the development of a general theory. This theory was developed by P. A. M. Dirac and G. P. Bergmann; Dirac gave the quantization procedure for such theories. For quite some time these results did not attract much attention from the experts. Appreciation came after the discovery of the unified theory of weak and electromagnetic interactions. The Yang–Mills fields (i.e. gauge theories with non-Abelian groups) were recognized as fundamental in nature. In the theory a small parameter still existed and, therefore, it seemed that nothing could prevent the application of perturbation theory.

However, it had earlier been established by R. Feynman that a straightforward extension of the standard QED formalism to the Yang–Mills theory would lead to catastrophe – the breaking of unitarity. This was a clear indication of the necessity for a general theory.

The situation became even more desperate after the discovery that strong interactions are also described by a non-Abelian gauge theory. Here the problem of a correct description was truly of paramount importance because the interaction could only be assumed to be small at short distances. The limitations of the trouble-free perturbation theory method used so far were emphasized by the phenomenon of quark confinement which, in particular, is characterized by a linearly rising potential. More to point, the very applicability of perturbation theory turned out to be questionable (because of the increase of the coupling constant with distance), the new theory of strong interactions, quantum chromodynamics (QCD), did not seem to provide even a qualitative explanation of confinement. The problem was made even worse by the fact that invoking new methods such as quasiclassical ones, not related to perturbation theory, was unsuccessful. Few results from the first ambush on the problem justified a subsequent siege.

It seems that some essential features of QCD escaped the attention of the physicists. The necessity to scrutinize the very foundations of the theory thus became evident and the starting point for such an analysis would be the fundamental works of Dirac and Bergmann.

This monograph, to a large extent, is devoted to studies of the foundations of the dynamics of gauge systems (constrained systems). As some of the most important, general features of gauge dynamics were not investigated thoroughly enough, this presentation is focused on the simplest models.[1] They are quite elementary, but at the same time, they possess all the characteristic features inherent in theories of this kind. Clearly, a good knowledge of simple gauge systems is very helpful for the understanding of gauge theories with infinitely many degrees of freedom, i.e. field theories. These models are Yang–Mills theories with matter fields in spacetime of the reduced dimension (0+1). Inspite of their maximal simplicity, but also maybe because of it, they have turned out to be quite useful. Studies of these models have led to the uncovering of a few peculiarities of gauge dynamics which were previously unnoticed. Even the simplest model (the scalar electrodynamics in spacetime (0+1)) allowed such a remarkable fact as physical phase space reduction to be revealed. It appears that the phase space of the only physical degree of freedom in this model is not a plane but rather a *cone* unfoldable into a half-plane. This drastically changes the dynamics of the system. For example, the frequency of the harmonic oscillator is doubled (consequently, the distance between its energy levels is also doubled in quantum theory). In models with an arbitrary gauge group, the structure of the physical

[1] It is said that, when introduced to a new phenomenon, W. K. Rentgen used to ask the question, "Where is the hydrogen atom here?"

phase space is determined by the Weyl group W, the discrete subgroup that characterizes the root pattern of the gauge group (W is the group of all compositions of the reflections in the hyperplanes orthogonal to simple roots). A remarkable synthesis of physics and rather abstract mathematics! Another example of the usefulness of studying simple gauge models is the gauge fixing problem. The insufficiency of the Lorenz gauge condition in eliminating gauge arbitrariness, discovered in non-Abelian gauge field theories, attracted much attention. Studies of this problem in simple models provide complete clarification of the issue, as well as a correct recipe for a quantum description in an arbitrary gauge.

The aforementioned examples already seem sufficient to indicate that the properties of gauge and non-gauge systems may be quite different, experience gained in working with ordinary systems might be insufficient, and without a good understanding of the mechanics of simple gauge systems, it would be difficult to figure out the peculiarities of gauge systems with many degrees of freedom. This is especially true for gauge field theories that have infinitely many degrees of freedom.

The years that have passed since the early studies on gauge theories have only shown that the importance of problems in the Hamiltonian formalism for gauge theories and the path integral method do not diminish. The results of both prominent and lesser known researchers scattered over the vast journal literature have not lost their significance for a new generation of theoretical physicists. The Hamiltonian mechanics resides in the very foundations of all physics. It seems more and more evident that the beauty of the world is based on chance. Namely, if one admits the existence of a fundamental source of random forces (a "thermostat"), then there naturally emerge the notions of the phase space, the Hamiltonian, Hamiltonian mechanics, quantum mechanics (the probability amplitudes, Planck constant, Schrödinger equation, and Fock space), the time arrow, and many other common features of our present physical understanding of nature. In this newly emerging picture, there is a place for the Nambu mechanics with several Hamiltonians. At the foundation of all this richness lies the Gibbs distribution which, in the simplest and most important case of the harmonic oscillator, is the consequence of the central limit theorem (the two-dimensional normal (Gauss) distribution). But the probability processes are described by the Wiener path integral. It turns out that the path integrals in quantum mechanics have the very same nature and emerge due to the very same reason. Recently a model of the "physical spacetime" has been proposed (first advocated by V. A. Fock). Its essence is that 3D space is modeled by a 3D network built of bosonic strings and placed into a thermostat. In this model there naturally emerge the Minkowski space, fermions, the idea of supersymmetry, and so on. At the same time, there appears a new aspect of Hamiltonian mechanics. The network is not a manifold, while Hamiltonian mechanics is formulated on an even-dimensional manifold. So there is a problem of developing the Hamiltonian mechanics on spaces which are not manifolds. This development is still very much in progress. Thus, it is evident that the problems of modern fundamental physics require a deep understanding and study of the mechanics of gauge systems.

Chapter 1 provides a survey of some classical aspects of Hamiltonian mechanics that are used throughout the book (Hamilton's least action principle, Hamiltonian equations of motion, canonical transformations, and so on). In addition to the traditional material, particular topics whose development was stimulated by the demands of modern quantum field theory are also included. First of all, one such topic is the mechanics with Grassmann (anticommuting) variables and the mechanics of mixed systems with bosonic and fermionic degrees of freedom. Noether's theorem is extended to Lagrangian systems on supermanifolds. It is illustrated with several examples including supersymmetry. A general formalism of the Hamiltonian mechanics on supermanifolds is completed with the Hamilton–Jacobi theory. Some extensions of Hamiltonian mechanics are discussed, e.g. Nambu mechanics. The question of non-canonical transformations, that are beginning to gain significance in modern physics, but whose general theory is still largely undeveloped, is also studied here. A classical example can be furnished by the holomorphic representation of the harmonic oscillator problem which is obtained by the transformation $(q, p) \to (a, a^*)$ which does not preserve the Poisson bracket. Another example is related to the so-called q-deformed systems (dynamical systems associated with quantum groups), the research area that became quite fashionable two decades ago. The last section of Chapter 1 contains a brief review of recent applications of Hamiltonian mechanics to the aforementioned problems in fundamental physics (the Gibbs distribution and the emergence of quantum mechanics). Overall, this chapter can be used as a modern supplement to a classical university course in theoretical mechanics.

Chapter 2 is devoted to the fundamentals of quantum theory, namely, those that concern the Hamiltonian path integral formalism. Although the basics of the formalism have already been established by Feynman, there are many aspects, important in practice, that have been studied much later. Among the most significant are the problem of changing variables and path integrals in curvilinear coordinates whose coordinate surfaces have a non-trivial topology, the behavior of the Hamiltonian path integral under canonical transformations, the operator ordering problem in the Hamiltonian, the path integral formalism for problems with boundaries, and others. The above listed problems are by no means far-fetched, they are encountered in the path integral formalism for the simplest constrained systems. The methods developed allow one to formulate the path integral formalism for problems with non-standard phase spaces. Many of the issues studied in this chapter are not usually included in standard textbooks on the path integral method.

Chapter 3 provides an elementary introduction to the theory of constrained systems. The presentation is illustrated by many examples. This chapter is not intended to give a comprehensive review of the vast subject of constrained dynamics. It merely offers the basic facts with an emphasis on what is relevant for the sequel.

Chapter 4 is devoted to the problem of quantization of constrained systems. In addition to the standard quantization recipes, the operator ordering problem in constraints is considered, and a particular non-standard case, when the "constraints" depend on velocities, but cannot be solved for these, is studied as well (Section 4.3).

In Chapter 5 dynamical systems with gauge symmetries are analyzed in detail. The physical phase space structure, the problem of eliminating non-physical degrees of freedom, and other peculiarities of the gauge dynamics are studied. Models with Abelian and non-Abelian gauge groups describing dynamics of both bosonic and fermionic degrees of freedom (mixed systems with commuting and anticommuting variables) are considered. The question of choosing a gauge fixing condition is thoroughly investigated as well as peculiarities of a gauge-fixed dynamics. In particular, with an example of Yang–Mills theories on cylindrical spacetime, such important issues as orbit space geometry, physical phase space structure, quantum theory on the orbit space, gauge-fixing ambiguity problem (Gribov problem), and their effects on a quantum description are presented. The emphasis is put on proving the equivalence of quantum theories where all non-physical degrees of freedom are eliminated (by, e.g. a gauge fixing) to explicitly gauge invariant quantum theories based on the Dirac quantization method. Some simple effects due to non-standard physical phase space structure are studied in gauge quantum theories. Most of the subject matter of this chapter can only be found in journal publications.

In Chapter 6, the Hamiltonian path integral formalism is developed for the models from Chapter 5. The main feature here is the non-standard structure of the physical phase space and its effects on the path integral formalism in gauge theories. The problem is tackled in two ways. First, the operator formalism for physical degrees of freedom with a non-standard phase space, developed in Chapter 5 for gauge theories, is applied to derive the corresponding path integral formalism by means of the methods of Chapter 2. In this way, the equivalence of the path integral formalism to the Dirac quantization method is established. Then a new method of developing the path integral formalism for gauge theories is presented. It is based on the projection method for constrained systems and no longer refers to the operator formalism. It offers a modification of the Kato–Trotter formula for an infinitesimal transition amplitude in the presence of a gauge symmetry. Its equivalence to the Dirac operator formalism is proved. Several examples are worked out in detail to demonstrate how the orbit space geometry and physical phase space structure are accounted for in the path integral formalism. The path integral formalism developed here solves the gauge fixing ambiguity problem. With an explicit example of Yang–Mills theories on the cylindrical spacetime, the modified path integral is shown to yield the correct spectrum of the theory that has been obtained by means of rigorous methods of axiomatic quantum field theory. A relation between the developed path integral formalism and the Morse theory used to study the orbit space structure in 4D Yang-Mills theory is established. The chapter is concluded with a study of the effects of a non-standard phase space of physical variables in gauge theories on their Green's functions. The material included in this chapter can only be found in the journal literature.

Chapter 7 is special. In the preceding chapters the simplest (soluble) gauge models are mostly studied, here the discussion concerns gauge field theories and, moreover, one of the most difficult problems in gauge field theories – confinement. This chapter is included in this monograph because its content is, in essence, a consistent implementation of the Dirac method developed for general constrained systems. An understanding of the key role of constraints and the extended group of gauge transformations allows one to approach the phenomenon from a somewhat more general point of view. The universal nature of the phenomenon becomes evident in this approach. It also makes it possible to separate features inherent only in QCD from those inherent in any gauge theory. From the viewpoint of fiber bundle theory, the path-ordered exponential (the operator of parallel transport) is the only basic building block in all gauge invariants. Coherent field excitations corresponding to the path-ordered exponential (under the assumption of their dynamical stability) are characterized by the energy rising linearly with the path length. In other words, objects with energies rising linearly with the distance between them are internally inherent to such theories and emerge at the very beginning of the quantum theory development. On the other hand, the relative simplicity in establishing the basic properties of confinement only emphasizes the real difficulties of the concrete problem of describing hadrons in QCD, even in the simplest limiting case of massive quarks and a large coupling constant. The problem of the dynamic stability of coherent excitations of quantum gauge fields generated by gauge invariant operators is illustrated with the soluble example of the decay of string-like excitations of quantum electromagnetic fields, followed by formation of the Coulomb field of static sources. This chapter also contains an analysis of gauge invariants in gauge theories with the Higgs phenomenon.

All supplementary material is included in Chapter 8. This chapter contains some basic facts on group theory, Lie algebras, Cartan–Weyl basis (which is used extensively in Chapters 5 and 6), a brief survey of Grassmann algebras and calculus on them, solving of equations of motion in dynamical systems on supermanifolds, standard gauge fixing procedures in the Hamiltonian path integral formalism, and some other technical facts that will be helpful when reading the book.

This book should be accessible to upper level undergraduate students and can certainly be used by graduate students specializing in theoretical and mathematical physics, although its content is based on recent studies which have not been reviewed in the monograph literature. Each of the chapters can be read independently. The introduction to each chapter is quite elementary and does not require special prerequisites beyond standard university courses on analytical mechanics and quantum theory. Exceptions to this may be Chapters 5, 6, and 7. For these the reader may need some basic knowledge of group theory and fiber bundle theory (although, the latter is not really necessary).

The authors are deeply indebted to Nathalie Shabanov for her substantial help with preparing a LaTeX electronic format of the manuscript and to S. L. Berdnikov for his assistance in production of some of the figures.

1
Hamiltonian formalism

1.1 Hamilton's principle of stationary action

A dynamical system with a finite number n degrees of freedom can be described by real functions of time $q^i(t)$ $(i = 1, 2, ..., n)$ which, together with the derivatives $\dot{q}^i(t)$, uniquely determine its state at any moment of time t. The collection of all values of q^i is called the configuration space M of the system. In the simplest case, M is a Euclidean space R^n. It can generally be an n-dimensional (differentiable) manifold. In a small neighborhood of each point, a manifold looks topologically like a Euclidean space R^n. This allows one to introduce local coordinates on the manifold M and develop calculus on M. Every point of M has a neighborhood U that is homeomorphic to R^n (there is a continuous and invertible map between points of U and some subset of R^n). Let q^i, $i = 1, 2, ..., n$, be rectangular coordinates in R^n. A neighborhood U with local coordinates in it (i.e. with the map $\phi : U \to R^n$) is called a chart. A manifold can be covered by a set of charts with transition maps between local coordinate systems in the overlap of any two charts. Thus, in a local coordinate system on M, the time evolution of a dynamical system is again described by n real functions $q^i(t)$, $i = 1, 2, ..., n$.

Consider trajectories of the system $q^i = q^i(t)$ on M through a point $q_0^i = q^i(0)$. The derivative $\dot{q}^i(0)$ is called a tangent vector to the curve at $q = q_0$. The tangent space T_q at $q = q_0$ is defined as the set of all tangent vectors $\dot{q}^i(0)$ at the point $q = q_0$. Thus, the velocity $\dot{q}$ is an element of T_q. The tangent bundle of a manifold M is a disjoint union $TM = \cup_{q\in M} T_q$. A state of a dynamical system is an element of TM as, at any moment of time t, it is defined by values of $q^i(t)$ and $\dot{q}^i(t)$. The quantities q^i and $\dot{q}^i$ are called generalized coordinates and velocities, respectively. It is worth noting that a configuration space can be infinite-dimensional if the system has infinitely many degrees of freedom. A field theory is an example of such a system.

Differential equations whose solution determines the trajectory of motion $q = q(t)$ in the system configuration space are called the equations of motion. For a sufficiently large class of dynamical systems, the equations of motion can be obtained from *Hamilton's principle of staionary action* (or, simply, Hamilton's principle) which states that the trajectory $q(t)$ with fixed end points $q(t_1)$ and $q(t_2)$ is an extremum of the action functional (see, e.g. [1–3])

$$S = S[q; t_1, t_2] = \int_{t_1}^{t_2} L(\dot{q}, q, t)\, dt. \tag{1.1}$$

The real function L of time $t \in R$, generalized coordinates, and velocities is called a Lagrangian of the dynamical system (i.e. $L: R \times TM \to R$).

Variations of a trajectory $q^i = q^i(t)$, $t_1 \le t \le t_2$, are a set of trajectories $q^i = q^i_\epsilon(t)$ parameterized by real ϵ so that

$$q^i_\epsilon(t) = q^i(t) + \epsilon\Phi^i(t) + O(\epsilon^2) \equiv q^i(t) + \delta q^i(t) + O(\epsilon^2), \tag{1.2}$$

and $\Phi^i(t_{1,2}) = 0$, where $\Phi^i(t)$ is called the variation field. The variational derivative $\delta S/\delta q$ is defined by

$$\frac{d}{d\epsilon} S[q_\epsilon; t_1, t_2]\bigg|_{\epsilon=0} = \int_{t_1}^{t_2} \frac{\delta S}{\delta q^i}\, \Phi^i(t)\, dt. \tag{1.3}$$

A curve $q^i = q^i(t)$ is an extremum of the action (1.1) if the variational derivative of the latter vanishes for this curve. The extremum condition

$$\frac{\delta S}{\delta q^i(t)} = 0\,, \qquad \delta q(t_1) = \delta q(t_2) = 0, \tag{1.4}$$

yields the Euler–Lagrange equation of motion

$$\frac{d}{dt}\frac{\partial}{\partial \dot{q}^i}\, L(\dot{q}, q, t) = \frac{\partial}{\partial q^i}\, L(\dot{q}, q, t). \tag{1.5}$$

Thus, all such dynamical systems are determined by their Lagrangians. A Lagrangian should satisfy some physical conditions, e.g. to be invariant under symmetry transformations of the dynamical system such as, for instance, the Lorentz symmetry transformations, etc. It is natural to demand that the Euler–Lagrange equations be self-consistent. For example, put $L = q$ (here $n = 1$). Equation (1.5) leads to an obvious contradiction, $0 = 1$. In other words, the corresponding action has no extremum at all.

It appears that most dynamical systems occurring in nature are described by second-order differential equations, that is, they do not involve higher-order derivatives with respect to time. However, there is no general law that prohibits theories with higher derivatives. As a rule, the existence of higher derivatives in the theory is associated with some internal structure of the object described by such a theory. For example, higher derivatives naturally emerge in an approximate description of elastic vibrations of a rod whose transverse dimensions are much smaller than its length [4]. In this case, the Lagrangian depends on higher time derivatives $q^{i(m)} = d^m q^i/dt^m$ ($m > 1$) so that the trajectory of motion is an extremum of the action

$$S = S[q; t_1, t_2] = \int_{t_1}^{t_2} L\left(q^{(m)}, q^{(m-1)}, ..., \dot{q}, q, t\right)\, dt.$$

The equations of motion resulting from the principle of stationary action (called the Hamilton–Ostrogradsky principle) are known as Ostrogradsky's equations [5]

$$\frac{\delta S}{\delta q^i(t)} = \sum_{k=0}^{m} (-1)^k \frac{d^k}{dt^k}\frac{\partial L}{\partial q^{i(k)}} = 0. \tag{1.6}$$

Note that the variational derivative of the action is evaluated under the condition that $\delta q^{(k)}(t_1) = \delta q^{(k)}(t_2) = 0$, for $k = 0, 1, ..., m-1$.

1.1.1 Poincaré equations

Hamilton's principle on manifolds has a peculiarity related to the choice of a basis in the tangent space. At any point $q \in M$ the operators (vector fields) $\hat{e}_i = \partial/\partial q^i$ define the rate of change of any function $F(q)$ along the trajectory $q^i = q^i(t)$: $\dot{F} = \dot{q}^i \hat{e}_i F$. The velocity $\dot{q}^i$ is an element of the tangent space T_q. In general, one can define n smooth linearly independent vector fields on M that serve as a new basis in T_q,

$$\hat{w}_i = w_i^j(q)\hat{e}_j, \quad [\hat{w}_i, \hat{w}_j] = c_{ij}^k(q)\hat{w}_k, \tag{1.7}$$

where $w_i^j(q)$ are functions of q and $[\,,\,]$ denotes the commutator. The rate of change of a function F along the trajectory can be written in the new basis:

$$\dot{F} = \dot{q}^i \hat{e}_i F = \omega^i \hat{w}_i F.$$

The quantities ω^i are position-dependent linear combinations of the velocities $\dot{q}^i$ and called *quasi-velocities* [6]. In differential geometry, the structure functions $c_{ij}^k(q)$ are called the *object of anholonomity* [7]. Note that under a change of local coordinates on M, $q \to Q$, the basis in T_q transforms as $\partial/\partial Q^i = J_i^j(q)\hat{e}_j$, where $J_i^j = \partial q^j/\partial Q^i$ is the Jacobian of the change of coordinates. Clearly, the new basis $\hat{w}_i = \partial/\partial Q^i$ is commutative, i.e. if $w_i^j = J_i^j$, then the structure functions vanish, $c_{ij}^k = 0$. When the structure functions are non-trivial, the basis vector fields are not integrable (there is no change of variables such that $\hat{w}_i = \partial/\partial Q^i$), and, hence, the quasi-velocity ω^i is not the time derivative of a local coordinate on M. Such bases in T_q are called *anholonomic*. Anholonomic bases turn out to be quite useful in constructing equations of motion for dynamical systems on manifolds.

Let $L(\dot{q}, q) = \tilde{L}(\omega, q)$ be the Lagrangian expressed as a function of local coordinates and quasi-velocities. Let $q_\epsilon^i(t)$ be a variation of the trajectory as defined in (1.2). Put

$$\frac{\partial}{\partial \epsilon} F(q_\epsilon)\Big|_{\epsilon=0} = \phi^i \hat{w}_i F, \tag{1.8}$$

where ϕ^i are the components of the variation vector field Φ in the new basis. They are now independent variations of the trajectory. From the commutativity of the derivatives, $[\partial/\partial t, \partial/\partial\epsilon]F(q_\epsilon(t)) = 0$, and the commutation relation (1.7), the variations of the quasi-velocities are obtained:

$$\frac{\partial \omega^i}{\partial \epsilon}\Big|_{\epsilon=0} = \frac{d\phi^i}{dt} + c_{kj}^i \omega^k \phi^j. \tag{1.9}$$

Computing the variation of the action $S = \int \tilde{L} dt$ by means of (1.8), (1.9), and the boundary conditions $\phi^i(t_{1,2}) = 0$ (as a consequence of $\Phi^i(t_{1,2}) = 0$), the equations of motion in the anholonomic basis are obtained:

$$\frac{d}{dt}\frac{\partial \tilde{L}}{\partial \omega^i} = \hat{w}_i \tilde{L} + c_{ik}^j \omega^k \frac{\partial \tilde{L}}{\partial \omega^j}. \tag{1.10}$$

Equations (1.10) are called the Poincaré equations [8]. Poincaré first obtained them in his studies of celestial mechanics. In a holonomic basis in T_q the Poincaré equations become the Euler–Lagrange equation (1.5).

To appreciate the Poincaré equations, consider a dynamics on a group manifold $M = G$. Let $\hat{w}_i$ be generators of left shifts on G, i.e. $g \to g_0 g$, $g_0, g \in G$ (see Section 8.1.1 for a summary of the group theory). Then the structure functions $c_{ij}^k(q)$ are constant and coincide with the structure constants of G. If, in addition, one imposes the condition that the Lagrangian is invariant under the left shifts, i.e. $\hat{w}_i \tilde{L} = 0$, then the Poincaré equations determine the time evolution of a "free" particle on the group manifold (in the form that does not depend on any particular choice of local coordinates on G) [6].

The simplest example of this kind is provided by the motion of a rigid body with one fixed point. The motion has the characteristic properties that the distance between any two points of the body remains constant, and one of its points always coincides with a fixed point in space. Clearly, all possible positions of the system are obtained from a particular one by rotations in space about a point. Thus, M is the group $SO(3)$ manifold. Let $\mathbf{r}_0$ be the position vector of a point of the body relative to the point about which it rotates. In time t the position vector is $\mathbf{r}(t) = U(t)\mathbf{r}_0$, where $U(t) \in SO(3)$ (i.e. $UU^T = U^T U = 1$ and $\det U = 1$, where U^T is the transpose of U). The velocity of this point is then $\dot{\mathbf{r}} = \dot{U}U^T\mathbf{r}$. Since the matrix $\dot{U}U^T$ is skew-symmetric (because $UU^T = 1$), there exists a vector $\boldsymbol{\Omega}$, called the angular velocity, such that $\dot{\mathbf{r}} = \boldsymbol{\Omega} \times \mathbf{r}$. In the frame moving with the body, the angular velocity vector is $\boldsymbol{\omega} = U^T\boldsymbol{\Omega}$. It can be shown that the kinetic energy of the system has the form (see, e.g. [1])

$$L = \frac{1}{2}\left(I_1\omega_1^2 + I_2\omega_2^2 + I_3\omega_3^2\right),$$

where the components ω_i are defined in the frame whose axes coincide with the principal axes of the tensor of momenta of inertia, and I_i $(i = 1, 2, 3)$ are principal values of the tensor. Apparently, ω_i are not the conventional generalized velocities of the dynamical system as there exist no coordinates q_i such that $\omega_i = \dot{q}_i$. So the Euler–Lagrange equation (1.5) would lead to incorrect equations for the rigid body dynamics, and the Poincaré equations must be applied. In the basis associated with the principal axes, $\hat{w}_i$ should generate rotations about the ith axis, and, hence, $[\hat{w}_i, \hat{w}_j] = \varepsilon_{ijk}\hat{w}_k$, where ε_{ijk} is the totally skew-symmetric unit tensor (the structure constants of $SO(3)$). According to (1.10), one obtains

$$\dot{\mathbf{L}} = \mathbf{L} \times \boldsymbol{\omega},$$

where $L_i = I_i\omega_i$ (no summation over i) are the components of the angular momentum vector $\mathbf{L}$. These are the well-known equations of a spinning top.

1.1.2 The existence of a Lagrangian for a dynamical system

Suppose that the time evolution of a dynamical system is described by a system of second-order differential equations. A natural question arises as to whether this system can always be cast in the form of the Euler–Lagrange equations. The answer is negative. There are so-called *non-Lagrangian* systems for which it is impossible to find a Lagrangian [9]. As an example, consider a dynamical system in $M = R^3$ whose time evolution is determined by the equations [10, 11]

$$\ddot{q}_i - \alpha\epsilon_{ijk}q^j\dot{q}^k = 0, \tag{1.11}$$

where α is a real constant, ϵ_{ijk} is the totally antisymmetric unit tensor, $\epsilon_{123} = 1$ (in [10], a more general system was studied where $\alpha = \alpha(q) = \lambda|\mathbf{q}|^{-3}$, $\lambda =$ const). The summation over repeated indices is assumed in (1.11). Clearly, the initial value problem for Eqs. (1.11) does have a solution. However, there is no Lagrangian such that the Euler–Lagrange equations coincide with (1.11).

In general, the existence of a Lagrangian for a dynamical system was studied by Helmholtz [12]. Given a system of second-order equations

$$G_i(\ddot{q},\dot{q},q) = H_{ij}(\dot{q},q)\ddot{q}^j + A_i(\dot{q},q) = 0,$$

where $i = 1, 2, ..., n$, the necessary and sufficient conditions on G_i as functions of q, $\dot{q}$, and $\ddot{q}$ in order for a Lagrangian L to exist are [12]

$$\frac{\partial G_i}{\partial \ddot{q}^j} = \frac{\partial G_j}{\partial \ddot{q}^i}, \tag{1.12}$$

$$\frac{\partial G_i}{\partial \dot{q}^j} + \frac{\partial G_j}{\partial \dot{q}^i} = \frac{d}{dt}\left(\frac{\partial G_i}{\partial \ddot{q}^j} + \frac{\partial G_j}{\partial \ddot{q}^i}\right), \tag{1.13}$$

$$\frac{\partial G_i}{\partial q^j} - \frac{\partial G_j}{\partial q^i} = \frac{1}{2}\frac{d}{dt}\left(\frac{\partial G_i}{\partial \dot{q}^j} - \frac{\partial G_j}{\partial \dot{q}^i}\right). \tag{1.14}$$

If these conditions are fulfilled, a Lagrangian L can be found by solving the following equations

$$\frac{\partial^2 L}{\partial \dot{q}^i \partial \dot{q}^j}\ddot{q}^j + \frac{\partial^2 L}{\partial \dot{q}^i \partial q^j}\dot{q}^j - \frac{\partial L}{\partial q^i} = G_i. \tag{1.15}$$

It is easy to verify that for G_i given in (1.11), conditions (1.12) and (1.13) are trivially satisfied, while the third Helmholtz condition (1.14) is violated.

However, the existence problem for a Lagrangian may be treated more broadly. Namely, one could take an *equivalent* set of equations $M_i^j G_j = 0$, where *integrating factors* M_i^j form a non-singular matrix and are functions of q, $\dot{q}$, and t. Then the Helmholtz conditions are viewed as equations for M_i^j. These equations can be recast in a geometrical form known as the Douglas theorem [13] (see also

Section 1.14.3). In particular, it should be noted that there are dynamical systems for which there exist many integrating factors leading to different Lagrangians.

As a simple example, consider a particle moving along the line under the friction force proportional to the particle velocity. The equation of motion has the form

$$\ddot{q} + \alpha\dot{q} = 0. \tag{1.16}$$

The Helmholtz condition (1.13) is not fulfilled for this equation. Consider an equivalent equation of the form

$$G \equiv M(\dot{q}, q)(\ddot{q} + \alpha\dot{q}) = 0.$$

The integrating factor M must be chosen so that G satisfies (1.13) (the other Helmholtz conditions are trivially satisfied for a one-dimensional motion):

$$\alpha\dot{q}\,\frac{\partial M}{\partial \dot{q}} + \alpha M = \dot{q}\,\frac{\partial M}{\partial q}. \tag{1.17}$$

This equation has many solutions and can easily be solved by separating variables $M(\dot{q}, q) = \Phi(\dot{q})\Psi(q)$. The corresponding Lagrangian is obtained by solving Eq. (1.15). Let $\beta = -\Psi'/\Psi \neq 0$ be the separation constant. Then $M(\dot{q}, q) = e^{-\beta q - \beta\dot{q}/\alpha}/\dot{q}$. Non-constant solutions ($\dot{q} \neq 0$) of (1.16) are extrema of the action

$$S = \int dtL, \qquad L = \dot{q}F(\dot{q})\exp(-\beta q), \tag{1.18}$$

where the function $F(\dot{q})$ satisfies the condition

$$\frac{dF}{d\dot{q}} = \frac{1}{\dot{q}^2}\exp\left(-\frac{\beta}{\alpha}\dot{q}\right).$$

If the separation constant in (1.17) is set to zero, i.e. $M = \Phi(\dot{q}) = 1/\dot{q}$ ($\beta = 0$), then the Lagrangian has the conventional form $L = T(\dot{q}) - V(q)$, where $V = \alpha q$ and $d^2T/d\dot{q}^2 = 1/\dot{q}$.

Alternatively, the equation of motion (1.16) can also be regarded as the Euler–Lagrange equation for the Lagrangian that explicitly depends on time:

$$L = \frac{1}{2}e^{\alpha t}\dot{q}^2. \tag{1.19}$$

The Helmholtz condition (1.13) is satisfied with the integrating factor explicitly depending on time, $M = e^{\alpha t}$. Furthermore, the dynamical system (1.16) can be viewed as a part of a larger Lagrangian dynamical system, e.g.

$$L(\dot{q}, q, \dot{Q}, Q) = \dot{Q}(\dot{q} + \alpha q).$$

The Euler–Lagrange equations of motion for the variables q and Q are decoupled, and, in particular, they coincide with (1.16) for $q(t)$. Such a possibility is not even included in the Helmholtz conditions.

The above analysis shows that the choice of the Lagrangian may not be unique, and additional physical principles should be invoked to limit it (e.g. to demand that the Hamiltonian coincides with the system energy (see Section 1.2), or that the Lagrangian has specific symmetries, etc.).

1.2 Hamiltonian equations of motion

Equations (1.5) are of the second order. Any system of second-order differential equations can be transformed to a system of first-order differential equations by increasing the number of independent functions. Indeed, setting

$$p_i = \frac{\partial L}{\partial \dot{q}^i}, \tag{1.20}$$

one finds from (1.5) that

$$\dot{p}_i = \frac{\partial L}{\partial q^i}. \tag{1.21}$$

The quantity p_i defined by (1.20) is called the canonical momentum conjugated to q^i. The matrix

$$T_{ij} = \frac{\partial^2 L}{\partial \dot{q}^i \partial \dot{q}^j}$$

is called the *Hessian matrix* (or simply, the *Hessian*) of the Lagrangian system. If it is not singular, $\det T \neq 0$, then by the implicit function theorem, the relation (1.20) defines the generalized velocities $\dot{q}^i$ as functions of q^i and p_i, that is,

$$\dot{q}^i = \dot{q}^i(p, q, t), \tag{1.22}$$

which must be substituted into (1.21). Then Eqs. (1.21) and (1.22) comprise the system of first-order differential equations for the generalized coordinates q^i and momenta p_i.

In Lagrangian systems with a singular Hessian matrix, the implicit function theorem does not apply, and the straightforward transformation to the corresponding first-order equations of motion is impossible. In this case, Eqs. (1.20) yield some relations $\phi_a(p, q, t) = 0$, called *constraints*, between the canonical variables. A general analysis of *constrained* systems will be given in Chapter 3. Here it is always assumed that the Hessian matrix is not singular.

Let f be a concave function in a Euclidean space. The matrix of its second derivatives, $\partial^2 f/\partial x^i \partial x^j$, is either positive or negative definite in the domain of f. By the implicit function theorem, the equation $\mathbf{y} = \partial f/\partial \mathbf{x}$ can be solved defining the function $\mathbf{x} = \mathbf{x}(\mathbf{y})$. The function

$$g(\mathbf{y}) = \mathbf{y} \cdot \mathbf{x}(\mathbf{y}) - f(\mathbf{x}(\mathbf{y}))$$

is called the Legendre transform of f [3].

The Hamiltonian of a Lagrangian system is the Legendre transform of $L(\dot{q}, q, t)$ with respect to the variable $\dot{q}$,

$$H(p, q, t) = p_i \dot{q}^i - L(\dot{q}, q, t), \tag{1.23}$$

where $\dot{q}^i = \dot{q}^i(p, q, t)$ is defined by (1.20) (i.e. $y_i = p_i$ and $x^i = \dot{q}^i$). The first-order differential equations (1.20) and (1.21) can also be obtained from the variational principle for the action written in the Hamiltonian form,

$$S_H = \int_{t_1}^{t_2} dt \left(p_i \dot{q}^i - H(p, q, t) \right), \tag{1.24}$$

with the same boundary conditions (1.4). It follows from (1.23) that the actions (1.24) and (1.1) are equivalent. In the Hamiltonian action (1.24), the independent variables are p_i and q^i. Therefore

$$\frac{\delta S_H}{\delta q^i} = -\dot{p}_i - \frac{\partial H}{\partial q^i} = 0, \qquad \frac{\delta S_H}{\delta p^i} = \dot{q}_i - \frac{\partial H}{\partial p^i} = 0,$$

or, after moving all the partial derivatives of the Hamiltonian to the right-hand side,

$$\dot{q}^i = \frac{\partial H}{\partial p_i}, \qquad \dot{p}_i = -\frac{\partial H}{\partial q^i}. \tag{1.25}$$

The system (1.25) is called the *Hamiltonian equations of motion.*

Thus, if $q^i(t)$ satisfies the Euler–Lagrange equations, then the pairs $(p_i(t), q^i(t))$ solve the Hamiltonian equations of motion and vice versa, meaning that the Euler–Lagrange and Hamiltonian systems of equations are equivalent. All the pairs (p_i, q^i) form the *phase space* of the system. A typical Lagrangian considered in what follows reads

$$L = \frac{1}{2} g_{ij}(q) \dot{q}^i \dot{q}^j + A_i(q) \dot{q}^i - V(q);$$

here $V(q)$ is a potential energy, the matrix g_{ij} is assumed to be non-singular and may depend on the coordinates. The Euler–Lagrange equations of this system are easy to obtain

$$g_{ij} \ddot{q}^j = -\Gamma_{nj,i} \dot{q}^n \dot{q}^j + F_{ij} \dot{q}^j - V_{,i}, \tag{1.26}$$

where $F_{ij} = A_{j,i} - A_{i,j}$, the index after the comma denotes the partial derivative with respect to the corresponding coordinate, for instance, $g_{nj,i} \equiv \partial g_{nj}/\partial q^i$, and $\Gamma_{nj,i} = 1/2(g_{ni,j} + g_{ji,n} - g_{nj,i}) \equiv [nj, i]$ are the Christoffel symbols.

Consider the Hamiltonian formalism for this system. By the definition (1.20) the canonical momenta are

$$p_i = \frac{\partial L}{\partial \dot{q}^i} = g_{ij} \dot{q}^j + A_i. \tag{1.27}$$

The generalized velocities are found as functions of the coordinates and momenta from (1.27),

$$\dot{q}^j = \dot{q}^j(p, q) = g^{ji}(q)(p_i - A_i(q)),$$

where g^{ij} is the inverse of the matrix g_{ij}, $g^{ij} g_{jk} = \delta^i_k$. The substitution of these relations into (1.23) yields the Hamiltonian of the system:

$$H = \frac{1}{2} g^{ij}(p_i - A_i)(p_j - A_j) + V. \tag{1.28}$$

It is straightforward to verify that, when $g_{ij} = \delta_{ij}$ and $A_i = 0$, this Hamiltonian turns into the sum of the kinetic and potential energies of a particle with unit mass whose position is given by the coordinates q^i.

Such a simple relation between the Hamiltonian and the system energy is not always possible. For example, a particle moving along an axis under the friction force is described by Eq. (1.16). If the Lagrangian is taken in the form (1.18), then the Hamiltonian obtained via the Legendre transform does not depend on time explicitly, and, therefore, is a conservative quantity. The Legendre transform for the time-dependent Lagrangian (1.19) leads to the Hamiltonian that also explicitly depends on time,

$$H = \frac{1}{2} e^{-\alpha t} p^2, \tag{1.29}$$

where $p = \partial L/\partial \dot{q} = e^{\alpha t}\dot{q}$. It follows from the Hamiltonian equations of motion (1.25) that $\dot{p} = 0$, i.e. the momentum is conserved, $p = p_0 = \text{const}$. Therefore,

$$H = e^{-\alpha t} E_0, \tag{1.30}$$

where $E_0 = p_0^2/2$ is the initial energy. Equation (1.30) shows that the Hamiltonian of the system is not conserved.

So, the same equations of motion can be obtained from different Lagrangians; the Hamiltonian depends on the choice of the Lagrangian. In Section 1.12 a particle with friction is also described as a system with a non-standard symplectic structure.

In addition, it is worth noting that there are dynamical systems for which the Hamiltonian equations of motion are impossible to construct. For example, put

$$p_i = \dot{q}_i, \qquad \dot{p}_i = \alpha \epsilon_{ijk} q_j p_k, \qquad i, j, k = 1, 2, 3. \tag{1.31}$$

Equations (1.31) are equivalent to the equations of motion (1.11) for which there exists no Lagrangian (see also a further discussion in Section 1.14.3).

1.3 The Poisson bracket

Consider the time evolution of a function $F(p, q, t)$ along a phase space trajectory $p = p(t)$ and $q = q(t)$. From the Hamiltonian equations of motion (1.25) one finds that

$$\begin{aligned} \frac{dF}{dt} &= \frac{\partial F}{\partial t} + \frac{\partial F}{\partial q^i}\dot{q}^i + \frac{\partial F}{\partial p_i}\dot{p}_i \\ &= \frac{\partial F}{\partial t} + \frac{\partial F}{\partial q^i}\frac{\partial H}{\partial p_i} - \frac{\partial F}{\partial p_i}\frac{\partial H}{\partial q^i} \equiv \frac{\partial F}{\partial t} + \{F, H\}. \end{aligned} \tag{1.32}$$

The symbol

$$\{A, B\} = \sum_{i=1}^{n} \left(\frac{\partial A}{\partial q^i} \frac{\partial B}{\partial p_i} - \frac{\partial A}{\partial p_i} \frac{\partial B}{\partial q^i} \right) \tag{1.33}$$

defined in (1.32) is called the *Poisson bracket* for functions A and B in the phase space of the system. The Poisson bracket has several remarkable properties. Namely, it is skew-symmetric,

$$\{A, B\} = -\{B, A\}, \tag{1.34}$$

satisfies both the Leibniz rule,

$$\{A, BC\} = \{A, B\}C + B\{A, C\}, \tag{1.35}$$

and the Jacobi identity,

$$\{\{A, B\}, C\} + \{\{B, C\}, A\} + \{\{C, A\}, B\} = 0 \tag{1.36}$$

for arbitrary A, B, and C. If the function F does not depend explicitly on time, then its evolution is determined by the equation $\dot{F} = \{F, H\}$. Taking the canonical coordinates and momenta as F, the Hamiltonian equations of motion are written in the symmetric form,

$$\dot{q}^i = \{q^i, H\}, \quad \dot{p}_i = \{p_i, H\}.$$

A quantity F is an *integral of motion* if $dF/dt = 0$ and, hence,

$$\frac{\partial F}{\partial t} + \{F, H\} = 0. \tag{1.37}$$

For integrals of motion that do not depend explicitly on time, Eq. (1.37) has a simpler form: $\{F, H\} = 0$. In particular, the Hamiltonian is an integral of motion if it does not depend explicitly on time because $dH/dt = \partial H/\partial t = 0$.

1.4 Canonical transformations

The Poisson bracket for the canonical coordinates and momenta is

$$\{q^i, p_j\} = \delta^i_j. \tag{1.38}$$

Consider functions

$$Q^i = Q^i(p, q)\,, \quad P_i = P_i(p, q) \tag{1.39}$$

such that

$$\{Q^i, P_j\} = \delta^i_j. \tag{1.40}$$

The functions (1.39) are said to define a *canonical transformation*, whereas Q^i and P_i are new generalized coordinates and momenta, respectively [1]. Put in (1.33),

$$\frac{\partial}{\partial q^i} = \frac{\partial Q^j}{\partial q^i}\frac{\partial}{\partial Q^j} + \frac{\partial P_j}{\partial q^i}\frac{\partial}{\partial P_j},$$
$$\frac{\partial}{\partial p_i} = \frac{\partial Q^j}{\partial p_i}\frac{\partial}{\partial Q^j} + \frac{\partial P_j}{\partial p_i}\frac{\partial}{\partial P_j}.$$

Then, by making use of (1.40), one finds

$$\{A, B\} = \frac{\partial A}{\partial Q^i}\frac{\partial B}{\partial P_i} - \frac{\partial A}{\partial P_i}\frac{\partial B}{\partial Q^i}.$$

Thus, canonical transformations do not change the Poisson bracket. This fact has an important consequence: the Hamiltonian equations of motion are form invariant under canonical transformations,

$$\dot{Q}^i = \{Q^i, \tilde{H}\} = \frac{\partial \tilde{H}}{\partial P_i}, \qquad \dot{P}_i = \{P_i, \tilde{H}\} = -\frac{\partial \tilde{H}}{\partial Q^i},$$

where $\tilde{H} = \tilde{H}(P, Q) = H(p(P, Q), q(P, Q))$ is the Hamiltonian in the new variables. A change of variables

$$q^i = q^i(Q) \quad \text{or} \quad Q^i = Q^i(q) \tag{1.41}$$

gives the simplest example of canonical transformations. The transformation (1.41) is called a *point transformation*. In order to find new canonical momenta, the condition (1.40) is invoked:

$$\{Q^i, P_j\} = \frac{\partial Q^i}{\partial q^k}\frac{\partial P_j}{\partial p_k} = \delta^i_j. \tag{1.42}$$

Together with the condition $\{P_j, P_i\} = 0$, Eq. (1.42) yields

$$P_j(q, p) = p_n \frac{\partial q^n}{\partial Q^j} = p_n (J^{-1})^n_j, \quad J^n_j \equiv \frac{\partial q^n}{\partial Q^j}. \tag{1.43}$$

where J^{-1} denotes the inverse of a matrix J. Equalities (1.43) define the new canonical momenta up to the gradient of an arbitrary function $f(q)$, $p_n \to p_n + \partial f(q)/\partial q^n$.

For example, consider a particle moving in a plane. The Lagrangian has the form,

$$L = \frac{1}{2}\dot{\mathbf{x}}^2 - V(\mathbf{x}),$$

where $\mathbf{x} \in R^2$. According to (1.20) and (1.23), the Hamiltonian reads

$$H = \frac{1}{2}\mathbf{p}^2 + V(\mathbf{x}).$$

The polar coordinates in the plane are defined by the point transformation

$$\mathbf{x} = (x_1, x_2) = (r\cos\varphi,\ r\sin\varphi). \tag{1.44}$$

Substitution of (1.44) into (1.43) gives the momenta

$$p_r = \left(\mathbf{p}, \frac{\partial \mathbf{x}}{\partial r}\right) = \frac{(\mathbf{p}, \mathbf{x})}{r}, \qquad p_\varphi = \left(\mathbf{p}, \frac{\partial \mathbf{x}}{\partial \varphi}\right) = (\mathbf{p}, T\mathbf{x}), \tag{1.45}$$

that are canonically conjugated to r and φ, respectively. Here $r = |\mathbf{x}|$, the 2×2 matrix T is skew-symmetric, $T_{ij} = -T_{ji}$, with $T_{12} = -1$ (T is a generator of $SO(2)$ rotations of the vector $\mathbf{x}$ about the origin), and $(\mathbf{p}, T\mathbf{x}) \equiv p_i T_{ij} x_j$ ($i, j = 1, 2$). It is not difficult to find the Hamiltonian in the new canonical variables:

$$\tilde{H} = \frac{1}{2}\, p_r^2 + \frac{1}{2r^2}\, p_\varphi^2 + V\left(\mathbf{x}(r, \varphi)\right).$$

The corresponding equations of motions can easily be obtained from this Hamiltonian.

1.5 Generating functions of canonical transformations

It is shown in Section 1.2 that the Hamiltonian equations of motion (1.25) determine stationary trajectories for the action:

$$\delta S_H = \delta \int_{t_1}^{t_2} dt L = \delta \int_{t_1}^{t_2} dt \left(p_i \dot{q}^i - H(p, q, t)\right) = 0. \tag{1.46}$$

On the other hand, the Hamiltonian equations of motion are proved to be invariant under canonical transformations. Therefore, new canonical variables should satisfy the equation

$$\delta \int_{t_1}^{t_2} dt \left(P_i \dot{Q}^i - \tilde{H}(P, Q, t)\right) = 0. \tag{1.47}$$

Thus, Eqs. (1.46) and (1.47) must hold simultaneously for any canonical transformation. It follows from this observation that the integrands in (1.46) and (1.47) may differ only by the total time derivative of a function F. The function F is called a *generating function* of a canonical transformation. Beside the time t, the generating function F may be a function of all the canonical variables involved, the old ones as well as the new ones, i.e. it may depend on $4n$ variables (n is the number of degrees of freedom). However, the old and new variables are related by the $2n$ conditions (1.39). Therefore there are only $2n$ independent arguments of the function F (safe for the time variable). Hence, the generating function can be written in one of the following four forms [1]:

$$F_1(q, Q, t), \quad F_2(q, P, t), \quad F_3(p, Q, t), \quad F_4(p, P, t). \tag{1.48}$$

The choice of generating function depends on the particular peculiarities of the problem under consideration. For instance, the variables q^i and Q^i are not independent for a point transformation (1.41), and the function F_1 is not suitable to generate the corresponding canonical transformation.

Let F_1 be chosen as a generating function. Then the integrands in (1.46) and (1.47) should differ by the total time derivative of F_1,

$$p_i \dot{q}^i - H = P_i \dot{Q}^i - \tilde{H} + \frac{d}{dt} F_1(q, Q, t). \tag{1.49}$$

Since the old and new variables are regarded as independent, Eq. (1.49) is satisfied only when the coefficients at $\dot{Q}^i$ and $\dot{q}^i$ in the right-hand side coincide with the corresponding coefficients in the left-hand side. This leads to the following relations:

$$p_i = \frac{\partial F_1}{\partial q^i}, \tag{1.50}$$

$$P_i = -\frac{\partial F_1}{\partial Q^i}, \tag{1.51}$$

$$\tilde{H} = H + \frac{\partial F_1}{\partial t}. \tag{1.52}$$

Relation (1.50) contains only p_i, q^i, Q^i, and t, therefore it defines Q^i as functions of p_i, q^i, and t. The substitution of $Q^i = Q^i(p, q, t)$ into the right-hand side of Eq. (1.51) determines $P_i = P_i(p, q, t)$. In this way the new canonical variables are defined as functions of the old ones.

Equations (1.51) and (1.50) show that the generating functions F_2 and F_3 can be obtained from F_1 by the Legendre transform for the variables Q^i and q^i, respectively (compare the Legendre transform of the Lagrangian to the Hamiltonian):

$$F_2(q, P, t) = P_i Q^i + F_1(q, Q, t), \tag{1.53}$$

$$F_3(Q, p, t) = -q^i p_i + F_1(q, Q, t). \tag{1.54}$$

The function F_4 is obtained from F_1 by the double Legendre transform for both of its arguments q^i and Q^i according to Eqs. (1.51) and (1.50):

$$F_4(p, P, t) = -p_i q^i + P_i Q^i + F_1(q, Q, t). \tag{1.55}$$

Substituting the solutions of Eqs. (1.53)–(1.55) into (1.49) and equating the coefficients at the time derivatives of the corresponding independent variables, one finds,

$$p_i = \frac{\partial F_2}{\partial q^i}, \qquad Q^i = \frac{\partial F_2}{\partial P_i}; \tag{1.56}$$

$$q^i = -\frac{\partial F_3}{\partial p_i}, \qquad P_i = -\frac{\partial F_3}{\partial Q^i}; \tag{1.57}$$

$$q^i = -\frac{\partial F_4}{\partial p_i}, \qquad Q^i = \frac{\partial F_4}{\partial P_i}. \tag{1.58}$$

The Hamiltonian $\tilde{H}$ is defined by Eq. (1.52) where F_1 is replaced by the corresponding generating function.

A simple example is provided by the generating function of the point transformation introduced in the preceding section. Put

$$F_2 = f^i(q,t)P_i; \tag{1.59}$$

then from Eq. (1.56) one infers that

$$Q^i = \frac{\partial F_2}{\partial P_i} = f^i(q,t).$$

It is easy to see that in this case Eq. (1.43) immediately follows from the first equation in (1.56). The canonical transformation becomes identical if $f^i = q^i$.

Consider a canonical transformation that is defined by the generating function

$$F_2^\epsilon(q,P,t) = q^i P_i + \epsilon G(q,P,t), \tag{1.60}$$

where ϵ is a parameter, $\epsilon \to 0$. Obviously, the canonical transformation is identical at $\epsilon = 0$. A canonical transformation generated by the function (1.60) is called *infinitesimal*. According to (1.56), one has,

$$p_i = P_i + \epsilon \frac{\partial G(q,P,t)}{\partial q^i}, \qquad Q^i = q^i + \epsilon \frac{\partial G(q,P,t)}{\partial P_i}. \tag{1.61}$$

The first equation in (1.61) can be solved in the first order of perturbation theory $P_i = p_i + O(\epsilon)$. Substituting the solution into the second equation in (1.61) and neglecting the terms $O(\epsilon^2)$, the new canonical variables as functions of the old ones under an infinitesimal canonical transformation are deduced:

$$P_i = p_i - \epsilon \frac{\partial G(q,p,t)}{\partial q^i} \qquad Q^i = q^i + \epsilon \frac{\partial G(q,p,t)}{\partial p_i}. \tag{1.62}$$

The function G is called the *generator* of an infinitesimal canonical transformation.

By means of relations (1.53)–(1.55), one can derive the infinitesimal functions $F_{1,3,4}^\epsilon$. For example, according to (1.53),

$$F_1^\epsilon(q,Q,t) = P_i(q^i - Q^i) + \epsilon G(q,P),$$

where $P_i = P_i(Q,q,t)$ is found by solving the second equation in (1.61).

Let $p_i = p_i(t)$ and $q^i = q^i(t)$ be a phase space trajectory of the system. Let us compare the values of the canonical variables at two instances separated by an infinitesimal time interval $\epsilon \to 0$,

$$\begin{aligned} P_i &\equiv p_i(t+\epsilon) = p_i + \epsilon \dot{p}_i + O(\epsilon^2), \\ Q^i &\equiv q^i(t+\epsilon) = q^i + \epsilon \dot{q}^i + O(\epsilon^2). \end{aligned}$$

Then in accord with the Hamiltonian equations of motion, it follows that

$$P_i = p_i - \epsilon \frac{\partial H(p,q,t)}{\partial q^i}, \tag{1.63}$$

$$Q_i = q^i + \epsilon \frac{\partial H(p,q,t)}{\partial p_i}, \tag{1.64}$$

where the terms $O(\epsilon^2)$ are neglected. A comparison of Eqs. (1.63) and (1.64) with Eqs. (1.62) shows that the canonical variables $q(t)$ and $p(t)$ are related to their initial values $q_0 = q(0)$ and $p_0 = p(0)$ by a canonical transformation generated by the Hamiltonian. The motion of a Hamiltonian system corresponds to the continuous unfolding of a canonical transformation that relates the current canonical variables q and p to their initial values $Q = q_0$ and $P = p_0$.

Finding the canonical transformation that relates the current canonical variables q and p to their initial values $Q = q_0$ and $P = p_0$ is equivalent to solving the equations of motion, since $p = p(p_0, q_0, t) = p(P, Q, t)$ and $q = q(p_0, q_0, t) = p(P, Q, t)$. The Hamiltonian in the new variables should be a constant function because $\dot{Q} = \dot{P} = 0$. The approach of solving the Hamiltonian equations of motion by means of canonical transformations is known as the Hamilton–Jacobi theory [1]. It is discussed in Section 1.10.1 for general theories with bosonic and fermionic degrees of freedom.

1.6 Symmetries and integrals of motion

1.6.1 Noether's theorem

Consider a map of the configuration space onto itself, $q_\epsilon : M \to M$, which smoothly depends on a parameter ϵ so that it becomes the identity map when $\epsilon = 0$. A trajectory $q = q(t)$ is mapped onto $q = q_\epsilon(t)$. Consider the action for the transformed trajectory,

$$S_\epsilon = \int_{t_1}^{t_2} L(\dot{q}_\epsilon, q_\epsilon, t)dt. \tag{1.65}$$

Noether's first theorem asserts that, if the action is invariant under the map q_ϵ, i.e. $S_\epsilon = S_0$, then the quantity

$$I = \frac{\partial L}{\partial \dot{q}^i} Q^i \tag{1.66}$$

is an integral of motion, where $Q^i = \left(\partial q^i_\epsilon / \partial \epsilon\right)_{\epsilon=0}$ [14, 15]. The proof is elementary. Since q_ϵ depends smoothly on ϵ, the condition of the theorem implies that $(dS_\epsilon/d\epsilon)_{\epsilon=0} = 0$. Put $q_\epsilon = q + \epsilon Q + O(\epsilon^2)$ for a trajectory $q = q(t)$ ($q(t)$ solves the Euler–Lagrange equations). Substituting $q_\epsilon(t)$ into (1.65), linearizing the latter in ϵ, integrating by parts the term with $\epsilon \dot{Q}$, making use of the Euler–Lagrange equations, and, finally, demanding that the term linear in ϵ vanishes, one infers that $I(t_2) - I(t_1) = 0$, where I is defined in (1.66). This means that I is indeed an integral of motion because t_2 and t_1 are arbitrary.

Suppose that L does not depend on the coordinates q. Then the action is invariant under translations $q^i \to q^i_\epsilon = q^i + \epsilon^i$, i.e. each coordinate can be translated independently. According to Noether's theorem, there exist $n = \dim M$ integrals of motion

$$I_i = \frac{\partial L}{\partial \dot{q}^i} = p_i;$$

here $Q^i = 1$. They coincide with the canonical momenta p_i.

There are integrals of motion associated with the invariance of the action under time transformations. Consider a parametric map $t = t_\epsilon(\tau)$ which depends smoothly on the parameter ϵ and turns into the identity map $t = \tau$ at $\epsilon = 0$. A trajectory $q = q(t)$ is mapped onto $q = q(t_\epsilon(\tau)) \equiv q_\epsilon(\tau)$. Let $q_1 = q(t_1)$ and $q_2 = q(t_2)$ be the initial and final points of the trajectory. Under the time transformations they do not change. Hence the range of $t_\epsilon(\tau)$ must be the same interval $t_1 \le t_\epsilon \le t_2$, and the range of the new time variable τ is determined by the inverse transform $\tau = t_\epsilon^{-1}(t)$. Put $\tau_{1\epsilon} = t_\epsilon^{-1}(t_1)$ and $\tau_{2\epsilon} = t_\epsilon^{-1}(t_2)$. Then the action for the transformed trajectory computed between the same initial and final points is defined by

$$S_\epsilon = \int_{\tau_{1\epsilon}}^{\tau_{2\epsilon}} L(\dot{q}_\epsilon(\tau), q_\epsilon(\tau), \tau) d\tau. \tag{1.67}$$

Noether's theorem claims that, if the action is invariant under the time transformation, that is, $S_\epsilon = S_0$, then the quantity

$$I = \left(\frac{\partial L}{\partial \dot{q}^i} \dot{q}^i - L \right) T \tag{1.68}$$

is an integral of motion, where $T = (\partial t_\epsilon / \partial \epsilon)_{\epsilon=0}$. Indeed, taking the derivative of the action (1.67) with respect to ϵ, making use of the Euler–Lagrange equation to transform the term containing $\partial \dot{q}_\epsilon / \partial \epsilon$, and setting the derivative to zero, one finds the condition

$$\frac{dS_\epsilon}{d\epsilon} = L(\dot{q}_\epsilon, q_\epsilon, \tau) \frac{\partial \tau}{\partial \epsilon} \Bigg|_{\tau_{1\epsilon}}^{\tau_{2\epsilon}} - \dot{q}_\epsilon \frac{\partial L}{\partial \dot{q}} (\dot{q}_\epsilon, q_\epsilon, \tau) \Bigg|_{\tau_{1\epsilon}}^{\tau_{2\epsilon}} = 0.$$

The conclusion results from this equation by setting $\epsilon = 0$ in it with the help of the following expansions. Put $t_\epsilon = \tau + T\epsilon + O(\epsilon^2)$. Hence, $q_\epsilon = q + \epsilon T \dot{q} + O(\epsilon^2)$ for any trajectory, $\tau_{1\epsilon} \approx t_1 - \epsilon T_1$ and $\tau_{2\epsilon} \approx t_2 - \epsilon T_2$, where $T_{1,2}$ are the values of T at $t_{1,2}$, for integration limits in (1.67). The expression in parentheses in (1.68) coincides with the canonical Hamiltonian so that $I = HT$. Note also that the time transformations can be generated by a function on TM, i.e. $T = T(\dot{q}, q, t)$, in general.

If the Lagrangian does not explicitly depend on time, then action (1.67) is invariant under time translations $t = t_\epsilon(\tau) = \tau + \epsilon$, which is proved by changing the integration variable in (1.67), $\tau = t - \epsilon$. Here $T = 1$ and the corresponding Noether integral (1.68) coincides with the Hamiltonian of the system:

$$I = \frac{\partial L}{\partial \dot{q}^i}\dot{q}^i - L = p_i\dot{q}^i - L = H$$

(see (1.20) and (1.23)).

Finally, consider transformations obtained by a composition of the maps q_ϵ, generated by Q (see (1.66)) and t_ϵ, generated by T, i.e. a trajectory $q = q(t)$, $t \in [t_1, t_2]$, is mapped onto $q = q_\epsilon(t_\epsilon(\tau)) \equiv q_\epsilon(\tau)$, $\tau \in [\tau_{1\epsilon}, \tau_{2\epsilon}]$. The action for the latter is given in (1.67). If it is invariant under the combined transformations, i.e. $S_\epsilon = S_0$, then the quantity

$$I = \left(\frac{\partial L}{\partial \dot{q}^i}\dot{q}^i - L\right)T + \frac{\partial L}{\partial \dot{q}^i}Q^i \tag{1.69}$$

is an integral of motion. The proof goes along the same line of arguments used to establish (1.66) and (1.68) by noting that the combined map is generated by $\dot{q}T + Q$. Indeed, $q_\epsilon \approx q(t_\epsilon(\tau)) + \epsilon Q^i(t_\epsilon(\tau)) \approx q + \epsilon(\dot{q}^i T + Q^i)$.

To illustrate (1.69), consider a dynamical system defined by the Lagrangian

$$L = \frac{1}{2}\dot{\mathbf{q}}^2 - V(\mathbf{q})\,, \quad V = \frac{g}{\gamma - 1}\left(q_1q_2q_3\right)^{1-\gamma}, \tag{1.70}$$

where $\mathbf{q} = (q_1, q_2, q_3) \in R^3$, and g and γ are numerical parameters. This dynamical system describes a self-similar compressible hydrodynamic flow [16, 17]. The Euler–Lagrange equations of motion can be written in the form,

$$\ddot{q}_i q_i = g\left(q_1q_2q_3\right)^{1-\gamma}; \tag{1.71}$$

no summation over i here. Let $\mathbf{q} = \mathbf{q}(t)$ be a solution to (1.71). Then it is easy to verify that the trajectory $\mathbf{q} = \mathbf{q}_\lambda(t) = c_\lambda \mathbf{q}(\lambda t)$, where $c_\lambda = \lambda^{2/(1-3\gamma)}$, is also a solution to (1.71). In other words, the equations of motion are covariant under the simultaneous scaling transformations of the coordinates and time:

$$\mathbf{q} \to \mathbf{q}_\lambda = c_\lambda \mathbf{q}, \quad t \to t_\lambda = \lambda t, \quad c_\lambda = \lambda^{2/(1-3\gamma)}. \tag{1.72}$$

This map smoothly depends on the parameter λ and turns into the identity map at $\lambda = 1$. Then according to (1.67) the action of the transformed trajectory reads,

$$S_\lambda = \int_{t'/\lambda}^{t''/\lambda} L(\dot{\mathbf{q}}_\lambda, \mathbf{q}_\lambda)dt = c_\lambda^2\lambda \int_{t'}^{t''} L(\dot{\mathbf{q}}, \mathbf{q})dt; \tag{1.73}$$

the latter equality is obtained by changing the integration variable $t \to \lambda t$. The action is invariant under the scaling transformations (1.72) if

$$c_\lambda^2\lambda = 1\,, \text{ or } \gamma = 5/3.$$

Note also that the Lagrangian (1.70) does not depend explicitly on time and, therefore, the energy $E = \dot{\mathbf{q}}^2/2 + V$ is an integral of motion. To find the integral of motion associated with the scaling symmetry, put $\lambda = e^\epsilon$ so that $c_\lambda = \lambda^{-1/2} = e^{-\epsilon/2}$, for $\gamma = 5/3$. Then $\mathbf{Q} = -\mathbf{q}/2$ and $T = t$. Hence, the integral (1.69) becomes,

$$I = Et - \frac{1}{2}\dot{\mathbf{q}} \cdot \mathbf{q} = \frac{1}{2}\frac{d}{dt}\left(Et^2 - \frac{1}{2}\mathbf{q}^2\right). \tag{1.74}$$

This equation allows one to completely determine the evolution of the radial variable $|\mathbf{q}|$. Indeed, by integrating (1.74), one infers that

$$W = \frac{1}{2}\mathbf{q}^2 - Et^2 + 2It$$

remains constant on the trajectory of the system. The constants W, E, and I are fixed by the initial positions and velocities. In the spherically symmetric case, $r = |\mathbf{q}|$, the Noether integrals are sufficient to find an explicit solution of the Euler–Lagrange equations. In the cylindrically symmetric case, the system is integrable thanks to the two Noether integrals (E and W) [18].

As a final remark, it is worth noting that the conclusion of Noether's theorem would not generally hold if the hypothesis that the action is invariant under symmetry transformations is replaced by a weaker condition of the covariance of the Euler–Lagrange equations under the symmetry transformations. For example, the Euler–Lagrange equations (1.71) are covariant under the transformations (1.72) for any γ, while the action is invariant only if $\gamma = 5/3$. It is not hard to verify that the quantity (1.69) with $\mathbf{Q} = 2\mathbf{q}/(1-3\gamma)$ and $T = t$ is an integral of motion if and only if $\gamma = 5/3$.

1.6.2 Integrals of motion and symmetry groups

Let the configuration space of the system be a vector space in which a Lie group G has a non-trivial representation. Suppose that the action is invariant with respect to transformation from G (a brief introduction to the group theory can be found in Section 8.1):

$$q^i \to q^i_\omega = \Omega^i_j q^j, \qquad \Omega = \exp(\omega_a T^a), \quad a = 1, 2, ..., N = \dim G, \tag{1.75}$$

where ω_a are parameters independent of time, the matrices T^a are generators of G,

$$\left[T^a, T^b\right] = F^{abc}T^c, \tag{1.76}$$

and F^{abc} are the structure constants. According to Noether's first theorem, the system has N integrals of motion (1.66). Since $Q^{ia} = \left(\partial q^i_\omega/\partial\omega_a\right)_{\omega=0} = T^{ai}_j q^j$, the quantities

$$L^a = \frac{\partial L}{\partial \dot{q}^i} T_j^{ai} q^j = p_i T_j^{ai} q^j \equiv p T^a q \tag{1.77}$$

are integrals of motion, $dL^a/dt = 0$. The Poisson bracket of the Noether integrals (1.77) is,

$$\{L^a, L^b\} = p_i \left[T^a, T^b\right]_j^i q^j = F^{abc} p_i T_j^{ci} q^j = F^{abc} L^c. \tag{1.78}$$

Thus, the quantities L^a can be viewed as generators of the group G if the Poisson bracket is identified with the Lie bracket in the Lie algebra of G. The latter is always possible because the Poisson bracket is skew-symmetric and satisfies the Jacobi identity. It follows from this observation that the operators $\hat{L}^a$, defined by

$$\hat{L}^a \Phi(p, q) = \{L^a, \Phi\}, \tag{1.79}$$

are generators of the symmetry transformations of smooth functions Φ on the phase space. Indeed, with the definition (1.79),

$$\left[\hat{L}^a, \hat{L}^b\right] \Phi = F^{abc} \hat{L}^c \Phi. \tag{1.80}$$

Since Φ is arbitrary, Eq. (1.80) implies the operator equality

$$\left[\hat{L}^a, \hat{L}^b\right] = F^{abc} \hat{L}^c, \tag{1.81}$$

which means that the operators (1.79) can be identified with the generators of G. Then $\hat{\Omega} = \exp(\omega_a \hat{L}^a)$ is the operator of the symmetry transformations of smooth functions on the phase space. In particular,

$$\hat{\Omega} q^i = [\exp(\omega_a T^a)]_j^i q^j, \qquad \hat{\Omega} p_i = [\exp(\omega_a T^a)]_i^j p_j. \tag{1.82}$$

Thus, any function Φ of the canonical variables that is invariant with respect to transformations from G should satisfy the equation

$$\hat{L}^a \Phi = 0. \tag{1.83}$$

If L^a are integrals of motion, then their Poisson bracket with the Hamiltonian vanishes, meaning that the Hamiltonian satisfies equation (1.83),

$$\hat{L}^a H = \{L^a, H\} = 0,$$

and, therefore, is invariant with respect to the symmetry transformations (1.82) from the group G.

In accordance with (1.78), the Noether integrals L_a can be viewed as generators of the Lie algebra if the Lie product is identified with the Poisson bracket because the latter is bilinear, skew-symmetric and satisfies the Leibnitz rule and Jacobi identity. The Lie algebra structure of integrals of motion holds in general. Let I_1 and I_2 be integrals of motion. Then $I_3 = \{I_1, I_2\}$

is an integral of motion. Indeed, by the Jacobi identity, $\dot{I}_3 = \{H, I_3\} = \{H, \{I_1, I_2\}\} = \{I_1, \{H, I_2\}\} + \{I_2, \{I_1, H\}\} = 0$ because $\{H, I_{1,2}\} = 0$. Thus, the integrals of motion form a Lie algebra (see Section 8.1.2). Furthermore, the algebra of operators (1.79) associated with integrals of motion is also a Lie algebra where the Lie product is the conventional commutator of operators, $[\hat{I}_1, \hat{I}_2] \equiv \hat{I}_1\hat{I}_2 - \hat{I}_2\hat{I}_1 = \hat{I}_3$ for any integrals of motion such that $I_3 = \{I_1, I_2\}$ (a proof is straightforward and based on the Jacobi identity). Since polynomials of integrals of motion are integrals of motion, the operators $\hat{I}$ associated with all integrals of motion form an associative algebra with the standard operator addition and multiplication. This algebra is a *universal enveloping algebra* [21] of the Lie algebra (the algebra of all polynomials of the independent integrals of motion).

The algebraic structure associated with integrals of motion can be used to reduce the original Hamiltonian dynamics to that with fewer degrees of freedom, or even to integrate the system completely. The goal may be reached in two steps. First, one should identify all independent integrals of motion I_α (labeled by α) whose Poisson bracket with each other vanishes, $\{I_\alpha, I_{\alpha'}\} = 0$. This problem is equivalent to a purely algebraic problem of finding all independent, mutually commuting elements $\hat{I}_\alpha$ (they generate the maximal Abelian subalgebra of the universal enveloping algebra). Second, a canonical transformation can be constructed after which some of the new canonical momenta coincide with the Poisson commuting integrals I_α [3, 19]. The identification of new momenta with I_α is possible precisely because of their Poisson commutativity $\{I_\alpha, I_{\alpha'}\} = 0$. After this canonical transformation, the new Hamiltonian is independent of the coordinates Φ_α canonically conjugated to the conserved momenta I_α (because $\dot{I}_\alpha = \{H, I_\alpha\} = \partial H/\partial\Phi_\alpha = 0$), and their equations of motion are easily integrated. If the number of I_α coincides with the number of degrees of freedom in the system, then the Hamiltonian is independent of all new canonical coordinates, and their evolution is linear in time because $\dot{\Phi}_\alpha = \partial H/\partial I_\alpha = \text{const}$ (they are called *cyclic* coordinates). The system is completely integrated.

While the first of the above two problems is the standard problem of representation theory [21], the second problem amounts to solving a partial differential equation for a generating function of the canonical transformation in question. For example, the independent mutually commuting operators for the Lie algebra (1.81) are the Casimir operators (independent invariant symmetric polynomial of $\hat{L}_a$; see Section 8.1.2) and the operators forming a basis in the Cartan subalgebra of the Lie algebra [21]. It is known that eigenvalues of the independent Casimir operators uniquely define a representation of the group G, while eigenvalues of independent elements of the Cartan subalgebra label basis vectors in this representation [20, 21].

To illustrate the treatment of the second problem, let us turn to a simple example. Consider a system in the three-dimensional Euclidean space. Let the symmetry group of the Hamiltonian be the group of orthogonal rotations $SO(3)$. If $\mathbf{r}$ and $\mathbf{p}$ are canonical coordinates and momenta, respectively, then the Hamiltonian depends only on the combinations, $\mathbf{r}^2$, $\mathbf{p}^2$, and $(\mathbf{p}, \mathbf{r})$ (defined via the Euclidean inner product). The components of the angular momentum $\mathbf{L} = \mathbf{r} \times \mathbf{p}$ are the Noether integrals of motion. Clearly, $\{L_a, L_b\} = \varepsilon_{abc} L_c$, where the components of the unit totally skew-symmetric tensor, ε_{abc}, are the structure constants of $SO(3)$ ($a, b, c = 1, 2, 3$ and $\varepsilon_{123} = 1$). The Poisson commuting integrals of motion are $\mathbf{L}^2$ (the Casimir operator of $SO(3)$ [21]) and L_3 (the generator of the Cartan subalgebra of $SO(3)$). To reduce the number of equations of motion, consider the canonical transformation associated with the point transformation to spherical coordinates, $x^1 = r \sin\theta \cos\varphi$, $x^2 = r \sin\theta \sin\varphi$, and $x^3 = r \cos\theta$, where x^i ($i = 1, 2, 3$) are rectangular components of the position vector $\mathbf{r}$. According to (1.43), the new canonical momenta are: $p_r = (\mathbf{p}, \mathbf{e}_r)$, $p_\theta = (\mathbf{p}, \mathbf{e}_\theta)$, and $p_\varphi = (\mathbf{p}, \mathbf{e}_\varphi)$, where $\mathbf{e}_r = \partial \mathbf{x}/\partial r$, $\mathbf{e}_\theta = \partial \mathbf{x}/\partial \theta$, and $\mathbf{e}_\varphi = \partial \mathbf{x}/\partial \varphi$ are orthogonal vectors so that $\mathbf{p} = p_r \mathbf{e}_r + r^{-1} p_\theta \mathbf{e}_\theta + (r \sin\theta)^{-1} p_\varphi \mathbf{e}_\varphi$. Then $\mathbf{L}^2 = \mathbf{x}^2 \mathbf{p}^2 - (\mathbf{x}, \mathbf{p})^2 = r^2 \left(\mathbf{p}^2 - p_r^2\right) = p_\theta^2 + p_\varphi^2 / \sin^2\theta$ and $L_3 = -p_\varphi$. Next, consider the canonical transformation $(\theta, p_\theta; \varphi, p_\varphi) \to (\Theta, P_\theta; \Phi, P_\varphi)$ in which the new momenta coincide with the Poisson commuting integrals of motion, $P_\theta = p_\theta^2 + p_\varphi^2/\sin^2\theta = \mathbf{L}^2$ and $P_\varphi = p_\varphi$. It can be defined by the generating function $F_2(\varphi, P_\varphi, \theta, P_\theta)$ that should satisfy the equation

$$\mathbf{L}^2 = \left(\frac{\partial F_2}{\partial \theta}\right)^2 + \frac{1}{\sin^2\theta}\left(\frac{\partial F_2}{\partial \varphi}\right)^2 = P_\theta, \tag{1.84}$$

in accord with the first equation in (1.56). As it is demanded that $p_\varphi = \partial F_2/\partial \varphi = P_\varphi$, the solution to (1.84) should be sought in the form $F_2 = \varphi P_\varphi + S(\theta, P_\theta, P_\varphi)$. The equation for S is easily integrated with the result

$$S(\theta, P_\theta, P_\varphi) = \int \sqrt{P_\theta - P_\varphi^2/\sin^2\theta}\; d\theta. \tag{1.85}$$

The new canonical variables Θ and Φ as functions of the spherical angles and their momenta are found according to (1.56). Since $\mathbf{p}^2 = p_r^2 + P_\theta/r^2$, $(\mathbf{p}, \mathbf{r}) = r p_r$, and $\mathbf{x}^2 = r^2$, the Hamiltonian depends only on r, p_r, and P_θ. Therefore, the Hamiltonian equations of motion are furnished by $\dot{\Phi} = \dot{P}_\phi = \dot{P}_\theta = 0$, and $\dot{\Theta} = \partial H/\partial P_\theta$, so that Φ, P_φ, and P_θ are constants, while $\Theta = \int (\partial H/\partial P_\theta) dt$ is fully determined by the solutions of the equations for r and p_r. The latter are the only remaining non-trivial equations of motion. Thus, using the Noether's integrals associated with the group $SO(3)$, the dynamical description has been reduced to a Hamiltonian system with a single degree of freedom.

If the Hamiltonian does not explicitly depend on time, then it is an integral of motion. Similarly to (1.84) and (1.85), one can construct a canonical transformation $(r, p_r; \Theta, P_\theta) \to (R, P_r; \Theta', P_\theta')$ such that $P_r = H(r, p_r, P_\theta)$ and

$P'_\theta = P_\theta$. Its generating function can be taken in the form $F_2(r, P_r, \Theta, P'_\theta) = \Theta P'_\theta + S(r, P_r, P'_\theta)$, where S is a solution of the ordinary differential equation $H(r, (\partial S/\partial r), P'_\theta) = P_r$ with constant parameters P_r and P'_θ. The canonical coordinate R becomes cyclic, $\dot{R} = \partial H/\partial P_r = 1$. The system is completely integrated.

1.7 Lagrangian formalism for Grassmann variables

Grassmann variables are used to describe fermionic degrees of freedom. By definition, these variables anticommute:

$$\xi^i \xi^j = -\xi^j \xi^i, \qquad i, j = 1, 2, ..., n; \tag{1.86}$$

here n is the number of degrees of freedom. The quantities ξ^i are called *generating elements* of the Grassmann algebra (or simply Grassmann numbers). It follows from (1.86) that $\xi^i \xi^i = 0$ (no summation over i). Elements of the Grassmann algebra (or functions of Grassmann variables) are elements of a vector space (over the field of real or complex numbers) which is the span of monomials $\xi^{i_1}\xi^{i_2}\cdots\xi^{i_k}$, $k = 0, 1, ..., n$. The case $k = 0$ corresponds to ordinary (commuting) numbers (c-numbers). The independent monomials form a basis in the Grassmann algebra and their number is the dimension of the Grassmann algebra. It follows from condition (1.86) that the dimension equals $\sum_{k=0}^{n} C_k^n = 2^n$, where $C_k^n = n!/[(n-k)!k!]$ are the binomial coefficients. Two elements of the Grassmann algebra coincide if their expansion coefficients over the basis are the same. In Section 8.2 the calculus (differentiation and integration) is developed for functions of Grassmann variables.

As in the case of bosonic degrees of freedom, the time evolution of a fermionic system is determined by functions of time $\xi^i = \xi^i(t)$ that take their values in the Grassmann algebra. The generating elements of this algebra are the initial values $\xi^i(t=0) = \xi_0^i$, $\dot{\xi}^i(t=0) = \dot{\xi}_0^i$, and so on, depending on the order of differential equations which the functions $\xi^i(t)$ satisfy. This interpretation requires some clarification because of the anticommutativity of $\xi^i(t)$ (for any moment of time t).

By analogy with the Dirac equation, first-order differential equations of motion are used to define the classical mechanics of fermionic degrees of freedom. However, as shown below, nothing prevents us from considering a dynamics with anticommuting variables that is described by the second- or even higher-order differential equations. If the functions satisfy first-order differential equations, then they are uniquely defined only by the initial Grassmann numbers ξ_0^i. The functions $\xi^i(t)$ are odd elements of the Grassmann algebra with generating elements ξ_0^i and the expansion coefficients depending on time:

$$\xi^i(t) = \sum_{k=1}^{n'} z^i_{i_1 i_2 \cdots i_{2k-1}}(t)\; \xi_0^{i_1}\xi_0^{i_2}\cdots\xi_0^{i_{2k-1}}, \tag{1.87}$$

where $n' = [n/2]$ is the next smallest integer to $n/2$. Because of the anticommutativity of ξ_0^i, the expansion coefficients z^i are totally skew-symmetric. The number of independent components in a totally skew-symmetric tensor of rank $2k-1$ is C_{2k-1}^n. Therefore the time evolution of a fermionic system with n degrees of freedom is determined by $n\sum_{k=1}^{n'} C_{2k-1}^n = n2^{n-1}$ real functions of time. The initial conditions $\xi^i(t=0) = \xi_0^i$ induce the initial conditions for the tensors z^i:

$$z_{i_1}^i(0) = \delta_{i_1}^i; \tag{1.88}$$

$$z_{i_1 i_2 \dots i_{2k-1}}^i(0) = 0, \quad k \geq 2. \tag{1.89}$$

If $\xi^i(t)$ satisfy second-order differential equations, then they are odd elements of the Grassmann algebra with generating elements ξ_0^i and $\dot{\xi}_0^i$. The representation (1.87) should be changed to,

$$\xi^i(t) = \sum_{k,l=1}^{n} z_{i_1 i_2 \cdots i_k i_1' i_2' \cdots i_l'}^i(t) \;\; \xi_0^{i_1} \xi_0^{i_2} \cdots \xi_0^{i_k} \dot{\xi}_0^{i_1'} \dot{\xi}_0^{i_2'} \cdots \dot{\xi}_0^{i_l'}, \tag{1.90}$$

where the summation is carried out over k and l whose sum is an odd number, $k + l = 2m + 1$, $m = 0, 1, \dots, n-1$. The initial conditions for the functions $z^i(t)$ have the following form:

$$z_{i_1}^i(0) = \delta_{i_1}^i, \quad \dot{z}_{i_1}^i(0) = 0; \quad z_{i_1'}^i(0) = 0, \quad \dot{z}_{i_1'}^i(0) = \delta_{i_1'}^i, \tag{1.91}$$

and the other tensors vanish at $t = 0$. In contrast to the first-order theory, here z^i and $\dot{z}^i$ must be fixed at the initial time moment because the functions $z^i(t)$ satisfy second-order differential equations. In higher-order theories, $\xi^i(t)$ is viewed as an odd element of the Grassmann algebra with generating elements ξ_0^i, $\dot{\xi}_0^i$, $\ddot{\xi}_0^i$, and so on.

Equations of motion can be defined by means of the principle of stationary action,

$$\delta S[\xi] = \delta \int_{t_1}^{t_2} L(\dot{\xi}, \xi, t)dt = 0, \qquad \delta\xi(t_1) = \delta\xi(t_2) = 0, \tag{1.92}$$

where the Lagrangian L is usually assumed to be an even element of the Grassmann algebra generated by $\dot{\xi}^i$ and ξ^i. The variational derivative with respect to ξ^i is defined similarly to (1.2) and (1.3) with the variation parameter ϵ being a Grassmann number independent of generating elements ξ^i. The variational derivative satisfies the Leibniz rule for the derivatives with respect to Grassmann variables (see Section 8.2) and, in particular,

$$\frac{\vec{\delta}}{\delta\xi^i(t)}\,\xi^j(t') = \xi^j(t')\,\frac{\overleftarrow{\delta}}{\delta\xi^i(t)} = \delta_i^j\delta(t-t').$$

The arrow indicates the direction in which the differentiation operation acts (the left or right derivative (see Section 8.2)). As a result, one finds,

$$\delta S[\xi] = \int_{t_1}^{t_2} dt \left(\delta\dot{\xi}^i \frac{\vec{\partial} L}{\partial \dot{\xi}^i} + \delta\xi^i \frac{\vec{\partial} L}{\delta \xi^i} \right) = 0. \tag{1.93}$$

As noted the variations $\delta\xi^i$ are independent generating elements that anticommute with ξ^i. So the ordering in the products in the right-hand side of Eq. (1.93) is essential. After integrating by parts, the equations of motion follow

$$\frac{d}{dt}\frac{\vec{\partial} L}{\partial \dot{\xi}^i} = \frac{\vec{\partial} L}{\partial \xi^i} \tag{1.94}$$

because $\delta\xi^i$ are arbitrary. Equations (1.94) are similar to the Euler–Lagrange equations for bosonic systems. If the Lagrangian in (1.92) depends on higher-order time derivatives of $\xi^i(t)$, the corresponding equations of motion would have the form (1.6) where all the partial derivatives should be replaced by the left partial derivatives. Then these equations define a theory with higher derivatives for Grassmann dynamical variables as well.

Substituting the decomposition (1.87) (or (1.90) depending on the order of the equations of motion) of $\xi^i(t)$ over a basis in the Grassmann algebra into the Euler–Lagrange equations (1.94) and equating the coefficients at the same basis elements in the left- and right-hand sides, the system of equations for the skew-symmetric tensors z^i is obtained which has to be solved with the initial conditions (1.88) and (1.89) (or (1.91)). Thus, the dynamics of n real Grassmann degrees of freedom is equivalent to the dynamics of all real totally skew-symmetric tensors of odd ranks that do not exceed n (or $2n$). In other words, for any dynamical system with n Grassmann variables, one can construct an equivalent dynamics with $n2^{n-1}$ (or $n2^{2n-1}$ for second-order equations) real commuting variables.

Example. Let $n = 3$. Put

$$L = \frac{i}{2}\alpha_{ij}\dot{\xi}^i\xi^j - iV(\xi),$$

where $\alpha_{ij} = \delta_{ij}$ for simplicity. Since the Lagrangian is an even element of the Grassmann algebra, the most general form of the (time-independent) potential V reads,

$$V = V_0 + \frac{1}{2}\epsilon_{ijk}V_k\xi^i\xi^j,$$

where V_k is a constant vector, and a constant V_0 is irrelevant for equations of motion and can be omitted. Equations (1.94) are

$$\delta_{ik}\dot{\xi}^k = \epsilon_{ijk}V_k\xi^j.$$

The decomposition (1.87) is written as,

$$\xi^i(t) = z^i_j(t)\xi^j_0 + z^i(t)\xi^1_0\xi^2_0\xi^3_0.$$

Its substitution into the equations of motion yields the equations for the decomposition coefficients z^i_j and z^i:

$$\dot{z}^i_j(t) = V_n\epsilon_{ikn}\, z^k_j(t); \tag{1.95}$$
$$\dot{z}^i(t) = V_k\epsilon_{ijk}z^j(t). \tag{1.96}$$

Equations (1.95) are three independent equations labeled by the index $j = 1, 2, 3$. Therefore all the equations of motion can be viewed as equations for four three-dimensional vectors $\mathbf{z}_j$ and $\mathbf{z}$. Because of the skew-symmetry of the matrix $\epsilon_{ijk}V_k$ the norm of these vectors is an integral of motion. A general solution of the equations of motion describes rotations of the vectors $\mathbf{z}_j$ and $\mathbf{z}$ about the origin:

$$z^i_\mu = \left[\exp(-\hat{V}t)\right]^{i\cdot}_{\cdot k} z^k_\mu(0) \equiv \Omega^i_k(t)z^k_\mu(0), \qquad \mu = 0, 1, 2, 3, \tag{1.97}$$

where $\mathbf{z}_0 \equiv \mathbf{z}$, $\hat{V}$ is the matrix $(\hat{V})_{ik} = \epsilon_{ikn}V_n$, and the matrix $\Omega(t)$ is an element of the group $SO(3)$. The components $z^j_\mu(t)$ oscillate about the origin. With the initial conditions (1.88) and (1.89), one obtains

$$z^i(t) = 0, \quad z^i_j(t) = \Omega^i_j(t),$$

or for the Grassmann variables,

$$\xi^i(t) = \Omega^i_j(t)\xi^j_0.$$

Some other examples of dynamical systems with Grassmann variables can be found in [22–24]. A general structure of the configuration space in such systems is studied in [23].

1.8 Hamiltonian formalism for Grassmann variables

As noted above, the first- and second-order formalism can be used to describe dynamics of fermionic degrees of freedom. In the first case, a typical Lagrangian is

$$L = \frac{i}{2}\alpha_{ij}\dot{\xi}^i\xi^j - V(\xi), \tag{1.98}$$

where $\alpha_{ij} = \alpha_{ji}$. If ξ^i are complex generating elements of the Grassmann algebra (see Section 8.2.2), then the Lagrangian should be written in the form,

$$L = \frac{i}{2}\alpha_{ij}\left(\dot{\xi}^{i*}\xi^j - \xi^{i*}\dot{\xi}^i\right) - V(\xi, \xi^*), \tag{1.99}$$

where α_{ij} is a symmetric real matrix. In the latter case, the Lagrangian must be real, $L^* = L$. The complex conjugation reverses the order in the product of complex generating elements similarly to the Hermitian conjugation of the matrix product (see Section 8.2), $(\xi^i\xi^j\cdots\xi^k)^* = \xi^{k*}\cdots\xi^{j*}\xi^{i*}$.

The Euler–Lagrange equations for the Lagrangian (1.98) have the form,

$$i\alpha_{ij}\dot{\xi}^j = -\frac{\vec{\partial} V}{\partial \xi^i}. \tag{1.100}$$

In the theory with complex generating elements, ξ^i and their complex conjugates ξ^{i*} are independent variables, and the equations of motion for the Lagrangian (1.99) are written as,

$$i\alpha_{ij}\dot{\xi}^j = -\frac{\vec{\partial} V}{\partial \xi^{i*}}, \tag{1.101}$$

$$i\alpha_{ij}\dot{\xi}^{i*} = -\frac{\vec{\partial} V}{\partial \xi^{i}}. \tag{1.102}$$

Equations (1.100)–(1.102) are of the first order, whereas Eq. (1.102) is obtained from (1.101) by complex conjugation (this can easily be verified by means of the relation $\vec{\partial}V/\partial\xi^{i*} = -(\vec{\partial}V/\partial\xi^i)^*$ which holds for any real function $V^* = V$).

A dynamics of Grassmann degrees of freedom in the second-order formalism can be defined by the following Lagrangian:

$$L = -\frac{1}{2}\alpha_{ij}(\xi)\dot{\xi}^i\dot{\xi}^j + \frac{1}{2}\left(A_i(\xi)\dot{\xi}^i - \dot{\xi}^i A_i(\xi)\right) - V(\xi), \tag{1.103}$$

where $\alpha_{ij} = -\alpha_{ij}$ is a skew-symmetric real matrix with elements being even elements of the Grassmann algebra, and where $A_i(\xi)$ are odd elements of the Grassmann algebra. The Lagrangian (1.103) can also be used in the case of complex generating elements if the indices i and j are replaced by a and b, and set $\xi^a = (\xi^i, \xi^{i*})$ (i.e. the indices a and b label all generating elements). Naturally, the matrix α_{ab} and functions A_a should be chosen so that the Lagrangian remains real, $L^* = L$. The Lagrangian (1.103) leads to the second-order equations of motion,

$$\alpha_{ij}\ddot{\xi}^j = -\Gamma_{nj,i}\dot{\xi}^n\dot{\xi}^j - F_{ij}\dot{\xi}^j + V_{,\vec{i}}\,, \tag{1.104}$$

where the Christoffel symbols on the Grassmann algebra are introduced, $\Gamma_{nj,i} = (\alpha_{ni,\vec{j}} - \alpha_{ji,\vec{n}} - \alpha_{nj,\vec{i}})/2$, the index with an overarrow after the comma denotes the left derivative with respect to the corresponding Grassmann variable, e.g. $\alpha_{ni,\vec{j}} = \vec{\partial}\alpha_{ni}/\partial\xi^j$, and $F_{ij} = A_{i,\vec{j}} + A_{j,\vec{i}}$. When deriving the equations of motion (1.104), the order in which ξ^i and their time derivatives appear in the products should be taken into account, for example, $\partial_t\alpha_{ni} = \dot{\xi}^j\alpha_{ni,\vec{j}} = -\alpha_{ni,\vec{j}}\dot{\xi}^j$, or $\partial_t A_i = \dot{\xi}^j A_{i,\vec{j}} = A_{i,\vec{j}}\dot{\xi}^j$, because α_{ij} and A_i are even and odd, respectively. As they appear, Eqs. (1.104) coincide with those describing a similar dynamical system with bosonic (commuting) variables (compare Eq. (1.26)); the difference in the definition of symbols $\Gamma_{nj,i}$ and F_{ij} is related to anticommutativity of the Grassmann variables ξ^i.

The canonical momentum conjugated to the variable ξ^i is defined as the right derivative of the Lagrangian with respect to $\dot{\xi}^i$:

$$p_i = L\frac{\overleftarrow{\partial}}{\partial\dot{\xi}^i}. \tag{1.105}$$

By definition, the canonical momenta are odd elements of the Grassmann algebra. The Hamiltonian is obtained by the corresponding Legendre transform of the Lagrangian for the velocity $\dot{\xi}^i$,

$$H = p_i\dot{\xi}^i - L. \tag{1.106}$$

The order in the product of p_i and $\dot{\xi}^i$ is relevant because the canonical momenta and velocities are odd elements and, hence, anticommute. It is set by the definition (1.105). The Hamiltonian is an even element of the Grassmann algebra generated by the canonical variables p_i and ξ^i. Here it is assumed that equations (1.105) have a unique solution for $\dot{\xi}^i = \dot{\xi}^i(p,\xi)$, so that the generating elements $\dot{\xi}^i$ and ξ^i can be changed to new generating elements p_i and ξ^i in the Grassmann algebra. The latter is not always possible (e.g. in the first-order formalism, see Eq. (1.108)). The Hamiltonian equations of motion can be obtained from the principle of stationary action in the Hamiltonian form,

$$\delta S[p,\xi] = \delta\int_{t_1}^{t_2} Ldt = \delta\int_{t_1}^{t_2} dt\left(p_i\dot{\xi}^i - H\right) = 0\ ,\quad \delta\xi^i(t_a) = 0,\quad a = 1,2,$$

where the generating elements p_i and ξ^i are viewed as independent variables. Carrying out the variation, one infers that

$$\begin{aligned}\delta S &= \int_{t_1}^{t_2} dt\left(\delta p_i\dot{\xi}^i - \dot{p}_i\delta\xi^i - \delta\xi^i\frac{\vec{\partial}}{\partial\xi^i}H - \delta p^i\frac{\vec{\partial}}{\partial p_i}H\right)\\ &= \int_{t_1}^{t_2} dt\left[\delta p_i\left(\dot{\xi}^i - \frac{\vec{\partial}H}{\partial p_i}\right) + \delta\xi^i\left(\dot{p}_i - \frac{\vec{\partial}H}{\partial\xi^i}\right)\right] = 0.\end{aligned}$$

The variations δp_i and $\delta\xi^i$ are odd elements of the Grassmann algebra and anticommute with p_i and ξ^i. As δp_i and $\delta\xi^i$ are arbitrary, the following equations of motion should hold:

$$\dot{\xi}^i = \frac{\vec{\partial}H}{\partial p_i};\quad \dot{p}_i = \frac{\vec{\partial}H}{\partial\xi^i}. \tag{1.107}$$

There is a subtlety in the Hamiltonian formalism for anticommuting variables. Equation (1.105) defines the canonical momentum as the right derivative of the Lagrangian with respect to the velocity. There would have been no inconsistency if it were defined by the left derivative $\vec{\partial}L/\partial\dot{\xi}^i = -L\overleftarrow{\partial}/\partial\dot{\xi}^i$, i.e. both the definitions differs only in the sign. The definition of the Hamiltonian also contains an ambiguity related to the anticommutativity of p_i and $\dot{\xi}^i$. The replacement of $p_i\dot{\xi}^i$ by $\dot{\xi}^i p_i = -p_i\dot{\xi}^i$ in (1.106) leads to the sign change in the right-hand sides of the Hamiltonian equations of motion (1.107) or, alternatively, the left derivatives are replaced by the right ones in them. This sign ambiguity can always be removed by additional physical conditions.

When comparing the Hamiltonian formalisms for the first- and second-order theories, it is easy to see that the rule (1.106) cannot be applied to the Lagrangian (1.98) because the equations

$$p_i = L\frac{\overleftarrow{\partial}}{\partial \dot{\xi}^i} = -\frac{1}{2}\alpha_{ij}\xi^j \tag{1.108}$$

do not define the velocities as functions of canonical momenta, and, hence, the latter can no longer be independent generating elements. Apparently, the same holds for the Lagrangian (1.99). Relations between canonical momenta and coordinates, like (1.108), are called *constraints.* The Hamiltonian formalism for *constrained systems* is studied in Chapter 3. It is shown there that the first-order formalism can be used to describe systems with commuting (bosonic) variables as well as with anticommuting (fermionic) ones, and it is equivalent to a Hamiltonian formalism where some of the canonical coordinates play the role of canonical momenta (see also [25, 26]). As a point of fact, any second-order system (with commuting or anticommuting variables) can be reformulated as a first-order Lagrangian system by going over to the Hamiltonian formalism and identifying the Hamiltonian with the new Lagrangian (i.e. by viewing the phase space of the system as a configuration space of the corresponding Lagrangian first-order system). Thus, bosonic and fermionic systems can always be described on an equal footing.

Consider the Hamiltonian formalism for the Lagrangian (1.103). The canonical momenta are written in the form,

$$p_i = L\frac{\overleftarrow{\partial}}{\partial \dot{\xi}^i} = \alpha_{ij}\dot{\xi}^j + A_i\,.$$

After some algebraic transformations, the Hamiltonian can be represented as,

$$H = \frac{1}{2}\alpha^{ij}(p_i - A_i)(p_j - A_j) + V,$$

which appears to have the same form as its bosonic analog (1.28) (this is why the canonical momenta are defined by the right derivative in (1.105) so that the Hamiltonian has the form (1.106)); here α^{ij} is the inverse of the matrix α_{ij}, $\alpha^{ik}\alpha_{kj} = \delta^i_j$.

Consider a function $F = F(p, \xi, t)$ on the phase space of the systems, which is now a Grassmann space, and let us study its time evolution along the trajectory of motion defined by the Hamiltonian equations of motion (1.107). One has,

$$\frac{dF}{dt} = \frac{\partial F}{\partial t} + F\frac{\overleftarrow{\partial}}{\partial \xi^i}\dot{\xi}^i + F\frac{\overleftarrow{\partial}}{\partial p^i}\dot{p}^i$$
$$= \frac{\partial F}{\partial t} + F\left(\frac{\overleftarrow{\partial}}{\partial \xi^i}\frac{\vec{\partial}}{\partial p_i} + \frac{\overleftarrow{\partial}}{\partial p_i}\frac{\vec{\partial}}{\partial \xi^i}\right)H \equiv \frac{\partial F}{\partial t} + \{F, H\}. \tag{1.109}$$

If the function F does not explicitly depend on time, then its time evolution is fully determined by the quantity $\{F, H\}$. In particular, choosing F to coincide with the canonical variables, the Hamiltonian equations of motion (1.107) can be written in the following form:

$$\dot{\xi}^i = \{\xi^i, H\}, \quad \dot{p}_i = \{p_i, H\}.$$

For any two functions A and B on the phase space of a Grassmann system, the quantity

$$\{A, B\} = A\left(\frac{\overleftarrow{\partial}}{\partial \xi^i}\frac{\vec{\partial}}{\partial p_i} + \frac{\overleftarrow{\partial}}{\partial p_i}\frac{\vec{\partial}}{\partial \xi^i}\right)B \tag{1.110}$$

is called the *Poisson bracket for Grassmann variables*. It has the following permutation property:

$$\{A, B\} = -(-1)^{\epsilon_A \epsilon_B}\{B, A\}, \tag{1.111}$$

where $\epsilon_{A,B}$ are the parities of the Grassmann algebra elements A and B ($\epsilon_A = 0$ or $\epsilon_A = 1$ if A is even or odd, respectively). Apparently, a generic element of the Grassmann algebra does not have any specific parity, for it is the sum of odd and even elements. So the permutation rule for the Poisson bracket of generic elements of the Grassmann algebra is established by means of the bilinearity of the Poisson bracket (1.110) and the rule (1.111). The Poisson bracket also satisfies the Leibniz rule,

$$\{A, BC\} = \{A, B\}C + (-1)^{\epsilon_B \epsilon_A} B\{A, C\} = \{A, B\}C + (-1)^{\epsilon_B \epsilon_C}\{A, C\}B, \tag{1.112}$$

and the Jacobi identity,

$$(-1)^{\epsilon_A \epsilon_C}\{\{A, B\}, C\} + (-1)^{\epsilon_C \epsilon_B}\{\{C, A\}, B\} + (-1)^{\epsilon_B \epsilon_A}\{\{B, C\}, A\} = 0. \tag{1.113}$$

If the functions A, B, and C do not possess a specific parity, then the Leibniz rule and Jacobi identity hold separately for their odd and even components.

A proof of relations (1.111)–(1.113) is based on the permutation rule for two elements of the Grassmann algebra with a specific parity,

$$AB = (-1)^{\epsilon_A \epsilon_B} BA. \tag{1.114}$$

The validity of this relation is obvious. If one of the two elements is even ($\epsilon = 0$), then A and B commute, and they anticommute if both the elements are odd ($\epsilon_A = \epsilon_B = 1$). The parity of the derivative of A with respect to any generating element equals $\epsilon(\partial A) = 1 - \epsilon_A$: if A is odd, then its derivative is even and vice versa. The following chain of equalities follows from the definition (1.110):

$$\begin{aligned}\{A, B\} &= (-1)^{\epsilon_{\partial A}\epsilon_{\partial B}} \left[\left(\frac{\vec{\partial}}{\partial p_i} B \right) \left(A \frac{\overleftarrow{\partial}}{\partial \xi^i} \right) + \left(\frac{\vec{\partial}}{\partial \xi^i} B \right) \left(A \frac{\overleftarrow{\partial}}{\partial p_i} \right) \right] \\ &= (-1)^{\epsilon_{\partial A}\epsilon_{\partial B} + \epsilon_A + \epsilon_B} \left[B \left(\frac{\overleftarrow{\partial}}{\partial \xi^i} \frac{\vec{\partial}}{\partial p_i} + \frac{\overleftarrow{\partial}}{\partial p_i} \frac{\vec{\partial}}{\partial \xi^i} \right) A \right] \\ &= -(-1)^{\epsilon_A \epsilon_B} \{B, A\},\end{aligned}$$

where the obvious relation,

$$\frac{\vec{\partial}}{\partial \theta_i} A = (-1)^{\epsilon_A + 1} A \frac{\overleftarrow{\partial}}{\partial \theta_i}, \tag{1.115}$$

has been used in which θ_i stands for any of the generating elements ξ^i and p_i.

The Leibniz rule for the Grassmann Poisson bracket is easily derived from the product rule for the Grassmann derivatives (see Section 8.2),

$$\frac{\vec{\partial}}{\partial \theta_i}(AB) = \left(\frac{\vec{\partial}}{\partial \theta_i} A \right) B + (-1)^{\epsilon_A} A \frac{\vec{\partial}}{\partial \theta_i} B, \tag{1.116}$$

$$(AB) \frac{\overleftarrow{\partial}}{\partial \theta_i} = (-1)^{\epsilon_B} A \frac{\overleftarrow{\partial}}{\partial \theta_i} B + A \left(B \frac{\overleftarrow{\partial}}{\partial \theta_i} \right), \tag{1.117}$$

where θ_i is any of the generating elements ξ^i and p_i. A proof of the Jacobi identity is based on the observation that the left-hand side of Eq. (1.113) is a linear homogeneous function of the second derivatives $\partial^2/\partial\theta_i\partial\theta_j$ of A, B, and C. For this reason, it is sufficient to show that the coefficients at the second derivatives vanish. The latter is not difficult to do by making use of the rules (1.114)–(1.117) and the anticommutativity of the derivatives,

$$\frac{\partial}{\partial \theta_i} \frac{\partial}{\partial \theta_j} = -\frac{\partial}{\partial \theta_j} \frac{\partial}{\partial \theta_i}.$$

By analogy with the bosonic case, one can define canonical transformations for Grassmann variables [22, 23], construct the corresponding generating functions, and introduce the notion of a symmetry of a Grassmann system and associated (Noether's) integrals of motion. This will be done when considering systems of a general type with commuting and anticommuting degrees of freedom (see Sections 1.9 and 1.10).

Example. Consider the Hamiltonian mechanics of the system (1.103) when $i, j = 1, 2$, and $A_i = 0$. In this case, $\alpha_{ij} = m\epsilon_{ij}$, where $\epsilon_{ij} = -\epsilon_{ji}$ is the unit skew-symmetric tensor, $\epsilon_{12} = 1$, and

$$V(\xi) = V_0 + \frac{1}{2} K \epsilon_{ij} \xi^i \xi^j.$$

Since $\epsilon_{ij}\epsilon_{jk} = -\delta_{ik}$, the inverse is $\alpha^{ij} = m^{-1}\epsilon^{ij} = -m^{-1}\epsilon_{ij}$. Therefore the Hamiltonian has the form,

$$H = \frac{1}{2m}\epsilon^{ij} p_i p_j + \frac{1}{2} K \epsilon_{ij} \xi^i \xi^j + V_0.$$

With (1.107) the equations of motion are obtained:

$$\dot{p}_i = K\epsilon_{ij}\xi^j, \qquad \dot{\xi}^i = \frac{1}{m}\epsilon^{ij} p_j. \tag{1.118}$$

They have to be supplemented by initial conditions,

$$p_i(t=0) = p_{0i}, \qquad \xi^i(t=0) = \xi_0^i, \tag{1.119}$$

in order for the motion to be uniquely determined. Equations (1.118) are equivalent to the following equations:

$$\ddot{\xi}^i - \frac{K}{m}\xi^i = 0, \tag{1.120}$$

with the initial conditions $\xi^i(0) = \xi_0^i$ and $\dot{\xi}^i(0) = m^{-1}\epsilon^{ij} p_{0j}$. Equations (1.120) describe a harmonic oscillator with frequency $\omega = (-K/m)^{1/2}$, if $K/m < 0$. The corresponding solution reads,

$$\xi^i(t) = \xi_0^i \cos(\omega t) + \frac{1}{m\omega}\epsilon^{ij} p_{0j} \sin(\omega t); \tag{1.121}$$

$$p_i(t) = p_{0i} \cos(\omega t) - m\omega\epsilon_{ij}\xi_0^j \sin(\omega t). \tag{1.122}$$

As noted above, solving the Hamiltonian equations of motion for Grassmann variables is equivalent to integrating a system of differential equations for real functions (see also Section 1.9 where equations of motion on supermanifolds are discussed). Indeed, the generating elements $p_i(t)$ and $\xi^i(t)$ can be decomposed similarly to (1.90) where the generating elements ξ_0^i should be replaced by p_{0i}. Substituting this decomposition into the Hamiltonian equations of motion (1.107) and equating the coefficients at the same basis elements of the Grassmann algebra, the equations for the expansion coefficients are obtained, while the conditions (1.119) induce the corresponding initial conditions. In particular, for the model discussed the decomposition (1.90) has the form,

$$\boldsymbol{\xi}(t) = z_1(t)\boldsymbol{\xi}_0 + z_2(t)\mathbf{p}_0 + z_3(t)\boldsymbol{\xi}_0 p_{01}p_{02} + z_4(t)\mathbf{p}_0\xi_0^1\xi_0^2\,, \tag{1.123}$$

$$\mathbf{p}(t) = Z_1(t)\mathbf{p}_0 + Z_2(t)\boldsymbol{\xi}_0 + Z_3(t)\mathbf{p}_0\xi_0^1\xi_0^2 + Z_4(t)\boldsymbol{\xi}_0 p_{01}p_{02}, \tag{1.124}$$

where the boldface letters denote two–vectors, z and Z are 2×2 matrices. The initial conditions follow from (1.119),

$$z^j_{1i}(0) = Z^j_{1i}(0) = \delta^j_i \,, \qquad (1.125)$$

$$z^j_{3i}(0) = Z^j_{3i}(0) = z^{ij}_2(0) = z^{ij}_4(0) = Z_{2ij}(0) = Z_{4ij}(0) = 0, \qquad (1.126)$$

The substitution of (1.123) and (1.124) into equations (1.118) yields the equations for the matrices z and Z. Their particular solutions satisfying the initial conditions (1.125) and (1.126) are,

$$z^i_{1j}(t) = \delta^i_j \cos(\omega t), \qquad z^{ij}_2(t) = \frac{1}{m\omega}\epsilon^{ij}\sin(\omega t); \qquad (1.127)$$

$$Z^i_{1j}(t) = \delta^i_j \cos(\omega t), \qquad Z_{2ij}(t) = -m\omega\epsilon_{ij}\sin(\omega t). \qquad (1.128)$$

The other matrices vanish. Obviously, the solutions (1.127) and (1.128) are equivalent to (1.121) and (1.122).

Thus, equations of motion in theories with Grassmann degrees of freedom can be formulated and solved either in the algebraic form (like (1.118), (1.121), and (1.122)) or in the c-number form (for the expansion coefficients (1.127) and (1.128)).

1.9 Hamiltonian dynamics on supermanifolds

Consider a system which contains both bosonic and fermionic degrees of freedom. As has been shown, a dynamical description of systems with commuting and anticommuting variables can be done on equal footing in the framework of the first-order formalism. Therefore, from the very beginning the Hamiltonian formalism may be used to describe the evolution of such a system. With this purpose, consider an even-dimensional Euclidean superspace Γ spanned by $2n$ commuting variables, being generalized canonical coordinates and momenta of the bosonic degrees of freedom, and by $2n'$ anticommuting variables, being generalized canonical coordinates and momenta of the fermionic degrees of freedom. Let θ^a, where $a = 1, 2, \ldots, 2(n + n')$, denote canonical coordinates in Γ, and the first $2n$ values of the index a correspond to the bosonic degrees of freedom, while the remaining $2n'$ values correspond to the Grassmann ones. The parities of the bosonic and fermionic variables are equal to 0 and 1, respectively. The left and right derivatives, $\vec{\partial}/\partial\theta^a$ and $\overleftarrow{\partial}/\partial\theta^a$, coincide with the ordinary derivative for commuting variables, and are defined as usual for the Grassmann ones.

Functions on Γ are elements of the Grassmann algebra generated by the anticommuting elements θ^a, i.e. they are polynomials in the anticommuting (fermionic) variables θ^a with coefficients being functions of the commuting (bosonic) variables θ^a. Apparently, each function can be uniquely decomposed into the sum of odd and even functions, and parity can be determined in the same way as in the Grassmann algebra: the product of either two even or odd functions is even, while the product of odd and even functions is odd. Hence, the parities ϵ^a of the generating elements determine the parity of any function on Γ. For functions which have a specific parity, the commutation rule is then defined by (1.114). Hamiltonian dynamics of any system is defined by the phase space, the Poisson bracket, the Hamiltonian as a function on the phase space. Then the evolution of any function F on the phase space is determined by the Hamiltonian equation of motion (1.32). The Hamiltonian is chosen to be an even function on Γ, $H = H(\theta)$. The Poisson bracket on Γ is, by definition, a natural generalization of (1.33) and (1.110), namely,

$$\{A, B\} = A\frac{\overleftarrow{\partial}}{\partial\theta^a}\omega^{ab}\frac{\vec{\partial}}{\partial\theta^b}B, \tag{1.129}$$

where the matrix ω^{ab} is skew-symmetric for bosonic variables and symmetric for the fermionic ones, and has the standard block form,

$$\left(\omega^{ab}\right) = \left(\overset{\circ}{\omega}{}^{ab}\right) \equiv \begin{pmatrix} 0 & \delta_{ij} & 0 & 0 \\ -\delta_{ij} & 0 & 0 & 0 \\ 0 & 0 & 0 & \delta_{i'j'} \\ 0 & 0 & \delta_{i'j'} & 0 \end{pmatrix}, \tag{1.130}$$

where δ_{ij} and $\delta_{i'j'}$ denote the unit matrices, $i, j = 1, 2, \ldots, n$ and $i', j' = 1, 2, \ldots, n'$ in the bosonic and fermionic sectors, respectively, while the zeros stand for the corresponding zero block-matrices. Clearly, the Poisson bracket (1.129) is nothing but the sum of the bosonic and fermionic Poisson brackets. Therefore the time evolution of any function F on Γ is described by Eq. (1.109) with the Poisson bracket (1.129). Thus, the Hamiltonian dynamics is defined.

For Γ being a generic supermanifold [27, 28], the symplectic structure can be introduced as follows. Every point of Γ has a neighborhood U homeomorphic to the Euclidean superspace of the same dimension. Therefore in every neighborhood $U \subset \Gamma$, one can introduce local coordinates (generating elements) θ^a as generating elements of the Euclidean superspace. A supermanifold Γ is then covered by a set of neighborhoods (charts) with the transition (parity-preserving) maps defined on their intersections (the relations between local coordinates in each chart at the points of intersection of the charts; see also Section 1.10 for a detailed algebraic description of parity-preserving coordinate transformations in a superspace). The Poisson bracket in each chart U is still defined by (1.129), but the matrix ω^{ab} now may depend on the local coordinates θ^a,

$$\omega^{ab} = \omega^{ab}(\theta), \quad \theta \in U. \tag{1.131}$$

The matrix ω^{ab} is assumed to be non-singular, i.e. it has the inverse for all $\theta \in U$. The Poisson bracket (either for bosonic or fermionic systems) has three characteristic properties: it satisfies the commutation rule, the Leibniz rule, and the Jacobi identity (compare (1.34)–(1.36) and (1.111)–(1.113)). These three properties are postulated for the Poisson bracket (1.129) defined on a supermanifold Γ. A generic matrix in Eq. (1.131) would not satisfy these three identities. However, Eq. (1.131) defines the Poisson bracket with its characteristic properties (the commutation and Leibniz rules, and the Jacobi identity) if the matrix ω^{ab} obeys the following relations [29]:

$$\epsilon(\omega^{ab}) = (\epsilon^a + \epsilon^b) \bmod 2, \tag{1.132}$$

$$\omega^{ab} = -(-1)^{\epsilon^a\epsilon^b}\omega^{ba}, \tag{1.133}$$

$$\omega^{ad}\frac{\vec{\partial}}{\partial\theta^d}\omega^{bc}(-1)^{\epsilon^a\epsilon^c} + \text{cycle}(a,b,c) = 0, \tag{1.134}$$

where a, b, and c are fixed, whereas cycle(a,b,c) denotes the sum over cyclic permutations of the indices a, b, and c.

To prove that the characteristic properties hold under these conditions on the matrix ω^{ab}, the following properties of the derivative are invoked:

$$\begin{aligned}\frac{\vec{\partial}}{\partial\theta^a}A &= (-1)^{(\epsilon_A+\epsilon^a)\epsilon^a}A\frac{\overleftarrow{\partial}}{\partial\theta^a},\\ \epsilon(\partial_a A) &= (\epsilon^a + \epsilon_A) \bmod 2,\end{aligned} \tag{1.135}$$

which are a trivial generalization of the corresponding relations in the pure Grassmann case. With the commutation rule (1.114), the derivatives $\vec{\partial}_b B$ and $A\overleftarrow{\partial}_a$ can be moved over to the left and to the right, respectively, in the right-hand side of Eq. (1.129). Then the left (right) derivative of B (A) is replaced by the right (left) derivative according to the rule (1.135). The commutation rule (1.111) for the Poisson bracket (1.129) follows from (1.133). The Leibniz rule is deduced directly from the product rule for the left and right derivatives,

$$\begin{aligned}\vec{\partial}_b(BC) &= (\vec{\partial}_b B)C + (-1)^{\epsilon_B\epsilon^b}B\vec{\partial}_b C,\\ (BC)\overleftarrow{\partial}_b &= B(C\overleftarrow{\partial}_b) + (-1)^{\epsilon_C\epsilon^b}B\overleftarrow{\partial}_b C.\end{aligned}$$

The proof of the Jacobi identity is based on Eq. (1.134) which the matrix ω^{ab} is required to satisfy. The technicalities are similar to those in the pure bosonic and fermionic cases, i.e. one should show that the coefficients at the second derivatives $\partial^2 A/\partial\theta^a\partial\theta^b$, $\partial^2 B/\partial\theta^a\partial\theta^b$, and $\partial^2 C/\partial\theta^a\partial\theta^b$ vanish; the corresponding rule of changing the order of differentiation is written as,

$$\frac{\partial}{\partial\theta^a}\frac{\partial}{\partial\theta^b} = (-1)^{\epsilon^a\epsilon^b}\frac{\partial}{\partial\theta^b}\frac{\partial}{\partial\theta^a}.$$

The difference is that, in addition to the linear combination of the second derivatives, there are terms containing the derivatives of ω^{ab}. The latter are cancelled out thanks to condition (1.134).

Thus, the Hamiltonian dynamics is defined on a supermanifold Γ. The equations of motion have the canonical form,

$$\dot{\theta}^a = \{\theta^a, H\} = \omega^{ab} \frac{\vec{\partial}}{\partial \theta^b} H. \tag{1.136}$$

The matrix ω^{ab} is called a symplectic structure on Γ. The (super)manifold Γ equipped with a symplectic structure is called a symplectic (super)manifold. Solutions of the equations of motion (1.136) are elements $\theta^a = \theta^a(t, \theta_0^a)$ of the *superalgebra* generated by the initial variables $\theta^a(t = t_0) = \theta_0^a$, with the fixed parities $\epsilon(\theta^a) = \epsilon(\theta_0^a) = \epsilon^a$, i.e. the time evolution can be viewed as a transformation from one set of generating elements to another.

By definition, the time evolution of bosonic and fermionic degrees of freedom is described, respectively, by even and odd functions of generating elements of the superalgebra. This means, in particular, that the initial values of bosonic variables are not necessarily to be chosen as real c-numbers. Such a choice would correspond to a particular subset in all possible solutions of the equations of motion and would also disagree with the theory of canonical transformations. The point being that under canonical transformations the Hamiltonian equations of motion (1.136) retain their form, while the new bosonic canonical variables (and, of course, their initial values) become even functions of the old bosonic and fermionic canonical variables (see below (1.174)–(1.176) in Section 1.10). Since a priori there is no restriction on the choice of local coordinates on Γ when solving the equations of motion, the initial conditions must be generic elements of the superalgebra with specific parities. The same interpretation of the classical dynamics also follows from quantum theory in a superspace (see Section 5.7.1).

Equations of motion (1.136) can be obtained from the principle of stationary action. If the symplectic structure has the canonical form (1.130), i.e. the phase space is a Euclidean superspace, then Hamiltonian equations of motion (1.136) define extrema of the action,

$$S = \int_{t_1}^{t_2} dt \left(\frac{1}{2} \overset{\circ}{\omega}_{ab} \theta^a \dot{\theta}^b - H(\theta) \right), \tag{1.137}$$

where $\overset{\circ}{\omega}_{ab}$ is the inverse of $\overset{\circ}{\omega}{}^{ab}$, $\overset{\circ}{\omega}{}^{ab} \overset{\circ}{\omega}_{bc} = \delta_c^a$. However, if $\omega^{ab} \neq \overset{\circ}{\omega}{}^{ab}$, then the extrema of the functional (1.137), where $\overset{\circ}{\omega}_{ab}$ is replaced by ω_{ab} which is the inverse of ω^{ab}

$$\omega^{dd'} \omega_{d'c} = \delta_c^d, \tag{1.138}$$

no longer satisfy Eqs. (1.136).

Before finding the correct action for the variational principle in the general case, some algebraic and geometric properties of the symplectic structure should be studied. The following consequences of relations (1.132)–(1.134) can be established:

$$\epsilon(\omega_{ab}) = (\epsilon^a + \epsilon^b) \bmod 2, \tag{1.139}$$

$$\omega_{ab} = \omega_{ba}(-1)^{(\epsilon^a+1)(\epsilon^b+1)}, \tag{1.140}$$

$$\vec{\partial}_a \omega_{bc} (-1)^{(\epsilon^a+1)\epsilon^c} + \text{cycle}(a, b, c) = 0. \tag{1.141}$$

Indeed, equalities (1.139) and (1.140) are a trivial consequence of the definition (1.138) of the covariant matrix elements ω_{ab}. To prove (1.141), one should differentiate equality (1.138) with respect to θ^a, multiply the resulting equality by ω_{bd}, and make the cyclic permutation of the indices a, b, and c. As a result, one infers,

$$(-1)^{(\epsilon^a+1)\epsilon^c}\omega_{bd}\vec{\partial}_a\omega^{dd'}\omega_{d'c}+\vec{\partial}_a\omega_{bc}(-1)^{(\epsilon^a+1)\epsilon^c}+\text{cycle}(a,b,c)=0. \tag{1.142}$$

The sum of three terms in the cyclic permutation of the indices in the first term of (1.142) vanishes thanks to the identity (1.134). This could easily be verified by solving (1.134), for example, for $\vec{\partial}_a\omega^{dd'}$ and substituting the solution into one of the terms of the cyclic permutation sum in (1.142). Thus, relations (1.132)–(1.134) imply relations (1.139)–(1.141). The converse is also true. Differentiating (1.138), multiplying the result by the inverse matrices $\omega^{bb'}$ to raise the indices (a,b,c), and carrying out the cyclic permutation for them, the following relations are established:

$$\omega^{ad}\vec{\partial}_d\omega^{bc}(-1)^{\epsilon^a\epsilon^c}+(-1)^{\epsilon^a\epsilon^c+\epsilon^b\epsilon^{b'}}\omega^{aa'}\omega^{cc'}\omega^{bb'}\partial_{c'}\omega_{a'b'}+\text{cycle}(a,b,c)=0. \tag{1.143}$$

A simple algebraic transformation shows that the sum generated by the cyclic permutation in the second term of (1.143) is proportional to the left-hand side of Eq. (1.141) if all the indices in the latter are replaced by the primed ones. Thereby Eq. (1.143) implies (1.134).

Equality (1.141) has a simple geometrical interpretation. The non-degenerate two-form $\omega=\omega_{ab}d\theta^a\wedge d\theta^b$ is closed on the supermanifold Γ,

$$d\omega=0. \tag{1.144}$$

Equality (1.141) follows immediately from Eq. (1.144) with the permutation rule in the basis of two-forms,

$$d\theta^a\wedge d\theta^b=(-1)^{(\epsilon^a+1)(\epsilon^b+1)}d\theta^b\wedge d\theta^a; \tag{1.145}$$

the latter is a consequence of the definition of ω and the property (1.140). Consider any two-form on Γ that satisfies the conditions (1.144) and (1.145). Then its coefficients should satisfy the conditions (1.139)–(1.141) (or (1.132)–(1.134)). Thus, with every closed non-degenerate two-form on Γ one can associate a symplectic structure on Γ (the Poisson bracket) and, hence, define a Hamiltonian dynamics [30].

To formulate the principle of stationary action for the Hamiltonian equations of motion (1.136), consider the matrix $\bar{\omega}_{ab}$ that is constructed from the matrix elements of the symplectic structure by the rule [29],

$$\bar{\omega}_{ab}(\theta)=(\theta^c\vec{\partial}_c+2)^{-1}\omega_{ab}(\theta)=\int_0^1\omega_{ab}(\alpha\theta)\alpha d\alpha. \tag{1.146}$$

It satisfies the equation

$$(\theta^c\vec{\partial}_c+2)\bar{\omega}_{ab}(\theta)=\omega_{ab}(\theta) \tag{1.147}$$

and the identities (1.139)–(1.141). The Hamiltonian equations of motion (1.136) define the extrema of the action,

$$S = \int_{t_1}^{t_2} dt \left(\theta^a \bar{\omega}_{ab} \dot{\theta}^b - H(\theta) \right). \tag{1.148}$$

This statement is proved by a straightforward calculation. By varying the action functional (1.148), one finds,

$$\delta S = \int_{t_1}^{t_2} dt \delta\theta^a \left[2\bar{\omega}_{ab}\dot{\theta}^b - (-1)^{\epsilon^b(\epsilon^c+1)}\theta^c \left((-1)^{(\epsilon^a+1)\epsilon^c} \bar{\omega}_{bc,\vec{a}} \right.\right.$$
$$\left.\left. + (-1)^{(\epsilon^b+1)\epsilon^a} \bar{\omega}_{ca,\vec{b}} \right) \dot{\theta}^b - H_{,\vec{a}} \right], \tag{1.149}$$

where the integration by parts has been done for the term with $\delta\dot{\theta}^a$. The boundary term vanishes if $\delta\theta^a(t_1) = \delta\theta^a(t_2) = 0$. All the variations $\delta\theta^a(t)$, $t_1 < t < t_2$, are regarded as generating elements of the superalgebra with the same parity as θ^a, and they are moved over to the very left according to their commutation rule with the matrix elements $\bar{\omega}_{ab}$. The expression in the parenthesis in δS can be replaced by $-(-1)^{\epsilon^b(\epsilon^c+1)}\bar{\omega}_{ab,\vec{c}}$ because the matrix $\bar{\omega}_{ab}$ satisfies the identity (1.141). As a result, the extrema of the action are found to be determined by the condition,

$$\begin{aligned} \delta S &= \int dt \delta\theta^a \left[(\theta^c \vec{\partial}_c + 2)\bar{\omega}_{ab}\dot{\theta}^b - H_{,\vec{a}} \right] \\ &= \int dt \delta\theta^a \left[\omega_{ab}\dot{\theta}^b - H_{,\vec{a}} \right] = 0, \end{aligned} \tag{1.150}$$

which leads to the Hamiltonian equations of motion (1.136). In the simplest case when $\omega_{ab} = \overset{\circ}{\omega}_{ab}$, one has $\bar{\omega}_{ab} = (1/2)\, \overset{\circ}{\omega}_{ab}$, and the action (1.148) turns into (1.137). Note that the variation (1.150) implies that the end points of the trajectory are viewed as parameters of the action (1.149) because the condition $\delta\theta^a(t_1) = \delta\theta^a(t_a) = 0$ has been used, while the intermediate points $\theta^a(t)$, $t_1 < t < t_2$, (the open interval) are the variables for which the necessary condition for extrema is sought. In this sense, Eqs. (1.136) provide a weaker condition for the extrema of the action (1.149) as no variation (minimization) with respect to the end points is done. The actual extrema are then found by varying $\theta^a(t_1)$, or $\theta^a(t_2)$, or both (i.e. by restricting $\theta^a(t_{1,2})$) so that Eqs. (1.136) have a solution. In this way the covariance of Hamilton's variational principle under general coordinate transformations in the phase space is achieved.

1.10 Canonical transformations on symplectic supermanifolds

Consider a neighborhood $U \subset \Gamma$ of a point θ in Γ. In the Hamiltonian mechanics, there is a theorem due to Darboux [3] which asserts that there exists a coordinate chart for U such that the symplectic two-form has the canonical form $\omega = \overset{\circ}{\omega}$

(or $\omega^{ab} = \overset{\circ}{\omega}{}^{ab}$). Such a chart is called the Darboux chart around the point θ. In other words, there are coordinates, called the Darboux coordinates,

$$\Theta^a = \Theta^a(\theta) \tag{1.151}$$

such that

$$\tilde{\omega}^{ab}(\Theta) = \Theta^a \frac{\overleftarrow{\partial}}{\partial\theta^{a'}} \omega^{a'b'}(\theta) \frac{\vec{\partial}}{\partial\theta^{b'}} \Theta^b = \overset{\circ}{\omega}{}^{ab}. \tag{1.152}$$

Let local coordinates θ^a in a chart U of the symplectic manifold Γ be the Darboux coordinates so that the symplectic structure has the canonical form. A change of variables $\Theta^a = \Theta^a(\theta)$ is said to be a canonical transformation if it preserves the canonical symplectic structure $\overset{\circ}{\omega}{}^{ab}$. It is easy to see that this requirement is equivalent to the following algebraic condition on the change of variables:

$$\overset{\circ}{\omega}{}^{ab} = \{\Theta^a, \Theta^b\}, \tag{1.153}$$

which is a generalization of (1.40) to the case of a generic symplectic supermanifold. Indeed, the symplectic structure in the new variables coincides with the Poisson bracket $\{\Theta^a, \Theta^b\}$ according to the transformation law of the coefficients of the two-form (1.152). Thanks to the property (1.153), the Poisson bracket remains invariant under canonical transformations,

$$\begin{aligned}\{A, B\} &= A\frac{\overleftarrow{\partial}}{\partial\theta^a} \overset{\circ}{\omega}{}^{ab} \frac{\vec{\partial}}{\partial\theta^b} B \\ &= \left(A\frac{\overleftarrow{\partial}}{\partial\Theta^a}\right)\left(\Theta^a \frac{\overleftarrow{\partial}}{\partial\theta^{a'}}\right) \overset{\circ}{\omega}{}^{a'b'} \left(\frac{\vec{\partial}}{\partial\theta^{b'}}\Theta^b\right)\left(\frac{\vec{\partial}}{\partial\Theta^b}B\right) = A\frac{\overleftarrow{\partial}}{\partial\Theta^a} \overset{\circ}{\omega}{}^{ab} \frac{\vec{\partial}}{\partial\Theta^b} B,\end{aligned}$$

in accordance with the composition rule for the left and right derivatives,

$$\frac{\vec{\partial}}{\partial\theta^a} A = \left(\frac{\vec{\partial}}{\partial\theta^a}\Theta^{a'}\right)\left(\frac{\vec{\partial}}{\partial\Theta^{a'}}A\right), \quad A\frac{\overleftarrow{\partial}}{\partial\theta^a} = \left(A\frac{\overleftarrow{\partial}}{\partial\Theta^{a'}}\right)\left(\Theta^{a'}\frac{\overleftarrow{\partial}}{\partial\theta^a}\right).$$

As a consequence of the invariance of the Poisson bracket, the Hamiltonian equations of motion (1.136) are also invariant under canonical transformations.

Let a set of neighborhoods U_i be a cover of the manifold Γ, i.e. $\Gamma = \cup_i U_i$, and let the local coordinates in each chart be the Darboux coordinates. The transition maps on the intersections $U_i \cap U_j$ should be canonical transformations because the symplectic structure has the canonical form in each neighborhood. Canonical transformations in each U_i together with the transition maps define a canonical transformation in the whole Γ.

Similarly to the bosonic case, one can define generating functions of the canonical transformation (1.151). Since the equations of motion retain their form under canonical transformations, the action in the new variables can differ from (1.137) only by a boundary term, i.e. by the integral of the total time derivative of some function F,

$$S = \int_{t_1}^{t_2} dt \left(\frac{1}{2} \Theta^a \overset{\circ}{\omega}_{ab} \dot{\Theta}^b - \tilde{H}(\Theta) + \frac{d}{dt} F(\theta, \Theta, t) \right), \tag{1.154}$$

where $\tilde{H} = H(\theta(\Theta))$. The function F can only depend on $2(n+n')$ variables because the old and new variables are related by $2(n+n')$ conditions (1.151) (n and n' are the numbers of bosonic and fermionic degrees of freedom, respectively). So any $2(n+n')$ of the $4(n+n')$ variables Θ^a and θ^a can be chosen as independent arguments of F. In general, given a function F, the problem of finding the functions (1.151) is equivalent to solving a system of differential equations. Thanks to the specific form of the canonical symplectic structure $\omega = \overset{\circ}{\omega}$, there are particular choices of the independent arguments of F for which this problem is algebraic, similarly to the pure bosonic case studied in Section 1.5.

Consider these cases. The action (1.154) can be written in the following form:

$$S = \int_{t_1}^{t_2} dt \left(\Theta^a \overset{\circ}{\omega}_{++ab} \dot{\Theta}^b - \tilde{H} + \frac{dF}{dt} \right), \tag{1.155}$$

where $\overset{\circ}{\omega}_{++ab}$ is the lower triangular part of the matrix $\overset{\circ}{\omega}_{ab}$,

$$\overset{\circ}{\omega}_{++ab} = \begin{pmatrix} 0 & 0 & 0 & 0 \\ \delta_{ij} & 0 & 0 & 0 \\ 0 & 0 & 0 & 0 \\ 0 & 0 & \delta_{i'j'} & 0 \end{pmatrix}. \tag{1.156}$$

The action (1.155) is obtained from (1.154) by integration by parts in terms containing the elements of the upper triangular part of the matrix $\overset{\circ}{\omega}_{ab}$, which is equivalent to changing the function F. The matrix (1.156) is then represented as the sum of its bosonic ($a, b = i, j$) and fermionic ($a, b = i', j'$) parts: $\overset{\circ}{\omega}_{++ab} = \overset{\circ}{\omega}{}^B_{+ab} + \overset{\circ}{\omega}{}^F_{+ab}$, i.e. $\overset{\circ}{\omega}{}^B_{+i'j'} = 0$ and $\overset{\circ}{\omega}{}^F_{+ij} = 0$. In the matrix notations, $(\overset{\circ}{\omega}{}^{-1})_{++} = (\overset{\circ}{\omega}{}^{-1})^B_+ + (\overset{\circ}{\omega}{}^{-1})^F_+$ where, as before, the contravariant components $\overset{\circ}{\omega}{}^{ab}$ are elements of the inverse matrix $\overset{\circ}{\omega}{}^{-1}$. The column Θ is composed of the columns Q, P, and Ψ, P^ψ (from top to bottom) where Q and P are the columns of new bosonic canonical coordinates and momenta, respectively, and similarly, Ψ and P^ψ are the columns of new fermionic canonical variables. Let the function F in (1.155) depend on one of the following 16 sets of arguments:

$$F_{bf} = F\left((\overset{\circ}{\omega}{}^{-1})_{\pm\pm}\Theta, (\overset{\circ}{\omega}{}^{-1})_{\pm\pm}\theta, t \right), \tag{1.157}$$
$$(\overset{\circ}{\omega}{}^{-1})_{\pm\pm} = (\overset{\circ}{\omega}{}^{-1})^B_\pm + (\overset{\circ}{\omega}{}^{-1})^F_\pm,$$

where $(\overset{\circ}{\omega}{}^{-1})^B_- = -(\overset{\circ}{\omega}{}^{-1T})^B_+$ and $(\overset{\circ}{\omega}{}^{-1})^F_- = (\overset{\circ}{\omega}{}^{-1T})^F_+$; the index $b = 1, 2, 3, 4$ in F_{bf} indicates the choice of bosonic arguments of F, while the index $f = 1, 2, 3, 4$

does so for fermionic arguments. The pair b and f fixes a particular sequence of the plus and minus signs in the right-hand side of Eq. (1.157). For example,

$$F_{11} = F(Q, q, \Psi, \psi, t), \tag{1.158}$$

which corresponds to the sequence of all (four) plus signs in the right-hand side of Eq. (1.157). From comparison of actions (1.155) and (1.137), with the latter written via $(\overset{\circ}{\omega}{}^{-1})_{++}$ just as (1.155), it follows that,

$$p_i = \frac{\partial}{\partial q^i} F_{11}, \quad p^{\psi}_{i'} = -\frac{\vec{\partial}}{\partial \psi^{i'}} F_{11}, \tag{1.159}$$

$$P_i = -\frac{\partial}{\partial Q^i} F_{11}, \quad P^{\psi}_{i'} = \frac{\vec{\partial}}{\partial \Psi^{i'}} F_{11}, \tag{1.160}$$

$$\tilde{H} = H + \frac{\partial}{\partial t} F_{11}, \tag{1.161}$$

in a full analogy with (1.50)–(1.52). The sign difference in the definitions of the bosonic and fermionic canonical momenta is associated with the anticommutativity of variables ψ and Ψ. Relations (1.159) furnish Q^i and $\Psi^{i'}$ as functions of the old canonical variables. Substituting these into (1.160), momenta P_i and $P^{\psi}_{i'}$ are obtained as functions of q, p, ψ, and p^{ψ}.

As in the pure bosonic case, one can show that all the functions (1.157) are related to one another by suitable Legendre transformations. There are peculiarities in solving Eqs. (1.159) for Q and Ψ because of the nilpotency of the Grassmann variables. Suppose first that there is just one Grassmann degree of freedom. In this case, function (1.158) should have the following representation:

$$F = F_0(q, Q, t) + F_1(Q, q, t)\psi\Psi, \tag{1.162}$$

because $\psi^2 = \Psi^2 = 0$. Then Eq. (1.159) implies that

$$p_i = \frac{\partial F_0}{\partial q^i} + \frac{\partial F_1}{\partial q^i}\psi\Psi, \quad p^{\psi} = -F_1\Psi. \tag{1.163}$$

The second equality in (1.163) defined Ψ as a function of p^{ψ}, q, and Q. Its substitution into the first equality in (1.163) yields an equation for the new bosonic coordinates Q^i. As Q^i are even elements of the Grassmann algebra, they should have the following decomposition in the Grassmann algebra basis:

$$Q^i = Q^i(q, p, \psi, p^{\psi}, t) = Q^i_0(q, p, t) + Q^i_1(q, p, t)\psi p^{\psi}, \tag{1.164}$$

i.e. two bosonic functions $Q^i_{1,0}$ are to be found. Substituting (1.164) into (1.163) one infers that,

$$p_i = \left.\frac{\partial F_0}{\partial q^i}\right|_{Q=Q_0} + \left.\frac{\partial^2 F_0}{\partial q^i \partial Q^j}\right|_{Q=Q_0} Q^j_1 \psi p^{\psi} - \left.\frac{\partial F_1}{\partial q^i}\frac{1}{F_1}\right|_{Q=Q_0} \psi p^{\psi}. \tag{1.165}$$

In the Grassmann algebra, this equality is equivalent to two relations,

$$p_i = \left.\frac{\partial F_0}{\partial q^i}\right|_{Q=Q_0}, \tag{1.166}$$

$$\left.\frac{\partial^2 F_0}{\partial q^i \partial Q^j}\right|_{Q=Q_0} Q_1^j = \left.\frac{\partial F_1}{\partial q^i}\frac{1}{F_1}\right|_{Q=Q_0}, \tag{1.167}$$

the first of which defines the functions $Q_0^j = Q_0^j(q,p,t)$, whereas the second one is a linear equation for Q_1^j. The new canonical momenta are found from (1.160) by substituting into them the explicit functions (1.164) defined by Eqs. (1.166) and (1.167),

$$P_i = -\left.\frac{\partial F_0}{\partial Q^i}\right|_{Q=Q_0} - \left(\left.\frac{\partial^2 F_0}{\partial Q^i \partial Q^j}\right|_{Q=Q_0} Q_1^j - \left.\frac{\partial F_1}{\partial Q^i}\frac{1}{F_1}\right|_{Q=Q_0}\right)\psi p^\psi, \tag{1.168}$$

$$P^\psi = -F_1(Q_0, q, t)\psi.$$

In a general case, the generating function (1.158), as an even element, is expanded over the standard basis in the Grassmann algebra with $2n'$ generating elements being $\psi^{i'}$ and $\Psi^{i'}$. After the differentiation, $\Psi^{i'}$ in the second equation in (1.159) should be replaced by its expansion, as an odd element, over the standard basis in the Grassmann algebra with generating elements $\psi^{i'}$ and $p^\psi_{i'}$. Equating the coefficients at the same basis elements in the resulting equality, a system of equations for the expansion coefficients of $\Psi^{i'}$ is obtained. Its solution furnishes the functions $\Psi^{i'} = \Psi^{i'}(Q, q, \psi, p^\psi, t)$. These functions are then substituted into the first equation in (1.159) in order to find Q^i. The resulting equation is solved in the same fashion by expanding Q^i, as an even element, over the standard basis in the Grassmann algebra with generating elements $\psi^{i'}$ and $p^\psi_{i'}$ and equating the coefficients at the same basis elements, which yields a system of equations for the expansion coefficients of Q^i. The substitution of the explicit functions $Q^i = Q^i(p, q, \psi, p^\psi, t)$ into $\Psi^{i'} = \Psi^{i'}(Q, q, \psi, p^\psi, t)$ gives $\Psi^{i'}$ as functions of the old canonical variables and time. Finally, the substitution of Q^i and $\Psi^{i'}$ into (1.160) produces the new canonical momenta as functions of the old canonical variables. A similar procedure applies to any of the sixteen generating functions (1.157); the generating elements of the Grassmann algebra have to be changed according to the choice of the arguments in the generating function.

As an example, consider a canonical transformation associated with a change of variables in the configuration superspace,

$$Q^i = Q^i(q,\psi), \quad \Psi^{i'} = \Psi^{i'}(q,\psi), \quad i = 1, 2, \ldots, n, \quad i' = 1, 2, \ldots, n'. \tag{1.169}$$

The functions Q^i and $\Psi^{i'}$ are, respectively, generic even and odd elements of the Grassmann algebra generated by $\psi^{i'}$. For each i and i', they contain $2^{n'-1}$ functions of q which are the coefficients in the expansion of Q^i and $\Psi^{i'}$ over the standard basis in the Grassmann algebra, because the dimensions of the even and odd subspaces coincide and equal $2^{n'-1}$. Thus, the change of variables in the superspace (1.169) is determined by $(n+n')2^{n'-1}$ scalar functions of q. To be a genuine change of variables, relations (1.169) must be invertible, i.e. they must admit the solution for q^i and $\psi^{i'}$,

$$q^i = q^i(Q,\Psi), \qquad \psi^{i'} = \psi^{i'}(Q,\Psi).$$

To find its explicit form, one should expand functions q^i and $\psi^{i'}$ over the even and odd basis elements, respectively, in the Grassmann algebra generated by $\Psi^{i'}$ and substitute these decompositions into relations (1.169) to obtain a system of linear equations for the decomposition coefficients of q^i and $\psi^{i'}$. Let Θ^a and θ^a denote the whole collections of the new and old variables, respectively. Consider the matrix $J_a^b = \vec{\partial}\Theta^b/\partial\theta^a$ which is the Jacobian of the change of variables. A solution to the system of equations which furnishes the explicit form of the inverse functions q^i and $\psi^{i'}$ is shown to exist if the superdeterminant of the matrix J, also known as the Berezian (see Section 8.2.2), does not vanish. This implies, in particular, that the Jacobian of the bosonic mapping $Q^i = Q^i(q,0)$ is non-singular, and that the numbers of old and new Grassmann variables must coincide, which demands that the matrix $\vec{\partial}\Psi^{i'}/\partial\psi^{j'}$ at $\psi = 0$ be non-singular, as well.

It is convenient to choose the function F_{--++} as the generating function of the canonical transformation associated with the change of variables (1.169). It is related to F_{11} by the Legendre transform similarly to (1.53),

$$F_{22}(q,P,\psi,P^\psi,t) = P_iQ^i + P^\psi_{i'}\Psi^{i'} + F_{11}(Q,q,\Psi,\psi,t)\,.$$

Therefore the canonical transformation is defined by the following relations:

$$p_i = \frac{\partial}{\partial q^i}F_{22}, \qquad p^\psi_{i'} = -\frac{\vec{\partial}}{\partial\psi^{i'}}F_{22}\,; \tag{1.170}$$

$$Q^i = \frac{\partial}{\partial P_i}F_{22}, \qquad \Psi^{i'} = \frac{\vec{\partial}}{\partial P^\psi_{i'}}F_{22}\,. \tag{1.171}$$

The Hamiltonian in the new variables has the form (1.161) where the generating function is to be changed accordingly. Equalities (1.171) should be equivalent to the change of variables in superspace (1.169) and, therefore, similarly to the bosonic case (1.59), one has,

$$F_{22} = P_iQ^i(q,\psi) + P^\psi_{i'}\Psi^{i'}(q,\psi),$$

for a point transformation in the superspace. Then the relations,

$$p_i = P_j\frac{\partial Q^j}{\partial q^i} + P^\psi_{j'}\frac{\partial\Psi^{j'}}{\partial q^i}, \qquad p^\psi_{i'} = -P_j\frac{\vec{\partial}Q^j}{\partial\psi^{i'}} + P^\psi_{j'}\frac{\vec{\partial}\Psi^{j'}}{\partial\psi^{i'}}, \tag{1.172}$$

which follow from (1.170), determine the new canonical momenta. Equations (1.172) can be written in the matrix form,

$$\begin{pmatrix} p \\ p^\psi \end{pmatrix} = \begin{pmatrix} \frac{\partial Q}{\partial q} & -\frac{\partial\Psi}{\partial q} \\ -\frac{\vec{\partial}Q}{\partial\psi} & -\frac{\Psi\overleftarrow{\partial}}{\partial\psi} \end{pmatrix}\begin{pmatrix} P \\ P^\psi \end{pmatrix}. \tag{1.173}$$

The diagonal blocks of the matrix in (1.173) are even elements of the Grassmann algebra, whereas the off-diagonal ones are odd. The determinant of such a matrix is called the superdeterminant or Berezian. If the superdeterminant does not vanish, then there exists the inverse matrix (see Section 8.2). By multiplying both sides of Eq. (1.173) by this, new canonical momenta P and P^{ψ} are obtained as functions of the old canonical variables p, q, p^{ψ}, and ψ.

To illustrate a point transformation in a superspace, consider the following generalization of the polar coordinates (1.44) to the superspace case [31]:

$$\mathbf{x} \equiv \begin{pmatrix} x^1 \\ x^2 \end{pmatrix} = \exp(T\phi) \begin{pmatrix} r \\ 0 \end{pmatrix}, \quad \boldsymbol{\psi} \equiv \begin{pmatrix} \psi^1 \\ \psi^2 \end{pmatrix} = \exp(T\phi) \begin{pmatrix} \xi^1 \\ \xi^2 \end{pmatrix} \equiv e^{T\phi}\boldsymbol{\xi}, \tag{1.174}$$

where the 2×2 matrix T is a generator of $SO(2)$ rotations in the plane (see (1.45)). The relation between the old and new bosonic variables (the rectangular coordinates x^i and the polar coordinates r and ϕ) coincides with (1.44). A peculiarity of the change of variables (1.174) is that the relation between the Grassmann variables depends on the new bosonic variable, the polar angle ϕ. For this reason the quantity $(\mathbf{p}, T\mathbf{x})$ is no longer the momentum canonically conjugated to ϕ because $\{(\mathbf{p}, T\mathbf{x}), \boldsymbol{\xi}\} = \{(\mathbf{p}, T\mathbf{x}), e^{-T\phi}\boldsymbol{\psi}\} = T\boldsymbol{\xi} \neq 0$.

The solution of Eq. (1.173) for the point transformation (1.174) defines the new canonical momenta,

$$\begin{pmatrix} p_1^{\xi} \\ p_2^{\xi} \end{pmatrix} = \exp(-T\phi) \begin{pmatrix} p_1^{\psi} \\ p_2^{\psi} \end{pmatrix}, \tag{1.175}$$

$$p_r = \frac{(\mathbf{x}, \mathbf{p})}{r}, \qquad p_\phi = (\mathbf{p}, T\mathbf{x}) + (\mathbf{p}^{\psi}, T\boldsymbol{\psi}), \tag{1.176}$$

where $(\mathbf{p}^{\psi}, T\boldsymbol{\psi}) = p_i^{\psi} T_{ij} \psi_j$. The Grassmann momenta are related in the same way as the Grassmann coordinates (compare the second equality in (1.174) and (1.175)). Note that the momentum p_ϕ depends also on the Grassmann variables. Although the equality $\{\phi, p_\phi\} = 1$ is furnished only by the first term in p_ϕ, the second term is needed to ensure that $\{p_\phi, \boldsymbol{\xi}\} = \{(\mathbf{p}, T\mathbf{x}) + (\mathbf{p}^{\psi}, T\boldsymbol{\psi}), e^{-T\phi}\boldsymbol{\psi}\} = Te^{-T\phi}\boldsymbol{\psi} - Te^{-T\phi}\boldsymbol{\psi} = 0$ (the Poisson bracket is calculated in the old variables).

Let the generating function have the form,

$$F_{22}^{\epsilon} = P_i q^i + P_{i'}^{\psi} \psi^{i'} + \epsilon G(q, P, \psi, P^{\psi}, t), \tag{1.177}$$

where ϵ is an infinitesimal parameter, $\epsilon \to 0$. Obviously, the function (1.177) defines the identical canonical transformation at $\epsilon = 0$ (as one might infer from (1.170) and (1.171)). Therefore the canonical transformation generated by the function (1.177) is infinitesimal. The new canonical variables differ from the old ones by infinitesimal variations, $P_i = p_i + \delta p_i + O(\epsilon^2)$, $P_{i'}^{\psi} = p_{i'}^{\psi} + \delta p_i^{\psi} + O(\epsilon^2)$, and similarly for the canonical coordinates. Substituting (1.177) into (1.170) and (1.171) and neglecting the terms $O(\epsilon^2)$, the new infinitesimally transformed canonical variables are obtained as functions of the old ones,

$$P_i = p_i - \epsilon\frac{\partial}{\partial q^i}G(q,p,\psi,p^\psi,t)\,,\; P^\psi_{i'} = p^\psi_{i'} + \epsilon\frac{\vec{\partial}}{\partial\psi^{i'}}G(q,p,\psi,p^\psi,t)\,, \tag{1.178}$$

$$Q^i = q^i + \epsilon\frac{\partial}{\partial p^i}G(q,p,\psi,p^\psi,t)\,,\; \Psi^{i'} = \psi^{i'} + \epsilon\frac{\vec{\partial}}{\partial p^\psi_{i'}}G(q,p,\psi,p^\psi,t)\,. \tag{1.179}$$

These equalities generalize Eqs. (1.62) to the case of systems with bosonic and fermionic variables.

If, in particular, the function G in (1.178) and (1.179) is set to be the Hamiltonian, then the infinitesimal canonical transformation can be written in the compact form,

$$\Theta^a = \theta^a + \epsilon\{\theta^a, H\} = \theta^a + \epsilon\dot{\theta}^a\,,$$

i.e. $\Theta^a(t) = \theta^a(t+\epsilon)$. Hence, just as in the bosonic case, the time evolution of a Hamiltonian system with bosonic and fermionic degrees of freedom is the continuous unfolding of a canonical transformation generated by the Hamiltonian.

1.10.1 Hamilton–Jacobi theory

Let θ^a_0 be the initial values of the canonical variables, and $\theta^a = \theta^a(\theta_0, t)$ be the system trajectory in phase space. Since time evolution is the continuous unfolding of a canonical transformation, there should exist a canonical transformation $\theta^a \to \Theta^a = \theta^a_0$ after which the canonical variables become constant in time. Clearly, finding such a canonical transformation is equivalent to solving the Hamiltonian equation of motion because $\theta^a = \theta^a(\Theta, t)$. The equations of motion become trivial, $\dot{\Theta}^a = 0$, and, as explained at the end of Section 1.5, the Hamiltonian in the new canonical variables is a constant which can always be assumed to vanish,

$$\tilde{H} = H(q,p,\psi,p^\psi,t) + \partial_t F_{22} = 0, \tag{1.180}$$

where the generating function of the canonical transformation is chosen to be $F_{22}(q,P,\psi,P^\psi,t)$. Equations (1.170) may be used so that Eq. (1.180) becomes

$$H\left(q, \frac{\partial S}{\partial q}, \psi, -\frac{\vec{\partial} S}{\partial \psi}, t\right) + \frac{\partial S}{\partial t} = 0, \tag{1.181}$$

where the conventional notation $F_{22} = S$ has been adopted. Equation (1.181) is a generalization of the Hamilton–Jacobi equation to systems with both bosonic and fermionic degrees of freedom. Its solution $S = S(q,\psi,t)$ is called Hamilton's principal function. The solution depends on n real integration constants α_i, $i=1,2,...,n$, and n' real Grassmann constants, $\xi_{i'}$, $i' = 1,2,...,n'$, where n and n' are the numbers of bosonic and fermionic degrees of freedom, respectively. By analogy with the bosonic case [1], these constants (or suitable functions of them) become new conservative canonical momenta,

$$P_i = \alpha_i, \quad P^{\psi}_{i'} = \xi_{i'}.$$

The conservative canonical coordinates are then determined by (1.171),

$$Q^i = \frac{\partial S}{\partial \alpha_i}, \qquad \Psi^{i'} = \frac{\vec{\partial} S}{\partial \xi_{i'}}.$$

To illustrate peculiarities of the Hamilton–Jacobi equation associated with the nilpotency of the Grassmann variables, consider the simplest case when the system has one bosonic and one fermionic degree of freedom. A generic Hamiltonian reads,

$$H = H_0(q,p) + H_1(q,p)\psi p^{\psi}.$$

Consequently, the solution to (1.181) is sought in the form,

$$S = S_0(q,\alpha,t) + S_1(q,\alpha,t)\psi\xi.$$

Substituting H and S into (1.180) and collecting the coefficients at the basis elements, 1 and $\psi\xi$, of the Grassmann algebra, the following system of equations is obtained:

$$H_0\left(q, \frac{\partial S_0}{\partial q}\right) + \frac{\partial S_0}{\partial t} = 0, \tag{1.182}$$

$$H'_{0p}\left(q, \frac{\partial S_0}{\partial q}\right)\frac{\partial S_1}{\partial q} - H_1\left(q, \frac{\partial S_0}{\partial q}\right) S_1 + \frac{\partial S_1}{\partial t} = 0, \tag{1.183}$$

where $H'_{0p}(q,p) = \partial H_0/\partial p$. For systems with several Grassmann degrees of freedom, Hamilton's principal function is a general even element of the Grassmann algebra generated by $\psi^{i'}$ and $\xi_{i'}$. An explicit example of a canonical transformation that nullifies the canonical Hamiltonian is constructed for a supersymmetric dynamics in the next section.

1.11 Noether's theorem for systems on supermanifolds

Noether's theorem has a natural extension to systems with fermionic and bosonic degrees of freedom. Here a slightly different formulation of Noether's theorem than that in Section 1.6.1 is given. It uses transformation properties of the Lagrangian rather than the invariance of the action functional. This formulation is more convenient for the discussion of supersymmetric dynamics below.

Let a supermanifold M be a configuration space. Let θ^α denote the set q^i and $\psi^{i'}$ of local coordinates in M. Then $\theta^\alpha = \theta^\alpha(t)$ defines a trajectory (curve) in M and derivative $\dot{\theta}^\alpha(0)$ is, by definition, an element of the tangent space T_θ at $\theta^\alpha = \theta^\alpha(0)$ which is a Euclidean superspace. The tangent bundle TM is defined similarly to the bosonic case. A Lagrangian is an even function on TM,

$L = L(\dot{\theta}, \theta, t)$. The Euler–Lagrange equations are obtained from the variational principle for the action $S = \int L dt$. When calculating the variation of the action with respect to θ^α, the parities of the latter should be taken into account (similarly to the case of the first-order action (1.149)). The Euler–Lagrange equations have the form,

$$\frac{d}{dt}\frac{\vec{\partial} L}{\partial \dot{\theta}^\alpha} = \frac{\vec{\partial} L}{\partial \theta^\alpha}. \tag{1.184}$$

A generalization of Ostrogradsky's equations (1.6) to dynamics on supermanifolds is also straightforward (details are omitted). Here discussion is limited to Lagrangian dynamics with at most second-order Euler–Lagrange equations.

Let the trajectory of a system be infinitesimally transformed,

$$\theta^\alpha_\epsilon(t) = \theta^\alpha(t) + \epsilon\Theta^\alpha(\dot{\theta}, \theta, t) + O(\epsilon^2), \tag{1.185}$$

where the transformation parameter ϵ is now allowed to be either a real or Grassmann variable with the corresponding commutation rule in its product with a function on TM. Obviously, for a Grassmann ϵ, the transformation is linear because $\epsilon^2 = 0$. Clearly, the trajectory transformation must be parity preserving and, therefore, given the parity of ϵ, the parity of the functions Θ^α is uniquely fixed. Suppose that the Lagrangian function has the following property:

$$L_\epsilon = L(\dot{\theta}_\epsilon, \theta_\epsilon, t) = L(\dot{\theta}, \theta, t) + \epsilon\frac{d}{dt}F(\dot{\theta}, \theta, t) + O(\epsilon^2) \tag{1.186}$$

for some function F on TM. Then the quantity

$$I = \Theta^\alpha \frac{\vec{\partial} L}{\partial \dot{\theta}^\alpha} - F \tag{1.187}$$

is an integral of motion. The proof is similar to the purely bosonic case. Calculating the variation of the Lagrangian under variation (1.185) of the trajectory, one finds from the supposition (1.186) that,

$$\frac{dF}{dt} = \dot{\Theta}^\alpha \frac{\vec{\partial} L}{\partial \dot{\theta}^\alpha} + \Theta^\alpha \frac{\vec{\partial} L}{\partial \theta^\alpha}. \tag{1.188}$$

Differentiation of (1.187) yields,

$$\frac{dI}{dt} = \dot{\Theta}^\alpha \frac{\vec{\partial} L}{\partial \dot{\theta}^\alpha} + \Theta^\alpha \frac{d}{dt}\frac{\vec{\partial} L}{\partial \dot{\theta}^\alpha} - \frac{dF}{dt} = \dot{\Theta}^\alpha \frac{\vec{\partial} L}{\partial \dot{\theta}^\alpha} + \Theta^\alpha \frac{\vec{\partial} L}{\partial \theta^\alpha} - \frac{dF}{dt} = 0,$$

where the equations of motion (1.184) and then relation (1.188) have been used to establish the Noether's theorem conclusion.

Naturally, the formulation of Noether's theorem given in Section 1.6.1 can also be extended to dynamical systems on supermanifolds. If the action functional is invariant under a transformation in M with $\Theta^\alpha = \left(\vec{\partial}\theta^\alpha_\epsilon/\partial\epsilon\right)_{\epsilon=0}$, and a simultaneous time transformation $t \to t_\epsilon$ with $T = (\vec{\partial} t_\epsilon/\partial\epsilon)_{\epsilon=0}$, then, similarly to the bosonic case, Noether's integral of motion has the form,

$$I = T\left(\dot{\theta}^\alpha \frac{\vec{\partial} L}{\partial \dot{\theta}^\alpha} - L\right) + \Theta^\alpha \frac{\vec{\partial} L}{\partial \dot{\theta}^\alpha}. \tag{1.189}$$

It is clear from comparison of the integrals (1.187) and (1.189) that there exists a time transformation such that, when combined with the coordinate transformation (1.185) with the property (1.186), it leaves the action invariant, i.e. the boundary terms arising in the variation of the action induced by (1.186) are compensated for by the time transformation.

1.11.1 Supersymmetry

If the symmetry transformation mixes the bosonic and fermionic variables, it is usually referred as to a supersymmetric transformation. Consider a general system with one bosonic and one fermionic degree of freedom. It is customary to use the first-order formalism for fermionic degrees of freedom. So, in order for the fermionic degree of freedom to have a non-trivial dynamics, it should be described by a complex Grassmann variable ψ. A generic Lagrangian for such a system has the form,

$$L = \frac{1}{2}\dot{q}^2 - \frac{i}{2}\dot{\psi}^*\psi + \frac{i}{2}\psi^*\dot{\psi} - V_0(q) - V_1(q)\psi^*\psi, \tag{1.190}$$

where V_0 is a potential for the bosonic degree of freedom, and function V_1 describes a coupling between the bosonic and fermionic degrees of freedom. Without loss of generality, an infinitesimal transformation that mixes the commuting and anticommuting variables is written as,

$$q_\epsilon = q + \epsilon\psi^* + \psi\epsilon^* + O(\epsilon^*\epsilon)\,, \tag{1.191}$$

$$\psi_\epsilon = \psi + i\epsilon A(\dot{q}, q), \quad \psi^*_\epsilon = \psi^* - i\epsilon^* A^*(\dot{q}, q), \tag{1.192}$$

where A is a complex function of $\dot{q}$ and q, and ϵ is a complex Grassmann variable.

The question arises whether there exist functions V_0, V_1, and A such that the Lagrangian (1.190) satisfies condition (1.186) of Noether's theorem. The answer is affirmative. The resulting dynamical system provides a simple example of supersymmetric systems. It is straightforward to verify that the variation of the Lagrangian (1.190) under transformations (1.191) and (1.192) has the following form:

$$L_\epsilon - L = \epsilon\psi^*\left(-\ddot{q} + \dot{A} - V_0' - iV_1 A\right) + \epsilon\frac{d}{dt}\left(\dot{q}\psi^* - \frac{1}{2}\psi^* A\right) + \text{c.c.}, \tag{1.193}$$

where the abbreviation "c.c." means "complex conjugation". Since ψ^* is an independent variable, the condition of Noether's theorem is satisfied, provided the expression in parenthesis in the first term of the right-hand side of (1.193) vanishes identically. This necessarily requires that

$$A(\dot{q}, q) = \dot{q} + iW(q) \tag{1.194}$$

to cancel the second derivative $\ddot{q}$, while the other terms are canceled out if, in addition, potentials V_0 and V_1 have a specific form,

$$V_0 = \frac{1}{2} W^2, \quad V_1 = W', \tag{1.195}$$

for any function $W(q)$. A well-known example of the supersymmetric harmonic oscillator corresponds to the choice $W = q$. In quantum mechanics the potential W must satisfy additional asymptotic conditions ($\int W dq \to \infty$ (or $-\infty$) as $|q| \to \infty$) otherwise the supersymmetry is spontaneously broken (the ground state is not supersymmetric, while the Hamiltonian is) [32, 33].

Since ϵ and ϵ^* are two independent parameters, there are two corresponding Noether's integrals. They are obtained from (1.187) by substituting in both the explicit form of $F = \dot{q}\psi^* - \psi^* A/2$, established in (1.193), and the transformation laws (1.191) and (1.192) (where the Grassmann variable ϵ^* should be moved over to the very left to comply with the ordering in the Noether's theorem conditions (1.185) and (1.186)),

$$\psi^* \frac{\partial L}{\partial \dot{q}} + iA \frac{\vec{\partial} L}{\partial \dot{\psi}} - F = \psi^* A \equiv \mathcal{Q}, \tag{1.196}$$

$$-\psi \frac{\partial L}{\partial \dot{q}} - iA^* \frac{\vec{\partial} L}{\partial \dot{\psi}^*} + F^* = -\psi A^* \equiv -\mathcal{Q}^*. \tag{1.197}$$

The Noether integrals are odd elements of the Grassmann algebra because parameters ϵ and ϵ^* are Grassmann variables. If the number of fermionic and bosonic degrees of freedom increase, it is possible to find systems with several (complex) supersymmetry parameters. The required restrictions on the potentials and couplings of the degrees of freedom can be deduced in the same way as presented above.

The existence of Grassmann integrals of motion, $\mathcal{Q}$ and $\mathcal{Q}^*$, also allows for partial or even complete integrability just as in the pure bosonic case studied in Section 1.6. This is illustrated below with the example of supersymmetric classical dynamics. First, the Hamiltonian formalism for complex Grassmann degrees of freedom described by the first-order equations (1.101) and (1.102) should be established. A simple way to do this is to calculate the total time derivative of a function $F(\xi, \xi^*, t)$. Similarly to (1.109), one infers from (1.101) and (1.102) that $\dot{F} = \partial_t F + \{F, H\}$, where $H = V$ and the Poisson bracket is defined by,

$$\{A, B\} = iA \left(\frac{\overleftarrow{\partial}}{\partial \xi^i} \alpha^{ij} \frac{\vec{\partial}}{\partial \xi^{*j}} + \frac{\overleftarrow{\partial}}{\partial \xi^{*i}} \alpha^{ij} \frac{\vec{\partial}}{\partial \xi^j} \right) B,$$

where α^{ij} is the inverse of α_{ij} which is a real symmetric matrix (compare (1.99)). The Poisson bracket satisfies the Leibniz rule (1.112) and the Jacobi identity (1.113). It is then straightforward to obtain the Hamiltonian formalism for the model (1.190) for which $\alpha^{ij} = \alpha_{ij} = 1$. Indeed, the supersymmetric Hamiltonian reads,

$$H = \frac{1}{2}p^2 + \frac{1}{2}W^2(q) + W'(q)\psi^*\psi, \tag{1.198}$$

where $p = \partial L/\partial \dot{q} = \dot{q}$, $\{q, p\} = 1$ and $\{\psi, \psi^*\} = \{\psi^*, \psi\} = -i$.

Because quantities (1.196) and (1.197) are integrals of motion ($A = p+iW(q)$), their Poisson bracket with the Hamiltonian vanishes. The integrals of motion form an algebra with respect to the Poisson bracket

$$\{Q, Q^*\} = \{Q^*, Q\} = -2iH, \qquad \{Q, H\} = \{Q^*, H\} = 0. \tag{1.199}$$

One of the peculiarities of this algebra is that the Lie bracket for the Grassmann generators Q and Q^* are symmetric. A Lie algebra in which the Lie bracket has the properties (1.111) and (1.113) is called a *superalgebra* (or a Z_2–graded Lie algebra). After the canonical quantization ($i\{\ ,\ \} \to [\ ,\ ]$), the integrals of motion becomes operators $\hat{Q}, \hat{Q}^\dagger$, and $\hat{H}$. They are generator of the *supersymmetry algebra* with the commutation relations that follow from (1.199):

$$[\hat{Q}, \hat{Q}^\dagger]_+ = 2\hat{H}, \qquad [\hat{Q}, \hat{H}] = [\hat{Q}^\dagger, \hat{H}] = 0,$$

where $[\ ,\ ]_+$ denotes the anticommutator [33].

To show the integrability of the supersymmetric mechanical system, a canonical transformation should be constructed such that all new canonical momenta are integrals of motion and the new canonical coordinates become cyclic. Consider first the following canonical transformation of the Grassmann canonical variables:

$$\Psi^* = \frac{\mathcal{Q}}{\sqrt{2H}} = \psi^* \frac{p + iW(q)}{\sqrt{p^2 + W^2(q)}} \equiv \psi^* e^{i\phi}, \quad \Psi = \psi e^{-i\phi}, \tag{1.200}$$

where $p = \dot{q}$ has been used. Evidently, $\{\Psi^*, \Psi\} = i$. The new canonical Grassmann variables are integrals of motion because they depend only on the supersymmetric integrals of motion and the Hamiltonian. It is easy to verify that a general solution of the equation

$$\{B, \Psi^*\} = 0$$

has the form $B(p, q, \psi^*, \psi) = B_0(p, q) + \{\phi, B_0\}\psi^*\psi$, in the class of even elements of the Grassmann algebra. Therefore the new canonical bosonic variables must have the form,

$$P = P_0 + \{\phi, P_0\}\psi^*\psi, \quad Q = Q_0 + \{\phi, Q_0\}\psi^*\psi. \tag{1.201}$$

Demanding that $\{Q, P\} = 1$ and making use of the Jacobi identity, functions P_0 and Q_0 are shown to be a canonical transformation of p and q, i.e. $\{Q_0, P_0\} = 1$.

Put $Q_0 = q$ and $P_0 = p$. Then, by inverting relations (1.201) and (1.200), the old canonical variables are obtained as the functions of the new ones,

$$p = P - \frac{W'(Q)P}{P^2 + W^2(Q)}\,\Psi^*\Psi, \quad q = Q - \frac{W(Q)}{P^2 + W^2(Q)}\,\Psi^*\Psi, \tag{1.202}$$

$$\psi^* = \frac{P - iW(Q)}{\sqrt{P^2 + W^2(Q)}}\,\Psi^*, \qquad \psi = \frac{P + iW(Q)}{\sqrt{P^2 + W^2(Q)}}\,\Psi, \tag{1.203}$$

where the explicit form of the Poisson brackets $\{\phi, p\}$ and $\{\phi, q\}$ have been substituted. Using Eqs. (1.202) and (1.203), the Hamiltonian in the new variables is found,

$$H = \frac{1}{2}P^2 + \frac{1}{2}W^2(Q). \tag{1.204}$$

It no longer depends on the Grassmann canonical variables (they are integrals of motion). The supersymmetry algebra commutation relations hold in the new canonical variables because $\mathcal{Q} = \Psi^*\sqrt{2H}$ and, hence, $\{\mathcal{Q}, \mathcal{Q}^*\} = -2iH$, thanks to the invariance of the Poisson bracket under canonical transformations. Finally, a canonical transformation $(P, Q) \to (\tilde{P}, \tilde{Q})$ is carried out which is generated by the function

$$F_2(Q, \tilde{P}) = \int_{Q_0}^{Q} \sqrt{2[\tilde{P} - W^2(q)]}\; dq,$$

where Q_0 is a constant. In accordance with the first equation in (1.56) the new momentum coincides with the Hamiltonian $\tilde{P} = H(P, Q)$ and, hence, the canonical coordinate $\tilde{Q}$ is cyclic, $\tilde{Q}(t) = t + \text{const}$. Integration of the supersymmetric system is complete. Due to the Grassmann nature of the supersymmetric integrals of motion, the time-independent canonical transformation constructed to integrate the system eliminates the fermionic canonical variables altogether from the Hamiltonian rather than making one of them cyclic as in the bosonic case (the Hamiltonian H cannot depend on, e.g. the new "momentum" Ψ^* only). If so desired, one can carry out one more (time-dependent) canonical transformation after which the Hamiltonian vanishes. This transformation is easy to construct as the Hamiltonian coincides with the canonical momentum. The composition of this and the other two canonical transformations constructed above provides the solution of the Hamilton–Jacobi equation for the supersymmetric dynamics.

1.12 Non-canonical transformations

In practice, non-canonical transformations might also be useful to study Hamiltonian dynamics. The simplest example is provided by the harmonic oscillator theory defined by the Hamiltonian

$$H = \frac{1}{2}(p^2 + q^2)\,. \tag{1.205}$$

Instead of the canonical variables p and q, consider the complex variables,

$$a = \frac{1}{\sqrt{2}}(p - iq), \qquad a^* = \frac{1}{\sqrt{2}}(p + iq). \tag{1.206}$$

In these variables the Hamiltonian (1.205) has a simple form,

$$H = a^* a\,,$$

and the Poisson bracket is induced by the Poisson bracket in the old canonical variables,

$$\{A, B\} = \frac{\partial A}{\partial q}\frac{\partial B}{\partial p} - \frac{\partial A}{\partial p}\frac{\partial B}{\partial q} = -i\left(\frac{\partial A}{\partial a}\frac{\partial B}{\partial a^*} - \frac{\partial A}{\partial a^*}\frac{\partial B}{\partial a}\right). \tag{1.207}$$

In particular,

$$\{a, a^*\} = -i\,.$$

The time evolution of any function on the phase space is determined by its Poisson bracket (1.207) with the Hamiltonian. The Hamiltonian equations of motion of the harmonic oscillator in the new variables are,

$$\dot{a} = \{a, H\} = -ia, \qquad \dot{a}^* = \{a^*, H\} = ia^*. \tag{1.208}$$

Thus, all the attributes of the Hamiltonian mechanics, the phase space of the Hamiltonian, and the Poisson brackets satisfying the Jacobi identity are retained, however the transformation $(q, p) \to (a, a^*)$ does not preserve the canonical symplectic structure, $\overset{\circ}{\omega}{}^{ab} \to -i\,\overset{\circ}{\omega}{}^{ab}$, $a, b = 1, 2$, i.e. this transformation is not canonical.

In order to study the class of non-canonical transformations that preserve the Hamiltonian form of the equations of motion, consider a general change of variables in a symplectic supermanifold,

$$\theta^a = \theta^a(\Theta, t). \tag{1.209}$$

Suppose that variables θ^a are the Darboux variables, and, hence, the symplectic two-form has the canonical form $\omega = \overset{\circ}{\omega}$ (such variables can always be found according to the Darboux theorem). In the new variables the Poisson bracket is defined by the matrix,

$$\omega^{ab} = \Theta^a_{\overleftarrow{,c}}\,\overset{\circ}{\omega}{}^{cd}\Theta^b_{,\vec{d}}, \tag{1.210}$$

where the index after the comma denotes the corresponding derivative with respect to the old variables θ^a. Matrix (1.210) satisfies the identities (1.132)–(1.134). The validity of identities (1.132) and (1.133) follows directly from the explicit form (1.210) of the matrix ω^{ab}. In order to prove identity (1.134), consider the inverse matrix,

$$\omega_{ab} = \theta^c_{,\vec{a}}\,\overset{\circ}{\omega}{}_{cd}\theta^d_{\overleftarrow{,b}}, \tag{1.211}$$

where the indices after the comma now denote the corresponding derivative with respect to the new variables Θ^a. It defines a two-form ω which is obtained from the canonical one $\overset{\circ}{\omega}$ by the change of variables (1.209). Therefore $d\overset{\circ}{\omega} = d\omega = 0$, because the closedness of a differential form does not depend on the choice of local coordinates on a supermanifold. Hence, the matrix (1.211) should satisfy identity (1.141), which is equivalent to identity (1.134) as has been shown in Section 1.9.

The new symplectic structure may depend explicitly on time, which generally leads to a breaking of the Hamiltonian form of the equations of motion. To illustrate this assertion, consider the dynamics of the system (1.29) in new variables Q and P:

$$q = Q\,, \quad p = e^{\alpha t/2}P. \tag{1.212}$$

The new phase space coordinates Q and P are no longer canonical because $\{Q, P\} = \exp(-\alpha t/2) \neq 1$, and the symplectic structure now depends explicitly on time,

$$\omega^{ab} = \exp\left(-\frac{\alpha}{2}t\right)\overset{\circ}{\omega}{}^{ab}. \tag{1.213}$$

Let us show that there exists no Hamiltonian for which the Hamiltonian equations of motion with the symplectic structure (1.213) coincide with the original equations of motion of the system. If the Hamiltonian form of the equations of motion were preserved, then the equations of motion should have been,

$$\dot{Q} = \{Q, \tilde{H}\} = \exp\left(-\frac{\alpha}{2}t\right)\frac{\partial \tilde{H}}{\partial P}, \tag{1.214}$$

$$\dot{P} = \{P, \tilde{H}\} = -\exp\left(-\frac{\alpha}{2}t\right)\frac{\partial \tilde{H}}{\partial Q}, \tag{1.215}$$

where $\tilde{H}$ is the Hamiltonian in the new variables. On the other hand, the substitution of (1.212) into the Hamiltonian equations of motion in the old variables ($\dot{q} = \partial H/\partial p = \exp(-\alpha t)p$ and $\dot{p} = -\partial H/\partial q = 0$, with H given in (1.29)) yields,

$$\dot{Q} = \exp\left(-\frac{\alpha}{2}t\right)P, \tag{1.216}$$

$$\dot{P} = -\frac{\alpha}{2}P. \tag{1.217}$$

Comparison of Eqs. (1.214) with (1.216) and (1.215) with (1.217) shows that the Hamilton function $\tilde{H}$ must satisfy the following equations:

$$\frac{\partial \tilde{H}}{\partial P} = P, \quad \frac{\partial \tilde{H}}{\partial Q} = \frac{\alpha}{2}\exp\left(\frac{\alpha}{2}t\right)P. \tag{1.218}$$

Apparently, no such function exists because it follows from (1.218) that $\tilde{H}$ does not satisfy Clairaut's theorem, $\partial^2\tilde{H}/\partial P\partial Q \neq \partial^2 H/\partial Q\partial P$, i.e. Eqs. (1.218) are not integrable. Thus, after the change of variables (1.212), the equations of motion (1.216) and (1.217) can no longer have the Hamiltonian form.

To study this phenomenon in general, consider the principle of stationary action in the new variables (1.209). The following relation can be proved:

$$\frac{1}{2}\theta^a \overset{\circ}{\omega}_{ab}\frac{d\theta^b}{dt} - H(\theta) = \Theta^a\bar{\omega}_{ab}(\Theta)\frac{d\Theta^b}{dt} - \tilde{H}(\Theta,t) + \frac{dF(\Theta,t)}{dt}, \qquad (1.219)$$

where $\bar{\omega}_{ab}(\Theta)$ is obtained from the matrix $\omega_{ab}(\Theta)$, defined in (1.211), by the rule (1.146), and

$$\tilde{H}(\Theta,t) = H(\theta(\Theta,t)) + \int_0^1 d\alpha \frac{\partial\theta^a(\alpha\Theta,t)}{\partial\alpha} \overset{\circ}{\omega}_{ab} \frac{\partial\theta^b(\alpha\Theta,t)}{\partial t}, \qquad (1.220)$$

$$F(\Theta,t) = \frac{1}{2}f(t) + \frac{1}{2}\int_0^1 d\alpha\theta^a(\alpha\Theta,t) \overset{\circ}{\omega}_{ab}\frac{\partial\theta^b(\alpha\Theta,t)}{\partial\alpha},$$

$$\frac{df}{dt} = \theta^a(0,t) \overset{\circ}{\omega}_{ab}\partial_t\theta^b(0,t).$$

Indeed, first, equality (1.219) is multiplied by 2 and the explicit expression for $\bar{\omega}_{ab}$ is substituted into the first term in its right-hand side. Next, this term is transformed with the help of the following identities:

$$\Theta^c\frac{\vec{\partial}}{\partial(\alpha\Theta^c)}\theta^a(\alpha\Theta,t) = \frac{\partial}{\partial\alpha}\theta^a(\alpha\Theta,t),$$

$$\alpha\dot{\Theta}^d\frac{\vec{\partial}}{\partial(\alpha\Theta^d)}\theta^a(\alpha\Theta,t) = \frac{d\theta^a(\alpha\Theta,t)}{dt} - \frac{\partial\theta^a(\alpha\Theta,t)}{\partial t}.$$

In doing so, all the terms containing $\partial_t\theta^b(\alpha\Theta,t)$ are canceled out in (1.219) after the substitution of (1.220) for the second term in the right-hand side of Eq. (1.219). Finally, the remaining part of the first term in the right-hand side of (1.219) is represented as the sum of two identical terms (just as $A = A/2 + A/2$) in one of which the integration by parts is carried out with respect to α. As a result, the right-hand side becomes identical to the left-hand side; the proof of Eq. (1.219) is complete.

It follows from (1.219) that the equations of motion in the new variables can be obtained by the variational principle for the action,

$$S = \int_{t_1}^{t_2} dt\left[\Theta^a\bar{\omega}_{ab}(\Theta,t)\dot{\Theta}^b - \tilde{H}(\Theta,t)\right].$$

The difference with the principle of stationary action for (1.148) is that the symplectic structure may now explicitly depend on time, which changes calculations (1.149) and (1.150). Namely, the variational derivative of the action depends on the partial derivative of the matrix $\bar{\omega}_{ab}$ with respect to time (compare (1.150)),

$$\delta S = \int_{t_1}^{t_2} dt \delta\Theta^a \left[\omega_{ab}\dot{\Theta}^b - \tilde{H}_{,\vec{a}} + (\partial_t \bar{\omega}_{ab})\Theta^b\right] = 0. \tag{1.221}$$

The derivative $\partial_t \bar{\omega}_{ab}$ occurs through the integration by parts in the term containing $\delta\dot{\Theta}^a$. According to (1.221), the extrema of the action satisfy the equations,

$$\dot{\Theta}^a = \omega^{ab}(\tilde{H}_{,\vec{b}} - (\partial_t \bar{\omega}_{bc})\Theta^c) \tag{1.222}$$

or

$$\nabla^a_{tb}\Theta^b = \omega^{ab}\tilde{H}_{,\vec{b}},$$

where $\nabla^a_{tb} = \delta^a_b \partial_t + A^a_b(\theta, t)$ is the "covariant" time derivative with the "connection" $A^a_b = \omega^{ac}\partial_t \bar{\omega}_{cb}$. The second term in the right-hand side of Eq. (1.222), which contains $\partial_t \bar{\omega}_{ac}$, breaks the Hamiltonian structure of the equations of motion. Indeed, if the right-hand side of Eq. (1.222) is written in the form $\omega^{ab}\tilde{H}'_{,\vec{b}}$ where

$$\begin{aligned} \tilde{H}' &= \tilde{H} + H_\omega, \\ H_{\omega,\vec{a}} &= -\partial_t \bar{\omega}_{ab}\Theta^b, \end{aligned} \tag{1.223}$$

then, in order for the function $H_\omega = H_\omega(\Theta, t)$ to exist, it is necessary and sufficient that

$$H_{\omega,\vec{a}\vec{b}} = (-1)^{\epsilon^a\epsilon^b} H_{\omega,\vec{b}\vec{a}}. \tag{1.224}$$

The substitution of (1.223) into (1.224) together with the identity (1.141) yields the condition on the matrix ω_{ab},

$$\partial_t \omega_{ab}(\Theta, t) = 0, \tag{1.225}$$

which implies that $H_\omega = 0$.

Thus, the Hamiltonian form of the equations of motion is preserved under general coordinate transformations on the phase space if and only if the symplectic structure in the new variables does not depend explicitly on time.

This does not mean, however, that the change of variables (1.209) is not allowed to have an explicit time dependence in order to preserve the Hamiltonian form of the equations of motion. As an example, consider the Hamiltonian

$$H(p, q, t) = \frac{1}{2m}p^2 + \frac{\nu}{2}q^2,$$

in which $m = m(t)$ and $\nu = \nu(t)$. Define the new variables by,

$$P = p^2/(2m), \quad Q = \nu q^2/2. \tag{1.226}$$

The symplectic structure in the new variables has the form,

$$\omega_{ab} = \frac{1}{2\Omega\sqrt{PQ}}\overset{\circ}{\omega}_{ab}, \quad a, b = 1, 2, \tag{1.227}$$

where $\Omega = (\nu/m)^{1/2}$ is the oscillator frequency. The condition (1.225) is satisfied, provided that frequency Ω (or the ratio of the functions $\nu(t)$ and $m(t)$) does not depend on time. The Hamiltonian in the new variables is defined by relation (1.220) in which one should set $\theta^1 = q$, $\theta^2 = p$, and, similarly, $\Theta^1 = Q$, $\Theta^2 = P$. This reads,

$$\tilde{H} = P + Q - \frac{1}{2\Omega}\sqrt{PQ}\left(\frac{\dot{\nu}}{\nu} + \frac{\dot{m}}{m}\right) = P + Q - \frac{1}{\Omega}\sqrt{PQ}\,\frac{\dot{m}}{m}; \tag{1.228}$$

the latter equality follows from $\partial_t \Omega = 0$. The Hamiltonian (1.228) together with the symplectic structure (1.227) determine the Hamiltonian dynamics of the system despite the change of variables in the phase space (1.226) depending explicitly on time.

1.13 Examples of systems with non-canonical symplectic structures

1.13.1 A particle with friction

One of the simplest dynamical systems that can be described by Hamiltonian dynamics with a non-canonical symplectic structure is a particle moving along a line subject to a friction force. The equation of motion is given by (1.16),

$$\ddot{q} + \alpha \dot{q} = 0. \tag{1.229}$$

Its general solution reads,

$$q(t) = q_0 + \frac{v_0}{\alpha}\left(1 - e^{-\alpha t}\right),$$

where $q(0) = q_0$ and $\dot{q}(0) = v_0$.

Consider the dynamics defined by the Hamiltonian

$$H = H_0(p)$$

(H_0 is a function of the momentum p) and the symplectic structure

$$\{q, p\} = 1 - \frac{qp}{a}, \tag{1.230}$$

where a is a constant of the action dimension. The phase space plane should be restricted to the region where the symplectic structure does not vanish, e.g. $\{q, p\} > 0$. The Hamiltonian equations of motion have the form,

$$\dot{p} = \{p, H_0\} = 0\,, \quad \dot{q} = \{q, H_0\} = \frac{\partial H_0}{\partial p}\left(1 - \frac{qp}{a}\right). \tag{1.231}$$

It follows from the first equation in (1.231) that the momentum is an integral of motion, $p(t) = p_0 = \text{const}$. Therefore the equation for coordinate q is linear,

$$\dot{q} + \gamma q = f_0, \quad f_0 = \frac{\partial H_0}{\partial p_0}, \tag{1.232}$$

$$\gamma = \frac{f_0 p_0}{a}. \tag{1.233}$$

Equations (1.232) and (1.229) are equivalent if one sets $f_0 = v_0 + \alpha q_0$, and quantity (1.233) is identified with the friction coefficient α.

To prove the self-consistency of the Hamiltonian dynamics, one should show that the symplectic two-form associated with the Poisson bracket (1.230) is closed. Since the closedness of the form does not depend on the coordinate choice, it is sufficient to construct the Darboux coordinates. These are [34, 35],

$$q = q(Q,P) = Q, \quad p = p(Q,P) = \frac{a}{Q}\left[1 - \exp\left(-\frac{QP}{a}\right)\right]. \tag{1.234}$$

A simple calculation shows that (1.230) holds provided that P and Q are canonically conjugated variables, $\{Q,P\} = 1$.

Equations of motion (1.231) can be obtained from the principle of stationary action for (1.148) where $\theta^1 = q$, $\theta^2 = p$, and

$$\bar{\omega}_{ab}(\theta) = -\frac{a}{2pq}\ln\left(1 - \frac{pq}{a}\right)\overset{\circ}{\omega}_{ab} = \frac{P(q,p)}{2p}\overset{\circ}{\omega}_{ab}. \tag{1.235}$$

When calculating the matrix (1.235), the integral representation (1.146) has been used.

It is curious that the canonical quantization of the Poisson bracket (1.230) (i.e. the replacement of q and p by the corresponding Hermitian operators with the commutation relation $[\,,\,] = i\hbar\{\,,\,\}$) leads to the so-called quantum q-particle theory (or a q-particle on the line [36]) with the deformation parameter depending on the ratio $\hbar/\alpha$ [34, 35]. Thus, quantum mechanics of a particle with friction is the simplest example of a q-deformed system [34, 35]. A modification of the Hamiltonian mechanics for the harmonic oscillator with friction has been studied in [37].

1.13.2 *q*-Oscillator

In the literature possible modifications of the Heisenberg algebra have been discussed for many years [38–42]. Interest in the problem was revived after the discovery of *quantum groups* [43–45]. In particular, the idea that the Lorentz symmetry of the spacetime is to be replaced by the quantum deformation of the Lorentz group at short distances was appealing from the physical point of view because it could provide a natural ultraviolet cut-off in quantum field theory. Similarly to ordinary classical groups, it appeared that representations of quantum groups can be realized with the help of the creation $\hat{b}_j^\dagger$ and annihilation $\hat{b}_j$ operators which obey the q-deformed Heisenberg–Weyl algebra [46],

$$\hat{b}_i\hat{b}_j^\dagger - q^{2\delta_{ij}}\hat{b}_j^\dagger\hat{b}_i = \hbar\delta_{ij}; \tag{1.236}$$

here, following the widely accepted convention, the deformation parameter (a real number) is denoted by q (not to be confused with the above notations for canonical coordinates). In the limit $q \to 1$, the algebra (1.236) becomes the conventional Heisenberg–Weyl algebra. It turns out that the algebra (1.236) can be obtained by the canonical quantization of a non-standard symplectic structure [47–50]. Here the consideration is limited to the case of a single q-oscillator. A general case can be found in [49].

Consider the following symplectic structure:

$$\{x, p\} = 1 - \frac{p^2 + \omega^2 x^2}{2E}, \tag{1.237}$$

where E and ω are some constants; the canonical coordinate is denoted here by x to avoid potential confusion with the deformation parameter q. In the complex (holomorphic) variables defined by,

$$x = \frac{i}{\sqrt{2\omega}}\,(b - b^*)\,, \quad p = \sqrt{\frac{\omega}{2}}\,(b + b^*)\,,$$

the symplectic structure has the form,

$$\{b, b^*\} = -i\left(1 - \frac{\omega b^* b}{E}\right). \tag{1.238}$$

The q-deformed Heisenberg–Weyl algebra (1.236) is obtained by the canonical quantization of (1.237) or (1.238) with the subsequent renormalization of the operators $\hat{b} \to (1 + \hbar\omega/(2E))^{-1/2}\hat{b}$, and similarly for $\hat{b}^\dagger$. The deformation parameter is $q = (1-\gamma)/(1+\gamma)$, where $\gamma = \hbar\omega/(2E)$.

Thus, the q-oscillator is obtained from the conventional harmonic oscillator by suitable deformation of the symplectic structure. The self-consistency of the Hamiltonian dynamics is established similarly to the case of the q-particle by finding the Darboux coordinates [48–50],

$$b = l\sqrt{\frac{\omega}{2}}\left(e^{-2iX/l} - e^{-iX/l}e^{-P/(\omega l)}\right), \tag{1.239}$$

$$b^* = l\sqrt{\frac{\omega}{2}}\left(e^{2iX/l} - e^{iX/l}e^{-P/(\omega l)}\right), \tag{1.240}$$

where the new constant l is defined by $E = \omega^2 l^2/2$ and $\{X, P\} = 1$. In the limit $l \to \infty$ ($E \to \infty$), b^* and b become the conventional holomorphic variables. The parameter l has the dimension of length. Since the dependence on the canonical coordinate X appears only in the form of the phase factor $e^{2iX/l}$, it is natural to associate the deformation with a compactification of the configuration space R into a circle of radius l.

Note that the canonical quantization could be carried out in the Darboux variables X and P instead of x and p. In this case, the operators $\hat{b}^\dagger$ and $\hat{b}$ are viewed as functions of the operators $\hat{X}$ and $\hat{P}$ satisfying the standard Heisenberg algebra $[\hat{X}, \hat{P}] = i\hbar$. They would also obey the deformed Heisenberg–Weyl commutation relations (1.236) but with a different deformation parameter, $q = \exp(-\hbar/(\omega l^2))$.

This means, in particular, that the transition to the Darboux variables and canonical quantization are non-commutative operations [50].

The Hamiltonian equations of motion for the q-oscillator are obtained in the usual way. Let the Hamiltonian be a function of b and b^*, $H = H_q(b, b^*)$. Then,

$$\dot{b} = \{b, H\} = -i\left(1 - \frac{\omega b^* b}{E}\right)\frac{\partial H_q}{\partial b^*}, \tag{1.241}$$

$$\dot{b}^* = \{b^*, H\} = i\left(1 - \frac{\omega b^* b}{E}\right)\frac{\partial H_q}{\partial b}. \tag{1.242}$$

If, for simplicity, the Hamiltonian is restricted to a special form $H_q = H_q(h)$, where $h = \omega b^* b$, then the equations of motion (1.241) and (1.242) describe the harmonic oscillator with the energy depending frequency [50],

$$\dot{b} = -i\Omega b, \quad \dot{b}^* = i\Omega b^*, \quad \Omega = \omega\left(1 - \frac{E_\omega}{E}\right)H_q'(E_\omega),$$

where $h = E_\omega = \mathrm{const}$ is an integral of motion, for $\{h, H_q\} = 0$.

So the q-deformation of dynamical systems may be described by a modification of the symplectic structure via a special non-canonical transformation. Examples of such transformations are provided by (1.234) (a q-particle), and by (1.239) and (1.240) (a q-oscillator). Equations (1.239) and (1.240) can be generalized to the case of any number of oscillators. The corresponding Hamiltonian dynamics is a special case of Hamiltonian systems on Kähler manifolds. This generalization allows one to construct representations of quantum groups in the space of functions on a phase space with a non-standard symplectic structure [50], similarly to that for the classical groups in Section 1.6.

1.14 Some generalizations of the Hamiltonian dynamics

1.14.1 Nambu Mechanics

Nambu mechanics [51] is defined on an n-dimensional phase space manifold by $n-1$ Hamiltonians H_a, $a = 1, 2, ..., n-1$ (n may be odd). If x^i, $i = 1, 2, ..., n$, are local coordinates, then the time evolution of any function $F = F(x)$ is determined by the Jacobian of the transformation $(x^1, x^2, ..., x^n) \to (F, H_1, ..., H_{n-1})$ [52],

$$\dot{F} = \frac{\partial(F, H_1, ..., H_{n-1})}{\partial(x^1, x^2, ..., x^n)}. \tag{1.243}$$

In particular,

$$\dot{x}^i = \frac{\partial(x^i, H_1, ..., H_{n-1})}{\partial(x^1, x^2, ..., x^n)}. \tag{1.244}$$

All the Hamiltonians are integrals of motion, $\dot{H}_a = 0$ as follows from (1.243). The dynamics is superintegrable because it has $n-1$ integrals of motion (the maximal number of integrals of motion for an n-dimensional dynamical system described by first-order equations). Recall that the integrability of the Hamiltonian dynamics with N degrees of freedom ($2N$ is the dimension of the phase space space) requires only N integrals of motion. The motion occurs along the trajectory that is the intersection of the level surfaces $H_a(x) = E_a$, where E_a are uniquely determined by the initial position $x^i(0) = x_0^i$. The system moves along this curve with a speed equal to the magnitude of the vector given in the right-hand side of Eq. (1.244). Nambu equation (1.243) is invariant under general coordinate transformations $x^i \to y^i$ whose Jacobian equals one, $\partial(x^1, ..., x^n)/\partial(y^1, ..., y^n) = 1$. This is the analogy of canonical transformations in the Hamiltonian mechanics. If the set of Hamiltonians is transformed to another set by a constant non-singular linear map whose matrix has the unit determinant, then the Nambu equations also remain invariant. So, a priori there is no particular preference or significance of one Hamiltonian over the other.

Nambu mechanics is related to the initial value problem for a generic first-order system, $\dot{x}^i = A^i(x)$, $x^i(0) = x_0^i$. Suppose that the vector field A^i is continuously differentiable and $A^i(x_0) \neq 0$. Then according to the flowbox theorem [53], there exists a neighborhood U of x_0^i in which one can construct a continuously differentiable diffeomorphism (a change of variables $x^i \to x^i(y)$) such that the vector field A^i is transformed to $A^j(\partial y^i/\partial x^j) = \delta^{i1}$. This implies that, in the neighborhood U, the trajectory is an intersection of the level surfaces, $y^j(x) = y^j(x_0)$, $j = 2, 3, ..., n-1$, and, therefore, the tangent vector $\dot{x}^i$ is orthogonal to the gradients $\partial_i y^j(x)$, $j \neq 1$. Put $H_a(x) = y^{a-1}(x)$, $a = 2, 3, ..., n-1$, and let s denote a parameter of the trajectory $x^i = x^i(s)$. Then the vector $\dot{x}^i$ is perpendicular to the gradients $\partial_i H_a$ and, therefore, must be proportional to the Jacobian $\partial(x^i, H_1, ..., H_{n-1})/\partial(x^1, ..., x^n)$. Modulo a reparameterization, this yields the Nambu equations (1.244). If the above diffeomorphism exists globally, the first-order system becomes the Nambu mechanics.

Due to the superintegrability, the Nambu mechanics can be related to a Hamiltonian mechanics in which one of the integrals of motion H_a serves as the Hamiltonian, say, $H_1 = H$. Consider new local coordinates $y^1 = q(x)$, $y^2 = p(x)$, and $y^{a+1} = H_a(x) \equiv h_{a-1}$, $a = 2, 3, ..., n-1$, where the functions $p(x)$ and $q(x)$ satisfy the condition,

$$\frac{\partial(q, p, h_1, h_2, ..., h_{n-2})}{\partial(x^1, x^2, ..., x^n)} = 1. \tag{1.245}$$

As noted above, the Nambu equations are invariant under general coordinate transformations with the unit Jacobian. Therefore in the new variables, the Nambu equations (1.244) become,

$$\dot{p} = -\frac{\partial H}{\partial q}, \quad \dot{q} = \frac{\partial H}{\partial p}, \quad \dot{h}_a = 0\,. \tag{1.246}$$

Clearly, this is nothing but the canonical Hamiltonian dynamics in which the Hamiltonian H also depends on the integrals of motion h_a, $H = H(p, q, h)$. The Hamiltonian equations of motion (1.246) are invariant under canonical transformations of p and q. This freedom is associated with the fact that a general solution of Eq. (1.245) is defined modulo general canonical transformation of p and q when the h_a remain fixed.

Equations (1.246) may be obtained from the stationary action principle for

$$S_N = \int \left[p\dot{q} + \lambda^a \dot{h}_a - H(q,p,h)\right] dt, \tag{1.247}$$

where the auxiliary variables or Lagrange multipliers λ^a serve as the momenta canonically conjugated to h_a. Their evolution is determined by the Hamiltonian equation of motion, $\dot{\lambda}^a = -\partial H/\partial h_a$, in addition to (1.246). So the Nambu equations (1.244) can be obtained from Hamilton's principle for the action (1.247) written via the original variables and the Lagrange multipliers λ^a,

$$S_N = \int \left[A_i \dot{x}^i + \lambda^a \partial_i h_a \dot{x}^i - H(x)\right] dt, \tag{1.248}$$

where the vector field, $A_i(x) = p(x)\partial_i q(x)$ is defined by a solution of (1.245). Different solutions of (1.245) are related by a canonical transformation. So the use of another solution in the action (1.247) would lead to modifying only the boundary term, which immediately follows from the canonical one-form transformation, $A_i dx^i = pdq = PdQ + dF$, for any canonical transformation $(p,q) \to (P,Q)$. A variational principle for the Nambu mechanics in terms of only the original variables x^i can also be formulated. The action is the analog of the Poincaré–Cartan integral invariant for a special $(n-1)$-differential form [54] rather than the one-form as in (1.248).

The above analysis shows that a Hamiltonian dynamics can always be cast in the form of the Nambu mechanics if the Hamiltonian contains some parameters (physical constants). The idea is to declare the physical parameters of the Hamiltonian as additional Nambu Hamiltonians h_a. For example, put $H = \left(x_1^2 + x_2^2 x_3^3\right)/2$ and $h_1 = x_3$. Since h_1 is an integral of motion $x_3 = \omega = \text{const}$, the dynamics of the other two variables coincides with the Hamiltonian dynamics of the harmonic oscillator with $H = (p^2 + \omega^2 q^2)/2$, where $p = x_1$ and $q = x_2$, as follows from (1.246). For odd n, the Nambu mechanics can also be interpreted as the Hamiltonian dynamics with a Lie–Poisson symplectic structure (instead of the conventional symplectic structure). This also relates the Nambu mechanics to a Hamiltonian dynamics on phase spaces of a non-trivial topology (e.g. a spherical phase space: $h_1 = \delta_{ij} x^i x^j$, $j, i = 1, 2, 3$). Relations between the Nambu mechanics and dynamics with quantum group symmetries can be established as well (see next section).

1.14.2 Lie–Poisson symplectic structure

Let X be a Lie algebra with a basis λ_a, $a = 1, 2, ..., n$, $[\lambda_a, \lambda_b] = F_{ab}^c \lambda_c$, with F_{ab}^c being the structure constants. Consider the space of analytic real functions on X.

Put $\nabla A(x) = \lambda_a \partial A/\partial x^a$ for a function $A(x)$ on X. The Lie–Poisson bracket is defined by,

$$\{A, B\}_{LP} = \langle x, [\nabla A, \nabla B] \rangle = F_c^{ab}\, x^c \, \frac{\partial A}{\partial x^a} \frac{\partial B}{\partial x^b}, \tag{1.249}$$

where $x = \lambda_a x^a$, and the basis is orthonormal with respect to the inner product, $\langle \lambda_a, \lambda_b \rangle = \delta_{ab}$. It is not difficult to verify that the Lie–Poisson bracket is skew-symmetric and satisfies both the Leibniz rule and the Jacobi identity. The latter follows from the Jacobi identity in the Lie algebra. In contrast to the canonical variables in the Hamiltonian dynamics, the phase space may have an odd dimension, the canonical symplectic structure has the form,

$$\{x^a, x^b\}_{LP} = F_c^{ab}\, x^c. \tag{1.250}$$

Given a Hamiltonian $H = H(x)$, the dynamics is defined by,

$$\dot{B} = \{B, H\}_{LP},$$

for any $B = B(x)$. In particular,

$$\dot{x}^a = F_c^{ab}\, x^c \, \frac{\partial H}{\partial x^b}. \tag{1.251}$$

A peculiarity of Lie–Poisson systems is that there are *absolute* integrals of motion, i.e. functions of x that remain integrals of motion for any choice of the Hamiltonian H. Indeed, the bracket (1.250) can be identified with the Lie product, while the local phase space coordinates x^a are viewed as generators of X. Hence, Casimir polynomials $C_i(x)$ of x (constructed by means of invariant symmetric tensors (see Section 8.1)) have the vanishing bracket with any x^a,

$$\{x^a, C_i(x)\}_{LP} = 0,$$

so that $\{C_i, H\}_{LP} = 0$ for any H. Therefore the phase space of the system is foliated into manifolds defined by the algebraic equations $C_i(x) = C_{i0} = \text{const.}$, because the system never leaves any such manifold regardless of the choice of the Hamiltonian. For example, the dynamics on the spherical phase space can be constructed by choosing $X = so(3) \sim su(2) \sim R^3$, i.e. $F_{abc} = \varepsilon_{abc}$. The Casimir invariant polynomial is $C(x) = \delta_{ab} x^a x^b$. Its level surfaces $C(x) = r_0^2$ are spheres of radius r_0. The canonical quantization of the Lie–Poisson symplectic structure for $X = su(2)$ (i.e. the replacement of x^a by Hermitian operators $\hat{x}^a$, with the commutation relation defined by the rule $[\,,\,] = i\hbar\{\,,\,\}_{LP}$, i.e. $[\hat{x}^a, \hat{x}^b] = i\hbar\varepsilon^{abc}\hat{x}^c$) leads to the quantum mechanics of the angular momentum (spin).

Consider the three-dimensional Nambu mechanics with the second Hamiltonian $H_2 = \delta_{ij} x^i x^j/2$. Then Eqs. (1.244) coincide with (1.251) for the algebra $su(2)$, where $F_{abc} = \varepsilon_{abc}$. A generic H_2 may be associated with a Casimir operator in a deformation of $su(2)$. Indeed, put $\{x^i, x^j\} = \omega^{ij}(x)$, where the symplectic structure ω^{ij} is to be chosen so that it is skew-symmetric, satisfies the Leibniz rule and the Jacobi identity. The skew-symmetry requires that $\omega^{ij} = \varepsilon^{ijk}\omega_k$ for some vector ω_k. The Leibniz rule holds naturally for the Poisson bracket defined by (1.129) (the fact that the phase space is odd-dimensional is irrelevant here). Finally, the Jacobi identity,

$$\{x^i, \{x^j, x^k\}\} + \{x^j, \{x^k, x^i\}\} + \{x^k, \{x^i, x^j\}\} = 0\,,$$

is equivalent to the following condition, $\varepsilon^{ijk}\omega_i\partial_j\omega_k = 0$. The simplest solution of this equation $\omega_i = \partial_i H_2$ leads to the Nambu mechanics (1.244).

The variables x^i may be viewed as generators of the "deformed" $SU(2)$ where the Lie bracket is defined by the Poisson bracket with the factor i. Then the function $H_2(x)$ is the Casimir invariant of the deformed group,

$$\{x^i, x^j\} = \varepsilon^{ijk}\,\partial_k H_2(x), \tag{1.252}$$

because $\{x^i, H_2\} = 0$. In particular, with the choice,

$$H_2(x) = x_1^2 + x_2^2 + \frac{\sinh^2(qx_3)}{q^2}$$

(q being a "deformation" parameter), relations (1.252) (with the aforementioned factor of i in the right-hand side) coincide with the commutation relations for generators of the quantum group $SU_q(2)$ [55]. The canonical quantization of this system leads to quantum mechanics of the so-called q-spin. When the deformation parameter tends to zero ($q \to 0$), the Casimir invariant H_2 of $SU_q(2)$ becomes the Casimir invariant of $SU(2)$, and the Poisson bracket (1.252) turns into the Lie–Poisson bracket for the $su(2)$ algebra.

1.14.3 Non-symplectic structures

Another natural extension of the Hamiltonian mechanics is obtained by relaxing the condition that the symplectic two-form is closed [37, 56], i.e. $d\omega \neq 0$. The Jacobi identity no longer holds and, hence, the Darboux theorem cannot be applied to such dynamics. The Liouville theorem (the phase space volume conservation) still holds. Indeed, the volume element is $V = (\wedge\omega)^n$, where the phase space dimension is $2n$. Then $dV = 0$, as the volume form V has the largest degree possible in the phase space. Evidently, the Poincaré invariants associated with lower-degree forms are no longer conserved quantities for a generic ω.

Non-symplectic dynamics may be used to describe physical systems such as those with anholonomic constraints [9]. Another example is provided by system (1.11) or (1.31). Consider a six-dimensional phase space spanned by three momenta p_i and three coordinates q_i. Consider a 6×6 matrix [56],

$$\omega^{\mu\nu} = \begin{pmatrix} g\varepsilon_{ijk}p_k & \delta_{is} \\ -\delta_{rj} & f\varepsilon_{rsk}q_k \end{pmatrix}, \tag{1.253}$$

where ε_{ijk} is a totally skew-symmetric unit tensor, $\varepsilon_{123} = 1$, $g = g(p)$, and $f = f(q)$ $(i, j, k, r, s = 1, 2, 3)$. Let the Hamiltonian be $H = p_i^2/2$ (a "free particle"). Here $p_i^2 \equiv p_i p_i$ and, similarly, $q_i^2 = q_i q_i$. Then the Hamiltonian equations of motion, $\dot{x}^\mu = \omega^{\mu\nu}(\partial H/\partial x^\nu)$, where x^μ denotes the pair (q_k, p_j), coincide with (1.31) if $f(q) = \alpha$. However, the two-form, $\omega = \omega_{\mu\nu} dx^\mu \wedge dx^\nu$ associated with the matrix (1.253), $\omega_{\mu\sigma}\omega^{\sigma\nu} = \delta_\mu^\nu$, is not closed. If $f = -\lambda/|q|^3$, where λ is a constant and $|q| = \left(q_i^2\right)^{1/2}$, then Eqs. (1.31) describe the motion of a charged particle in the magnetic monopole field,

$$\ddot{q}_i = \lambda \frac{\varepsilon_{ijk} q_j \dot{q}_k}{|q|^3}.$$

These equations were first considered by Poincaré [10]. In modern notation, $\lambda = e g_e/(mc)$, where e is the electric charge, m is the particle mass, g_e is the magnetic charge, and c is the speed of light in a vacuum.

1.15 Hamiltonian mechanics. Recent developments

Hamiltonian and Lagrangian mechanics are traditionally assumed to provide equivalent, but different in appearance, descriptions of the time evolution of mechanical systems. For the simplest systems, this is indeed so. More elaborate studies show, however, that the Hamiltonian mechanics is a more general theory. This conclusion is based, in particular, on the analysis of relations between the classical and quantum theories.

A Hamiltonian mechanics is defined by an even-dimensional manifold, Γ^{2n}, a non-degenerate closed two-form, $\omega = \omega_{ij} dx^i \wedge dx^j$ $(\det \omega_{ij} \neq 0$ and $d\omega = 0)$ on it, and a Hamiltonian $H = H(x)$ as a real function on Γ^{2n}; here x^i are local coordinates on Γ^{2n}. The Poisson bracket is then defined by the two-form,

$$\{f, g\} = \omega^{ij} \frac{\partial f}{\partial x^i} \frac{\partial g}{\partial x^j}, \quad \omega^{ik}\omega_{kj} = \delta_j^i, \quad i, j, k = 1, 2, ..., 2n, \tag{1.254}$$

for any smooth functions f and g on Γ^{2n}. Finally, one has to "postulate" the existence of time, as an evolution parameter, and the equation of motion,

$$\dot{f} = \{f, H\}, \quad \dot{f} = \frac{df}{dt}. \tag{1.255}$$

The Jacobi identity is equivalent to the closedness of the two-form. There are two obvious distinctions of the Hamiltonian mechanics over the Lagrangian. First, the former requires an additional structure, the closed two-form, to define the time evolution. Second, in the variational approach for systems with n degrees of freedom, the equations of motion are obtained by varying $2n$ independent variables in the Hamiltonian formalism, while in the Lagrangian formalism only n independent variables are needed.

The first, less trivial, observation follows from the very construction of the Hamiltonian mechanics: the Hamiltonian equations of motion (1.255) are not invariant under the time reversal $t \to -t$. They are invariant only under simultaneous transformations, $t \to -t$ and $\omega \to -\omega$. That the two-form ω is not degenerate makes the phase space Γ^{2n} orientable, which implies the existence of the time arrow. For example, a phase space trajectory of the harmonic oscillator is an ellipse. For any choice of initial conditions, the ellipse is always traced out *clockwise*. The clockwise orientation is reversed only if $t \to -t$. This resolves an old paradox that, even if the Euler–Lagrange equations of motion of a point-like particle in an external force field happen to be invariant under the time inversion $t \to -t$, the traveling back in time is not possible.

As noted, the covariance of Hamilton's variational principle (1.150) under canonical transformations demands that the end points of the varied phase space trajectory $q^i(t)$, $p_i(t)$ remain fixed, i.e. the variation is carried out only with respect to $q^i(t)$ and $p_i(t)$ with $t_1 < t < t_2$ (an open interval), where $t_{1,2}$ are the initial and final times. The $2n$ Hamiltonian equations of motion do not have solutions under the generic choice of $4n$ initial and final conditions. Since variations of the final (or initial) point are still allowed to extremize the action, one can always find the initial and final configurations for which the Hamiltonian equations of motion have solutions which are the actual extrema. This is in contrast to the least action principle for the Lagrangian action where the extremum trajectory exists if both the final and initial points are fixed in the *configuration* space. It turns out to be possible to find a covariant variational principle in which the variation with respect to the final point is carried out simultaneously with all the intermediate points of the trajectory, while only the initial point remains fixed [57]. It is worth noting that it follows from the quantum mechanics that the initial value problem can be resolved by considering two action functionals. Namely, one should add the integration with respect to q into the representation (2.4) of the quantum transition amplitude, and introduce two time parameters ϵ_1 and ϵ_2 (corresponding to the two evolution operators $\hat{U}_{\epsilon_1}$ and $\hat{U}_{\epsilon_2}$). Then the two functionals,

$$S_{1H} = \int (p_i \dot{q}^i - H)dt, \qquad S_{2H} = \int (-q^i \dot{p}_i - H)dt$$

should be varied, the first one with respect to p_i only, the second one with respect to q^i. This solves the initial value problem. Apparently, the variational principle motivated by quantum mechanics is not covariant under general coordinate transformations in the phase space, which is natural as the quantization and coordinate transformations are not commutative, in general (see Chapter 2).

Remarkably, the Hamiltonian formalism allows one to find a possible answer to the fundamental question in quantum mechanics: while being causal, why does the Schrödinger equation only allow for predictions of probabilities of events [58]? Suppose there is a source of random forces at the Planck distances (e.g. a thermal bath). Then one can only define a probability that a system is in a particular state. The probability is determined by the Gibbs distribution,

$G = h^{-1}\exp[-\beta H(p,q)]$, where $\beta = (kT)^{-1}$, k is the Boltzman constant, h is a normalization constant, and T is the temperature (of the thermal bath). Both the Hamiltonian equations of motion and the Schrödinger equation follow from this hypothesis. Let x denote the pair (p,q) of canonical variables. A variation δx can be decomposed into the sum,

$$\delta x = \delta x_\perp + \delta x_\parallel, \tag{1.256}$$

where the variations $\delta x_\perp$ do not change the Gibbs distribution (i.e. the Hamiltonian is not changed), while $\delta x_\parallel$ do ($\delta x_\parallel$ and $\delta x_\perp$ are variations along the level surfaces of the Hamiltonian and perpendicular to them, respectively). Consider first the variations $\delta x = \delta x_\perp$. The classical equations of motion follow from the condition that H remains constant along the trajectory,

$$\delta H(x) = \partial_i H(x)\delta x^i = 0. \tag{1.257}$$

A solution of Eq. (1.257) can be written in the form, $\delta x^i = \omega^{ij}\partial_j H\delta t$, where $\omega^{ij} = -\omega^{ij}$ and t is a parameter. If the matrix ω^{ij} is dimensionless, then t has the dimension of time. Putting $\dot{x}^i = \delta x^i/\delta t$, the equations of motion are obtained, $\dot{x}^i = \omega^{ij}\partial_j H$. If a non-degenerate closed two-form can be associated with the matrix w^{ij} (which is not always possible), then these equations define Hamiltonian dynamics (see (1.254) and (1.255)).

Next, consider the harmonic oscillator $H = \omega(p^2+q^2)/2$. It turns out that the evolution of small deviations from the equilibrium distribution is described by the probability amplitudes. In order to demonstrate this, complex canonical variables (1.206) are introduced. The Gibbs distribution determines the measure,

$$d\mu(a^*,a) = \frac{-i}{2\pi} da^* \wedge da\, \beta\omega e^{-\beta\omega a^* a}, \qquad \int d\mu(a^*,a) = 1. \tag{1.258}$$

Consider small deviations from the equilibrium distribution: $a^* \to a^* + \epsilon^*$, $a \to a + \epsilon$. One has,

$$d\mu \to d\mu_\epsilon = d\mu |f_\epsilon(a)|^2, \quad f_\epsilon(a) = e^{-\beta\omega\epsilon^* a} e^{-\beta\omega\epsilon^*\epsilon/2}, \qquad \int d\mu_\epsilon = 1. \tag{1.259}$$

Non-equilibrium distributions tend towards equilibrium ones with time. The law according to which the measure evolves is determined by solution of the classical equations of motion for the complex function f_ϵ (neglecting terms $O(1/t_r)$ where t_r is the relaxation time), which is the dynamical variable. It can therefore be concluded that:

(1) The Gibbs distribution determines the probability density in the phase space Γ^2;
(2) Its normalization constant $h = 2\pi/(\beta\omega)$ has the dimension of the action and can be identified with the Planck constant;
(3) The measure (1.259) determines the probability distribution of the harmonic oscillators in a non-equilibrium state;

(4) The integral

$$\int \frac{da^* \wedge da}{2\pi i\hbar}\, e^{-a^* a/\hbar} |f_\epsilon(a)|^2 \equiv \|f_\epsilon\|^2 , \quad \hbar = \frac{1}{\beta\omega}, \tag{1.260}$$

is the square of the norm of the function f_ϵ in the Fock space.

The latter observation allows one to define the inner product,

$$(g, f) = \int d\mu(a^*, a)\overline{g(a)} f(a), \tag{1.261}$$

where the overbar denotes the complex conjugation. With this inner product, any change in $f(a)$ (e.g. multiplication by a) can be described as the action of an operator, for example,

$$\hat{a}^\dagger\, \overline{f(a)} = a^* \overline{f(a)}\,, \quad \hat{a}\, \overline{f(a)} = \hbar \frac{d}{da^*}\, \overline{f(a)}, \quad [\hat{a}, \hat{a}^\dagger] = \hbar. \tag{1.262}$$

This can be verified by integration by parts in (1.261). The complex functions f and g are dynamical variables and possess all of the properties of probability amplitudes. For example, one can show that (g, f) is also a probability amplitude, and $|(f, g)|^2$ is the probability of finding the system characterized by the state vector f in the state g. Since the Hamiltonian contains both variables, a^* and a, when deriving the equations of motion, the non-commutativity of the corresponding operators should be taken into account. The operator ordering rule can already be established from classical theory,

$$H = \frac{\omega}{2} \begin{pmatrix} a^* \\ a \end{pmatrix} \begin{pmatrix} 0 & 1 \\ 1 & 0 \end{pmatrix} \begin{pmatrix} a^* \\ a \end{pmatrix} = \frac{\omega}{2} \left(a^* a + a a^*\right).$$

If the damping is neglected, then distribution (1.259) evolves according to the classical equations of motion for the variable $f(a)$,

$$\dot{f}(a) = \{f, H\} = -i\omega a\, \frac{df(a)}{da}, \tag{1.263}$$

where $\{A, B\} = i(\partial_{a^*} A \partial_a B - \partial_a A \partial_{a^*} B)$ is the Poisson bracket in Γ^2. By multiplying Eq. (1.263) by the Planck constant $\hbar$, the Schrödinger equation for the harmonic oscillator is obtained. The second classical Hamilton equation is deduced by complex conjugation. The absence of the "vacuum energy" term $\hbar\omega/2$ in (1.263) is explained by the fact that the non-commutativity of the operators has been neglected (see [59] for details). This is a consequence of the influence of the thermostat. Changing to polar coordinates, $a = \rho e^{i\varphi}$, in Γ^2, one infers that,

$$\{A, B\} = -\frac{1}{2\rho} \left(\partial_\varphi A \partial_\rho B - \partial_\rho A \partial_\varphi B\right),$$

$H = \omega\rho^2$, and the classical equations of motion become,

$$\dot{\varphi} = -\omega\,, \quad \dot{\rho} = 0. \tag{1.264}$$

The first equation in (1.264) describes a uniform motion along a circle in the phase space, while the second one is a consequence of the stationarity of the Gibbs distribution. The circular motion must be periodic, which leads to the energy quantization, $\omega_n = n\omega$ [59]. It should be emphasized that the classical motion leads to the energy quantization. The normalizability of the probability distributions eliminates the states with negative energies ($n \geq 0$, compare (1.263) and (1.259) for $f_n = a^n$).

It is worth noting that taking into account the damping of the oscillator motion leads to massless fields becoming massive, and, in particular, to a non-vanishing cosmological constant [59].

Thus, both classical Hamiltonian mechanics and quantum mechanics appear as different aspects of the existence of the Gibbs distribution. The significance of harmonic oscillators is that, first, both the strings and fields are their ordered sets, and, second, their Gibbs distribution is normal, i.e. predicted by the central limit theorem in the probability theory [59]. It is interesting that placement of a bosonic string into a thermostat leads to the emergence of fermions and the idea of supersymmetry [59]. The cosmological constant $\lambda = 1/(2t_r^2)$ also appears.

From the point of view of a motion with fixed energy, Maupertuis' principle can be analyzed [37]. This asserts that the abbreviated action,

$$S_M = \int p_i dq^i$$

attains its minimum on the trajectory of the actual motion. At first glance, this action does not seem to contain any information about the dynamics of the system. But there is a tacit assumption that the motion conserves the energy, i.e. that the Hamiltonian is an integral of motion, $H(p,q) = E$. Bearing this fact in mind, the action should be written in the form,

$$S_M = \int dt\, \left[p_i(t)\dot{q}^i(t) - \lambda(t)(H - E)\right], \tag{1.265}$$

where $\lambda(t)$ is the Lagrange multiplier which is also varied. Rewriting the action (1.265) as,

$$S_M = \int d\tau\, \left[p_i(\tau)q'^i(\tau) - (H - E)\right],$$

where $d\tau = \lambda(t)dt$, $q'^i = dq^i(\tau)/d\tau$, one infers that the extrema of S_M are attained at the trajectories that solve equations

$$q'^i(\tau) = \frac{\partial H}{\partial p_i(\tau)}, \quad p'_i(\tau) = -\frac{\partial H}{\partial q^i(\tau)}.$$

In contrast to the Hamiltonian equations of motion in physical time t, following from Hamilton's principle, these equations determine only the shape of the trajectory in the phase space. The variation of action (1.265) with respect to λ yields the energy conservation condition. This demonstrates the meaning of Maupertuis' principle (see also the end of Section 3.1.1).

2
Hamiltonian path integrals

2.1 Introduction

2.1.1 Preliminary remarks

The path integral formalism is one of the most important tools in modern theoretical physics. Integration in functional spaces was considered for the first time by Norbert Wiener in his studies of diffusion processes [60, 61]. The corresponding integration measure is now known as the Wiener measure, i.e. the so-called Gaussian path integrals were studied. In quantum mechanics, functional integrals were introduced by R. Feynmann [62, 63] who was inspired by important observations due to P. A. M. Dirac concerning a relation between wave function and classical action [64, 65].

In modern physics the path integral formalism is applied not only to developing a quantum description for systems with infinitely many degrees of freedom as, for example, in quantum field theories or condensed matter physics, but also to ordinary quantum mechanical problems. Moreover there is a tendency to reformulate standard quantum mechanical problems (such as a particle in a "box" or potential well [66, 67], or a particle in the Coulomb field [68]) within the framework of the path integral method. The reason for doing this is rather obvious: physicists develop a new mathematical tool, a new language and, in order to comprehend all of its subtleties, it is necessary to investigate as many concrete examples as possible.

Use of the path integral formalism is widespread in quantum field theory because it is the most adequate mathematical tool for a wide range of problems which allows, not only for elegant formulation of problems, but also for a great deal of simplification of their theoretical analysis. In addition to these important, but still rather technical advantages, the path integral method opens up new possibilities for applications – it is natural for developing a quasiclassical approximation which allows us to carry out a theoretical analysis beyond conventional perturbation theory. In quantum mechanics path integrals have turned out to be useful not only because they provided a new insight into old problems or a better understanding of the method itself. The new approach allows us to solve several important problems: (i) quantization in curvilinear coordinates [69], (ii) quantization in curved spaces [69], as well as the operator ordering problem in constraints from a sufficiently large class (constraints at most quadratic in canonical momenta [70]).

Path integrals in which the integration is carried over trajectories in the configuration or phase space are called *Lagrangian* or *Hamiltonian* path integrals, respectively. The former were developed first [62]. Path integrals in the Hamiltonian formulation were introduced later [71–74] although the Hamiltonian path integral formalism is closer to quantum mechanics and, in fact, naturally follows from the latter. The Lagrangian path integral is derived from the Hamiltonian one by carrying out the integration over canonical momenta.

2.1.2 Quantization

Although the path integral formalism allows for generalization of standard (canonical) quantization [65], the standard quantization methods are used for its derivation because they are well developed and tested in experiments. Generally, a quantization is said to be a set of rules that define the quantum mechanical description of a given classical system, i.e. a description of the system by means of the probability amplitudes. The canonical quantization [65] means that all canonical variables q^k and p_k, where k labels degrees of freedom, and, hence, all functions of them are turned into operators that satisfy the canonical commutation relations,

$$[\hat{q}^k, \hat{p}_j] = i\hbar\{q^k, p_j\} = i\hbar\delta^k_j. \tag{2.1}$$

The operator algebra (2.1) is known as the Heisenberg algebra ($i^2 = -1$). The operators act in a Hilbert space whose elements are called state vectors. The canonical operators $\hat{q}^k$ and $\hat{p}_k$ are self-adjoint in the Hilbert space. The time evolution of state vectors is determined by the Hamiltonian operator which generates time shifts. The above recipe is by no means universal. It is applicable firstly only in Cartesian coordinates and secondly, when there is no operator ordering problem in the Hamiltonian and other functions of canonical variables that might be of interest. Note that upon canonical quantization an operator corresponding to a classical function $H(p, q)$ may not be uniquely defined because the arguments are no longer commutative. The aforementioned restrictions are troublesome. Clearly, even the simplest gauge theories may turn out be quite problematic for canonical quantization, since the physical variables are often associated with curvilinear coordinates. The path integral formalism appears to be more flexible and points to a closer connection between classical and quantum mechanics than one usually assumes. Here the quantization problem discussion is limited to the above remarks. Further and more elaborate studies of the quantization problem can be found in the textbooks [65, 75].

2.2 Hamiltonian path integrals in quantum mechanics

2.2.1 Definition of the Hamiltonian path integral

Let $\hat{H} = H(\hat{q}, \hat{p})$ be the Hamiltonian operator (the operator ordering problem is assumed to be somehow resolved so that $\hat{H}$ is Hermitian). Consider the matrix element (or the kernel) of the evolution operator, $\hat{U}_{t-t'} = \exp[-i\hat{H}(t-t')/\hbar]$. By partitioning the time interval $t-t'$ into N segments $\varepsilon = (t-t')/N$, the evolution operator kernel can be written as the convolution,

$$\langle q|\hat{U}_{t-t'}|q'\rangle = \int dq_1 \cdots dq_{N-1} \langle q|\hat{U}_\varepsilon|q_{N-1}\rangle \cdots \langle q_1|\hat{U}_\varepsilon|q'\rangle. \tag{2.2}$$

The kernel (2.2) defines the amplitude for the system to be in a state $|q\rangle$ at time t if it was in state $|q'\rangle$ at time $t' \le t$ (here $\hat{q}|q\rangle = q|q\rangle$). If the system were in a state $|\psi_0\rangle$, then in a time t its wave function would be $\psi_t(q) = \langle q|\psi_t\rangle = \langle q|\hat{U}_t|\psi_0\rangle = \int dq' U_t(q,q')\psi_0(q')$. The path integral is defined as the limit of (2.2) when $N \to \infty$ ($\varepsilon \to 0$) if it exists. For what follows the key question is: with what accuracy should the matrix element of the operator $\hat{U}_\varepsilon$ be computed as a function of ε when calculating the limit? From the structure of the product of kernels in (2.2), it is clear that in the decomposition of $\hat{U}_\varepsilon$ in the power series over ε only the linear terms should be retained. The reason is simple. Since $\hat{U}_\varepsilon \to \hat{I}$ as $\varepsilon \to 0$ (here $\hat{I}$ is the identity operator), the expression (2.2) is of the following type:

$$(1 + a_1/N) \cdots (1 + a_N/N) = \exp \sum \ln(1 + a_i/N), \qquad N \to \infty,$$

for a sequence a_i, where terms $o(1/N)$ in parentheses can be neglected in the limit. Here the problem of the existence of the limit in (2.2) will not be discussed. The interested reader may refer to the mathematical literature [76]–[95]. It is assumed that the operator $\hat{H}$ is such that the limit exists. In what follows the infinitesimal kernels $\langle q|\hat{U}_\epsilon|q'\rangle$ are computed with the accuracy determined above unless otherwise specified.

To begin with, consider the simplest case of a system with n degrees of freedom for which

$$\hat{H} = H(\hat{q}, \hat{p}) = \frac{\hat{p}^2}{2} + V(\hat{q}). \tag{2.3}$$

Here, for brevity, $\hat{p}^2 \equiv \hat{p}_i^2$. In what follows the index enumerating the degrees of freedom is suppressed whenever this is possible without confusion. The Hamiltonian path integral is obtained from (2.2) by taking the limit $\varepsilon \to 0$ after substituting the following expression for $U_\varepsilon(q,q')$:

$$\begin{aligned}\langle q|\hat{U}_\epsilon|q'\rangle &= \langle q| \exp\left(-\frac{i}{\hbar}\hat{H}\epsilon\right)|q'\rangle \approx \left\langle q\left|1 - \frac{i\epsilon}{\hbar}H(\hat{q},\hat{p})\right|q'\right\rangle \\ &= \left\langle q\left|1 - \frac{i\epsilon}{\hbar}H(q,p)\right|p\right\rangle \langle p|q'\rangle \approx \int dp \exp\left(-\frac{i}{\hbar}H(q,p)\epsilon\right)\langle q|p\rangle\langle p|q'\rangle\,.\end{aligned} \tag{2.4}$$

Here $\hat{p}|p\rangle = p|p\rangle$ so that

$$\langle q|H(\hat{q},\hat{p})|p\rangle = H(q,p)\langle q|p\rangle. \tag{2.5}$$

When transforming the transition amplitude in (2.4), the terms of order $O(\varepsilon^2)$ are omitted. Taking into account that $\langle q|p\rangle = (2\pi\hbar)^{-n/2}\exp(ipq/\hbar)$, where $pq \equiv p_i q^i$, one infers that,

$$U_\varepsilon(q,q') \approx \int\limits_{-\infty}^{\infty} \frac{dp}{(2\pi\hbar)^n} \exp\left\{\frac{i}{\hbar}\left[p(q-q') - \varepsilon H(q,p)\right]\right\}. \tag{2.6}$$

The approximation (2.2) of the Hamiltonian path integral has the form,

$$\langle q|\hat{U}_{t-t'}|q'\rangle \approx \int\limits_{-\infty}^{\infty} \prod_{k=1}^{N-1} \frac{dp_k dq_k}{(2\pi\hbar)^n} \frac{dp_0}{(2\pi\hbar)^n} \exp\left\{\frac{i}{\hbar}S_N\right\}, \tag{2.7}$$

$$S_N = \sum_{k=1}^{N} \left[p_{k-1}(q_k - q_{k-1}) - \varepsilon H(q_k, p_{k-1})\right],$$

where $q_N = q$ and $q_0 = q'$. The variables q_k, $k = 1, 2, ..., N-1$ can be interpreted as the values of the configuration space coordinate taken along a trajectory $q = q(t)$, $q(k\varepsilon) = q_k$. In the formal continuum limit $N \to \infty$ ($\varepsilon \to 0$), the difference $q_k - q_{k-1}$ is interpreted as $q_k - q_{k-1} = \varepsilon\dot{q}(k\varepsilon) + o(\varepsilon)$, so that the sum in S_N turns into the Riemann sum for the corresponding integral, and the conventional representation of the transition amplitude $U_{t-t'}(q,q')$ by the path integral follows,

$$\langle q|\hat{U}_{t-t'}|q'\rangle = \int \prod_{\tau\in(t',t)} \frac{dq(\tau)dp(\tau)}{(2\pi\hbar)^n} \exp\left\{\frac{i}{\hbar}\int\limits_{t'}^{t} d\tau\left[p\dot{q} - H(q,p)\right]\right\}. \tag{2.8}$$

Three remarks are in order for this well-known formal derivation of the path integral. First, the path integral (2.8) is a symbol (a convenient notation). Its meaning is determined by the pre-limit expression (2.7). The limit of (2.7) as $N \to \infty$ serves as the definition of the symbol (2.8). Other definitions would make sense as long as they do not contradict (2.7) and (2.8), because the latter follow from quantum mechanics. Second, Eq. (2.8) does not seem useful, for it assumes the carrying out of an infinite number of integrations (uncountably many, in fact). It turns out, however, that the integral (2.8) can be viewed as an iterated integral. Moreover, in some important particular cases (when, e.g. $H(p,q)$ is bilinear), it can be explicitly calculated. The fruitfulness of the path integral method largely depends on the latter property. Third (by number, not by importance), the question about the very existence of the limit of (2.7) is crucial. There is a theorem due to Kato and Trotter [94, 95] that addresses this question. In the present case the theorem is stated as [89]: *Let $\hat{A}$ and $\hat{B}$ be self-adjoint operators in a separable Hilbert space with domains being $D(\hat{A})$ and $D(\hat{B})$ such that the operator $\hat{A}+\hat{B}$ is self-adjoint in the intersection $D(\hat{A})\cap D(\hat{B})$. Then there exists the limit (in the operator sense), $[\exp(it\hat{A}/N)\exp(it\hat{B}/N)]^N \to \exp[i(\hat{A}+\hat{B})t]$ as $N \to \infty$.*

Let us discuss Eqs. (2.2)–(2.8) in detail.

It is clear from (2.2) that the matrix element (2.6) is the fundamental quantity of the formalism that is being developed, and the path integral is, in fact, fully defined by the infinitesimal kernel $U_\varepsilon(q,q')$. The factor at $i/\hbar$ in the exponential (2.6) can be written as,

$$\varepsilon\left[p(q-q')/\varepsilon - H(q,p)\right] \approx \int_0^\epsilon dt[p\dot{q} - H] \equiv S_H, \tag{2.9}$$

i.e. it is nothing but the classical Hamiltonian action of the system. Classical Hamiltonian equations of motion (1.25) follow from application of the least action principle. Equations (2.6) and (2.9) are the key to understanding the relation between classical and quantum mechanics: the transition amplitude (2.6) is determined by the classical action (S_H should be viewed as a functional of $q(t)$ and $p(t)$). When $\hbar \to 0$ in (2.8), the contribution from the classical trajectory is dominant in the amplitude (2.8). Conversely, this explains why the Hamiltonian equations of motion describe the motion of the classical system (if the amplitude (2.8) is postulated, then the classical mechanics is a limiting case of the quantum mechanics).

2.2.2 Lagrangian path integrals

Substitution of the Hamiltonian (2.3) into (2.6) and subsequent integration over the momentum variables yields,

$$\begin{aligned} U_\varepsilon(q,q') &= (2\pi i\varepsilon\hbar)^{-n/2}\exp\left\{\frac{i}{\hbar}\left[\frac{(q-q')^2}{2\varepsilon} - \varepsilon V(q)\right]\right\} \\ &\approx (2\pi i\varepsilon\hbar)^{-n/2}\exp\left\{\frac{i}{\hbar}\int_0^\varepsilon d\tau L\right\}. \end{aligned} \tag{2.10}$$

The exponential in (2.10) is proportional to the Lagrangian action $L(q,q') = \int d\tau L(q,\dot{q})$ of a classical trajectory as a function of the initial and final positions. In the case of the Hamiltonian (2.3) the trajectory between points q and q' can be approximated by a straight line segment in the integral $\int dtL$ when $\varepsilon \to 0$ (see [62]). After substituting (2.10) into (2.2) and taking the limit $N \to \infty$, the transition amplitude is written in the form,

$$\begin{aligned} &\langle q|\hat{U}_{t-t'}|q'\rangle \\ &= \lim_{N\to\infty}\int\prod_{k=1}^{N-1}\frac{dq_k}{(2\pi i\varepsilon\hbar)^{n/2}}\exp\left\{\frac{i}{\hbar}\sum_{k=1}^{N}\left[\frac{(q_k-q_{k-1})^2}{2\varepsilon} - \varepsilon V(q_k)\right]\right\} \\ &\equiv \int_{q(t')=q'}^{q(t)=q} Dq(\tau)\exp\left\{\frac{i}{\hbar}\int_{t'}^{t} d\tau\left[\frac{1}{2}\dot{q}^2 - V(q)\right]\right\}. \end{aligned} \tag{2.11}$$

The representation of the transition amplitude by the Lagrangian path integral (2.11) is called the Feymann–Kac formula [87]. In the case of a free particle ($V = 0$), the transition amplitude $U_\varepsilon^{(0)}$ depends only on the difference $q - q'$, $U_\varepsilon^{(0)}(q, q') = U_\varepsilon(q - q')$, and the following relation holds, $\int dq'' U_{\varepsilon_1}(q - q'')U_{\varepsilon_2}(q'' - q') = U_{\varepsilon_1+\varepsilon_2}(q - q')$. Hence, the evolution operator kernel for a finite time interval $t - t'$ reads,

$$U_{t-t'}(q - q') = [2\pi i(t - t')\hbar]^{-n/2} \exp\left[\frac{i}{\hbar}\frac{(q - q')^2}{2(t - t')}\right]. \tag{2.12}$$

Path integrals with quadratic functions in the exponential are called *Gaussian*. Note that Eq. (2.12) is obtained from the pre-limit integral in (2.11) only if the integration is carried out in the infinite limits (over $(-\infty, \infty)$).

Expression (2.10) is derived by calculating the ordinary n-dimensional integral (2.6). In order to work with the new symbol, i.e. the path integral introduced in (2.8) or (2.11), one should develop basic rules to handle and manipulate it in practical applications. This goal is achieved by studying computational techniques for multiple n-dimensional integrals and their properties, because the rules to handle the symbols (2.8) and (2.11) as well as their properties are determined by admissible transformations of the corresponding pre-limit expressions (2.7) and (2.11) and by the properties of the latter. The transformation rules are based on the following expression for an N-dimensional Gaussian integral:

$$G_N = \int \prod_{i=1}^{N} dx_i e^{-xAx+bx} = \pi^{N/2}(\det A)^{-1/2} e^{\frac{1}{4}bA^{-1}b}, \tag{2.13}$$

where $xAx \equiv x_i A_{ij} x_j$ is positive definite, $bx \equiv b_i x_i$. Formula (2.13) is proved by changing variables via an orthogonal transformation so that the form xAx becomes diagonal in the new variables. Some important facts resulting from expression (2.13) for the Gaussian integrals and other related formulas, that are often used, can be found in Section 8.3.1.

The above derivation of the Hamiltonian path integral gives rise to several questions [96].

(1) When deriving (2.4), the resolution of unity $1 = \int dp|p\rangle\langle p|$ is inserted to the right of operator $\hat{H}$. Had the resolution of unity been inserted to the left of $\hat{H}$, one would have obtained a different result because,

$$\langle p|H(\hat{q}, \hat{p})|q'\rangle = H(q', p)\langle p|q'\rangle. \tag{2.14}$$

Even for the simplest Hamiltonian (2.3), when the functions $H(q, p)$ in (2.5) and (2.14) coincide with the classical Hamiltonian, one should answer the question: are these two ways of deriving the path integral equivalent, i.e. can the difference $\varepsilon[H(q, p) - H(q', p)] \approx \varepsilon V_{,i}(q^i - q'^i)$ be neglected in (2.6)? Here $V_{,i} = \partial_i V$. As shown below, the answer is affirmative for the Hamiltonian (2.3). In a general case, when the Hamiltonian contains products of non-commuting operators, for example, the sum $\hat{q}\hat{p} + \hat{p}\hat{q}$, neither (2.5) nor (2.14) are applicable, because the classical Hamiltonian functions $H(q,p)$ and $H(q',p)$ they contain are obtained by replacing the canonical operators by the corresponding classical variables q and p or q' and p in $H(\hat{q},\hat{p})$, while the matrix element $\langle q|\hat{q}\hat{p} + \hat{p}\hat{q}|q'\rangle$ equals $\langle q|p\rangle p\langle p|q'\rangle(q+q')$. Is the replacement $q' \to q$ or $q \to q'$ allowed in the latter matrix element, i.e. can one neglect the terms $p\Delta q$ ($\Delta q \equiv q - q'$)? The answer is negative. In practice, there are Hamiltonians with more complicated combinations of non-commuting operators. So this question is just a particular case of a more general one. The quantum mechanical amplitudes (2.8) and (2.11) contain only classical quantities (q, p, S_H, S). A natural question arises: how is the non-commutativity of canonical operators reflected in the path integral method?

(2) The functions of coordinates in the pre-limit expressions are taken only at the boundary points of the interval Δq. How can one go over to the kernels (2.6) with the classical functions taken at internal points of the interval, for instance, $H((q+q')/2,p)$?

(3) How can one change integration variables in the path integral?

(4) Equation (2.8) cannot be used in curvilinear coordinates because of, in particular, the amplitude $\langle q|p\rangle$ change. What is the Hamiltonian path integral in this case?

(5) In curvilinear coordinates the canonical operators may have quite different spectra (e.g. discrete, lying in an interval or in a semi-axis as, for example, in polar coordinates). Is the general form of the Hamiltonian path integral preserved under the change of variables?

(6) The range of physical coordinates may be restricted due to boundary conditions (for example, a particle in an infinite potential well). What are the integration limits in this case?

(7) Finally, the last question (again, by number, not by its importance): when neglecting the terms $(\varepsilon p^2)^2$, etc., as $\varepsilon \to 0$ in (2.4), it has been ignored that integration over the momentum p is carried to infinite limits and, hence, there always exist values of p such that εp^2 is not small. Is the above limiting procedure justified? The same question concerns the terms $(\varepsilon V)^2$, etc., because $V(q)$ may not be bounded (e.g. $V \to \infty$ as $q \to \infty$), while the integration with respect to q in (2.7) is also extended to the whole real line. The question of what can be neglected and what cannot is usually addressed by examining soluble problems which are "close" or similar (in some sense) to the one studied. In many such cases, the solution (2.12) of the problem of a free particle can be investigated with this purpose. It is easily seen that function (2.12) cannot be expanded into a power series over $\varepsilon = t - t'$ in a neighborhood of $\varepsilon = 0$, in contrast to what has been assumed in (2.4). In other words, it seems that the exponential $\exp(-i\varepsilon p^2/(2\hbar))$ must be taken into account exactly i.e. without its linearization. This question is discussed in Section 2.3.3. The question about the class of admissible potentials

$V(q)$, for which the pre-limit representation (2.11) is valid, is studied in Section 2.7.

Answers to the above questions essentially comprise the content of this chapter.

2.3 Non-standard terms and basic equivalence rules

2.3.1 Non-standard terms

The discussion here is limited to studying dynamical systems defined by a Lagrangian of the form,

$$L = \frac{1}{2}\, g_{ij}(q)\dot{q}^i\dot{q}^j + A_i(q)\dot{q}^i - V_L(q), \tag{2.15}$$

where the metric tensor g_{ij}, vector field A_i, and the scalar function V_L are assumed to be independent of time. This Lagrangian determines a rather general class of dynamical systems that includes almost all physically interesting cases, and which, on the other hand, has a relatively simple form and can be studied by rather elementary means. The corresponding canonical Hamiltonian reads,

$$H = \frac{1}{2}g^{ij}(p_i - A_i)(p_j - A_j) + V_L, \tag{2.16}$$

where g^{ij} is the inverse of the metric tensor, $g^{ik}g_{kj} = \delta^i_j$. Upon canonical quantization of the system, the operator ordering problem in the Hamiltonian arises immediately. For the time being, let this question be left with no specific answer by assuming that one or other operator ordering has been made, and let us investigate instead the effects of operator ordering choice on the path integral form. Consider two particular choices:

$$\hat{H}^s = \frac{1}{2}g^{ij}(\hat{q})\hat{p}_i\hat{p}_j + f^j(\hat{q})\hat{p}_j + V_H(\hat{q}), \quad V_H = V_L + \frac{1}{2}A_iA^i, \tag{2.17}$$

$$\hat{H}^a = \frac{1}{2}\hat{p}_i\hat{p}_j g^{ij}(\hat{q}) + p_j f^j(\hat{q}) + V_H(\hat{q}), \quad f^i = -A^i,\ \hat{H}^a \neq \hat{H}^s. \tag{2.18}$$

The indices s and a stand for a *standard* (the $\hat{p}$'s are to the right of all the $\hat{q}$'s) and *anti-standard* (the $\hat{q}$'s are to the right of all the $\hat{p}$'s) operator ordering. The difference between the two corresponding path integrals will be studied. Consider the infinitesimal evolution operator kernel (2.6). The fundamental character of kernels $U_\varepsilon(q, q')$ is worth emphasizing once again: they are the simplest elements of the theory used to obtain the path integral and to determine its form and properties. This is the reason for starting with their investigation. So, let us find $U_\varepsilon(q, q')$ for $\hat{H}^s$ and $\hat{H}^a$. Making use of Eq. (2.5) in the former case and (2.14) in the latter, one finds that the exponentials in (2.6) differ by the term $(-i\varepsilon/\hbar)(H(q,p) - H(q',p)) \equiv (-i\varepsilon\hbar)\Delta H$ where,

$$\Delta H = \frac{1}{2} p_i p_j \left(g^{ij}_{,k}(q)\Delta^k - \frac{1}{2} g^{ij}_{,kl}(q)\Delta^k \Delta^l + \cdots \right)$$
$$+ p_j \left(f^j_{,k}(q)\Delta^k + \cdots \right) + V_{,k}(q)\Delta^k + \cdots ,$$
$$\Delta^k \equiv (q - q')^k = \Delta q^k. \tag{2.19}$$

Should these terms be taken into account in (2.7)? Obviously, one should with at least the first three of them. If the difference (2.19) could be neglected, this would mean that the operator ordering of non-commuting operators in $\hat{H}$ is irrelevant in the path integral formalism, and the path integral representation (2.8) is the same for the essentially distinct Hamiltonians $\hat{H}^s$ and $\hat{H}^a$. Apparently, this is not so, and the difference (2.19) (apart from the last term) should yield a contribution to the path integral which must be accounted for.

The terms that are to be added to the leading term $p\Delta$ in the exponential of the pre-limit expression (2.7) and which contain the products $(p)^n(\Delta)^m$ ($n \geq 0$ and $m \geq 1$, except for $n = m = 1$) are said to be *extra terms* or *non-standard terms.* These terms have complete information about the operator ordering in the Hamiltonian [97]. In order to solve problems by the path integral method for Hamiltonians of the form (2.16), one should develop rules for how to handle the extra terms. It turns out that one can get rid of the extra terms (i.e. one can turn a Hamiltonian with extra terms into an equivalent Hamiltonian without them [97, 98]) by means of the so-called *equivalence rules.*

2.3.2 Basic equivalence rules

The simplest way of deriving equivalence rules for the difference (2.19) is to move the operators $\hat{p}$ in $\hat{H}^s$ to the left by means of canonical commutation relations. The result would be the operator $\hat{H}^a$ plus terms containing the commutators of $\hat{p}$ with functions of $\hat{q}$. The latter are naturally related to the extra terms (2.19). Here a different approach is proposed. Since the aim of this section is to develop the path integral formalism, the equivalence rules will be derived within the Hamiltonian path integral method [97]–[101].

Consider the matrix element,

$$\begin{aligned} M_{qq'} &= \langle q|[\hat{q}^{i_1}, [\cdots, [\hat{q}^{i_k}, \hat{p}_{j_1} \cdots \hat{p}_{j_l}] \cdots]]|q'\rangle \\ &= \Delta^{i_1} \ldots \Delta^{i_k} \langle q|\hat{p}_{j_1} \cdots \hat{p}_{j_l}|q'\rangle \\ &= \int \frac{dp}{(2\pi\hbar)^n} \exp\left(\frac{i}{\hbar} p_i \Delta^i\right) \Delta^{i_1} \ldots \Delta^{i_k} p_{j_1} \cdots p_{j_l}. \end{aligned} \tag{2.20}$$

Using the obvious relations,

$$[\hat{q}^i, \hat{p}_j] = i\hbar\delta^i_j, \quad [\hat{q}^i, \hat{p}_{j_1} \cdots \hat{p}_{j_l}] = i\hbar\partial(\hat{p}_{j_1} \cdots \hat{p}_{j_l})/\partial\hat{p}_i, \quad \partial\hat{p}_i/\partial\hat{p}_j = \delta^j_i, \tag{2.21}$$

the matrix element (2.20) can be written in the form,

$$
\begin{aligned}
M_{qq'} &= \langle q|(i\hbar)^k \partial^{i_1} \cdots \partial^{i_k} (\hat{p}_{j_1} \cdots \hat{p}_{jl})|q'\rangle \\
&= \int \frac{dp}{(2\pi\hbar)^n} \exp\left(\frac{i}{\hbar} p_i \Delta^i\right) (i\hbar)^k \partial^{i_1} \cdots \partial^{i_k} (p_{j_1} \cdots p_{jl}).
\end{aligned}
\tag{2.22}
$$

Here $\partial^i = \partial/\partial p_i$. Combining (2.20) and (2.22), one obtains,

$$
\int \frac{dp}{(2\pi\hbar)^n} \exp\left(\frac{i}{\hbar} p_i \Delta^i\right) \left[\Delta^{i_1} \cdots \Delta^{i_k} p_{j_1} \cdots p_{jl} \right.
$$

$$
\left. -(i\hbar)^k \partial^{i_1} \cdots \partial^{i_k} p_{j_1} \cdots p_{jl}\right] = 0. \tag{2.23}
$$

The basic equivalence rules (2.23) form a foundation for the Hamiltonian equivalence rules [97, 98]. According to (2.23) the products $\Delta^{i_1} \cdots \Delta^{i_k} p_{j_1} \cdots p_{j_l}$ in the infinitesimal evolution operator kernel (2.6) can be replaced by the expressions $(i\hbar)^k \partial^{i_1} \cdots \partial^{i_k} p_{j_1} \cdots p_{j_l}$ which do not contain the extra terms. It follows from (2.23) that, if $k > l$, then integral (2.20) vanishes. Equation (2.23) is generalized to a generic function $F(q, p, \Delta)$,

$$
\int \frac{dp}{(2\pi\hbar)^n} \exp\left(\frac{i}{\hbar} p_i \Delta^i\right) \left[F(q, p, \Delta) - F^a(q, p, i\hbar\partial_p) \cdot 1\right] = 0, \tag{2.24}
$$

where $F^a \cdot 1$ means that all operators ∂_p in $F(q, p, i\hbar\partial_p)$ are placed to the left of all the momenta p ("anti-standard" ordering, reflected by the index a) and then the resulting operator is applied to the unity.

Remark. The rule (2.24) holds for any q.

Relations (2.23) and (2.24) are the basic equivalence rules. They follow from the fundamental postulates of quantum mechanics. By their means one can easily derive equivalence rules for generic Hamiltonians [100, 101].

It follows from the above analysis that the explicit form of the transition amplitude (2.6) is fully determined by peculiarities of the quantum mechanical description. Indeed, according to (2.20) and (2.21), expressions of the form $p^n \Delta^m$ in the path integral (more accurately, in its pre-limit form) correspond to the multiple commutator $[\hat{q}, [\cdots, [\hat{q}, \hat{p} \cdots \hat{p}] \cdots]]$. According to (2.5) and (2.14), the choice of the function $g_{ij}(q)$ or $g_{ij}(q')$ in H (2.16) is related to standard or anti-standard operator ordering in the Hamiltonian.

The end points of the interval $[q, q']$ do not exhaust all possibilities for the choice of argument of the functions in (2.6). One can take a generic *base* point $q_\alpha = q - \alpha\Delta q$ ($0 \le \alpha \le 1$) where the index α determines the position of the base point within the interval. What is the operator ordering that corresponds to such a choice of argument in (2.6)? In other words, to which operator does the choice of the function,

$$
H(q_\alpha, p) \equiv H_\alpha = \frac{1}{2} g^{ij}(q_\alpha) p_i p_j + f^j(q_\alpha) p_j + V_H(q_\alpha) \tag{2.25}
$$

in (2.6) correspond? In order to answer this question, let us introduce the α-bracket (in the case of a one-dimensional space, $n = 1$) [99],

$$\{\hat{p}^m, f(\hat{q})\}_\alpha \equiv \alpha\hat{p}\,\{\hat{p}^{m-1}, f(\hat{q})\}_\alpha + (1-\alpha)\,\{\hat{p}^{m-1}, f(\hat{q})\}_\alpha\,\hat{p}, \tag{2.26}$$

whereas $\{\hat{p}^0, f(\hat{q})\}_\alpha \equiv f(\hat{q})$. For $m = 1$ expression (2.26) is written in the form, $\{\hat{p}, \hat{f}\}_\alpha = \alpha\hat{p}\hat{f} + (1-\alpha)\hat{f}\hat{p}$. In particular, $\{\hat{p}, \hat{f}\}_{1/2} = (\hat{p}f(\hat{q}) + f(\hat{q})\hat{p})/2$ corresponds to symmetric operator ordering. If, by definition, $\{\hat{f}, \hat{p}\}_\alpha = \alpha\hat{f}\hat{p} + (1-\alpha)\hat{p}\hat{f}$, then it is easy to see that $\{\hat{p}^m, \hat{f}\}_{1-\alpha} = \{\hat{f}, \hat{p}^m\}_\alpha$. It is not hard to verify that the following relation holds:

$$\langle q|\{\hat{p}^m, f(\hat{q})\}_\alpha|q'\rangle = \int \frac{dp}{2\pi\hbar}\, e^{\frac{i}{\hbar}p\Delta}\, p^m f(q_\alpha). \tag{2.27}$$

For $m = 0$ it is trivial (according to (2.23) its right-hand side is independent of α). Next, one infers that,

$$\langle q|\{\hat{p}, f(\hat{q})\}_\alpha|q'\rangle = \int \frac{dp}{2\pi\hbar}\, e^{\frac{i}{\hbar}p\Delta}\, p\,[\alpha f(q') + (1-\alpha)f(q)],$$

from which the validity of (2.27) for $m = 1$ is established by taking into account that $q = q_\alpha + \alpha\Delta$, $q' = q_\alpha - (1-\alpha)\Delta$, and that the terms $p\Delta^r$ ($r > 1$) do not contribute to the integral. For $m > 1$, relation (2.27) can be proved by mathematical induction. First, the α-bracket is replaced by the sum (2.26). Second, the resolution of unity $\int |q''\rangle\langle q''|dq''$ is inserted between the operators in each product. Third, by making use of the the induction hypothesis (2.27), all the p's can be replaced by the derivatives of the corresponding exponentials. Finally, integration by part is carried out. The derivatives of $f(q_\alpha)$ are canceled out. A generalization of the α-bracket to an n-dimensional space is obvious ($\hat{p}^m \to \hat{p}_{i_1}\cdots\hat{p}_{i_m}$ and so on).

Thus, function (2.25) appears in the exponentials (2.6) when α-ordering is chosen in the Hamiltonian,

$$H_\alpha(\hat{q}, \hat{p}) = \frac{1}{2}\,\{\hat{p}_i\hat{p}_j, g^{ij}(\hat{q})\}_\alpha + \{\hat{p}_i, f^i(\hat{q})\}_\alpha + V_H(\hat{q}). \tag{2.28}$$

The potential energy $V_H(q)$ may be taken at any point q_α. As is shown below, this does not affect the value of the path integral. Apparently, the choice of the mid-point, $q_{1/2} = (q + q')/2$ as the base point corresponds to symmetric *Weyl operator ordering.*

The technical tools given above are sufficient to derive the Hamiltonian equivalence rules. But before turning to details, it is instructive to find out how the basic equivalence rules (2.20) and (2.23) are modified in curvilinear coordinates, as well as to establish the basic equivalence rules for the Lagrangian formalism.

2.3.3 Basic integrals in curvilinear coordinates. Lagrangian basic equivalence rules

Here only topologically trivial curvilinear coordinates are discussed, i.e. the case when coordinate surfaces are topologically equivalent to a hyper-planc (obtained by continuous deformation from the hyper-plane). In other words, curvilinear coordinates like spherical or cylindrical ones are excluded. These are discussed separately (see Section 2.7).

In curvilinear coordinates the basic formulas of quantum mechanics should be modified in the following way. *The scalar product*:

$$\langle\psi_1|\psi_2\rangle \equiv \langle\psi_1|q\rangle\langle q|\psi_2\rangle \equiv \int d^n q\sqrt{g}\psi_1^*(q)\psi_2(q), \quad g = \det g_{ij}(q), \tag{2.29}$$

where g_{ij} is the metric tensor. *The momentum operators*:

$$\hat{P}_j = \frac{\hbar}{i} g^{-1/4}\frac{\partial}{\partial q^j} g^{1/4}; \quad [\hat{q}^k, \hat{P}_j] = i\hbar\delta_j^k. \tag{2.30}$$

These operators are self-adjoint relative to the scalar product (2.29) in the Hilbert space of states $|\psi\rangle$. Their eigen-functions are,

$$\langle q|p\rangle = (2\pi\hbar)^{-n/2} g^{-1/4}\exp(ipq/\hbar), \tag{2.31}$$

and the identity operator kernel (the δ-function) is given by,

$$\int dp\,\langle q|p\rangle\langle p|q'\rangle = (gg')^{-1/4}\delta^{(n)}(q-q') \equiv \delta(q,q'), \quad g' = g(q'). \tag{2.32}$$

The *Laplace–Beltrami* operator (which is proportional to the kinetic energy operator) takes the form,

$$\Delta_{LB} = \frac{1}{\sqrt{g}}\frac{\partial}{\partial q^i}\sqrt{g}g^{ij}\frac{\partial}{\partial q^j} = -\frac{1}{\hbar^2} g^{-1/4}\hat{P}_i\sqrt{g}g^{ij}\hat{P}_j g^{-1/4}. \tag{2.33}$$

It is also often written as $g^{-1/2}\partial_i \circ g^{1/2}g^{ij}\partial_j$; the symbol $\circ$ indicates that the partial derivative ∂_i should be understood as an operator acting after all of the operators to its right (not just as the derivative of $g^{1/2}g^{ij}$).

It is convenient to define here the *Christoffel symbols*,

$$[ij,k] = (g_{ik,j} + g_{jk,i} - g_{ij,k})/2, \tag{2.34}$$

the *Riemann curvature tensor*,

$$\begin{aligned} R_{ikjl} = &\frac{1}{2}(g_{ij,kl} - g_{il,kj} - g_{kj,il} + g_{kl,ij}) \\ &+g^{mn}([ij,m][kl,n] - [kj,m][il,n]), \end{aligned} \tag{2.35}$$

the *Ricci tensor*, and the *scalar curvature*,

$$R_{ij} = g^{kl}R_{iklj}, \quad R = g^{ij}R_{ij} \tag{2.36}$$

($R > 0$ for a sphere in a Euclidean space). Here and in what follows the indices of the same variance are contracted with the help of the metric tensor, $R = R_{jj}$.

It is clear from (2.31) and (2.32) how to modify equations (2.6)–(2.8) when going over to curvilinear coordinates. For example,

$$\left\langle q \left| \exp\left(-\frac{i}{\hbar}\hat{H}_\alpha\varepsilon\right) \right| q' \right\rangle \approx (gg')^{-1/4} \int_{-\infty}^{\infty} \frac{dp}{(2\pi\hbar)^n} \exp\left\{\frac{i}{\hbar}[p\Delta - \varepsilon H(q_\alpha, p)]\right\}, \tag{2.37}$$

i.e. the factor $(gg')^{-1/4}$ appears in the amplitude. The integral (2.20) also acquires the same factor $(gg')^{-1/4}$, while the basic equivalence rules (2.23) do not change. As the integration in the resolution of unity is carried out with the measure $\sqrt{g}dq \equiv \bar{d}q$, a similar modification occurs in the multiple integral (2.7) and the path integral (2.8). The latter has the form,

$$\langle q|\hat{U}_{t-t'}|q'\rangle = (gg')^{-1/4} \int \prod_t \frac{dq(t)dp(t)}{(2\pi)^n} \exp\left\{\frac{i}{\hbar}\int_{t'}^{t} dt[p\dot{q} - H(q,p)]\right\}. \tag{2.38}$$

In examining (2.37) an important observation is that the non-standard terms can also occur in the factor at the exponential in the kernel (2.37). For instance, when changing the argument q' to q, $g(q') = g(q-\Delta) = g - g_{,k}\Delta^k + \cdots$. Here questions about treatment of the non-standard terms as well as about their smallness also arise. Let us study this problem in more detail.

The key point is as follows. As argued in Section 2.2, it is important to establish the smallness hierarchy of the quantities being considered. The scale is set by ε; all other quantities are to be compared with this. In the basic equivalence rules, $\exp(ip\Delta/\hbar)$ is viewed as a quantity of order $O(1)$. When studying the dynamics, one should be able to accurately describe the motion of a free particle, i.e. the kernel (2.12) (as $t \to t'$) should be regarded as the $O(1)$ order quantity. Let us find out whether the equivalence rules should be changed under such a condition.

Thus, integrals of the form,

$$(gg')^{-1/4} \int \frac{dp}{(2\pi\hbar)^n} \exp\left[\frac{i}{\hbar}(p\Delta - \varepsilon p^2/2)\right] \Delta^{i_1}\cdots\Delta^{i_k} p_{j_1}\cdots p_{j_l}, \tag{2.39}$$

where $p^2 = g^{ij}p_ip_j$, are subject to study. In contrast to (2.20), the integral (2.39) does not vanish when $k > l$ because of the Gaussian exponential in the integrand. First of all let us prove that this exponential can be expanded in powers of ε in the integral (2.39). Put $k = l = 0$. The integration with respect to p yields,

$$U_\varepsilon(q, q') = (g/g')^{1/4}(2\pi i\varepsilon\hbar)^{-n/2} \exp\left(\frac{i}{\hbar}g_{ij}\frac{\Delta^i\Delta^j}{2\varepsilon}\right). \tag{2.40}$$

The infinitesimal evolution operator (2.40) is applied to a wave function,

$$\psi_\varepsilon(q) = \int \bar{d}q' U_\varepsilon(q, q')\psi(q'), \quad \bar{d}q = g^{1/2}d^nq. \tag{2.41}$$

When expanding the wave function into the power series, $\psi(q') = \psi(q-\Delta) = \psi(q) - \Delta^j\psi_{,j}(q) + \frac{\Delta^j\Delta^k}{2}\psi_{,jk}(q)\cdots$, the problem of smallness of the terms $(\Delta)^m$ arises. A qualitative estimate follows from the representation (2.40): the leading contribution to the integral (2.41) is given by those values of Δ for which the exponential in (2.40) is $\Delta^2/\varepsilon = O(1)$, i.e. $\Delta \sim \varepsilon^{1/2}$. The exact relation is given by the following *Lagrange equivalence rules* which confirm the aforementioned qualitative arguments (see their derivation below):

$$\int_{-\infty}^{\infty} \frac{d\Delta}{(2\pi i\varepsilon\hbar)^{n/2}} \exp\left(\frac{i}{\hbar}g_{ij}\frac{\Delta^i\Delta^j}{2\varepsilon}\right)\left[\Delta^{j_1}\cdots\Delta^{j_{2k}} - (i\varepsilon\hbar)^k g^{j_1\cdots j_{2k}}\right] = 0. \tag{2.42}$$

Here $g^{j_1\cdots j_{2k}}$ stands for the tensor in brackets in Eq. (2.47) below; the product of an odd number of Δ's is equivalent to zero. According to (2.42), the product $\Delta^i\Delta^j$ in the integrand can be replaced by $i\varepsilon\hbar g^{ij}$. Suppose that g_{ij} is independent of q, then (2.41) leads to,

$$\psi_\varepsilon \approx \psi_0 + \frac{1}{2}(i\varepsilon\hbar)g^{ij}\psi_{0,ij}, \tag{2.43}$$

from which, in particular, the Schrödinger equation for a free particle follows. On the other hand, by approximating the exponential function in (2.39) at $k = l = 0$, $\exp(-i\varepsilon p^2/2\hbar) \approx 1 - i\varepsilon g^{ij}p_ip_j/2\hbar$, making use of the relation $p_i\exp(ip\Delta/\hbar) = -i\hbar\partial\exp(ip\Delta/\hbar)/\partial\Delta^i$, and applying the result of integration with respect to p to the wave function $\psi(q')$, the expansion (2.43) is obtained again. This proves that the expansion of the exponential in (2.39) into the power series in ε is indeed admissible. It follows from the latter that, after expansion into the power series in ε, the equivalence rules for integrals of the type (2.39) are reduced to the basic equivalence rules (2.23). Thereby the aforementioned problem of non-standard terms in the pre-exponential factor is solved. The explicit form of the corresponding equivalence rules is established in Section 2.4

The discussion is completed with the derivation of Eq. (2.42) (of course, the constant factor $(2\pi i\epsilon\hbar)^{-n/2}$ may be omitted). Differentiation of the obvious equality,

$$\int_{-\infty}^{\infty} \frac{d^n\Delta}{(2\pi i\epsilon\hbar)^{n/2}} \exp\left(\frac{i}{\hbar}g_{ij}\frac{\Delta^i\Delta^j}{2\varepsilon}\right) = g^{-1/2} \tag{2.44}$$

with respect to g^{ij} yields,

$$\int_{-\infty}^{\infty} \frac{d^n\Delta\ \Delta^k\Delta^l}{(2\pi i\varepsilon\hbar)^{n/2}} \exp\left(\frac{i}{\hbar}g_{ij}\frac{\Delta^i\Delta^j}{2\epsilon}\right) = g^{-1/2}g^{kl}(i\varepsilon\hbar), \tag{2.45}$$

where the relation

$$dg = gg^{ij}dg_{ij}$$

(which follows from $\ln\det\hat{A} = \operatorname{Tr}\ln\hat{A}$) has been used to calculate the derivative. With the help of the equality

$$dg^{ij} = -g^{ik}g^{jl}dg_{kl}, \tag{2.46}$$

which is a consequence of the identity $d(g^{ik}g_{kj}) = 0$, the higher derivatives with respect to g_{ij} can be calculated with the result,

$$\int\limits_{-\infty}^{\infty} \frac{d^n\Delta\,\Delta^{j_1}\dots\Delta^{j_{2k}}}{(2\pi i\varepsilon\hbar)^{n/2}} \exp\left(\frac{i}{\hbar}g_{ij}\frac{\Delta^i\Delta^j}{2\varepsilon}\right) = g^{-1/2}(i\varepsilon\hbar)^k[g^{j_1j_2}\dots g^{j_{2k-1}j_{2k}}+$$

$$+\text{perm.}]_{(2k-1)!!} \equiv g^{-1/2}(i\varepsilon\hbar)^k g^{j_1\cdots j_{2k}}, \tag{2.47}$$

where the brackets contain the sum over all possible distinct products of tensors g^{ij}; all together $(2k)!/(k!2^k) = (2k-1)!!$ terms (indicated by the subscript of the bracket). After replacing $g^{-1/2}$ in (2.47) by the integral (2.44), the desired relation (2.42) follows. Sometimes a generalized relation turns out to be useful:

$$\int\limits_{-\infty}^{\infty} \frac{d^n\Delta}{(2\pi i\varepsilon\hbar)^{n/2}} \exp\left(\frac{i}{\hbar}g_{ij}\frac{\Delta^i\Delta^j}{2\varepsilon}\right)$$

$$\times\left[\Delta^{j_1}\dots\Delta^{j_{2k}} - (i\varepsilon\hbar)^l\sum g^{j_1\cdots j_{2l}}\Delta^{j_{2l+1}}\dots\Delta^{j_{2k}}\right] = 0, \tag{2.48}$$

where the summation is carried out over all $(2k-1)!!/(2l-1)!!(2k-2l-1)!!$ permutations of indices between the sets $(j_1,\dots,j_{2l})$ and $(j_{2l+1},\dots,j_{2k})$ that correspond to permutations in the identity $g^{j_1\cdots j_{2k}} = \sum g^{j_1\cdots j_{2l}}g^{j_{2l+1}\cdots j_{2k}}$ (where $1 \le l < k$ is an iteger). Thus, $(\Delta)^{2m} \sim \varepsilon^m$. It follows from the latter, in particular, that the term $V_{,k}\Delta^k$ in (2.19) can be safely neglected because, when multiplied by ε, it becomes of order $\varepsilon^{3/2}$.

Remark. If the maximal power of the momentum in the Hamiltonian equals n ($H \sim p^n/n$), then the leading contribution is due to trajectories for which $\Delta \sim \varepsilon^{1/n}$. This can be demonstrated by various means. The simplest is related to the Lagrangian formalism: $\partial H/\partial p = \dot{q}$, i.e. $p^{n-1} = \dot{q}$ and $L = p\dot{q} - H = (1-1/n)\dot{q}^{n/(n-1)}$; hence, the infinitesimal action $\varepsilon L \sim (\Delta^n/\varepsilon)^{1/(n-1)}$, that is, $\varepsilon L = O(1)$ when $\Delta^n/\varepsilon \sim 1$. The statement holds even for $n = 1$.

2.4 Equivalence rules

2.4.1 Hamiltonian equivalence rules

From the basis equivalence rules (2.23) it is not hard to derive the equivalence rules for systems described by Lagrangians of the form (2.15) [100, 101]. The most general expression for the infinitesimal evolution operator kernel reads (modulo an irrelevant factor),

$$U_\epsilon(q,q') = \int \frac{dp}{(2\pi\hbar)^n}\Big(1+\alpha(q,\Delta)\Big)\exp\left\{\frac{i}{\hbar}\Big[p_i\Big(\Delta^i + c^i(q,\Delta)\Big)\right.$$
$$\left. - \varepsilon\Big(a(p,q,\Delta)+V(q)\Big)\Big]\right\}, \tag{2.49}$$

$$a(p,q,\Delta) = \frac{1}{2}p_ip_ja^{ij}(q,\Delta) + p_if^i(q,\Delta). \tag{2.50}$$

Representation (2.49) covers all practically interesting cases. The following terms of the power series in Δ of the functions α, c, a, and f are sufficient:

$$\alpha(q,\Delta) = \alpha_k\Delta^k + \alpha_{kl}\Delta^k\Delta^l + O(\Delta^3), \tag{2.51}$$
$$c^i(q,\Delta) = c^i_{jk}\Delta^j\Delta^k + c^i_{jkl}\Delta^j\Delta^k\Delta^l + O(\Delta^4), \tag{2.52}$$
$$a^{ij}(q,\Delta) = g^{ij} + a^{ij}_k\Delta^k + a^{ij}_{kl}\Delta^k\Delta^l + O(\Delta^3), \tag{2.53}$$
$$f^i(q,\Delta) = f^i + f^i_{,j}\Delta^j + O(\Delta^2). \tag{2.54}$$

The power series truncation in (2.51)–(2.54) is determined by the estimates obtained in Section 2.3, namely, $p\Delta \sim i\hbar$ and $\Delta \sim \varepsilon^{1/2}$, and also by the possibility of neglecting terms that are small compared to ε. Note that the expansion of c^i begins with terms quadratic in Δ because the linear term is explicitly written in (2.49).

To derive the effective Hamiltonian without extra terms, the generalized equivalence rule (2.24) can be used. One infers,

$$U_\varepsilon(q,q') = \int \frac{dp}{(2\pi\hbar)^n}\exp\left\{\frac{i}{\hbar}[p_i\Delta^i - \varepsilon V(q)]\right\}$$
$$\times\left\{[1+\alpha(q,i\hbar\partial_p)]\left[1-\frac{i\epsilon}{\hbar}a(p,q,i\hbar\partial_p)\right]\right.$$
$$\left.\times\exp\left[\frac{i}{\hbar}p_ic^i(q,i\hbar\partial_p)\right]\right\}^a\cdot 1 + O(\varepsilon^2), \tag{2.55}$$

where superscript a indicates that all the operators ∂_p are placed to the left of all the p's, and the operator $\{\ \}^a$ is applied to the unit constant function. After carrying out the necessary differentiations, the kernel (2.55) in the standard form without extra terms reads,

$$U_\varepsilon^{\rm eff}(q,q') = \int \frac{dp}{(2\pi\hbar)^{n/2}}\exp\left\{\frac{i}{\hbar}[p_i\Delta^i - \varepsilon H_{\rm eff}(q,p)]\right\}, \tag{2.56}$$

$$H_{\rm eff} \equiv \frac{1}{2}p_ip_jg^{ij}(q) + f^i_{\rm eff}(q)p_i + V_{\rm eff}(q)$$
$$= V(q) + \frac{i\hbar}{\varepsilon}\left\{[1+\alpha(q,i\hbar\partial_p)]\left[1-\frac{i\varepsilon}{\hbar}a(p,q,i\hbar\partial_p)\right]\right.$$
$$\left.\times\left[1+\frac{i}{\hbar}p_ic^i(q,i\hbar\partial_p) + \frac{1}{2}\left(\frac{i}{\hbar}p_ic^i(q,i\hbar\partial_p)\right)^2\right] - 1\right\}^a\cdot 1. \tag{2.57}$$

Calculation of the necessary derivatives in (2.57) in which the functions α, c, a, and f are defined in (2.50)–(2.54) is not difficult and yields the result,

$$f^i_{\text{eff}} = f^i + i\hbar\left(a^{ik}_k + g^{ik}\alpha_k - 2c^k_{kl}g^{il} - c^i_{kl}g^{kl}\right), \tag{2.58}$$

$$\begin{aligned} V_{\text{eff}} &= V(q) + i\hbar\left[f^k_{,k} + \alpha_k f^k - 2c^k_{kl}f^l\right] \\ &+\hbar^2\Big[3g^{ij}c^k_{ijk} + 2a^{ij}_j c^k_{ki} + a^{ij}_k c^k_{ij} + 2c^i_{ij}\alpha^j + g^{ij}c^k_{ij}\alpha_k \\ &- 2\left(c^k_{ki}c^l_{lj} + c^l_{ki}c^k_{lj} + c^k_{kl}c^l_{ij}\right)g^{ij} - a^{ij}_{ij} - a^{ki}_k\alpha_i - g^{ij}\alpha_{ij}\Big]. \end{aligned} \tag{2.59}$$

Equations (2.58) and (2.59) are derived for the base point being q (i.e. all functions in them are taken at the point q). However, the equivalence rules (2.23) and (2.24) are independent of the choice of base point (see the remark in Section 2.3.2). Therefore the expressions found are also valid if the argument of the functions α, c, a, and f is changed to $q_\alpha = q - \alpha\Delta$, $0 \le \alpha \le 1$, [102], that is, these functions can be taken at any point of the interval $[q, q']$. It is also worth noting that rule (2.24) would hold for Hamiltonians that are not necessarily bilinear in momenta.

2.4.2 Lagrangian equivalence rules

The Hamiltonian equivalence rules derived above allow one to obtain analogous Lagrangian equivalence rules. Derivation of the Lagrangian equivalence rules (see [96]) can be simplified substantially [99]. Let the matrix element of the infinitesimal evolution operator have the form (here all the factors not relevant for our purposes are omitted as well),

$$\bar{U}_\varepsilon(q, q') = [1 + \alpha(q, \Delta)]\exp\left\{\frac{i}{\hbar}S(q, q')\right\}, \tag{2.60}$$

where,

$$\begin{aligned} S(q, q') = &\ \frac{1}{\varepsilon}\left[\frac{1}{2}g_{ij}(q)\Delta^i\Delta^j + B_{ijk}(q)\Delta^i\Delta^j\Delta^k + C_{ijkl}(q)\Delta^i\Delta^j\Delta^k\Delta^l\right] \\ &+ A_i(q)\Delta^i + D_{ij}(q)\Delta^i\Delta^j - \varepsilon V(q). \end{aligned} \tag{2.61}$$

The latter is the most general form of $S(q, q')$ which gives contributions of order ε to (2.60). Apparently, the tensors B, C, and D in (2.61) are symmetric. The exponential in (2.60) can be written as an integral with respect to the momenta,

$$\exp\left[\frac{i}{\hbar}S(q, q')\right] = \frac{(2\pi i\epsilon\hbar)^{n/2}}{\sqrt{g}}\int\frac{dp}{(2\pi\hbar)^n}\exp\left\{\frac{i}{\hbar}\left[S_H(q, p, \Delta) + D_{ij}\Delta^i\Delta^j\right]\right\}, \tag{2.62}$$

where,

$$S_H(q,p,\Delta) = p_i\bar{c}^i(q,\Delta) - \varepsilon\left[\frac{1}{2}g^{ij}(q)p_ip_j + \bar{f}^i(q,\Delta)p_i + \bar{V}_H(q,\Delta)\right],$$
$$\bar{c}^i = \Delta^i + B^i_{jk}\Delta^j\Delta^k + \left(C^i_{jkl} - \frac{1}{2}B^{im}_j B_{mkl}\right)\Delta^j\Delta^k\Delta^l,$$
$$\bar{f}^i = -A^i + B^i_{jk}A^j\Delta^k,$$
$$\bar{V}_H = V_L + \frac{1}{2}A_iA^i.$$

In order to make use of the equivalence rules (2.57)–(2.59), the exponential $\exp(iD\Delta^2/\hbar)$ in (2.62) is expanded into the power series which is then truncated, preserving only the term linear in D. This procedure leads to the replacement $\alpha_{ij} \to \alpha_{ij} + (i/\hbar)D_{ij}$ in (2.60). By applying the equivalence rules (2.58) and (2.59), in which it is taken into account that the tensors a^{ij}_k and a^{ij}_{kl} vanish, the kernel (2.60) can be represented in the form (2.56) (also modulo an irrelevant constant factor). The integration with respect to the momenta within it leads to the sought-for equivalent expression for the kernel (2.60),

$$\bar{U}^{\text{eff}}_\varepsilon(q,q') = \exp\left\{\frac{i}{\hbar}\left[g_{ij}\frac{\Delta^i\Delta^j}{2\varepsilon} + A^{\text{eff}}_i\Delta^i - \varepsilon V^{\text{eff}}\right]\right\}, \tag{2.63}$$
$$A^{\text{eff}}_i = -f^{\text{eff}}_i = A_i + i\hbar\left[B_{ikk} + 2B^k_{ki} - \alpha_i\right], \tag{2.64}$$
$$V^{\text{eff}} = V + \hbar^2\left[C_{ijkl}g^{ijkl} - B^{ijk}B_{\{ijk\}} + \frac{1}{2}\alpha_i\alpha^i - \alpha^i_i\right] - i\hbar D^i_i, \tag{2.65}$$

where $B_{\{ijk\}} = B_{ijk}+B_{jki}+B_{kij}$. The essence of the Lagrangian equivalence rules is contained in Eqs. (2.63)–(2.65). Note that for their derivation a particular case of the Hamiltonian equivalence rules (2.58) and (2.59) when $a^{ij}_k = a^{ij}_{kl} = 0$ appears to be sufficient. Derivation of (2.63)–(2.65) in the framework of the Lagrangian formalism only is technically much more involved.

A few concluding *remarks* are in order.

(1) The basis equivalence rules (2.23), (2.24), (2.42), and (2.48) hold exactly, while the kernels (2.49) and (2.56) (or (2.60) and (2.63)) are equivalent up to terms of order $O(\varepsilon^2)$,

$$\int dq'\left[U_\varepsilon(q,q') - U^{\text{eff}}_\varepsilon(q,q')\right]\psi(q') = O(\varepsilon^2), \quad \varepsilon \to 0.$$

(2) The equivalence rules (2.57)–(2.59) allow one to transfer extra terms from $a^{ij}(q,\Delta)$ to $c^i(q,\Delta)$ or $\alpha(q,\Delta)$ and back. For example, one can get rid of the extra terms in the "Hamiltonian" (2.50) by redefining c^i and α in (2.49). The corresponding explicit formulas are given in [99].

(3) Terms α_i and α_{ij} are not contracted with the other tensors which implies the possibility of getting rid of extra terms in the pre-exponential factor and

those in the exponential independently, in particular, from the former by the replacements $A_i \to A_i - i\hbar\alpha_i$ and $V \to V + \hbar^2(\alpha_i\alpha^i/2 - \alpha_{ii})$.

(4) In the pre-exponential factor, the following terms are to be retained:

$$\Delta,\ \Delta^2,\ \varepsilon p^n\Delta^m,\ p^n\Delta^k,\ 0 \le m \le n,\ 1 \le k \le n+2,\ n = 0, 1, \ldots. \tag{2.66}$$

(5) The operator nature of the canonical variables emerging in the path integral formalism was already understood by R. Feynman (see [62]). A comprehensive analysis of the problem in the Lagrangian formalism was given in [69] (see also [103]). In the framework of the Hamiltonian formalism, the problem is discussed in [73, 104–107]. The Lagrangian equivalence rules were first given in [69] (the simplest case) and in [107, 108]. In their most general and complete form, they were found in [96]. The Hamiltonian equivalence rules for a one-dimensional space were formulated in the publications [97, 109], while for Hamiltonians of the form (2.16) and (2.17) in spaces of an arbitrary dimension this was done in [98]. A general approach to the study of Hamiltonian equivalence rules was developed in [99–101].

2.5 Rules for changing the base point

2.5.1 Ambiguities of the formal expression (2.8)

One of the virtues of the path integral formalism is its direct relation to classical mechanics: the classical Hamiltonian appears in the exponential (2.8). As demonstrated above, this happens only for the simplest functions of the form (2.3). Already for dynamical systems defined by the functions (2.15) and (2.16), the operator ordering problem in the Hamiltonian emerges upon quantization. Path integrals are not exempt from this problem either. It reveals itself in two ways: first, as the problem of extra terms and, second, related to the latter, as the problem of choosing the base point (i.e. choosing the point in the interval $[q_i, q_{i-1}]$ at which the values of the functions g^{ij} and f^i are taken in the pre-limit expressions like (2.7)); see Section 2.3.2. In the framework of the operator approach, a change of the base point implies a change of the operator ordering in the Hamiltonian. Thus, the path integrals corresponding to distinct base points but with the same classical Hamiltonians are generally different. In particular, it follows from this observation that, first, the simplest expression (2.8) is not suitable for Hamiltonian functions of the form (2.16), and second, the formal symbol of the path integral should be supplied by a parameter α, $0 \le \alpha \le 1$, to indicate the base point q_α at which the values of the functions involved are taken in the pre-limit expression; for systems with the operator ordering problem in the Hamiltonian, the formal symbol of the path integral is meaningless unless the base point is indicated. In what follows, all "classical" functions like Hamiltonian and Lagrangian functions or action are supplied by an index specifying the base point, i.e. $H_\alpha(q, p)$ and $L_\alpha(q, \dot{q})$ or S_α, respectively. The operator ordering problem in the Hamiltonian is studied in Section 2.8.

2.5.2 Rules for changing the base point

Here the rules for changing the base point are formulated for the Hamiltonian and Lagrangian path integrals. The Hamiltonian formalism is studied first. According to (2.28), in order to define the transformation $q_\alpha \to q_\beta$, it is sufficient to find a relation between the α- and β-brackets (2.26) (i.e. between the brackets with distinct indices). This can be accomplished in two different ways. By making use of the relation,

$$q_\alpha = q_\beta - \delta\Delta, \quad \delta = \alpha - \beta, \quad \Delta = q - q', \tag{2.67}$$

the function $f(q_\beta - \delta\Delta)$ in Eq. (2.27) can be expanded into a power series over Δ to which the equivalence rules (2.23) are then applied. The resulting series obviously contains only a finite number of terms, because the power of Δ cannot exceed that of p according to (2.23). All the functions in the right-hand side would depend on q_β. Therefore, by reading Eq. (2.27) backwards, from right to left, a relation between the α- and β-brackets is inferred. The second way is purely algebraic. Consider an obvious equality,

$$\{\hat{p}, \hat{f}\}_\alpha = \{\hat{p}, \hat{f}\}_\beta + \delta[\hat{p}, \hat{f}], \quad \delta = \alpha - \beta. \tag{2.68}$$

This can also be written in the form,

$$\{\hat{p}, \hat{f}\}_\alpha = \{\hat{p} + \delta[\hat{p},\ ], \hat{f}\}_\beta, \tag{2.69}$$

where $\{[\hat{p},\ ], \hat{f}\}_\beta = \{1, [\hat{p}, \hat{f}]\}_\beta$. In fact, Eq. (2.68) defines the right-hand side of (2.69). The sought-for relation between the brackets with distinct indices follows immediately,

$$\{\hat{p}^m, f(\hat{q})\}_\alpha = \{(\hat{p} + \delta[\hat{p},\ ])^m, f(\hat{q})\}_\beta. \tag{2.70}$$

For direct verification of (2.70) the following relation might be useful:

$$\{\hat{p}^m, [\hat{p}, \hat{f}]\}_\alpha = [\hat{p}, \{\hat{p}^m, \hat{f}\}_\alpha] \qquad m = 0, 1, \ldots, \tag{2.71}$$

which is a trivial consequence of the linearity of the α-bracket in $\hat{f}$ and the commutativity of the momentum operators amongst each other.

Application of the generalized equation (2.70), where $\hat{p}^m \to \hat{p}_{i_1} \cdots \hat{p}_{i_m}$, to the Hamiltonian (2.28) yields,

$$\hat{H}_\alpha = \hat{H}_\beta - i\hbar\delta\left\{\hat{p}_i, \hat{g}^{ij}_{,j}\right\}_\beta - \frac{\hbar^2}{2}\delta^2\hat{g}^{ij}_{,ij} - i\hbar\delta\hat{f}^i_{,i}. \tag{2.72}$$

According to (2.72) there is a relation between the Hamiltonian functions in the exponential (2.6) which is given by,

$$H_\alpha(q,p) = H_\beta(q,p) - i\hbar\delta\left[p_i g^{ij}_{,j}(q_\beta) + f^j_{,j}\right] - \frac{\hbar^2}{2}\delta^2 g^{ij}_{,ij} \equiv H_\beta + H'_\beta. \tag{2.73}$$

For the functions in (2.73) whose argument is not shown, the choice of the base point is not essential: its change does not create any relevant terms. Thus, in the Hamiltonian path integral the following rule for changing the base point holds [96]:

$$\int \prod_t \frac{dpdq}{(2\pi\hbar)^n} \exp\left[\frac{i}{\hbar}\int_{t'}^{t}(p\dot{q} - H_\alpha)dt\right]$$
$$= \int \prod_t \frac{dpdq}{(2\pi\hbar)^n} \exp\left[\frac{i}{\hbar}\int_{t'}^{t}\left(p\dot{q} - H_\beta - H'_\beta\right)dt\right], \qquad (2.74)$$

where H'_β is defined in (2.73). In addition, Eq. (2.74) represents a sample of the correct expression for the Hamiltonian path integral for non-trivial Hamiltonians; the indices α and β indicate the points of intervals at which the values of the functions are taken.

The rules for changing the base point in the Lagrangian path integrals may be derived by two methods. In the framework of the Lagrangian formalism, the goal is achieved by changing the argument (2.67) in all functions, the subsequent expansion of the latter into power series over Δ in which only relevant extra terms are retained, and, finally, applying the Lagrangian equivalence rules (2.64) and (2.65). In doing so, the pre-exponential factor, which emerges when changing q_α to q_β in the measure $\sqrt{g}\,dq$, must also be taken into account. The above procedure leads to the following equivalence rules for changing the base point in the Lagrangian path integral:

$$\int D_\alpha q(t) \exp\left[\frac{i}{\hbar}\int_{t'}^{t} L_\alpha dt\right] = \int D_\beta q(t) \exp\left[\frac{i}{\hbar}\int_{t'}^{t}\left(L_\beta + L'_\beta\right)dt\right], \qquad (2.75)$$

where,

$$L'_\beta = -i\hbar\delta g_{ik,k}\dot{q}^i - i\hbar A_{j,j} - \frac{\hbar^2\delta^2}{2}\left(g_{ij,ij} - g_{kj,i}g_{ki,j}\right) \qquad (2.76)$$

for the Lagrangian of the form (2.15), and the index in $D_\alpha q$ refers to the argument of the density $\sqrt{g}$ in the measure (see (2.29)).

A faster method is based on integration over the momenta in (2.74). In this approach the base point in the measure $\sqrt{g}dq$ changes automatically. One should only bear in mind that the term H'_β is linear in momenta, i.e. $f^i : \ f^i \to \tilde{f}^i = f^i - i\hbar\delta g^{ij}_{,j}$ in the right-hand side of (2.74). Because of this and the equality $V_L = V_H - f^2/2$, the difference,

$$-\frac{1}{2}\left(\tilde{f}^2 - f^2\right) = i\hbar\delta f_i g^{ij}_{,j} + \frac{\hbar^2\delta^2}{2} g^{ik}_{,k} g^{ij}_{,j}$$

should be added to V_L (in addition to the terms already written in (2.74)). As a result, the effective Lagrangian L^{eff}_β is obtained in the form (2.15) in which,

$$A^{\text{eff}}_i = -\tilde{f}_i = -f_i + i\hbar\delta g_{ik} g^{kj}_{,j}, \quad \delta = \alpha - \beta, \qquad (2.77)$$

$$V^{\text{eff}}_L = V_L + i\hbar\delta f_i g^{ij}_{,j} + \frac{\hbar^2\delta^2}{2} g^{ik}_{,k} g^{ij}_{,j} - i\hbar\delta f^j_{,j} - \frac{\hbar^2\delta^2}{2} g^{ij}_{,ij}. \qquad (2.78)$$

To see the equivalence of the answers provided by Eq. (2.76) and Eqs. (2.77) and (2.78), one should go over to the contravariant tensors in the latter by means of $f^i = -g^{ij}A_j$, $A_{j,j} = f_i g^{ij}_{,j} - f^j_{,j}$ and the identities following from (2.46),

$$g^{ij}_{,j} g^{ik}_{,k} = g_{ij,j} g_{ik,k} \; ; \quad g^{ij}_{,ij} = g_{ki,i} g_{kj,j} + g_{ki,j} g_{kj,i} - g_{ij,ij} .$$

Recall that the indices of the same variance are contracted with the help of the metric tensor. Equations (2.75) and (2.76) solve the problem of changing the base point in the Lagrangian path integral for Lagrangians of the form (2.15).

In conclusion, it should be emphasized that here and in the preceding section the very choice of the Hamiltonian has not been the subject of the discussion, but rather the analysis of different operator orderings within it in relation to technical peculiarities of the path integral formalism. The question about choosing the Hamiltonian operator is discussed in Section 2.8.

Back in the 1970s, questions about the uniqueness of the very definition of path integrals were debated (see [105]), and the ambiguity of the standard formulation was interpreted as a defect of their definition via the discretization procedure (2.7) and (2.11) ([110], p.60). As shown above, the ambiguities associated with path integrals are two-fold. First, they reflect the dependence of the result of calculations on operator ordering in the Hamiltonian. However, this is not a drawback by any means. A real drawback would have been exactly the converse where the pre-limit expressions like (2.2) and (2.7) turned out be insensitive to operator ordering. Second, expressions like (2.8) do not define the path integral unambiguously, which is also not a defect of the very notion of the path integral defined by the limiting procedure, but rather due to the inadequacy of notation. As explained, functions in equations like (2.8) must be supplied by indices that determine the base point at which these functions are calculated in the pre-limit expressions (see (2.74) and (2.75)). The rules for changing the base point were formulated in [96].

2.6 Canonical transformations and Hamiltonian path integrals

2.6.1 Preliminary remarks

When working with Hamiltonian path integrals, it is important to know their properties under canonical transformations, in particular, one should be able to change variables within them. This rather difficult point is related to two problems: first, the general problem of the correspondence of canonical and unitary transformations (the terms *canonical transformation* and *unitary transformation* refer to transformations in the classical and quantum theories, respectively); second, it concerns the peculiarities of changing the Hamiltonian path integral under canonical transformations (a similar problem arises when changing variables in the Lagrangian path integral).

The importance of the first problem is evident. Since the path integral formalism is a purely quantum formalism, it is clear that one should have a general theory for the behavior of probability amplitudes under such transformations. There is a well-known statement due to Dirac [65] that, for systems having classical analogs, unitary transformations are analogous to canonical ones. For infinitesimal transformations, this analogy is straightforward [65]. In general, this problem is far from being solved [111–113]. The types of difficulties that occur may be gleaned from the simplest example of the coordinate transformation $Q = q^2$; the ranges of the old and new coordinates turn out to be different, while unitary transformations do not change the spectrum of the operators (see [111, 112] for more details).

As to the second problem, here the difficulties are related to the necessity of taking into account extra terms, additional to the classical action of the system. As noted, when describing the simplest system, the path integral contains only the classical Hamiltonian (2.3). Thereby an illusion is created that the problem of canonical transformations can be resolved only by a formalism of classical Hamiltonian mechanics. However, the point is that the extra terms, disappearing in the classical limit, must be retained in the path integral. If canonical transformations do not explicitly depend on time, the difference between the old and new infinitesimal actions is given by the expression (see Section 1.5),

$$p\Delta q - P\Delta Q = \frac{\partial F_1(q,Q)}{\partial q}\Delta q + \frac{\partial F_1(q,Q)}{\partial Q}\Delta Q + O(\Delta^2), \tag{2.79}$$

where $F_1(q,Q)$ is the generating function. If Eq. (2.79) is divided by ε and we let ε go to zero, the well-known equation of classical mechanics is obtained, $p\dot{q} - P\dot{Q} = dF_1/dt$, because terms like $(\Delta q)^2/\varepsilon \approx \dot{q}\Delta q$ tend to zero (along classical trajectories). These terms cannot, however, be neglected in the quantum description and, hence, in path integrals; the relevant extra terms must be retained in Eq. (2.79) (in particular, terms $O(\Delta^2)$). This is the key to the problem.

Yet another difficulty is related to the technical problem of constructing a unitary operator for a given generating function of a canonical transformation. No unique solution of this problem exists. So, in what follows, only canonical transformations determined by a generator are considered. A final remark: it is useful to distinguish between the change of variables in the path integral and a canonical transformation. The change of integration variables does not affect the boundary values of coordinates (e.g. q and q' in (2.7)) with respect to which no integration is carried out; when changing variables in the pre-limit expression, one should only make sure that all relevant extra terms are retained in the limit $N \to \infty$. Even though new and old integration variables can be related by a canonical transformation, such a change of variables does not imply any unitary transformation of vectors in the Hilbert space. The new variables may even be non-canonical. In other words, only the actual calculation method of the kernel $U_t(q,q')$ changes. Under unitary transformations corresponding to canonical transformations, the boundary values of coordinates q and q' are transformed as well as the wave functions, in addition to the change in the integration variables.

The problem of retaining relevant extra terms under a change of integration variables is considered first. The main points are easier to elucidate in the Lagrangian path integral formalism. Above all, the question about changing variables in the Lagrangian path integral is of interest on its own.

2.6.2 Change of variables in Lagrangian path integrals. Coordinates topologically equivalent to Cartesian coordinates

A coordinate system in a Euclidean space is specified by its coordinate surfaces. A particular coordinate surface is the surface on which this coordinate has a fixed value. Every point is then viewed as the point of intersection of specific coordinate surfaces. Since the coordinate surfaces are embedded into a Euclidean space, they are described by suitable algebraic equations. When changing variables in quantum mechanics, the topology of coordinate surfaces plays a significant role. Two sets of coordinates whose coordinate surfaces can be continuously deformed into one another are called *topologically equivalent* or simply *equivalent*. Topologically equivalent coordinates can be obtained, for example, by a small continuous deformation of a given net of coordinate surfaces. Coordinates topologically equivalent to Cartesian ones are said to be *trivial* (their coordinate surfaces can be continuously deformed to hyper-planes). The polar and Cartesian coordinates in the plane provide an example of topologically non-equivalent coordinates (circles, being coordinate curves of the radial variable, cannot be continuously deformed into straight lines). The change of integration variables associated with a change in topology of coordinate surfaces creates specific problems in the path integral formalism and will be studied separately in Section 2.7.

For systems described by a Lagrangian of the form (2.15) the standard infinitesimal action reads,

$$S(q,\Delta) = g_{ij}\frac{\Delta^i\Delta^j}{2\varepsilon} + A_i\Delta^i - \varepsilon V. \tag{2.80}$$

It depends on the difference of the coordinates $\Delta = q - q'$ which, upon change in variables $q^i = q^i(Q)$, is replaced by the truncated power series,

$$\begin{aligned}\Delta^i = q^i(Q) - q^i(Q') = q^i_{,j}(Q)\Delta^j_Q - \frac{1}{2}q^i_{,jk}(Q)\Delta^j_Q\Delta^k_Q \\ +\frac{1}{6}q^i_{,jkl}(Q)\Delta^j_Q\Delta^k_Q\Delta^l_Q - \cdots ,\end{aligned} \tag{2.81}$$

where $\Delta_Q = Q - Q'$. All terms explicitly written in (2.81) must be taken into account (the last term $\sim \Delta_Q^3$ makes a contribution $\sim \Delta_Q^4/\varepsilon$ when (2.81) is substituted into the first term in (2.80)).

Let d be the matrix with elements $d^i_j = q^i_{,j}(Q)$. Put

$$c^i_{jk} = -\frac{1}{2}\left(d^{-1}\right)^i_{i'} q^{i'}_{,jk}(Q)\ ; \quad c^i_{jkl} = \frac{1}{6}(d^{-1})^i_{i'}q^{i'}_{,jkl}(Q), \tag{2.82}$$

where d^{-1} is the inverse matrix. Functions $q(Q)$ are assumed to be smooth. Then,

$$\Delta^{i'} = d_i^{i'}\left(\Delta_Q^i + c_{jk}^i \Delta_Q^j \Delta_Q^k + c_{jkl}^i \Delta_Q^j \Delta_Q^k \Delta_Q^l + \cdots\right) \equiv d_i^{i'} \tilde{c}^i(Q, \Delta_Q). \tag{2.83}$$

After substitution of (2.83) into (2.80), the latter assumes a form identical to that of (2.61) if $g_{ij} \to \tilde{g}_{ij}$, $A_i \to \tilde{A}_i$, and the tensors $\tilde{g}_{ij}, \tilde{A}_i, B_{ijk}$, and so on are expressed via the tensors d_j^i, c_{jk}^i, and c_{jkl}^i:

$$\tilde{g}_{ij}(Q) = g_{i'j'}(q(Q)) d_i^{i'} d_j^{j'}\ ; \qquad B_{ijk} = \tilde{g}_{ii'} c_{jk}^{i'}\ ;$$
$$C_{ijkl} = \tilde{g}_{ii'} c_{jkl}^{i'} + \tilde{g}_{i'j'} c_{ij}^{i'} c_{kl}^{j'}/2\ ; \qquad \tilde{A}_i = d_i^j A_j\ ; \quad D_{ij} = \tilde{A}_k c_{ij}^k. \tag{2.84}$$

Now the general formulas (2.64) and (2.65) can be used. After some algebraic transformations the following result is inferred: the infinitesimal effective action S^{eff} is the sum in the square brackets in (2.63) in which g_{ij} is replaced by $\tilde{g}_{ij}$ defined in (2.84) and,

$$A_j^{\text{eff}} = A_j + i\hbar\left(\tilde{g}_{jl} c_{kk}^l + 2c_{jk}^k\right);$$
$$V^{\text{eff}} = V + \hbar^2\left[3c_{jkk}^j + c_{jj}^l c_{kk}^l/2 - 2c_{ij}^l c_{lj}^i\right] + \frac{i\hbar}{2} A_k q_{,jj}^k. \tag{2.85}$$

Summation over the repeated lower (upper) indices is carried out with the help of the tensor $\tilde{g}^{ij}(\tilde{g}_{ij})$, for example, $c_{kk}^l = c_{ik}^l \tilde{g}^{ik}$. Equations (2.85) determine the infinitesimal evolution operator kernel in the new variables so that the rule to change variables in (2.41) is given by,

$$\psi_\varepsilon \approx \int \frac{\sqrt{g'}dq'}{(2\pi i\varepsilon\hbar)^{n/2}} \exp\left[\frac{i}{\hbar} S(q, \Delta)\right] \psi(q')$$
$$\approx \int \frac{\sqrt{\tilde{g}'}dQ'}{(2\pi i\varepsilon\hbar)^{n/2}} \exp\left[\frac{i}{\hbar} S^{\text{eff}}(Q, \Delta_Q)\right] \psi(q(Q')). \tag{2.86}$$

Here $\tilde{g}' = g' J^2(Q')$, $J = \det q_{,j}^i$ (the Jacobian of the change of variables), and S is taken at the base point q, i.e. $\alpha = 0$. For Lagrangians of the particular form, $L = \dot{q}^2/2 - V$, the effective action was derived in [114], and for general Lagrangians this was done in [96]. Let us briefly discuss point transformations in the Hamiltonian path integral. It $q^i = q^i(Q)$, $P_i = d_i^j p_j$, and $d_i^j = q_{,i}^j$, then the transformation $q, p \to Q, P$ is linear in momenta. The extra terms emerge from $p_i \Delta^i$ in the Hamiltonian infinitesimal action in the exponential. In the new canonical variables this term turns into $P_i \tilde{c}^i(Q, \Delta_Q)$, where $\tilde{c}^i$ is given by (2.83). At this point the Hamiltonian equivalence rules (2.52), (2.58), and (2.59) should be applied ($\alpha = 0$, $a^{ij}(Q, \Delta_Q) = \tilde{g}^{ij}$, and $f^i(Q, \Delta_Q) \equiv \tilde{f}^i(Q) = (d^{-1})_j^i f^j$).

2.6.3 Canonical and unitary transformations

The path integrals allow us to elucidate the relation between canonical and unitary transformations. Infinitesimal and finite transformations from old variables

q^i, p_i to new ones Q^i, P_i $(i = 1, 2, ..., n)$ are considered separately. For simplicity of notation the indices are omitted in what follows. The classical and quantum theories are discussed simultaneously.

Infinitesimal transformations. As pointed out in Chapter 1, canonical transformations can be defined by one of the four generating functions $F_1(q, Q)$, $F_2(q, P)$, $F_3(p, Q)$, and $F_4(p, P)$. For the present study, it is most convenient to use F_2. In this case the transformation from old to new variables is determined by the equations,

$$\partial F_2(q, P)/\partial q = p, \quad \partial F_2(q, P)/\partial P = Q. \tag{2.87}$$

An *infinitesimal canonical transformation* is defined by the function,

$$F_2^\omega = qP + G(q, P)\omega, \tag{2.88}$$

where ω is a parameter, $\omega \to 0$, and the function $G(q, P)$ is the *generator* of the canonical transformation. According to (2.87) one has,

$$p = P + \frac{\partial G(q, P)}{\partial q}\,\omega, \quad Q = q + \frac{\partial G(q, P)}{\partial P}\,\omega. \tag{2.89}$$

As $P = p + O(\omega)$, neglecting terms $O(\omega^2)$, these relations can also be written in the form,

$$P = p - \frac{\partial G(q, p)}{\partial q}\,\omega, \quad Q = q + \frac{\partial G(q, p)}{\partial p}\,\omega. \tag{2.90}$$

The infinitesimal action in the new variables differs from that in the old variables by the differential,

$$dt\,[p\dot{q} - H(q, p)] - dt\left[P\dot{Q} - \tilde{H}(Q, P)\right] = pdq - PdQ = dF_1(q, Q), \tag{2.91}$$

where $\tilde{H}(Q, P) = H(q(Q, P), p(Q, P))$ (see Chapter 1), and $pdq \equiv p_i dq^i$ and so on. The function F_1 can be found from F_2 (and vice versa),

$$F_1(q, Q) = F_2(q, P) - QP; \tag{2.92}$$

here $P = P(q, Q)$ is the solution of the second equation in the system (2.87). For infinitesimal transformations,

$$F_1(q, Q) = P(q - Q) + G(q, P)\omega. \tag{2.93}$$

The *unitary transformation* corresponding to the canonical transformation (2.89) is generated by the operator,

$$\hat{G} = G(\hat{q}, \hat{p}), \tag{2.94}$$

which is obtained by replacing the arguments of the function $G(q, p)$ in (2.88) by the canonical operators, $q, p \to \hat{q}, \hat{p}$, $[\hat{q}, \hat{p}] = i\hbar$. The replacement is assumed to be carried out in Cartesian coordinates, and the operator order problem in $\hat{G}$ is resolved in one way or the other so that $\hat{G}$ is self-adjoint. A finite unitary transformation generated by a self-adjoint operator is determined by the operator,

$$\hat{U}_\tau = \exp[-(i/\hbar)\hat{G}\tau] \tag{2.95}$$

according to the rule,

$$\hat{A}_\tau = \hat{U}_\tau^\dagger \hat{A} \hat{U}_\tau, \qquad |\tau, A\rangle = \hat{U}_\tau^\dagger |A\rangle, \tag{2.96}$$

where τ is a parameter of the transformation, $\hat{A}$ is an operator, and $|\tau, A\rangle$ is the eigenvector of the operator $\hat{A}_\tau$ corresponding to the eigenvalue A. For an infinitesimal transformation ($\tau = \omega \to 0$) one has,

$$\hat{q}_\omega = \hat{q} + \frac{i}{\hbar}[\hat{G}, \hat{q}]\omega = \hat{q} + \frac{\partial \hat{G}(\hat{q}, \hat{p})}{\partial \hat{p}}\omega, \tag{2.97}$$

$$\hat{p}_\omega = \hat{p} + \frac{i}{\hbar}[G, \hat{p}]\omega = \hat{p} - \frac{\partial \hat{G}(\hat{q}, \hat{p})}{\partial \hat{q}}\omega. \tag{2.98}$$

In order to emphasize the analogy with the classical expressions (2.90), the symbols of derivatives with respect to operators are introduced here; their definition is evident. Equations (2.97) and (2.98) define explicitly infinitesimal unitary transformations of the canonical operators $\hat{q}$ and $\hat{p}$ that correspond to the infinitesimal canonical transformation (2.90). Clearly, if the procedure of constructing the operator $\hat{G}$ from the function $G(q,p)$ is not specified, equations (2.95) and (2.96) are purely formal. The path integral method allows one to obtain such a procedure in quite a few cases (see Section 2.8).

Finite transformations. In classical and quantum theories these are determined by the generating functions F_i and the operator $\hat{U}_\tau$, respectively. Let us show how the functions $F_1(q,Q)$ and $F_2(q,P)$ can be obtained from the function $G(q,P)$, and find the path integral representation for the kernel of the operator $\hat{U}_\tau$.

Canonical transformations. If τ is a parameter of the finite canonical transformation, then equations (2.90) can be written as differential equations,

$$\overset{\circ}{p}(\tau) = -\partial G(q,p)/\partial q(\tau), \quad \overset{\circ}{q}(\tau) = \partial G(q,p)/\partial p(\tau), \tag{2.99}$$

in which the derivative $df(\tau)/d\tau$ is denoted by $\overset{\circ}{f}(\tau)$ (to distinguish it from the time derivative denoted by the over-dot). Equations (2.99) can be viewed as the initial value problem. Put $q(0) = q$, $p(0) = p$, $q(\tau) = Q$, and $p(\tau) = P$. Solution of the initial value problem (2.99), $Q = Q(q,p,\tau)$ and $P = P(q,p,\tau)$, gives an explicit relation between the old and new canonical variables. To find the functions F_1 and F_2, note that according to (2.93),

$$F_1^\omega(q,Q) = [-P\overset{\circ}{q} + G(q,P)]\omega \equiv -I(\overset{\circ}{q}, q)\omega, \tag{2.100}$$

where $\overset{\circ}{q}\,\omega = Q - q$, $q = q(\tau)$, $Q = q(\tau+\omega)$, $P = p(\tau+\omega)$, and the solution of the system (2.99) is substituted for P. It is clear from (2.100) that there is an analogy with classical mechanics in which $G(q,p)$ can be viewed as the Hamiltonian function, $-F_1^\omega$ as the infinitesimal action, and the parameter τ as the time variable, i.e. the function $I(\overset{\circ}{q}, q)$ may be interpreted as the Lagrangian function in the analytic mechanics. It is not hard to verify that the solution of equations

$$\frac{d}{d\tau}\frac{\partial I(\overset{\circ}{q}, q)}{\partial \overset{\circ}{q}} - \frac{\partial I(\overset{\circ}{q}, q)}{\partial q} = 0, \tag{2.101}$$

together with the condition $\partial I/\partial \overset{\circ}{q} = p$, define the very same canonical transformation as the solution of (2.99). The proof goes along the same lines as the proof of the equivalence of the Lagrangian and Hamiltonian equations of motion. The generating function of a finite transformation $F_1(q, Q)$ is the extremum (minimum) of the functional $\int I(\overset{\circ}{q}, q)d\tau'$, with the fixed boundary values $q(0) = q$ and $q(\tau) = Q$:

$$-F_1(q, Q) \equiv R(q, Q) = \text{extremum} \int_0^\tau I(\overset{\circ}{q}, q)d\tau'. \tag{2.102}$$

The extremum is attained at the solutions of Eq. (2.101). It is also clear that the equations,

$$\partial F_1(q, Q)/\partial q = p, \qquad \partial F_1(q, Q)/\partial Q = -P, \tag{2.103}$$

which follow from the definition (2.91), and Eqs. (2.99) or (2.101) define the same canonical transformation. Indeed, according to (2.102), i.e. if Eq. (2.101) is fulfilled, $\delta F_1 = -(\partial I/\partial \overset{\circ}{q})\delta q|_0^\tau$ from which it follows that $\delta F_1 = -P\delta q(\tau) + p\delta q(0)$, after taking into account the relations $\partial I/\partial \overset{\circ}{q} = p(\tau)$, $p(0) = p$, and $p(\tau) = P$. The derived expression for δF_1 is equivalent to (2.103). Thereby the equivalence of Eqs. (2.101) and (2.103) is proved. Equation (2.102) does define the function $F_1(q, Q)$ in (2.91). The function $F_2(q, P)$ is related to $F_1(q, Q)$ by (2.92). Equations (2.100) and (2.102) solve the problem of constructing a finite canonical transformation if its generator is known.

Unitary transformations. The matrix element $\langle Q|\hat{U}_\tau|q\rangle$ can be represented by the path integral by means of the standard procedure given in Section 2.2,

$$\langle Q|\hat{U}_\tau|q\rangle = \int_{q(0)=q}^{q(\tau)=Q} \prod_\tau \frac{dqdp}{(2\pi\hbar)^n} \exp\left\{\frac{i}{\hbar}\int_0^\tau \left[p\,\overset{\circ}{q} - G(q, p)\right] d\tau\right\}. \tag{2.104}$$

The representation makes sense only if the operator ordering is specified in the generator $\hat{G}$. When the standard ordering is used (all the operators $\hat{p}$ are positioned to the right of all the $\hat{q}$'s), the function $G(q, p)$ in (2.104) is defined by the equality $G(q, p)\langle q|p\rangle = \langle q|\hat{G}|p\rangle$. Suppose that the function G is at most quadratic in momenta. Then the integration with respect to the momenta can be done in (2.104). The result reads,

$$\langle Q|\hat{U}_\tau|q\rangle = \int Dq(\tau) \exp\left\{\frac{i}{\hbar}\int_0^\tau I(\overset{\circ}{q}, q)d\tau\right\}. \tag{2.105}$$

This is the desired representation of the kernel of the unitary operator $\hat{U}_\tau$. In the limit $\hbar \to 0$, the well-known quasi-classical approximation for this kernel is obtained [75, 115]:

$$\langle Q|\hat{U}_\tau|q\rangle \approx \frac{D^{1/2}}{(2\pi i\hbar)^{n/2}} \exp\left[\frac{i}{\hbar}R(q,Q)\right] \ ; \quad D = \det\left|-\frac{\partial^2 R}{\partial q^j \partial Q^k}\right|, \tag{2.106}$$

for functions $I(\overset{\circ}{q}, q)$ bilinear in $\overset{\circ}{q}$. The asymptote (2.106) can actually be derived directly from (2.105) (see [110], p.113 and also [116]). It is easier however to make use of the Schrödinger equation, $i\hbar\partial U/\partial\tau = G(Q, -i\hbar\partial/\partial Q)U$ for the kernel $U = \langle Q|\hat{U}_\tau|q\rangle$ written in the form (2.106). The equations derived for the functions R and D are analogous to Eqs. (2.101) and (2.103). The method due to Van Vleck [115] for calculating the pre-exponential factor in (2.106) is given in Section 8.3.4.

An example. Consider an infinitesimal point transformation $Q(q) = q + \phi(q)\omega$, where ϕ is an arbitrary (smooth) function. It is defined by the generator $G(q, P) = P_i\phi^i(q)$, i.e. according to (2.90),

$$P_j = p_j - \phi^i_{,j}(q)p_i\omega, \quad Q^j = q^j + \phi^j(q)\omega. \tag{2.107}$$

The quantum operator corresponding to (2.107) reads,

$$\hat{G} = \frac{1}{2}\left[\phi^j(\hat{q})\hat{p}_j + \hat{p}_j\phi^j(\hat{q})\right] = \phi(\hat{q})\hat{p} + \frac{\hbar}{2i}\phi^j_{,j}(\hat{q}). \tag{2.108}$$

Thanks to the linearity of $G(q, P)$ in momenta the operator ordering problem is solved by symmetrization, i.e. by demanding the hermiticity of $\hat{G}$ (no other possibility exists). The unitary operator is,

$$\hat{U}_\omega = \exp[-(i/\hbar)\hat{G}\omega].$$

The wave functions are transformed according to the relation,

$$\begin{aligned}\tilde{\psi} = \hat{U}^\dagger_\omega\psi &\approx \left(1 + \omega\phi^i\frac{\partial}{\partial q^i} + \frac{\omega}{2}\phi^j_{,j}\right)\psi(q) \\ &\approx \left(1 + \frac{\omega}{2}\phi^j_{,j}(q)\right)\psi(q + \phi(q)\omega) \approx J^{1/2}\psi(Q(q)),\end{aligned} \tag{2.109}$$

where $J = \det Q^i_{,j} \approx 1 + \phi^j_{,j}\omega$ is the Jacobian of the transformation. If the original coordinates are Cartesian, then $J = \sqrt{g}$ with g_{ij} being the metric tensor in the new coordinates. In particular, it follows from (2.109) that the formal transformation $\psi(q) \to \psi(Q(q))$ is not unitary (the factor $J^{1/2} = g^{1/4}$ is missing); but it is "unitary" with respect to the scalar product $(\psi_1, \psi_2) = \int dq\psi_1^*(q)\psi_2(q)$, because here $dq \to JdQ$ under the change of variables, i.e. effectively, $\psi_i(q) \to J^{1/2}\psi_i(Q(q))$ $(i = 1, 2)$.

The path integral representation of a unitary operator (see (2.104) and (2.105)) was obtained in [118], but, in contrast to [117], without discussing the problem of the extra terms.

2.6.4 Canonical transformations of the Hamiltonian path integrals

In classical mechanics the Hamiltonian formalism is invariant under canonical transformations. Does a similar statement hold for the Hamiltonian path integrals? Or, in other words, are the basic formulas of the classical Hamiltonian mechanics sufficient to use when making canonical transformations in the Hamiltonian path integrals? The answer is negative in general. The reason is that the extra terms, that are omitted in the classical theory, must be taken into account. Let us turn to a more detailed study of this problem, limiting the discussion to the case of infinitesimal canonical transformations [96, 117]. Consider the basic formula,

$$\psi_\epsilon(q) \approx \int \frac{dq''dp''}{2\pi\hbar} \exp\left\{\frac{i}{\hbar}\left[p''(q-q'') - \varepsilon H(q'',p'')\right]\right\}\psi(q'') \, ; \tag{2.110}$$

for the sake of simplicity a system with a single degree of freedom ($n = 1$) and the standard Hamiltonian (2.3) is considered. There are two ways to tackle the problem. The first is the "operator" one. The Hamiltonian is transformed $\hat{H} \to \hat{H}_\omega$ by means of $\hat{U}_\omega$ in accordance with the rule (2.96),

$$\hat{H}_\omega = \hat{U}_\omega^\dagger \hat{H} \hat{U}_\omega, \tag{2.111}$$

then the standard Hamiltonian path integral formalism is developed for the new Hamiltonian. The second way is to go over to the new canonical variables in (2.110) using (2.90). The first approach turns out to be the simplest because it avoids lengthy and rather exhausting calculations associated with accurate accounting of the extra terms.

An infinitesimally small canonical transformation $q'',p'' \to q',p'$ (2.90) (here $Q = q''$ and $P = p''$, i.e. the inverse transformation is considered) is merely a change of integration variables in (2.110). Carrying a similar canonical transformation of the arguments $q,p \to \tilde{q},\tilde{p}$ in (2.110) and neglecting terms of order $O(\omega^2)$, Eq. (2.110) can be written in the form,

$$\begin{aligned}\psi_\varepsilon \approx \int \frac{dq'dp'}{2\pi\hbar} \exp\Bigg\{&\frac{i}{\hbar}p'\frac{\partial G(\tilde{q},\tilde{p})}{\partial\tilde{p}}\omega + \frac{i}{\hbar}\left[p'(\tilde{q}-q') - \varepsilon H_\omega(q',p')\right] \\ &-\frac{i}{\hbar}p'\frac{\partial G(q',p')}{\partial p'}\omega - \frac{i}{\hbar}(\tilde{q}-q')\frac{\partial G(q',p')}{\partial q'}\omega\Bigg\}\psi\left(q' + \frac{\partial G(q',p')}{\partial p'}\omega\right),\end{aligned} \tag{2.112}$$

where

$$H_\omega(q,p) = H(q + \omega\partial G/\partial p, p - \omega\partial G/\partial q). \tag{2.113}$$

To advance further, it is necessary to specify the generator G. The discussion is limited to functions of the form,

$$G(q,p) = \nu p^2/2 + U(q), \tag{2.114}$$

which comprise a sufficiently large class of canonical transformations. Since $\partial G/\partial p = \nu p$ and $\partial G/\partial q = U'(q) = dU/dq$, for the Hamiltonian corresponding to (2.3) one has,

$$H_\omega(q,p) \approx H(q,p) - pY(q)\omega, \quad Y = U'(q) - \nu V'(q), \tag{2.115}$$

while in the quantum case the variables q and p should be replaced by the canonical operators $\hat{q}$ and $\hat{p}$ in (2.113). In doing so, the operator ordering problem arises in the Hamiltonian (2.111),

$$\hat{H}_\omega \approx H(\hat{q},\hat{p}) - \frac{1}{2}\left(\hat{p}Y(\hat{q}) + Y(\hat{q})\hat{p}\right)\omega. \tag{2.116}$$

For instance, if anti-standard ordering is used (the $\hat{p}$s are to the left of the $\hat{q}$s), the Hamiltonian reads,

$$\hat{H}^a_\omega \approx H(\hat{q},\hat{p}) - \hat{p}Y(\hat{q})\omega - \frac{i\hbar}{2}Y'(\hat{q})\omega. \tag{2.117}$$

As compared to (2.115), $\hat{H}^a_\omega$ contains a term proportional to $\hbar$ and, hence, specific for quantum theory. Since the Hamiltonian path integral formalism comprises a quantum description of dynamical systems, the terms omitted in the classical limit must be taken into account in it; in this particular case this is the term proportional to Y'. The above observation is a characteristic feature of canonical transformations in the Hamiltonian path integral formalism.

Even though the problem considered is clear enough, it is desirable to have direct verification of the equivalence of the two approaches to solving it because their starting points are quite distinct. Unfortunately, the corresponding technicalities are very involved and not presented here. The interested reader can find them in the Appendix of [117].

It is instructive to elucidate the equivalence of expression (2.112) to that obtained from (2.110) by a unitary transformation generated by $\hat{G}$,

$$\psi_\varepsilon(q) = \langle q|\exp(-i\varepsilon\hat{H}/\hbar)|\psi\rangle \equiv \left\langle q\left|\hat{U}_\omega\exp(-i\varepsilon\hat{H}_\omega/\hbar)\hat{U}^\dagger_\omega\right|\psi\right\rangle. \tag{2.118}$$

Inserting the resolution of unity $1 = \int dq|q\rangle\langle q|$ between all the operators and $|\psi\rangle$ in the right-hand side of (2.118) and making use of the explicit form of the kernels of the operators involved (they have a form analogous to (2.6)), the wave function (2.118) is represented as,

$$\psi_\varepsilon(q) \approx \int \frac{dPdQ}{2\pi\hbar}\frac{dP'dQ'}{2\pi\hbar}\frac{dp'dq'}{2\pi\hbar}\exp\left\{\frac{i}{\hbar}\left[P(q-Q) - \omega G(q,P)\right.\right.$$
$$\left.\left. + P'(Q-Q') - \varepsilon H^a_\omega(Q',P') + p'(Q'-q') + \omega G(q',p')\right]\right\}\psi(q'). \tag{2.119}$$

The result of the unitary transformation (2.119) appears to be quite different from the analogous result (2.112) obtained by the canonical transformation of the argument of the wave function (2.110). Nevertheless both the results are identical. The proof of this statement turns out to be technically quite tedious but rather instructive. First, in the course of calculations, Eq. (2.79) is illustrated, and the necessity for taking into account the terms $O(\Delta^2)$ within it becomes clear. Second, a new feature of the formalism appears where extra terms of higher orders $O(\Delta^4)$ than stated in (2.66) should be accounted for. The peculiarity of the example studied is related to both the existence of two small parameters ε and ω and the necessity to retain terms of order $O(\varepsilon\omega)$.

The necessary calculations are carried out as follows. The integration with respect to Q in (2.119) yields $\delta(P-P')$, and the subsequent integration with respect to P gives the unit factor. Omitting in the resulting expression the primes at the integration variables Q' and P', one obtains,

$$\psi_\epsilon(q) \approx \int \frac{dPdQ}{2\pi\hbar}\frac{dp'dq'}{2\pi\hbar}\exp\left\{\frac{i}{\hbar}[-G(q,P)\omega + P(q-Q) - \varepsilon H^a_\omega(Q,P) + p'(Q-q') + G(q',p')\omega]\right\}\psi(q'). \tag{2.120}$$

The next step is the integration with respect to P in (2.120). Put $q - Q = \Delta_\omega$. The coefficient at P^2 in the exponential in (2.120) is $\varepsilon + \nu\omega \equiv \Omega$ (not just ε). This fact as well as the necessity to retain terms of order $O(\varepsilon\omega)$ does allows for direct application of the equivalence rules from Section 2.4, but the general approach presented there is suitable here, too. The integration with respect to P results in,

$$\psi_\varepsilon(q) \approx \int \frac{dQdp'dq'}{2\pi\hbar(2\pi i\Omega\hbar)^{1/2}}\exp\left\{\frac{i}{\hbar}\left[\frac{[\Delta_\omega + \varepsilon\omega Y(Q)]^2}{2\Omega} - p'\Delta_\omega - U(q)\omega - V(Q)\varepsilon + \frac{i\hbar}{2}Y'(Q)\varepsilon\omega + p'(q-q') + G(q',p')\omega\right]\right\}\psi(q'). \tag{2.121}$$

Now the standard procedure can be used to calculate relevant terms of the integral (2.121) in the limit $\varepsilon,\omega \to 0$. The integrand is decomposed into a power series of Δ_ω and then the integration with respect to Q is carried out, i.e., Eq. (2.42) is used again where $n = 1$, $g_{ij} \to 1$, $g^{j_1\cdots j_{2k}} \to (2k-1)!!$, and $\varepsilon \to \Omega$. In the decompositions of Y and V, the linear and quadratic terms, respectively, must be retained, $Y(Q) \approx Y(q) - Y'(q)\Delta_\omega$ and $V(Q) \approx V(q) - V'(q)\Delta_\omega + V''(q)\Delta_\omega^2/2$, while the argument of $Y'(Q)$ may immediately be replaced by q as it comes with the coefficient $\varepsilon\omega$. A striking difference from the equivalence rules of Section 2.4 is that the terms $\varepsilon\Delta_\omega^2$, equivalent to $\varepsilon\Omega \approx \nu\varepsilon\omega$ in accord with (2.42), are taken into account. For this very reason, terms up to order $\Delta_\omega^4 \sim \Omega^2 \approx 2\nu\varepsilon\omega$ in the expansion of $\exp(p'\Delta_\omega(i\hbar)$ have to be retained; the terms $\varepsilon\omega\Delta_\omega/\Omega$ must also be taken into account. After carrying out the specified calculations, Eq. (2.121) becomes

$$\psi_\varepsilon(q) \approx \int \frac{dp'dq'}{2\pi\hbar} \exp\left\{\frac{i}{\hbar}\left[-G(q,p')\omega + p'(q-q') - \varepsilon\left(\frac{p'^2}{2} + V(q)\right.\right.\right.$$
$$\left.\left.\left. - U'(q)p'\omega + \frac{i\hbar\omega}{2}U''(q)\right) + G(q',p')\omega\right]\right\} \psi(q'). \qquad (2.122)$$

It is now not hard to transform the obtained expression to the form comparable to Eq. (2.112). The coefficient at $p'^2/2$ in the exponential (2.122) is ε. Therefore the results of Section 2.4 can be used. The proof is completed by carrying out the following steps:

1) replace $q \to q'$ in $V(q)$ and $U''(q)$;
2) substitute $p'U'(q) \approx p'U'(q') + p'U''(q')\Delta$, $\Delta = q - q'$;
3) replace q by $\tilde{q} + \omega\partial G(\tilde{q},\tilde{p})/\partial\tilde{p}$ (here $(q,p) \to (\tilde{q},\tilde{p})$ is the canonical transformation generated by G);
4) change the integration variable $q' \to q' + \omega\partial G(q',p')/\partial p'$.

The validity of the above operations is evident (in step 1, the replacement yields a difference only in terms of order $\varepsilon\Delta$, while in step 4 the derivative $\partial G/\partial p'$ is independent of q' for G given in (2.114)). Making use of Eqs. (2.54), (2.58), and (2.59) in which all the tensors α, a, and c vanish and $f^i_{,k} = -U''(q')$, the term $U''(q')p'\Delta$ is replaced by $i\hbar U''(q')$ after step 2. Finally, taking into account that $H(q' + \omega\partial G/\partial p', p' - \omega\partial G/\partial q') \approx H(q' + \omega\partial G/\partial p', p') - p'U'(q')\omega \equiv H_\omega(q',p')$, Eq. (2.122) can be rewritten in the form,

$$\psi_\varepsilon(q) \approx \int \frac{dp'dq'}{2\pi\hbar} \exp\left\{\frac{i}{\hbar}\left[p'\frac{\partial G(\tilde{q},\tilde{p})}{\partial\tilde{p}}\omega - G(\tilde{q},p')\omega + p'(\tilde{q}-q') - \varepsilon H_\omega(q',p')\right.\right.$$
$$\left.\left. + \frac{i\hbar}{2}U''(q')\varepsilon\omega - p'\frac{\partial G(q',p')}{\partial p'}\omega + G(q',p')\omega\right]\right\}\psi\left(q' + \frac{\partial G}{\partial p'}\omega\right). \quad (2.123)$$

This equation still differs from (2.112). Consider the following approximation of the difference in (2.123):

$$G(q',p') - G(\tilde{q},p') \approx (q' - \tilde{q})\frac{\partial G(q',p')}{\partial q'} - \frac{1}{2}(q'-\tilde{q})^2\frac{\partial^2 G(q',p')}{\partial q'^2}. \qquad (2.124)$$

It is not hard to be convinced by means of (2.42) again that the replacement $(q'-\tilde{q})^2 \to i\varepsilon\hbar$ is justified in (2.124) (and, hence, in (2.123)). As a result all the terms of order $\varepsilon\omega$ are canceled out and the final expression becomes identical to (2.112).

In conclusion let us discuss Eq. (2.79) in somewhat more detail. The terms outside the square brackets in the exponential (2.112) are nothing but the difference $\delta = p''(q-q'') - p'(\tilde{q}-q')$ (compare the text before (2.112)). In classical mechanics, it is the differential dF_1 if $\Delta t = \varepsilon \to 0$ (see (2.91)). As pointed out before, when working with Hamiltonian path integrals, such an approximation is far from being sufficient. Taking into account that $p'' = p' - \omega\partial G(q',p')/\partial q'$ and

$q - q'' = \tilde{q} - q' + O(\omega)$, the difference can be brought to the form,

$$\delta \approx -\left[p'(\tilde{q} - q) - p'(q' - q'') + (\tilde{q} - q')\frac{\partial G(q', p')}{\partial q'}\omega\right],$$

where terms $O(\omega^2)$ are neglected. A further transformation is done by invoking (2.124) and the definition (2.93) of the function $F_1^\omega(q', q'') = p'(q' - q'') + G(q', p')\omega$ (here p' must be substituted by the solution of the equation $q'' = q' + \omega\partial G(q', p')/\partial p'$). The final result reads,

$$\delta = -\left[F_1^\omega(\tilde{q}, q) - F_1^\omega(q', q'') - \frac{(\tilde{q} - q')^2}{2}\frac{\partial^2 G(q', p')}{\partial q'^2}\omega\right]. \tag{2.125}$$

In classical theory, the ratio δ/ε tends to $-dF_1^\omega/dt$ as $\varepsilon \to 0$, because the last term in (2.125) vanishes in this limit, $(\tilde{q} - q')^2/\varepsilon \to 0$. In the Hamiltonian path integral the last term cannot be neglected as proved above. A "non-classical" term $i\hbar U''(q')\varepsilon\omega/2$ corresponds to the last term in (2.123) which occurs through the non-commutativity of the canonical operators $\hat{q}$ and $\hat{p}$ when converting (2.116) into (2.117). Thus, here, just as before, the operator nature of canonical variables reveals itself in the extra terms.

So, the Hamiltonian path integral does change its form under canonical transformations, since the above analysis proves that it is insufficient to replace the classical Hamiltonian by the transformed one because of a possible occurrence of non-classical terms like $(-i\hbar\omega/2)[U''(q) - \alpha V''(q)]$. In this sense the Hamiltonian path integral is not "covariant" under canonical transformations. Strictly speaking, it is not covariant even under linear canonical transformations, when the function $U(q)$ is quadratic in (2.114), because in this case the term proportional to $\hbar$ in (2.117) does not vanish either, $Y' = \text{const.} \neq 0$.

Here the discussion has been limited to generators of the form (2.114). An addition of terms linear in p would not change the calculations in any significant way. If the function G has higher powers of p, it would be impossible to repeat the calculations, since the integrals with respect to the momentum cannot be computed analytically. However, even without explicit integration, all the formulas remain valid because the analogous expressions are obtained for finite canonical transformations with generators of the form (2.114).

The problem of canonical transformations in path integrals has also been studied in [119].

2.7 Problems with non-trivial boundary conditions

2.7.1 A particle in an infinite well

So far only problems without boundary conditions have been considered in which a natural requirement of decreasing the wave function at the spatial infinity automatically follows from the general formulas (2.2) or (2.41). There are problems where the configuration space has boundaries. The simplest and a well-known example is provided by the problem of a quantum particle in an infinite potential well [66, 67],

$$i\hbar\partial_t\psi(q,t) = -\frac{\hbar^2}{2}\frac{\partial^2\psi(q,t)}{\partial q^2}, \quad 0 \le q \le L, \quad \psi(0,t) = \psi(L,t) = 0, \qquad (2.126)$$

for $t \ge 0$ and $\psi(x,0) = \psi_0(x)$. Similar problems also arise when changing coordinates (e.g. the angular variables [96, 120]) or for quantum dynamics on compact manifolds (group manifolds [121–123]) and so on. In all these cases a straightforward application of the standard scheme [62, 63] does not lead to a correct path integral representation. The reason is clear. The solution must satisfy the required boundary conditions, whereas the formalism developed so far guarantees only a decrease of the solution at the spatial infinity. Thus, a modification of the formalism is needed.

In order to illustrate the kind of difficulties in the standard scheme of the path integral construction, consider the example (2.126). A naive application of the convolution formula for N infinitesimal evolution operator kernels for a free particle,

$$U_t(q,q') = \lim_{N\to\infty}(2\pi i\varepsilon\hbar)^{-1/2}\int_0^L \prod_{j=1}^{N-1}\frac{dq_j}{(2\pi i\varepsilon\hbar)^{1/2}}\exp\left[\frac{i}{\hbar}\sum_{j=0}^{N-1}\frac{(q_{j+1}-q_j)^2}{2\varepsilon}\right] \qquad (2.127)$$

($\varepsilon = t/N$, $q_N = q$, $q_0 = q'$) is not applicable even when $t = \varepsilon$, $\varepsilon \to 0$. Although the evolved wave function,

$$\psi_\epsilon(q) \approx \int_0^L \frac{dq'}{(2\pi i\epsilon\hbar)^{1/2}}\exp\left[\frac{i}{\hbar}\frac{(q-q')^2}{2\epsilon}\right]\psi_0(q') \equiv \int_0^L U_\varepsilon(q-q')\psi_0(q')dq' \quad (2.128)$$

satisfies the Schrödinger equation for $q \in (0,L)$, it does not fulfill the boundary condition $\psi_\varepsilon(0) = \psi_\varepsilon(L) = 0$, even if the initial wave function does so, $\psi_0(0) = \psi_0(L) = 0$. Hence, the representation (2.127) does not solve the problem. Moreover, the multiple integral (2.127) cannot be explicitly computed despite the simplicity of the integrand. Thus, the problem is associated with the fact that the fundamental solution of (2.126) used in the convolution (2.127) does not satisfy the boundary condition.

A correct approach has already been pointed out in [66]. The function (2.127) satisfies the Schrödinger equation (2.126). So one could try to make a linear combination of such solutions that would fulfill the boundary conditions. The problem resembles the problem concerning the electrostatic potential of a point-like charge between parallel perfectly conducting planes. In order to satisfy the zero boundary conditions at points $q = 0$ and $q = L$, one should subtract, respectively, $\exp[i(q+q')^2/2\varepsilon\hbar]$ and $\exp[i(q-2L+q')^2/2\varepsilon\hbar]$ from the exponential in the integrand in (2.128). These terms correspond to the motion from the points

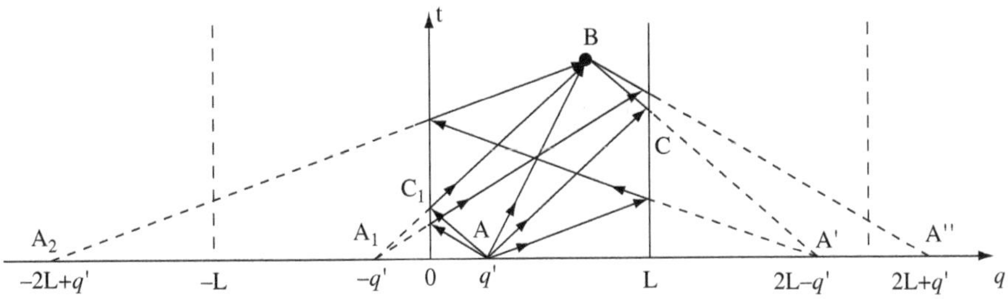

Fig. 2.1. World lines of a particle in an infinite square well

obtained by "mirror" reflections of the point q' about the boundary points, i.e. from $-q'$ and $2L - q'$ (compare the method of images in electrostatics). The procedure is depicted in Fig. 2.1. The contributions of the new trajectories (the world lines) A_1C_1B and $A'CB$ exactly equal contributions from trajectories with reflections, AC_1B and ACB, respectively. However, the first function (the first difference) fails to satisfy the zero boundary condition at $q = L$, while the second one fails to do so at $q = 0$. So, the process of adding contributions of "mirror" points needs to be continued: one should add the exponentials corresponding to the motion from the points that are mirror images of $-q'$ and $2L - q'$ relative to the boundaries $q = L, 0$, and so on. As a result of the method of images (Pauli, [66], p.171), the sum,

$$K'(q,q';\varepsilon) = \sum_{n=-\infty}^{\infty} [U_\varepsilon(q - q' + 2Ln) - U_\varepsilon(q + q' + 2Ln)], \qquad (2.129)$$

appears in the integrand in (2.128) in place of $U_\varepsilon(q,q')$. It is not difficult to prove that the function (2.129) is nothing but the evolution operator kernel (the fundamental solution) in the problem (2.126),

$$K(q,q';t) = \sum_{n=1}^{\infty} \psi_n(q)\psi_n^*(q') \exp(-iE_nt/\hbar), \quad t > 0, \qquad (2.130)$$

if $t = \varepsilon$; here $\psi_n(q) = (2/L)^{1/2} \sin(\pi nq/L)$ are normalized eigenfunctions of the free particle Hamiltonian that satisfy the zero boundary conditions and correspond to the eigenvalues $E_n = \pi^2\hbar^2n^2/2L^2$. The proof can be done via the Poisson summation formula [120] (see Section 8.3.2),

$$\sum_{n=-\infty}^{\infty} f(n\hbar)e^{in(q-q'')} = \sum_{n=-\infty}^{\infty} \int_{-\infty}^{\infty} \frac{dpdq'}{\hbar} f(p)e^{ip(q-q')/\hbar}\delta(q' - q'' + 2\pi n), \quad (2.131)$$

or by means of some properties of the elliptic functions [66]. The solution is found, but the derivation of the path integral is associated with an awkward procedure of integrating the Gaussian exponentials in the kernel K' in finite limits. Equation

(2.128) with the kernel (2.129) substituted for $U_\varepsilon(q-q')$ can be written in the form,

$$\psi_\varepsilon(q) \approx \sum_{n=-\infty}^{\infty} \left[\int_{2Ln}^{L+2Ln} dq' U_\varepsilon(q-q')\psi_0(q'-2Ln) \right.$$
$$\left. - \int_{2Ln-L}^{2Ln} dq' U_\varepsilon(q-q')\psi_0(-q'+2Ln) \right]. \tag{2.132}$$

Let Ψ_0 be a function defined in the whole real axis so that it coincides with ψ_0 in the interval $[0, L]$ and the following properties hold:

$$\Psi_0(-q) = -\Psi_0(q), \quad \Psi_0(q+2Ln) = \Psi_0(q), \quad \Psi_0(0) = \Psi(L) = 0. \tag{2.133}$$

Relations (2.133) define the values of Ψ_0 outside the interval $[0, L]$. Equation (2.132) can now be represented in a simpler form,

$$\psi_\varepsilon(q) \approx \int_{-\infty}^{\infty} dq' U_\varepsilon(q-q')\Psi_0(q'). \tag{2.134}$$

Note that this equation also defines an extension of ψ_ε to the whole real axis if $q \in (-\infty, \infty)$ in the right-hand side. It is straightforward to verify that function (2.134) is a solution of the Schrödinger equation and satisfies the conditions (2.133). The advantage of the representation (2.133) is that one can integrate over the whole real axis in the convolution of the infinitesimal kernels $U_\varepsilon(q-q')$, thus obtaining the conventional representation of the path integral, if the resulting kernel is applied to the function Ψ_0 which is defined in the whole real axis via the function ψ_0 (in this case, Ψ_0 is odd and periodic with the period being $2L$). The problem of the path integral construction is solved.

Thus, there are two possibilities:

(1) one can use the evolution operator kernel K' (2.129) and integrate in the finite limits (i.e. over the physical configuration space) when calculating the action of the evolution operator;
(2) one can use the evolution operator kernel $U_\varepsilon(q-q')$ instead and integrate in the infinite limits.

In the latter case $U_\varepsilon(q-q')$ is to be applied to the function Ψ_0, the extension of the initial wave function ψ_0 to the whole real axis in accordance with (2.133).

It is convenient to write the results obtained above in a form that is suitable for generalization to a large set of similar problems (periodic boundary conditions, residual discrete gauge groups, boundaries with special geometries, etc.). Define an operator $\hat{Q}$ by the kernel,

$$Q(q, q') = \sum_{n=-\infty}^{\infty} \left[\delta(q-q'+2Ln) - \delta(q+q'+2Ln)\right], \tag{2.135}$$

where $q \in (-\infty, \infty)$ and $q' \in [0, L]$. Then,

$$\Psi_0(q) = \int_0^L dq' Q(q, q')\psi_0(q') \tag{2.136}$$

and, according to (2.132) and (2.134),

$$K(q, q'; t) = \int_{-\infty}^{\infty} dq'' U_t(q - q'')Q(q'', q'). \tag{2.137}$$

If the sum (2.130) is substituted into (2.137) for K and letting t go to zero, the following representation of the kernel of the operator $\hat{Q}$ is derived:

$$Q(q, q') = \sum_n \psi_n(q)\psi_n^*(q'); \tag{2.138}$$

here $q, q' \in [0, L]$ because, by (2.4) or (2.6),

$$\langle q|\hat{U}_t|q'\rangle \to \langle q|q'\rangle = \delta(q - q'), \quad t \to 0. \tag{2.139}$$

The eigenfunctions $\psi_n(q)$ may be extended to the whole real axis so that the extension satisfies the conditions (2.133). This allows for the extension $q \in (-\infty, \infty)$ in (2.138). It follows from Eqs. (2.135)–(2.139) that the operator $\hat{Q}$ plays a twofold role. First, its action on ψ_0 gives Ψ_0, an odd and $2L$-periodic extension of ψ_0 into the whole real axis. Second, it provides the fulfillment of the boundary conditions for the physical evolution operator kernel (2.137), i.e. in the operator form $\hat{K}_t = \hat{U}_t\hat{Q}$. The operator $\hat{Q}$ will be called an *extension operator*.

At this point it is already not too difficult to establish a general structure of the solution of the initial value problem for quantum problems with boundaries of a similar type [124]. Let $\hat{H} = -\hbar^2\Delta/2 + V$ be a Hamiltonian describing the motion of a particle in an n-dimensional Euclidean space, and let V be an analytic function of each variable q^i in some neighborhood of the real axis. Let Γ denote the boundary of a region Ω to which the motion is restricted. The boundary is assumed to be sufficiently smooth and, for the sake of discussion, the region Ω to be compact. Let $U_t(q, q')$ be defined by (2.11) where the integration with respect to q is done in the infinite limits. According to the aforementioned general guideline one should be able to construct either the kernel $K_t(q, q')$ or the (extended) function Ψ_0. The formal solutions are given by,

$$K_t(q, q') = \int_{-\infty}^{\infty} U_t(q, q'')Q(q'', q')dq'', \; t > 0, \tag{2.140}$$

$$\Psi_0(q) = \int_{q' \in \Omega} Q(q, q')\psi_0(q')dq', \tag{2.141}$$

where,

$$Q(q,q') = \sum_{(k)} \psi_{(k)}(q)\psi^*_{(k)}(q'), \tag{2.142}$$

and $\psi_{(k)}$ form a complete orthonormal set of the eigenfunctions of the operator $\hat{H}$, $\hat{H}\psi_{(k)} = E_{(k)}\psi_{(k)}$, satisfying the boundary condition,

$$\psi_{(k)}(q)\big|_{\Gamma} = 0\ ; \tag{2.143}$$

(k) denotes the complete set of quantum numbers that label the vectors $\psi_{(k)}$. Because of the analyticity of the potential, the functions $\psi_{(k)}(q)$ can be analytically extended to the complement of Ω, and thereby the kernel $Q(q,q')$ and Ψ_0 are defined for all values of q in the Euclidean space. Obviously, for $q, q' \in \Omega$, the kernel $Q(q,q')$ coincides with the delta-function $\delta(q-q')$. If q is not in Ω, then Q is an extension of the δ-function outside Ω. Equations (2.140)–(2.142) establish a general structure of the path integral representation of the kernel K.

The above general statements are not too hard to prove. Making use of the linear approximation of the infinitesimal evolution operator kernel (see (2.4)),

$$\int_{-\infty}^{\infty} U_\varepsilon(q,q')\Psi_0(q')dq' \approx \left(1 - \frac{i\varepsilon}{\hbar}\hat{H}\right)\Psi_0(q), \quad \varepsilon \to 0, \tag{2.144}$$

and taking into account (2.142), Eq. (2.140) can be transformed in the following way:

$$K_\varepsilon(q,q') \approx \left(1 - \frac{i\varepsilon}{\hbar}\hat{H}\right) Q(q,q') \approx \sum_{(k)} \exp\left(\frac{-i\varepsilon E_{(k)}}{\hbar}\right) \psi_{(k)}(q)\psi^*_{(k)}(q'). \tag{2.145}$$

This equation shows immediately that $K_t(q,q')$ is indeed the evolution operator kernel of the boundary value problem: it clearly satisfies the Schrödinger equation and the boundary conditions; it converges to the δ-function as $t \to 0$. It is necessary to integrate in the infinite limits in (2.140) to establish the first equality in (2.145). When applying this general scheme to concrete examples, one should make sure that the integrals involved converge or, more broadly, that a meaningful interpretation of them does exist. However, integrals like (2.134) with the kernel (2.12) are easy to make meaningful even for functions Ψ_0 growing rather fast at spatial infinity (e.g. as a Gaussian exponential); the function ψ_0 is normalized in the physical region Ω. In [124] this approach was applied to the problem of a particle in a circular infinite well. In this particular case all the calculations can be explicitly carried out.

In quantum mechanics there are problems where the physical coordinate ranges in an interval of length L, but instead of zero boundary conditions at the end points of the interval, the periodic ones are imposed (a particle in a circle). In this case, only the periodicity of the kernel K is required. Suppose that the function $U_t(q,q')$ satisfies the Schrödinger equation in the whole space and has the property,

$$U_t(q+nL,q') = U_t(q,q'-nL), \tag{2.146}$$

where n is an integer and $U_t(q,q') \to \delta(q-q')$ as $t \to 0$. Then kernel K is the sum,

$$K_t(q,q') = \sum_{n=-\infty}^{\infty} U_t(q+nL,q') = \int_{-\infty}^{\infty} dq'' U_t(q,q'')Q(q'',q'), \tag{2.147}$$

where,

$$Q(q,q') = \sum_{n=-\infty}^{\infty} \delta(q-q'+nL), \tag{2.148}$$

and, hence,

$$\psi_t(q) = \int_0^L dq' K_t(q,q')\psi_0(q') \tag{2.149}$$

is the solution of the problem because $\psi_t(q+nL) = \psi_t(q)$. After substitution of (2.147) into (2.149), the latter assumes the form,

$$\psi_t(q) = \int_{-\infty}^{\infty} dq' U_t(q,q')\Psi_0(q'), \qquad \Psi_0 = \hat{Q}\psi_0, \tag{2.150}$$

which is convenient for the path integral representation.

The meaning of the sum (2.147) is just as simple as that of the sum (2.129). The difference in the summand of (2.129) is necessitated by the zero boundary conditions at the points $q = 0$ and $q = L$, and the summation accounts for contributions of trajectories connecting q' and q with multiple reflections from the boundaries, while here the sum in (2.147) ensures that the amplitude satisfies the periodic boundary conditions. It takes into account the fact that, in addition to a straight trajectory that connects q' and q, there are other extremum trajectories with non-trivial winding numbers which go about the circle either clockwise or counterclockwise multiple times before reaching the end point q, and their contributions to the transition amplitude cannot be neglected.

It is not hard to figure out how the solution of similar problems will look. For instance, for a particle in the half-line $q \geq 0$ (zero boundary condition at $q = 0$), one has,

$$Q(q,q') = \delta(q-q') - \delta(q+q'); \tag{2.151}$$

for a particle in a sector in a plane with the angle π/n $(n = 2, 3, \ldots)$,

$$Q(\mathbf{x}, \mathbf{x}') = \sum_k \left[\delta(\mathbf{x} - c_k\mathbf{x}') - \delta(\mathbf{x} - \tilde{c}_k\mathbf{x}'\right], \tag{2.152}$$

where c_k and $\tilde{c}_k$ are 2×2 matrices, elements of the (finite) group of all reflections in the lines forming the boundaries of the sector (c_k corresponds to an even number of reflections, while $\tilde{c}_k$ to an odd number). This group is a particular case of the *Coxeter groups* [125–127]. A Coxeter group is an abstract group that admits a formal description in terms of mirror symmetries. Its elements are defined as distinct products of abstract generators (basis reflections) $g_1, g_2, \ldots, g_p$ that satisfy the condition $(g_ig_j)^{m_{ij}} = 1$, where the integer $m_{ii} = 1$ and $m_{ij} \geq 2$ for $i \neq j$; the condition $m_{ij} = \infty$ means that no condition $(g_ig_j)^{m_{ij}} = 1$ should be imposed. The finite Coxeter groups are precisely the finite Euclidean reflection groups; the symmetry groups of regular polyhedra are an example [126, 127]. Consider p hyper-planes h_i, $i = 1, 2, \ldots, p$, in an n-dimensional Euclidean space through the point $\mathbf{x} = 0$. The equations of the hyper-planes have the form $\mathbf{x}\cdot\mathbf{n}_i = 0$, where $\mathbf{n}_i$ is the unit normal of the hyper-plane h_i, and mutual orientations of the normals are fixed by,

$$\mathbf{n}_i \cdot \mathbf{n}_j = -\cos\frac{\pi}{m_{ij}}, \qquad m_{ij} = 2, 3, \ldots. \tag{2.153}$$

The generators are defined by the conditions: $g_i\mathbf{n}_i = -\mathbf{n}_i$ and $g_i\mathbf{x} = \mathbf{x}$, if $\mathbf{x} \in h_i$. The set of possible distinct products of the generators form the Coxeter group C. The convex region D bounded by the planes h_i such that $\mathbf{x}\cdot\mathbf{n}_i > 0$ for any $\mathbf{x} \in D$ is called *the fundamental domain.* The union of all regions cD where $c \in C$, together with the planes h_i, coincides with the whole Euclidean space. Clearly, the problem of a particle inside the fundamental domain with zero boundary conditions is also solved by applying equations like (2.151) and (2.152). For example, let the first (positive) quadrant in the plane be the fundamental domain ($n = m_{12} = 2$). In Fig. 2.2 the direct trajectory (the straight line segment (AB)) and trajectories reflected from the boundaries of the fundamental domain are shown, which contribute to the sum in (2.152). Solutions for problems with the corresponding periodic boundary conditions are obtained just as easily: the difference in (2.152) should be replaced by the sum. Another particular case of the Coxeter group is given by Weyl groups of Lie algebras (the symmetry group of the root system of a Lie algebra). The fundamental region is the so-called *Weyl chamber* [129]. The problem of constructing the path integral for a system in the Weyl chamber appears in studies of some gauge models and discussed in Chapter 6 in detail. Finally, when the potential V is not zero, the problem can be formulated in the whole space by extending V outside the fundamental domain by means of a suitable operator $\hat{Q}$:

$$Q(\mathbf{x}, \mathbf{x}') = \sum_{c\in C} \delta(\mathbf{x} - c\mathbf{x}') \tag{2.154}$$

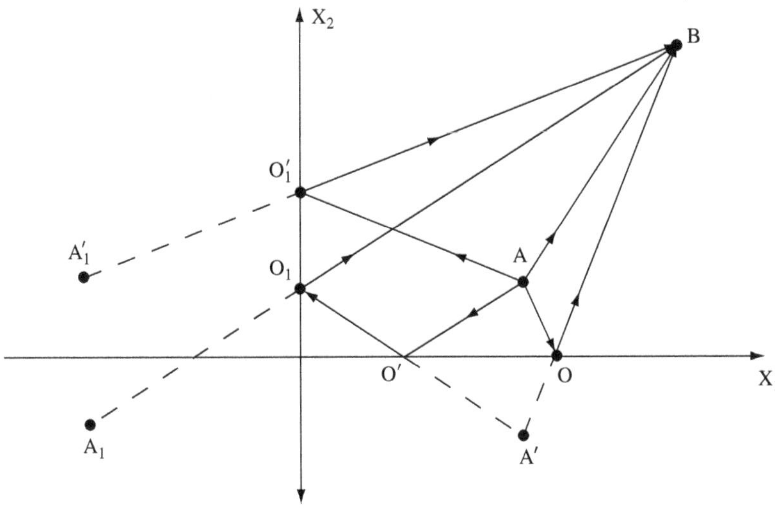

Fig. 2.2. Reflected trajectories for a particle confined in the positive quadrant

(the extension symmetric with respect to the Coxeter group). The problem of the path integral description of a particle dynamics in a sector has also been studied in [130, 131].

It should be mentioned that not all Coxeter groups are finite (e.g. the Weyl groups of infinite-dimensional Kac–Moody algebras used in string theory; the affine Weyl group arising in the path integral formalism for topological Yang–Mill theories studied in detail in Chapter 6), and not all can be described in terms of symmetries and Euclidean reflections (e.g. hyperbolic Coxeter groups [126–128]). Not all physically interesting boundary value problems can be related to Coxeter groups (e.g. the above sector problem, but with the angle being π/μ, where $\mu > 1$ is irrational).

2.7.2 A particle in a disk

Consider the problem of a particle in an infinite circular potential well of radius R [124]. The kernel,

$$U_\varepsilon(\mathbf{x}, \mathbf{x}') = (2\pi\varepsilon\hbar)^{-1} \exp\left[\frac{i}{\hbar}\frac{|\mathbf{x}-\mathbf{x}'|^2}{2\varepsilon}\right], \quad \mathbf{x} = (x_1, x_2), \tag{2.155}$$

is a solution of the Schrödinger equation. In the spirit of the above approach, the kernel (2.155) should be used to find the fundamental solution of the problem that also satisfies the zero boundary condition at $|\mathbf{x}| = R$. Suppose that the function (2.155) is defined in the entire plane. An extension Ψ_0 of the initial wave function ψ_0 to the non-physical region (outside the disk) is sought such that the evolved function,

$$\psi_t(\mathbf{x}) = \int \frac{d\mathbf{x}'}{2\pi i t\hbar} \exp\left[\frac{i}{\hbar}\frac{|\mathbf{x}-\mathbf{x}'|^2}{2t}\right]\Psi_0(\mathbf{x}'), \quad t \geq 0, \tag{2.156}$$

solves the initial value problem and satisfies the zero boundary condition at $|\mathbf{x}| = R$; the integration region in (2.156) coincides with the entire plane. It turns out that the extension with required properties does exist and the initial value problem solution can indeed be given by (2.156).

To prove the statements as asserted, consider the orthonormal set of eigenfunctions of the Hamiltonian (in polar coordinates),

$$\psi_{km}(r,\phi) = \tilde{c}_{km} J_m(\lambda_{km} r) e^{im\phi}, \quad k, |m| = 0, 1, 2, ..., \tag{2.157}$$

where $\tilde{c}_{km}$ are normalization constants, J_m are the Bessel functions, and λ_{km} are found from the boundary condition $J_m(\lambda_{km} R) = 0$; they determine eigenvalues of the Hamiltonian: $\hat{H}\psi_{km} = E_{km}\psi_{km}$ and $E_{km} = \hbar^2\lambda_{km}^2/2$. The initial wave function is expanded over the basis functions (2.157),

$$\psi_0(r,\phi) = \sum_{k,m} c_{km}\psi_{km}(r,\phi), \quad r \in [0,R], \quad \phi \in [0, 2\pi). \tag{2.158}$$

The Bessel functions can be extended analytically to the region $r > R$. The function ψ_0 is extended to the non-physical region $r > R$ by means of the series (2.158), i.e. it is postulated that outside the disk $r \leq R$, ψ_0 is defined by the series (2.158) in which the functions ψ_{km} are taken at $r > R$. In essence, this is a way to extend a non-analytic function (ψ_0 is an element of the Hilbert space). It should be noted that, depending on the problem, the very same function is extended differently, i.e. Eq. (2.158) defines an extension *natural* for the problem studied. Let Ψ_0 be the natural extension of ψ_0 defined by the series (2.158).

Next it is proved that function (2.156) gives a solution of the initial value problem, that is, for all finite $t > 0$, ψ_t satisfies both the Schrödinger equation in the disk and the zero boundary condition. In polar coordinates, Eq. (2.156) has the form,

$$\psi_t(\mathbf{x}) = \sum_{k,m} C_{km} \int_0^\infty \int_0^{2\pi} \frac{r'dr'd\phi'}{2\pi i\hbar t} \exp\left\{\frac{i}{\hbar}\frac{[r^2 + r'^2 - 2rr'\cos(\phi - \phi')]}{2t}\right\} \times J_m(\lambda_{km} r') e^{im\phi'}, \tag{2.159}$$

where $C_{km} = c_{km}\tilde{c}_{km}$. Making use of the expansion,

$$e^{iz\cos\phi} = \sum_{n=-\infty}^{\infty} i^n e^{in\phi} J_n(z)$$

and the relation,

$$\int_0^\infty dr r e^{i\gamma r^2} J_m(\alpha r) J_m(\beta r) = \frac{i}{2\gamma} e^{-\frac{\beta\alpha i}{2} m} e^{\frac{\alpha^2+\beta^2}{4i\gamma}} J_m\left(\frac{\alpha\beta}{2\gamma}\right)$$

(a consequence of Eq. 6.633.2 in [132]), Eq. (2.159) is transformed to,

$$\psi_t(\mathbf{x}) = \sum_{k,m} C_{km} J_m(\lambda_{km} r) e^{im\phi} e^{-iE_{km}t/\hbar}, \tag{2.160}$$

from which it is clear that ψ_t is indeed the solution of the problem. Thus, the extension Ψ_0 of ψ_0 to the entire plane allows us to use (2.156), i.e. to integrate over trajectories in the whole space.

The above solution can also be restated in terms of the extension operator $\hat{Q}$. Put

$$Q(\mathbf{x}, \mathbf{x}') = \sum_{k,m} \psi_{km}(\mathbf{x}) \psi_{km}(\mathbf{x}'). \tag{2.161}$$

The series obviously converges to the δ-function in the physical region $|\mathbf{x}| \le R$ and $|\mathbf{x}'| \le R$, while for $|\mathbf{x}| > R$ it defines a natural extension of the δ-function to the non-physical region. The extended function Ψ_0 is now viewed as the result of the action of the operator $\hat{Q}$ with the kernel (2.161),

$$\Psi_0(\mathbf{x}) = \hat{Q}\psi_0(\mathbf{x}) = \int\limits_{|\mathbf{x}'| \le R} d\mathbf{x}' Q(\mathbf{x}, \mathbf{x}') \psi_0(\mathbf{x}'). \tag{2.162}$$

Substituting (2.162) into (2.156), the latter can be written as,

$$\psi_t(\mathbf{x}) = \int\limits_{|\mathbf{x}'| \le R} d\mathbf{x}' K_t(\mathbf{x}, \mathbf{x}') \psi_0(\mathbf{x}'), \tag{2.163}$$

$$K_t(\mathbf{x}, \mathbf{x}') = \int d\mathbf{x}'' U_t(\mathbf{x}, \mathbf{x}'') Q(\mathbf{x}'', \mathbf{x}'), \tag{2.164}$$

where $t > 0$ and the arguments $\mathbf{x}$ and $\mathbf{x}'$ range over the physical region. Equation (2.164) determines a superposition of the kernels (2.155) that satisfies the required boundary condition; the kernel (2.155) has a standard representation by the path integral.

2.7.3 General problems with zero boundary conditions

Let $\Omega \subset R^n$ be the physical configuration space. The wave functions satisfy the zero boundary condition,

$$\psi|_\Gamma = 0, \quad \Gamma = \partial\Omega. \tag{2.165}$$

The boundary Γ divides the space into two disconnected parts, Ω and its complement Ω^c. The way the path integral problem has been treated so far requires natural extension of the initial wave function into the non-physical region Ω^c. For generic boundaries, there are no symmetries that could be used to construct this extension, while an explicit solution of the eigenvalue problem for the Hamiltonian would also hardly be possible and, hence, representation (2.141) of the extension operator kernel becomes practically useless. It is therefore desirable to find an alternative relation between the evolution operator kernels for the same system (the same Hamiltonian) but with and without the boundary condition (2.165) imposed, avoiding an explicit use of the spectral representation (2.141) of the extension operator. Such a relation does exist [133], although its practical application for simple boundaries is rather technical [134]. The problem is studied for Euclidean path integrals; the evolution operator kernel is first obtained for the imaginary time $\beta = it \geq 0$ and then an analytic continuation $\beta \to it$ is carried out, where t is the physical time.

Consider the fundamental solution of the problem,

$$\begin{aligned} -\partial_\beta U_\beta &= \hat{H} U_\beta, \quad U_0(q, q') = \delta(q - q'), \quad q, q' \in \Omega \\ U_\beta\big|_{q\in\Gamma} &= U_\beta\big|_{q'\in\Gamma} = 0, \quad \beta > 0, \end{aligned} \tag{2.166}$$

subject to the boundary condition (2.165). Suppose that $\hat{H}$ has a natural extension (e.g. an analytic extension) to the whole R^n so that the fundamental solution of the problem (without boundaries)

$$-\partial_\beta U^0_\beta = \hat{H} U^0_\beta, \quad U^0_0(q, q') = \delta(q - q'), \quad q, q' \in R^n \tag{2.167}$$

can be written via the Euclidean path integral,

$$U^0_\beta(q, q') = \int\limits_{q(0)=q'}^{q(\beta)=q} Dq\, e^{-S[q]},$$

where $S[q]$ is the Euclidean action corresponding to the Hamiltonian $\hat{H}$ and the integration is carried out over paths $q = q(\tau)$, $\tau \in [0, \beta]$, with pinned end points $q(0) = q'$ and $q(\beta) = q$. The problem is to find a relation between this path integral and the fundamental solution of the boundary problem (2.166).

Let $\delta_\Gamma(q)$ be the δ-function with support on Γ, i.e. for any $\psi(q)$,

$$\int dq\ \delta_\Gamma(q)\ \psi(q) = \int_\Gamma d\sigma_\xi \psi(q_\xi),$$

where $q = q_\xi$ are parametric equations of Γ, and $d\sigma_\xi$ is the corresponding measure to integrate over Γ (the surface Γ is assumed to be piece-wise smooth). Consider the fundamental solution $U^\lambda_\beta(q, q')$ of the problem (2.167) in which the Hamiltonian is modified by adding a singular potential,

$$\hat{H} \to \hat{H}_\lambda = \hat{H} + \lambda\delta_\Gamma(q). \tag{2.168}$$

This has the formal path integral representation,

$$U^\lambda_\beta(q, q') = \int\limits_{q(0)=q'}^{q(\tau)=q} Dq \exp\left\{-S[q] - \lambda \int_0^\beta \delta_\Gamma(q(\tau))d\tau\right\}. \tag{2.169}$$

The last term in the exponential contributes only if the path crosses the surface Γ. An accurate definition of the amplitude (2.169) as a solution to an integral equation is given below (see (2.172)).

The following relation holds which solves the posed problem:

$$U_\beta(q, q') = \lim_{\lambda\to\infty} U_\beta^\lambda(q, q'). \tag{2.170}$$

Before proving this relation, a further remark is in order. Intuitively, the statement can be understood from the following observation. Let q_ξ be a point on Γ and $\mathbf{n}_\xi$ be a normal to Γ at q_ξ. Consider a small solid cylinder C_h with the axis through q_ξ and parallel to $\mathbf{n}_\xi$. The cylinder extends to both sides of Γ by an infinitesimal length h. Let $\Delta\Gamma$ be the portion of Γ that lies inside C_h. Integrating both sides of the Schrödinger equation, $\hat{H}_\lambda\psi = E\psi$, over C_h, one infers,

$$\int_{\Delta\Gamma} d\sigma_\xi \psi = -\frac{1}{\lambda}\int_{C_h} dq(\hat{H} - E)\psi. \tag{2.171}$$

If the integral in the right-hand side of this equation remains finite in the limit $h \to 0$, then the solution ψ must satisfy the boundary condition (2.165) in the limit $\lambda \to \infty$ because $\Delta\Gamma$ is arbitrary. For example, for a free particle $\hat{H} \sim \Delta$ and the right-hand side of (2.171) is proportional to the jump of the normal derivative of ψ at q_ξ. It is also noteworthy that this approach can equally well be used for the complementary region Ω^c.

Consider the integral equation for the fundamental solution (2.169),

$$U_\beta^\lambda(q, q') = U_\beta^0(q, q') - \lambda \int_\Gamma d\sigma_\xi \int_0^\beta d\tau U_{\beta-\tau}^0(q, q_\xi) U_\tau^\lambda(q_\xi, q'). \tag{2.172}$$

It can be derived from the integral form of the equation (2.166),

$$e^{-\beta(\hat{H}+\hat{V})} = e^{-\beta\hat{H}} + \int_0^\beta d\tau e^{-(\beta-\tau)\hat{H}}\hat{V}e^{-\tau(\hat{H}+\hat{V})}, \tag{2.173}$$

for $\hat{V} = \lambda\delta_\Gamma$. Relation (2.173) is obtained as follows. The operator solution $\hat{U}_\beta = e^{-\beta(\hat{H}+\hat{V})}$ of Eq. (2.166) is written in the form $\hat{U}_\beta = e^{-\beta\hat{H}}(1 + \hat{W}_\beta)$ where the operator $\hat{W}_\beta$ obeys the initial condition $\hat{W}_{\beta=0} = 0$. The substitution of this representation into Eq. (2.166) yields the equation $\partial_\beta \hat{W}_\beta = e^{\beta\hat{H}}\hat{V}\hat{U}_\beta$. Relation (2.173) is then deduced by integrating both the sides of this equation over the interval $[0, \beta]$. Carrying out the Laplace transform of (2.172) yields,

$$\bar{U}_s^\lambda(q, q') = \bar{U}_s^0(q, q') - \lambda \int_\Gamma d\sigma_\xi \bar{U}_s^0(q, q_\xi)\bar{U}_s^\lambda(q_\xi, q'), \tag{2.174}$$

where

$$\bar{U}_s^\lambda = \int_0^\infty d\beta e^{-s\beta} U_\beta^\lambda$$

is the Laplace transform of U_β^λ. Equation (2.174) is local in the parameter s and can be solved. Let $\hat{K}_s^\lambda$ be the operator defined by the kernel $\bar{U}_s^\lambda(q, q')$, where

$q \in \Gamma$ and $\hat{K}_s^\Gamma$ is the operator defined by the kernel $\bar{U}_s^0(q, q')$ where both the arguments are restricted to Γ, i.e. $q, q' \in \Gamma$. Then it follows from (2.174) that,

$$\hat{K}_s^\lambda = \hat{K}_s^0 - \lambda \hat{K}_s^\Gamma \hat{K}_s^\lambda .$$

Hence,

$$\hat{K}_s^\lambda = \left[1 + \lambda \hat{K}_s^\Gamma\right]^{-1} \hat{K}_s^0 .$$

This solution establishes an explicit dependence of the kernel $\bar{U}_s^\lambda$ on λ in the right-hand side of (2.174). Thus, after its substitution into the right-hand side of (2.174), the limit $\lambda \to \infty$ can be explicitly taken with the result,

$$\bar{U}_s(q, q') = \bar{U}_s^0(q, q') - \int_\Gamma d\sigma_\xi d\sigma_{\xi'} \bar{U}_s^0(q, q_\xi) M_s(q_\xi, q_{\xi'}) \bar{U}_s^0(q_{\xi'}, q'), \tag{2.175}$$

where $\hat{M}_s K_s^\Gamma = K_s^\Gamma \hat{M}_s = 1$ or,

$$\int_\Gamma d\sigma_{\xi''} M_s(q_\xi, q_{\xi''}) \bar{U}_s^0(q_{\xi''}, q_{\xi'}) = \int_\Gamma d\sigma_{\xi''} \bar{U}_s^0(q_\xi, q_{\xi''}) M_s(q_{\xi''}, q_{\xi'}) = \delta_{\xi, \xi'}, \tag{2.176}$$

with $\delta_{\xi,\xi'}$ being the unit operator kernel with respect to the measure $d\sigma_\xi$.

If either q or q' is restricted to Γ in (2.175), then its right-hand side vanishes because of (2.176). Hence, the inverse Laplace transform of (2.175) satisfies the zero boundary conditions at Γ. The Laplace transform of (2.175) is also proportional to U_β^0, which satisfies Eq. (2.167). Thus, the inverse Laplace transform of the kernel (2.175) is the fundamental solution of the boundary problem (2.165) expressed via the path integral (2.169).

Specific examples of the above formalism can be found in [134]. For simple boundaries Γ, like an infinite rectangular well, boundaries of a modular domain in R^n relative to actions of finite reflection groups, Eq. (2.175) can be further simplified to the form (2.140) and (2.152). The method can be applied to more complicated examples where Ω is the modular domain relative to affine Weyl groups [135]. Problems with moving boundaries can also be studied in the same way [134].

2.7.4 A particle in the potential q^k

Regarding Eq. (2.41) with the kernel (2.10), R. Feynman made a remark that this is valid only for potentials growing no faster than q^2 at infinity [62]. Pauli devoted more attention to this issue [66], giving a concrete example of the insufficiency of Eqs. (2.41) and (2.10), namely, the problem of a particle in an infinite well. The latter can be viewed as a quantum mechanical problem of describing a particle moving in a force field with the potential

$$V(q) = \left|1 - \frac{2q}{L}\right|^\infty , \tag{2.177}$$

which grows faster than any power of $|q|$. The physical content of the proposed solution (2.135) and (2.137) for the problem with the potential (2.177) has been clarified: operator $\hat{Q}$ takes into account the necessary contributions of trajectories reflected from the boundaries. It is therefore natural to investigate the class of potentials for which the equation

$$\psi_\varepsilon(q) \approx \int_{-\infty}^{\infty} \frac{dq'}{(2\pi i\varepsilon\hbar)^{n/2}} \exp\left\{\frac{i}{\hbar}\left[\frac{(q-q')^2}{2\varepsilon} - \varepsilon V(q')\right]\right\} \psi_0(q'), \tag{2.178}$$

$\varepsilon \to 0$, becomes invalid. Clearly, this should be related to whether the operator $\hat{H} = \hat{p}^2/2 + V(\hat{q})$ is self-adjoint. A detailed study of this problem for power potentials is due to P. Choquard [136].

Let us elaborate this question in more detail (see also [96]). The key point here is as follows. As emphasized in Sections 2.2–2.4, in the basic equations of the path integral formalism, like (2.178), all the terms of order ε should be retained. Therefore the value of the path integral is fully determined by the asymptotic behavior of (2.178) which can be investigated by the stationary phase approximation [137]. The leading term in the exponential is usually $(\Delta q)^2/\varepsilon$. This is indeed so for potentials that grow no faster than q^2 at spatial infinity. But for faster growing potentials the term $\varepsilon V(q')$ at large q' cannot be neglected, for it is no longer small in comparison with $(\Delta q)^2/\varepsilon$.

This statement can be illustrated with an example of the potential $V(q) = \lambda q^k$ in one-dimensional space ($n = 1$). The following technical trick is useful for this purpose (suggested by S. Yu. Slavyanov). In (2.178) a change of variables is made, $q = y/\varepsilon^\alpha$ and $q' = y'/\varepsilon^\alpha$, where the parameter α is fixed by the condition that the powers of both the terms in the square brackets coincide, $\varepsilon^{-1-2\alpha} = \varepsilon^{1-\alpha k}$, and, hence,

$$\alpha = 2/(k-2). \tag{2.179}$$

Equation (2.178) assumes the form,

$$\psi_\varepsilon\left(\frac{y}{\varepsilon^\alpha}\right) \approx \frac{1}{\varepsilon^\alpha}\int_{-\infty}^{\infty} \frac{dy'}{(2\pi i\varepsilon\hbar)^{1/2}} \exp\left\{\frac{i}{\hbar}\frac{1}{\varepsilon^{2\alpha+1}}\left[\frac{(y-y')^2}{2} - \lambda y'^k\right]\right\} \psi_0\left(\frac{y'}{\varepsilon^\alpha}\right). \tag{2.180}$$

If $k > 2$, then $2\alpha + 1 > 1$ because of (2.179) and the asymptote of the integral in (2.180) as $\varepsilon \to 0$ is found by the standard method. The stationary points are obtained by solving the algebraic equation,

$$y' - y - k\lambda y'^{k-1} = 0, \tag{2.181}$$

which has $k-1$ solutions. One of them for $y \to 0$ (as $\epsilon \to 0$, while q remains fixed, and, hence, $y = O(\epsilon^\alpha)$) is found by perturbation theory,

$$y_0' \approx y + k\lambda y^{k-1}, \tag{2.182}$$

and corresponds to the assumption that the leading term is indeed $(\Delta q)^2/\varepsilon$. The other solutions are small; they give contributions of the form,

$$\sum_m c_m \exp(iA_m/\hbar\varepsilon^{2\alpha+1})\psi_0\left(y'_m/\varepsilon^\alpha\right), \tag{2.183}$$

where y'_m are roots of Eq. (2.181) enumerated by m, and $A_m = \left(y - y'_m\right)^2/2 - \lambda y'^k_m$. Similar results are obtained when the integration in (2.178) is carried out over an interval or semi-axis. For intervals the amplitude ψ_0 is taken at the boundary points, while here the asymptote of (2.183) as $\varepsilon \to 0$ is determined by the behavior of ψ_0 at infinity. It is clear that, for a sufficiently fast decreasing ψ_0 ($\psi_0 \sim 1/q^{(1/\alpha+\delta)}$, $\delta > 0$), contributions of all other stationary points can be neglected. In order for their contributions to be negligible in the path integral, the wave function ψ_t is required to decrease just as fast at a finite t. For states with a finite energy (even k) $\langle\psi_0|\hat{H}|\psi_0\rangle < \infty$, $\psi \sim 1/q^{(k+1+2\delta)/2}$, $\delta > 0$. But $(k+1+2\delta)/2 > 1/\alpha + \delta = (k-2)/2 + \delta$, i.e. the above condition on the decrease of ψ_0 is always satisfied. Thus, it is indeed sufficient to take into account only solution (2.182) in (2.180); the additional stationary points occurring in the case of potentials of the form q^k $(k > 2)$ can be discarded.

P. Choquard also investigated contributions of the trajectories connecting $q(0)$ and $q(t)$ to the asymptote of (2.178) (see [136]). For the potentials $V \sim q^{2k}$, it was found that, beside the main trajectory, there are reflected trajectories with extremal properties. The conclusion of [136] is similar to the one given above: the contribution of reflected trajectories may be neglected at a sufficiently small time interval.

2.7.5 Topologically nontrivial coordinates

The cylindrical and spherical coordinates provide an example of topologically nontrivial coordinates (i.e. curvilinear coordinates in a Euclidean space whose coordinate surfaces cannot be continuously deformed to those of the Cartesian coordinates (hyper-planes)). As explained in Section 2.6, coordinate transformations that preserve the (trivial) topology of the Cartesian coordinate surfaces do not cause any substantial problem in the path integral formalism, and can be handled only by taking into account the necessary extra terms. In contrast, coordinate transformations under which the topology of coordinate surfaces changes create additional subtleties that need be clarified. The problem is illustrated with an example of cylindrical coordinates in a three-dimensional Euclidean space. The spectra of the canonical operators $\hat{x}_i$ and $\hat{p}_i$, associated with the Cartesian coordinates and containing the entire real axis, are changed to the following spectra:

$$\hat{z} = \hat{x}_3 \ : (-\infty, \infty); \quad \hat{p}_z = -i\hbar\partial_z \quad : \ (-\infty, \infty); \tag{2.184}$$

$$\hat{r} = \left(\hat{x}_1^2 + \hat{x}_2^2\right)^{1/2} : [0, \infty); \quad \hat{p}_r = -i\hbar r^{-1/2}\partial_r r^{1/2} : (-\infty, \infty); \tag{2.185}$$

$$\hat{\phi} = \tan^{-1}\left(\hat{x}_2\hat{x}_1^{-1}\right) : [0, 2\pi); \quad \hat{p}_\phi = -i\hbar\partial_\phi \quad :0, \pm\hbar, \pm 2\hbar, \ldots \tag{2.186}$$

Thus, even for a rather simple coordinate transformation, four different kinds of spectra occur. The eigenfunctions of the momentum operators $\hat{p}_z$, $\hat{p}_r$, and $\hat{p}_\phi$ are,

$$\frac{c}{\sqrt{r}}\exp\left[\frac{i}{\hbar}\left(p_z z + p_r r + m\hbar\phi\right)\right], \quad m = 0, \pm 1, \ldots. \tag{2.187}$$

Basic equations like (2.7) would contain integrations in various limits and even summations. Moreover, $\hat{p}_r$ is not self-adjoint in $[0,\infty)$ [138, 139] and its eigenfunctions $\langle r|p\rangle \sim \exp(irp/\hbar)/\sqrt{r}$ are not orthogonal,

$$\begin{aligned}\langle p|p'\rangle &= \int_0^\infty drr\,\langle p|r\rangle\langle r|p'\rangle \\ &= \frac{1}{2\pi\hbar}\int_0^\infty drr\,\frac{1}{r}\exp\left[\frac{i}{\hbar}(p'-p)r\right] = \frac{i}{2\pi}\frac{1}{p'-p+i\epsilon},\end{aligned} \tag{2.188}$$

i.e. $\langle p|p'\rangle$ is not a δ-function. Nevertheless, in the set of physical wave functions,

$$\psi(p) = \langle p|r\rangle\langle r|\psi\rangle = (2\pi\hbar)^{-1/2}\int_0^\infty dr\sqrt{r}\exp(-ipr/\hbar)\psi(r), \tag{2.189}$$

analytic for $\operatorname{Im} p < 0$ and decreasing as $\operatorname{Im} p \to -\infty$, the expression (2.188) is the unit operator kernel,

$$\langle p|p'\rangle\langle p'|\psi\rangle = \frac{i}{2\pi}\int_{-\infty}^{\infty}\frac{dp'\psi(p')}{p'-p+i\epsilon} = \langle p|\psi\rangle. \tag{2.190}$$

This observation gives hope that there is the possibility of satisfactory formulation of the path integrals in cylindrical (spherical) coordinates. This turns out to be so. By making use of the methods proposed in [96, 120], it will be shown that, despite the diversity of spectra (2.184)–(2.186) in (2.187), the path integral can be written in the standard form when the integration over all variables is carried out in the infinite limits. To begin with, consider the simplest case of polar coordinates. Because of the importance of the problem, which remained unsolved for a long time, the solution is obtained in two different ways.

Polar coordinates; Lagrangian path integrals. Consider the problem of the path integral formalism in polar coordinates for a free particle of unit mass moving in a plane. The evolution operator kernel reads,

$$\begin{aligned}\langle \mathbf{x}|\hat{U}_t|\mathbf{x}'\rangle &= \frac{1}{2\pi i t\hbar}\exp\left[\frac{i}{\hbar}\frac{(\mathbf{x}-\mathbf{x}')^2}{2t}\right] \\ &= \frac{1}{2\pi i t\hbar}\exp\left[\frac{i}{2t\hbar}(r^2 + r'^2 - 2rr'\cos\Delta_\phi)\right] = K_t(r, r', \Delta_\phi),\end{aligned} \tag{2.191}$$

where $\Delta_\phi = \phi - \phi'$; it is periodic in the angular variable ϕ: $K_t(r, r', \Delta_\phi + 2n\pi) = K_t(r, r', \Delta_\phi)$ as it should be. However, when constructing the path integral representation of the kernel (2.191), only the asymptote of K as $t \to 0$ is used, which is determined by critical points of the expression in the exponential, i.e. by solutions of the equations,

$$\sin\Delta_\phi = 0\ ; \qquad r - r'\cos\Delta_\phi = 0. \tag{2.192}$$

Solutions of (2.192) can be written in the form,

$$\phi_0' = \phi + 2\pi n, \quad r_0' = r\ ; \tag{2.193}$$
$$\phi_0' = \phi + \pi + 2\pi n, \quad r_0' = -r, \quad n = 0, \pm 1, \ldots. \tag{2.194}$$

Expanding the expression in the exponential (2.191) into a power series over Δ_ϕ in the neighborhoods of points (2.193) and (2.194), and limiting the expansion to only extra terms relevant in the limit $\varepsilon \to 0$ [69, 103], the following representation of the asymptote K_ε of the kernel (2.191) ($t = \varepsilon \to 0$) is obtained [96]:

$$K_\varepsilon = \frac{1}{2\pi i\varepsilon\hbar}\sum_{n=-\infty}^{\infty}\left\{\exp\left[\frac{i}{2\varepsilon\hbar}\left(\Delta_r^2 + rr'\left(\Delta_{n\phi}^2 - \frac{\Delta_{n\phi}^4}{12}\right)\right)\right]\right.$$
$$\left. + \begin{pmatrix} r' \to -r' \\ \phi' \to \phi' + \pi \end{pmatrix}\right\}, \tag{2.195}$$

where $\Delta_r = r - r'$ and $\Delta_{n\phi} = \phi - \phi' + 2\pi n$. Equation (2.195) can be rewritten as,

$$K_\varepsilon(r, r'', \phi, \phi'') = \int_{-\infty}^{\infty}\int_{-\infty}^{\infty}\frac{dr'd\phi'}{2\pi i\varepsilon\hbar}$$
$$\times \exp\left\{\frac{i}{2\varepsilon\hbar}\left[\Delta_r^2 + rr'\left(\Delta_\phi^2 - \frac{\Delta_\phi^4}{12}\right)\right]\right\} Q_2(r', \phi'; r'', \phi''), \tag{2.196}$$

where the variables r'' and ϕ'' range over the physical region $[0, \infty)$ and $[0, 2\pi)$, respectively, and

$$Q_2(r, \phi; r', \phi') = \delta(r - r')Q(\phi - \phi') + \delta(r + r')Q(\phi - \phi' + \pi), \tag{2.197}$$
$$Q(\phi - \phi') = \sum_{n=-\infty}^{\infty}\delta(\phi - \phi' + 2\pi n). \tag{2.198}$$

The kernel (2.196) defines the action of $\hat{K}_\varepsilon$ on a wave function: $\hat{K}_\varepsilon\psi_0(r, \phi) = \int dr'r'd\phi' K_\varepsilon(r, r', \phi, \phi')\psi_0(r', \phi')$, where the integral is evaluated over the physical region. Substituting $r' = r - \Delta_r$ into (2.196) and eliminating all the extra terms by means of the equivalence rules (2.61), (2.63)–(2.65), the following representation of the infinitesimally evolved wave function is obtained:

$$\psi_\varepsilon \approx \int_{-\infty}^{\infty}\frac{\sqrt{rr'}dr'd\phi'}{2\pi i\varepsilon\hbar}\exp\left\{\frac{i}{\hbar}\left[\frac{\Delta_r^2 + r^2\Delta_\phi^2}{2\varepsilon} + \varepsilon\frac{\hbar^2}{8r^2}\right]\right\}\Psi_0(r', \phi'), \tag{2.199}$$

where the function Ψ_0 is defined in the entire *polar plane*, $(r,\phi) \in R^2$, by means of the extension operator $\hat{Q}_2$,

$$\Psi_0(r,\phi) = \int_0^\infty dr' \int_0^{2\pi} d\phi' Q_2(r,\phi;r',\phi')\psi_0(r',\phi'), \tag{2.200}$$

i.e. $\Psi_0(r,\phi+2\pi n) = \Psi_0(r,\phi)$, $\Psi_0(-r,\phi) = \Psi_0(r,\phi+\pi)$, and $\Psi_0 = \psi_0$ in the physical region. It is straightforward to be convinced that ψ_ε satisfies the Schrödinger equation and is extended into the non-physical region in the same way as Ψ_0. Therefore the whole standard machinery of the path integral formalism can now be easily employed.

Polar coordinates; Hamiltonian path integral. There is an alternative derivation of the kernel (2.191) in polar coordinates which allows one to obtain it directly in the Hamiltonian form. The kernel (2.191) is invariant with respect to the symmetry transformations,

$$\phi \to \phi + 2\pi, \tag{2.201}$$

$$r \to -r, \qquad \phi \to \phi + \pi, \tag{2.202}$$

and similarly for ϕ' and r'. As $t = \varepsilon \to 0$, the approximation $\cos\Delta_\phi \approx 1 - \Delta_\phi^2/2$ is valid (in this case the argument of the exponential in (2.191) turns into $i\varepsilon L/\hbar$ where $L = (\dot{r}^2 + r^2\dot{\phi}^2)/2$ is the classical Lagrangian). However this approximation, as well as the more accurate one that also includes the extra term $\Delta_\phi^4/24$, breaks the fundamental symmetry (2.201) and (2.202). This difficulty can be avoided in the following way. The part of the exponential (2.191) that contains the cosine function is expanded into a Bessel series [140],

$$\exp(-iz\cos\Delta_\phi) = \sum_{m=-\infty}^{\infty} J_m(z) \exp\left[i\left(\Delta_\phi + \frac{3\pi}{2}\right)\right], \quad z = \frac{rr'}{\varepsilon\hbar} > 0, \tag{2.203}$$

where J_m is the Bessel function. When $\varepsilon \to 0$ the Bessel function has the asymptotic form [140],

$$J_m(z) = \sqrt{\frac{2}{\pi z}} \cos\left[z - \left(m + \frac{1}{2}\right)\frac{\pi}{2} + \frac{m^2 - 1/4}{2z}\right] + O(z^{-5/2}). \tag{2.204}$$

Next Eq. (2.131) is used, which is written in the form,

$$\sum_{m=-\infty}^{\infty} f(m\hbar)e^{im(\phi-\phi'')} = \int_{-\infty}^{\infty} \frac{dp\,d\phi'}{\hbar} f(p) e^{\frac{i}{\hbar}p(\phi-\phi')} Q(\phi' - \phi''). \tag{2.205}$$

Taking into account (2.204) and (2.205) the following expression for the asymptotic form of the exponential (2.203) is derived:

$$e^{-iz\cos(\phi-\phi'')} = \sqrt{\frac{2\pi i}{z}} \int_{-\infty}^{\infty} \frac{dpd\phi'}{2\pi\hbar} e^{\frac{i}{\hbar}p(\phi-\phi')} \left\{ \exp\left[-iz - \frac{i}{\hbar}A_z(p)\right] \right.$$

$$\left. \times Q(\phi'-\phi'') + \exp\left[iz + \frac{i}{\hbar}A_z(p) - \frac{i\pi}{2}\right] Q^{\pi}(\phi'-\phi'') \right\} + O(z^{-5/2}); \quad (2.206)$$

where $A_z(p) = (p^2 - \hbar^2/4)/(2\hbar z)$ and

$$Q^{\pi}(\phi) = Q(\phi+\pi)\ ; \quad (2.207)$$

the argument ϕ in the kernels $Q(\phi-\phi')$ and $Q^{\pi}(\phi-\phi')$ ranges over the whole real line even if ϕ' lies in the physical interval $[0, 2\pi)$. Substitution of (2.206) into the equation,

$$\psi_{\varepsilon}(r,\phi) = \int_0^{\infty}\int_0^{2\pi} \frac{r''dr''d\phi''}{2\pi i\varepsilon\hbar} \exp\left[\frac{i}{\hbar}\frac{r^2 + r''^2 - 2rr''\cos(\phi-\phi'')}{2\varepsilon}\right] \psi_0(r'',\phi''), \quad (2.208)$$

yields,

$$\psi_{\varepsilon} \approx \int\int_{-\infty}^{\infty} \frac{dp_{\phi}d\phi'}{2\pi\hbar} \int_0^{\infty}\int_0^{2\pi} \frac{r''dr''d\phi''}{2\pi i\varepsilon\hbar} \sqrt{\frac{2\pi i\varepsilon\hbar}{rr''}}$$

$$\times \left\{ \exp\left[\frac{i}{\hbar}\left(\frac{(r-r'')^2}{2\varepsilon} - \varepsilon\frac{p_{\phi}^2 - \hbar/4}{2rr''}\right)\right] Q(\phi'-\phi'') \right.$$

$$\left. + \frac{1}{i}\exp\left[\frac{i}{\hbar}\left(\frac{(r+r'')^2}{2\varepsilon} + \varepsilon\frac{p_{\phi}^2 - \hbar^2/4}{2rr''}\right)\right] Q^{\pi}(\phi'-\phi'') \right\} \psi_0(r'',\phi''). \quad (2.209)$$

After moving the factor $1/i$ under the square root, $1/\left(i\sqrt{r''}\right) = 1/\sqrt{-r''}$, the first and second terms in the curly brackets differ only in the sign of r'' (disregarding the difference between Q and Q^{π}). Next the identities,

$$1 = \int_{-\infty}^{\infty} dr'\delta(r' \mp r'') \quad (2.210)$$

are inserted into these terms (the one with the upper sign is for the first term). The factors depending on r and r'' at the exponentials are represented in the form,

$$\frac{r''}{\sqrt{\pm rr''}} = \frac{r'}{\sqrt{rr'}}\sqrt{\frac{r''}{\pm r'}}. \quad (2.211)$$

Now all the variables r'' are replaced by $\pm r'$ in all functions (safe for ψ_0) in accordance with (2.210). As a result,

$$\psi_\varepsilon(r,\phi) \approx \int\limits_{-\infty}^{\infty} \frac{r'dr'}{(2\pi i\varepsilon\hbar)^{1/2}} \int\limits_{-\infty}^{\infty} \frac{dp_\phi d\phi'}{2\pi\hbar} \frac{1}{\sqrt{rr'}} \exp\left\{\frac{i}{\hbar}\left[\frac{\Delta_r^2}{2\varepsilon} + p_\phi\Delta_\phi - \varepsilon\frac{p_\phi^2 - \hbar^2/4}{2rr'}\right]\right\} \Psi_0(r',\phi'), \tag{2.212}$$

where Ψ_0 is defined in (2.200). The sought-for representation follows immediately from (2.212),

$$\psi_\varepsilon \approx \int\limits_{-\infty}^{\infty} \frac{dp_r dr' dp_\phi d\phi'}{(2\pi\hbar)^2} \frac{r'}{\sqrt{rr'}} \exp\left\{\frac{i}{\hbar}\left[p_r\Delta_r + p_\phi\Delta_\phi - \varepsilon\left(\frac{p_r^2}{2} + \frac{p_\phi^2 - \hbar^2/4}{2rr'}\right)\right]\right\} \Psi_0(r',\phi') \tag{2.213}$$

(the integration with respect to p_r turns (2.213) into (2.212)). Note that there is no problem in defining the square root $(rr')^{-1/2}$ when r' is negative, because the contribution to the integral comes only from $r' \approx r$ and the square root is positive. Making use of the equivalence rules (2.57)–(2.59), the factor $(r'/r)^{1/2}$ at the exponential can be included into the Hamiltonian by modifying the potential and the terms linear in momentum. After integration with respect to p_ϕ, Eq. (2.212) becomes (2.199).

It is not difficult to verify that ψ_ε in (2.213) satisfies the same equation as function (2.208) and is defined for all real r and ϕ; it is periodic in ϕ and $\psi_\varepsilon(-r,\phi) = \psi_\varepsilon(r,\phi+\pi)$. The latter allows us to transform (2.208) to the form that is valid for a finite time t,

$$\psi_t(r,\phi) = \int\limits_{-\infty}^{\infty} r'dr'd\phi' U_t(r,\phi;r',\phi')\Psi_0(r',\phi'), \tag{2.214}$$

where the function Ψ_0 defined in (2.200) and the kernel U_t is represented by the path integral [120],

$$U_t(r,\phi;r',\phi') = \frac{1}{\sqrt{rr'}} \int\limits_{-\infty}^{\infty} \prod_{\tau=0}^{t} \frac{dp_r dr dp_\phi d\phi}{(2\pi\hbar)^2} \exp\left\{\frac{i}{\hbar}\int\limits_0^t \left[p_r\dot{r} + p_\phi\dot{\phi} - \frac{p_r^2}{2} - \frac{p_\phi^2 - \hbar^2/4}{2r^2}\right] d\tau\right\}; \tag{2.215}$$

here $r(0) = r'$, $\phi(0) = \phi'$, $r(t) = r$, and $\phi(t) = \phi$. This is the representation of the evolution operator kernel by the path integral in polar coordinates.

Spherical coordinates in n-dimensional Euclidean space. The above construction of the path integral can be generalized to the case of n-dimensional Euclidean space when the evolution operator kernel differs from (2.191) only in the pre-exponential factor $(2\pi it\hbar)^{-n/2}$, and the squared norm of the vector in the exponential is written as,

$$|\mathbf{x}-\mathbf{x}'|^2 = r^2 + r'^2 - 2\sum_{j=1}^{n-2}\left[h_j(q)h_j(q')\cos\theta_j\cos\theta'_j\right] \\ - 2h_{n-1}(q)h_{n-1}(q')\cos\left(\theta_{n-1}-\theta'_{n-1}\right). \tag{2.216}$$

Here θ_j are the angles in the n-dimensional spherical coordinate system (θ_{n-1} is usually denoted by ϕ as the polar angle, $h_j = r\sin\theta_1\cdots\sin\theta_{j-1}$ are the Lamé coefficients, $j = 2, 3, \ldots, n-1$, and $h_1 = r$; the variable q stands for the set of spherical coordinates $(\theta_1, \ldots, \theta_{n-1}, r)$; all the angles range over the interval $[0, \pi]$ apart from ϕ for which $0 \le \phi < 2\pi$. Relations between Cartesian and spherical coordinates have the form,

$$\begin{aligned} x_1 &= r\cos\theta_1\ ; \quad x_2 = r\sin\theta_1\cos\theta_2; \ \ldots; \\ x_{n-1} &= r\sin\theta_1\cdots\sin\theta_{n-2}\cos\theta_{n-1}; \\ x_n &= r\sin\theta_1\cdots\sin\theta_{n-1}, \end{aligned} \tag{2.217}$$

i.e. $x_j = h_j\cos\theta_j$ for $j < n$ and $x_n = h_{n-1}\sin\theta_{n-1}$. Put

$$z'_j = h_j(q)h_j(q')/(\varepsilon\hbar) \equiv h_jh'_j/(\varepsilon\hbar). \tag{2.218}$$

The sequel is essentially based on the repeated application of Eq. (2.206), first, to $\exp\left(-iz'_{n-1}\cos\Delta\theta_{n-1}\right)$ (which comes from the last term in (2.216)), then, because of the factors $\exp(\pm iz)$ (in the present case, $\exp\left(\pm iz'_{n-1}\right)$) in (2.206) and the equality,

$$z'_{j-1}\cos\theta_{j-1}\cos\theta'_{j-1} \pm z'_j = z'_{j-1}\cos\left(\theta_{j-1}\mp\theta'_{j-1}\right), \tag{2.219}$$

to the exponential $\exp\left[-iz'_{n-2}\cos\left(\theta_{n-2}\mp\theta'_{n-2}\right)\right]$, and so on. Before turning to the technical details of this procedure, it should be emphasized again that, as in the case of polar coordinates, the expansion of $|\mathbf{x}-\mathbf{x}'|^2$ in powers of $\Delta\theta_j$ (see (2.219)) in the argument of the first exponential (2.191), cannot be truncated and limited to a finite number of terms in the limit $\varepsilon \to 0$. When constructing the path integral, the asymptote that preserves the discrete symmetries of the infinitesimal evolution operator kernel in spherical coordinates should be used, i.e. relations (2.203)–(2.206) are to be used. It follows from (2.216) that the kernel $U_\varepsilon(\mathbf{x}, \mathbf{x}')$ of the problem studied is invariant under the following transformations:

$$\theta_i \to \theta_i + 2\pi, \qquad i = 1, 2, \ldots, n-1, \tag{2.220}$$

$$\theta_i \to -\theta_i, \qquad \theta_{i+1} \to \theta_{i+1} + \pi, \qquad i = 1, 2, \ldots, n-2, \tag{2.221}$$

$$r \to -r, \qquad \theta_1 \to \theta_1 + \pi, \tag{2.222}$$

which is the set of all transformations preserving the right-hand sides of (2.217).

Let us now turn to the path integral derivation. Unfortunately, all the technicalities are too lengthy to present here in detail. Therefore, only the necessary, most important parts of the derivation are given. To begin with, suitable, more compact notations are introduced. Equation (2.206) is rewritten in the form,

$$\exp\left[-iz_j''\cos\left(\theta_j\mp\theta_j''\right)\right]=c_j\int d\sigma_j'\left[E\left(h_j''\right)e^{-iz_j''}Q\left(\theta_j\mp\theta_j''\right)\right.$$
$$\left.+E\left(-h_j''\right)e^{iz_j''}Q^{\pi}\left(\theta_j'\mp\theta_j''\right)\right]+\cdots,\quad(2.223)$$

where,

$$E\left(h_j''\right)=\frac{1}{\sqrt{h_j''}}\exp\left\{\frac{i}{\hbar}\left[p_j\left(\theta_j-\theta_j'\right)-\varepsilon\frac{p_j^2-\hbar^2/4}{2h_jh_j''}\right]\right\},$$
$$d\sigma_j'=\frac{dp_jd\theta_j'}{2\pi\hbar},\qquad c_j=\left(\frac{2\pi i\varepsilon\hbar}{h_j}\right)^{1/2},\quad(2.224)$$

$h(q')=h'$, $h(q'')=h''$, $z''=hh''/\varepsilon\hbar$, and $Q^{\pi}(\theta)=Q(\theta+\pi)$ (compare (2.207) and (2.218)). Relation (2.223), in which the upper sign is taken and $j=n-1$, is applied to transform the exponential that contains $\cos\left(\theta_{n-1}-\theta_{n-1}''\right)$ in the kernel $U_\varepsilon(\mathbf{x},\mathbf{x}'')$, in accordance with (2.216) and (2.224). The resulting expression is then multiplied by $\exp\left(-iz_{n-2}''\cos\theta_{n-2}\cos\theta_{n-2}''\right)$ according to (2.216) and (2.217). Taking account of both of the signs in the arguments $\mp iz_{n-1}''$ of the exponentials in (2.223) and equation (2.219), this yields that the factors of $E\left(h_{n-1}''\right)$ and $E\left(-h_{n-1}''\right)$ have a form similar to those in the left-hand side of Eq. (2.223), with the signs of θ_{n-2}'' being plus and minus, respectively. The subsequent application of (2.223) to transform these factors leads to the following representation:

$$\exp\left[-iz_{n-2}''\cos\theta_{n-2}\cos\theta_{n-2}''-iz_{n-1}''\cos\left(\theta_{n-1}-\theta_{n-1}''\right)\right]=c_{n-1}c_{n-2}$$
$$\times\int d\sigma_{n-1}'d\sigma_{n-2}'\left\{E\left(h_{n-1}''\right)Q\left(\theta_{n-1}'-\theta_{n-1}''\right)\left[E_z\left(h_{n-2}''\right)Q\left(\theta_{n-2}'-\theta_{n-2}''\right)\right.\right.$$
$$\left.+E_z\left(-h_{n-2}''\right)Q^{\pi}\left(\theta_{n-2}'-\theta_{n-2}''\right)\right]+E\left(-h_{n-1}''\right)Q^{\pi}\left(\theta_{n-1}'-\theta_{n-1}''\right)$$
$$\left.\times\left[E_z\left(h_{n-2}''\right)Q\left(\theta_{n-2}'-\theta_{n-2}''\right)+E_z\left(-h_{n-2}''\right)Q^{\pi}\left(\theta_{n-2}'-\theta_{n-2}''\right)\right]\right\},\quad(2.225)$$

where $E_z\left(\pm h_j''\right)=E\left(\pm h_j''\right)\exp\left(\mp iz_j''\right)$. A further transformation of (2.225) is based on the fact that,

$$E\left(-h_{n-1}\left(r'',\theta_1'',\ldots,\theta_{n-2}''\right)\right)Q\left(\theta_{n-2}'+\theta_{n-2}''\right)$$
$$=E\left(h_{n-1}\left(r'',\theta_1'',\ldots,\theta_{n-3}'',\theta_{n-2}'\right)\right)Q\left(\theta_{n-2}'+\theta_{n-2}''\right),\quad(2.226)$$
$$E\left(h_{n-1}\left(r'',\theta_1'',\ldots,\theta_{n-2}''\right)\right)Q^{\pi}\left(\theta_{n-2}'-\theta_{n-2}''\right)$$
$$=E\left(-h_{n-1}\left(r'',\theta_1'',\ldots\theta_{n-3}'',\theta_{n-2}'\right)\right)Q^{\pi}\left(\theta_{n-2}'-\theta_{n-2}''\right).\quad(2.227)$$

Making use of these relations, Eq. (2.225) is rewritten in the form,

$$\exp[\cdots] = c_{n-1}c_{n-2}\int d\sigma'_{n-1}d\sigma'_{n-2}\left[E\left(h''_{n-1}\left(\theta'_{n-2}\right)\right)E_z\left(h''_{n-2}\right)Q_2(\theta';\theta'')\right.$$
$$\left.+E\left(-h''_{n-1}\left(\theta'_{n-2}\right)\right)E_z\left(-h''_{n-2}\right)Q_2^{\pi}(\theta';\theta'')\right], \tag{2.228}$$

where the new notation, $h_{n-1}\left(r'',\theta''_1,\ldots,\theta''_{n-3},\theta'_{n-2}\right) \equiv h''_{n-2}\left(\theta'_{n-2}\right)$ is introduced to emphasize that the argument in h''_{n-1} is changed, $\theta''_{n-2}\to\theta'_{n-2}$, and,

$$Q_2(\theta';\theta'') = Q\left(\theta'_{n-2}-\theta''_{n-2}\right)Q\left(\theta'_{n-1}-\theta''_{n-1}\right)$$
$$+Q\left(\theta'_{n-2}+\theta''_{n-2}\right)Q^{\pi}\left(\theta'_{n-1}-\theta''_{n-1}\right) \equiv Q_2\left(\theta'_{n-2},\theta'_{n-1};\theta''_{n-2},\theta''_{n-1}\right), \tag{2.229}$$
$$Q_2^{\pi}(\theta';\theta'') = Q_2\left(\theta'_{n-2}+\pi,\theta'_{n-1};\theta''_{n-2},\theta''_{n-1}\right). \tag{2.230}$$

Equation (2.228) has the same structure as (2.223). Therefore the procedure can be repeated but now with the multiplier, $\exp\left(-iz''_{n-3}\cos\theta_{n-3}\cos\theta''_{n-3}\right)$ which corresponds to the next term in the sum (2.216). All of the calculations repeat those already carried out above up to the term with θ_1. A deviation occurs for the radial variable r. The radial variable should be treated in the same way as in the case of polar coordinates, i.e. one should substitute the identities (2.210) into the corresponding terms and take into account that all the h_j's change the sign when r does also. Finally, the pre-exponential factors and the Jacobian $\sqrt{g''} = h''_1\cdots h''_n \equiv h''$ ($h_n = 1$ for the spherical coordinates) must be taken into account in the expressions analogous to (2.209). Making use of the identity (2.211) written for h,

$$\frac{h''}{\sqrt{\pm hh''}} = \frac{h'}{\sqrt{hh'}}\sqrt{\frac{h''}{\pm h'}}, \tag{2.231}$$

and taking into account the sign change of h'' under $\theta''\to\theta'$ (as in (2.226) and (2.227)), the following analogy of Eq. (2.212) is obtained in the case of spherical coordinates in n-dimensional Euclidean space:

$$\psi_\varepsilon(q) \approx \int_{-\infty}^{\infty}\frac{dr'}{(2\pi i\varepsilon\hbar)^{1/2}}\int\prod_{j=1}^{n-1}d\sigma'_j\frac{\sqrt{g'}}{(gg')^{1/4}}$$
$$\times\exp\left\{\frac{i}{\hbar}\left[\frac{\Delta_r^2}{2\varepsilon}+\sum_{j=1}^{n-1}\left(p_j\Delta\theta_j-\varepsilon\frac{p_j^2-\hbar^2/4}{2h_jh'_j}\right)\right]\right\}\Psi_0(q'), \tag{2.232}$$

where,

$$\Psi_0(q) = \int_S d^nq'Q_n(q;q')\psi_0(q'), \tag{2.233}$$

S denotes the physical range of the spherical coordinates $\theta_1,\ldots,\theta_{n-1},r$ (in which (2.217) is a one-to-one mapping), and the kernel Q_n is defined by the recurrence relation that generalizes (2.229) and (2.230),

$$Q_j\left(\theta_{n-j},\ldots,\theta_{n-1};\theta'_{n-j},\ldots,\theta'_{n-1}\right)=Q\left(\theta_{n-j}-\theta'_{n-j}\right)Q_{j-1} +Q\left(\theta_{n-j}+\theta'_{n-j}\right)Q^{\pi}_{j-1}, \tag{2.234}$$

$$Q^{\pi}_j=Q_j\left(\theta_{n-j}+\pi,\theta_{n-j+1},\ldots,\theta_{n-1};\theta'_{n-j},\ldots,\theta'_{n-1}\right),\quad Q_1\equiv Q, \tag{2.235}$$

if $j<n$, and for $j=n$,

$$Q_n\left(r,\theta_1,\ldots,\theta_{n-1};r',\theta'_1,\ldots,\theta'_{n-1}\right)=\delta(r-r')Q_{n-1}+\delta(r+r')Q^{\pi}_{n-1}. \tag{2.236}$$

Direct calculation shows that Q_n, defined by (2.234)–(2.236) and functions Ψ_0 and ψ_ε, given in (2.233) and (2.232), respectively, do not change under the transformations (2.220)–(2.222). Next, it is straightforward to verify that Q_n is the unit operator kernel in the set of functions invariant under (2.220)–(2.222),

$$\Psi_0(q)=\int_S dq'Q_n(q,q')\Psi_0(q').$$

The listed properties are sufficient to develop the path integral representation of the evolution operator kernel from (2.232). The Hamiltonian path integral is readily obtained by adding the integration with respect to p_r (see the transformation of (2.212) to (2.213)). The result is [120],

$$\psi_t(q)=\int dq'\sqrt{g'}U_t(q,q')\Psi_0(q'), \tag{2.237}$$

$$U_t(q,q')=(gg')^{-1/4}\int_{-\infty}^{\infty}\prod_{\tau=0}^{t}\frac{dpdq}{(2\pi\hbar)^n}\exp\left\{\frac{i}{\hbar}S_H\right\}, \tag{2.238}$$

$$S_H=\int_0^t\left[p_i\dot q^i-H(p,q,\hbar)\right]d\tau,$$

$$H(p,q,\hbar)=\frac{p_r^2}{2}+\sum_{j=1}^{n-1}\frac{p_j^2-\hbar^2/4}{2h_j^2(q)}; \tag{2.239}$$

here, $dq\equiv drd\theta_1\cdots d\theta_{n-1}$, $dp\equiv dp_rdp_1\cdots dp_{n-1}$, and p_j are momenta canonically conjugated to the angles θ_j. Integration with respect to all of the canonical variables is carried out over $(-\infty,\infty)$. Just as in the case of polar coordinates, when deriving (2.238) and (2.239) from (2.232), h'_j is replaced by h_j. According to the equivalence rules (2.58) and (2.59), this does not lead to any change in the potential because h_j does not contain θ_j, the variable canonically conjugated to p_j. In the pre-limit expression (or in the time slicing approximation) of (2.238), the function (2.239) is taken at the left boundary points of the partition intervals, although here this is not relevant.

Recall that the Laplace operator in n-dimensional spherical coordinates can be written in the form,

$$\Delta = -\hbar^{-2}\left[\hat{P}_r^2 + \sum_{j=1}^{n-1}\frac{1}{h_j^2}\left(\hat{P}_j^2 - \frac{\hbar^2}{4}\right)\right],$$

where (compare (2.30)),

$$\hat{P}_k = \frac{\hbar}{i}\frac{1}{g^{1/4}}\frac{\partial}{\partial q^k}g^{1/4}, \quad q^n = r, \quad q^j = \theta_j, \quad j \neq n, \quad k = 1, 2, \dots, n, \tag{2.240}$$

while the classical Hamiltonian of a free particle of unit mass reads,

$$H = \frac{1}{2}\left(p_r^2 + \sum_{j=1}^{n-1}\frac{p_j^2}{h_j^2}\right). \tag{2.241}$$

The validity of equations (2.237) and (2.238) may be established directly by, e.g. finding the Schrödinger equation for the function (2.232). Integration with respect to the momenta (integration in the infinite limits is crucial here) gives the Lagrangian path integral. Its validity is established by constructing the corresponding Schrödinger equation for it [62, 63], which is impossible to do without infinite integration limits (for the spherical coordinates, in this particular case).

A transformation of the path integral to general coordinates q^k topologically equivalent to the spherical ones (i.e. to such q^k that the equations $q^k = q^k(r, \theta_1, \dots, \theta_{n-1})$ have only one solution; q^k are smooth functions with the range being the entire real line when $r, \theta_1, \dots, \theta_{n-1}$ have the same range) is carried out according to the general rules developed in Section 2.6.

General curvilinear coordinates (see [120] as well as Section 6.6.2). The problem of changing variables in the path integral has been solved above for spherical coordinates and topologically equivalent to them. However a generalization of the proposed approach to the case of general curvilinear coordinates is not obvious. Let us discuss this question.

In the Hamiltonian path integral one can always make a change of variables from Cartesian coordinates to new specified curvilinear coordinates (see Section 2.6). The new canonical momenta would range over the entire real line (as a consequence of this, a change of variables is a point transformation). Thereby the problem of discrete spectra of the canonical momenta is avoided. As far as integration with respect to the coordinates is concerned, it is closely related to the problem of constructing the operator $\hat{Q}$.

Let the Cartesian coordinates x^i $(i = 1, 2, ..., n)$ be analytic functions $x^i = x^i(q)$ of the new coordinates q^i in the whole R^n. Consider the form $|\mathbf{x} - \mathbf{x}'|^2$, which appears in the evolution operator kernel for a free particle (2.191), in the new coordinates. Suppose that the mapping defined by $x^i(q)$ is one-to-one if the range of q is restricted to a subset $S \subset R^n$. In other words, the inverse functions

$q^i = q^i(\mathbf{x})$ have the range S and the domain R^n. Furthermore, let there exist a discrete group G of transformations of q with elements g_k such that the form $|\mathbf{x}(q) - \mathbf{x}(q')|^2$ is invariant under transformations from G of either q or q', and the transformed variables $g_k q$ ($g_k \neq 1$) do not belong to S. The group G is said to be *a covering group* if the sets $g_k S$ do not intersect and their union covers R^n (except, maybe, for a zero measure set), and S is said to be *a modular domain* in R^n relative to G. An example of such a group is served by the set of transformations (2.220)–(2.222). Now the necessity to integrate with respect to each q^i in the infinite limits can be found in two ways. For instance, one could note that the exponential (2.191) does not have just one stationary point, $q = q_0' \in S$, as the points $q_k = g_k q_0'$ ($g_k \in G$) are also stationary. All of them must be taken into account because, even if just one point is neglected, the G symmetry of the evolution operator kernel becomes broken. According to the assumption that G is a covering group ($\cup_k g_k S = R^n$ (modulo a set of zero measure)), the integration with respect to q is to be done over R^n. On the other hand, the case $S \neq R^n$ is similar to the problem of a particle in a region with boundaries, when one should make sure that the evolution operator kernel satisfies the boundary conditions, which effectively leads to the extension of the integration region to the whole R^n (see Section 2.7.1). More accurately, instead of boundary conditions, here, more or less complicated symmetry properties must hold for the evolution operator kernel. The latter is achieved by the contributions of additional trajectories. For example, the periodicity of the wave function in the angular variable of the polar coordinates is achieved by taking into account not only the shortest trajectory connecting the points (r, ϕ) and (r', ϕ'), but also trajectories different by angles that are integer multiples of 2π. (Here, stationary trajectories are meant.)

Let us turn to the problem of constructing the kernel of the operator $\hat{Q}$. This also arises when solving problems with boundaries by the path integral method (see Section 2.7.1). Earlier the kernel of $\hat{Q}$ emerged in a natural way in the method based on specific properties of spherical coordinates. In a general case, the following observation helps to find it. It is known that the limit of the kernel $U_\varepsilon(q, q')$ as $\varepsilon \to 0$ is the delta-function in S. The existence of the symmetry group G allows one to *(i)* extend the function $U_\varepsilon(q, q')$ to the whole R^n, and *(ii)* construct the kernel of $\hat{Q}$, because the latter is nothing but the unit operator kernel (in the set S) extended by the symmetry properties to R^n [96, 117, 124]. In other words, the kernel of the operator $\hat{Q}$ is the limit of the kernel of the evolution operator $\hat{U}_t$, extended to beyond the physical region S as $t \to 0$, i.e.

$$Q(q, q') = \lim_{\varepsilon \to 0} U_\varepsilon(q, q') \, (g g')^{1/4} . \tag{2.242}$$

For example, the kernel $Q_2/(rr')^{1/2}$ (see (2.197) and (2.198)) is the limit of the kernel $K_t(r, r', \Delta\phi)$ in (2.191), analytically extended to all real r and ϕ, as $t \to 0$.

A direct derivation of the kernels $Q_n(r, \theta_1, \ldots; r', \theta_1', \ldots)$ in (2.236) for $n = 2, 3$ from the corresponding δ-functions $\delta(\mathbf{x} - \mathbf{x}')$ is given in Appendix 8.3.

A list of basic rules for constructing the path integral for a free particle in generic curvilinear coordinates is as follows:

(1) the argument of the exponential in the first line of (2.191) is written in the new coordinates with a given topology that are defined by analytic functions;
(2) the Hamiltonian operator is expressed via the new canonical operators, $\hat{q}$ and Hermitian momenta (see (2.33)), with a specified operator ordering, e.g. by moving all the momentum operators to the right of all the $\hat{q}$'s;
(3) the extension operator kernel is constructed as the limit (2.242) by making use of the explicit form of the kernel (2.191) in the new variables (or by any other means); the kernel $Q(q, q')$ is defined for all real q.

The final result has the following form:

$$\psi_t(q) = \int dq' \sqrt{g'} \int_S dq'' U_t(q, q') Q(q', q'') \psi_0(q''), \tag{2.243}$$

$$U_t(q, q') = \frac{1}{(gg')^{1/4}} \int \prod_{\tau=0}^{t} \frac{dq dp}{(2\pi\hbar)^n} \exp\left\{ \frac{i}{\hbar} \int_0^t [p_i \dot{q}^i - H_{\text{eff}}(q, p)] d\tau \right\}, \tag{2.244}$$

where integration with respect to q and p is carried out over R^n, and the symbol $H_{\text{eff}}(p, q)$ means that all quantum corrections to the classical Hamiltonian $H(p, q)$, which are associated with the non-commutativity of the canonical operators, are included, i.e. all relevant extra terms are taken into account. Moreover, the chosen operator ordering should be specified in (2.243) (by indicating the corresponding base point in the pre-limit expression (2.7)). The Lagrangian formalism is obtained by integrating with respect to p. An explicit form of Q in the case of a gauge symmetry is given in Appendix 8.5 (see also Sections 6.5 and 6.7). A generalization to curvilinear coordinates in a superspace (systems with commuting and anti-commuting variables) is given in Section 6.6.2 (see also [31]).

A non-vanishing potential. Here the analogy with the problem of a particle in a region with boundaries also appears useful in finding a general form of the path integral. Let the potential and old variables be real analytic functions of the new variables q. Then the Schrödinger equation can be analytically extended outside S to the whole R^n apart, maybe, from a zero measure set where the Jacobian of the change of variables vanishes. Clearly, a solution to this equation satisfying specific physical conditions (like, for example, boundary conditions, regularity conditions, periodicity, etc.) would also be the solution of the original problem. The path integral representation of the evolution operator kernel for this new Schrödinger equation is not hard to develop in which integration with respect to all canonical variables is carried out in the infinite limits. The kernel obtained allows us to find the wave function satisfying the "extended" Schrödinger equation at any time t, and thereby to determine the wave function that solves the original problem. The only difference from the original problem is that the initial wave function in the latter ψ_0 is defined in the physical region S, while in the auxiliary (extended) problem it should be defined in the whole space. Hence, the question arises how to extend ψ_0 outside the physical region S so that the wave function resulting from the action of the auxiliary evolution operator on the extended ψ_0 satisfies the boundary conditions and other physical conditions for all t. The problem is reduced to finding the kernel of the operator $\hat{Q}$ which can be defined as the limit (2.242) because the kernel $U_\varepsilon(q, q')$ is defined in the whole R^n (analyticity), while its limits as $\varepsilon \to 0$ coincides with the δ-function in the physical region S. In each particular case one should verify that the wave function obtained in such a way indeed solves the original problem. Technical details of this procedure are identical to those given in Section 2.7.1. An alternative procedure for constructing the path integral is based on extension of the potential $V(q)$ to the non-physical region by means of the group G: $V = \hat{Q}V$ where the kernel of $\hat{Q}$ reads,

$$Q(q, q') = \sum_k \delta(q - g_k q'), \quad g_k \in G \tag{2.245}$$

(compare (2.154)).

Example: parabolic coordinates [120]. The above general procedure is illustrated with an example of parabolic coordinates [141] in R^2,

$$z = \frac{1}{2} w^2, \quad dx_1^2 + dx_2^2 - \left(q_1^2 + q_2^2\right)\left(dq_1^2 + dq_2^2\right),$$

where $z = x_1 + ix_2$, $w = q_1 + iq_2$, $q_1 \geq 0$, $\infty > q_2 > -\infty$, and the Lamé coefficients are equal, $h_1 = h_2 = \left(q_1^2 + q_2^2\right)^{1/2}$. The quadratic form in (2.191) reads,

$$|\mathbf{x} - \mathbf{x}'|^2 = \frac{1}{4}\left[(q_1 + q_1')^2 + (q_2 + q_2')^2\right]\left[(q_1 - q_1')^2 + (q_2 - q_2')^2\right]. \tag{2.246}$$

The covering symmetry group of the form (2.246) is Z_2 whose non-trivial element acts as,

$$q_1 \to -q_1, \qquad q_2 \to -q_2. \tag{2.247}$$

The kernel of $\hat{Q}$ is immediately derived from (2.247),

$$Q(q, q') = \delta\left(q_1 - q_1'\right)\delta\left(q_2 - q_2'\right) + \delta\left(q_1 + q_1'\right)\delta\left(q_2 + q_2'\right).$$

The Hamiltonian operator of a free particle reads,

$$\hat{H} = -\frac{\hbar^2}{2}\frac{1}{h_1 h_2}\left(\partial_1^2 + \partial_2^2\right) = \frac{1}{2\sqrt{g}}\hat{P}_j^2 + \frac{1}{2g}(2i\hbar\hat{q}_j\hat{P}_j - \hbar^2), \tag{2.248}$$

where $\hat{P}_j = -i\hbar g^{-1/4}\partial_j g^{1/4}$, $g = h_1^2 h_2^2$, $j = 1, 2$. The final equations are identical to (2.243) and (2.244); the right-hand side of (2.248) can be taken as $H_{\text{eff}}(p, q)$, in which all the canonical operators are replaced by the corresponding c-numbers. The validity of the procedure can be established directly by making use of the invariance of the action with respect to the inversion (2.247), and a peculiar feature of the model that the asymptotic expression of (2.246) as $\Delta_q \to 0$ coincides with the exact one because the terms Δ_q^3/ε and Δ_q^4/ε are to be retained. Note that the effective action in which extra terms are omitted would no longer be the invariant under the inversion (2.247).

2.8 Quantization by the path integral method

2.8.1 Lagrangian formalism

The canonical quantization method implies, as noted in Section 2.1, replacement of canonical variables by operators obeying the commutation relations (2.1). It is also pointed out there that its applicability is rather limited (Cartesian coordinates, the operator ordering problem) and, because of that, quantum mechanics in curvilinear coordinates usually begins with canonical quantization in Cartesian coordinates, after which the Laplace operator is written in the new variables.

In the framework of the path integral method it is possible not only to obtain the evolution operator kernel, but also to change integration variables, i.e. this method is not "weaker" than the operator one. The path integral method even seems to be preferable because it deals directly with the most important object for physics, the evolution operator kernel $U_t(q, q')$. It is then natural to ask: is it possible to construct the kernel $U_t(q, q')$ using only classical mechanics? Let us elaborate on this issue.

For simple Lagrangians which appear in (2.11), the infinitesimal kernel $U_\varepsilon(q, q')$ can be represented in the form [66, 69, 142],

$$U_\varepsilon(q, q') = (2\pi i\hbar)^{-n/2} D^{1/2} \exp\left[\frac{i}{\hbar} S(q, q')\right], \tag{2.249}$$

where,

$$S(q, q') \approx \varepsilon L = (q - q')^2/2\varepsilon - \varepsilon V(q), \quad \varepsilon \to 0, \tag{2.250}$$

is the action computed for solution of the classical equations of motion, expressed via the initial and final values of the coordinates, q' and q, and

$$D = \det \| - \partial^2 S/\partial q^i \partial q'^j \| = \varepsilon^{-n}. \tag{2.251}$$

In (2.249) the classical action is approximated by (2.250); the classical trajectory in it is approximated by a straight line through the points q and q' [62, 63]. This is well justified for dynamical systems with the Hamiltonian (2.3) because the kernel (2.249) satisfies the proper Schrödinger equation. However, the recipe (2.249)–(2.251) is not applicable to Lagrangians of the form,

$$L = g_{ij}(q)\dot{q}^i\dot{q}^j/2 + A_i(q)\dot{q}^i - V(q) \tag{2.252}$$

(here and in the sequel the functions g_{ij} and A_i are assumed to be independent of time).

What is a criterion for the validity of equations like (2.10) and (2.249)? Function (2.249) is the evolution operator kernel, i.e. it describes the propagation of a particle and must have the following properties:

(1) to satisfy the Schrödinger equation,

$$i\hbar\partial_t\langle q|\hat{U}_t|q'\rangle = \hat{H}\langle q|\hat{U}_t|q'\rangle, \quad t > 0, \tag{2.253}$$

where $\hat{H}$ is the Hamilton operator acting on the first argument;

(2) to tend to the unit operator kernel as $t \to 0$, and

$$\langle q|\hat{U}_t|q'\rangle \to \langle q|q'\rangle, \quad t \to 0\ ; \tag{2.254}$$

(3) to provide for the fulfilling of boundary (or equivalent to them) conditions.

As becomes clear from what follows, the function (2.249) for the Lagrangian (2.252) does not satisfy the proper Schrödinger equation. In order to find a correct expression for $U_\varepsilon(q, q')$, the following should be noted. It is not yet established with what accuracy the action $S(q, q')$ in (2.249) should be calculated for a classical trajectory. According to Dirac (see [64, 65]), $S(q, q')$ is the classical action that is calculated for the true classical trajectory connecting points q' and q. As already noted, the Feynman recipe [62, 63] (an approximation of the classical trajectory by a straight line segment) is not appropriate for more complicated systems with the Lagrangian (2.252). That expression (2.249) for the kernel $U_\varepsilon(q, q')$ formally coincides with the quasi-classical representation of the wave function [75, 115] motivates conjecture that the corresponding function should look similar, even in more general cases. In Appendix 8.3, the suggested form of the kernel is found by different means. Thus, it is *postulated* that the kernel $U_\varepsilon(q, q')$ has the form,

$$U_\varepsilon(q, q') \approx (2\pi i\hbar)^{-n/2}(gg')^{-1/4}D^{1/2}\exp\left[\frac{i}{\hbar}S(q, q')\right], \tag{2.255}$$

$$D = \det\| - \partial^2 S/\partial q^i \partial q'^j\|\ ; \tag{2.256}$$

the factor $(gg')^{-1/4}$ is included in accordance with Eqs. (2.31) and (2.37).

The next aspect to be studied is the accuracy with which the action $S(q, q')$ should be calculated in (2.255). According to [64, 65], the action,

$$S(q, q') = \int_0^\varepsilon L(q, \dot{q})dt, \quad q = q(\varepsilon), \quad q' = q(0), \tag{2.257}$$

should be computed for the true classical trajectory connecting points q' and q. It is known, however, that it is sufficient to retain only relevant extra terms, given in (2.61), in order to develop the path intgeral formalism. Let us find out the consequences when the Dirac proposition is literally followed for the action $S(q, q')$ in (2.255).

Let us calculate the contribution of all relevant extra terms to $S(q,q')$. The Euler–Lagrange equations of motion for the Lagrangian (2.252) are,

$$g_{ij}\ddot{q}^j + [jk,i]\dot{q}^j\dot{q}^k - F_{ij}\dot{q}^j + V_{,j} = 0, \quad F_{ij} = A_{j,i} - A_{i,j}, \tag{2.258}$$

where the Christoffel symbols $[jk,i]$ are defined in (2.34). Their solution, expressed as a function of q and q', is to be substituted into integral (2.257). With this purpose, the initial value problem solution,

$$q(t) = q' + \dot{q}'t + \ddot{q}'t^2/2 + \dddot{q}'\, t^3/6 + \cdots \tag{2.259}$$

is expanded in a power series over $\Delta = q - q'$; here $q' = q(0)$, $\dot{q}' = \dot{q}(0)$, etc. To achieve this goal, it is necessary to determine the initial values of the derivatives of $q(t)$ (up to the third order) as power series of Δ. Setting $t = \varepsilon$ in (2.259), one obtains $\dot{q}'_{(0)} = \Delta/\varepsilon$ in the leading order (the index (0) indicates the leading order approximation). To calculate $\dot{q}'$ in higher-orders, the higher order derivatives $\ddot{q}'$, $\dddot{q}'$, ..., should be determined. The second derivative $\ddot{q}$ can be expressed via $\dot{q}$ and q using the equations of motion (2.258). The derived expression is then differentiated with respect to time to obtain $\dddot{q}$ as a function of $\ddot{q}$, $\dot{q}$, and q into which the expression for $\ddot{q}$ is substituted. As a result, $\dddot{q}$ becomes a function of $\dot{q}$ and q. If the found values $\ddot{q}(0) = \ddot{q}' \equiv \ddot{q}'(\dot{q}',q')$ and $\dddot{q}(0) = \dddot{q}' \equiv \dddot{q}'(\dot{q}',q')$ are substituted into (2.259), setting also $t = \varepsilon$, then the following non-linear relation is obtained:

$$\dot{q}' = \frac{\Delta}{\varepsilon} - \ddot{q}'(\dot{q}',q')\varepsilon/2 - \dddot{q}'(\dot{q}',q')\varepsilon^2/6 - \cdots; \tag{2.260}$$

recall $\Delta = q - q'$. Equation (2.260) is solved iteratively for $\dot{q}'$ as a power series in Δ. As a result, the following expressions for $\dot{q}'$, $\ddot{q}'$, and $\dddot{q}'$ in (2.259) are derived (with the required accuracy):

$$\begin{aligned}\dot{q}'^j \approx{}& \frac{1}{\varepsilon}\Big\{\Delta^j + \frac{1}{2}g^{jn}\Big[[ik,n]\Delta^i\Delta^k + \frac{1}{3}\Big([ik,n]_{,l} - g^{mm'}[lk,m][in,m']\Big)\\ &\times\Delta^i\Delta^k\Delta^l\Big]\Big\} + \frac{1}{2}g^{ij}F_{ki}\Delta^k;\end{aligned} \tag{2.261}$$

$$\begin{aligned}\ddot{q}'^j \approx{}& -\frac{1}{\varepsilon^2}g^{jk}[mn,k]\Big\{\Delta^m\Delta^n + g^{mm'}[il,m']\Delta^i\Delta^l\Delta^n\Big\}\\ &+ g^{ij}F_{ik}\frac{\Delta^k}{\varepsilon};\end{aligned} \tag{2.262}$$

$$\begin{aligned}\dddot{q}'^j \approx{}& -\frac{1}{\varepsilon^3}\Big\{g^{jk}_{,l}[mn,k] + g^{jk}[mn,k]_{,l}\\ &- 2g^{jk}[in,k]g^{im'}[ml,m']\Big\}\Delta^m\Delta^n\Delta^l.\end{aligned} \tag{2.263}$$

All the functions in the right-hand sides of these equations are taken at the point q'. Note that the approximation of the trajectory by a straight line segment would correspond to retaining just the first term Δ/ε in the expansion (2.261). Equations (2.260)–(2.263) take into account that the trajectory is curved. Next one has,

$$S(q,q') = \int_0^\varepsilon L dt \approx L_0\varepsilon + \frac{\varepsilon^2}{2}\frac{dL_0}{dt} + \frac{\varepsilon^3}{6}\frac{d^2L_0}{dt^2}. \tag{2.264}$$

Here L_0, dL_0/dt, and so on are the Lagrangian and its total time derivatives at $t = 0$, calculated after substituting the solution (2.259) into L. It turns out that the contribution of the last term in (2.264), which is of order Δ^4/ε, vanishes, while the second term contributes $(A_{m,n} - A^k[mn,k])\Delta^m\Delta^n/2$. The first one is easy to calculate by substituting (2.261) into L_0. Gathering all the terms together, the following expression for the action is inferred:

$$\begin{aligned} S(q,q') \approx \frac{1}{2\varepsilon}\Big[& g_{ij}\Delta^i\Delta^j + \frac{1}{2}g_{ij,k}\Delta^i\Delta^j\Delta^k \\ & + \frac{1}{6}\left(g_{ij,kl} - \frac{1}{2}[ij,m]g^{mn}[kl,n]\right)\Delta^i\Delta^j\Delta^k\Delta^l\Big]_{q'} \\ & + \left[A_i\Delta^i + \frac{1}{2}A_{j,i}\Delta^i\Delta^j - \varepsilon V\right]_{q'} \equiv \tilde{S}(q',\Delta); \end{aligned} \tag{2.265}$$

the index q' at the square brackets means that all the functions are taken at the point q'.

Thus, the action has been found. It remains to calculate the determinant (2.256) of the matrix,

$$-\frac{\partial^2 S(q,q')}{\partial q^i \partial q'^j} \equiv -\phi_{ij} \qquad \left(\phi^{jk}\phi_{ki} = \phi^{kj}\phi_{ik} = \delta^j_i;\ \phi_{ij} \neq \phi_{ji}\right). \tag{2.266}$$

Making use of (2.265) one deduces,

$$\begin{aligned} \phi_{ij} \approx -\frac{1}{\varepsilon}\Big\{ & g_{ij} + [jk,i]\Delta^k + \frac{1}{6}\Big[[kl,i]_{,j} + [jl,i]_{,k} + [jk,i]_{,l} \\ & - g^{mn}\Big([ij,m][kl,n] + [jk,m][il,n] + [jl,m][ik,n]\Big)\Big]\Delta^k\Delta^l - \frac{\varepsilon}{2}F_{ij}\Big\}_{q'}. \end{aligned} \tag{2.267}$$

As before, only the relevant terms are retained. The determinant $D = \det(-\phi_{ij})$ of the matrix (2.267) is calculated by means of the decomposition,

$$\begin{aligned} \det(A+B) = \det A\Big\{ & 1 + \mathrm{Tr}(A^{-1}B) + \frac{1}{2}[\mathrm{Tr}(A^{-1}B)]^2 \\ & - \frac{1}{2}\mathrm{Tr}(A^{-1}BA^{-1}B) + \cdots\Big\}, \end{aligned} \tag{2.268}$$

for operators A and B [69]. The result (with $A_{ij} = g_{ij}/\varepsilon$) reads,

$$D = \frac{1}{\varepsilon^n} g' \left\{ 1 + \frac{1}{2g'} g_{,k}\Delta^k + \frac{1}{2} g^{ij} g^{mn} \Big([jk,i][ml,n] - [jk,m][nl,i] \Big) \Delta^k \Delta^l \right.$$
$$+ \frac{1}{6} \left[g^{ij} \left(g_{ij,kl} + g_{il,kj} - \frac{1}{2} g_{kl,ij} \right) - g^{ij} g^{mn} \Big([ij,m][kl,n] \right.$$
$$\left. \left. + 2[ik,m][jl,n] \Big) \right] \Delta^k \Delta^l \right\}. \tag{2.269}$$

Next the expansion,

$$g^{1/2} \approx g'^{1/2} \left\{ 1 + \frac{1}{2g'} g_{,k}\Delta^k + \frac{1}{4} \Big[g^{ij} g^{mn} \left(2[jk,i][ml,n] \right. \right.$$
$$\left. \left. - g_{jn,l} g_{im,k} \right) + g^{ij} g_{ij,kl} \Big] \Delta^k \Delta^l \right\}_{q'},$$

and the definition of the Ricci tensor R_{kl} (2.36) are employed to transform (2.269) to the following compact form:

$$D = \frac{(gg')^{1/2}}{\varepsilon^n} \left[1 + \frac{1}{6} R_{kl} \Delta^k \Delta^l \right] ; \tag{2.270}$$

the choice of the argument of R_{kl} (or the base point) is not relevant in (2.270).

The infinitesimal evolution operator kernel is determined by Eqs. (2.255), (2.265), and (2.270). In order to assess the obtained result, let us derive the equation which the kernel (2.255) satisfies. The fastest way to achieve the goal was proposed by Pauli (see [66]) for a much simpler case when $g_{ij,k} = 0$. The idea is to make use of both the Hamilton–Jacobi equation (see Section 1.10.1),

$$\frac{\partial S}{\partial t} + H \left(\frac{\partial S}{\partial q}, q \right) = 0, \tag{2.271}$$

where,

$$H(p,q) = g^{ij}(p_i - A_i)(p_j - A_j)/2 + V, \tag{2.272}$$

and the conservation law,

$$\frac{\partial D}{\partial t} + \frac{\partial}{\partial q^i} \left[g^{ij} \left(\frac{\partial S}{\partial q^j} - A_j \right) D \right] = 0. \tag{2.273}$$

Equation (2.273) is easy to derive. It follows from the identity $\ln\det\phi = \mathrm{Tr}\ln\phi$ and the definitions (2.256) and (2.266) that,

$$\frac{\partial D}{\partial t} = D\phi^{ji} \frac{\partial \phi_{ij}}{\partial t} ; \quad \frac{\partial D}{\partial q^k} = D\phi^{ji} \frac{\partial \phi_{ij}}{\partial q^k} \tag{2.274}$$

(compare (2.45)). Acting with the operator $\partial^2/\partial q^i \partial q'^j$ on equality (2.271), multiplying the result by ϕ^{ji}, contracting the indices, and taking into account (2.266),

(2.272), and (2.274), equation (2.273) is inferred. For the derivation of the Schrödinger equation, the following relations are invoked for the kernel (2.255):

$$\frac{\hbar}{i}\frac{\partial U}{\partial t} = \left[\frac{\partial S}{\partial t} + \frac{1}{2}\frac{\hbar}{i}\frac{1}{D}\frac{\partial D}{\partial t}\right] U, \tag{2.275}$$

$$\left(\frac{\hbar}{i}\frac{\partial}{\partial q^j} - A_j\right) U = \left[\frac{\partial S}{\partial q^j} - A_j + \frac{\hbar}{2i}\frac{1}{Dg^{-1/2}}\frac{\partial}{\partial q^j}\left(Dg^{-1/2}\right)\right] U, \tag{2.276}$$

$$\frac{1}{2\sqrt{g}}\left(\frac{\hbar}{i}\frac{\partial}{\partial q^i} - A_i\right)\sqrt{g}g^{ij}\left(\frac{\hbar}{i}\frac{\partial}{\partial q^j} - A_j\right) U$$

$$= \frac{1}{2}g^{ij}[\;]_i[\;]_j U + \frac{\hbar}{2i\sqrt{g}}\frac{\partial}{\partial q^i}\left\{\sqrt{g}g^{ij}[\;]_j\right\} U, \tag{2.277}$$

where the symbol $[\;]_j$ denotes the expression in square brackets in (2.276). Summing the left- and right-hand sides of Eqs. (2.275) and (2.277) and adding to both sides of the resulting equality the term VU, one infers,

$$\left(\frac{\hbar}{i}\frac{\partial}{\partial t} + \hat{H}\right) U = -\frac{\hbar^2}{2}\frac{1}{\sqrt{Dg^{-1/2}}}\Delta_{LB}\sqrt{Dg^{-1/2}}\,U, \tag{2.278}$$

where,

$$\hat{H} = \frac{1}{2\sqrt{g}}\left(\frac{\hbar}{i}\frac{\partial}{\partial q^i} - A_i\right)\sqrt{g}g^{ij}\left(\frac{\hbar}{i}\frac{\partial}{\partial q^j} - A_j\right) + V, \tag{2.279}$$

and Δ_{LB} is the Laplace–Beltrami operator (2.33); the terms, which do not contain $\hbar$ and those proportional to $\hbar$, are canceled out because of relations (2.271) and (2.273), respectively. The terms proportional to $\hbar^2$ are expressed via the scalar curvature R. Making use of (2.33) and (2.270) and neglecting contributions $O(\Delta)$, Eq. (2.278) is written in the form [69],

$$i\hbar\frac{\partial U}{\partial t} = \left(\hat{H} + \frac{\hbar^2 R}{12}\right) U. \tag{2.280}$$

An alternative derivation of this equation is based on the Lagrangian equivalence rules (2.64) and (2.65) for S and D, given in (2.265) and (2.270) with subsequent differentiation of U (i.e. calculating the action of the operator $-i\hbar\partial_t + \hat{H}$ defined in (2.279) on U). It is noteworthy that, if the factor $D^{1/2}$ is neglected in (2.255), the coefficient at R is doubled in (2.280).

The result (2.280) is remarkable in several aspects. First, Eq. (2.280) gives the correct Schrödinger equation in curvilinear coordinates (a Euclidean space, $R_{ijkl} = 0$) – the Hamiltonian has exactly the form (2.279) where $A_j = 0$; moreover, the wave equation coincides with (2.280) even if $A_i \neq 0$. Therefore the method is appropriate for quantization in curvilinear coordinates, which is a great advantage in comparing with canonical quantization (2.1). Second, when deriving (2.280), no assumption has been made that the space is flat. More to the point, Eq. (2.280) implies that the Schrödinger equation requires corrections when the scalar curvature does not vanish (there is a pure quantum correction $\hbar^2 R/12$ to the potential – it is beautiful!). However, in this latter part, the postulate (2.255) requires practical (experimental) verification, although there is little doubt about its validity. In this regard, it should be noted that the canonical quantization in a curved space may be carried out by embedding the latter into a Euclidean space of a higher dimension (see, e.g., [143] and references therein). The system necessarily has second-class constraints (see Chapters 3 and 4). An explicit form of the quantum potential (like the curvature term) depends on the way the second class constraints are taken into account when quantizing the theory (see, e.g., [59], [143], [144]). However the very existence of the embedding space, as a physical reality of some sort, implies an addition hypothesis, which is not required at all in the path integral quantization method. Finally, from Eq. (2.280) the question arises: how does it turn out to be possible to solve a purely quantum problem of operator ordering in the Hamiltonian by only making use of the classical equations of motion?

The answer requires a careful analysis. Here the discussion is limited to brief remarks. The argument of the exponential in (2.255) is nothing but the extremal action, i.e. the action computed for a solution of the classical equations of motion. The peculiarity of the quantum problem as compared to the classical one is that in the former the main contribution is given by continuous *nowhere differentiable* trajectories that are characterized by the property $\Delta^2 \sim \varepsilon$, while for classical trajectories $\Delta \sim \varepsilon$. This is exactly what determines the accuracy with which the extremal value of S in (2.255) is to be computed as $\varepsilon \to 0$,

$$S = S_0 + O(\Delta^2), \tag{2.281}$$

where $S_0 = \varepsilon L(q, \Delta/\varepsilon)$; the terms $O(\Delta^2)$ are extra terms giving contributions of order $\sim \Delta^2$ and $\sim \Delta^4/\varepsilon$ (i.e. $\sim i\hbar\varepsilon$ and $\sim \hbar^2\varepsilon$) that must be taken into account. These very terms determine the operator ordering in the Hamiltonian (they affect the form of the Schrödinger equation). From this point of view, the statement that the classical equations of motion contain information about operator ordering in the Hamiltonian seems natural. The path integral quantization shows that there exist more intimate relations between the classical and quantum description than the commonly accepted ones. The necessity of taking into account the "curvature" of the trajectory (more accurately, the insufficiency of its approximation by a straight line segment, as in (2.10) and (2.12)) is related to the fact that non-differentiable trajectories ($\Delta/\varepsilon \sim \varepsilon^{-1/2} \to \infty$) give the main contribution to the path integral (2.11). The relation $\varepsilon \sim \Delta^2$, characteristic for Brownian motion [61], means that the *fractal dimension* of the trajectory equals two [145, 146]. It looks as though the trajectories would broaden, turning into two-dimensional objects. The question deserves deeper study (see also Section 1.15).

2.8.2 Hamiltonian formalism

The above consideration cannot be directly generalized to the Hamiltonian formalism [99, 100]. Quantization in the framework of the Hamiltonian path integral method is a separate important problem. Its significance is understood already from the practical problems which occur when working in the Hamiltonian formalism. For example, when a canonical transformation is defined by a generator $G(p,q)$ (see Section 2.6), its extension to quantum mechanics requires development of a procedure to construct the corresponding operator $\hat{G}$, in which the operator ordering problem may occur. Of course, given G, an analogous Lagrangian formalism can be developed (see (2.105)), but this looks rather artificial. A similar problem may arise when quantizing dynamical systems with constraints (the operator ordering problem in constraints, see Chapter 4). However, regardless of practical reasonings, the question is of interest on its own.

For the simplest Hamiltonians such as (2.3), the infinitesimal evolution operator kernel U_ε has the form (2.6),

$$U_\varepsilon(q,q') \approx \int \frac{dp}{(2\pi\hbar)^n} \exp\left\{\frac{i}{\hbar}\left[p\Delta - \varepsilon H(q,p)\right]\right\}, \quad \Delta = q - q'. \tag{2.282}$$

In order to generalize (2.282) to the case of a curved space when H is given by (2.25),

$$H = \frac{1}{2}g^{ij}(q)p_ip_j + f^i(q)p_i + V_H \quad \left(V_H = V_L + \frac{1}{2}A_iA^i\right), \tag{2.283}$$

the problem is to determine S_H in the argument of the exponential (the expression in the square brackets in (2.282)) and the pre-exponential factor. The function S_H is found first. It turns into the action for a classical trajectory connecting the points q and q' after replacing p by the solution of the equation $\partial S_H/\partial p = 0$. With this observation and by analogy with the Lagrangian formalism (see (2.257)), the Hamiltonian action is replaced by the expression,

$$S_H = \int_0^\varepsilon dt\left[p\dot{q} - H(q,p)\right], \tag{2.284}$$

which is calculated taking into account all relevant extra terms. Substitution of solutions of the Hamiltonian equations of motion,

$$\dot{q}^i = p^i + f^i, \quad \dot{p}_i = [ij,k]p^jp^k - f^j_{,i}p_j - V_{,i}, \tag{2.285}$$

into (2.284) with subsequent use of the expansion, analogous to (2.264), yields the action,

$$S_H(q,p,q') = \varepsilon[p\dot{q} - H]_0 + \frac{\varepsilon^2}{2}\frac{d}{dt}[p\dot{q}]_0 + \frac{\varepsilon^3}{6}\frac{d^2}{dt^2}[p\dot{q}]_0 + \cdots, \tag{2.286}$$

where the curvature of the trajectory has been taken into account. The derivatives are evaluated at $t = 0$ in (2.286); that the Hamiltonian H is an integral of motion is accounted for in (2.286), also.

The peculiarity of the problem as compared with the Lagrangian formalism is in the following. On the one hand, the action is to be calculated for the classical trajectory, i.e. to substitute the solutions of (2.285), $q(t) = q(t, q_0, p_0)$ and $p(t) = p(t, q_0, p_0)$, depending on the initial values, $q_0 = q(0) = q'$ and $p_0 = p(0) = p$, into (2.284), while the first equation in (2.285) allows us to get rid of p altogether. On the other hand, p is an independent variable in respect of which the integration is carried out (see (2.282)), i.e. its presence is mandatory in (2.284). In view of this, the following approach is chosen. In the first term, p is retained as it is, while the solutions (2.259)–(2.263), expressed via q and q', are substituted for $q(t)$ and $\dot{q}(t)$. In doing so, the stationary point of the argument of the exponential in (2.282) is still determined by the Hamiltonian equations of motion $\dot{q}^i = \partial H/\partial p_i$. The remaining terms in (2.286) do not affect the classical equations of motion and, therefore, one may go over from p to $\dot{q}$ in them. When calculating their explicit form, only the relevant extra terms should be kept; it is sufficient to treat only the terms explicitly written in (2.286) (for example, the first term among those omitted in (2.286) is of order $\varepsilon^4 p d^4 q/dt^4 \sim \varepsilon^4 p \Delta^4/\varepsilon^4 \sim \Delta^3 = O(\varepsilon^{3/2})$). The contribution of the last term in (2.286), whose order is $\Delta^4/\varepsilon \sim \varepsilon$, vanishes, while the second one gives,

$$\frac{\varepsilon^2}{2}\frac{d}{dt}[p_i \dot{q}^i] = \left(A_{i,j} - [ij,k]A^k\right)\frac{\Delta^i \Delta^j}{2}, \qquad A_i = -g_{ij} f^j. \tag{2.287}$$

Gathering all the contributions, the action can be written as,

$$S_H = \varepsilon p_i \dot{q}'^i(\Delta) - \varepsilon \left[\frac{1}{2} g^{ij} p_i p_j + f^i p_i + V_H\right] + A_{k;l}\Delta^k \Delta^l, \tag{2.288}$$

where $A_{k;l} = A_{k,l} - [kl,j]A^j$ is the covariant derivative, and $\dot{q}'$ is given in (2.261). All the functions are taken at the end point $q' = q(0)$. Of course, the integration with respect to the momentum p in (2.282) after substitution of the expression (2.288) into the exponential gives the action (2.265). It is then left to find the pre-exponential factor. This should be expressed via the second derivatives of S_H like D in (2.255). The difficulty is that, in contrast to S, the action S_H depends on the three variables: q, q', and p. So, this factor should be guessed. Therefore the answer is given right away, and then proved. For the Hamiltonian of the form (2.283), the kernel $U_\varepsilon(q, q')$ is given by the expression,

$$U_\varepsilon(q, q') = \int \frac{dp}{(2\pi\hbar)^n} \bar{D}_H^{1/2} (gg')^{-1/4} \exp\left[\frac{i}{\hbar} S_H(q, p, q')\right], \tag{2.289}$$

in which S_H is defined by the equalities (2.288) and (2.261). The function $\bar{D}_H$ is defined as follows. Put

$$D_H(q, p, q') = \det \begin{Vmatrix} \partial^2 S_H/\partial q^i \partial q'^j & \partial^2 S_H/\partial q^i \partial p_j \\ \partial^2 S_H/\partial p_i \partial q'^j & \partial^2 S_H/\partial p_i \partial p_j \end{Vmatrix}. \tag{2.290}$$

Substituting the solution, $p = \bar{p}(q, q')$ of the classical equations of motion into D_H, the function $\bar{D}_H$ is obtained

$$\bar{D}_H(q, q') \equiv D_H(q, \bar{p}(q, q'), q') = D(q, q'). \tag{2.291}$$

To establish the validity of the recipe (2.289)–(2.291), it is sufficient to prove that the kernel (2.289) turns into the kernel (2.255) after integration with respect to p. As already noted, this is indeed so for the exponential – the verification is not difficult. If the determinant D_H were independent of the momenta, the transformation $D_H \to \bar{D}_H$ would not have been necessary. However D_H does depend on p and, when calculating $D_H^{1/2}$ as a power series over Δ, one should use $\bar{D}_H$. That the choice of $\bar{D}_H$ in (2.289) is correct is proved as follows. The action S_H in (2.288) is written in the form,

$$S_H(q, p, q') = \frac{1}{2} \frac{\partial^2 S_H}{\partial p_i \partial p_j} (p_i - \bar{p}_i)(p_j - \bar{p}_j) + \bar{S}_H(q, q'), \tag{2.292}$$

where $\bar{p}(q, q') = \dot{q}' - f(q')$ is the solution of the classical equations of motion $p(q, q', t)$ at $t = 0$ and expressed via the initial and final points of the trajectory, q and q': $\bar{p}(q, q') = p(q, q', 0)$ (see (2.285) and (2.261)). Clearly, in Eq. (2.292), $\bar{S}_H(q, q') = S_H(q, \bar{p}(q, q'), q') = S(q, q')$, where S is given in (2.265). Put

$$A_{ij} = \frac{\partial^2 S_H}{\partial q^i \partial q'^j}, \quad B_i^l = \frac{\partial^2 S_H}{\partial q^i \partial p_l}, \quad C_j^k = \frac{\partial^2 S_H}{\partial q'^j \partial p_k}, \quad D_p^{kl} - \frac{\partial^2 S_H}{\partial p_k \partial p_l}. \tag{2.293}$$

The determinant D_H in (2.290) can be represented in block form via the matrices (2.293),

$$D_H = \det \begin{pmatrix} \hat{A} & \hat{B} \\ \hat{C} & \hat{D}_p \end{pmatrix} = \det \hat{D}_p \ \det \left(\hat{A} - \hat{B} \hat{D}_p^{-1} \hat{C} \right). \tag{2.294}$$

Here $(\hat{A})_{ij} \equiv A_{ij}$ and similarly for the matrices $\hat{B}, \hat{C}$, and $\hat{D}_p$. The determinant of the matrix $\hat{D}_p$ may be ignored in (2.294) because it is canceled out by the corresponding determinant resulting from the integration with respect to p in (2.289) (save for some factors, irrelevant for now, which get combined with $(2\pi\hbar)^{-n}$)). The matrices $\hat{A}$, $\hat{B}$, and $\hat{C}$ computed at the solutions $p = \bar{p}$ have the following form:

$$A_{ij}\Big|_{\bar{p}} = \frac{\partial^2 \bar{S}_H}{\partial q^i \partial q'^j} + D_p^{kl} \frac{\partial \bar{p}_k}{\partial q^i} \frac{\partial \bar{p}_l}{\partial q'^j}, \quad B_i^l\Big|_{\bar{p}} = -D_p^{kl} \frac{\partial \bar{p}_k}{\partial q^i}, \quad C_j^k\Big|_{\bar{p}} = -D_p^{lk} \frac{\partial \bar{p}_l}{\partial q'^j}.$$

Substituting these expressions into the second determinant in (2.294), one infers,

$$\det \left(\hat{A} - \hat{B} \hat{D}_p^{-1} \hat{C} \right) \Big|_{\bar{p}} = \det \left\| \frac{\partial^2 \bar{S}_H}{\partial q^i \partial q'^j} \right\| = \det \left\| \frac{\partial^2 S}{\partial q^i \partial q'^j} \right\| = (-1)^n D,$$

where D is the determinant (2.256); the factor $(-1)^n$ is included into an overall constant factor. The proof of the equivalence of the Hamiltonian quantization recipe, ((2.289) and (2.291)) and the Lagrangian one ((2.255) and (2.256)) is complete.

Remark. The symmetric operator ordering (more accurately, the Weyl ordering) is sometimes offered as a recipe to order operators when quantizing in curvilinear coordinates. The procedure does not sustain even a simple test. The Hamilton function in the spherical coordinates contains products of canonical variables whose Poisson brackets vanish, i.e. formal replacement of the canonical variables by the corresponding operators has no operator ordering ambiguity whatsoever. But, as a point of fact, such a formal procedure yields an incorrect Hamiltonian (compare (2.33)). Thus, its acceptance as a quantization recipe would lead, for instance, to an incorrect spectrum of the hydrogen atom (see also Section 4.2).

3
Dynamical systems with constraints

3.1 Introduction

In practice one often deals with problems whose dynamical variables, i.e. coordinates and momenta q^i, p_i, are subject to additional conditions, called constraints, that are specific for the dynamical system in question. The constraints can be explicitly specified at the very beginning as a consequence of essential dynamical features of the problem. For example, if a point particle moves in a plane along a circle of radius R, then its coordinates must satisfy the condition $\mathbf{x}^2 \equiv x_1^2 + x_2^2 = R^2$. In general, the constraints may also include velocities. In the framework of the Lagrangian formalism, the constraints are incorporated into the dynamical description by means of the *Lagrange multiplier method* [9, 15, 147]. The Lagrangian of a dynamical system is modified by adding a linear combination of the constraints. The coefficients are then regarded as additional dynamical variables. In particular for the motion constrained to a circle, this implies that

$$L(\mathbf{x}, \dot{\mathbf{x}}) \to \tilde{L}(\mathbf{x}, \dot{\mathbf{x}}, \lambda) = L(\mathbf{x}, \dot{\mathbf{x}}) + \frac{1}{2}\lambda(t)(R^2 - \mathbf{x}^2), \tag{3.1}$$

where the Lagrangian L describes the motion in the absence of constraints, and λ is the Lagrange multiplier. By demanding that the first variations of the new action with respect to $\mathbf{x}$ and λ vanish, the new equations of motion and the constraint are obtained.

A different approach to constrained dynamics was proposed by d'Alambert [6, 9, 15]. In his method, the equations of motion for constrained dynamics are obtained from the stationary action principle for the original action $S = \int L dt$, in which variations of the dynamical variables are no longer independent but rather required to be compatible with the constraints. For instance, in the above example of a particle on a circle, the variations $\delta\mathbf{x}$ must satisfy the condition $(\mathbf{x}, \delta\mathbf{x}) = 0$, i.e. allowed virtual displacements are those that occur along the constraint surface (the circle in this case). The d'Alambert equations are also supplemented by the constraints, which define the complete system of equations of motion. The two approaches are not generally equivalent, although both of them are used to describe actual physical systems. A comparison of the Lagrange and d'Alambert methods is given in the next section.

The Lagrangian (3.1) admits an alternative interpretation. The parameter λ can be viewed as a new dynamical variable in addition to x_1, x_2. From the latter it becomes clear that constraints can be introduced implicitly, meaning that their very existence can be established by analyzing the Lagrangian. Denoting λ by x_3 in (3.1), one has

$$\tilde{L}(\mathbf{x}, \dot{\mathbf{x}}, x_3, \dot{x_3}) = L(\mathbf{x}, \dot{\mathbf{x}}) - \frac{1}{2} x_3(\mathbf{x}^2 - R^2), \qquad \mathbf{x} = (x_1, x_2). \tag{3.2}$$

Formally, the Lagrangian (3.2) seems to have no Lagrange multipliers as all the variables x_1, x_2, x_3 enter into the formalism on an equal footing. But in fact x_3 plays the role of λ. Thus, in classical Lagrangian theory constraints do not lead to any difficulties and can be naturally included into the formalism by means of the Lagrange multiplier method.

The situation is a little different in the Hamiltonian formalism used to introduce a quantum description. According to the Dirac recipe, the theory is formulated in terms of the canonical variables q, p, which are then promoted to operators.[1] So, before going over to a quantum description, the Hamiltonian formalism for systems with constraints should be developed and studied. The constraints can now be imposed on both the canonical coordinates and momenta. For example, by going over to the Hamiltonian formalism one infers that $p_3 = \partial\tilde{L}/\partial\dot{x}_3 \equiv 0$ for the Lagrangian (3.2). Hence $p_3 = 0$ is a constraint imposed on the canonical momentum p_3.

A general reason for constraints occuring is a degeneracy of the Hessian matrix (the matrix of second derivatives of the Lagrangian with respect to generalized velocities). If, for example, in (3.2) the term L has the form $L(\mathbf{x}, \dot{\mathbf{x}}) = \frac{1}{2}\sum_{i,k} T_{ik}\dot{x}_i\dot{x}_k - V(\mathbf{x})$, $i, k = 1, 2$, so that $\det(T_{ik}) \neq 0$, then the kinetic energy in $\tilde{L}$ can be written as a quadratic form $\frac{1}{2}\sum_{i,k=1}^{3} \tilde{T}_{ik}\dot{x}_i\dot{x}_k$. Obviously, here $\tilde{T}_{3k} = \tilde{T}_{k3} = 0$, and, hence, $\det(\tilde{T}_{ik}) = 0$, or, equivalently, the Hessian matrix is singular. In accordance with the implicit function theorem, the equations $p_i = \partial L/\partial\dot{x}_i$, which are used to eliminate velocities when going over from a Lagrangian formalism to a Hamiltonian one, cannot be solved to find $\dot{x}_i$ when the Hessian matrix is singular, for some of the equations turn out to be conditions on coordinates and momenta (constraints). The Lagrangians for which

$$\det(T_{ik}) = 0, \quad T_{ik} = \frac{\partial^2 L}{\partial\dot{q}^i\partial\dot{q}^k}, \tag{3.3}$$

are called *degenerate*. Sometimes they are also called, not quite adequately, *singular*. In the case of multiple constraints it is necessary to verify whether they are not contradictory to one another as well as to the equations of motion.

Another characteristic example is provided by the Lagrangian of a relativistic particle,

$$L = -m\sqrt{\dot{x}_\mu^2}, \tag{3.4}$$

[1] This recipe is correct only in Cartesian coordinates, see Chapter 2

where $\dot{x}_\mu = dx_\mu/d\tau$, x_μ is the generalized coordinate (a four-vector), and τ is an invariant parameter viewed as time. In this case the canonical momenta $p_\mu = \partial L/\partial \dot{x}_\mu = m^2\dot{x}_\mu/L$ are not independent variables. As easily seen, they satisfy the equation $p^2 = m^2$ (a more comprehensive analysis of the system is given in Chapter 4). The constraint appears here as a result of the desire to retain a Lorentz-invariant description of the system. In the physical framework, the coordinates $\mathbf{x}$ are functions of time $t = x_0$, i.e. the physical nature of the variables x_0 and $\mathbf{x}$ is quite different. But for the sake of obtaining a covariant formalism they are treated on the same footing as independent dynamical variables in the Lagrangian formalism. This can only be done at the price of having an auxiliary parameter τ viewed as time in the Lagrangian formalism so that $x_\mu = x_\mu(\tau)$ describes the particle trajectory. Thus, in order to have covariant equations of motion, an additional degree of freedom $x_0(\tau)$ is introduced. The occurrence of constraints in this and similar cases is evident – thanks to the constraints, the new dynamical system remains equivalent to the old one.

The addition of new (non-physical) degrees of freedom in order to obtain an explicitly Lorentz-invariant description and, as a consequence, the occurrence of constraints, are typical for massless fields with higher spins ($J > 1/2$). Note, however, that theories with Fermi fields ($J = 1/2$) also have constraints (see Section 3.3.3). The well-known examples are provided by the electromagnetic and Yang–Mills fields ($J = 1$) and the gravitational field ($J = 2$). It is these theories that stimulated the study of Hamiltonian dynamics of constrained systems and their quantization. The characteristic property of the aforementioned (gauge) theories is the degeneracy of their Lagrangians. For instance, the Lagrangian of a free electromagnetic field is given by $L = -(1/4)F_{\mu\nu}^2$, where $F_{\mu\nu} = \partial_\mu A_\nu - \partial_\nu A_\mu$. The Hessian matrix

$$T^{\mu\nu} = \partial^2 L/\partial \dot{A}_\mu \partial \dot{A}_\nu = g^{0\mu}g^{0\nu} - g^{00}g^{\mu\nu}$$

is diagonal with diagonal elements being (0,1,1,1), and, hence, $\det(T^{\mu\nu}) = 0$, that is, the electromagnetic field Lagrangian is degenerate. In the framework of quantum field theory, this leads to an additional difficulty in perturbation theory: besides conventional divergences inherent to local field theories, pure kinematic ones appear, associated with the non-existence of the inverse of a singular matrix.

Finally, a general remark concerning the Lagrangian and Hamiltonian formalisms in relativistic quantum theory. A quantization is usually carried out in the Hamiltonian formalism. In doing so it is difficult to maintain explicit Lorentz-invariance of the resulting quantum theory. The Hamiltonian formalism not being explicitly covariant can lead to the breaking of the Lorentz-invariance of the theory as a consequence of an inappropriate choice of Hamiltonian. On the contrary, if a quantum system is described in the framework of the Lagrangian formalism, e.g. by the Feynman integral, the relativistic invariance is easily achieved by choosing a Lorentz-invariant Lagrangian. However, in such an approach, it is not straightforward to establish the probability conservation, i.e. unitarity of the S-matrix. This is why one first chooses a Lorentz-invariant Lagrangian from which a quantum Hamiltonian is obtained. If the Hamiltonian happens to be self-adjoint, the corresponding relativistic theory has a unitary S-matrix. This is the reason for having the Lagrangian formalism as a starting point in our study of constrained systems.

3.1.1 Comparison of the Lagrange and d'Alambert methods for constrained dynamics

As noted, the Lagrange treatment of constrained systems is based on the stationary action principle for the extended action

$$\tilde{S} = \int dt \tilde{L}(\dot{q}, q, t) = \int dt \left[L(\dot{q}, q, t) + \lambda^s \phi_s(\dot{q}, q) \right] \equiv S + \int dt \lambda^s \phi_s,$$

where L is the Lagrangian of the corresponding unconstrained system, ϕ_s are the constraints (enumerated by s), and λ^s are the Lagrange multipliers viewed as auxiliary dynamical variables. The equations of motion follow from the stationary action principle,

$$\delta \tilde{S} = \int dt \left\{ \left[-\frac{d}{dt} \frac{\partial \tilde{L}}{\partial \dot{q}^i} + \frac{\partial \tilde{L}}{\partial q^i} \right] \delta q^i + \phi_s \delta \lambda^s \right\} = 0. \tag{3.5}$$

The variations δq^i and $\delta \lambda^s$ are independent. So the equations of motion consist of two sets of equations: the Euler–Lagrange equations for the Lagrangian $\tilde{L}$ and the constraints $\phi_s = 0$.

In the d'Alambert approach [6, 9, 15], the constraints $\phi_s = 0$ are supplemented by the equations of motion that follow from the stationary action principle for the unconstrained system,

$$\delta S = \int dt \left[-\frac{d}{dt} \frac{\partial L}{\partial \dot{q}^i} + \frac{\partial L}{\partial q^i} \right] \delta_c q^i = 0, \tag{3.6}$$

where the variations of dynamical variables $\delta_c q^i$ are no longer independent but rather must be compatible with the constraints (the trajectory is varied only within the constraint surface in the tangent bundle on which the Lagrangian is defined (see Section 1.1)), which is indicated by the subscript c. Constraints $\phi_s(\dot{q}, q, t) = 0$ are called *holonomic* if there exist functions $\sigma_s(q, t)$ (i.e. independent of the velocities $\dot{q}^i$) such that the system $\phi_s = 0$ is equivalent to $d\sigma_s/dt = 0$. Otherwise the constraints $\phi_s = 0$ are called *anholonomic*. In other words, constraints $\phi_s(\dot{q}, q) = 0$ are holonomic if they are integrable. It can be proved that the Lagrange and d'Alambert approaches are equivalent only for holonomic constraints [6]. There are actual physical systems for which the correct dynamical description can only be obtained by the d'Alambert method (see below).

As an example of a holonomic constrained system, consider the free particle of unit mass on a circle of radius R, i.e. $L = \dot{\mathbf{x}}^2/2$ in (3.1). The Lagrange approach (3.5) yields the following equations of motion:

$$\ddot{\mathbf{x}} = -\lambda \mathbf{x}\,, \quad \mathbf{x}^2 = R^2. \tag{3.7}$$

The constraint solution reads $x^1 = R\cos\varphi$ and $x^2 = R\sin\varphi$ where φ is the dynamical variable. The constraint also implies that $(\dot{\mathbf{x}}, \mathbf{x}) = 0$. Hence, the first vector equation in (3.7) is equivalent to two scalar equations $(\ddot{\mathbf{x}}, \mathbf{x}) = -\lambda R^2$ and $(\ddot{\mathbf{x}}, \dot{\mathbf{x}}) = 0$. The latter is easily solved $\varphi(t) = \omega t + \varphi_0$, where ω and φ_0 are constants, while the former determines $\lambda(t)$. Thus, the motion is a revolution of the particle along the circle with the constant rate ω.

In the d'Alambert approach (3.6), the variations compatible with the constraint are $\delta\mathbf{x} = T\mathbf{x}\delta\eta$, where $\delta\eta$ is an independent variation parameter and T is the unit skew-symmetric 2×2 matrix, $T_{12} = 1$ (the generator of $SO(2)$ rotations). The complete system of equations reads

$$(\ddot{\mathbf{x}}, T\mathbf{x}) = 0, \qquad \mathbf{x}^2 = R^2. \tag{3.8}$$

Introducing the same dynamical variable φ to solve the constraint, one can see that the first equation in (3.8) gives $\ddot{\varphi} = 0$ and, hence, $\varphi(t) = \omega t + \varphi_0$. So the dynamics is equivalent in both the approaches.

As an example of an anholonomic system, consider an ice skater on an inclined plane. The skater is described by three variables q^i $(i = 1, 2, 3)$ where $q^1 \equiv x$ and $q^2 \equiv y$ are rectangular coordinates on the plane (the position of the skater on the plane), and $q^3 \equiv \varphi$ is the angle of the skate with the positive x-semi-axis counted counterclockwise (it describes the rotation of the skater about the axis normal to the plane). The skater's mass and moment of inertia are set to one for simplicity. The skater moves on the plane without friction. Suppose that the x-axis is set to be perpendicular to the acceleration of free fall. If the angle of the plane with the horizontal plane is α, then the gravity force acting on the skater has only the y-component of the magnitude $-g\sin\alpha \equiv -k$, where g is the free fall acceleration. Thus, the Lagrangian has the form,

$$L = \frac{1}{2}\left(\dot{x}^2 + \dot{y}^2 + \dot{\varphi}^2\right) - V, \quad V = ky. \tag{3.9}$$

The first two terms are the kinetic energy of the center of mass, the third one is the rotation energy, and V is the gravitational potential energy. The skater's motion is constrained by the condition that the skater cannot slide sidewise, i.e. the skater's velocity remains parallel to the skate orientation all the time. Let $\mathbf{e} = (\cos\varphi, \sin\varphi)$ be the unit vector parallel to the skate. Then the constraint is

$$\phi = \dot{x}\sin\varphi - \dot{y}\cos\varphi = 0; \tag{3.10}$$

the velocity component normal to $\mathbf{e}$ remains zero in due course. The constraint (3.10) is anholonomic. Indeed, let $M = M(x, y, \varphi)$ be an integrating factor for

Eq. (3.10) so that $d\sigma/dt = M\phi$ for some $\sigma = \sigma(x, y, \varphi)$. It is then straightforward to verify that the function σ must satisfy the equations

$$\frac{\partial \sigma}{\partial x} = M \sin\varphi, \qquad \frac{\partial \sigma}{\partial y} = -M \cos\varphi, \qquad \frac{\partial \sigma}{\partial \varphi} = 0,$$

which have no solution for any choice of $M \neq 0$. Qualitative physical features of a skater's motion are easy to anticipate. For example, if the skater is not rotating at the initial moment of time, $\dot{\varphi}(0) = 0$, the motion will occur along a straight line, and the skater is not expected to develop any rotation in due course. So the predictions of the d'Alambert and Lagrange methods for this system can be compared and related to the anticipated physical picture.

Consider first the d'Alambert approach. The allowed variations of x, y, and φ must be such that no displacement of the skater occurs in the direction perpendicular to $\mathbf{e}$. Therefore the variations δx and δy must satisfy the condition $\delta x \sin\varphi - \delta y \cos\varphi = 0$, and $\delta\varphi$ is not restricted. Hence, $\delta x = \delta\eta \cos\varphi$ and $\delta y = \delta\eta \sin\varphi$, where $\delta\eta$ is an independent variation parameter. These variations are substituted into (3.6) and the coefficients at $\delta\eta$ and $\delta\varphi$ are set to zero to obtain the d'Alambert equations of motion,

$$\ddot{x} \cos\varphi + (\ddot{y} + k) \sin\varphi = 0, \qquad \ddot{\varphi} = 0, \tag{3.11}$$

which are supplemented by the constraint equation (3.10). The second equation is easily integrated $\varphi(t) = \omega t + \varphi_0$. A general solution to the constraint (3.10) is $\dot{x} = v \cos\varphi$ and $\dot{y} = v \sin\varphi$, where the unknown function of time v is to be determined by the first equation in (3.11), which becomes $\dot{v} + k \sin\varphi = 0$ and, therefore, $v(t) = v_0 + (k/\omega)[\cos\varphi(t) - \cos\varphi_0]$. It is convenient to introduce the complex variable $z = x + iy$ in the plane. Then $\dot{z} = ve^{i\varphi}$ and, therefore, $z(t) = z_0 + Z(t) - Z(0)$ where

$$Z(t) = -iAe^{i\varphi(t)} - iB\left(e^{2i\varphi(t)} + 2i\varphi(t)\right), \; A = \frac{v_0}{\omega} - \frac{k\cos\varphi_0}{\omega^2}, \; B = \frac{k}{4\omega^2}, \tag{3.12}$$

and z_0 is the initial position of the skater.

The system has two dynamical features to be noted. First, if the skater has no initial rotation $\omega = 0$, then $v(t) = v_0 - tk\sin\varphi_0$ and $z(t) = z_0 + e^{i\varphi_0}(v_0 t - t^2 k \sin\varphi_0/2)$ from which it follows that

$$\frac{y(t) - y_0}{x(t) - x_0} = \tan\varphi_0,$$

i.e. the skater moves along a straight line and develops no rotation in due course. This is anticipated by physical reasoning. Second, if $\omega \neq 0$, the skater always remains in the strip

$$|y(t) - y_0 + Y(0)| \leq \sqrt{A^2 + B^2}, \qquad Y(0) = \operatorname{Im} Z(0),$$

i.e. the skater cannot slide down along the inclined plane below a certain threshold.

In contrast, the Lagrange approach predicts a drastically different motion of the skater, not compatible with the actual physical picture. Indeed, let λ be the Lagrange multiplier for the constraint (3.10). Using the same solution of the constraint (3.10), $\dot{x} = v\cos\varphi$ and $\dot{y} = v\sin\varphi$, the Euler–Lagrange equations for the Lagrangian $\tilde{L}$ (see (3.5)) are transformed to the following form

$$\ddot{\varphi} = \lambda v, \quad \dot{v} = -\lambda\dot{\varphi} - k\sin\varphi, \quad \dot{\lambda} = v\dot{\varphi} + k\cos\varphi. \tag{3.13}$$

The last two equations in (3.13) can be combined into a single equation for the complex variable $\xi = \lambda - iv$, so that $\dot{\xi} = i\dot{\varphi}\xi + ke^{i\varphi}$. Its general solutions reads,

$$\xi = \lambda - iv = e^{i(\varphi-\varphi_0)}\xi_0 + e^{i\varphi}kt, \quad \xi_0 = \lambda_0 - iv_0.$$

It is then substituted into the first equation in (3.13) to obtain the equation for φ. The initial value λ_0 of the Lagrange multiplier can be fixed by the initial angular velocity $\dot{\varphi}(0) \equiv \omega_0$ using the general solution for $\varphi(t)$.

First, note that, even if initially $\omega_0 = 0$, the skater generally develops rotation $\dot{\varphi}(t) \neq 0$, $t > 0$ (apart from the static solution for all variables, the constant solution $\varphi = \varphi_0$ is only possible for $\varphi_0 = \pm\pi/2$, i.e. when the skater slides along the y-axis). Second, a further analysis shows that the skater can drift downhill any distance, in contrast to the prediction of the d'Alambert method. This is established by studying the asymptotic solution of $\dot{y} = v\sin\varphi \sim -kt\sin^2\varphi$ as $t \to \infty$, where φ is the asymptotic solution of the first equation in (3.13) (technical details are omitted).

Constraints can be used to eliminate some (non-physical) variables in the original (unconstrained) Lagrangian. The Euler–Lagrange equations for the resulting physical Lagrangian describe the same dynamical picture as the method of Lagrange multipliers or the d'Alambert approach, provided the constraints are holonomic. For instance, the physical Lagrangian of a particle on a circle is $L = R^2\dot{\varphi}^2/2$. For anholonomic constraints, the d'Alambert equations cannot be cast in the form of the Euler–Lagrange equations for the Lagrangian in "physical" variables. For example, the constraint (3.10) can be solved for $\dot{x}$. The solution can then be used to eliminate the x degree of freedom from the Lagrangian (3.9). The Euler–Lagrange equations of motion for the "physical" variables y and φ lead to the Lagrange description of the skater, which is different from the d'Alambert approach.

A simple and instructive example of the equivalence of the d'Alambert and Lagrange approaches when constraints are holonomic is provided by the Maupertuis' principle discussed at the end of Section 1.15. Let θ^i, $i = 1, 2, \ldots, 2N$, be canonical variables in the phase space $\Gamma = R^{2N}$ so that $\{\theta^i, \theta^j\} = \omega^{ij}$ is the canonical (Darboux) symplectic structure on Γ. Let ω_{ij} be the inverse of ω^{ij}, i.e. $\omega_{ik}\omega^{kj} = \delta_i^j$. Maupertuis' principle asserts that the abbreviated action $S_M = \frac{1}{2}\int \omega_{ij}\theta^i\theta^j\,dt$ attains its minimum on the trajectory of the actual motion, provided the motion conserves the energy, i.e. the equation $H(\theta) = E$ is a (holonomic) constraint. By the d'Alambert's principle, the variations $\delta\theta^j$ are not independent and subject to the condition $\delta H = h_j \delta\theta^j = 0$ where $h_j = \partial H/\partial\theta^j$. Its general solution reads

$$\delta\theta^i = P^i_j \delta\eta^j, \quad P^i_j = \delta^i_j - \frac{h^i h_j}{h^2},$$

where $\delta\eta^j$ are independent variations, and $h^2 = h_i h^i = \delta^{ij} h_i h_j$. The matrix P^i_j is the projector onto a hyperplane orthogonal to the gradient of H. Therefore the stationary action principle $\delta S_M = 0$ yields the equations of motion $P^i_j \omega_{ik} \dot{\theta}^k = 0$. As the matrix P^i_j has a one-dimensional null space spanned by the vector h^i, these equation can be written in the equivalent form $\omega_{ik}\dot{\theta}^k = \lambda h_i$ where λ is an arbitrary function of time. Defining a new parameter τ on the trajectory by the relation $d\tau = \lambda\, dt$, the equations of motion assume the Hamiltonian form

$$\frac{d\theta^k}{d\tau} = \omega^{kj} \frac{\partial H}{\partial\theta^j}.$$

Since λ is arbitrary and, hence, τ is not the physical time, they determine only the shape of a phase space trajectory.

3.2 A general analysis of dynamical systems with constraints

3.2.1 The Hamiltonian formalism

In fundamental physics the Lagrangian contains the whole information about the system dynamics, including possible constraints on dynamical variables. As noted, a general reason for constraints existing in the Lagrangian formalism is the singularity of the Hessian matrix. The latter leads to some peculiarities of the Hamiltonian formalism that are analyzed here. The discussion follows the work of Dirac [148]. Let a Lagrangian $L(q, \dot{q})$ describe a dynamical system with N degrees of freedom. To change to the Hamiltonian formalism, the velocities $\dot{q}$ should be expressed as functions of the canonical momenta defined by

$$p_i = \frac{\partial L(q, \dot{q})}{\partial \dot{q}^i}, \qquad i = 1, 2, \ldots, N. \tag{3.14}$$

These N equations must be solved for $\dot{q}^i$. As noted above, the latter is possible if the rank of the matrix $T_{ik} = \partial^2 L/\partial\dot{q}^i \partial\dot{q}^k$ is N. The Hamiltonian formalism is then developed in a standard fashion via the Legendre transformation. Here the case of interest is when the rank of the Hessian matrix is less than N.

Suppose that the rank of T equals R, $R < N$. Then the matrix T_{ik} has $N - R$ zero eigenvalues and, hence, only R equations of the system (3.14) can be solved for $\dot{q}^k$. Without loss of generality they are assumed to be the first R equations. With this assumption one has

$$\dot{q}^k = f^k(q, p_l, \dot{q}^r), \quad k, l = 1, 2, \ldots, R, \quad r = R + 1, \ldots, N. \tag{3.15}$$

Here and in what follows the following abbreviation is used:

$$f^k(q,p_l,\dot{q}^r) = f^k(q_1,\ldots,q_N,p_1,\ldots,p_R,\dot{q}^{R+1},\ldots,\dot{q}^N).$$

The inequality $R < N$ means that, after inserting (3.15) in the remaining $N-R$ equations (3.14), the dependence upon $\dot{q}^r$ disappears, and they turn into relations between q and p only,[2]

$$\phi_s(q,p) = 0, \qquad s = 1,2,\ldots,N-R \equiv S. \tag{3.16}$$

In the Hamiltonian formalism these are conditions to be imposed on the canonical variables q^i and p_i. They are called *constraints*. It is noteworthy that the functions ϕ_s themselves are also called constraints. For brevity such a terminological liberty is also adopted here. The existence of constraints means that only a subspace of the total $2N$-dimensional phase space is physical. Equations (3.16) can be rewritten in another form. Set

$$\varphi_r(q,p_k) = \left.\frac{\partial L(q,\dot{q}^k,\dot{q}^r)}{\partial \dot{q}^r}\right|_{\dot{q}^k=f^k(q,p_k,\dot{q}^r)} = p_r. \tag{3.17}$$

Here the dependence of the Lagrangian on the velocities $\dot{q}^r = \{\dot{q}^{R+1},\ldots,\dot{q}^N\}$ and $\dot{q}^k = \{\dot{q}^1,\ldots,\dot{q}^R\}$ is explicitly indicated, and the fact that $\varphi_r(q,p_k)$ do not depend on $\dot{q}^r$ is taken into account. Instead of (3.16) one obtains

$$\phi_s(q,p) = p_{R+s} - \varphi_{R+s}(q,p_k) = 0, \quad s = 1,2,\ldots,S. \tag{3.18}$$

In doing so a more specific form of the functions ϕ_s is established. Namely, they are linear in p_{R+s}. The Hamiltonian is defined in the standard way,

$$\tilde{H} = p_i\dot{q}^i - L = p_k f^k(q,p_l,\dot{q}^r) + p_r\dot{q}^r - L(q,f^k,\dot{q}^r). \tag{3.19}$$

The velocities $\dot{q}^{R+1},\ldots,\dot{q}^N$ cannot be expressed in terms of the momenta and, therefore, are retained in $\tilde{H}$. The latter is not, strictly speaking, a canonical Hamiltonian. Let us prove the following:

(1) With the constraints (3.18) enforced, $\tilde{H}$ does not depend on $\dot{q}^r$;
(2) Without the constraints (3.18), $\tilde{H}$ depends linearly on $\dot{q}^r$ and p_r.

Indeed, differentiating (3.19) with respect to $\dot{q}^r$ one infers that

$$\frac{\partial \tilde{H}}{\partial \dot{q}^r} = p_k\frac{\partial f^k}{\partial \dot{q}^r} + p_r - \frac{\partial L}{\partial f^k}\frac{\partial f^k}{\partial \dot{q}^r} - \frac{\partial L}{\partial \dot{q}^r} = p_r - \varphi_r. \tag{3.20}$$

The first and third terms cancel each other out because of the first R equations in (3.14). The first proposition, $\partial\tilde{H}/\partial\dot{q}^r = 0$, is proved by combining (3.20) and (3.18). Let $H(q,p_k)$ be the right-hand side of (3.19) after taking into account the constraints. As the right-hand side of (3.20) does not depend on $\dot{q}^r$, the integration with respect to $\dot{q}^r$ yields

$$\tilde{H} = H(q,p_k) + \dot{q}^r\left[p_r - \varphi_r(q,p_k)\right], \tag{3.21}$$

[2] There are more complicated cases in which this is not always possible (see Chapter 4).

which completes the proof of the second proposition. Equation (3.21) establishes a general form of the Hamiltonian in the presence of constraints.

Now the question arises: how is this Hamiltonian to be used? If the constraints are ignored, then $\tilde{H}$ depends upon the velocities $\dot{q}^r$, which is quite an unusual property of the Hamiltonian. If the constraints are enforced, then the canonical variables q and p are no longer independent variables. What is to be done? The way out was suggested by Dirac [148]. Consider the variation of the function $p_i\dot{q}^i - L(q,\dot{q})$ under an infinitesimal change of all its arguments,

$$\delta(p_i\dot{q}^i - L) = \delta p_i\dot{q}^i + p_i\delta\dot{q}^i - \frac{\partial L}{\partial \dot{q}^i}\delta\dot{q}^i - \frac{\partial L}{\partial q^i}\delta q^i.$$

By taking into account the equations of motion $d/dt\partial L/\partial\dot{q}^i = \partial L/\partial q^i$ and the definition (3.14) it follows that

$$\delta(p_i\dot{q}^i - L) = \dot{q}^i\delta p_i - \dot{p}_i\delta q^i \equiv \delta H(q,p). \tag{3.22}$$

Equation (3.22) shows that $p_i\dot{q}^i - L$ is a function of q and p. If q^i and p_i were independent variables, the Hamiltonian equations of motion would have followed from (3.22) by equating the coefficients at the variations δp_i and δq^i in the relation[3]

$$\dot{q}^i\delta p_i - \dot{p}_i\delta q^i = \delta H(q,p) = \frac{\partial H}{\partial p_i}\delta p_i + \frac{\partial H}{\partial q^i}\delta q^i. \tag{3.23}$$

However, this cannot be done because the variations δp_i and δq^i are no longer independent. According to Eqs. (3.16),

$$\frac{\partial \phi_s}{\partial q^i}\delta q^i + \frac{\partial \phi_s}{\partial p_i}\delta p_i = 0, \qquad s = 1,2,\ldots,S. \tag{3.24}$$

In order to derive the Hamiltonian equations of motion, the method of Lagrange multipliers is applied. Multiplying equations (3.24) by arbitrary functions of time u_s and adding the result to the right-hand side of (3.22), one obtains the equality

$$\dot{q}^i\delta p_i - \dot{p}_i\delta q^i = \delta H + u_s\left(\frac{\partial \phi_s}{\partial q^i}\delta q^i + \frac{\partial \phi_s}{\partial p_i}\delta p_i\right), \tag{3.25}$$

from which the Hamiltonian equations of motion are inferred,

$$\begin{aligned} \dot{q}^i &= \frac{\partial H}{\partial p_i} + u_s\frac{\partial \phi_s}{\partial p_i}; \\ \dot{p}_i &= -\frac{\partial H}{\partial q^i} - u_s\frac{\partial \phi_s}{\partial q^i}. \end{aligned} \tag{3.26}$$

[3] It is not surprising that equations (3.23) are equivalent to the Hamiltonian equations of motion as the Lagrangian equations of motion are used when deriving (3.22).

Equations (3.26) must also be supplemented by the constraints (3.16). Thus, for $2N + S$ unknown functions q^i, p_i, and u_s there are $2N$ equations (3.26) and S constraints (3.16).

The system (3.26) can be rewritten in terms of the Poisson brackets,

$$\dot{q}^i = \{q^i, H_T\}, \quad \dot{p}_i = \{p_i, H_T\}, \tag{3.27}$$

where

$$H_T = H(q, p_k) + u_s\phi_s(q, p). \tag{3.28}$$

Equations (3.26) and (3.27) are identical if the Poisson brackets are defined so that $\{q, u_s\} = \{p, u_s\} = 0$. According to Dirac [148], it is not necessary to define the brackets explicitly, rather it is sufficient to assume that they have all the properties of the Poisson brackets. In particular, for an arbitrary function A of q and p, the relations $\{u_s\phi_s, A\} = u_s\{\phi_s, A\} + \{u_s, A\}\phi_s$ hold. Then, for example,

$$\{q^i, H_T\} = \{q^i, H\} + u_s\{q^i, \phi_s\} + \{q^i, u_s\}\phi_s.$$

Thanks to the constraints (3.16), the last term vanishes here regardless of any specific definition of $\{q, u_s\}$. If the linearity of ϕ_s in the momenta p_r (compare (3.18)) is taken into account together with the fact that the Hamiltonian $H(q, p_k)$ in (3.28) does not depend on p_r, then the reason to include the terms $u_s\phi_s$ into H_T becomes apparent. According to (3.27) and (3.18)

$$\dot{q}^r = \{q^r, H_T\} = u_{r-R}, \tag{3.29}$$

that is, these terms, or, more accurately, the coefficients u_s, determine the rate of change of the $N - R$ non-physical coordinates q^r with time.

Thus, the Hamiltonian equations of motion (3.27) define the evolution of canonical variables with time. Now the self-consistency of the dynamics has to be verified. The constraints $\phi_s = 0$ must hold at every moment of time, which requires that $\dot{\phi}_s = 0$ or, equivalently,

$$\{\phi_s, H_T\} = \{\phi_s, H\} + u_{s'}\{\phi_s, \phi_{s'}\} = 0, \tag{3.30}$$

where the constraints are taken into account. The following possibilities arise:

(1) Equations (3.30) are identically satisfied;
(2) They give rise to new conditions

$$\psi_m(q, p) = 0, \qquad m = 1, 2, \ldots, M, \tag{3.31}$$

which cannot be reduced to (3.16) and, hence, must be regarded as new constraints;

(3) Equations (3.30) are conditions on u_s;
(4) Equations (3.30) are inconsistent.

The first case needs no comment as the theory is self-consistent. In the second case, the self-consistency of the theory requires further restrictions on the canonical variables. The second case is possible only if

$$\{\phi_s, \phi_{s'}\} = C_{ss'}^{s''}(q,p)\phi_{s''}, \tag{3.32}$$

where $C_{ss'}^{s''}$ are some functions of q and p, i.e. when the terms containing u_s disappear after imposing constraints in Eqs. (3.30). If the conditions (3.32) are not fulfilled, meaning that $\{\phi_s, \phi_{s'}\}$ differs from zero for some of s and s' even after taking into account the constraints (for example, if it is a nonzero constant), then the third possibility is realized. In this case some of the coefficients u_s or all of them can be found from Eqs. (3.30). Finally, it may happen that the conditions (3.30) cannot be satisfied with any choice of functions q, p, and u_s. The theory is inconsistent. Here are some examples of pathological Lagrangians: (1) Set $L = q$. The equation of motion is $\partial L/\partial q = 0$, i.e. $1 = 0$ (nonsense). (2) Set $L = \dot{q}$, the equation of motion reads $(d/dt)\partial L/\partial \dot{q} = 0$ or $0 = 0$ (a trivial identity; the Lagrangian is the total time derivative).

In the second case the self-consistency analysis must be carried out for the new constraints ψ_m in the same way as it was for constraints ϕ_s until all further constraints are found. The procedure is terminated at a stage when the self-consistency conditions yield no more new constraints.

Before turning to examples the following helpful terminology is introduced [148]. The constraints (3.16) and (3.31) are called *primary* and *secondary* constraints, respectively. The definition indicates only the stage of the self-consistency analysis at which the constraints appear; there is no distinction between them otherwise. Yet, all further constraints resulting from (3.31) are viewed as secondary.

When dealing with constraints there are two types of equalities that occur. Besides the conventional ones there are those that hold only after taking into account the constraints, i.e. they are satisfied only on a subspace of the phase space defined by the constraints. Such equalities are called *weak*. To distinguish the equalities that hold in the weak sense, the sign $\approx$ is used in place of $=$. For example, (3.32) can be written as $\{\phi_s, \phi_{s'}\} \approx 0$. Equalities which hold in the entire phase space are called *strong* and the conventional sign $=$ is used in these.

Any dynamical quantity in a constrained theory can be attributed to one of the following two classes [148]. A function $A(q,p)$ is said to be a dynamical quantity of the *first class* if its Poisson brackets with all the constraints vanish in the weak sense,

$$\{A, \phi_j\} \approx 0, \quad j = 1, 2, \ldots, J. \tag{3.33}$$

From now and on all the constraints, both the primary and secondary ones, are denoted by ϕ_j unless specified otherwise. The relation (3.33) can be written in the equivalent form,

$$\{A, \phi_j\} = d_{jj'}(q,p)\phi_{j'}. \tag{3.34}$$

In other words, any quantity that vanishes in the weak sense is a linear combination of constraints in the strong sense. Of course, the coefficients $d_{jj'}$ can also depend on ϕ_j. All quantities that are not of the first class are called dynamical quantities of the *second class*. The self-consistency of the theory requires the equality (3.30) to hold, that is,

$$\{H_T, \phi_j\} = h_{jk}\phi_k \approx 0, \tag{3.35}$$

with h_{jk} being functions of q, p. Hence, *the Hamiltonian is always a first-class quantity.*

Constraints are also classified in the same way. If for some of the constraints the condition (3.32) is satisfied, then the corresponding constraints are called *the first-class constraints*. In this case, the constraints are said to be *in involution*. Thus, the Poisson brackets of the first-class constraints are in involution. Constraints whose Poisson brackets are not in involution are called *second-class constraints*. It should be stressed that constraints cannot be used before calculating the Poisson brackets as in the definition of the latter all the canonical variables q and p are assumed to be independent.

3.2.2 Examples of systems with constraints

1. *A simple constrained system* is defined by the Lagrangian

$$L = \frac{1}{2}(\dot{x}_1 + \dot{x}_2)^2. \tag{3.36}$$

The canonical momenta are $p_1 = \dot{x}_1 + \dot{x}_2$ and $p_2 = \dot{x}_1 + \dot{x}_2$ so that the primary constraint reads $\phi = p_2 - p_1 = 0$. The "Hamiltonian" (3.19) is then $\tilde{H} = \frac{1}{2}p_1^2 + \dot{x}_2(p_2 - p_1)$. The Lagrangian (3.36) is invariant under (gauge) transformations $\delta x_1 = \omega(t)$ and $\delta x_2 = -\omega(t)$, with $\omega(t)$ being an arbitrary function of time. In the new variables, $X_1 = x_1 + x_2$ and $X_2 = \frac{1}{2}(x_2 - x_1)$, the Lagrangian assumes the form $L = \frac{1}{2}\dot{X}_1^2$ from which it becomes clear that this is a dynamical system with just one degree of freedom. The physical phase space of the system has a smaller dimension than the total phase space. This is the essence of the effect of constraints on dynamics. The equations of motion define only the evolution of X_1, while that of X_2 remains completely arbitrary. Clearly, X_2 is a non-physical degree of freedom. In particular, it changes under the gauge transformations, while X_1 does not. Let us now elucidate the meaning of the Hamiltonian H_T,

$$H_T = \frac{1}{2}p_1^2 + u(p_2 - p_1). \tag{3.37}$$

The theory is obviously self-consistent because $\{\phi, H_T\} = 0$. From the equations of motion, $\dot{x}_1 = p_1 - u$, $\dot{x}_2 = u$, $\dot{p}_1 = 0$, and $\dot{p}_2 = 0$ and the constraint $p_1 = p_2$, it follows that the function $u(t)$ is arbitrary. Let us find X_1 and X_2 as functions of time. Because of $\{X_1, \phi\} = 0$ and $\{X_2, \phi\} = 1$ we have $\dot{X}_1 = \{X_1, H_T\} = p_1$ and $\dot{X}_2 = -\frac{1}{2}p_1 + u$, from which the meaning of the terms $u_s\phi_s$ in H_T becomes evident. They do not affect the evolution of physical variables and determine the evolution of unphysical variables. Their presence makes the theory more general. Observe that variations of the function u yield the gauge transformations. As the physical variables are invariant under the gauge transformations, the function u may be fixed at will, e.g. $u = 0$, or $u = 1$, or set $u = p_1/2$ so that $X_2 = \text{const}$. In doing so, the evolution of non-physical degrees of freedom becomes fixed. The procedure is called a *gauge fixing*. The Hamiltonian H_T is invariant under gauge transformations in the weak sense.

2. *A system with first-class constraints.* A less trivial example of a constrained system is provided by the Lagrangian

$$L(\mathbf{x}, \dot{\mathbf{x}}, \mathbf{y}, \dot{\mathbf{y}}) = \frac{1}{2}(\dot{\mathbf{x}} - [\mathbf{y}, \mathbf{x}])^2 - V(\mathbf{x}^2), \tag{3.38}$$

where $\mathbf{x}$ and $\mathbf{y}$ are three-dimensional vectors and $[\mathbf{x}, \mathbf{y}]$ denotes the cross product, $[\mathbf{x}, \mathbf{y}]_i = \varepsilon_{ijk} x_j y_k$, where ε_{ijk} is the totally skew-symmetric unit tensor, $\varepsilon_{123} = 1$. The canonical momenta for $\mathbf{y}$ form the primary constraints, $\boldsymbol{\pi} = \partial L/\partial \dot{\mathbf{y}} = 0$. The other momenta are $\mathbf{p} = \partial L/\partial \dot{\mathbf{x}} = \dot{\mathbf{x}} - [\mathbf{y}, \mathbf{x}]$. The Hamiltonian reads

$$H_T = H + (\mathbf{u}, \boldsymbol{\pi}), \quad H = \frac{1}{2}\mathbf{p}^2 + V(\mathbf{x}^2) + (\mathbf{y}, [\mathbf{x}, \mathbf{p}]), \tag{3.39}$$

where $(\cdot, \cdot)$ denotes the Euclidean scalar product. Next the self-consistency of the system must be investigated. From $\{y_i, \pi_j\} = \delta_{ij}$, $i, j = 1, 2, 3$, it follows that $\dot{\mathbf{y}} = \mathbf{u}$ where $\mathbf{u}$ is an arbitrary function, while

$$\{\boldsymbol{\pi}, H_T\} = -[\mathbf{x}, \mathbf{p}] \equiv -\mathbf{M} = 0 \tag{3.40}$$

yields three secondary constraints. They are nothing but the components of the angular momentum and, hence, only two of them are independent. It is easy to see that

$$\{\pi_i, M_j\} = 0 \quad \{M_i, M_j\} = \varepsilon_{ijk} M_k, \tag{3.41}$$

i.e. the constraints are in involution. With the relation

$$\{M_i, H_T\} = \varepsilon_{ijk} y_j M_k \tag{3.42}$$

the theory is proved to be self-consistent, and H_T is a first-class quantity. The Lagrangian (3.38) is invariant under the group of gauge transformations

$$\delta \mathbf{x} = [\boldsymbol{\omega}(t), \mathbf{x}], \quad \delta \mathbf{y} = \dot{\boldsymbol{\omega}}(t) + [\boldsymbol{\omega}(t), \mathbf{y}], \tag{3.43}$$

where $\boldsymbol{\omega}(t)$ is an arbitrary infinitesimal function of time. Note that the first equation in (3.43) shows that the gauge transformations of $\mathbf{x}$ are rotations about the origin, i.e. they form the group $SO(3)$.

It is instructive to analyze solutions of the equations of motion. The canonical variables $\mathbf{x}$ and $\mathbf{p}$ satisfy the constraint,

$$[\mathbf{x}, \mathbf{p}] = 0, \tag{3.44}$$

and the Hamiltonian equations of motion are,

$$\dot{\mathbf{x}} = \mathbf{p} + [\mathbf{y}, \mathbf{x}], \quad \dot{\mathbf{p}} = -\nabla V + [\mathbf{y}, \mathbf{p}]. \tag{3.45}$$

The equation $\dot{\mathbf{y}} = \mathbf{u}$, as one will see, contains no new information – the vector $\mathbf{y}$ turns out to be arbitrary. Equation (3.44) implies that the vectors $\mathbf{x}$ and $\mathbf{p}$ must be parallel, $\mathbf{p} = \lambda(t)\mathbf{x}$ where λ is an arbitrary function of time. That the function λ remains arbitrary is a consequence of the fact that only two of three equations (3.44) are independent. Substituting $\mathbf{p}$ into (3.45) and taking for simplicity $V = \frac{1}{2}\mathbf{x}^2$, one finds,

$$\dot{\mathbf{x}} = \lambda\mathbf{x} + [\mathbf{y}, \mathbf{x}], \qquad \dot{\lambda}\mathbf{x} + \lambda\dot{\mathbf{x}} = -\mathbf{x} + \lambda[\mathbf{y}, \mathbf{x}]. \tag{3.46}$$

Multiplying the first of these equations by λ and subtracting from it the second one the equation for λ is obtained, $\dot{\lambda}\mathbf{x} = -(\lambda^2 + 1)\mathbf{x}$, i.e.

$$\lambda = -\tan(t + c) \tag{3.47}$$

with c being a constant. The first equation in (3.46) can be rewritten in the matrix form,

$$\dot{\mathbf{x}} = A(t)\mathbf{x}, \quad A(t) = \lambda(t) - \hat{A}(t), \quad (\hat{A})_{ij}(t) = \varepsilon_{ijk}y_k(t). \tag{3.48}$$

The solution of this equation is

$$\mathbf{x}(t) = e^{\Lambda(t)}T\exp\left(-\int_0^t \hat{A}(t')dt'\right)\mathbf{x}(0), \tag{3.49}$$

where $\Lambda(t) = \int \lambda dt$ such that $\Lambda(0) = 0$ and T in front of the exponential stands for time ordering (by definition, the time-ordered exponential is the solution of the homogeneous differential equation (3.48) with $A = \hat{A}$ being an arbitrary time-dependent matrix). The momentum $\mathbf{p}$ is obtained by multiplying (3.49) by $\lambda(t)$, while the vector $\mathbf{y}(t)$ remains arbitrary. From the nine unknown quantities, $\mathbf{x}$, $\mathbf{p}$, and $\mathbf{y}$ only six are determined, i.e. three of nine equations (3.44) and (3.45) turn out to be consequences of the others. In fact, this illustrates the second Noether theorem [14].

Let us discuss the solution. A gauge can be fixed by setting $\mathbf{y} = 0$. Then, as follows from (3.49), the particle moves along a straight line through the origin that is parallel to $\mathbf{x}(0)$. This is the physical content of the system dynamics. If $\mathbf{y}(t) \neq 0$, then taking into account that the matrices $\hat{G}^k$, $\hat{G}^k_{ij} = \varepsilon_{ijk}$, are generators of rotations in the three-dimensional Euclidean space and $\hat{A} = \hat{G}^k y_k(t)$, it is concluded that the time-ordered exponential in (3.49) describes a (non-uniform in time) rotation of the initial position vector about the origin. Thus, the particle moves along a line through the origin, while the line itself arbitrarily rotates with time about the origin. The rotations have no effect on the physical motion along the line which is solely determined by the potential V.

3. *Second-class constraints.* Let the theory be given by the Lagrangian (3.2) where $L = (1/2)\dot{\mathbf{x}}^2$. Then

$$\mathbf{p} = \frac{\partial \tilde{L}}{\partial \dot{\mathbf{x}}} = \dot{\mathbf{x}}, \quad p_3 = 0, \tag{3.50}$$

where $p_3 \equiv \phi_1 = 0$ is the primary constraint. The Hamiltonian is

$$H_T = H + u\phi_1, \qquad H = \frac{1}{2}\mathbf{p}^2 + x_3\phi_2(\mathbf{x}), \tag{3.51}$$

where $\phi_2 \equiv \frac{1}{2}(\mathbf{x}^2 - R^2)$. The self-consistency condition reads,

$$\{\phi_1, H_T\} = -\phi_2(\mathbf{x}) = 0, \tag{3.52}$$

which is the secondary constraint. It is necessary to verify that the new constraint does not contradict the equations of motion,

$$\{\phi_2(\mathbf{x}), H_T\} = (\mathbf{x}, \mathbf{p}) \equiv \tilde{\phi}_3 = 0. \tag{3.53}$$

This is yet another constraint. By calculating the Poisson bracket,

$$\{\phi_2, \tilde{\phi}_3\} = \mathbf{x}^2 \approx R^2, \tag{3.54}$$

the constraints ϕ_2 and $\tilde{\phi}_3$ are proved to be of the second class. The self-consistency analysis must be continued. In doing so, one more constraint is found,

$$\{\tilde{\phi}_3, H_T\} = \mathbf{p}^2 - x_3\mathbf{x}^2 \equiv \phi_4 = 0. \tag{3.55}$$

This is not in involution with the primary constraint,

$$\{\phi_4, \phi_1\} = -\mathbf{x}^2 \approx -R^2. \tag{3.56}$$

Therefore ϕ_1 and ϕ_4 are another pair of second-class constraints. Note that ϕ_1 is found to be the second-class constraint only at the final stage of the analysis. It is easy to see that

$$\{\phi_4, H_T\} = -(4x_3\tilde{\phi}_3 + u\mathbf{x}^2) \approx -uR^2 = 0, \tag{3.57}$$

which is not a new constraint, rather it imposes a condition on u, namely, $u = 0$. The analysis is terminated as all the self-consistency conditions have been exhausted. It remains to point out that the constraints $\tilde{\phi}_3$ and ϕ_4 are also not in involution, $\{\tilde{\phi}_3, \phi_4\} = 2\phi_4 + 4x_3\mathbf{x}^2$, unlike ϕ_2 and ϕ_4 for which $\{\phi_2, \phi_4\} = 2\tilde{\phi}_3$. However, the linear combination of the constraints $\phi_3 = \tilde{\phi}_3 - 4x_3\phi_1$ is in involution with ϕ_4, $\{\phi_3, \phi_4\} = 2\phi_4$. The Lagrangian equations of motion, $\ddot{\mathbf{x}} = -x_3\mathbf{x}$ and $\mathbf{x}^2 = R^2$, are easily solved in the polar coordinates, $r = (\mathbf{x}^2)^{1/2}$ and $\varphi = \tan^{-1}(x_2/x_1)$. The solution reads $\varphi = \nu t + \varphi_0$, $r = R$, and $x_3 = \nu^2 = \text{const}$. This contains no arbitrariness despite the presence of the primary constraint. Thus, even though the Lagrangian (3.2) is singular, it is not invariant under any gauge transformations (otherwise, according to the second Noether theorem [14], the solutions would have contained arbitrary functions of time).

An attempt to generalize the Hamiltonian by adding to it a linear combination of all constraints,

$$H_E = H + u_\alpha \phi_\alpha \ , \quad \alpha = 1, 2, 3, 4, \tag{3.58}$$

results in nothing new. Indeed,

$$\begin{aligned}\{\phi_\alpha, H_E\} = \{\phi_\alpha, H\} + \{\phi_\alpha, \phi_\beta\} u_\beta \approx \{\phi_\alpha, \phi_\beta\} u_\beta \approx \\ \approx R^2 M_{\alpha\beta} u_\beta = 0,\end{aligned}$$

where $M_{\alpha\beta}$ is a skew-symmetric matrix with nonzero elements being $M_{14} = M_{23} = -M_{41} = -M_{32} = 1$. Thus, all the coefficients u_α are fixed: $u_\alpha = 0$. The system does not admit a more general description.

Finally, the question may arise whether the standard Poisson bracket formalism is admissible to obtain new constraints. This question seems rather natural as the formalism assumes the variables q and p to be independent, while in due course for the self-consistency analysis they are demanded to satisfy some relations (constraints). Recall that the very existence of primary constraints requires introducing arbitrary functions u_s into the Hamiltonian formalism (Section 3.2.1). The difference with the subsequent self-consistency analysis is that the primary constraints appear when carrying out the Legendre transform to obtain the Hamiltonian formalism in which it is supposed that all the canonical variables are independent. Once the Hamiltonian formalism is established, the occurrence of new constraints does not require its modification because the constraints themselves appear by using the Poisson brackets and, hence, enter the formalism on an equal footing with the Hamiltonian equations of motion.

4. *Classes of primary constraints.* Can second-class constraints occur amongst the primary constraints? The following example answers this question. In the theory with the Lagrangian,

$$L = \frac{1}{2}\dot{x}_1^2 - \dot{x}_2 x_3 \tag{3.59}$$

the primary constraints are $p_2 = -x_3$, $p_3 = 0$. Put $\phi_1 = p_2 + x_3$ and $\phi_2 = p_3$. Their Poisson bracket does not vanish weakly, $\{\phi_1, \phi_2\} = 1$. Hence, the constraints are of the second class. The Lagrangian (3.59) is a special case of

$$L = \frac{1}{2}\dot{x}_1^2 - (\dot{x}_2 x_3 + \kappa \dot{x}_3 x_2).$$

It is not hard to verify that, if $\kappa = +1$, the primary constraints are of the first class, while for $\kappa = -1$ they appears to be of the second class.

5. *An inconsistent dynamical system with constraints.* Put

$$L = \frac{1}{2}\dot{x}_1^2 + x_2. \tag{3.60}$$

As $p_1 = \dot{x}_1$ and $p_2 = 0$, the Hamiltonian is $H_T = H + up_2$ where $H = \frac{1}{2}p_1^2 - x_2$. The self-consistency condition $\{p_2, H_T\} = 0$ is impossible to satisfy because $\{p_2, H_T\} = 1$. The theory (3.60) is inconsistent.

6. *A system with a fictitious degree of freedom.* A theory with the Lagrangian $L = \frac{1}{2}\dot{x}_1^2 + \dot{x}_2$ though being self-consistent, is empty for x_2 because the equation of motion for this degree of freedom is fulfilled identically, $0 = 0$.

3.2.3 The Lagrangian formalism

Let us elucidate characteristic features of constrained systems in the Lagrangian formalism. As pointed out in Section 3.2.1 such systems are described by Lagrangians whose Hessian matrix, $T_{ik} = \partial^2 L/\partial\dot{q}^i\partial\dot{q}^k$, is singular (see (3.3)). The equations of motion are,

$$\frac{d}{dt}\frac{dL}{d\dot{q}^i} - \frac{\partial L}{\partial q^i} = T_{ik}\ddot{q}^k + \frac{\partial^2 L}{\partial\dot{q}^i\partial q^k}\dot{q}^k - \frac{\partial L}{\partial q^i} = 0, \quad i = 1, 2, \ldots, N. \tag{3.61}$$

If the rank R of the matrix T_{ik} is less then N, then the Euler–Lagrange equations (3.61) are second-order differential equations only for R functions $q^k(t)$, i.e. only R of them can be solved for $\ddot{q}^k$,

$$\ddot{q}^k = F^k(q, \dot{q}, \ddot{q}^{R+s}), \quad k = 1, 2, \ldots, R, \quad s = 1, 2, \ldots, N - R. \tag{3.62}$$

The remaining $N - R$ relations are not equations of motion because they do not contain the second derivatives $\ddot{q}^{R+s}$ after substituting the functions (3.62) into them (in particular, some of the equalities (3.61) may follow from the others). In other words, the $N - R$ Euler–Lagrange equations turn out to be conditions on coordinates and velocities,

$$\Theta_s(q, \dot{q}) = 0, \quad s = 1, 2, \ldots, N - R. \tag{3.63}$$

The number of independent conditions (3.63) cannot exceed the number of primary constraints. The very existence of relations (3.63) implies that the initial conditions, i.e. coordinates and velocities,

$$q^i(0) = \alpha^i, \quad \dot{q}^i(0) = \beta^i, \quad i = 1, 2, \ldots, N,$$

cannot be arbitrary and must satisfy (3.63), otherwise the equations of motion have no solution.

In the Lagrangian formalism, the self-consistency analysis amounts to verifying whether the conditions (3.63) are compatible with the time evolution, meaning that their time derivative must vanish as well,

$$\dot{\Theta}_s = \frac{\partial\Theta_s}{\partial\dot{q}^i}\ddot{q}^i + \frac{\partial\Theta_s}{\partial q^i}\dot{q}^i = 0. \tag{3.64}$$

If the matrix $\partial\Theta_s/\partial\dot{q}^i$ has nontrivial left eigenvectors $\eta_a^s = \eta_a^s(q,\dot{q})$ with zero eigenvalue, i.e. $\eta_a^s\partial\Theta_s/\partial\dot{q}^i = 0$, then Eq. (3.64) may yield new restrictions on the velocities and coordinates which are not linear combinations of the previously established ones and whose time derivatives must be required to vanish too, and so on until all independent conditions are found and their mutual compatibility is established. Such an analysis of "constraints" in the Lagrangian approach (they are called "the Lagrangian constraints") is carried out in detail in [149–151].

3.3 Physical variables in systems with constraints

3.3.1 The extended group of gauge transformations

Constraints should be satisfied at every moment of time. As the system evolves in time, it must remain in the surface of constraints defined by the equations $\phi_\alpha(q,p) = 0$ in the total phase space. This means that some of the canonical variables p and q are redundant or non-physical. It is natural to reformulate the theory only in terms of physical degrees of freedom or, in other words, to go over to a Hamiltonian dynamics in the reduced or *physical* phase space. Realization of this idea is different for theories with first- and second-class constraints. As shown earlier, the key difference between systems with first- and second-class constraints is that the dynamics of first-class constrained systems is more general as the solutions of equations of motion contain arbitrary functions of time (see (3.28)). The number of such functions equals the number of independent first-class constraints. The existence of such an arbitrariness may mean that the theory possesses a local (gauge) symmetry with the first-class constraints being its generators.

Indeed, consider the change of a function $F = F(p,q)$ in an infinitesimal interval of time Δt in the framework of the generalized Hamiltonian dynamics. One has [148],

$$F_{\Delta t} \approx F_0 + \dot{F}_0\Delta t = F_0 + \{F, H_E\}\Delta t = F_0 + (\{F,H\} + u_\alpha\{F,\phi_\alpha\})\,\Delta t, \quad (3.65)$$

where H is the Hamiltonian of the system, $F_0 = F_{t=0}$, u_α are arbitrary functions of time, and ϕ_α are the first-class constraints (both primary and secondary ones). The functions u_α are arbitrary and may be changed at will to other functions $u'_\alpha = u_\alpha + \delta u_\alpha$. Then the quantity (3.65) changes as

$$\delta F_{\Delta t} = -\delta u_\alpha\{\phi_\alpha, F\}\Delta t \equiv -\Delta t\,\delta u_\alpha \hat{\phi}_\alpha F, \quad (3.66)$$

where the operators $\hat{\phi}_\alpha$ are defined in (1.79). Since ϕ_α are in involution, it follows that

$$[\hat{\phi}_\alpha, \hat{\phi}_\beta] \approx c^\gamma_{\alpha\beta}\hat{\phi}_\gamma. \quad (3.67)$$

If $c^\gamma_{\alpha\beta}$ do not depend on the canonical variables, the operators $\hat{\phi}_\alpha$ are generators of a Lie algebra, and the commutation relation holds in the strong sense otherwise the algebra (3.67) is called *open*. Equation (3.66) shows that the first-class constraints are generators of transformations induced by variations of arbitrary functions of time that determine the time evolution of unphysical degrees of freedom in the theory (compare (3.29)). As the variations $u_\alpha \to u'_\alpha$ cannot change the physical state of the system, the transformations generated by the first-class constraints are gauge transformations. The gauge transformations form a group as its generators fulfill (3.67).

The evolution of physical quantities cannot depend on choice of u_α. Thus any physical quantity should satisfy the condition

$$\delta F_{\Delta t} \approx 0,$$

which means that F must be a gauge-invariant function on the phase space,

$$\hat{\phi}_\alpha F = \{\phi_\alpha, F\} \approx 0 \ , \quad \alpha = 1, 2, ..., M. \tag{3.68}$$

In order to obtain solutions of Eq. (3.68) which are non-trivial in the weak sense, i.e. $F \not\approx 0$, one has to find a canonical transformation such that the constraints become non-singular linear combinations of M new independent canonical momenta [152]. Solutions to (3.68) are then given by functions independent of these momenta and M coordinates canonically conjugated to them.

Thus, unlike theories with second-class constraints, the physical phase space for first-class constrained systems does not coincide with the surface $\phi_\alpha = 0$. There are gauge transformations acting in the surface of constraints, which are generated by the operators $\hat{\phi}_\alpha$. The physical phase space is therefore the surface of constraints modulo the gauge transformations. The geometry of the physical phase space is studied in Chapter 5. It must be emphasized that *all* the first-class constraints (*primary and secondary*) generate gauge transformations. So the algebra (3.67) may be larger than the gauge symmetry algebra of the Lagrangian. Transformations generated by all first-class constraints are called *extended gauge transformations* [148]. The notion of the extended gauge group is further elaborated with the following example.

The Lagrangian (3.38) is invariant under the gauge transformations (3.43) which are rotations of the position vector $\mathbf{x}$ about the origin. The orthogonal group $G = SO(3)$ has three generators. But the algebra of the first-class constraints (3.41) contains six linearly independent generators. The operators $\hat{M}_i$ generate $SO(3)$-rotations of $\mathbf{x}$ and $\mathbf{p}$, while leaving the other canonical variables untouched, $\hat{M}_i \pi_j = \hat{M}_i y_j = 0$. The operators $\hat{\pi}_i$ commute with each other and generate only translations of y_i. Evidently, the operators $\hat{\boldsymbol{\pi}}$ and $\hat{\mathbf{M}}$ are generators of a Lie algebra that is larger than $SO(3)$. The Dirac extended gauge group, $\tilde{G}$, is the direct product of the group of three-dimensional translations of the vector $\mathbf{y}$ and the group of $SO(3)$-rotations of vectors $\mathbf{x}$ and $\mathbf{p}$,

$$\tilde{G} = T_3(\mathbf{y}) \times SO(3)(\mathbf{x}, \mathbf{p}).$$

Equations (3.68) for the model (3.38) are decoupled into two sets,

$$\hat{\boldsymbol{\pi}} F(\mathbf{y}, \boldsymbol{\pi}, \mathbf{x}, \mathbf{p}) = \frac{\partial}{\partial \mathbf{y}} F \approx 0, \tag{3.69}$$

$$\hat{\mathbf{M}}F(\mathbf{y},\boldsymbol{\pi},\mathbf{x},\mathbf{p}) \approx 0. \tag{3.70}$$

It follows from Eq. (3.69) that physical quantities do not depend on $\mathbf{y}$, and, because of the constraint $\boldsymbol{\pi} = 0$, they are also independent of $\boldsymbol{\pi}$. Thus, the variables $\boldsymbol{\pi}$ and $\mathbf{y}$ are unphysical. Because $\hat{\mathbf{M}}$ are generators of the group $SO(3)$, solutions to (3.70) are given by functions of $SO(3)$-invariants that can be composed of the vectors $\mathbf{p}$ and $\mathbf{x}$. If $F = F(\mathbf{x},\mathbf{p})$ is an analytic function, then F can only depend on three combinations $\mathbf{p}^2$, $\mathbf{x}^2$, and $(\mathbf{x},\mathbf{p})$ [153]. On the surface of constraints, $\mathbf{M} = 0$, the vector $\mathbf{p}$ is proportional to $\mathbf{x}$ because there are only two independent constraints (see (3.44) and what follows). Hence, the invariants are not independent. In fact, only two of them are independent. To see this, the theory is formulated in new canonical variables associated with spherical coordinates in the configuration space, $x_1 = r\sin\theta\cos\varphi$, $x_2 = r\sin\theta\sin\varphi$, and $x_3 = r\cos\theta$. Making use of (1.43) new canonical momenta are found,

$$\begin{aligned}
p_r &= \frac{(\mathbf{x},\mathbf{p})}{r}, \quad r = |\mathbf{x}|,\\
p_\theta &= \sin\varphi M_1 - \cos\varphi M_2 \approx 0,\\
p_\varphi &= -M_3 \approx 0,
\end{aligned}$$

that are conjugated to r, θ, and φ, respectively. The invariants $\mathbf{x}^2$ and $(\mathbf{x},\mathbf{p})$ are functions of r and p_r. Finally, the remaining invariant on the constraint surface is

$$\mathbf{p}^2 = p_r^2 + \frac{1}{r^2}\left(p_\theta^2 + \frac{1}{\sin^2\theta}p_\varphi^2\right) \approx p_r^2 \,.$$

Thus, p_r and r can be identified with the physical canonical variables. A general solution of Eqs. (3.69) and (3.70) reduced on the surface of constraints reads $F = F(p_r, r)$. Alternatively, one could use new canonical variables associated with the spherical coordinates in the momentum space. Of course, the physical canonical variables resulting from this approach are related to p_r and r by a canonical transformation in the physical phase space.

The model has *five* unphysical degrees of freedom $(\mathbf{y}, \theta, \psi)$, although there are *six* first-class constraints. The reason is that the vector $\mathbf{x}$ has a stationary group $SO(2)$ (rotations about the vector $\mathbf{x}$ itself), which is a subgroup of the gauge group $SO(3)$. Thus, among the three constraints $\mathbf{M}$, only two are independent.

3.3.2 Elimination of unphysical variables in second-class constrained systems. The Dirac bracket

Mechanical systems with second-class constraints do not allow for a generalization of dynamics. There is no gauge arbitrariness, and the physical phase space is fully defined by the constraints themselves. However, the constraints cannot be solved *before* evaluating the Poisson brackets in the Hamiltonian equations of motion, for it would lead to either a contradiction or a modification of the original

(Lagrangian) dynamics. Indeed, consider the example (3.51). In Section 3.2.2 it is shown that the Lagrangian equations of motion have the solution (see the text between (3.57) and (3.58)),

$$x_1 = R\cos(\nu t + \varphi_0)\ ,\quad x_2 = R\sin(\nu t + \varphi_0)\ ,\quad x_3 = \nu^2 = \text{const.} \tag{3.71}$$

On the other hand, using the constraints $\phi_\alpha = 0$, the variables p_3, x_3, p_1 and x_2 can be eliminated from the Hamiltonian (3.51), thus selecting x_1 and p_2 as the physical variables. Then the Hamiltonian in the physical phase space has the form,

$$H_{ph} = \frac{R^2}{2x_1^2}\, p_2^2\ . \tag{3.72}$$

The corresponding Hamiltonian equations of motion are trivial,

$$\dot{p}_2 = \{p_2, H_{ph}\} = 0\,,\quad \dot{x}_1 = \{x_1, H_{ph}\} = 0$$

because $\{p_2, x_1\} = 0$. The latter seems to imply that there is no symplectic structure on the physical phase space! Not to mention that a general solution of these equations of motion obviously does not coincide with (3.71).

This "contradiction" between the Hamiltonian and Lagrangian formalisms is resolved by means of the *Dirac bracket* [148]. Let $\{\phi_\alpha, \phi_\beta\}^{-1}$ denote the inverse to the matrix $\{\phi_\alpha, \phi_\beta\}$, where ϕ_α are independent second-class constraints. The Dirac bracket is defined by

$$\{F_1, F_2\}_D = \{F_1, F_2\} - \{F_1, \phi_\alpha\}\{\phi_\alpha, \phi_\beta\}^{-1}\{\phi_\beta, F_2\}. \tag{3.73}$$

It is easy to show that the Dirac bracket is skew-symmetric and satisfies the Leibnitz rule and Jacobi identity [148]. Hence, it defines a symplectic structure on the phase space. The Dirac bracket does not change the equations of motion,

$$\dot{F} = \{F, H\}_D = \{F, H\} - \{F, \phi_\alpha\}\{\phi_\alpha, \phi_\beta\}^{-1}\{\phi_\beta, H\} \approx \{F, H\},$$

because all the terms proportional to $\{\phi_\alpha, H\}$ are equal to zero in the weak sense. The generalized Hamiltonian $H_E = H + u_\alpha\phi_\alpha$ does not lead to any new dynamics as the Dirac bracket of any function of canonical variables and any second-class constraint vanishes,

$$\{F, \phi_\alpha\}_D = \{F, \phi_\alpha\} - \{F, \phi_\gamma\}\{\phi_\gamma, \phi_\beta\}^{-1}\{\phi_\beta, \phi_\alpha\} = 0.$$

Thus, if the Dirac bracket is used in place of the Poisson bracket, the constraints $\phi_\alpha = 0$ can be solved *before* calculating the Dirac bracket in the equations of motion, and the aforementioned contradiction does not occur.

For the example (3.51), one finds,

$$\{\phi_\alpha, \phi_\beta\}^{-1} = \frac{1}{R^2}\begin{pmatrix} 0 & 0 & 0 & -1 \\ 0 & 0 & -1 & 0 \\ 0 & 1 & 0 & 0 \\ 1 & 0 & 0 & 0 \end{pmatrix}.$$

Then, taking $p_2 \equiv p$ and $x_1 \equiv q$ as the physical variables, the physical Hamiltonian coincides with (3.72), and the symplectic structure on the physical phase space is given by,

$$\{q,p\}_D = \frac{1}{R^2} q\sqrt{R^2 - q^2}. \tag{3.74}$$

Hamiltonian equations of motion with the symplectic structure (3.74) are equivalent to the Lagrangian equations of motion in the model (3.2) with $L = \dot{\mathbf{x}}^2/2$.

It is easy to verify that the equalities

$$q = R\cos\varphi\ , \quad p = -\frac{1}{R}p_\varphi \cos\varphi\ , \quad \{\varphi, p_\varphi\} = 1$$

define the Darboux variables for the symplectic structure (3.74). Here $\varphi \in [0, 2\pi)$ and p_φ ranges over the whole real line. The symplectic structure defined by the Dirac bracket depends on the choice of local coordinates in the physical phase space. For example, the choice $p_1 \equiv p'$ and $x_1 \equiv q'$ as the physical variables leads to

$$\{q', p'\}_D = 1 - \frac{q'^2}{R^2}\ , \quad H_{ph} = \frac{R^2 p'^2}{2(R^2 - q'^2)}. \tag{3.75}$$

The Darboux variables in this case are

$$q' = R\cos\varphi\ , \quad p' = -\frac{1}{R}p_\varphi \sin\varphi. \tag{3.76}$$

However, the corresponding symplectic two-form is invariant under the choice of local coordinates in the physical phase space (the constraint's surface),

$$\{q,p\}_D^{-1} dq \wedge dp = \{q'p'\}_D^{-1} dq' \wedge dp', \tag{3.77}$$

where

$$q' = q, \quad p' = \frac{p}{q}\sqrt{R^2 - q^2}.$$

So, the elimination of non-physical variables in systems with second-class constraints generally leads to Hamiltonian dynamics with a non-standard symplectic structure. This structure generates an invariant, non-degenerate, and closed two-form on the physical phase space defined by the constraints. The Poisson bracket and the Hamiltonian depend on the choice of physical variables (the choice of local coordinates on the surface of second-class constraints). Different coordinates in the physical phase space are related by transformations that leave the symplectic two-form invariant.

Another important example of a theory with second-class constraints is provided by the so-called first-order systems whose Lagrangian is linear in velocities.

3.3.3 The first-order formalism and Hamiltonian mechanics

In theories with Fermi fields, a Lagrangian is generally linear in velocities. To elucidate peculiarities of the Hamiltonian formalism for such theories [26], consider a model with the Lagrangian

$$L = \frac{1}{2}\alpha^{ij}\xi_i\dot{\xi}_j - V(\xi)\ , \quad i,j = 1,2,\ldots,2n, \quad \alpha^{ij} = -\alpha^{ji}. \tag{3.78}$$

Without loss of generality, the discussion can be limited to the simplest case of $n = 1$ and $\alpha^{12} = -1$. Then, when going over to the Hamiltonian formalism $\pi_1 = \partial L/\partial\dot{\xi}_1 = \xi_2/2$, $\pi_2 = \partial L/\partial\dot{\xi}_2 = -\xi_1/2$, two primary second-class constraints occur: $\phi_1 = \pi_2 + \xi_1/2 = 0$, $\phi_2 = \pi_1 - \xi_2/2 = 0$, $\{\phi_1, \phi_2\} = 1$. Apparently, ϕ_1 and ϕ_2 are just a pair of canonically conjugated variables. Another pair reads $\xi = -(\pi_2 - \xi_1/2)$ and $\pi = (\pi_1 + \xi_2/2)$ so that $\{\xi, \phi_i\} = \{\pi, \phi_i\} = 0$.

The constraints can be solved, i.e. the non-physical variables can be eliminated from the theory, if the Poisson bracket $\{\ ,\ \}$ is simultaneously replaced by the Dirac one (see Section 3.3.2),

$$\{f,g\}_D = \{f,g\} - \{f,\phi_i\}\{\phi_i,\phi_j\}^{-1}\{\phi_j,g\}.$$

The weak equalities holds $\xi \approx \xi_1$, $\pi \approx \xi_2$ and $\{\xi_1, \xi_2\}_D = 1$, through which the Hamiltonian is found,

$$H(\xi,\pi) = \pi_1\dot{\xi}_1 + \pi_2\dot{\xi}_2 - \frac{1}{2}(-\xi_1\dot{\xi}_2 + \xi_2\dot{\xi}_1) + V(\xi_1,\xi_2) \approx V(\xi_1,\xi_2). \tag{3.79}$$

The action $A = S_H = \int L dt$ for the Lagrangian (3.78) is nothing but the action in the Hamiltonian form where V is the Hamiltonian. Thus, ξ_1 and ξ_2 are canonically conjugated variables. The conclusion is easily extended to any integer n if one takes into account that there always exists a linear transformation of ξ by which the matrix α can be transformed to the standard block-diagonal form with the blocks equal to $\tau \equiv -i\tau_2$, where τ_2 is the Pauli matrix. Thus, the first-order formalism is in fact the Hamiltonian formalism. If the matrix α is brought to the standard block-diagonal form, then $\xi_{2i-1} = q_i$, $\xi_{2i} = p_i$, $i = 1,2,\ldots,n$, $V(\xi) = H(p,q)$, and the action $A = \int L(\xi,\dot{\xi})dt$ is of the Hamiltonian form,

$$A = S_H = \int dt\left[\frac{1}{2}(p_i\dot{q}_i - q_i\dot{p}_i) - H(p,q)\right].$$

The equations of motion in the theory with the Lagrangian (3.78) are identical to the Hamiltonian equations of motion $\dot{q}_i = \partial H/\partial p_i$, $\dot{p}_i = -\partial H/\partial q_i$.

3.4 Nonlinear Poisson brackets and systems with constraints

3.4.1 Dynamics with a nontrivial symplectic structure and systems with second-class constraints

It is shown in Section 3.3.2 that the elimination of unphysical variables in theories with second-class constraints by means of the Dirac bracket generally leads to a Hamiltonian formalism with a non-standard symplectic structure on the physical phase space (a nonlinear Poisson bracket). It is natural to investigate the inverse problem: Given a symplectic structure, find a Hamiltonian dynamics with second-class constraints in an extended phase space equipped with the standard symplectic structure such that the symplectic structure induced by the Dirac bracket on the physical phase space coincides with the given one [29].

Consider a Hamiltonian mechanics with a non-standard symplectic structure (see Section 1.9),

$$\{\theta^a, \theta^b\}_\theta = \omega^{ab}(\theta), \quad a = 1, 2, \dots 2N. \tag{3.80}$$

Here for the sake of simplicity a pure bosonic system is studied, i.e. $\epsilon^a = 0$ and the action is given in (1.148). The phase space is extended by adding new variables π_a while the old variables θ^a are viewed now as generalized coordinates canonically conjugated to π_a. In other words, the standard symplectic structure is defined on the extended phase space,

$$\{\theta^a, \theta^b\} = \{\pi_a, \pi_b\} = 0, \quad \{\theta^a, \pi_b\} = \delta^a_b. \tag{3.81}$$

The theory in the extended phase space is postulated to have the constraints

$$\varphi_a(\theta, \pi) = \pi_a + \bar{\omega}_{ab}(\theta)\theta^b = 0, \tag{3.82}$$

where the matrix $\bar{\omega}_{ab}$ is defined in (1.146). It is then proved that equalities (3.82) are second-class constraints, and the corresponding Dirac bracket reproduces the symplectic structure (3.80) on the constraint surface $\varphi_a = 0$, i.e. the extended theory is equivalent to the original theory.

Indeed, using the identity (1.141) for the matrix $\bar{\omega}_{ab}$ and equality (1.147), it is easy to verify that

$$\Delta_{ab} = \{\varphi_a, \varphi_b\} = \omega_{ab}(\theta). \tag{3.83}$$

The determinant of this matrix is not zero by the assumption that the symplectic structure (3.80) is not degenerate. So the constraints (3.82) are of the second class. The Dirac bracket of θ^a and θ^b on the constraint surface (3.82) reads,

$$\{\theta^a, \theta^b\}_D = -\{\theta^a, \varphi_c\}\Delta^{cd}\{\varphi_d, \theta^b\} = \omega^{ab}(\theta) = \{\theta^a, \theta^b\}_\theta. \tag{3.84}$$

Thus, according to Eqs. (3.83) and (3.84), the system with the action (1.148) is equivalent to the dynamical system with the action

$$S_\varphi = \int dt \left[\pi_a \dot{\theta}^a - H(\theta) - \lambda^a \varphi_a(\theta, \pi)\right], \tag{3.85}$$

where λ^a are the Lagrange multipliers. It is instructive to prove this statement using the equations of motion. The stationary action principle yields the equations,

$$\delta S_\varphi / \delta\lambda^a = \varphi_a = 0, \tag{3.86}$$

$$\delta S_\varphi / \delta\pi_a = \dot{\theta}^a - \lambda^b \partial\varphi_b / \partial\pi_a = \dot{\theta}^a - \lambda^a = 0, \tag{3.87}$$

$$\delta S_\varphi / \delta\theta^a = -\dot{\pi}_a - \partial_a H - \lambda^b \partial_a \varphi_b = 0. \tag{3.88}$$

Differentiation of the constraints (3.82) with respect to time furnishes the relation $\dot{\pi}_a = -\bar{\omega}_{ab}\dot{\theta}^b - \theta^c \partial_b \bar{\omega}_{ac} \dot{\theta}^b$. Substituting it and (3.87) into (3.88) and then making use of identities (1.141) and (1.147) for the matrix $\bar{\omega}_{ab}$, the Hamiltonian equations of motion are obtained for the original dynamical system with the symplectic structure (3.80),

$$\omega_{ab}\dot{\theta}^b = \partial_a H. \tag{3.89}$$

Note that due to Eqs. (3.86) and (3.87) only the variables θ^a are independent.

3.4.2 Abelian conversion of the second-class constraints

Suppose that the phase space of the system (3.85) is extended even further by adding a new set of canonical variables ϕ^a for which the canonical symplectic structure is postulated,

$$\{\phi^a, \phi^b\}_\phi = \overset{\circ}{\omega}{}^{ab}, \quad \{\phi^a, \theta^b\}_\phi = \{\phi^a, \pi_b\}_\phi = 0, \tag{3.90}$$

while the Poisson brackets for the variables θ^a and π_b is given by (3.81). Here the index ϕ is used to denote the Poisson bracket on the doubly extended phase space. With the second-class constraints (3.82) a set of first-class constraints $\sigma_a(\pi, \theta, \phi) = 0$ is associated such that

$$\{\sigma_a, \sigma_b\}_\phi = 0, \tag{3.91}$$

and

$$\sigma_a(\pi, \theta, \phi = 0) = \varphi_a(\pi, \theta). \tag{3.92}$$

Equations (3.91) form a system of nonlinear first-order differential equations for the functions σ_a with the initial conditions (3.92). Thanks to the regularity of the functions (3.82) in the phase space, the initial value problem (3.91) and (3.92) has a unique solution in the neighborhood of $\phi^a = 0$ in the extended phase space. This solution defines the first-class constraints $\sigma_a = 0$.

The procedure of converting second-class constraints into first-class ones by extending the phase space is called an *Abelian conversion* of second-class constraints. The conversion is called *non-Abelian* if the right-hand side of (3.91) is a linear combination of the constraints σ_a. The conversion formalism was introduced in [154] and further developed in [29, 155, 156].

A solution of (3.91) may be sought in the form [157],

$$\sigma_a(\pi,\theta,\phi) = \varphi_a(\pi,\theta) + K_a(\theta,\phi)\ , \quad K_a(\theta,\phi=0) = 0. \tag{3.93}$$

A substitution of (3.93) into (3.91) yields the following equation for the functions K_a:

$$\partial_a K_b - \partial_b K_a - \{K_a, K_b\}_\phi = \omega_{ab}(\theta). \tag{3.94}$$

Its solution can be found as a power series in ϕ^a [158]. An alternative approach is discussed in Section 3.4.3.

Consider the action

$$S_\sigma = \int dt \left[\pi_a \dot{\theta}^a + \frac{1}{2}\phi^a \overset{\circ}{\omega}_{ab}\dot{\phi}^b - \tilde{H}(\theta,\phi) - \lambda^a \sigma_a(\pi,\theta,\phi) \right], \tag{3.95}$$

where λ^a are the Lagrangian multipliers, and the new Hamiltonian is defined as a solution of the linear equation

$$\{\tilde{H}, \sigma_a\}_\phi = \frac{\partial \tilde{H}}{\partial \phi^c} \overset{\circ}{\omega}{}^{cd} \frac{\partial K_u}{\partial \phi^b} + \frac{\partial \tilde{H}}{\partial \theta^a} = 0 \tag{3.96}$$

with the initial condition

$$\tilde{H}(\theta, \phi = 0) = H(\theta). \tag{3.97}$$

The system with the action (3.95) is equivalent to the system with the action (1.148), i.e. the equations of motion for physical degrees of freedom coincide with (3.89). Indeed, the stationary action principle for (3.95) furnishes the equations of motion,

$$\dot{\pi}_a = \{\pi_a, \tilde{H}_E\}_\phi = -\partial\tilde{H}/\partial\theta^a - \lambda^b \partial\sigma_b/\partial\theta^a, \tag{3.98}$$

$$\dot{\theta}^a = \{\theta^a, \tilde{H}_E\}_\phi = \lambda^a, \tag{3.99}$$

$$\dot{\phi}^a = \{\phi^a, \tilde{H}_E\}_\phi = \overset{\circ}{\omega}{}^{ab}\partial\tilde{H}/\partial\phi^b + \lambda^b \overset{\circ}{\omega}{}^{ac}\partial\sigma_b/\partial\phi^c, \tag{3.100}$$

$$\sigma_a = 0, \tag{3.101}$$

where $\tilde{H}_E = \tilde{H} + \lambda^a \sigma_a$. Because of (3.91) and the definition (3.96), equations (3.101) are the first-class constraints. They generate gauge transformations,

$$q \to q + \delta q\ , \quad \delta q = \varepsilon^a \hat{\sigma}_a q \equiv \varepsilon^a \{\sigma_a, q\}_\phi\ , \quad \lambda^a \to \lambda^a - \dot{\varepsilon}^a. \tag{3.102}$$

Here q runs over the set (π, θ, ϕ) and $\varepsilon^a = \varepsilon^a(t)$ are arbitrary infinitesimal functions of time. With the Poisson bracket calculated in (3.102), one finds

$$\delta\pi_a = \varepsilon^b \partial\sigma_b/\partial\theta^a = \varepsilon^b \partial K_b/\partial\theta^a\ , \quad \delta\theta^a = -\varepsilon^b\partial\sigma_b/\partial\pi_a = -\varepsilon^a, \tag{3.103}$$

$$\delta\phi^a = \varepsilon^c \partial\sigma_c/\partial\phi^b \overset{\circ}{\omega}{}^{ba} = \varepsilon^c \partial K_c/\partial\phi^b \overset{\circ}{\omega}{}^{ba}. \tag{3.104}$$

The action (3.95) and equations of motion (3.98)–(3.101) are invariant under the transformations (3.102)–(3.104).

To eliminate the gauge arbitrariness, i.e. to go over to the dynamics in the physical phase space, one has to fix arbitrary functions λ^b (to fix a gauge) and to solve the constraints (3.101). Set λ^b so that $\dot{\phi}^a = 0$, and put $\phi^a(t=0) = 0$, which corresponds to the gauge condition $\phi^a = 0$.[4] Using equations (3.101) the momenta π_a can be excluded from the equations of motion. It follows from (3.96) and (3.97) that

$$\left.\frac{\partial \tilde{H}}{\partial \phi^c} \overset{\circ}{\omega}{}^{cb} \frac{\partial K_a}{\partial \phi^b}\right|_{\phi=0} = -\left.\frac{\partial \tilde{H}}{\partial \theta^a}\right|_{\phi=0} = -\frac{\partial H}{\partial \theta^a}. \tag{3.105}$$

Since $K_a(\theta, \phi = 0) = 0$ for all values of θ^a, $\partial K_a/\partial\theta^b|_{\phi=0} = 0$. The latter relation is then used (3.94) to infer that

$$\{K_a, K_b\}|_{\phi=0} = \left.\frac{\partial K_a}{\partial \phi^{a'}} \overset{\circ}{\omega}{}^{a'b'} \frac{\partial K_b}{\partial \phi^{b'}}\right|_{\phi=0} = -\omega_{ab}(\theta). \tag{3.106}$$

Equation (3.100) is solved for λ^b under the gauge condition $\dot{\phi}^a = 0$. The solution is substituted into (3.98) and (3.99). By differentiating the constraint (3.101) with respect to time, an equation for $\dot{\pi}_a$ as a function of θ^a and $\dot{\theta}^a$ ($\phi^a = 0$) is obtained. Solving this, substituting $\dot{\pi}_a$ into (3.98), and using the identities (3.105) and (3.106), Eqs. (3.89) are derived. This completes the proof of the equivalence of the two theories. There also exists an alternative proof based on a gauge-invariant approach, discussed in Section 3.4.3.

Thus, the systems with the actions (1.148), (3.85), and (3.95) have identical dynamics in the physical phase space. The dynamical equivalence of systems with non-standard symplectic structures and systems with first- (or second-) class constraints is used in quantization [29], [154–158]. The canonical "quantization" rule, $[,] = i\hbar\{,\}$ (with the subsequent replacement θ^a by operators $\hat{\theta}^a$), may lead to the breaking of the associativity of the operator algebra of $\hat{\theta}^a$, when formally applied to the Poisson bracket (3.80), meaning that the Jacobi identity may no longer hold in the resulting quantum theory. A nontrivial problem here is to find an operator ordering in $\omega^{ab}(\hat{\theta})$ which would guarantee the Jacobi identity after the quantization. In the conversion formalism, the quantization problem is reduced to the canonical quantization of the constrained system (3.85), while the operator ordering problem does not occur in the constraints (3.82) (compare (3.81)). The operators of the second-class constraints (3.82) are then converted into the corresponding first-class constraint operators by adding extra canonical operators $\hat{\phi}_a$ [157–159]. All the relations (3.90)–(3.94), (3.96), and (3.97) have the

[4] A gauge-invariant approach is considered in Section 3.4.3.

same form in quantum theory if $\{\ ,\ \}_\phi$ is replaced by $(i\hbar)^{-1}[\ ,\ ]$. The problem of ordering of the operators $\hat{\theta}^a$ in $\omega^{ab}(\hat{\theta}$ is now equivalent to finding the operators $\hat{K}_a = \hat{K}_a(\hat{\theta}, \hat{\phi})$, where the operators $\hat{\phi}^a$ satisfy the canonical commutation relations. The operators $\hat{K}_a$ can be sought as a power series in $\hat{\phi}^a$ [157].

Remark. Generally, the program of quantization based on the conversion of constraints (generalized canonical quantization), is self-consistent only in a local approach, i.e. without taking into account the topology of the phase space [159]. Quantization has only been performed for a few systems with nonlinear Poisson brackets and a nontrivial topology of the phase space (see, e.g. [160]).

3.4.3 The Darboux variables and conversion of constraints

If the Darboux variables are known for the symplectic structure, then the Abelian conversion equations can be explicitly solved. A symplectic structure can locally be written in the form (see Section 1.11),

$$\omega_{ab} = \frac{\partial \zeta^{a'}}{\partial \theta^a} \overset{\circ}{\omega}_{a'b'} \frac{\partial \zeta^{b'}}{\partial \theta^b}, \tag{3.107}$$

where $\zeta^a = \zeta^a(\theta)$ are the Darboux variables, $\{\zeta^a, \zeta^b\}_\zeta = \overset{\circ}{\omega}{}^{ab} = \{\zeta^a(\theta), \zeta^b(\theta)\}_\theta$. Equation (3.107) is substituted into (3.94), and the latter is then multiplied by the matrices $\partial\theta^a/\partial\zeta^c$ from the left and by $\partial\theta^b/\partial\zeta^d$ from the right. By taking into account the equality $\partial^2\theta^a/\partial\zeta^b\partial\zeta^c = \partial^2\theta^a/\partial\zeta^c\,\partial\zeta^b$, Eq. (3.94) is transformed to the following form:

$$\partial\mathcal{K}_d/\partial\zeta^c - \partial\mathcal{K}_c/\partial\zeta^d - \{\mathcal{K}_c, \mathcal{K}_d\}_\phi = \overset{\circ}{\omega}_{cd}, \tag{3.108}$$

where $\mathcal{K}_a = K_b\partial\theta^b/\partial\zeta^a$. Note that (3.108) is invariant under canonical transformations of variables ϕ^a because the bracket $\{\ ,\ \}_\phi$ does not change under such transformations. It is invariant under canonical transformations of the variables ζ^a as well,

$$\zeta^a \to \zeta'^a(\zeta), \quad \{\zeta'^a, \zeta'^b\}_\zeta = \overset{\circ}{\omega}{}^{ab}, \quad \mathcal{K}_a \to \frac{\partial\zeta^b}{\partial\zeta'^a}\mathcal{K}_b, \tag{3.109}$$

because the Darboux variables are defined up to a general canonical transformation. Hence, any solution of (3.108) is defined up to two independent canonical transformations of the variables ζ^a and ϕ^a, respectively. Similarly, equation (3.94) is invariant under canonical transformations of the variables ϕ^a and any transformation $\theta^a \to \theta'^a(\theta)$ that preserves the symplectic structure, that is,

$$\{\theta'^a, \theta'^b\}_\theta = \omega^{ab}(\theta'), \quad K_a \to \frac{\partial\theta^b}{\partial\theta'^a}K_b.$$

The simplest solution of (3.108) is given by a linear function,

$$\mathcal{K}_a = \overset{\circ}{\omega}_{ab}\phi^b.$$

In this particular case, the Abelian constraints have the form

$$\sigma_a = \varphi_a(\theta,\pi) + \partial\zeta^b/\partial\theta^a \overset{\circ}{\omega}_{bc}\phi^c, \tag{3.110}$$

from which is not hard to find a solution of (3.96). Substituting (3.110) into (3.96) and carrying out some simple algebraic transformations, one infers that

$$\frac{\partial\tilde{H}}{\partial\zeta^a} - \frac{\partial\tilde{H}}{\partial\phi^a} = 0. \tag{3.111}$$

The solution of (3.111) satisfying the initial condition (3.97) reads,

$$\begin{aligned} &\tilde{H}(\theta,\phi) = H(\vartheta(\theta,\phi)), \\ &\vartheta^a(\theta,\phi) = \theta^a(\zeta(\theta)+\phi)\,, \quad \vartheta^a(\theta,\phi=0) = \theta^a, \end{aligned} \tag{3.112}$$

where the function $\theta^a = \theta^a(\zeta)$ defines the Darboux variables for the symplectic structure (3.80). Note that the functions (3.112) are only defined up to independent canonical transformations of variables ζ^a and ϕ^a. They are also gauge invariant because $\{\vartheta^a,\sigma_b\}_\phi = 0$. Hence, their evolution describes the evolution of physical degrees of freedom in gauge-invariant theory (3.95). By a straightforward calculation it can be verified that

$$\{\vartheta^a,\vartheta^b\}_\phi = \omega^{ab}(\vartheta).$$

Therefore,

$$\dot{\vartheta}^a = \{\vartheta^a,\tilde{H}_E\}_\phi = \omega^{ab}(\vartheta)\partial H(\vartheta)/\partial\vartheta^b,$$

which implies that the dynamics of gauge-invariant variables in the extended theory is equivalent to that of the initial theory with the action (1.148).

4
Quantization of constrained systems

4.1 The Dirac method

The problem of quantization of dynamical systems with constraints arose more then half a century ago in relation to the quantization problem of electromagnetic fields. Already in the 1920s E. Fermi [161, 162], W. Heisenberg, and W.Pauli [163] (details are reviewed in [164]) found a correct recipe for quantization of such systems. Later, when studying quantum theories of gravity and Yang–Mills fields, this problem became more significant. P. Dirac [165–167] (see also [168, 169]) developed a general approach to the problem, classified constraints, and proposed a general recipe for their quantization. Following Dirac, the problem of quantizing constrained systems attracted lots of attention. There now exists a vast literature devoted to the subject (see, e.g. [170–181]).

When developing a quantum theory for dynamical systems with constraints, the following questions arise:

(1) Is it possible to convert first-class constraints into operator equalities?
(2) Is it possible to eliminate non-physical variables before quantization in the case of first-class constraints?
(3) How does one treat second-class constraints?

These and other related questions will be considered in this chapter.

4.1.1 Systems with first-class constraints

In a sufficiently small neighborhood of any point in the physical phase space, independent first-class constraints can be viewed as some conditions on generalized momenta [152]. Clearly, they cannot be promoted into quantum operator equalities because this would violate the corresponding canonical commutation relations (in the equality $\hat{q}\hat{p} - \hat{p}\hat{q} = i$ one cannot set $\hat{p} = 0$). Thus, if a quantum theory is sought whose classical limit automatically reproduces the first-class constraints, one can only restrict the (Hilbert) space of state vectors by demanding that admissible vectors be annihilated by the operators of the constraints. In so doing, the above inconsistency with the commutation relations is avoided because the operators of first-class constraints are in involution. The corresponding subspace of the Hilbert space is called a physical subspace. Thus, the quantum first-class constraints $\hat{\phi}_a$ vanish when applied to vectors Φ from the physical Hilbert subspace,

$$\hat{\phi}_a \Phi = 0. \tag{4.1}$$

The founders of quantum mechanics clearly understood this [161–163]. Meanwhile some objections to this scheme have been offered (they are not often discussed, as a rule). One of the main arguments is as follows. If $p_a = 0$ is a constraint then physical state vectors Φ ($\hat{p}_a \Phi = 0$) are not normalizable, and, hence, they do not belong to the Hilbert space. It is not difficult to parry this argument. Let us distinguish between the constraints being generators of shifts in compact spaces (such as angular momentum $\mathbf{M}$, see (3.40) and Section 3.3) and non-compact ones (such as $\boldsymbol{\pi}$, see Section 3.3). It is clear that in the first case the problem of non-normalizability does not occur at all because the normalization integral is taken over a compact space and cannot diverge. The second case cannot be considered as a drawback or failure of the recipe, either. The occurrence of a constraint means that the corresponding variable is not physical. It belongs to a non-physical configuration space which is orthogonal to the physical one and, therefore, cannot influence any physical processes. Divergence of the normalization integral appears just when integrating over the non-physical variables. Thus, in the case of Cartesian coordinates, the latter may be ignored or, equivalently, the integration over them may be omitted. A mathematically sound renormalization procedure for physical states has also been developed (see, e.g. [182]). A quantization recipe alternative to the Dirac method [171], where non-physical degrees of freedom are eliminated before quantization, is considered in Section 4.1.4 (see also Section 8.4).

4.1.2 Systems with second-class constraints

In suitable local coordinates in the phase space, second-class constraints can be viewed as conditions on generalized canonically conjugated variables, i.e. as conditions like $q^\alpha = 0$, $p_\alpha = 0$. Of course, they cannot be promoted to the corresponding operator equalities. Moreover, they cannot be used as conditions on the state vectors – there are no such vectors satisfying the equalities $\hat{q}\Phi = \hat{p}\Phi = 0$ because the latter contradicts the commutation relations $[\hat{q}, \hat{p}] = i$.

Therefore one has the following possibilities:

(1) To eliminate non-physical degrees of freedom before quantization;
(2) To reformulate the Hamiltonian formalism (to change the classical Poisson brackets [148, 165]).

Practical applications show that the elimination of non-physical variables generally makes the problem even more complicated. The second way seems more appropriate (see also the remark at the end of this section). Its principal points are as follows.

Let ϕ_α be second-class constraints with nonzero Poisson brackets (even in the weak sense). The Poisson bracket is replaced by the Dirac bracket in the phase space [148, 165],

$$\{f, g\}_D = \{f, g\} - \{f, \phi_\alpha\}\{\phi_\alpha, \phi_\beta\}^{-1}\{\phi_\beta, g\}, \tag{4.2}$$

where $\{\phi_\alpha, \phi_\beta\}^{-1}$ stands for the inverse of the matrix $\{\phi_\alpha, \phi_\beta\}$, that is, $\{\phi_\alpha, \phi_\gamma\}\{\phi_\gamma, \phi_\beta\}^{-1} = \delta_\alpha^\beta$. The meaning of going over to the Dirac bracket is that the generalized canonically conjugated variables corresponding to the second-class constraints turn into independent variables that are no longer canonically conjugated. In other words, they become generalized coordinates (or momenta) because $\{\phi_\alpha, \phi_\beta\}_D = 0$. Their number is twice as many as the number of corresponding non-physical coordinates (or momenta). As a result, instead of $\{q, p\} = 1$ it can be postulated that $\{q, p\}_D = 0$, i.e. the variables $q = Q$, $p = Q'$ are declared as new generalized coordinates Q and Q'. No explicit introduction of variables canonically conjugated to them is necessary. In fact, the new canonical variables are introduced by means of a non-canonical transformation because the Poisson bracket is changed. The contradiction with the canonical commutation relation can be avoided altogether if the Dirac bracket is used in the quantization recipe (2.1) in place of the Poisson bracket, $\{,\} \to \{,\}_D$, because the canonically conjugated operators corresponding to the second-class constraints become commutative. In this approach, the condition $\hat{\phi}_\alpha \Phi = 0$ means that physical state vectors are independent of the non-physical variables. More accurately, the solution of the equation $q\Phi = 0$ is given by $\Phi = \delta(q)\Phi'$, where Φ' depends only on the remaining variables. In this sense both the approaches being discussed appear equivalent: Whether the non-physical variables corresponding to the second-class constraints are eliminated before quantization or ignored after quantization according to Eqs. (2.1), (4.1), and (4.2) (i.e. the states Φ' in the reduced configuration space are used), the result is the same. It should be noted, however, that the elimination of non-physical canonical variables can be carried out in many ways. Consequently, the physical Hamiltonian may acquire quantum corrections, associated with a particular procedure of eliminating the non-physical degrees of freedom, which disappear in the classical limit (they are proportional to the Planck constant). This is further discussed in Section 4.4.

Remark. It is possible to develop a quantization procedure, equivalent to the Dirac treatment of constraints, in which, in contrast to the Dirac method, the first- and second-class constraints are treated on the same footing [182]. The main advantage of this approach is that first- and second-class constraints are treated on the same footing. An explicit and universal form of the projection operator that projects the total Hilbert space onto the physical subspace can be established in the framework of the coherent state formalism [182]. In particular, the projection operator formalism offers a universal renormalization procedure for the norm of the Dirac states in the total Hilbert space even if gauge orbits in a first-class constrained system are not compact. A disadvantage of this approach is that the universal projection operator is rather cumbersome to use in practical applications.

4.1.3 First-class constraints. Can non-physical degrees of freedom be eliminated before quantization?

With regard to the aforementioned critique (non-normalizability of the physical states) of the Dirac recipe (4.1), it is often suggested that supplementary (gauge fixing) conditions are imposed on the canonical variables [165] (Dirac introduced them in the classical theory),

$$\chi_\alpha(q, p) = 0, \tag{4.3}$$

such that $\{\chi_\alpha, \phi_\beta\} \neq 0$ (even in the weak sense). The conditions $\chi_\alpha = 0$ together with the constraints $\phi_\alpha = 0$ are then viewed as second-class constraints. By solving $\phi = \chi = 0$ before quantization, all the non-physical degrees of freedom can be eliminated from the theory. In this case the problem of non-normalizability of state vectors Φ satisfying (4.1) is solved in a somewhat radical way – the non-physical variables are absent from the very beginning. This recipe stems from the rule for eliminating non-physical variables in the Hamiltonian path integral quantization formulated in [171] (see also [180, 181]). When applied to simple systems with Hamiltonians like (2.3) the formal Hamiltonian path integral method creates an illusion that the problem of non-commutativity of canonically conjugated quantities is totally avoided and one may proceed entirely in the framework of the classical Hamiltonian formalism [74, 171]. Naturally, in such an approach there is no difference between non-physical degrees of freedom associated with the first- and second-class constraints. Their elimination by means of substituting the δ-functions of the constraints and supplementary conditions into the path integral measure makes the theory equivalent to that obtained by quantizing after the elimination of non-physical degrees of freedom. Is such a recipe equivalent to the Dirac recipe? The answer is no. Generally speaking, quantization and elimination of non-physical degrees of freedom are not commutative operations [183]. The following example proves this statement [96, 183].

Consider a dynamical system with the Lagrangian

$$L(\mathbf{x}, \dot{\mathbf{x}}, y, \dot{y}) = \frac{1}{2}\left[\left(\frac{d}{dt} - yT\right)\mathbf{x}\right]^2; \tag{4.4}$$

here $\mathbf{x}$ is a two-dimensional vector, $\mathbf{x} = (x_1, x_2)$; $T = -i\tau_2$, τ_2 is the Pauli matrix, and $(T\mathbf{x})_i = T_{ij}x_j$. For $y \equiv 0$, it describes a "free" particle of unit mass in a two-dimensional Euclidean space (more details can be found in Chapter 5). The Lagrangian is invariant under the Abelian group of gauge transformations, $\delta y = \dot{\omega}$, $\delta \mathbf{x} = \omega T\mathbf{x}$, with $\omega(t)$ being an arbitrary infinitesimal function of time. There are two first-class constraints, $p_y = \partial L/\partial \dot{y} = 0$ and $\sigma = (\mathbf{p}, T\mathbf{x}) = x_1p_2 - x_2p_1 = 0$. The constraint σ generates rotations in the plane (x_1, x_2) (compare Section 3.3). In Cartesian coordinates the classical Hamiltonian is

$$H = \mathbf{p}^2/2 + y\sigma, \tag{4.5}$$

while in the polar coordinates, it reads,

$$H = \frac{1}{2}\left(p_r^2 + p_\phi^2/r^2\right) + yp_\phi, \quad p_\phi = \sigma. \tag{4.6}$$

After elimination of the non-physical variables ($y = p_y = p_\phi = \phi = 0$), a one-dimensional problem is obtained with the Hamiltonian,

$$H = p_r^2/2. \tag{4.7}$$

There are no difficulties in quantizing the theory if one sets aside the problem with both the range of variations of the physical variables (does r belong to half of the axis or to the entire axis?) and, consequently, the integration measure in the scalar product of the Hilbert space (dr or rdr?). In this approach answers to these questions are not evident at all. Note that there is no necessity to use polar coordinates to eliminate the non-physical degrees of freedom. Suppose the supplementary condition (4.3) is chosen so that $\chi = x_2$, $\{\chi, \sigma\} = x_1$. Then x_2 and p_2 can be eliminated to obtain $H = p_1^2/2$, where x_1 runs over the whole real axis and $||\psi||^2 = \int dx_1 |\psi(x_1)|^2$. Note that $\{\chi, \sigma\} = x_1 \neq 0$ only if $x_1 \neq 0$ so that the condition $\chi = x_2 = 0$ becomes inadmissible when $x_1 = 0$. This subtlety is discussed in detail at the end of Section 8.4. Ignoring it at this point, the canonical quantization of the reduced theory gives,

$$\hat{H} = -\frac{\hbar^2}{2}\frac{\partial^2}{\partial x_1^2}. \tag{4.8}$$

There is no integration over the non-physical variables (ϕ or x_2) in the normalization integrals of state vectors. The classical theory defined by the Hamiltonian (4.7) can also be quantized. Does thus obtained theory coincide with the one defined by (4.8)? Or, more generally, are theories corresponding to different supplementary conditions (unitary) equivalent?

The Dirac recipe (see [148, 165]) gives the result different from (4.8). Carrying out the canonical quantization for the Hamiltonian (4.5), we obtain (in the polar coordinates),

$$\hat{H} = -\frac{\hbar^2}{2}\left(\frac{\partial^2}{\partial r^2} + \frac{1}{r}\frac{\partial}{\partial r} + \frac{1}{r^2}\frac{\partial^2}{\partial \phi^2}\right) + \hat{y}\hat{p}_\phi, \quad \hat{p}_\phi = -ih\partial_\phi. \tag{4.9}$$

When restricted to the physical subspace composed of the vectors annihilated by the constraint, $\hat{p}_\phi \Phi = 0$, the Hamiltonian becomes,

$$\hat{H} = -\frac{\hbar^2}{2}\left(\frac{\partial^2}{\partial r^2} + \frac{1}{r}\frac{\partial}{\partial r}\right). \tag{4.10}$$

Clearly, the Hamiltonians (4.8) and (4.10) do not coincide. Besides, the operator (4.9) acts on functions normalized according to the equality

$$\int_0^\infty rdr \int_0^{2\pi} d\phi |\Psi|^2 = 1. \tag{4.11}$$

For physical functions independent of the angle ϕ ($\partial_\phi \Phi = 0$), the normalizing condition reads,

$$\int_0^\infty r dr |\Phi|^2 = 1. \tag{4.12}$$

The quantum descriptions obtained by the two recipes differ. The latter has been noted by many authors (besides [183] see also [175], [184–186]).

This discrepancy, when encountered, is often greeted with an air of surprise. As a point of fact, the reason for this phenomenon is quite simple, well documented in textbooks on quantum mechanics and, hence, should not be surprising at all. The discrepancy occurs only when non-physical variables are associated with curvilinear coordinates (in the case of non-physical variables described by Cartesian coordinates, like y and p_y, both approaches give an identical result). As noted by Dirac [65], the canonical quantization, $q, p \to \hat{q}, \hat{p}$ so that $[\hat{q}, \hat{p}] = i\hbar$ and $f(q,p) \to f(\hat{q}, \hat{p})$, works only in Cartesian coordinates. In curvilinear coordinates it generally fails. Therefore, an elimination of non-physical variables by means of suitable curvilinear coordinates with subsequent quantization implies the use of the canonical quantization recipe in curvilinear coordinates which is not generally consistent. When non-physical variables are eliminated prior to quantization the very information about the "curvilinear" nature of the remaining physical variables can be lost.

A priori both the approaches seems logically admissible when first-class constraints are present:

(1) One may first eliminate all the non-physical variables from the Hamiltonian and, then, quantize;
(2) One may quantize and use (4.1) to eliminate the non-physical variables.

In the framework of the abstract, pure mathematical approach (i.e. ignoring the physics), there seems to be no logical reason to prefer one of these methods to the other. But for dynamical systems that occur in nature the equivalence disappears. That the non-physical variables are invoked in gauge theories is not just a fancy way of formulating gauge-invariant Lagrangians – there is deep physics behind this. The power of the gauge principle is universally appreciated and accepted; as for its origin see [187]. That is why one should seriously accept the existence of non-physical variables and preserve the structures associated with them. And if one has to deal with non-physical variables and to take them into account in classical theory, undoubtedly they should be preserved in quantum theory because the former is a limiting case of the latter. But, if in quantum theory there are non-physical operators, then the only way to take them into account and to get rid of the corresponding unphysical variables is to restrict the total Hilbert space to the physical subspace where they disappear (compare (4.1)).

In favor of the Dirac method there are more prosaic arguments. A gauge symmetry may change the phase space of physical degrees of freedom (see Chapter 5).

This property of classical theory has serious physical consequences. For example, it leads to a doubling of the frequency of a harmonic oscillator. Meanwhile a simpleminded elimination of unphysical variables by solving the first-class constraints with the corresponding supplementary (gauge) conditions absolutely ignores this fact. And, most importantly, subsequent quantization of the (gauge-fixed) theory generally leads to the dependence of physical quantities on the supplementary condition $\chi = 0$, i.e. on the choice of the gauge (for more details see Section 5.3.2, the last paragraph of Section 5.3, and Section 8.4).

In summary:

(1) In the Dirac method, dynamical systems with first- and second-class constraints are quantized differently;
(2) For first- and second-class constraints the operations of eliminating the non-physical variables and quantization do not generally commute.

4.1.4 First-class constraints. The effective Hamiltonian

First-class constraints generate translations in the space of non-physical variables that are canonically conjugated to the constraints. So, the first-class constraints are generators of gauge transformations. The condition (4.1) guarantees that physical state vectors are independent of non-physical variables. Any physical operator $A(\hat{q}, \hat{p})$ should also be independent of non-physical variables, which implies that the following condition must hold:

$$[\hat{\phi}_\alpha, \hat{A}] = 0. \tag{4.13}$$

The condition (4.13) may be understood in the weak sense. In other words, the physical operators and state vectors must be invariant under the gauge transformations generated by all first-class constraints. The latter often form a Lie algebra (in the strong sense). Then the problem of finding physical operators is equivalent to the problem of classifying all invariants of the corresponding group (see also Section 3.3).

According to the above remarks, there are two possible approaches to study gauge systems: the invariant and non-invariant ones. In the invariant approach one uses only manifestly gauge-invariant variables, while the non-invariant approach allows for gauge non-invariant quantities to be employed. The latter must however be constructed so that it is equivalent to the invariant approach. Studies of various models show that, though being rather attractive at first sight, the invariant approach may lead to considerable technical complications even for relatively simple models (see Chapter 5). For the model (4.4) it amounts to using the canonical variables $r = (\mathbf{x}^2)^{1/2}$ and $p_r = (\mathbf{x}, \mathbf{p})/r$ that are the only independent invariants of the group of rotations in the plane (the gauge group $SO(2)$). But already for the adjoined representation of a generic simple gauge group, the classification problem of invariants and the corresponding calculations become extremely involved (see, e.g., Section 5.4.4). The non-invariant approach, in which the physical Hamiltonian and other physical quantities are expressed via some of the canonical variables that are not explicitly gauge invariant (such as, e.g. x_1 and p_1 in the model (4.4)), turns out to be more practical.

Let us show how to construct the effective (physical) Hamiltonian in the non-invariant approach for constraints linear in canonical momenta [96, 188]. Put $q^i = (q^a, q^\alpha)$, where q^a denote physical variables, while q^α stands for the non-physical ones, $i = 1, 2, \ldots, n$; $a = 1, 2, \ldots, n-s$; $\alpha = n-s+1, \ldots, n$. The Greek indices always numerate "non-invariant physical variables". Suppose that the constraints are linear in the canonical momenta,

$$\phi_\alpha(q,p) = \phi^i_\alpha(q)p_i = \phi^\beta_\alpha p_\beta + \phi^a_\alpha p_a. \tag{4.14}$$

The constraints (4.14) are also assumed to form the set of all independent constraints, and they have no operator ordering problem (the operator ordering problem in constraints is studied in Section 4.2). Any of the $n-s$ variables q^a varying under the gauge transformations generated by the constraints can be taken as non-invariant physical variables. The supplementary (gauge) conditions [165] are chosen in the form,

$$\chi^\alpha(q) \equiv (\phi^{-1})^\alpha_\beta q^\beta = 0, \quad \det \left\|\phi^\beta_\alpha\right\| \neq 0, \tag{4.15}$$

where $(\phi^{-1})^\alpha_\beta$ is the inverse of the matrix ϕ^β_α. Equality (4.15) defines a parameterization of the physical configuration space. If two functions f and g in the total configuration space are equal on the physical configuration space (i.e. on the subset defined by (4.15)), they are said to be equivalent; the equivalence relation is denoted as $f \simeq g$. This is to be contrasted with the weak equality of functions coinciding on the constraint surface in the total phase space (denoted by $\approx$). The Poisson bracket of χ^α and ϕ_β can then be written as,

$$\{\chi^\alpha, \phi_\beta\} \simeq \delta^\alpha_\beta\,. \tag{4.16}$$

The physical variables are also assumed to be *weakly stationary* or *weakly invariant* by which it is meant that

$$\{q^a, \phi_\alpha\} \simeq 0\,, \tag{4.17}$$

or

$$\phi^a_\alpha(q) \simeq 0. \tag{4.18}$$

Geometrically, the weak invariance implies that the coordinate curves of the physical and non-physical coordinates (set by (4.15)) are locally orthogonal. The effective (physical) Hamiltonian is obtained by eliminating all the non-physical variables from the total Hamiltonian only with the help of the condition (4.1). The total Hamiltonian is assumed to be bilinear in canonical momenta. The momentum operators are ordered in the standard way (by means of the commutation relations, any Hamiltonian polynomial in the momenta can be brought to the standard form),

$$\hat{H} = \frac{1}{2} g^{ij}(\hat{q})\hat{p}_i\hat{p}_j + \tilde{f}^j(\hat{q})\hat{p}_j + \tilde{V}_H(\hat{q}). \tag{4.19}$$

Making use of (4.1) and (4.14) the action of the non-physical momentum operators $\hat{p}_\alpha$ on physical state vectors is computed,

$$\hat{p}_\alpha \Phi = -(\phi^{-1})^\beta_\alpha \phi^a_\beta \hat{p}_a \Phi. \tag{4.20}$$

Relation (4.20) is applied to eliminate all the non-physical momenta $\hat{p}_j$ in the Hamiltonian (4.19) acting on a physical vector. In doing so, the right-hand side of Eq. (4.20) does not generally appear as a physical vector because ϕ^{-1} and ϕ^a_β may depend on the non-physical variables. Therefore, when computing the action of the Hamiltonian (4.19) on a physical vector, Eq. (4.20) cannot be used again to eliminate all the non-physical momenta in the second momentum $\hat{p}_i$ in the bilinear term of (4.19). The canonical commutation relations should be invoked to move $\hat{p}_i$ to the very right so that it acts on the physical vector Φ and only then Eq. (4.20) applies again. As a result the Hamiltonian acting on a physical vector can contain terms that depend on the unphysical variables q^α, but not on $\hat{p}_\alpha$. The dependence on q^α is eliminated by means of the supplementary conditions (4.15) and Eq. (4.18). Thus for the terms in the Hamiltonian bilinear in p_α one infers,

$$\hat{p}_\alpha \hat{p}_\beta \to -\hat{p}_\alpha (\phi^{-1})^\gamma_\beta \phi^a_\gamma \hat{p}_a = -(\phi^{-1})^\gamma_\beta \phi^a_\gamma \hat{p}_a \hat{p}_\alpha + i\hbar \frac{\partial}{\partial q^\alpha}\left[(\phi^{-1})^\gamma_\beta \phi^a_\gamma\right]\hat{p}_a. \tag{4.21}$$

The first term in the right-hand side of Eq. (4.21) as well as that of Eq. (4.20) vanishes thanks to (4.18) and (4.15), i.e. when $q^\alpha = 0$. A nontrivial contribution comes from the second term because the derivative $\partial_\alpha \phi^a_\gamma$ does not vanish in the limit $q^\alpha \to 0$. Thus, the operator (4.19) acts on physical state vectors Φ satisfying Eq. (4.1) as the following effective Hamiltonian:

$$\hat{H}_{\text{eff}} = \frac{1}{2} g^{ab}\hat{p}_a\hat{p}_b + \left\{\tilde{f}^a + \frac{i\hbar g^{\alpha\beta}}{2}\partial_\alpha\left[(\phi^{-1})^\gamma_\beta \phi^a_\gamma\right]\right\}\hat{p}_a + \tilde{V}_H. \tag{4.22}$$

There is a more convenient form of (4.22) which involves only the constraints (4.14) and supplementary conditions (4.15),

$$\hat{H}_{\text{eff}} = \left\{\hat{H} + \frac{i\hbar}{2} g^{ij}\partial_i\left(\chi^\gamma_{,j}\hat{\phi}_\gamma\right)\right\}_{\chi^\alpha = p_\beta = 0}. \tag{4.23}$$

Here the summation is carried out over the indices i and j instead of α and β as in the last term (4.22). The validity of this representation is easy to verify by using the identity

$$g^{ij}\partial_i\left(\chi^\gamma_{,j}\phi_\gamma\right) = g^{\alpha\beta}\partial_\alpha\left(\chi^\gamma_{,\beta}\phi_\gamma\right) + g^{a\beta}\partial_a\left(\chi^\gamma_{,\beta}\phi_\gamma\right) + g^{\alpha b}\partial_\alpha\left(\chi^\gamma_{,b}\phi_\gamma\right) + g^{ab}\partial_a\left(\chi^\gamma_{,b}\phi_\gamma\right) \tag{4.24}$$

and the equalities following from (4.14)–(4.18),

$$\chi^\gamma_{,\alpha} \simeq (\phi^{-1})^\gamma_\alpha, \quad \chi^\gamma_{,a} \simeq 0, \quad p_\alpha = 0.$$

All the terms in the right-hand side of Eq. (4.24), beside the first one, vanish on the physical phase space determined by the conditions $q^\alpha = p_\alpha = 0$.

The procedure of constructing the effective Hamiltonian (4.23) is illustrated with the model (4.4). The constraint and supplementary condition,

$$\phi = x_1p_2 - x_2p_1, \quad \chi = x_2/x_1,$$

are used to eliminate the variables x_2, p_2, and y. Applying (4.23) to the Hamiltonian H given in (4.5) we find,

$$\begin{aligned}\hat{H}_{\text{eff}} &= \left\{\frac{1}{2}\hat{\mathbf{p}}^2 + \frac{i\hbar}{2}\partial_j\left[(\hat{x}_2/\hat{x}_1)_{,j}(\hat{x}_1\hat{p}_2 - \hat{x}_2\hat{p}_1)\right] + \hat{y}\hat{\phi}\right\}_{y=x_2=p_2=0} \\ &= \frac{1}{2}\left(\hat{p}_1^2 - \frac{i\hbar}{x_1}\hat{p}_1\right) = -\frac{\hbar^2}{2}\left(\frac{\partial^2}{\partial x_1^2} + \frac{1}{x_1}\frac{\partial}{\partial x_1}\right).\end{aligned} \tag{4.25}$$

This is nothing but the Hamiltonian (4.10) in the "non-invariant" form. Furthermore the condition $\chi = \phi^{-1}x_2 = x_2/x_1$ makes sense only if $\phi = x_1 \neq 0$, that is, the physical range of x_1 should be restricted, e.g., to $x_1 > 0$. The Hamiltonian (4.25) is hermitian on the half-axis if the inner product contains the measure dx_1x_1 (compare with (4.12)). In this way the one-dimensional theory defined by the hermitian Hamiltonian (4.25) retains the information that the only physical variable of the theory is described by a curvilinear coordinate. This subtlety has an intrinsic relation with the fact that the physical phase space in this theory has a non-standard (conic) structure as shown in Chapter 5. Note that the Hamiltonians (4.22) and (4.23) are obtained in the form convenient for the (Hamiltonian) path integral presentation of the evolution operator kernel.

4.2 The operator ordering problem in constraints

4.2.1 Quantization of a system with zero Hamiltonian

In applications one often encounters dynamical systems with the vanishing canonical Hamiltonian (e.g. general relativity [148]), i.e. the Hamiltonian turns out to be a first-class constraint. In order to learn how to handle such theories, the operator ordering problem in constraints must be solved first. Any non-relativistic system with the Lagrangian

$$L = \frac{1}{2}g_{ij}(q)\dot{q}^i\dot{q}^j - f_i(q)\dot{q}^i - V(q) \tag{4.26}$$

can be transformed into a system with the vanishing Hamiltonian by treating the time t as an independent dynamical variable, while the role of time is played by an auxiliary independent parameter τ [148]. The corresponding new Lagrangian L^* and new velocities $\overset{\circ}{q}$ are furnished by the following expressions:

$$\int L(q,\dot{q})dt = \int L^*(q\,,\overset{\circ}{q})d\tau, \qquad \overset{\circ}{q}{}^i \equiv \frac{dq^i}{d\tau} = \dot{q}^i\frac{dt}{d\tau}; \qquad t \equiv q^0(\tau), \tag{4.27}$$

$$L^* = \left(\frac{1}{2}g_{ij}\,\overset{\circ}{q}{}^i\,\overset{\circ}{q}{}^j/\,\overset{\circ}{t}{}^2 - f_i\,\overset{\circ}{q}{}^j/\,\overset{\circ}{t} - V\right)\overset{\circ}{t}. \tag{4.28}$$

The new theory has one constraint. Indeed, the canonical momenta are

$$p_i = \frac{\partial L^*}{\partial\,\overset{\circ}{q}{}^i} = g_{ij}\,\overset{\circ}{q}{}^j/\,\overset{\circ}{t} - f_i, \qquad p_0 = \frac{\partial L^*}{\partial\,\overset{\circ}{t}} = -\frac{1}{2}g_{ij}\,\overset{\circ}{q}{}^i\,\overset{\circ}{q}{}^j/\,\overset{\circ}{t}{}^2 - V. \tag{4.29}$$

The first set of equations in (4.29) is solved for $\overset{\circ}{q}{}^j$,

$$\overset{\circ}{q}{}^j/\,\overset{\circ}{t} = g^{ij}(p_i + f_i)\,,$$

and the solution is substituted into the second one. The result reads,

$$\phi(q,p) \equiv p_0 + H(q,p) = 0, \tag{4.30}$$

$$H = \frac{1}{2}g^{ij}(p_i + f_i)(p_j + f_j) + V. \tag{4.31}$$

Equality (4.30) is a primary constraint. The corresponding canonical Hamiltonian

$$H^*(q,p) = p_i\,\overset{\circ}{q}{}^i + p_0\,\overset{\circ}{t} - L^* = \overset{\circ}{t}\left[p_0 + \frac{1}{2}g^{ij}(p_i + f_i)(p_j + f_j) + V\right] \approx 0 \tag{4.32}$$

is proportional to the constraint (4.30) and, hence, vanishes in the weak sense. When quantizing the system, the following two questions arise:

(1) How are the operators ordered in the constraint (4.30)?
(2) How are such systems treated?

The first question is fully investigated in the next section. The second is answered following the recipe proposed in [70].

The constraint (4.30) is a consequence of a gauge invariance associated with the time reparameterization symmetry,

$$\int L^*(q, dq/d\tau)d\tau = \int L^*(q, dq/d\tau')d\tau', \quad \tau' = f(\tau), \tag{4.33}$$

where $f(\tau)$ is an arbitrary function. The problem of the gauge arbitrariness is usually solved by fixing the gauge $t = \tau$, i.e. by going over to the familiar dynamics. But it is more convenient to preserve the gauge invariance and, following Dirac [148, 165], to describe the theory by the generalized Hamiltonian

$$\hat{H}_T = \hat{H}^* + v\hat{\phi}, \quad \phi = p_0 + H, \tag{4.34}$$

where v is an arbitrary function of τ. Here $H^* = 0$. It is also assumed that the operators in $\hat{\phi}$ are somehow ordered. Consider the infinitesimal evolution operator kernel ($\omega = \Delta\tau \to 0$),

$$U_\omega(q, q') = \left\langle q, q^0 \left| \exp\left[-\frac{i}{\hbar} v \left(\hat{p}_0 + \hat{H}\right) \omega\right] \right| q', q'^0 \right\rangle. \tag{4.35}$$

As $\hat{p}_0$ commutes with $\hat{H}$, the matrix element (4.35) is factorized,

$$U_\omega(q, q') = \left\langle q^0 \left| \exp\left[-\frac{i}{\hbar} v\omega\hat{p}_0\right] \right| q'^0 \right\rangle \left\langle q \left| \exp\left[-\frac{i}{\hbar} v\omega\hat{H}\right] \right| q' \right\rangle. \tag{4.36}$$

The first factor in (4.36) equals $\delta(q^0 - q'^0 - v\omega)$. Thus, the application of the kernel (4.35) to a wave function $\psi(q', q'^0)$ yields

$$\begin{aligned} \psi(q, q^0) &= \int dq' dq'^0 U_\omega(q, q') \psi(q', q'^0) \\ &= \int dq' \left\langle q \left| \exp\left[-\frac{i}{\hbar} v\omega\hat{H}\right] \right| q' \right\rangle \psi(q', q^0 - v\omega). \end{aligned} \tag{4.37}$$

With $q^0 \equiv t$ it follows that a correct evolution law has been established for wave functions (see (2.2)); evidently, $v = \overset{\circ}{t}$. In this approach there is no need to fix a gauge or choose supplementary conditions – the correct answer emerges automatically. The method can be applied to generic Hamiltonians, not quadratic in momenta. In Section 4.3 it is used to develop a quantum description of a relativistic particle.

4.2.2 The operator ordering in constraints

The rule proposed below to solve the operator ordering problem in constraints is based on the following arguments. The Hamiltonian generates shifts in time. The rules for ordering operators in Hamiltonians are formulated in Chapter 2. They are established by utilizing only the information contained in the classical theory. But the first-class constraints are also generators of shifts in the space of non-physical variables. Therefore it seems reasonable to attempt to extend the recipe of Section 2.8 to the present case, although some doubts might be expressed due to the fact that in the classical theory the constraints vanish. However in Section 4.2.1 it is argued that even in this case general recommendations due to Dirac (see [148, 165]) can still be applicable. A close examination of the above example of the theory with the vanishing Hamiltonian shows that all doubts can be dismissed and the correctness of the recipe is affirmed.

Indeed, a correct quantum description of the system with the Lagrangian (4.26) can be obtained by the operator ordering recipe developed in Chapter 2 and applied to the Hamiltonian (4.31). Its validity is also verified for Hamiltonians in curvilinear coordinates. On the other hand, the dynamical system with the constraint (4.30) is equivalent to that with the Lagrangian (4.26). That is why application of the recipe from Section 2.8 to the constraint (4.30) should give the result whose validity has already been established. Thus, the operator ordering recipe of Chapter 2 is applicable at least to constraints quadratic in momenta.

Example. The efficiency of the recipe is demonstrated with one more example [100]. For (4.31), consider the limiting case $g^{ij} \to 0$, $V \to 0$, and $g^{ij} f_j \to f^i \neq 0$. Then H can be identified with a generator $G(q, p) = p_i f^i(q)$ of point transformations of the canonical variables in classical theory. Following the procedure of Section 2.8 the kernel of the corresponding unitary operator is constructed. This example is interesting in two aspects. First, it is impossible to go over to the Lagrangian theory because the equation $\partial G/\partial p = \dot{q}$ cannot be solved for p to express it as a function of $\dot{q}$. Second, there is a clear criterion of validity of the recipe, which is the well-known formula for changing variables in the integral (or the symmetrization rule: $G(q,p) \to g(\hat{q}, \hat{p}) = (\hat{p}\hat{f} + \hat{f}\hat{p})/2$). In order to calculate the infinitesimal action S_H, defined in (2.284), one should first find $\dot{q}$ as a function of Δ. Note that, according to the remark at the end of Section 2.3, the main contribution to the path integral comes from those trajectories for which $\Delta \sim \epsilon$. For this reason, in the expansion

$$q(t) = q(0) + \dot{q}(0)t + \ddot{q}(0)t^2/2 + \cdots , \tag{4.38}$$

it is sufficient to retain only the terms explicitly given in (4.38) (unlike Eq. (2.259)). Setting $t = \epsilon$ in (4.38) and substituting for $\ddot{q}$ the result of differentiation of the first of the "equations of motion" with respect to time,

$$\dot{q}^j = \frac{\partial G}{\partial p_j} = f^j(q), \quad \dot{p}_i = -\frac{\partial G}{\partial q^i} = -f^k_{,i} p_k, \tag{4.39}$$

i.e. $\ddot{q}^j = f^j_{,k}(q')\dot{q}^k(0)$ (here, just as in Chapter 2, $q(0) = q'$), one infers

$$\dot{q}^j(0) \approx \frac{(q-q')^j}{\epsilon} - \frac{\epsilon}{2}\ddot{q}(0) = \frac{\Delta^j}{\epsilon} - \frac{\epsilon}{2} f^j_{,k}(q')\dot{q}^k(0), \tag{4.40}$$

which is the analog of Eq. (2.260). The solution of (4.40) for $\dot{q}(0)$ is

$$\dot{q}^j \approx \frac{1}{\epsilon}\left(\Delta^j - \frac{\epsilon}{2} f^j_{,k}(q')\Delta^k\right).$$

As pointed out earlier, in the case of $G = p_i f^i(q)$ the terms $O(\Delta^2)$ give contributions of order $O(\epsilon^2)$ which should be neglected. The two last terms in Eq. (2.286) turn out to be zero. For instance, for the second term, according to (4.39), the following chain of equalities holds,

$$d(p\dot{q})/dt = \dot{p}_i \dot{q}^i + p_i \ddot{q}^i = -f^k_{,i} p_k \dot{q}^i + p_i f^i_{,k} \dot{q}^k = 0.$$

Hence, the sought-for infinitesimal action reads,

$$S_H(q,p,q') = p_j\left(\Delta^j - \frac{\epsilon}{2}f^j_{,k}(q')\Delta^k\right) - \epsilon f^j(q')p_j.$$

The remaining problem is to calculate the determinant D_H:

$$\frac{\partial^2 S_H}{\partial q^i \partial q'^j} = \frac{\partial^2 S_H}{\partial p_k \partial p_l} = 0, \quad \frac{\partial^2 S_H}{\partial p_l \partial q^i} = \delta^l_i - \frac{\epsilon}{2}f^l_{,i}(q'),$$
$$\frac{\delta^2 S_H}{\partial p_k \partial q'^j} = -\left(\delta^k_j - \frac{\epsilon}{2}f^k_{,j} + \epsilon f^k_{,j}\right) + O(\epsilon\Delta),$$

and, therefore,

$$D_H = \det\left\|\delta^l_i - \frac{\epsilon}{2}f^l_{,i}\right\| \det\left\|\delta^k_j + \frac{\epsilon}{2}f^k_{,j}\right\| = 1 + O(\epsilon^2).$$

Now almost everything is in place to find the kernel $U_\epsilon(q,q')$. The only obstacle left is that, if $g^{ij} \to 0$, then $\det\|g_{ij}\| = g \to \infty$, whereas the final expression (2.289) contains the factor $(gg')^{-1/4}$ and the sought-for kernel seems to vanish. However, the point is that the action of the kernel (2.289) on a function $\psi(q')$ implies an integration with the measure $\sqrt{g'}dq'$, that is, the factors containing g and g' in front of the exponential and the measure are combined to yield the overall factor $(g'/g)^{1/4}$ which tends to 1 when $g \to \infty$. Hence, the factor $(gg')^{-1/4}$ in (2.289) should be omitted for the present case. The result is

$$U_\epsilon(q,q') \approx \int \frac{dp}{(2\pi\hbar)^n} \exp\left\{\frac{i}{\hbar}p_i\left[\Delta^i - \frac{\epsilon}{2}f^i_{,k}(q')\Delta^k - \epsilon f^j(q')\right]\right\}. \tag{4.41}$$

Let us change the base point q' to q in (4.41). Since $f^j(q') \approx f^j(q) - f^j_{,k}(q)\Delta^k$, it is necessary to change the sign in the second term in the exponential. Integrating then over p and using the following property of a multi-dimensional δ-function:

$$\delta^{(n)}(Ax) = \delta^{(n)}(x)/|\det A|,$$

the kernel (4.41) is expressed in the form,

$$U_\epsilon(q,q') = \delta^{(n)}\left(\Delta + \frac{\epsilon}{2}f_{,k}(q)\Delta^k - \epsilon f(q)\right) \approx \frac{\delta^{(n)}(\Delta - \epsilon f(q))}{\det\left\|\delta^i_j + \frac{\epsilon}{2}f^i_{,j}(q)\right\|}$$
$$\approx \delta^{(n)}(\Delta - \epsilon f)\left(1 - \frac{\epsilon}{2}f^j_{,j}\right),$$

from which it follows that

$$\psi_\epsilon(q) = \int dq' U_\epsilon(q,q')\psi_0(q') \approx \left(1 - \frac{\epsilon}{2}f^j_{,j}\right)\psi_0(q - \epsilon f). \tag{4.42}$$

Equation (4.42) coincides, actually, with Eq. (2.109) (in Section 2.6 the inverse transformation is considered). With the help of the basic equivalence rules (2.23), $\epsilon p_j \Delta^k = \epsilon i\hbar\delta^k_j$, the exponential in (4.41) can be written in the form

$p_i\Delta^i - \epsilon\left[p_j f^j(q') + \frac{i\hbar}{2} f^k_{,k}\right]$. The sum in the brackets is nothing but the "matrix element" $\langle p|\hat{G}|q'\rangle/\langle p|q'\rangle$ of the operator $\hat{G} = (\hat{p}\hat{f} + \hat{f}\hat{p})/2 = \hat{p}_j f^j(\hat{q}) + \frac{i\hbar}{2} f^k_{,k}(\hat{q})$ that plays the role of the Hamiltonian in Eqs. (2.4) and (2.14). This completes the proof of validity of the operator ordering recipe from Chapter 2 for generators (constraints) linear in momenta.

4.3 Relativistic particle

4.3.1 Classical theory

The quantization problem for a relativistic particle with the action

$$S = -m\int \sqrt{1-\dot{\mathbf{x}}^2}dt \tag{4.43}$$

is well known. An explicitly invariant form of the action (4.43) is often preferable,

$$S = -m\int \sqrt{\dot{x}^2}d\tau, \quad \dot{x} = dx/d\tau, \quad \dot{x}^2 \equiv \dot{x}^\mu \dot{x}^\mu, \quad \mu = 0,1,2,3, \tag{4.44}$$

where the auxiliary parameter τ is called the invariant time. Similarly to (4.33), the action (4.44) is invariant under reparameterization transformations, $\tau = f(\tau')$, with f being an arbitrary function. Consequently, the system has a constraint. The constrained dynamics here appears to be of a new type [189] where the "constraints" depend on the velocities but cannot be solved for them.

Let us investigate the Hamiltonian formalism in detail. The Lagrangian has the form,

$$L = -m(\dot{x}^2)^{1/2}, \qquad x^0 = t. \tag{4.45}$$

The momenta canonically conjugated to x^μ are

$$p_\mu = \partial L/\partial \dot{x}^\mu = -mg_{\mu\nu}\dot{x}^\nu/\sqrt{\dot{x}^2}. \tag{4.46}$$

The first three equations of the following four relating the velocities and momenta,

$$p_i = m\dot{x}^i/\sqrt{\dot{x}^2}, \quad i = 1,2,3, \quad p_0 = -m\dot{x}^0/\sqrt{\dot{x}^2}, \tag{4.47}$$

are solved for $\dot{x}^i$,

$$(\dot{x}^i)^2 = (\dot{x}^0)^2 p_i^2/E_p^2, \qquad E_p = (\mathbf{p}^2 + m^2)^{1/2} \tag{4.48}$$

(no summation over i). Equations (4.48) have two solutions,

$$\dot{x}^i = \pm|\dot{x}^0|p_i/E_p. \tag{4.49}$$

Note that in Eqs. (4.47) the arithmetical value of the square root is taken, $(\dot{x}^2)^{1/2} = m|\dot{x}^0|/E_p$. Therefore the first three equations in (4.47) are satisfied only by $\dot{x}^i = p_i|\dot{x}^0|/E_p$. Its substitution into the fourth equation in (4.47) gives the primary "constraint",

$$\phi_1 = p_0 + E_p\epsilon(\dot{x}^0) = 0, \tag{4.50}$$

where $\epsilon(x)$ is the sign function of x (it equals 1, if x is positive, and -1 otherwise). The Hamiltonian of the system,

$$H = p_\mu \dot{x}^\mu - L = \dot{x}^0(p_0 + E_p\epsilon(\dot{x}^0)) \approx 0 \tag{4.51}$$

vanishes when the constraint (4.50) is taken into account (compare (4.32)).

An interesting feature of the Hamiltonian formalism is that, strictly speaking, one cannot really call (4.50) a constraint and (4.51) a Hamiltonian because they retain the dependence on the velocity $\dot{x}^0$. The peculiarity of the system with the Lagrangian (4.45) is that, even though the "constraint" (4.50) explicitly contains the velocity $\dot{x}^0$, it cannot be solved for this. It is noteworthy that the sign change $\dot{x}^0 \to -\dot{x}^0$ is equivalent to changing the sign of the energy ($E_p \to -E_p$), i.e. equations (4.50) and (4.51) formally admit negative energies. All these subtleties are usually omitted (see, e.g. [190]), and the velocity $\dot{x}^0$ is eliminated by squaring both sides of (4.46), which yields the constraint,

$$\phi_2 = p^2 - m^2 = 0 \tag{4.52}$$

or $p_0^2 = (E_p\epsilon(\dot{x}_0))^2 = E_p^2$. The same relation follows from (4.50). Although the constraint (4.52) looks reasonable – the four-momentum of a free relativistic particle does satisfy it – it contradicts the statement that all primary constraints are linear in non-physical momenta (see Chapter3). Furthermore, even though the total Dirac Hamiltonian [148] $H_T = H + \tilde{v}\phi_2$ is now quadratic in the momenta, this does not seem to be advantageous any more because the energy of a relativistic particle is E_p, and in such a theory the evolution operator kernel and wave function (explicitly known: $\exp(-ipx)$) have a different dependence on the momenta. Let us show that the "constraint" (4.50) allows one to develop a consistent quantum theory.

4.3.2 Quantum theory

Taking into account (4.51) the total Dirac Hamiltonian H_T is

$$H_T = H + v\phi_1 = v(p_0 + E_p\epsilon(\dot{x}^0)), \tag{4.53}$$

where v should normally be viewed as an arbitrary function of τ. In the present case the problem does not appear as a regular one because H_T depends on the sign of $\dot{x}^0$. So, it is not surprising that, as will be shown, the function v is also not quite arbitrary – a satisfactory solution of the problem turns out to be possible only if it is positive, $v > 0$.

Thus, in the present theory it is impossible to completely eliminate the velocity $\dot{x}^0$ from the constraint and the Hamiltonian. How should the Hamiltonian formalism be developed in such a case? The key observation that turns out to be helpful is that only the sign of $\dot{x}^0$ actually enters into (4.50) and (4.51), and, therefore, if the motion is restricted so that $\dot{x}^0 > 0$, then the velocity $\dot{x}^0$ drops out from ϕ_1 and H_T. In doing so, the aforementioned obstacle on the way to a quantum description disappears. One should only verify self-consistency of the resulting quantum theory, i.e. one has to convince oneself that the enforced restrictions are consistent with the quantum laws and do not lead to contradictions.

The quantization rule for systems with a vanishing Hamiltonian [70] established in Section 4.2 is applied. Put $\{x^\mu, p_\nu\} = \delta^\mu_\nu$, $[x^\mu, p_\nu] = i\delta^\mu_\nu$, and $\langle x|p\rangle = (2\pi)^{-2}\exp(ipx)$, where $px = p_\mu x^\mu$. The kernel of the infinitesimal evolution operator reads,

$$U_\omega(x, x') \equiv \langle x|e^{-i\omega \hat{H}_T}|x'\rangle = \int \frac{d^4p}{(2\pi)^4} \exp\{i\left[p_\mu \Delta x^\mu - \omega v(p_0 + E_p)\right]\}. \quad (4.54)$$

Here $\omega = \Delta\tau > 0$, $\Delta x = x - x'$, and $\dot{x}^0 > 0$. The evolution operator is assumed to act only on functions from the "physical" subspace, that is, satisfying the condition $\theta(-\hat{\dot{x}}^0)\psi = 0$. Hence, one should verify that the action of the operator with the kernel (4.54) does not take state vectors outside of the physical space. The integration with respect to p_0 in (4.54) yields,

$$U_\omega(x, x') = \int \frac{d^3p}{(2\pi)^3} \exp\left\{i\left[p_i \Delta x^i - \omega v E_p\right]\right\} \delta\left(x_0 - x_0' - \omega v\right). \quad (4.55)$$

The δ-function in (4.55) guarantees that $(x^0 - x'^0)/\omega$ (i.e. $\dot{x}^0$) is positive provided $v(\tau) > 0$. The latter condition is consistent with the demand of a monotonous growth of time t and, in fact, equivalent to it.

Thus, in this problem the Lagrangian multiplier $v(\tau)$ is not quite arbitrary – its sign turns out to be fixed. Under this condition the conventional method of developing the Hamiltonian formalism and the resulting quantum theory appear to be self-consistent. Inserting the kernel (4.55) into the equation,

$$\tilde{\psi}(x, \tau + \Delta\tau) = \int d^4x' U_\omega(x, x')\tilde{\psi}(x', \tau)$$

and replacing $v\omega$ by Δx^0 it can also be written in the form,

$$\psi(x^0, \mathbf{x}) = \int \frac{d^3p d^3x'}{(2\pi)^3} \exp\left\{i\left[p_i \Delta x^i - \Delta x^0 E_p\right]\right\} \psi(x'^0, \mathbf{x}'), \quad (4.56)$$

where $x^0 = x'^0 + v\omega$. The variables x^0 and τ are not independent in (4.56). It is then sufficient to indicate only the dependence on x^0. In what follows the argument τ is omitted.

By construction, all wave functions are Lorentz-invariant. However, at first glance the measure in (4.56) does not seem to be so, even though the starting formulas are. The point is that d^3p and d^3x transform as zero components of four-vectors, thus the measure in (4.56) is actually invariant as being the scalar product of two time-like vectors in the rest frame of one of them (in which the space-like component vanishes). This explanation, however, does not completely settle the matter. For Lorentz-invariant functions $\psi(x)$ the scalar product differs from that in non-relativistic quantum mechanics. As is known [191], the scalar product for invariant state vectors with positive energies is given by,

$$(\psi_1, \psi_2) = i \int d^3x \psi_1^* \overleftrightarrow{\partial_0} \psi_2, \quad \psi_1^* \overleftrightarrow{\partial_0} \psi_2 = \psi_1^* \partial_t \psi_2 - \partial_t \psi_1^* \psi_2, \tag{4.57}$$

while the very derivation of (4.56) implies the other form of the scalar product. Let us show that representation (4.56) is consistent with (4.57). First the identity $1 = (2E_p)^{-1}(2E_p)$ is inserted into the integrand in (4.56). In the coordinate representation for functions satisfying the Klein–Fock–Gordon (KFG) equation,[1] the multiplication by $2E_p$ is equivalent to the action of $i \overleftrightarrow{\partial_0}$. Therefore (4.56) is identical to

$$\psi(x) = i \int d^3x \tilde{U}_\epsilon(\mathbf{x}, \mathbf{x}') \overleftrightarrow{\partial_{t'}} \psi(x'), \tag{4.58}$$

where $x^0 = t$, $t - t' = \epsilon$, and

$$\tilde{U}_\epsilon(\mathbf{x}, \mathbf{x}') = \int d\mu(p) \exp\left[-i(E_p \Delta t - \mathbf{p}\Delta\mathbf{x})\right], \quad d\mu(p) \equiv \frac{d^3p}{(2\pi)^3 2E_p}. \tag{4.59}$$

Equation (4.59) gives the correct expression for the kernel of the infinitesimal evolution operator. An iterated convolution of the kernel (4.59) with itself by means of the rule (4.58) yields the following representation of the kernel $\tilde{U}_{t-t'}$ by the path integral:

$$\begin{aligned} \tilde{U}_{t-t'}(\mathbf{x}, \mathbf{x}') = & \int \prod_{\tau=t'}^{t} \frac{d^3p(\tau) d^3x(\tau)}{(2\pi)^3} \frac{1}{2E_p(t')} \exp\left\{ i \int_{t'}^{t} d\tau [\mathbf{p}\dot{\mathbf{x}} - E_{p(\tau)}] \right\} \\ & \times \delta(\mathbf{x}(t') - \mathbf{x}') = \int d\mu(p) \exp\left[-ip(x - x')\right], \quad px = E_p t - \mathbf{p}\mathbf{x}. \end{aligned} \tag{4.60}$$

This can be rewritten in an equivalent form by replacing the path integral measure in (4.60) by

$$\prod_{\tau=t'}^{t} d\mu(p(\tau)) d^3x(\tau) i \overleftrightarrow{\partial_\tau}. \tag{4.61}$$

[1] This equation is traditionally attributed to O. Klein and W. Gordon. It is little known that V. Fock established this equation independently [192], prior to W. Gordon (for historical remarks, see [193, 194]).

The meaning of the symbolic expression (4.61) is evident in view of the definition (4.58)–(4.60). The parameter τ in the first exponential in (4.60) is just an integration variable. Note that the action in (4.60) is not quadratic in the momenta. Nevertheless the path integral is calculated explicitly, i.e. reduced to an ordinary three-dimensional integral. This turns out to be possible thanks to the trivial dependence of the action on $\mathbf{x}(\tau)$. In contrast to the conventional procedure where the integration is first carried out with respect to $\mathbf{p}(\tau)$, here the integration with respect to $\mathbf{x}(\tau)$ can be done first.

Equation (4.60) allows one to find a path integral representation for the propagator of a scalar particle. To do so, both the positive and negative energy solutions of the KFG equation must be invoked. But the change of the sign of E_p is equivalent to changing the sign of $\dot{x}^0$ in (4.50) and (4.51). Therefore the introduction of particles with negative energies is (formally) equivalent to allowing them to move backward in time. The sought-for representation is thus given by the sum,

$$\Delta_F(x-x') = \theta(t-t')\tilde{U}_{t-t'}(\mathbf{x},\mathbf{x}') + \theta(t'-t)\tilde{U}_{t-t'}(\mathbf{x},\mathbf{x}')|_{E_p\to -E_p},$$

where the kernel $\tilde{U}_{t-t'}$ is defined by the first equality (4.60). The validity of the result becomes evident if the second equality in (4.60) is used,

$$\Delta_F(x-x') = \int d\mu(p)\, \{\theta(t-t')\exp\left[-i(E_p(t-t') - \mathbf{p}(\mathbf{x}-\mathbf{x}'))\right] \\ + \theta(t'-t)\exp\left[i(E_p(t-t') - \mathbf{p}(\mathbf{x}-\mathbf{x}'))\right]\}\,. \tag{4.62}$$

Equation (4.62) is the standard expression for the Stückelberg–Feynman propagator. This can be verified directly from the definition,

$$\Delta_F(x-x') \equiv \langle 0|\theta(t-t')\hat{\phi}(x)\hat{\phi}(x') + \theta(t'-t)\hat{\phi}(x')\hat{\phi}(x)|0\rangle,$$

where the free scalar field operator $\hat{\phi}$ should be expressed via the creation and annihilation operators.

4.4 Elimination of non-physical variables. The second-class constraints

It is shown in Section 4.1 that second-class constraints should be solved before quantization, otherwise the Poisson bracket is to be replaced by the Dirac bracket. It turns out that the resulting quantum theories depend on the way the non-physical variables are eliminated. The statement is illustrated with an example of the theory with the Lagrangian (3.2). The corresponding Hamiltonian formalism is developed in Section 3.2.2 (see Eqs. (3.50)–(3.58)).

Let the Cartesian coordinates $x_1 \equiv q$ and $p_1 \equiv p$ be chosen to span the physical phase space (the constraint surface). Then their Dirac bracket and the physical Hamiltonian are given by (3.75). In quantum theory the commutation relation for the corresponding operators $\hat{q}$ and $\hat{p}$ are induced by the Dirac bracket,

$$[\hat{q},\hat{p}] = i\hbar\{q,p\}_D|_{q=\hat{q},p=\hat{p}} = i\hbar\left(1-\frac{\hat{q}^2}{R^2}\right). \tag{4.63}$$

From Eq. (4.63) it follows that the operators $\hat{q}$ and $\hat{p}$ cannot be viewed as standard canonical variables because their commutation relation does not form the Heisenberg algebra. In particular, the uncertainty relation is modified. There are some other physical consequences resulting from the deformation of the Heisenberg algebra (for more details, see [38–42]).

Alternatively, one could choose local coordinates on the physical phase space with the standard symplectic structure. They should be chosen so that their Poisson bracket with the second-class constraints vanishes. In this case, the Dirac and Poisson brackets coincide for physical variables, and, hence, the standard Heisenberg algebra emerges upon quantization. Consider the canonical transformation $\mathbf{p},\mathbf{x} \to p_r, p_\varphi, r, \varphi$ associated with the polar coordinates in the configuration space. Then $p_r = (\mathbf{p},\mathbf{x})/r \approx 0$, $r = |\mathbf{x}| \approx R$, and, hence, the physical phase space is parameterized by p_φ and φ. Their Dirac bracket coincides with the Poisson one,

$$\{\varphi, p_\varphi\}_D = \{\varphi, p_\varphi\} = 1, \tag{4.64}$$

because $\{\varphi,\phi_\alpha\} = \{p_\varphi,\phi_\alpha\} = 0$. The Hamiltonian in the physical phase space has the form,

$$H_{ph} = \frac{1}{2R^2}p_\varphi^2. \tag{4.65}$$

Upon quantizing the theory, $\{\ ,\ \}_D \to [\ ,\ ]/(i\hbar)$, there is no ordering problem in the Hamiltonian (4.65) because $\hat{p}_\varphi = -i\hbar\partial_\varphi$.

It is instructive to compare the two approaches. To construct quantum theory, beside the commutation relations, the Hilbert space in which these operators are self-adjoint must be specified. For the theory defined by (4.64) and (4.65), the Hilbert space is the space of 2π-periodic functions with the standard scalar product. In the approach with the nontrivial symplectic structure the problem can be solved, e.g. by the conversion method (see Section 3.4.2) or by means of the Darboux variables (see Section 3.4). The Darboux variables for the symplectic structure (3.75) are given in (3.76). Therefore the operators satisfying the commutation relation (4.63) can be represented by the self-adjoint operators,

$$\hat{q} = R\cos\hat{\varphi}, \quad \hat{p} = -\frac{1}{R}(\sin\hat{\varphi})^{1/2}\hat{p}_\varphi \circ (\sin\hat{\varphi})^{1/2}\,, \quad [\hat{\varphi},\hat{p}_\varphi] = i\hbar,$$

where $[\hat{\varphi},\hat{p}_\varphi] = i\hbar$. In particular, in the Hilbert space of periodic functions where $\hat{\varphi}$ is diagonal, $\hat{p}_\varphi = -i\hbar\partial_\varphi$, and $\hat{p}_\varphi^\dagger = \hat{p}_\varphi$. The procedure implies that the Darboux variables are introduced *after* quantization. The quantum theory construction is completed by finding a self-adjoint Hamiltonian operator corresponding to the classical Hamiltonian (3.75). This amounts to specifying the ordering of the operators $\hat{q}$ and $\hat{p}$ or $\hat{\varphi}$ and $\hat{p}_\varphi$ in the Hamiltonian $\hat{H}_{ph}$ in (3.75). The condition

$\hat{H}_{ph} = \hat{H}_{ph}^{\dagger}$ is generally not sufficient to fix the ordering uniquely. For example, for a real parameter γ, the following representation provides a Hermitian Hamiltonian:

$$\begin{aligned}\hat{H}_{ph} &= \frac{R^2}{2}(R^2 - \hat{q}^2)^{-\gamma/2}\,\hat{p}\,(R^2 - \hat{q}^2)^{-1+\gamma}\,\hat{p}\,(R^2 - \hat{q}^2)^{-\gamma/2} \\ &= \frac{1}{2R^2}(\sin\hat{\varphi})^{1/2-\gamma}\hat{p}_{\varphi} \circ (\sin\hat{\varphi})^{2\gamma-1}\hat{p}_{\varphi} \circ (\sin\hat{\varphi})^{1/2-\gamma} \\ &\equiv \frac{1}{2R^2}\hat{p}_{\varphi}^2 + \Delta V_q(\hat{\varphi}) = \hat{H}_{ph}^{\dagger},\end{aligned} \tag{4.66}$$

where the quantum potential $\Delta V_q(\varphi)$ is,

$$\Delta V_q(\varphi) = \frac{\hbar^2}{2R^2}(\gamma - 1/2)\left[(\gamma - 3/2)\frac{\cos^2\varphi}{\sin^2\varphi} - 1\right]. \tag{4.67}$$

Only for $\gamma = 1/2$ does both approaches give the same quantum theory. The quantum potential (4.67) does not depend on φ if $\gamma = 3/2$.

Thus, the physical canonical variables in the quantum theory of systems with second-class constraints cannot be treated as conventional canonical variables if the symplectic structure generated by the Dirac bracket differs from the standard one (beside a possible nontrivial topology of the physical phase space). The construction of a quantum theory with the canonical variables satisfying the Heisenberg algebra requires finding the Darboux variables for the symplectic structure generated by the Dirac bracket. From the above example it follows that the operations of quantization and going over to the Darboux variables do not generally commute. In other words, the quantum theory may depend on the choice of local coordinates in the physical phase space which are used in quantization. The analogy can be made with a similar (but simpler) operator ordering problem in the canonical quantization in curvilinear coordinates in the configuration space. The same problem often occurs when quantizing first-class constrained systems in which solving the constraints requires the use of curvilinear coordinates. But overall, various operator orderings lead to Hamiltonians differing only in terms proportional to the Planck constant which disappear in the classical limit. In a pure mathematical framework, quantum theories that have the same classical limit are equally acceptable. Some additional physical principles must be invoked in order to favor a particular quantum theory.

5
Phase space in gauge theories

A key feature of gauge theories is the presence of non-physical degrees of freedom whose time evolution is not determined by equations of motion. It seems natural to eliminate such degrees of freedom from the theory and obtain an effective dynamics only for physical variables. In this chapter this problem is studied in detail. There are two ways to achieve this goal, invariant and non-invariant. In the invariant approach, only gauge-invariant variables are used, while the non-invariant route is associated with the use of quantities that are not gauge invariant. However, it only appears non-invariant, similarly to the use of local coordinates on a manifold to compute coordinate-invariant quantities, e.g. curvature or geodesic length. Both approaches describe the very same theory and, therefore, must be equivalent. Unfortunately, there is a shortage of rigorous studies to establish the equivalence of the invariant and non-invariant approaches for general gauge systems. For that reason the discussion is limited to concrete models. The analysis of the physical phase space structure is in the focus of the discussion since this problem has been studied only recently and is not discussed in the monograph literature and textbooks on gauge theories. It appears that for a large class of gauge theories the physical phase space may have a non-trivial structure, i.e. different from the conventional symplectic plane for each physical degree of freedom. This leads to several new phenomena in classical as well as quantum dynamics [183] (see reviews [26, 195]).

As there are two approaches to developing an effective theory for physical degrees of freedom, the physical phase space structure is analyzed in both of them, invariant and non-invariant. Suppose there is a dynamical system with N degrees of freedom and M independent first-class constraints (including primary, secondary, etc.), $\phi_a = 0$, $a = 1, 2, ..., M$. The constraints must be satisfied at any moment in time. Therefore, in the total phase space Γ of the system, the trajectory must lie on the constraint surface defined by the equations $\phi_a = 0$. However the physical phase space does not coincide with the constraint surface. It has been shown in Section 3.3 that a solution of the Hamilton equations of motion, that satisfies the constraints $\phi_a = 0$, depends on M arbitrary functions of time. Variations of these functions yield gauge transformations of the solution and, hence, do not change the physical state of the system. In Hamiltonian mechanics, each point of the phase space represents only one state of a dynamical system. Hence, in gauge theories, the physical phase space is obtained from the constraint surface by identifying all points related by gauge transformations,

$$\Gamma_{ph} = \Gamma|_{\phi_a=0}/\tilde{G}, \tag{5.1}$$

where $\tilde{G}$ is the *Dirac extended gauge group* which is generated by all independent constraints ϕ_a (compare Section 3.3). Relation (5.1) defines the physical phase space in the invariant approach as a factor space of the constraint surface in the total phase space with respect to the extended gauge group. Clearly, Γ_{ph} has the dimension $2(N-M)$.

In the non-invariant approach, for the sake of discussion, the total phase space is assumed to be Euclidean. In other words, the generalized coordinates q and the corresponding canonical momenta p are N-dimensional vectors, $\Gamma = R_q^N \oplus R_p^N \equiv \Gamma^{2N}$. By fixing a gauge, or by imposing additional conditions on the canonical variables in order to eliminate gauge freedom in the equations of motion, the total phase space is restricted to a subspace $\Gamma^{2(N-M)}$. The studies of simple gauge models have already shown that in the space of physical degrees of freedom there may exist *redundant* (discrete) gauge transformations which cannot reduce the number of the physical degrees of freedom but do lead to a further reduction of their phase space [183, 196].[1] The redundant gauge group S depends on the gauge chosen, and, in natural gauges, coincides with the Weyl (or affine Weyl) group [26, 195–198]. For generic gauges, the set S of redundant gauge transformations may not form a group [26, 195]. Thus,

$$\Gamma_{ph} = \Gamma^{2(N-M)}/S.$$

The key observation to note here is based on the well-known fact that points of the configuration or phase space related by a gauge transformation are physically indistinguishable. The redundant gauge transformation identifies some points of $\Gamma^{2(N-M)}$, which leads to its further reduction without changing its dimensions. For example, a phase plane, Γ^2, can be turned into a cone that is unfolded into a half-plane by a suitable cut [183]. The physical phase space reduction leads to dynamical effects.

5.1 A simple model

Consider a dynamical system defined by the Lagrangian [183],

$$L(\mathbf{x}, \dot{\mathbf{x}}, y, \dot{y}) = \frac{1}{2}\left[\left(\frac{d}{dt} - yT\right)\mathbf{x}\right]^2 - V(\mathbf{x}^2), \tag{5.2}$$

where the two-dimensional vector $\mathbf{x}$ and scalar y are dynamical variables, $T = -i\tau_2$ with τ_2 being the Pauli matrix, $(T\mathbf{x})_i = T_{ij}x_j$. The Lagrangian (5.2) describes a non-relativistic particle of unit mass moving in a two-dimensional

[1] This phenomenon has been rediscovered and is intensively studied in string theory [199].

Euclidean space. In the new variables, $\phi = (x_1 + ix_2)/\sqrt{2}$, the Lagrangian has the form,

$$L(\phi, \phi^*, \dot{\phi}, \dot{\phi}^*, y, \dot{y}) = \frac{1}{2}\left(\frac{d}{dt} - iy\right)\phi\left[\left(\frac{d}{dt} - iy\right)\phi\right]^* - V(2\phi\phi^*), \tag{5.3}$$

from which it follows that (5.2) is the Lagrangian of the scalar electrodynamics in the (0+1) spacetime. The variable $y = A_0(t)$ is nothing but the time component of the vector potential A_μ. The Lagrangian is invariant under the gauge group transformations,

$$\mathbf{x} \to \exp(T\omega)\mathbf{x}, \qquad y \to y + \dot{\omega}, \tag{5.4}$$

where $\omega = \omega(t)$ is an arbitrary function of time.

5.1.1 Classical Hamiltonian dynamics

The canonical momenta of the system are

$$\mathbf{p} = \partial L/\partial\dot{\mathbf{x}} = \dot{\mathbf{x}} - yT\mathbf{x}, \qquad \pi = \partial L/\partial\dot{y} = 0, \tag{5.5}$$

and the Hamiltonian of the model is given by,

$$H = (\mathbf{p}, \dot{\mathbf{x}}) + \pi\dot{y} - L = \frac{1}{2}\mathbf{p}^2 + V(\mathbf{x}^2) + y(\mathbf{p}, T\mathbf{x}). \tag{5.6}$$

Applying the Dirac formalism [148], it is easy to see that the system has two first-class constraints, $\pi = 0$ and $\sigma = (\mathbf{p}, T\mathbf{x}) = x_1p_2 - x_2p_2$, such that σ generates rotations about the origin in two-dimensional Euclidean space [183], and π is the generator of translations of the variable y. Indeed, let us compute the Poisson brackets of the constraints with all canonical variables,

$$\{\pi, \mathbf{x}\} = \{\pi, \mathbf{p}\} = 0, \quad \{\pi, y\} = -1 \tag{5.7}$$

$$\{\sigma, \pi\} = \{\sigma, y\} = 0, \quad \{\sigma, \mathbf{x}\} = -T\mathbf{x}, \quad \{\sigma, \mathbf{p}\} = -T\mathbf{p}, \tag{5.8}$$

from which it follows that the operators $\hat{g}_\pi = e^{-\omega_1\hat{\pi}}$ and $\hat{g}_\sigma = e^{-\omega_2\hat{\sigma}}$, where $\hat{\pi} = \{\pi, \cdot\}$ and $\hat{\sigma} = \{\sigma, \cdot\}$, generate the following gauge transformations (compare Section 1.6):

$$\hat{g}_\pi\pi = \pi, \qquad \hat{g}_\pi y = y + \omega_1 \tag{5.9}$$

$$\hat{g}_\sigma\mathbf{x} = \exp(\omega_2 T)\mathbf{x}, \qquad \hat{g}_\sigma\mathbf{p} = \exp(\omega_2 T)\mathbf{p}, \tag{5.10}$$

with $\omega_{1,2}$ being arbitrary functions of time. The variables $\mathbf{x}$ and $\mathbf{p}$ are invariant under the action of $\hat{g}_\pi$, while y and π under that of $\hat{g}_\sigma$. Note that the gauge group (5.4) of the Lagrangian is a subgroup of the extended gauge group defined by the condition $\omega_1 = \dot{\omega}_2$.

Thus, in the model studied, the extended gauge group of the system $\tilde{G}$, generated by all independent first-class constraints, is $T \times SO(2)$ where T is the group of translations on R. Thus, out of the three degrees of freedom, y, x_1, and x_2, only one is physical.

As pointed out above, there are two ways to eliminate non-physical degrees of freedom. The first one is explicitly gauge invariant. Consider a canonical transformation $\mathbf{x}$, $\mathbf{p} \to r, \theta, p_r, p_\theta$, where $r = (\mathbf{x}^2)^{1/2}$ and θ are the polar coordinates, $p_r = (\mathbf{p}, \mathbf{n}_r)$, $p_\theta = (\mathbf{p}, \mathbf{n}_\theta)r$ are the corresponding canonical momenta, with $\mathbf{n}_r = \mathbf{x}/r$ and $\mathbf{n}_\theta$ being the unit vectors tangent to the coordinate curves of the polar coordinate system. The variables r and p_r are invariant under the extended gauge transformations and, hence, can be taken as the invariant physical canonical variables.[2]

The second (non-invariant) method of eliminating non-physical degrees of freedom is as follows. Thanks to (extended) gauge symmetry, all points of a circle $\mathbf{x}^2 = \text{const}$ are physically indistinguishable. Concentric circles form gauge orbits in the space spanned by $\mathbf{x}$. Any of its points can be taken as representative of the gauge orbit. In particular, one can take a curve intersecting each concentric circle only once, then the variable changing along such a curve can serve as a physical degree of freedom. For example, consider the semi-axis $x_2 = 0$, $x_1 \geq 0$. It is more convenient, however, to let x_1 span the entire axis on which the points x_1 and $-x_1$ are identified. This is equivalent to a dynamical system with one degree of freedom and the gauge group Z_2 containing two elements:[3] the identity element and the reflection $\hat{P}x_1 = -x_1$. The equivalence of the non-invariant and invariant formalisms is obvious. For general gauge systems, the relations between the two approaches is not that straightforward.

Let us turn to the question of the physical phase space structure. In the invariant approach, $r \geq 0$ and $-\infty < p_r < \infty$, so the physical phase space seems to be a half-plane. However, more accurate consideration shows that points p_r and $-p_r$ on the line $r = 0$ must be identified, thus turning the half-plane into a cone [183]. The easiest way to see this is to consider trajectories in $\Gamma(r, p_r)$ for a simple potential, say, $V = \mathbf{x}^2/2$ in the Hamiltonian (5.6) (the variable y plays the role of a Lagrange multiplier).

A general solution of the constraint $\sigma = 0$ has the form $\mathbf{p} = \lambda\mathbf{x}$, where $\lambda = \lambda(t)$ is a function of time determined by the dynamics of the system. Substituting this into the Hamilton equations of motion,

$$\dot{\mathbf{x}} = \{\mathbf{x}, H\} = \mathbf{p} + yT\mathbf{x}, \quad \dot{\mathbf{p}} = \{\mathbf{p}, H\} = -\mathbf{x} + yT\mathbf{p}, \tag{5.11}$$

and then combining them, the equation for λ is inferred,

$$\mathbf{x}\dot{\lambda} = -(1 + \lambda^2)\mathbf{x}. \tag{5.12}$$

[2] Because of (5.9) and the second equation in (5.5), the variables y and π describe a purely non-physical degree of freedom (compare Section 3.3).

[3] Note that conventional (non-gauge) dynamics can be supplemented by discrete gauge symmetry only by "hand", that is, it cannot be defined by a Lagrangian without non-physical variables. Here Z_2 appears as a subgroup of the extended gauge group.

For $\mathbf{x} \neq \mathbf{0}$, one has $\lambda(t) = \tan(t + t_0)$. Without loss of generality, the constant t_0 can be set to zero, which corresponds to the zero initial condition for the canonical momentum $\mathbf{p}(0) = \mathbf{0}$. Substituting the explicit form of $\lambda(t)$ into a general solution of the first equation in (5.11),

$$\mathbf{x}(t) = e^{\int_0^t d\tau[\lambda(\tau)+y(\tau)T]}\,\mathbf{x}(0), \tag{5.13}$$

one infers,

$$\mathbf{x}(t) = -\cos(t)\, e^{\int_0^t d\tau y(\tau)T}\,\mathbf{x}(0), \qquad \mathbf{p}(t) = \tan(t)\mathbf{x}(t). \tag{5.14}$$

The time evolution of the physical variables is given by,

$$r(t) = |\mathbf{x}(t)| = |\cos(t)|\, r(0), \quad p_r(t) = \lambda(t)|\mathbf{x}(t)| = \tan(t)|\cos(t)|\, r(0). \tag{5.15}$$

Note that the solution (5.14) may also contain $|\cos(t)|$ in place of $-\cos(t)$. This happens if, after cancelation of $\mathbf{x}$ in (5.12), the resulting equation is assumed to be valid for all values of t. When substituting $\lambda = \tan(t)$ into (5.13), the factor $|\cos(t)|$ occurs in (5.14) instead of $-\cos(t)$. However, $\mathbf{x}(t)$ vanishes when $\cos(t)$ changes sign, and the differential equation for $\lambda(t)$ becomes inapplicable at the moments of time when $\cos(t) = 0$. The solution (5.14) is written under the requirement that it coincides with the corresponding solution of equation (5.11) for $y = 0$. As a matter of fact, the replacement of $-\cos(t)$ by its absolute value $|\cos(t)|$ in (5.14) is equivalent to a gauge transformation (see below Section 5.1.2) and, therefore, has no effect on the time evolution of the gauge invariant variables (5.15). From (5.15) it follows that at $t = \pi/2$ ($r = 0$) the trajectory on the phase plane (r, p_r) has a discontinuity because p_r changes its sign (see Fig. 5.1f). For physical reasons it is clear however that no sudden jumps in the momentum can occur in this mechanical system as the potential is regular at $r = 0$. Therefore the points $(r = 0, p_r)$ and $(r = 0, -p_r)$ should be identified to eliminate the artificial jump. This means that the phase space of the system is a cone that is unfolded into a half-plane [183] as shown in Fig. 5.1e.

The conclusion is, of course, independent of the choice of potential. To see this, let us use (5.1) to find the physical phase space. Obviously, $\Gamma = R^2(y,\pi) \times R^4(\mathbf{x},\mathbf{p})$. The constraint surface reads $R(y) \times R^4(\mathbf{x},\mathbf{p})|_{\sigma=0}$. As a result, we have

$$\begin{aligned}\Gamma_{ph} &= [R(y) \times R^4(\mathbf{x},\mathbf{p})|_{\sigma=0}]/[T \times SO(2)] \\ &= [R(y)/T] \times [R^4(\mathbf{x},\mathbf{p})|_{\sigma=0}/SO(2)] \\ &= R^4(\mathbf{x},\mathbf{p})|_{\mathbf{p}=\lambda\mathbf{x}}/SO(2),\end{aligned} \tag{5.16}$$

in accord with the action of the extended gauge group. By a suitable gauge transformation the vector $\mathbf{x}$ can be directed along the first axis $x_i = \delta_{i1}x$. Hence, on the constraint surface, the same gauge transformation brings the momentum to the form $p_i = \delta_{i1}p$, where $p = \lambda x$. However, the gauge freedom has not yet been exhausted. A transformation remains, a rotation through angle π, under which

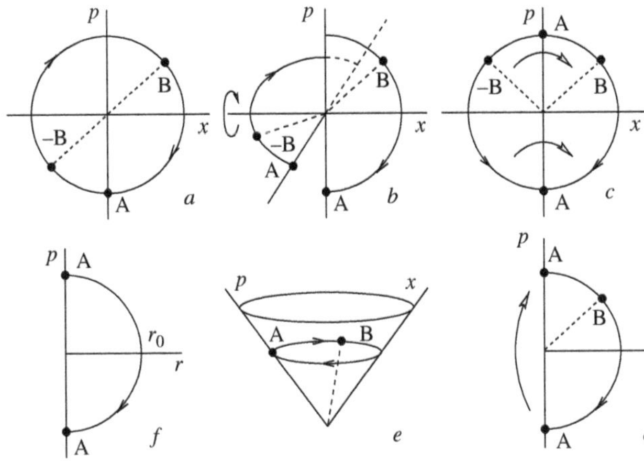

Fig. 5.1. *a*. The phase-space plane (p, x) and the oscillator trajectory on it. The states $\mathrm{B} = (p, x)$ and $-\mathrm{B} = (-p, -x)$ are gauge equivalent and are to be identified. The identification can be done in three steps of the following, convenient, cut-and-glue procedure shown in panels (b)–(d) (each step corresponds to no particular gauge transformation, while their composition does);
b. The phase-space plane is cut along the p-axis. The half-plane $x < 0$ is rotated about the x-axis through angle π (as though the plane is embedded into the three-dimensional space).
c. The resulting plane is then folded along the p-axis so that states B and $-B$ are identified;
d. Two copies of each state on the p-axis, which occur upon the cut (e.g. state A), are glued back to remove this doubling;
e. The resulting conic phase space. Each point corresponds to one physical state of the gauge system. The oscillator trajectory does not have any discontinuity;
f. The physical motion of the harmonic oscillator in the local gauge-invariant variables (p_r, r). The trajectory has a discontinuity at state A. The discontinuity occurs through the cut of the cone along a ray outgoing from the cone tip. In the particular parameterization of the physical phase space by (p_r, r), the cut coincides with the momentum axis $r = 0$. Adapted from [195].

x changes its sign: $x \to -x$. Under this gauge transformation the momentum p must also change its sign because $p = \lambda x$ on the constraint surface. Therefore,

$$\Gamma_{ph} = R^2(x, p)/Z_2 = \mathrm{cone}(\pi), \tag{5.17}$$

where the group Z_2 acts on the phase plane $R^2(x, p)$ as a reflection about the origin,

$$Z_2 : \quad (x, p) \to (\pm x, \pm p) \tag{5.18}$$

and the symbol cone(α) denotes a cone that is unfolded into a sector of the plane with angle α. In (5.17), $\alpha = \pi$, meaning that the cone is unfolded into a half-plane. In particular, cone$(2\pi) = R^2$. A geometrical illustration of (5.17) is given in Fig. 5.1.

Consider the dynamics of the physical degree of freedom in the non-invariant approach based on removing the gauge freedom by fixing a gauge condition for dynamical variables. The solution (5.13) is valid for any potential regular at the origin. It admits the following interpretation [183]. A particle moves along a straight line through the origin, while the line itself is rotated about the origin. Indeed, matrix T is a generator of rotations in the plane about the origin and $y(\tau)$ is an arbitrary function of time. Points of the plane that can be transformed to one another by rotations about the origin are physically indistinguishable. Hence, with a suitable choice of $y(\tau)$, i.e., by *choosing a gauge*, the dynamical problem can be reduced to studying the motion of the particle along a particular line, say, x_1, which corresponds to the *unitary gauge*, $x_2(t) = 0$.

However, condition $x_2 = 0$ is not sufficient to completely eliminate the gauge freedom. The redundant gauge transformations change the sign of x_1. Moreover, by combining the gauge condition with the solution of the constraint equation, one finds $p_1 = \lambda x_1$, and, hence, the redundant gauge transformations form the discrete group (5.18). The latter, as has been argued, leads to a further reduction of the phase space of the physical degree of freedom, turning the phase plane into a cone, $\Gamma_{ph}(x_1, p_1) = \text{cone}(\pi)$. For the model studied both the approaches, invariant and non-invariant, are equivalent. An advantage of the non-invariant approach is appreciated even more, when investigating general gauge models where finding canonical gauge-invariant variables produces substantial technical difficulties.

The discrete group Z_2 that turns the phase plane into a cone is a subgroup of the gauge group. Therefore it acts locally in time, that is, for any given moment of time $t > 0$, the sign of $x_1(t)$ can be changed. Given a trajectory of the system $x_1(t)$ in the unitary gauge, one can always construct a gauge transformation such that $x_1(t) \to |x_1(t)|$ (see details in [200]).

5.1.2 Symplectic structure on the physical phase space

The observed singularities of the phase-space trajectories are essentially artifacts of the *coordinate description* and, hence, depend on the parameterization of the physical phase space. For instance, the cone can be parameterized by another set of *canonical* gauge-invariant variables,

$$p_r = |\mathbf{p}| \geq 0, \quad r = \frac{(\mathbf{p}, \mathbf{x})}{p_r}, \quad \{r, p_r\} = 1. \tag{5.19}$$

It is easy to convince oneself that $r(t)$ would have discontinuities, rather than momentum p_r. This set of local canonical variables in the physical phase space is associated with the cut of the cone along the coordinate axis. In general, local canonical variables in the physical phase space are determined up to canonical transformations,

$$(p_r, r) \to (P_R, R) = (P_R(r, p_r), R(p_r, r)), \quad \{R, P_R\} = 1. \tag{5.20}$$

Coordinate singularities associated with arbitrary local canonical coordinates in the physical phase space may be tricky to analyze (compare Section 5.3.3).

It is also of interest to find out whether there could exist canonical variables in which the discontinuities in the classical phase space trajectories do not occur [195]. Let us return to the canonical variables where the momentum p_r changes sign as the trajectory passes through the origin $r=0$. The sought-for new canonical variables must be even functions of p_r when $r=0$ and be *regular* on the half-plane $r \geq 0$. Then the trajectory in the new coordinates would not suffer the discontinuity. In the vicinity of the origin, put

$$R = a_0\left(p_r^2\right) + \sum_{n=1}^{\infty} a_n(p_r)r^n, \qquad P_R = b_0\left(p_r^2\right) + \sum_{n=1}^{\infty} b_n(p_r)r^n. \tag{5.21}$$

The Poisson bracket $\{R, P_R\} = \sum c_n(p_r)r^n$ is a power series in r with coefficients being functions of p_r. It follows from the condition $\{R, P_R\} = 1$ that $c_0 = 1$ and $c_n = 0$, $n > 0$. Computing c_0 via the functions a_n and b_n, the equation $c_0 = 1$ becomes,

$$2p_r\left[a_1(p_r)b_0'\left(p_r^2\right) - a_0'\left(p_r^2\right)b_1(p_r)\right] = 1, \tag{5.22}$$

where a_0' and b_0' are the derivatives of a_0 and b_0 with respect to their argument. This has no solution for regular functions $a_{0,1}$ and $b_{0,1}$. Indeed, if functions a_n and b_n were regular, then so should the combination in brackets in (5.22) also have been, which cannot be true because $a_1b_0' - a_0'b_1 = 1/(2p_r)$, as follows from (5.22). A solution exists only for functions singular at $p_r = 0$. For instance, one can take $R = r/p_r$ and $P_R = p_r^2/2$, $\{R, P_R\} = 1$ which is obviously singular at $p_r = 0$. In these variables the evolution of the canonical momentum does not have abrupt jumps, however, the new canonical coordinate does have jumps as the system goes through the states with $p_r = 0$.

In general, the existence of singularities is due to the condition that a_0 and b_0 must be even functions of p_r. This condition leads to the factor $2p_r$ in the left-hand side of Eq. (5.22), thus making it impossible for b_1 and a_1 to be regular everywhere. Thus, although in the conic phase space the trajectories are regular, the motion always exhibits singularities when described in any local canonical coordinates on the physical phase space.

This analysis reveals an important and rather general feature of gauge theories. The physical phase space in gauge theories may be non-Euclidean. While phase-space trajectories are smooth in the physical phase space, they may appear to have singularities when described in local canonical coordinates. The singularities depend entirely on the choice of local canonical coordinates, even though this choice is made in a gauge-invariant way (see a further discussion of this in Section 5.3.3). What remains coordinate- and gauge-independent is the geometrical structure of the physical phase space which, however, reveals itself through the coordinate singularities occurring in any particular parameterization of the physical phase space by local canonical variables.

The absence of local canonical coordinates in which the dynamical description does not have singularities may seem rather disturbing. This is partly because of our custom of often identifying canonical variables with physical quantities which can be directly measured, such as, for instance, positions and momenta of particles in classical mechanics. In gauge theories canonical variables, that are defined through the Legendre transformation of the Lagrangian, cannot always be measured and, in fact, may not even be physical quantities. For example, in electrodynamics the vector potential and the electric field strength are fundamental canonical variables, while the electric and magnetic fields are measurable (gauge-invariant) quantities. However, they are not canonically conjugated variables. Nevertheless their Poisson bracket is well defined and may be used to construct a symplectic structure that determines the evolution of all gauge-invariant quantities (being functions of the electric and magnetic fields) via the Hamiltonian equation motion.

This analogy can also be extended to the mechanical model studied here. A simple choice of phase-space coordinates, that are not necessarily canonically conjugated, but in which the time evolution is regular, is

$$Q = \mathbf{x}^2, \quad P = (\mathbf{p}, \mathbf{x}). \tag{5.23}$$

The variables (5.23) are gauge invariant and in one-to-one correspondence with the canonical variables r, p_r parameterizing the physical (conic) phase space: $Q = r^2, P = p_r r, r \geq 0$. Due to analyticity in the original phase space variables, they also have a smooth time evolution $Q(t), P(t)$. However, their Poisson bracket is not canonical,

$$\{Q, P\} = 2Q. \tag{5.24}$$

Nevertheless it is acceptable to formulate Hamiltonian dynamics of physical degrees of freedom. The Hamiltonian assumes the form,

$$H = \frac{1}{2Q} P^2 + V(Q).$$

Therefore,

$$\dot{Q} = \{Q, H\} = 2P, \quad \dot{P} = \{P, H\} = \frac{P^2}{Q} - 2QV'(Q).$$

The solutions $Q(t)$ and $P(t)$ are regular for regular V, and there is no need to "remember" where the cut on the cone has been made.

Remark. The Poisson bracket (5.24) can be regarded as a skew-symmetric product (commutator) of two basis elements of the Lie algebra of the dilatation group. This observation allows one to quantize the symplectic structure: $[\hat{Q}, \hat{P}] = 2i\hbar\hat{Q}$. There are three irreducible representations of this algebra in which the spectrum of $\hat{Q}$ is either strictly positive, or zero, or strictly negative. So the case $\hat{Q} > 0$ may be used to describe quantum theory on the conic phase space. The corresponding representation can be realized by the so-called affine coherent states, and the coherent-state representation of the path integral formalism can also be developed [201].

5.1.3 Phase space in polar coordinates

It might seem that the phase space in polar canonical variables r and p_r is a half-plane even in the absence of gauge symmetry because of the inequality $r \geq 0$. A more thorough analysis shows that this is not so [26, 183]. The phase space Γ of each of the two degrees of freedom for a particle in a plane is a full plane, $\Gamma(x_i, p_i) = R^2$, $i = 1, 2$. In polar coordinates, the radial variable r and the polar angle θ take their values only in part of the real line. What are the phase spaces of the corresponding degrees of freedom? One might suggest that $\Gamma(r, p_r)$ is a half-plane and $\Gamma(\theta, p_\theta)$ is a strip of width 2π because $r \geq 0$ and θ ranges over the interval $0 \leq \theta < 2\pi$, while the corresponding canonical momenta can take any real values. However, a more accurate analysis shows that the phase spaces of the radial and angular degrees of freedom are full planes.

In order to be convinced that this is indeed so for r and p_r, consider the motion of a particle through the origin in the rectangular coordinate system. Suppose that the motion occurs along the x_1 axis. As long as the particle remains on the positive semi-axis, the equality $r = x_1$ holds and no paradox arises. When passing through the origin, the variable x_1 changes its sign, while r does not, and the canonical variables θ and p_r change abruptly by a jump, $\theta \to \theta + \pi$ and $p_r = |\mathbf{p}| \cos\theta \to -p_r$. The jumps are not related to the action of any forces and do not contradict the equations of motion. In fact, the kinematics of the system admits an interpretation that allows us to avoid these artificial discontinuities of the phase space trajectories. Indeed, relations between the rectangular and polar coordinates, $x_1 = r\cos\theta$, $x_2 = r\sin\theta$, are invariant under the discrete transformations, $\theta \to \theta + \pi$ and, simultaneously, $r \to -r$. This means that the motion of the system with the values of the polar coordinates being $\theta + \pi$ and $r > 0$ cannot be distinguished from that with values θ and $r < 0$. The phase space trajectories in the *plane* spanned by r, p_r and in the (x_1, p_1)-plane turn out to be the same as one would anticipate.

Stated in a different way, $\Gamma(r, p_r)$ can be viewed as a Riemann surface with two conic leaves that is obtained from the plane by the following cut-and-glue procedure. The (r, p_r)-plane is cut along the line $r = 0$, then the edge $(r = 0^+, p_r > 0)$ is glued to the edge $(r = 0^-, p_r < 0)$ and $(r = 0^+, p_r < 0)$ to $(r = 0^-, p_r > 0)$. Every continuous trajectory on the (x_1, p_1)-plane remains continuous on the Riemann surface $\Gamma(r, p_r)$ under the change of variables. The procedure is shown in Fig. 5.2. Note that no curvature is associated with the point $r = p_r = 0$, because the phase space vector (r, p_r) sweeps an angle 2π around two conic leaves before returning to the initial point. Each leaf corresponds to values of θ that differ by π and, by this reasoning, they describe physically distinct states. When the gauge symmetry is turned on, the leaves become physically indistinguishable as

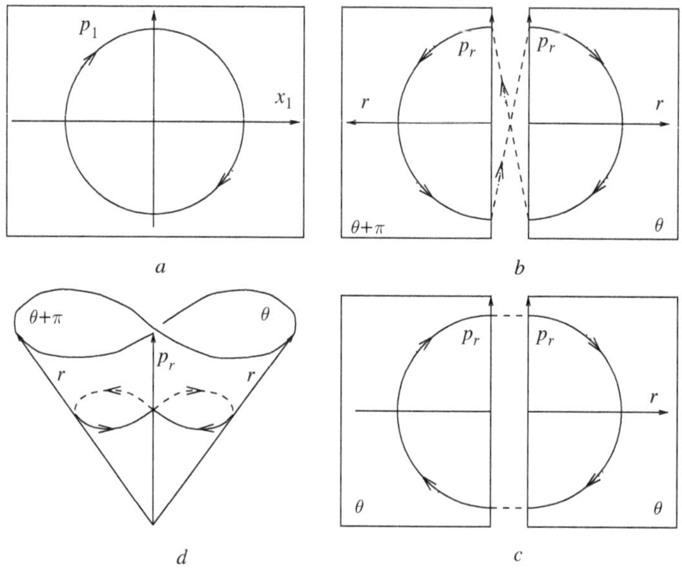

Fig. 5.2. *a*. A phase-space trajectory of the harmonic oscillator. The initial conditions are such that $x_2 = p_2 = 0$ for all moments of time. The system moves through the origin $x_1 = 0$;
b. The same motion is represented in the canonical variables associated with polar coordinates. When passing the origin $r = 0$, the trajectory suffers a discontinuity caused by a jump of the canonical momenta. The discontinuity can be removed in two ways:
c. One can convert the motion with values of canonical coordinates $(-p_r, r; p_\theta, \theta + \pi)$ into the equivalent motion $(p_r, -r; p_\theta, \theta)$, thus making a full phase-space plane out of two half-planes.
d. Another possibility is to glue directly the points connected by the dashed lines in Panel b. The resulting surface is the Riemann surface with two conic leaves. It has no curvature at the origin because the phase-space radius vector (p_r, r) sweeps the total angle 2π around the two conic leaves before returning to the initial state. Adapted from [195].

the polar angle is no longer a physical variable. The Riemann surface $\Gamma(r, p_r)$ obviously turns into cone(π).

Finally, studying the motion of the particle about the origin of the rectangular coordinate system, one sees that, for a complete description of the particle state, the angular variable θ should be allowed to take all real values in order to account for the number of full rotations about the origin. Hence, $\Gamma(\theta, p_\theta)$ is also a full plane. All of the above is in complete agreement with the Hamiltonian path integral formalism in polar coordinates given in Chapter 2.

5.1.4 Quantum theory. Invariant and non-invariant approaches

A canonical quantization of the system is carried out by replacing the canonical variables by the corresponding operators, $\mathbf{x}, \mathbf{p}, y, \pi \to \hat{\mathbf{x}}, \hat{\mathbf{p}}, \hat{y}, \hat{\pi}$, in the Hamiltonian and constraints. There is no operator ordering problem here. According to the commonly accepted Dirac quantization scheme of constrained systems presented in Chapter 4, the physical states are defined by the conditions

$$\hat{\pi}\Phi = 0, \qquad \hat{\sigma}\Phi = (\hat{\mathbf{p}}, T\hat{\mathbf{x}})\Phi = 0. \tag{5.25}$$

The first equation implies that the physical wave function is independent of y. In what follows, this variable will be ignored, e.g. when normalizing the physical states, the integration with respect to y is omitted. In fact, the non-physical variable y is eliminated *before* quantization, even though the elimination of non-physical variables and quantization do not commute. However, the latter typically occurs when non-physical variables are associated with curvilinear coordinates and, hence, is admissible for the Cartesian variable y.

The second condition in (5.25) means that Φ is invariant under the rotation group $SO(2)$, that is, in the coordinate representation it is independent of the polar angle φ. Therefore, $\Phi = \Phi(\mathbf{x}^2)$ because $\mathbf{x}^2$ is the only independent invariant of $SO(2)$ that can be obtained from the vector $\mathbf{x}$. The scalar product in the physical Hilbert space is defined by

$$\frac{1}{2\pi}\int d^2x \Phi_1^*(\mathbf{x}^2)\Phi_2(\mathbf{x}^2) = \int_0^\infty drr\Phi_1^*(r^2)\Phi_2(r^2).$$

The physical states are normalized as,

$$\frac{1}{2\pi}\int d^2x|\Phi|^2 = \int_0^\infty drr|\Phi(r^2)|^2 = 1. \tag{5.26}$$

Note that, while the resulting theory appears to be one-dimensional, the integration is carried out with a non-standard measure. This comprises the invariant approach. By going over to polar coordinates in the Hamiltonian (5.6),

$$\hat{H} = \tfrac{1}{2}\left[\hat{p}_r^2 + r^{-2}\hat{p}_\varphi^2 - (4r^2)^{-1}\right] + y\hat{p}_\varphi + V(r^2)$$

the energy operator is obtained in the physical subspace ($\hat{p}_\varphi\Phi = \hat{\pi}\Phi = 0$),

$$\hat{H}_{\text{ph}} = \frac{1}{2}\left(-\frac{\partial^2}{\partial r^2} - \frac{1}{r}\frac{\partial}{\partial r}\right) + V(r^2) = \frac{1}{2}\left(\hat{p}_r^2 - \frac{1}{4r^2}\right) + V(r^2). \tag{5.27}$$

Here $\hbar = 1$ and $\hat{p}_r\Phi = -ir^{-1/2}\partial(r^{1/2}\Phi)/\partial r$. The physical wave functions are normalized according to (5.26).

One can get rid of the non-physical variables before quantization. In the canonical variables r, p_r and φ, p_φ, the Hamiltonian (5.6) assumes a simple form, $H_{\text{ph}} = p_r^2/2 + V(r^2)$ when the constraints $\pi = p_\varphi = 0$ are taken into account. However, the corresponding quantum Hamiltonian obtained by the formal quantization rule, $p_r \to \hat{p}_r = -i\partial_r$, does not coincide with (5.27). This approach is, a priori, acceptable. But it has several significant drawbacks. First, the theory is formulated on the semi-axis, $r \geq 0$, where the operator $\hat{p}_r = -i\partial_r$ is not self-adjoint and cannot be extended to a self-adjoint operator [138, 139]. Second, in order for the Schrödinger equation to have meaning, it must be supplemented by a boundary condition at the point $r = 0$; that is, some additional principles must be invoked. Third, the choice of gauge-invariant variables is not unique. One can make a generic canonical transformation of r, p_r. The physical Hamiltonian certainly depends on the choice of physical canonical variables (for more details see Section 5.3.2). If the new canonical momentum is formally replaced by the corresponding derivative operator in the classical Hamiltonian, then it is clear that quantum theories associated with different choices of the physical canonical variables are not (unitary) equivalent, in general.

The Dirac quantization method [148] does not have the aforementioned drawbacks (compare Section 5.3.2). For this reason it is accepted as a "benchmark" when comparing different quantization recipes.

The non-invariant approach is based on the use of the condition $x_2 = p_2 = 0$ in the Hamiltonian (5.6). However, such a straightforward procedure is not admissible. It is incorrect to quantize the theory after solving the constraint $x_1 p_2 = x_2 p_1$ for p_2, substituting it into (5.6), and imposing the condition $x_2 = 0$. The reason is that the constraint must be fulfilled only on the physical states. Therefore, according to Section 4.1, the substitution $p_2 = (x_2/x_1)p_1$ in the term p_2^2 of (5.6) can be done only for one momentum [96, 188]: $p_2^2\Phi = p_2(x_2/x_1)p_1\Phi$, because the state $(x_2/x_1)p_1\Phi$ is no longer physical. The second momentum p_2 should be commuted with x_2 so that it acts on Φ. After that the above substitution is justified and can be carried out with the sequential elimination of the variable x_2 by means of the gauge condition $x_2 = 0$. As a result, the term p_2^2 turns into $(-i/x_1)p_1$. The sought-for Hamiltonian reads,

$$\hat{H}_{\text{ph}} = \frac{1}{2}\left(\hat{p}_1^2 - \frac{i}{x_1}\hat{p}_1\right) + V\left(x_1^2\right). \tag{5.28}$$

The operator (5.28) is defined on the whole real axis, and the residual gauge freedom associated with the group Z_2 (compare (5.18)) must be taken into account. The latter means that the physical states are subject to an additional gauge condition,

$$\hat{P}\Phi(x_1) = \Phi(-x_1) = \Phi(x_1), \quad \hat{P}\hat{x}_1 = -\hat{x}_1\hat{P}. \tag{5.29}$$

Recall that $\hat{P}$ is the only non-trivial element of Z_2. It follows from (5.29) that the physical states Φ are even functions of x_1, i.e. $\Phi = \Phi\left(x_1^2\right)$. The physical states are normalized according to (5.26),

$$\frac{1}{2}\int_{-\infty}^{\infty} dx_1 |x_1| \left|\Phi\left(x_1^2\right)\right|^2 = 1. \tag{5.30}$$

The choice of the integration measure requires explanation. In the non-invariant approach, the integration measure can be found by demanding that the Hamiltonian (5.28) be Hermitian. Integration over the whole real axis corresponds to the initial formulation of the problem: x_1 can attain all real values. But the physical range of x_1 is the semi-axis $x_1 \geq 0$. For this reason, the normalization integral (5.30) has the factor 1/2 and contains $|x_1|$ instead of x_1. The quantum theories obtained in the invariant and non-invariant approaches are obviously identical.

Equations (5.26), (5.29), and (5.30) clarify the role of the discrete gauge group in quantum theory. First, eigenfunctions of the Hamiltonian that are not invariant under the discrete group must be eliminated. Second, the normalization condition changes (the factor $|x_1|/2$ in (5.30)). Yet, there is one more feature worth noting. If the non-physical variables are eliminated before quantization, the resulting theory is described by the Hamiltonian $p_1^2/2 + V\left(x_1^2\right)$, which clearly differs from the theory defined by (5.28) and (5.30). All that has been stated about the invariant approach holds for the non-invariant method as well. Here, as in the invariant approach, in order to avoid possible dependence of the quantum theory on gauge choice, one has to apply the Dirac method (the non-invariant version of it, as shown above).

The example discussed illustrates the statement about non-commutativity of the canonical quantization and elimination of non-physical degrees of freedom in constrained systems [96, 175, 183, 186]. The resulting theories may have different Hamiltonians and scalar products. Of course, in particular cases, the two procedures may happen to be commutative. For example, variable y in the model (5.2) can be eliminated at any stage after going over to the Hamiltonian formalism. The non-commutativity noted above is also relevant for Yang–Mills theory [175, 202].

5.2 Harmonic oscillator with a conic phase space

The elimination of non-physical degrees of freedom in the classical theory for the model discussed in Section 5.1 is straightforward. Solving the constraint $p_2x_1 - p_1x_2 = 0$ for p_2, substituting the solution into (5.6), and setting then $x_2 = 0$, one obtains the Hamiltonian $p_1^2/2 + V\left(x_1^2\right)$. The effects of the redundant gauge symmetry on the system dynamics is taken into account by identifying the points (x_1, p_1) and $(-x_1, -p_1)$ in the phase space. In the case of a harmonic oscillator, $V = x_1^2/2$, a phase-space trajectory is a circle centered at the origin and whose radius is determined by the energy of the system. As the states (x_1, p_1) and $(-x_1, -p_1)$ on the circle are physically identical, the system returns to the initial state twice as fast. Hence, the physical period of oscillations is halved, or the physical frequency is doubled. The observed phenomenon should also appear in quantum theory, i.e. the distance between energy levels of the harmonic oscillator should double.

Let us illustrate the validity of this statement with an example of an n-dimensional isotropic harmonic oscillator with the gauge group $SO(n)$. The gauge transformations are rotations of the position vector $\mathbf{x}$ about the origin [183]. The Lagrangian is a natural generalization of (5.2),

$$L = \tfrac{1}{2}(\dot{\mathbf{x}} + y_a T_a \mathbf{x})^2 - \tfrac{1}{2}\mathbf{x}^2, \tag{5.31}$$

where T_a^{ij} are skew-symmetric matrices being the generators of $SO(n)$, $a=1,2,...,n(n-1)/2$. Going over to the Hamiltonian formalism is analogous to that for model (5.2). The canonical momenta $\mathbf{p} = \partial L/\partial \dot{\mathbf{x}} = \dot{\mathbf{x}} + y_a T_a \mathbf{x}$ and $\pi_a = \partial L/\partial \dot{y} = 0$, and the canonical Hamiltonian reads,

$$H = \tfrac{1}{2}\mathbf{p}^2 + \tfrac{1}{2}\mathbf{x}^2 - y_a \sigma_a, \tag{5.32}$$

where

$$\sigma_a = \{\pi_a, H\} = p_i T_a^{ij} x_j = 0 \quad (i,j = 1,2,\ldots,n), \tag{5.33}$$

are the secondary constraints. They generate the action of $SO(n)$ on the phase-space variables. All the constraints are in involution (the first-class constraints):

$$\{\sigma_a, H\} = f_{abc} y_b \sigma_c, \quad \{\pi_a, H\} = \sigma_a, \tag{5.34}$$

$$\{\sigma_a, \sigma_b\} = f_{abc}\sigma_c, \qquad \{\sigma_a, \pi_b\} = \{\pi_a, \pi_b\} = 0, \tag{5.35}$$

where f_{abc} are the structure constants of the group $SO(n)$, $[T_a, T_b] = f_{abc} T_c$.

It is easy to see that the system has only one physical degree of freedom and its phase space is cone(π). Indeed, the quantities σ_a are nothing but the components of the angular momentum of a particle moving in R^n. Therefore equalities (5.33) imply that only the radial motion is physical (the angular momentum vanishes). A general solution of the constraints (5.33) has the form $\mathbf{p} = \lambda \mathbf{x}$ as for the model in Section 5.1. By a suitable gauge transformation (by an n-dimensional rotation) the vector $\mathbf{x}$ and, hence, $\mathbf{p}$ can be directed along one of the coordinate axes, say, $x_i = \delta_{i1} x$. Moreover, by a gauge rotation through angle π one can change the sign of x at any moment of time (the redundant gauge symmetry Z_2). Accordingly, on the phase plane spanned by x and its canonical momentum p, the points (x, p) and $(-x, -p)$ have to be identified so that the plane turns into the cone, cone(π). Hence, the physical frequency of the oscillator doubles as in the model studied in Section 5.1.

5.2.1 Quantum theory. Coordinate representation

The quantum theory is constructed by the standard (canonical quantization) procedure. All canonical variables are replaced by operators and the Poisson brackets by commutators, $\{\,,\,\} \to -i[\,,\,]$. As in the simple $SO(2)$ model, the non-physical degrees of freedom y_a, π_a are eliminated before quantization. Physical states are defined by the conditions,

$$\hat{\sigma}_a \Phi_{ph}(\mathbf{x}) = 0. \tag{5.36}$$

The $n(n-1)/2$ constraints in the system are not all independent. As the stationary group of any nonzero vector in R^n is $SO(n-1)$, only

$n(n-1)/2-(n-1)(n-2)/2=n-1$ constraints are independent. The physical states are functions that are independent of the angular variables: $\Phi_{ph}(\mathbf{x})=\Phi(r)$. Thanks to the $n-1$ independent constraints, out of n degrees of freedom only one remains.

The Schrödinger equation for radial functions in the n-dimensional spherical coordinate system is written in the form,

$$\left[-\frac{d^2}{dr^2}-\frac{n-1}{r}\frac{d}{dr}+\frac{l(l+n-2)}{r^2}+r^2\right]\Phi(r)=2E\Phi(r), \tag{5.37}$$

where $l=0,1,...$ Substituting $\Phi(r)=r^l\exp(-r^2/2)f(r)$ into (5.37) and setting $z(t)=f(r)$ with $t=r^2$, the function $z(t)$ is shown to satisfy the equation,

$$tz''+(\alpha-t)z'-\beta z=0, \tag{5.38}$$

where $\alpha=l+n/2$ and $\beta=(\alpha-E)/2$. Its solution, regular at $r=0$, is given by the confluent hypergeometric function,

$$z(t)={}_1F_1(\beta,\alpha;t). \tag{5.39}$$

From the condition that $\Phi(r)$ should decrease as $r\to\infty$, or, equivalently, that $z(t)$ is a polynomial, which happens when $\beta=-k$ (a non-positive integer), the energy spectrum is found,

$$E_{kl}=2k+l+\frac{n}{2},\quad k=0,1,\ldots. \tag{5.40}$$

Taking into account the relation between the confluent hypergeometric function ${}_1F_1$ and the Laguerre polynomials L_k^α [132, Section 8.972]

$${}_1F_1(-k,\alpha+1;t)=L_k^\alpha(t)\Gamma(k+1)\Gamma(\alpha+1)/\Gamma(k+\alpha+1),$$

the solution of (5.37) can be written in the form,

$$\Phi_{kl}(r)=Cr^lL_k^{l-1+(n/2)}(r^2)e^{-r^2/2}, \tag{5.41}$$

with C being a normalization constant. The physical solutions are given by the spherically symmetric functions Φ_{k0}. From (5.40) and (5.41) it follows that at $l=0$ the distance between the energy levels is indeed doubled, and the physical states are even functions of r. The latter implies, in particular, that in the non-invariant approach the regular (at zero) eigenfunctions of the Hamiltonian (5.28) (with $V=x_1^2/2$) automatically satisfy the condition (5.29).

In conclusion, let us discuss the physical meaning of the variable x_1 (which describes the physical degree of freedom in the non-invariant approach). At the very beginning, it was clear that x_1 is not "quite" physical because it changes under the gauge transformation, $x_1\to -x_1$. From (5.29) and (5.30) it follows that all matrix elements of x_1 in the physical space vanish,

$$\langle\Phi_2|\hat{x}_1|\Phi_1\rangle=\frac{1}{2}\int_{-\infty}^{\infty}dx_1|x_1|^{n-1}\Phi_1^*\left(x_1^2\right)x_1\Phi_2\left(x_1^2\right)=0. \tag{5.42}$$

Equality (5.42) is equivalent to the following statement: the state with a specific value of x_1 cannot belong to the physical Hilbert space $\mathcal{H}_{ph}$. Indeed, if $\hat{x}_1\psi = x_1\psi$, then $\hat{P}\psi \neq \psi$, while $\hat{P}\Phi = \Phi$ for every physical state. The operators $\hat{P}$ and $\hat{x}_1$ do not commute, $\hat{P}\hat{x}_1 = -\hat{x}_1\hat{P}$, and therefore they cannot have common eigenstates except those for which $\hat{x}_1\psi = 0$. Only even powers of x_1 have a physical meaning, that is, the variables like $|x_1| = x_1\epsilon(x_1), x_1^2, x_1^4, ...$, where $\epsilon(x_1)$ is the sign function (by the Weierstrass theorem, $|x_1|$ can be viewed as an even power series). This observation is relevant for quantum field theory. Excitations described by the absolute value of the field cannot have the conventional propagator, i.e. they do not propagate [183], [200], [203–206] (see also Section 5.8.3).

Let us summarize the findings. After the elimination of non-physical degrees of freedom, there are two equivalent approaches to describe the studied feature of the dynamical system: (1) the system has a conical phase space; (2) it is invariant under a discrete gauge group (in this case, it is Z_2). The first approach is useful for studying classical Hamiltonian dynamics, while the second one is convenient for the Lagrangian path integral method in quantum theory. The physical consequence of this dynamical feature is a change of the physical spectrum of the Hamiltonian as well as a reduction of the Hilbert space. In particular, the physical frequency of the harmonic oscillator is doubled, and the Hilbert space consists only of parity invariant states. It should be stressed though that condition (5.29) is relevant only when the non-physical variables are eliminated (when a one-dimensional system equivalent to (5.2) is studied). In the framework of the Dirac formalism, all the information about the physical states is contained in the conditions (5.25) and (5.36).

5.2.2 The Fock space representation

It is instructive to solve the quantization problem for system (5.2) with the harmonic oscillator potential in the Fock space representation as the latter is a commonly used tool in quantum field theory. Setting $V = \mathbf{x}^2/2$ in (5.6) and introducing the annihilation and creation operators by $\mathbf{a} = (\mathbf{p} - i\mathbf{x})/\sqrt{2}$, $\mathbf{a}^\dagger = (\mathbf{p} + i\mathbf{x})/\sqrt{2}$, the Hamiltonian operator can be written in the form,

$$\hat{H} = (\hat{\mathbf{a}}^\dagger, \hat{\mathbf{a}}) + 1 + i\hat{y}\,(\hat{\mathbf{a}}^\dagger, T\hat{\mathbf{a}}), \quad \left[a_i, a_j^\dagger\right] = \delta_{ij}. \tag{5.43}$$

The states $\left(\hat{a}_1^\dagger\right)^m \left(\hat{a}_2^+\right)^n |0\rangle$, $m, n = 0, 1, \ldots$, form an orthogonal basis in the Hilbert space (excluding the y degree of freedom). From the condition that physical states are annihilated by the constraint operator $\hat{\sigma}\ |\Phi\rangle = 0$, where $\hat{\sigma} = (\hat{\mathbf{a}}^\dagger, T\hat{\mathbf{a}})$, an orthogonal basis of the physical Hilbert space is inferred,

$$|\Phi_k\rangle \sim [(\hat{\mathbf{a}}^\dagger)^2]^k\ |0\rangle. \tag{5.44}$$

The result (5.44) is evident if one notes that $\hat{\sigma}$ is nothing but a generator of rotations of the vector $\hat{\mathbf{a}}$, $[\hat{\sigma}, \hat{\mathbf{a}}^\dagger] = -T\hat{\mathbf{a}}^\dagger$, $[\hat{\sigma}, \hat{\mathbf{a}}] = -T\hat{\mathbf{a}}$, and it annihilates the vacuum state, $\hat{\sigma}\,|0\rangle = 0$. Any physical state can therefore be obtained by applying an invariant polynomial of $\mathbf{a}^\dagger$ to the vacuum state. Any such polynomial is a polynomial of the only independent rotation invariant $(\hat{\mathbf{a}}^\dagger)^2$, which proves (5.44). It then follows from the commutation relation

$$[(\hat{\mathbf{a}}^\dagger, \hat{\mathbf{a}}),\ (\hat{\mathbf{a}}^\dagger)^2] = 2(\hat{\mathbf{a}}^\dagger)^2 \tag{5.45}$$

that the distance between the energy levels is doubled. In the case of the n-dimensional harmonic oscillator, whose physical states satisfy $\hat{\sigma}_a|\Phi\rangle = 0$ (see (5.33)), the orthonormal physical basis reads [200, 207],

$$|\Phi_k\rangle = \left(\frac{4^k k!\Gamma(k+n/2)}{\Gamma(n/2)}\right)^{-1/2} [(\mathbf{a}^\dagger)^2]^k|0\rangle. \tag{5.46}$$

The distance between the physical energy levels is doubled as well, as one might deduce from (5.45).

It should be emphasized that any physical state contains excitations of all n harmonic oscillators and the physical picture that emerges is not reducible to a merely one-dimensional harmonic oscillator with a doubled frequency. Even in the case $n = 3$ in which Eq. (5.40) for $l = 0$ and $n = 3$ gives $rL_k^{1/2}(r^2) \sim H_{2k+1}(r)$ [132, Section 8.972] the normalization coefficients in (5.46) differ from those of the harmonic oscillator with a double frequency. Apparently, the coherent states obtained from the oscillator basis states (5.46) do not coincide with the conventional coherent states for the harmonic oscillator with a doubled frequency [207].

The Fock space representation turns out to be quite useful in quantizing gauge systems where dynamical variables realize the adjoint representation of generic Lie groups (see Sections 5.4 and 5.6).

5.3 The residual discrete gauge group and the choice of physical variables

One might get the impression from the above arguments that the discrete gauge group Z_2, being a subgroup of the gauge group, is a characteristic feature of the dynamical system like, e.g. the dimension of the physical phase space. In a way, this is true. An accurate statement is that the residual gauge symmetry is fully determined by the choice of a gauge fixing condition [208], while the group Z_2 is the simplest, from all the possibilities, as a consequence of the gauge condition $x_2 = 0$ being the simplest way to eliminate the non-physical degree of freedom in the model (5.2). The structure of the residual gauge group can be significantly complicated if unsuitable ways are used to eliminate non-physical degrees of freedom. This is similar to the problem of gauge fixing in Yang–Mills theories [209, 211], known as the Gribov problem. Let us examine the details.

5.3.1 A non-invariant approach

Suppose that in model (5.2) a physical variable is defined as a parameter spanning a curve in the (x_1, x_2) plane rather than the straight line $x_2 = 0$. There are several essentially distinct possibilities for the choice of curve, which are illustrated in Fig. 5.3 by curves 1–3. The concentric circles are gauge orbits. The gauge $x_2 = f_1(x_1)$ (curve 1) is not acceptable at all, for the physical region $0 \le r < r_0$ ($r = |\mathbf{x}|$) is eliminated. Curve 1 is tangential to the circle $S(r_0)$ of radius r_0 and does not intersect it. With the choice $x_2 = f_2(x_1)$ (curve 2), the orbits with radii $r < r_1$ and $r > r_2$ are intersected twice, while those with $r_1 < r < r_2$, four times. In contrast to $x_2 = 0$, transformations that relate the points of intersection of each orbit with curve 2 do not form a subgroup of the gauge group $SO(2)$. Indeed, the point $\mathbf{x}_2$ of the dashed circle $S(r)$ may be transformed to the point $\mathbf{x}_1$ by a suitable rotation about the origin, but under such a rotation point $\mathbf{x}_3$ is not moved to $\mathbf{x}_2$, $\mathbf{x}_4$ to $\mathbf{x}_3$, or $\mathbf{x}_1$ to $\mathbf{x}_4$. Thus a combination of two residual transformations does not generally give another residual transformation. Clearly, the residual gauge symmetry can be made as complicated as desired by choosing the curve accordingly. For the line $x_2 = 0$ the physical range of x_1 is given by non-negative real numbers. The physical range of x_1 for the curve $x_2 = f(x_1)$

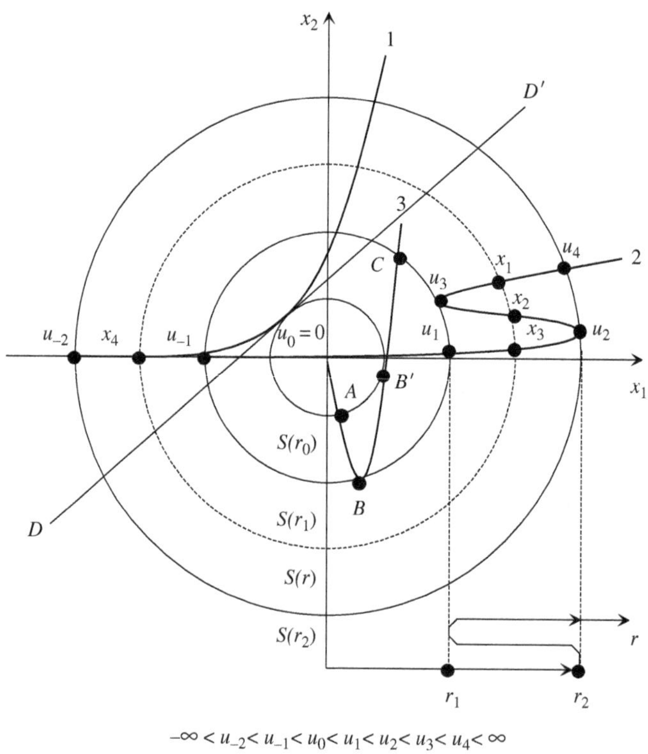

Fig. 5.3. An illustration of the change of variables associated with the gauge transformation law and the gauge condition chosen.

may even be a disconnected set. A simple example is provided by curve 3. The arc segments AB, BB' and $B'C$ represent gauge equivalent states. Therefore two of them should be eliminated by restricting the range of x_1, which implies that the physical range of x_1 becomes disjoint. Clearly, one should bear in mind all such potential complications when studying more realistic gauge theories where the gauge orbit structure is not that simple or even unknown.

The model discussed serves as a good illustration of the problem of gauge fixing ambiguity [209]. An inconvenient choice of a gauge fixing condition can make the problem meaningless (curve 1). Alternatively, it can enormously complicate the problem (an extension of curve 2 with an increasing amplitude of oscillation).

It should be noted that, although the origin of gauge fixing ambiguity discovered in [209] and in the model (5.2) is the same, namely, the physical variables are associated with curvilinear coordinates that span the orbit space, the work [209] deals with fixing the gauge freedom of the theory, that is, fixing y, while here elimination of unphysical degrees of freedom, for instance, x_2 after the elimination of y is studied. The gauge fixing ambiguity phenomenon is often attributed to a complexity of the gauge group because the problem emerged when studying non-Abelian gauge theories (see also [210]). From the above example, it is clear that a similar problem may occur even in an Abelian theory. In a non-Abelian gauge theory, invariant gauges are inappropriate (e.g. the Fermi gauge) because they do not admit finite residual groups [210]. This implies that relativistic invariant gauge conditions do not appear natural to such theories (see also [211]). A further discussion of Yang–Mills theories is given in Section 5.8.

In the model considered the gauge orbit structure is very simple, a set of concentric circles in a plane. In realistic gauge theories it is not that simple and often unknown. The physical configuration space is the orbit space. In particular, it can be shown that the orbit space in classical Yang–Mills field theories contains non-contractable spheres of any dimension [211]. A gauge fixing is nothing but parameterization of the orbit space by elements of the gauge fixing surface in the total configuration space. It might turn out to be suitable, like $x_2 = 0$ in the above mechanical model, or unsuitable like curve 2 in Fig. 5.3. If the orbit space structure is unknown, all gauges can a priori be used on an equal footing. Therefore one should be able to describe classical and quantum dynamics in a general case, taking into account all possible complications that may result from a not so suitable choice of a gauge condition. In addition, one should be able to select admissible gauge conditions. The gauge fixing surface of an admissible gauge intersects each orbit at least once. An example of an inadmissible gauge is given in Fig. 5.3 by curve 1. The problem is important for the path integral formalism in gauge theories. A method for obtaining a correct path integral formalism with an arbitrary admissible gauge condition is given in Chapter 6. Here an operator formalism of quantum theory is developed in the framework of the Dirac theory under a general choice of physical variables.

5.3.2 A description of physical variables by curvilinear coordinates

As argued in Section 5.1, the gauge $x_2 = 0$ is directly related to the gauge-invariant dynamical description of the model in terms of the invariant variable $r = (\mathbf{x}^2)^{1/2}$. How should more complicated gauge conditions, like those defined by curves 2 and 3 in Fig. 5.3, be treated in a gauge-invariant approach? Consider a generalization of polar coordinates in a plane defined by [24, 212, 213],

$$\begin{pmatrix} x_1 \\ x_2 \end{pmatrix} = e^{\theta T} \begin{pmatrix} f_1(u) \\ f_2(u) \end{pmatrix}, \tag{5.47}$$

where $f_i(u)$, $i=1,2$, are functions of a real parameter u. The case $f_1(u)=u$, $f_2=0$ corresponds to polar coordinates with $u^2 = \mathbf{x}^2$. Under the action of the matrix $e^{\theta T}$, the ray $x_2 = 0$, $x_1 = r \geq 0$ sweeps the entire plane when θ ranges from 0 to 2π. For arbitrary f_i, the relations $x_1 = f_1(u)$, $x_2 = f_2(u)$ are parametric equations of a curve in the (x_1, x_2) plane. The parameter u is gauge invariant by construction because $f_1^2(u) + f_2^2(u) \equiv r^2(u) = \mathbf{x}^2$.

Without loss of generality, the variable u may be assumed to range over the whole real line (e.g. a parameterization of the gauge fixing curve by the arclength counted from the origin). Then the mapping (5.47) is not generally one-to-one if pairs (u, θ) span the whole plane R^2 (there are points in the (u, θ) plane that are mapped onto the same point in the (x_1, x_2) plane), i.e. it does not define a change of variables. Let $\tilde{K} \subset R^2$ be the set of all (u, θ) such that the mapping (5.47) is one-to-one. So, if the domain of (u, θ) is restricted to $\tilde{K}$, relation (5.47) defines a change of variables. The very existence of $\tilde{K}$ imposes some conditions on f_i. The curve $x_i = f_i(u)$ should pass through the origin and go to infinity $|\mathbf{x}| \to \infty$. In what follows it is assumed, without loss of generality, that $u \in R$, $f_i(0) = 0$ and $r(u) \to \infty$ as $u \to \pm\infty$. In addition, it is also demanded that the curve be smooth (the functions f_i are at least twice continuously differentiable). The smoothness assumption simplifies the derivation of the physical Hamiltonian in quantum theory.

In order to find $\tilde{K}$ one should study symmetry transformations of the plane $(u, \theta) \to (u_s, \theta_s)$ under which $\mathbf{x}$ in (5.47) does not change. For example, for the polar coordinates ($f_1 = u \equiv r$, $f_2 = 0$) these transformations form a group that consists of two subgroups, $\tilde{S} = S_\theta \times S$, $S_\theta : \theta \to \theta + 2\pi n$, $r \to r$, with n being an integer, and $S : r \to -r$, $\theta \to \theta + \pi$ (see Section 2.7 and [120]). The points of the (r, θ) plane related by transformations from $\tilde{S}$ correspond to the same point of the (x_1, x_2) plane. Hence, $\tilde{K}$ is the fundamental modular domain of the plane $R^2(r, \theta)$ relative to the group $\tilde{S}$. Any point of $R^2(r, \theta)$ can be obtained from a point in $\tilde{K}$ by the action of a suitable element from $\tilde{S}$. For instance, $\tilde{K}$ can be chosen to be the strip $\theta \in [0, 2\pi)$, $r \in [0, \infty)$. By analogy, $\tilde{S}$ for the mapping (5.47) can be defined as $\tilde{S} = S_\theta \times S$, where $S_\theta : \theta \to \theta + 2\pi n$, $u \to u$ with n being an integer, and $S : u \to u_s = u_s(u)$, $\theta \to \theta + \theta_s(u)$. An important difference with the particular case of polar coordinates is that the transformations from S can be functions of u, and the number of independent elements of S (i.e. the number of functions u_s and θ_s) may also depend on u. An explicit form of these functions is determined by conditions under which the gauge fixing curve $x_i = f_i(u)$ intersects a gauge orbit (a circle centered at the origin). Indeed, from the definition of S it follows that,

$$\begin{pmatrix} x_1 \\ x_2 \end{pmatrix} = e^{\theta T} \begin{pmatrix} f_1(u) \\ f_2(u) \end{pmatrix} = e^{(\theta+\theta_s)T} e^{-\theta_s T} \begin{pmatrix} f_1(u) \\ f_2(u) \end{pmatrix}$$
$$= e^{(\theta+\theta_s)T} \begin{pmatrix} f_1(u_s) \\ f_2(u_s) \end{pmatrix}. \tag{5.48}$$

The last equality in (5.48) implies that a point $\mathbf{x} \in R^2$ is not changed under the transformation $(u, \theta) \to (u_s, \theta + \theta_s)$ in (5.47) and therefore this transformation belongs to S. It is also clear that the points $x_i = f_i(u)$ and $x_i^s = f_i(u_s)$ lie on the same circle (gauge orbit) because they can be obtained from one another by rotation through angle θ_s about the origin. Thus, the transformations $u \to u_s$ define other possible values of the parameter u at which the curve $\mathbf{x} = \mathbf{f}(u)$ intersects the same orbit (circle of radius $r(u)$). In particular, the latter means that all functions u_s can be found by solving the equation

$$r^2(u_s(u)) = r^2(u), \tag{5.49}$$

where $r^2(u) = \mathbf{f}^2(u) - \mathbf{x}^2(u)$.

Given a particular u, Eq. (5.49) may have many solutions. Moreover, the structure of solutions of (5.49) may be rather complicated. For instance, the gauge fixing curve can have self-intersections, countably many intersections with a particular gauge orbit, etc. All such cases are not discussed here. For the sake of simplicity, it is demanded that there be no self-intersections and Eq. (5.49) has a finite number of solutions for each value of $u \in R$, i.e. the curve $\mathbf{x} = \mathbf{f}(u)$ intersects each orbit finitely many times.

Under the above simplifying assumptions, there exist sets R_α such that Eq. (5.49) has a fixed number N_α of solutions if $u \in R_\alpha$. Let S_α be the set of transformations from S acting on R_α. Then $S = \prod_\alpha S_\alpha$. The fundamental modular domain $\tilde{K}$ is $\theta \in [0, 2\pi)$ and $u \in K$, where $K = \bigcup_\alpha K_\alpha$ and $K_\alpha = R_\alpha / S_\alpha$ is the fundamental modular domain of the set R_α relative to the action of S_α. Let $u \in K$ in (5.49). Then in every subset K_α, the solutions $u_s(u)$ are continuous mappings, $u_s : K_\alpha \to K_{\alpha s}$, where $K_{\alpha s}$ is the range of u_s. By the definition of S_α, there should be $R_\alpha = \bigcup_S K_{\alpha s}$. Note that S_α contains the identity mapping $K \to K$ which corresponds to a trivial solution $u_s = u$ of (5.49). The mappings u_s are also one-to-one and, hence, there exist the inverse mappings $u_s^{-1} : K_{\alpha s} \to K_\alpha$. If f_i are differentiable, then u_s^{-1} are differentiable too.

In what follows the mapping $u_s : K \to K_s = \bigcup_\alpha K_{\alpha s}$ is treated as a function $u_s = u_s(u)$, which, strictly speaking, requires clarification because each S_α has a different number of elements. So the symbol $u_s(u)$ denotes an element of the set

S_α whenever $u \in K_\alpha$. This should not lead to any confusion because u always belongs to one of the domains $K_\alpha \subset K$. When constructing an explicit form of u_s, the region K must be specified as a collection of intervals and a half-axis. The choice of K is obviously not unique. Once K is specified, the domains K_α as well as the functions $u_s(u)$, $u \in K_\alpha$, are found by solving (5.49). Note that (5.49) defines a mapping $R \to R_+$ where $u \in R$ and $r \in R_+$ (R_+ is isomorphic to the orbit space) with every choice of f_i. The modular domain K is determined by the condition that the mapping $r(u)$ becomes one-to-one when restricted on $K \subset R$. The mappings $u_s :\ K \to R$ are then uniquely defined by the condition $r(u_s) = r(u)$.

For example, for curve 2 in Fig. 5.3, the range of $u \in R$ is split into three regions, $R_1 = [u_{-1}, 0] \bigcup [0, u_1]$, $R_2 = [u_{-2}, u_{-1}] \bigcup [u_1, u_2] \bigcup [u_2, u_3] \bigcup [u_3, u_4]$, and $R_3 = (-\infty, u_{-2}] \bigcup [u_4, \infty)$. The points $f_i(u_a)$, $a = -2, -1, \ldots, 4$, are indicated by the symbols u_a in Fig. 5.3. The numbers of points of intersection of the curve with gauge orbits for each R_α are $N_1 = 2$, $N_2 = 4$, and $N_3 = 2$. One can take any of the intervals $[u_{-1}, 0]$ and $[0, u_1]$ as K_1, any of the half-axes in R_3 as K_3, and, finally, any of the four intervals in R_2 as K_2. Thus, there are $2 \cdot 4 \cdot 2 = 16$ ways to construct the fundamental modular domain K. Having fixed a particular choice of K, the functions $u_s(u)$ define one-to-one mappings, $K_\alpha \to K_{\alpha s}$, $\alpha = 1, 2, 3$. For instance, if $[0, u_1] = K_1 \subset K$, then the only non-trivial element u_s of S_1 maps K_1 onto $K_{1s} = [-u_1, 0]$ so that $r(u_s) = r(u)$ from which an explicit form of the function $u_s(u)$ can be found.

Another useful way to define the functions $u_s(u)$ is the following. A gauge fixing curve is a mapping $\mathbf{f} :\ R \to R^2$, admissible in the aforementioned sense, i.e. a mapping from a parameter (Euclidean) space into the total configuration space. Let $\hat{s}$ be elements of the group S_f of automorphisms of R subject to the condition that the points $\mathbf{f}(\hat{s}u)$ belong to the same gauge orbit for every fixed u, $r(\hat{s}u) = r(u)$. Then the factor space $R/S_f \subset R$ is isomorphic to the orbit space. Since every $\hat{s}(R/S_f)$ is also isomorphic to the orbit space, any of $\hat{s}(R/S_f)$ can be chosen as the modular domain K. The set of mappings S is then a restriction of S_f to K. By making a particular choice of K, a representation of S by functions $u_s(u)$ on K is obtained by restricting every $\hat{s}$ to K, that is, $\hat{s}u = u_s(u)$, $u \in K$. This notation is used below. Note that distinct elements of S_f may become identical when restricted to K. The points $\mathbf{f}(u)$ and $\mathbf{f}(\hat{s}u)$, for every $\hat{s} \in S_f$, are related to one another by a gauge transformation $\Omega_s \mathbf{f}(u) = \mathbf{f}(\hat{s}u)$, $\Omega_s \in SO(2)$. Since S_f is a group, $\hat{s}^2 \in S_f$, but generally $\Omega_s^2 \mathbf{f}(u) \neq \mathbf{f}(\hat{s}^2 u)$. Moreover, there may be no element $\hat{s}_1 \in S_f$ such that $\Omega_s^2 \mathbf{f}(u) = \mathbf{f}(\hat{s}_1 u)$. A simple example is a parabolic gauge fixing curve, $x_1 = u$ and $x_2 = u^2$. Note that $S_f = Z_2$ in this case ($S_f : u \to \pm u$). For a positive $u > 0$, there is $\Omega_s = \Omega_s(u) \in SO(2)$ such that $\Omega_s(u)\mathbf{f}(u) = \mathbf{f}(-u)$, i.e. a counterclockwise rotation in the (x_1, x_2)–plane that transforms the vector $\mathbf{f}(u) = (u, u^2)$ into the vector $\mathbf{f}(-u) = (-u, u^2)$. However, under this rotation, the vector $\mathbf{f}(-u)$ does not become $\mathbf{f}(u)$. Moreover, the point $\Omega_s(u)\mathbf{f}(-u)$ does not even lie on the curve $\mathbf{x} = \mathbf{f}(u)$. Hence, the rotation $\Omega_s^2(u)$ does not belong to the set of the residual gauge transformations. Thus, the set of residual gauge transformations may not form a group in general. Although the composition of two functions u_{s_1} and u_{s_2} from S is not defined, the composition $\hat{s}_2\hat{s}_1$ is an element of S_f. Since every physical quantity Φ is defined on the orbit space, it is convenient to talk about the residual gauge symmetry in the sense of the group S_f when a gauge fixing surface is chosen to parameterize the gauge orbit space. Every physical quantity is invariant under transformations from S_f, $\Phi(\hat{s}u) = \Phi(u)$.

The change of variables (5.47) does not have a global coordinate grid on the plane for generic $\mathbf{f}(u)$. However, the plane can be divided into rings so that the coordinate grid exists in each ring, i.e. the coordinate grid on the plane is obtained by gluing coordinate grids in each ring. As an example, consider the coordinate grid associated with curve 2 in Fig. 5.3. Set $K_1 = [0, u_1]$, $K_2 = [u_1, u_2]$, and $K_3 = [u_4, \infty)$. Then in the disk bounded by the circle $S(r_1)$, the grid consists of concentric circles spanned by the variable $\theta \in [0, 2\pi)$ and a bundle of curves $l_1(\theta)$ spanned by $u \in K_1$ which are obtained from the curve $\mathbf{x} = \mathbf{f}(u)$, $u \in K_1$, by rotating all its points through the same angle θ. The coordinate grid in the disk bounded by the circles $S(r_1)$ and $S(r_2)$ also contains concentric circles (gauge orbits) and the family $l_2(\theta)$ of curves $\mathbf{x} = \exp(T\theta)\mathbf{f}(u)$, $u \in K_2$, where $\theta \in [0, 2\pi)$ labels the curves. The coordinate grid in the rest of the plane (outside the disk bounded by the circle $S(r_2)$) is obtained similarly.

The coordinate grid in the plane depends on the choice of K. If in the above example $K_2 = [u_1, u_2]$ is replaced by the interval $[u_3, u_2]$, then the family $l_2(\theta)$ consists of different curves because the vector function $\mathbf{f}(u)$ traces out different curves when u ranges over the intervals $[u_1, u_2]$ and $[u_3, u_2]$. The gluing procedure is changed as well when K is changed. In other words, the coordinate curve $\theta = \text{const.}$ as a continuous function of u depends on the choice of K, but the very existence of such a curve is guaranteed by the definition of regions R_α. In the above example, $K = [0, u_1] \bigcup [u_1, u_2] \bigcup [u_4, \infty)$, the coordinate curve $\theta = 0$ is obtained by gluing the three pieces of $\mathbf{x} = \mathbf{f}(u)$ with $u \in [0, u_1]$, $u \in [u_1, u_2]$, and $u \in [u_4, \infty)$, which is done by suitable rotation of each curve, i.e. no rotation for the first two pieces and the third one is rotated so that point $\mathbf{f}(u_4)$ goes into point $\mathbf{f}(u_2)$ (point u_4 goes into u_2 in Fig. 5.3). A generic curve corresponding to a non-zero θ is obtained by rotating the curve $\theta = 0$ through any given angle.

Once the new coordinate grid is defined, the dynamics can be analyzed in the new variables (5.47). The momenta canonically conjugated to the variables θ and u are, respectively,

$$p_\theta = (\mathbf{p}, T\mathbf{x}) = \sigma, \quad p_u = \frac{1}{2}(\mathbf{p}, \mathbf{x})\frac{d}{du}\ln \mathbf{x}^2. \tag{5.50}$$

Note that for $u \in K$ there is a one-to-one correspondence between u and $r = |\mathbf{x}|$, and the derivative d/du can always be written as $r'(u)(d/dr)$; this simplifies calculations. By making use of (5.47) it is straightforward to verify that $\{p_\theta, p_u\} = 0$, $\{\theta, p_\theta\} = \{u, p_u\} = 1$. To obtain the latter equality, the derivative $\partial u/\partial x_i$ is found by differentiating the equality $\mathbf{x}^2 = f_1^2(u) + f_2^2(u)$ with respect to x_i. It follows from (5.50) that θ is a non-physical variable because $p_\theta = \sigma$ and σ

is the secondary constraint, while u is the physical variable. Equations (5.47) and (5.50) define a canonical transformation. By writing the canonical Hamiltonian in the new canonical variables and setting in it $p_\theta = \sigma = 0$, the Hamiltonian for the physical degree of freedom described by p_u and u is obtained and so is its Hamiltonian dynamics. The vectors $\mathbf{x}$ and $T\mathbf{x}$ are orthogonal ($(\mathbf{x}, T\mathbf{x}) = 0$ because T is skew-symmetric), and, therefore, the following decomposition holds,

$$\mathbf{p} = p_\theta \frac{T\mathbf{x}}{\mathbf{x}^2} + p_u \frac{\mathbf{x}}{\mu(u)}, \tag{5.51}$$

where $\mu(u) = (\mathbf{f}(u), \mathbf{f}\,'(u))$ with $\partial_u \mathbf{f}(u) \equiv \mathbf{f}\,'(u)$. Substituting (5.51) into (5.6) and setting $p_\theta = 0$, the physical Hamiltonian is inferred,

$$H_{\text{ph}} = \frac{1}{2} \frac{r^2(u)}{\mu^2(u)}\, p_u^2 + V(r^2(u)). \tag{5.52}$$

The Hamiltonian equations of motions are derived in the standard way.

Points of the phase plane (u, p_u) related by transformations from S: $u \to \hat{s}u = u_s$, $p_u \to \hat{s}p_u = p_{u_s} = (du_s/du)^{-1} p_u$ (here $d/du_s = (du_s/du)^{-1} d/du$) in (5.50) correspond to the states of the system which differ from one another only by the value of the angle, $\theta \to \theta + \theta_s(u)$, similarly to the states (r, p_r) and $(-r, -p_r)$ in the polar coordinates which differ by the value of the angle $\theta \to \theta + \pi$ (see Section 5.1.1). In the gauge theory in question, θ is a non-physical variable. Therefore all points (u_s, p_{u_s}) of the phase plane are indistinguishable and should be identified. This feature of the physical phase space $\Gamma(u, p_u)$ must be taken into account when describing the system dynamics. The latter observation establishes a relation between invariant and non-invariant descriptions of the classical dynamics. Both methods coincide if an additional reduction of the phase space due to the residual gauge symmetry in the non-invariant approach is taken into account. Indeed, the action of S on the invariant variable u coincides with the action of the residual discrete gauge symmetry in the gauge $x_i = f_i(u)$. The points of the gauge fixing curve $x_i^s = f_i(u_s)$ lie on the same gauge orbit (see (5.49)) and, hence, are related by gauge transformations, while the points u and u_s are related by transformations from S. Note, however, that S is a subset of symmetry transformations $\tilde{S}$ of the change of variables and not generally related to the discrete residual gauge symmetry in a gauge-fixed theory. The mappings from S induce residual gauge transformations if the change of variables is constructed so that the coordinate surfaces of some variables coincide with gauge orbits, while the other variables span a portion of the gauge fixing surface through which each gauge orbit passes only once. Such a change of variable is said to be *compatible with the gauge transformation law and the gauge fixing.*

Thus, a gauge condition is admissible if there exists a change of variables in the total configuration space that is compatible with it and with the gauge transformation law.

5.3.3 Peculiarities of Hamiltonian dynamics in invariant variables

As noted above, with any given gauge fixing condition one can associate a formally gauge-invariant description. Thus, all artifacts inherent in an inappropriate gauge fixing may well emerge in the formally gauge-invariant approach. In particular, the following observation is of interest [195]. Consider a regular trajectory in the total phase (or configuration) space that is a solution of the equation of motion. It is defined modulo gauge transformations. The latter can be used to gauge transform the trajectory to a gauge fixing surface. Thus obtained trajectory may have singularities associated with a particular gauge fixing which cannot be ignored in the dynamical description. A similar phenomenon has been found in lattice Yang–Mills theories where regular configurations of Yang–Mills fields acquire monopole-like singularities when transformed to the so-called maximal Abelian gauge [214, 392]. Interestingly enough, the Yang–Mills configurations that have such a property give a dominant contribution to the Wilson loop expectation value in the confinement phase (for a review see [393]). So, when describing Yang–Mills dynamics in the maximal Abelian gauge, the dynamics of monopole-like defects must be taken into account [396–399].

To illustrate the point, let us compare trajectories in the physical phase space spanned by canonical variables $r = |\mathbf{x}|$, $p_r = (\mathbf{x}, \mathbf{p})/r$, and u, p_u. They are related by the canonical transformation $r = r(u) = |\mathbf{f}(u)|$, $p_r = rp_u/\mu = p_u(dr/du)^{-1}$. Suppose that $\mathbf{f}$ is chosen so that $dr/du = 0$ only at two points $u = u'_{1,2}$ and $dr/du > 0$ as $u < u'_2$ and $u > u'_1$, while $dr/du < 0$ if $u \in (u'_2, u'_1)$. Under these assumptions the curve $\mathbf{x} = \mathbf{f}(u)$, $u \geq 0$, goes from the origin, crosses the circle $|\mathbf{x}| = r_1 = r(u_1)$ at $\mathbf{x} = \mathbf{f}(u_1)$ and reaches the circle $|\mathbf{x}| = r_2 = r(u'_2)$, touches it at $\mathbf{x} = \mathbf{f}(u'_2)$, returns back to the circle $|\mathbf{x}| = r_1$, and, after touching it at the point $\mathbf{x} = \mathbf{f}(u'_1)$, tends to infinity, after crossing the circle $|\mathbf{x}| = r_2$ at $\mathbf{x} = \mathbf{f}(u_2)$. An example of such a curve is given in Fig. 5.3 (curve 2) and in Fig. 5.4 (right)).

In the neighborhood of the origin, the physical phase space has a conic structure. This local structure is preserved upon the canonical transformation to the variables u, p_u because it is a smooth and one-to-one mapping of the strip $r \in (0, r_1)$ on $u \in (0, u_1)$. The same holds for the mapping of the half-plane $r > r_2$ onto the half-plane $u > u_2$. Troubles occur in the domain $r \in (r_1, r_2)$ where the inverse function $u = u(r)$ becomes multi-valued. It has three branches in this particular case. States belonging to the strips $u \in (u_1, u'_2)$, $u \in (u'_2, u'_1)$, and $u \in (u'_1, u_2)$ are physically equivalent as they are related by transformations from S. However the canonical variables $p_r, r \in (r_1, r_2)$ remain invariant under these transformations.

To investigate what happens to phase-space trajectories in the region $u \in (u_1, u_2)$, consider a motion with a constant momentum p_r and suppose that the particle is outgoing from the origin $r = 0$. Two such trajectories are depicted in the upper-left panel of Fig. 5.4. In the (p_u, u)-plane, the particle motion

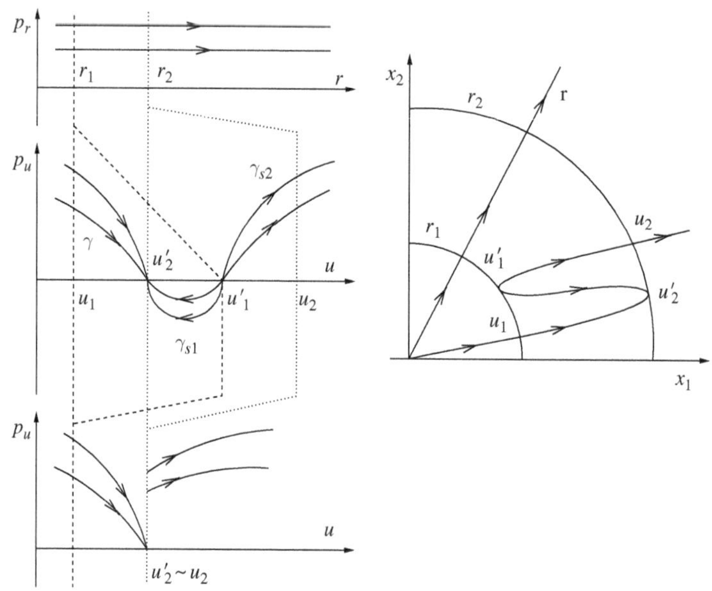

Fig. 5.4. Phantom trajectories caused by coordinate singularities occurring through a bad parameterization of the physical phase space. The arrows on the curves indicate the direction in which the phase space point moves with time. Adapted from [195].

corresponds to a point running along a curve so that u increases from 0 to u_1. As soon as the phase-space point crosses the line $u = u_1$, there appear two "phantom" phase-space trajectories outgoing from the point $p_u = 0, u = u'_1$. Indeed, the motion from $r = r_1$ to $r = r_2$ corresponds to three trajectories in the u–space: from $u = u_1$ to $u = u'_2$, from $u = u'_1$ to $u = u'_2$, and from $u = u'_1$ to $u = u'_2$ as one might see in the right panel of Fig. 5.4. Note also that $p_{u'_1} = p_{u'_2} = 0$ since $dr/du = 0$ at $u = u'_{1,2}$. The upper-left panel of Fig. 5.4 shows two trajectories $p_r =$ const. Their images in the new canonical variables are shown in the middle-left panel of Fig. 5.4. The reason for such an odd behavior of the image trajectories is that the interval (r_1, r_2) is represented by the three intervals (u_1, u'_2), (u'_2, u'_1), and (u'_1, u_2) in the u-parameterization. They are ranges of the three branches of the multi-valued function $u(r)$. The dashed and dotted lines in the figure show the "splitting" of the points $r_1 \to u_1, u'_1$ and $r_2 \to u_2, u'_2$, respectively. The trajectories outgoing from $u = u'_1$ appear at the time when the phase space trajectory crosses the line $u = u_1$.

If $\hat{s}_1$ and $\hat{s}_2$ map (u_1, u'_2) onto (u'_2, u'_1) and (u'_1, u_2), respectively, so that $r(u) = r(\hat{s}_{1,2}u)$, $u \in (u_1, u'_2)$, then the "phantom" trajectories, shown in Fig. 5.4 as curves γ_{s1} and γ_{s2}, are described by the pairs $(\hat{s}_{1,2}p_u, \hat{s}_{1,2}u)$, $\hat{s}_{1,2} \in S$, where the point (p_u, u) traces out the curve γ. The variable u increases along γ in the phase-space region $u \in (u_1, u'_2)$. Since $du_{s_1}/du < 0$ and $du_{s_2}/du > 0$, the variable u increases along the "phantom" trajectory $(\hat{s}_2 p_u, \hat{s}_2 u)$, while the point $(\hat{s}_1 p_u, \hat{s}_1 u)$ moves along γ_{s_1} so that u decreases. Note that the momentum $\hat{s}_1 p_u$ is negative for this trajectory since dr/du is negative in the interval (u_1, u_2). The points (p_u, u) and $(\hat{s}_1 p_u, \hat{s}_1 u)$ arrive at $(p_u = p_{u'_2} = 0, u = u'_2)$ at the same time and "annihilate" each other, whereas a "phantom" particle moving along the branch $(\hat{s}_2 p_u, \hat{s}_2 u)$ arrives at the line $u = u_2$. In the next moment of time the system leaves the interval $r \in (r_1, r_2)$ (or $u \in (u_1, u_2)$).

Such "branching" of classical phase-space trajectories is a pure artifact of an inappropriate parameterization of the physical phase space in the invariant approach or, alternatively, of a bad gauge fixing. It has to be removed by gluing all the "phantom" trajectories (branches) by identifying all points in the (u, p_u)-space that are related by transformations from S (see the text after (5.52)). In so doing, one cannot however avoid breaking the trajectories at the singular points $u = u'_{1,2}$. This feature will always be present when describing the dynamics in these variables (or in the corresponding gauge). Indeed, consider trajectories approaching the line $u = u_1$ with *different* momenta p_r from the origin and crossing it. Since the motion in phase space strips $u \in (u'_2, u'_1)$ and $u \in (u'_1, u_2)$ is physically equivalent to the one in the strip $u \in (u_1, u'_2)$, the former should be removed from the physical domain of the local canonical variables u and p_u. The state $(u = u'_2, p_u = 0)$ is equivalent to the state $(u = u_2, p_u = 0)$, so they can be glued together making just a point-like joint between two phase-space domains $u < u'_2$ and $u > u_2$. In general, the edges of the cut shown by the dotted line in Fig. 5.4 should be glued since the phase-space points in the vicinity of $u = u'_2$ are S-equivalent to those in the vicinity of $u = u_2$ and, therefore, correspond to the same physical states. This procedure restores the original *conic* structure of the physical phase space which certainly cannot depend on its parameterization. However, the continuity of the phase-space trajectories is lost. Every trajectory approaching the line $u = u'_2$ from the origin would fall into the point $p_u = 0$ on this line, because $p_u = dr/du p_r$ and (dr/du) vanishes at $u = u'_2$. So there is no trajectory that could cross this line with nonzero momentum p_u. On the other hand, trajectories approaching the line $u = u_2$ from infinity can have a non-zero momentum p_u. Therefore one always gains the discontinuity by gluing the lines $u = u'_2$ and $u = u_2$ (see the bottom left panel of Fig. 5.4).

It is interesting to note that in a non-invariant approach the artificial "attractor" at the phase-space point ($p_u = 0, u = u'_2$, i.e. the point through which all phase space trajectories outgoing from the origin must pass) corresponds to one of the zeros of the Faddeev–Popov determinant in the gauge chosen (see Section 8.4). It may obviously be absent in another gauge or, alternatively, in another parameterization of the physical phase space. Thus, the use of formally gauge-invariant canonical variables (i.e. those whose Poisson bracket with the constraints vanishes) may well exhibit the same type of singularities as the non-invariant approach based on gauge fixing. For this reason, it is important to study the geometrical structure of the physical phase space before introducing its explicit parameterization either via gauge fixing or by local formally gauge-invariant canonical coordinates, in order to avoid unnecessary (artificial) complications associated with a bad parameterization.

5.3.4 Quantum theory in invariant variables and gauge fixing

Let us now develop the quantum theory and, in particular, show that physical amplitudes are independent of the choice of function $\mathbf{f}$ if the structure of the physical configuration (or phase) space of the physical variable u is taken into account upon quantization. This goal is achieved by quantizing *before* elimination of the non-physical variable θ, and by defining the scalar product of physical states in accordance with the rule of changing variables.

After quantization of the system (see (5.25)), the change of variables (5.47) is carried out in all operators. The second equality in (5.25) is equivalent to the equation $-i\partial\Phi/\partial\theta = 0$, i.e. $\Phi = \Phi(u)$. Then, carrying out the change of variables in the Laplace operator and dropping all of its terms containing the derivative $\partial/\partial\theta$ on the right, the Schrödinger equation in the physical Hilbert space $\mathcal{H}^f_{\rm ph}$ (the index f indicates the choice of the function $\mathbf{f}$ in (5.47)) is obtained,

$$\left[-\frac{1}{2\mu(u)}\frac{d}{du}\circ\frac{r^2(u)}{\mu(u)}\frac{d}{du}+V(r^2(u))\right]\Phi_E(u)=E\Phi_E(u). \tag{5.53}$$

To compute the Laplace operator in the coordinates (5.47) it is convenient to make use the following identity:

$$d\mathbf{x} = d\theta T\mathbf{x} + \exp(\theta T)\mathbf{f}' du,$$

from which it follows that,

$$(\mathbf{x}, d\mathbf{x}) = (\mathbf{f}, \mathbf{f}')du = \mu du,$$
$$(T\mathbf{x}, d\mathbf{x}) = r^2 d\theta - (\mathbf{f}, T\mathbf{f}')du.$$

Hence,

$$\frac{\partial u}{\partial \mathbf{x}} = \frac{\mathbf{x}}{\mu}, \quad \frac{\partial \theta}{\partial \mathbf{x}} = \frac{T\mathbf{x}}{r^2} + \frac{(\mathbf{f}, T\mathbf{f}')}{r^2\mu}\mathbf{x}.$$

The eigenstates of the Hamiltonian in $\mathcal{H}^f_{\rm ph}$ are normalized as

$$\sum_\alpha \int_{K_\alpha} du|\mu(u)|\Phi^*_E(u)\Phi_{E'}(u) = \int_K du|\mu(u)|\Phi^*_E(u)\Phi_{E'}(u) = \delta_{EE'}. \tag{5.54}$$

Here the equality $dx_1 dx_2 = \mu(u)du d\theta$ has been used. Integration with respect to θ gives the factor 2π for states from $\mathcal{H}^f_{\rm ph}$ which is not included in the normalization of Φ_E. In general, K may be chosen so that the Jacobian is negative, $\mu(u) \le 0$, $u \in K_\alpha$ for some α. This is the reason for the absolute value of μ in (5.54). Without loss of generality, K is chosen so that $\mu(u) \ge 0$, $u \in K$, and the absolute value in the integration measure can be omitted. The first equality in (5.54) implies a certain rule to orient the integration regions K_α. It is assumed that in the integral $\int_{K_\alpha}$ the upper limit is always greater than the lower one, i.e. $\sum_\alpha \int_{K_\alpha} du = \int_K du$ is the sum of lengths of all intervals K_α.

Let us prove that the spaces $\mathcal{H}^f_{\rm ph}$ with different $\mathbf{f}$ are isomorphic. By changing variables in (5.53),

$$\frac{d}{du} = \frac{dr}{du}\frac{d}{dr} = \frac{\mu}{r}\frac{d}{dr},$$

the Hamiltonian operator in (5.53) is turned into (5.27). Relation (5.47) defines a change of variables and, therefore, there is a one-to-one correspondence between the variables u and r from which the relation between the integration measures follows:

$$\sum_\alpha \int_{K_\alpha} |\mu(u)| du = \int_0^\infty r dr,$$

that is, (5.54) is equivalent to (5.26). This means that all $\mathcal{H}^f_{\rm ph}$ are isomorphic to one another as they are isomorphic to the Hilbert space of the theory (5.27) and (5.26). According to the analysis of Section 5.2.1, the solutions Φ_E of (5.53) are regular functions of $r^2 = f_i^2(u)$. The scalar product of elements of $\mathcal{H}^f_{\rm ph}$, or the amplitudes, are independent of $\mathbf{f}$ because

$$\Phi_E(u) \equiv \tilde{\Phi}_E(r^2(u)) = \tilde{\Phi}_E(\mathbf{x}^2). \tag{5.55}$$

From the equality (5.55) the S-invariance of the physical states follows:

$$\Phi(u) = \Phi(u_s) \equiv \hat{s}\Phi(u)\ , \qquad \Phi \in \mathcal{H}^f_{\rm ph}. \tag{5.56}$$

This completes the proof of the following important fact. In the Dirac quantization method, the physical amplitudes are independent of parameterization of the physical configuration space (or, the choice of physical variables that span the orbit space, or, the choice of gauge fixing condition), while the Hamiltonian (5.53) as well as the scalar product (5.54) explicitly depends upon it.

For the sake of comparison, consider an alternative method when the elimination of non-physical variables is carried out before quantization. A naive replacement of p_u by the operators $-i\partial_u$ is impossible in the classical physical Hamiltonian (5.52) because of the operator ordering problem. Besides, the Hamiltonian operator turns out to be non-Hermitian. One has to postulate an operator ordering rule in order to make the Hamiltonian Hermitian, for example, by introducing a Hermitian momentum operator $p_u \to -i\mu^{1/2}\partial_u \circ \mu^{-1/2}$. There is however a more serious problem. Having ordered operators in a certain way to make the Hamiltonian Hermitian, one still can get a *gauge-dependent* quantum theory. The proof of the unitary equivalence of quantum theories for different $\mathbf{f}$'s relies on the specific form of the Hamiltonian operator (5.53), i.e. on the *specific* operator ordering. Only with this specific operator ordering is the theory independent of the choice of any particular parameterization of the orbit space (or the choice of gauge fixing condition). Thus, in the framework of the alternative quantization method, a special investigation is required in order to obtain a gauge-independent quantum theory. The latter implies, in fact, the use of the Dirac method (see Section 8.4). It should be noted that the same problem also occurs in the path integral quantization method (see Chapter 6).

5.4 Models with arbitrary simple compact gauge groups

In spite of an obvious simplicity the model discussed in Section 5.1 contains all the characteristic features inherent in gauge systems with generic gauge groups as well as with more degrees of freedom (e.g. gauge field theories). Here simple mechanical models with arbitrary compact gauge groups are considered. This problem is important in view of the face that, before studying the dynamics of physical degrees of freedom, one has to be able to identify them, which is not generally a simple task for a generic gauge group. However, the difficulties here are not so related to the gauge group itself but rather to its representations. In Section 5.2 the model with gauge group $SO(n)$ is studied which is a non-Abelian group of an arbitrary rank ($n \geq 2$). Nevertheless, the example is viewed as the simplest. This is so because the dynamical variables realize an elementary (vector) representation of $SO(n)$ in which there is only one physical degree of freedom. Here models are studied in which the dynamical variables realize the adjoint representation of an arbitrary compact gauge group. This is important for Yang–Mills theories where the fields belongs to the adjoint representation of the gauge group.

5.4.1 Classical theory. Selection of physical degrees of freedom

Consider the gauge model defined by the Lagrangian [196, 197],

$$L(x, \dot{x}, y, \dot{y}) = \frac{1}{2}\mathrm{Tr}\,(D_t x)^2 - V(x), \tag{5.57}$$

where $D_t = \partial_t + [y, \cdot\,]$, x, y are elements of a Lie algebra X of a simple compact group G. Without loss of generality, a matrix realization of X is used so that the trace in (5.57) and multiplication are defined in the sense of the matrix algebra. Let λ_a be an orthonormal basis in X, $\mathrm{Tr}\,\lambda_a\lambda_b = \delta_{ab}$ and $[\lambda_a, \lambda_b] = F^c_{ab}\lambda_c$ (the summation over repeated indices is assumed), where F^c_{ab} are the structure constants, $a, b, c = 1, 2, \ldots, N = \dim G$, so that $x = x_a\lambda_a$ (similarly for y). The real components x_a and y_a are the dynamical variables in the system. The Lagrangian is invariant under gauge transformations,

$$x \to \Omega x \Omega^{-1}, \quad y \to \Omega y \Omega^{-1} + \Omega\partial_t\Omega^{-1}, \quad \Omega = \Omega(t) \in G, \tag{5.58}$$

provided $V(\Omega x \Omega^{-1}) = V(x)$. To simplify the notations, set $xy = \mathrm{Tr}\,xy$ for any $x, y \in X$ (the trace notation will be restored if any possible confusion arises). The function L defines a system with N primary constraints, $\partial L/\partial \dot{y}_a = \pi_a = 0$, the Hamiltonian

$$H = \frac{1}{2}p^2 + V(x) + y_a p T_a x \tag{5.59}$$

(here $p = \lambda_a p_a$, $p_a = \partial L/\partial \dot{x}_a$), and N secondary constraints, $\sigma_a \equiv pT_a x = \mathrm{Tr}\,\lambda_a[p,x] = F_{abc}p_b x_c = 0$. All the constraints are of the first class, i.e. they are in involution. The relations (5.34) and (5.35) hold for them if f_{abc} are replaced by F_{abc}. In the total configuration space X spanned by x_a one should find physical variables and determine their time evolution. Two approaches, invariant and non-invariant, are also possible here.

First of all let us determine how many physical degrees of freedom the system has. In the invariant approach the answer is straightforward. One has to find all independent gauge-invariant quantities that can be constructed from $x \in X$. These are independent invariant polynomials of x and, hence, are uniquely determined by the symmetric invariant tensors in the adjoint representation of group G. The number of such tensors equals the rank l of G (see Section 8.1). Thus, the system has l physical degrees of freedom. This means, in particular, that among N constraints $\sigma_a = 0$ of the system there are only $N - l$ independent ones. However, a treatment of the theory in terms of invariant variables is not so easy because the invariants are homogeneous polynomials of x of various degrees. For groups of rank $l = 2$ this is done in Section 5.4.5 below (see also [195, 197, 200]).

A non-invariant approach appears easier. Its key element is the following representation of any element $x \in X$ [21]:

$$x = S(z)hS^{-1}(z), \tag{5.60}$$

where $h = h_i\lambda_i$, $i = 1, 2, \ldots, l$, and λ_i form a basis in the space of all mutually commutative elements of the algebra X, that is, h is an element of the Cartan subalgebra H of X [215, 216]. The group element $S(z)$ is obtained by the exponential mapping of an element $z = z_\alpha\lambda_\alpha \in X \setminus H$ into group G, $\alpha = l+1, l+2, \ldots, N$. The variables h_i and z_α are real. The dynamics of the system can therefore be formulated in their terms. Equation (5.60) shows that any $x \in X$ can be brought to the Cartan subalgebra by a suitable adjoint action of the group on x. Indeed, in accord with the gauge transformation law (5.58),

$$h = \Omega x \Omega^{-1} \in H, \text{ if } \Omega = S^{-1}(z). \tag{5.61}$$

Any element of X can be decomposed into the sum $x = h + z$, where $h \in H$ and $z \in X \setminus H$. By a suitable gauge transformation, the component z can always be nullified. Thus, the gauge $z = 0$ is compatible with the gauge transformation law. In this decomposition, the variables h_i are analogous to x_1 from Section 5.1, while z_a are to x_2. They can be eliminated by a gauge transformation.

So, any element of X is gauge equivalent to some element of the Cartan subalgebra [216]. The dimension of the maximal commutative subalgebra determines the number of physical degrees of freedom of the system. The latter can be described by functions $h_i(t)$. Equation (5.60) shows that among x_a there are $N - l$ non-physical variables that can be eliminated. Even though there are N secondary constraints, only $N - l$ of them are independent. This follows from a simple observation that $h \in H$ is invariant under the adjoint action of an

l-parametric subgroup of G that is formed by all mutually commutative elements of G, the Cartan subgroup. Note, if formally $z \in H$ in (5.60), then $ShS^{-1} = h$ because $[z, h] = 0$ for any two elements of H. Alternatively, the above observation can also be understood as follows. It is known that generators λ_a of a group have l zero eigenvalues in the adjoint representation [215]. Hence, there are l linearly independent vectors which are annihilated by l of N generators. In the example of Section 5.2 with the gauge group $SO(n)$, a generic $\mathbf{x} \in R^n$ has the stationary subgroup $SO(n-1)$ (rotations about $\mathbf{x}$) and its $(n-1)(n-2)/2$ generators nullify $\mathbf{x}$. Note that $\mathbf{x}$ is not in the adjoint representation of $SO(n)$, that is why the number of non-physical degrees of freedom is different, $\dim SO(n) - \dim SO(n-1) = n - 1 \neq \dim SO(n) - \operatorname{rank} SO(n)$ (compare the table in Section 8.1.2)).

The space of physical variables is now identified. Let us turn to studying its structure. Although the dimension of the physical space l cannot be decreased, its further reduction is still possible by means of gauge transformations. It is known [21] that a Lie group contains a finite discrete subgroup W (the Weyl group) whose adjoint action on the Cartan subalgebra H is not trivial, but leaves H invariant. The Weyl group is generated by reflections in the hyperplanes in H that are orthogonal to the roots of X (a Lie algebra can be viewed as a linear space with a scalar product [21]; see Section 8.1.2 for details). The group W is isomorphic to the group of all permutations of the roots, i.e. the root system is invariant under the action of W on H. The gauge $x = h$ is called an *incomplete global gauge with the residual symmetry group* $W \subset G$ (see [210]). The very existence of the residual discrete gauge group leads to a further reduction of the space spanned by variables h. Recall that the group Z_2 in the model of Section 5.1 contains the reflection $x_1 \to -x_1$, i.e. by its action the whole real line $x_1 \in R$ can be restored from the set $x_1 \geq 0$. Let $D_W(X)$ be the set of the hyperplanes in H containing the origin $h = 0$ and orthogonal to positive roots α, $(\alpha, h) = \alpha_i h_i = 0$. The complement of $D_W(X)$ in H consists of the set of open convex cones called Weyl chambers [216]. The Weyl group acts simply transitively on the set of Weyl chambers. The positive Weyl chamber K^+ is defined by the condition that $h \in K^+$ if $(h, \omega) = h_i\omega_i > 0$ for all simple roots ω. Hence, $(H \setminus D_W(X))/W \sim K^+$, i.e. any element of H which is not in $D_W(X)$ can be obtained from an element of K^+ by a transformation from W. The physical configuration space is isomorphic to the closure of K^+ (the set $D_W(X)$ has zero measure (volume) in H).

The residual discrete gauge group W acting in H leads also to a further reduction of the phase space spanned by h and its canonically conjugated momentum. Indeed, the total phase space is $\Gamma = X \times X \sim R^{2N}$ spanned by the components of x and their canonical momenta $p = \partial L/\partial \dot{x} = D_t x$. Note that the momentum $p \in X$ also transforms as an element of the adjoint representation of the gauge group. The constraint surface in the total phase space is defined by $\sigma = [p, x] = 0$. The physical phase space is $\Gamma_{ph} = \Gamma|_{\sigma=0}/G$ where G has the same adjoint action on x and p simultaneously (compare (5.1); the group of translations of y_a generated by the primary constraints $\pi_a = 0$ is already factored out). According to (5.61) there exists a gauge group element Ω such that $h = \Omega x \Omega^{-1} \in H$. The corresponding transformed momentum $p_h = \Omega p \Omega^{-1}$ must also be an element of H because on the constraint surface $\Omega[p,x]\Omega^{-1} = [p, h_h] = 0$. Hence the physical phase space Γ_{ph} is obtained from $\Gamma_{2l}(h, p_h) = H \times H \sim R^{2l}$, spanned by components of h and p_h, by identifying the pairs (whw^{-1}, wp_hw^{-1}) where $w \in W$, because the Weyl group is the set of all residual discrete gauge transformations. Thus, $\Gamma_{ph} = \Gamma_{2l}(h, p_h)/W$ which is a $2l$-dimensional hypercone. With a suitable cut, the hypercone can be unfolded into $K^+ \times H \sim K^+ \times R^l$ [195, 196].

For generic configurations $h \in K^+$ the physical phase space has no singularities and is locally flat. When h approaches a generic point on the boundary $(h, \omega) = 0$ of the Weyl chamber K^+, the physical phase space exhibits a conic singularity. Indeed, one can always make a linear canonical transformation such that one of the canonical coordinates, say $h^\perp$, varies along the line perpendicular to the boundary, while the others span hyperplanes parallel to hyperplane $(h, \omega) = 0$, being a part of the Weyl chamber boundary. In the new variables, the Weyl transformation that flips sign of the root ω will change signs for $h^\perp$ and its canonical momentum, while leaving the other canonical variables unchanged. Thus, at a generic point of the Weyl chamber boundary, the physical phase space has a *local* structure $R^{2(l-1)} \times \text{cone}(\pi)$.

The Weyl chamber boundary is not a smooth manifold and contains intersections of two hyperplanes, $(\omega_1, h) = 0$ and $(\omega_2, h) = 0$. At these points, the two local conic singularities of the physical phase space associated with simple roots $\omega_{1,2}$ would merge, forming locally a four-dimensional hyperconic singularity. This singularity cannot simply be described as a direct product of two cones (π). This would only be the case if the roots $\omega_{1,2}$ were orthogonal. In general, the tip of the hypercone would be "sharper" than that of $\text{cone}(\pi) \times \text{cone}(\pi)$, meaning that the hypercone can always be put inside $\text{cone}(\pi) \times \text{cone}(\pi)$ when their tips are at the same point. This can be understood again in local canonical variables where the coordinates h are split into a pair $h^\perp$ that spans a plane perpendicular to the intersection of two hyperplanes $(\omega_{1,2}, h) = 0$, and the others orthogonal to $h^\perp$. The root pattern in any plane containing at least two roots (e.g. a plane through the origin and parallel to the $h^\perp$–plane) is isomorphic to one of the root patterns of the groups of rank two, i.e. $SU(3)$, $Sp(4) \sim SO(5)$, G_2 or just $SU(2) \times SU(2)$. In the latter case the simple roots $\omega_{1,2}$ are orthogonal. A modular domain of $h^\perp$ coincides with the Weyl chamber of one of these groups and is contained in the positive quadrant being the Weyl chamber for $SU(2) \times SU(2)$. That is, a solid region bounded by the hypercone spanned by $p^\perp$ and $h^\perp$ and isomorphic to the quotient space R^4/W is contained in the solid region bounded by $\text{cone}(\pi) \times \text{cone}(\pi)$. The procedure is straightforward to generalize to the boundary points belonging to intersections of three hyperplanes $(\omega_{1,2,3}, h) = 0$, etc. At the origin, the physical phase space has the most singular point being the tip of a $2l$-dimensional hypercone which is "sharper" than $[\times\text{cone}(\pi)]^l$.

Just as in the case of the simplest models discussed above, the nontrivial (hyperconic) structure of the physical phase has drastic dynamical consequences. It is sometimes the case that studying a quantum mechanical system turns out to be easier than its classical limit. The present model is a good example of this observation. So, to investigate the dynamical peculiarities of this system, let us turn first to the quantum theory of a harmonic oscillator, $V = x^2/2$ in (5.59).

5.4.2 Quantum theory. Harmonic oscillator

A quantum treatment of the model is not hard to develop. Thanks to the skew-symmetry of the matrices T_a (i.e. the structure constants), there is no operator ordering problem in the canonical quantization of the constraints and the Hamiltonian. According to the Dirac method (see [148]), the physical Hilbert space is defined by the conditions,

$$\hat{\pi}_a \Phi = 0, \quad \hat{\sigma}_a \Phi = 0, \quad \Phi \in \mathcal{H}_{ph}, \; a = 1, 2, \dots, N. \tag{5.62}$$

The first set of equations in (5.62) is not physically interesting. They simply mean that Φ is independent of y_a in the coordinate representation of Φ (compare Section 5.1). The second set of equations is far less trivial because their solution requires suitable curvilinear coordinates. One can introduce local coordinates in X some of which span gauge orbits (the angular or orbital coordinates), while the others are transverse to the gauge orbits and, hence, gauge invariant. Transformation of the kinetic energy operator $\hat{p}_a^2$ to such curvilinear coordinates is a rather difficult technical problem [216]. It is solved in Sections 5.4.4 and 5.4.5. However this difficulty can be avoided in the case of the harmonic oscillator. To elucidate the effect of the constraints $\hat{\sigma}_a \Phi = 0$ on the quantum dynamics, here the harmonic oscillator is studied (a generic system is considered in the next section).

Introducing the operators $\hat{a}_b = (\hat{p}_b - i\hat{x}_b)/\sqrt{2}$ and $\hat{a}_b^\dagger$, the Hamiltonian (5.59) with $V = x_a^2/2$ and $\hat{\sigma}_b$ can be written in the form,

$$\hat{H} = \hat{a}^\dagger \hat{a} + \frac{N}{2} + i y_b \hat{a}^\dagger T_b \hat{a}, \quad \hat{\sigma}_b = \hat{a}^\dagger T_b \hat{a}, \tag{5.63}$$

where $\hat{a}^\dagger \hat{a} = \sum \hat{a}_b^\dagger \hat{a}_b$. The vacuum state is defined by $\hat{a}_b|0\rangle = 0$. The states

$$\left(\hat{a}_1^\dagger\right)^{n_1} \cdots \left(\hat{a}_N^\dagger\right)^{n_N} |0\rangle, \quad n_i = 0, 1, \dots, \tag{5.64}$$

form an orthogonal basis of the total Hilbert space. The operators $\hat{\sigma}_a$ obey the commutation relations,

$$[\hat{\sigma}_a, \hat{\sigma}_b] = F_{ab}^c \hat{\sigma}_c, \tag{5.65}$$

that is, they are generators of the gauge group. Hence, from (5.62)–(5.65) and the equality $\hat{\sigma}_a|0\rangle = 0$ it follows that elements of the physical Hilbert space are obtained by applying invariant polynomials of $\hat{a}_b^\dagger$ to the vacuum state $|0\rangle$. The number of independent invariant polynomials equals the rank l of the group [21]. Thus, the basis vectors of the physical Hilbert space have the form [196],

$$P_{r_1}^{n_1}(\hat{a}^\dagger)\cdots P_{r_l}^{n_l}(\hat{a}^\dagger)|0\rangle, \tag{5.66}$$

where P_{r_i} is an invariant homogeneous Casimir polynomial of degree r_i, $i=1,2,...,l$ and $n_i = 0,1,\ldots$. The degrees r_i for all simple groups are listed in the table of Section 8.1. From the homogeneity of the Casimir polynomials it follows that

$$[\hat{a}^\dagger\hat{a}, P_{r_i}(\hat{a}^\dagger)] = r_i P_{r_i}(\hat{a}^\dagger). \tag{5.67}$$

By making use of (5.67) the spectrum of physical excitations is obtained,

$$E = \sum_{i=1}^{l} r_i n_i + \frac{N}{2}. \tag{5.68}$$

In the corresponding system without gauge symmetry ($y = 0$ in (5.57)), the spectrum reads,

$$E = \sum_{i=1}^{N} n_i + \frac{N}{2}. \tag{5.69}$$

Gauge symmetry drastically changes the energy level pattern. Since r_i are positive integers, the degeneracy of each energy level E is determined by the number of solutions of the linear Diophantine equation (5.68) for a fixed integer $E-N/2$. The physical states (5.66) have an interesting feature that they can only be obtained by collective excitations of all oscillators. Indeed, the Casimir polynomial of the lowest degree for all groups is $P_2(\hat{a}^\dagger) = \sum \hat{a}_a^{\dagger 2}$. Physical states generated by this contain excitations of all N oscillators.

It is noteworthy that by eliminating non-physical degrees of freedom first (e.g. by fixing the gauge $x = h$) with subsequent canonical quantization, one obtains the quantum theory of l *decoupled* harmonic oscillators $\left(H = p_h^2/2 + h^2/2\right)$. The spectrum is then $E = \sum_{i=1}^{l} n_i + l/2$, which is obviously quite different from (5.68). By comparing it with the spectrum (5.69), the gauge fixed quantum theory seems to allow for independent excitations of "physical" oscillators (with the same unit frequency) described by h. The reason for such a discrepancy with gauge invariant treatment is that the structure of the physical phase space is not taken into account upon quantizing the gauge fixed theory. In fact, it will be shown next that the hyperconic structure of the physical phase space makes it impossible to have independent excitations of the components of h and thereby induces *kinematic coupling* between the physical degrees of freedom which, in turn, leads to the physical frequencies of independent vibrations of the l-dimensional isotropic oscillator becoming r_i, in full accordance with (5.68).

5.4.3 Classical theory. Analysis of dynamics for gauge groups of rank 2

Equation (5.68) shows that the spectrum of the system is determined by quite subtle characteristics of the gauge group, the degrees of independent invariant Casimir polynomials. Thanks to gauge symmetry, the number of physical degrees of freedom is reduced from $N = \dim G$ to $l = \dim H = \operatorname{rank} G$. The basis of physical excitations (5.66) and the physical spectrum (5.68) lead to the conclusion that the system contains l independent subsystems each of which has the spectrum $E_i = r_i n + \text{const}$, $n = 1, 2, \ldots$. The distance between the energy levels is increased by r_i times as compared to that of the original oscillator. What does this fact tell us?

To answer this question, let us analyze the dynamics in the case of gauge groups of rank 2 when the oscillator has only two physical degrees of freedom described by h_1 and h_2. The physical configuration space, the Weyl chamber K^+, is a sector in the plane spanned by $h_{1,2}$ bounded by two rays outgoing from the origin. For the groups $SU(3), SO(5) \sim Sp(4)$, and G_2, the angles α between the rays are, respectively, $\alpha = \pi/3, \pi/4, \pi/6$. A trajectory of the oscillator for group $SU(3)$ is shown in Fig. 5.5. The initial conditions are chosen so that the solutions of equations of motion have the form,

$$h_2(t) = A_2 \cos t, \qquad h_1(t) = A_1 \sin t,$$

and $A_2 > A_1$. The Weyl chamber is a sector with angle $\pi/3$ which is shown as a gray area in the figure. The ray OO' is its symmetry axis.

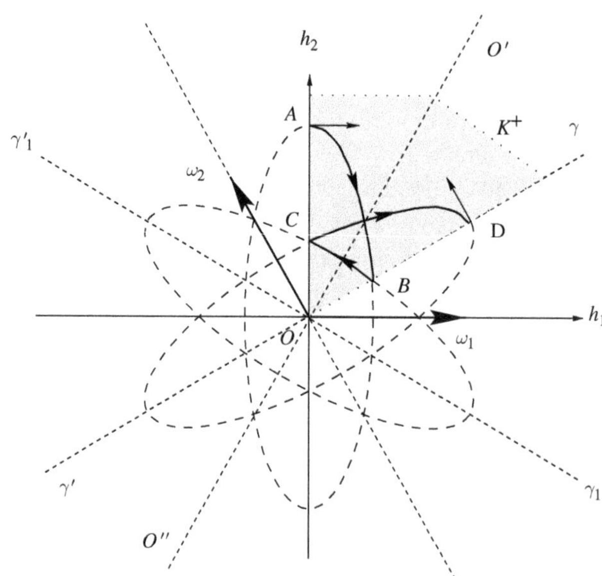

Fig. 5.5. Classical dynamics in the Weyl chamber. Adapted from [195] – the figure first appeared in the original Russian version of this book;

In the initial moment of time $t = 0$ the oscillator is located at point A. It then follows the elliptic trajectory extended along the axis h_2 and at $t = \pi/3$ reaches point B, i.e. the boundary of the Weyl chamber. Further motion along the ellipse in the sector bounded by rays $O\gamma$ and $O\gamma_1$ is gauge equivalent to the motion from B to C in the Weyl chamber K^+. It looks as though the oscillator hits the boundary, reflects from it and arrives at point C at time $t = 2\pi/3$. Although at point B the oscillator momentum abruptly changes direction, there is *no* force causing it because the oscillator potential is smooth on the entire plane. The momenta right before and after hitting the boundary wall are *gauge equivalent*. They are related by the Weyl transformation being the reflection relative to the line $\gamma'\gamma$ perpendicular to root ω_2. So there is no dramatic change in physical state of the system at the moment of reaching the boundary. Just like in the $SO(n)$ model of Section 5.2, the trajectory remains *smooth* on the hyperconic physical phase space. The momentum jump is a *coordinate artifact* occurring through a cut made on the physical phase space to parameterize it by a particular set of local canonical coordinates $p_h \in H$ and $h \in K^+$. Within the path integral formalism for the model, it will be shown that the phase of the transition amplitude does *not* change under such a reflection, in full contrast with the realistic reflection from an infinite potential wall, where the phase would be shifted by π (see Section 6.3).

At point C the oscillator hits the boundary of the Weyl chamber one more time and follows the elliptic segment $C \to D$. Again, at the very moment of collision, no abrupt change of the physical state occur. Finally, at $t = \pi$ the oscillator reaches point D, reflects from the boundary and goes back the same way to point A, returning there at $t = 2\pi$.

What are the independent frequencies of this two-dimensional isotropic oscillator? It is quite surprising that they do not equal just one (the frequency that enters into the Hamiltonian), but rather $r_1 = 2$ and $r_2 = 3$. By definition the angular frequency is $2\pi/T$, where T is the time in which the system returns into the initial state upon a periodic motion. The system state is specified by values of the momentum and position of the physical degree of freedom in question. Let us decompose the motion of the system into oscillations along axis $O'O''$ and the angular motion about the origin O. After the oscillator passes segments $A \to B \to C$, the angular variable attains its initial value, since points A and C lie on the same ray, while the equality of the corresponding (angular) canonical momentum at points A and C follows from its conservation law. The angular degree of freedom returns to the initial state two more times as the oscillator follows the path $C \to D \to C$, and then returns to the initial state after passing segments $C \to B \to A$. Thus, the period of the angular variable is three times less than that of the angular variable of an ordinary two-dimensional isotropic oscillator, i.e. the physical frequency is tripled. From Fig. 5.5 one can easily see that the states of the radial degree of freedom at points A and D are the same, so the physical frequency of the radial degree of freedom is doubled.

A similar analysis can also be carried out for groups $SO(5) \sim Sp(4)$ and G_2. The independent frequencies also coincide with the corresponding degrees of independent invariant polynomials, $r_1 = 2$ and $r_2 = 4, 6$, respectively. Note that the Weyl chamber is a sector with angle π/r_2. So, in classical theory the hyperconic structure of the physical phase space reveals itself through the effect of reflections of physical trajectories from the boundaries of the physical configuration space when the latter is parameterized by elements of the Cartan subalgebra. One should stress again that the effect of changing the physical frequencies of the oscillator does not depend on the choice of local canonical variables and is essentially due to the hyperconic structure of the physical phase space. To calculate the effect, a particular parameterization of the physical phase space has been used, $h \in K^+$ and $p_h \in H$. The choice of parameterization is a matter of convenience. Had one taken another set of local canonical coordinates, say, by making cuts of the hyperconic phase space such that the momentum variable p_h is restricted to the Weyl chamber, one would have arrived at the very same conclusion about the oscillator frequencies. The message is therefore: whatever local canonical variables are assumed, there are coordinate singularities associated with them which always occur and should be carefully taken into account when solving the dynamical problem, because they contain information about the geometry of the physical phase space.

Remark. The analysis of Section 5.1.2 can be extended to the hyperconic phase space with the same conclusion of the non-existence of phase space coordinates with standard symplectic structure in which the motion has no singularities. It is also possible to find local coordinates on the physical phase space with non-standard symplectic structure, similar to (5.23), such that the singularities are absent [195].

Another important observation is that the geometry of the physical phase space does not permit excitations of the Cartesian degrees of freedom h^i independently, even though the Hamiltonian does not contain any interaction between them. This effect can be anticipated from the fact that the residual Weyl transformations mix the h^i. It will be shown in Chapter 6 that such a *kinematic* coupling between the physical degrees of freedom appears to be crucial for constructing a correct path integral formalism for gauge systems. If the Hamiltonian does not contain any coupling between the degrees of freedom, then the transition amplitude in quantum mechanics is factorized over the degrees of freedom. This, however, is not the case for the model in question. This can readily be seen from the correspondence principle. For example, set $V = 0$ in the above model. In addition to the straight trajectory connecting the initial point $h' \in K^+$ and the final point $h \in K^+$, there are $2r_2 - 1$ trajectories which involve several reflections from the boundary of the Weyl chamber. Since no change of physical state occurs at the very moment of reflection, these trajectories would also be acceptable classical trajectories contributing to the semi-classical transition amplitude at the same footing as the straight one. The reflected trajectories inside K^+ are gauge equivalent to *straight* lines in H connecting points $wh'w^{-1} \in wK^+$, $w \in W$, and $h \in K^+$, as shown in Fig. 6.1 in Chapter 6. Hence, they satisfy the classical equations of motion. Using the Weyl symmetry they can be mapped into piecewise straight continuous trajectories inside the Weyl chamber. The contribution of these trajectories makes it impossible to factorize the semi-classical transition amplitude because the reflected trajectories cannot be associated with an excitation of any particular Cartesian degrees of freedom h^i.

5.4.4 Invariant canonical variables. Groups of rank 2

Analysis of the classical dynamics of the isotropic harmonic oscillator shows that independent excitations of the Cartesian degrees of freedom h^1 and h^2 are impossible due to the non-Euclidean structure of their physical phase space. If, say, $p_{h_2} = \dot{h}_2 = 0$ in the initial moment of time, then after hitting the boundary of the Weyl chamber, the momentum $p_{h_1} = \dot{h}_1$ will be re-distributed between both physical degrees of freedom, thus exciting the h_2-degree of freedom. Because of the kinematic coupling the only independent excitations possible are *collective* gauge-invariant excitations of the original degrees of freedom. Accordingly, in quantum theory their wave functions are explicitly gauge invariant.

For the Hamiltonian (5.59) and groups of rank $l = 2$, consider new gauge invariant variables [195, 197],

$$\Phi_1 = (\mathrm{Tr}\, x^2)^{1/2}, \qquad \Phi_2 = \Phi_1^{-r} \mathrm{Tr}\, x^r, \tag{5.70}$$

where r is the degree of the second independent Casimir polynomial. The use of a matrix representation is just a matter of technical convenience. The invariant *independent* (Casimir) polynomials can also be written via symmetric invariant *irreducible* tensors, $\mathrm{Tr}\, x^2 \sim \delta_{ab} x^a x^b$ and $\mathrm{Tr}\, x^r \sim d^{(r)}_{a_1 \cdots a_r} x^{a_1} \cdots x^{a_r}$. Every symmetric invariant tensor in a Lie algebra can be decomposed over the basis formed by irreducible symmetric invariant tensors. Ranks of irreducible tensors are degrees of independent Casimir polynomials in the Lie algebra (see Section 8.1.2). The canonical momenta conjugate to the new variables read,

$$\pi_i = \frac{\mathrm{Tr}\,(p e_i)}{\mathrm{Tr}\, e_i^2}, \qquad e_i = \frac{\partial \Phi_i}{\partial x^a} \lambda_a \equiv \frac{\partial \Phi_i}{\partial x}, \qquad i = 1, 2; \tag{5.71}$$

By straightforward computation one can convince oneself that the elements e_i have the following properties:

$$\mathrm{Tr}\,(e_1 e_2) = 0, \qquad [e_1, e_2] = 0.$$

Hence, with every $x \in X$ one can associate a Cartan subalgebra in X spanned by the (local) basis elements $e_{1,2}$. Next, straightforward computations yield,

$$\mathrm{Tr}\, e_1^2 = 1\,, \quad \mathrm{Tr}\, e_2^2 = \frac{r^2}{\Phi_1^2}\left(c_2 + c_1 \Phi_2 - \Phi_2^2\right) \equiv \frac{r^2}{\Phi_1^2}\left[a - (\Phi_2 - b)^2\right],$$

where $b = c_1/2$, $a = c_2 + c_1^2/4$, and the constants $c_{1,2}$ depend on the structure constants and are determined by the decomposition of the gauge-invariant polynomial $[\mathrm{Tr}\,(\lambda_a x^{r-1})]^2 = (c_1\Phi_2 + c_2)\Phi_1^{2(r-1)}$ over the basis polynomials $\mathrm{Tr}\,x^2$ and $\mathrm{Tr}\,x^r$. For example, for $SU(3)$, $r = 3$ and $c_1 = 0$, $c_2 = 1/6$.

Let us decompose the canonical momentum p over the basis e_i, $p = \pi_i e_i + \tilde{p}$ where, by construction, $\mathrm{Tr}\,(e_i\tilde{p}) = 0$. A solution to the constraint equation $[p, x] = [\tilde{p}, x] = 0$ is $\tilde{p} = 0$. That is, all the components of p orthogonal to the Cartan basis elements e_i must vanish since the commutator of $\tilde{p}$ and $x \sim e_1$ does not belong to the Cartan subalgebra according to (5.70) and (5.71). The physical Hamiltonian of an isotropic harmonic oscillator, $V = \mathrm{Tr}\,x^2/2 = \Phi_1^2/2$, assumes the form,

$$H_{ph} = \frac{1}{2}\pi_1^2 + \frac{r^2\pi_2^2}{2\Phi_1^2}\left[a - (b - \Phi_2)^2\right] + \frac{1}{2}\Phi_1^2. \tag{5.72}$$

From $\mathrm{Tr}\,e_2^2 \geq 0$ the range of Φ_2 can be determined,

$$-1 \leq (\Phi_2 - b)/\sqrt{a} \leq 1.$$

The Hamiltonian equations of motion are

$$\begin{aligned}
\dot{\pi}_1 &= \{\pi_1, H_{ph}\} = -\Phi_1 + \frac{r^2\pi_2^2}{\Phi_1^3}\left[a - (b - \Phi_2)^2\right], \\
\dot{\Phi}_1 &= \{\Phi_1, H_{ph}\} = \pi_1, \\
\dot{\pi}_2 &= \{\pi_2, H_{ph}\} = \frac{r^2\pi_2^2}{\Phi_1^2}\,(b - \Phi_2), \\
\dot{\Phi}_2 &= \{\Phi_2, H_{ph}\} = \frac{r^2\pi_2}{\Phi_1^2}\left[a - (b - \Phi_2)^2\right].
\end{aligned}$$

They admit the following oscillating solutions *independently* for each degree of freedom:

$$\begin{aligned}
&\Phi_2(t) = \pi_2(t) = 0; \\
&\Phi_1(t) = \sqrt{2E}\,|\cos t|, \quad \pi_1(t) = -\sqrt{2E}\,\sin t\,\varepsilon(\cos t),
\end{aligned} \tag{5.73}$$

where E is the energy, ε denotes the sign function, and

$$\begin{aligned}
&\Phi_1(t) = \sqrt{E}, \qquad \pi_1(t) = 0; \\
&\Phi_2(t) = b + \sqrt{a}\cos(rt), \quad \pi_2(t) = -\frac{\sqrt{E}}{r\sqrt{a}\sin(rt)}.
\end{aligned} \tag{5.74}$$

The absolute value in (5.73) is necessary because Φ_1 cannot be negative. Solutions (5.73) and (5.74) show that the independent frequencies are indeed the degrees of the independent Casimir polynomials, 2 and r. Clearly, the variables $\cos^{-1}[(\Phi_2 - b)\,/\sqrt{a}]$ and Φ_1 are associated with the angular and radial variables introduced in the previous section when studying the motion of the oscillator in the Weyl chamber.

Thanks to the gauge invariance of the new variables $\Phi_{1,2}$, the dynamics on the physical (hyperconic) phase space described by any local canonical variables is proved to be gauge invariant. For instance, if local canonical variables on the hyperconic physical phase space are $h \in K^+$ and $p_h \in H$ (see Section 5.4.3), then there is a canonical transformation that relates (h, p_h) to the gauge-invariant canonical variables (5.70) and (5.71). It is found by setting $x = ShS^{-1}$ and $p = Sp_hS^{-1}$ in the right-hand sides of (5.70) and (5.71). The kinematic coupling between $\Phi_{1,2}$ is absent because the new variables are invariant relative to the Weyl group. However, to excite either of $\Phi_{1,2}$ *independently*, excitations of *both* Cartesian degrees of freedom $h_{1,2}$ are needed (see Section 5.4.3). Thus, removal of the kinematic coupling via a suitable canonical transformation is *equivalent* to restoration of the explicit gauge invariance. A similar analysis can be carried out for groups of higher rank.

Let us now turn to the quantum theory and demonstrate that the kinematic coupling (or the nontrivial physical phase-space structure) associated with the the residual discrete gauge group W leads to the physical wave functions being invariant under the Weyl group transformations. In turn, the Weyl-invariant wave functions appear to have unique analytic gauge-invariant continuations to the whole configuration space X which coincide with the gauge-invariant states (5.66) in the coordinate representation.

5.4.5 Quantum theory. Coordinate representation

To construct the quantum theory of the system with the Hamiltonian (5.59) in coordinate representation, the constraint equations (5.62) for physical wave functions should be solved first. The trivial non-physical degrees of freedom y^a and π_a are eliminated from the very beginning and are not considered. The second equation in (5.62) can be solved in curvilinear coordinates (5.60). Under gauge transformations the variables z are transformed like the angles of the spherical coordinate system in the model of Section 5.2. As the operators $\hat{\sigma}_a$ are generators of gauge transformations, the physical wave functions must be independent of z (the operators of independent constraints generate shifts of z). Thus, $\Phi_{ph}(x) = \Phi(h)$.

In order to find the physical Hamiltonian, one should write the Laplace operator, $\Delta = -\hat{p}_a^2$, $\hat{p}_a = -i\partial/\partial x^a$ in curvilinear coordinates (5.60) and omit all terms that contain the derivative $\partial/\partial z$ on the right. From (5.60) it follows that $dx = dShS^{-1} + SdhS^{-1} + ShdS^{-1}$. Making use of the identity $dS^{-1} = -S^{-1}dSS^{-1}$, the metric tensor in the new variables is found, $\mathrm{Tr}\,(dx)^2 = ds^2$,

$$\begin{aligned} ds^2 &= \mathrm{Tr}\,\left\{(dh)^2 + 2\left[(hS^{-1}dS)^2 - h^2(S^{-1}dS)^2\right] + \left[S^{-1}dS, h\right]dh\right\} \\ &= \mathrm{Tr}\,\left[(dh)^2 + [h, S^{-1}dS]^2\right] = dh_i^2 + \tilde{g}_{\alpha\beta}(h,z)dz^\alpha dz^\beta. \end{aligned} \tag{5.75}$$

Here $\mathrm{Tr}\left(\left[S^{-1}dS,h\right]dh\right)=0$, because $dh\in H$ and $[S^{-1}dS,h]\in X\setminus H$ lie in the orthogonal subspaces of X. The metric tensor has a block diagonal structure, $g_{ij}=\delta_{ij}$, $g_{i\alpha}=g_{\alpha j}=0$, and $g_{\alpha\beta}=\tilde{g}_{\alpha\beta}$, where $i,j=1,2,\ldots,l$; $\alpha,\beta=l+1,\ldots,N$. In a real orthonormal basis in X, $S^{-1}dS=i\lambda_a F^a_\alpha(z)dz^\alpha$, where $F^a_\alpha(z)$ are some real functions of z (see Section 8.1.2). Taking into account that $[\lambda_i,\lambda_j]=0$ and $[\lambda_i,\lambda_\alpha]\in X\setminus H$, one infers $[h,S^{-1}dS]=i\lambda_\gamma h^j F_{j\delta\gamma}F^\delta_\alpha(z)dz^\alpha$. Here $F_{j\delta\gamma}$ are the structure constants. They are imaginary in the basis considered (see (8.15)). Hence,

$$\tilde{g}_{\alpha\beta}=F^\gamma_\alpha(z)g_{\gamma\delta}(h)F^\delta_\beta(z),\quad g_{\alpha\beta}=\omega_{\alpha\gamma}\omega_{\beta\gamma},\qquad \omega_{\alpha\beta}(h)=ih^jF_{j\alpha\beta},\tag{5.76}$$

and the Laplace–Beltrami operator can be written in the form,

$$\Delta=\frac{1}{g^{1/2}}\partial_a\circ g^{1/2}g^{ab}\partial_b=\kappa^{-2}\partial_i\circ\kappa^2\partial_i+\frac{1}{g^{1/2}}\partial_\alpha\circ g^{1/2}\tilde{g}^{\alpha\beta}\partial_\beta,\tag{5.77}$$

where $g^{1/2}=\det\omega(h)\det F(z)$, with $F^\alpha_\beta(z)$ being the elements of the matrix $F(z)$, $g^{ac}g_{cb}=\delta^a_b$, $\partial_i=\partial/\partial h^i$, and $\partial_\alpha=\partial/\partial z^\alpha$. In Section 8.1.3 an explicit form of the function $\det\omega(h)=\mu(h)$ is calculated,

$$\mu(h)=\kappa^2(h)\ ,\quad \kappa(h)=\prod_{\alpha>0}(\alpha,h)\,,$$

where the product is taken over all positive roots α of the algebra X. For any twice differentiable function Φ, the following identity holds, $\kappa^{-2}\partial_i(\kappa^2\partial_i\Phi)=\kappa^{-1}\partial_i^2(\kappa\Phi)-\kappa^{-1}\left(\partial_i^2\kappa\right)\Phi$. The function $\kappa(h)$ is harmonic, $\partial_i^2\kappa(h)=0$ (see the proof in Section 8.1) and therefore the so-called "quantum potential" vanishes, $V_q=\kappa^{-1}(\partial_i^2\kappa)/2=0$. Thus, the Hamiltonian in the physical subspace reads,

$$\hat{H}_{ph}=-\frac{1}{2}\kappa^{-1}\partial_i^2\circ\kappa+V(h).\tag{5.78}$$

The scalar product for physical states is given by,

$$(\Phi_1,\Phi_2)=\int_{K^+}dh\mu(h)\Phi_1^*(h)\Phi_2(h),\tag{5.79}$$

where the integration is carried out over the Weyl chamber. Here and in what follows, h is viewed as a vector in $R^l\sim H$ with components $(h_1,\ldots,h_l)$ in the standard basis, the Riemann measure d^lh is denoted dh, and $\Phi_{1,2}(h)$ stand for functions on R^l, etc. The weight $\mu(h)$ is proportional to the volume of the gauge orbit containing h. The equality (5.60) defines a change of variables, i.e. a one-to-one correspondence between $x\in X\sim R^N$ and the new variables h and z, provided $h\in K^+$ and $S(z)$ belongs to the coset space G/G_H, where G_H is the Cartan subgroup of G [216]. This explains the integration region in (5.79). Since $\sqrt{g}=\mu(u)\det F(z)$, $\det F(z)dz$ defines the measure on the coset space G/G_H, and the volume of the gauge orbit through an element $h\in K^+$ is $\mu(h)$ multiplied by the volume of G/G_H. The latter constant is included into the norm of physical states.

By making the substitution

$$\Phi_E(h) = \kappa^{-1}(h)\varphi_E(h), \tag{5.80}$$

the Schrödinger equation $\hat{H}_{ph}\Phi_E = E\Phi_E$ is transformed to the Schrödinger equation for a particle in the l-dimensional Euclidean space,

$$\left[-\tfrac{1}{2}\partial_i^2 + V(h)\right]\varphi_E = E\varphi_E. \tag{5.81}$$

However not all functions of the form (5.80), where φ_E are solutions of (5.81) normalized in R^l, are admissible as the basis elements of the physical Hilbert space $\mathcal{H}_{\rm ph}$. Indeed, the physical states must be gauge invariant,

$$\Phi_{\rm ph}(x) = \Phi_{\rm ph}(\Omega x \Omega^{-1}) \equiv \hat{\Omega}\Phi_{\rm ph}(x), \qquad \Omega \in G. \tag{5.82}$$

So physical states must be functions only of h, $\Phi_{\rm ph}(x) = \Phi_{\rm ph}(h)$. The Weyl group W is a subgroup of the gauge group G. Therefore the physical states should also be stationary relative to W, $\hat{w}\Phi_{\rm ph}(h) = \Phi_{\rm ph}(h_w) = \Phi_{\rm ph}(h)$ for any $w \in W$, where the action of $\hat{w}$ is defined by (5.82) and $h_w = whw^{-1}$ stands for the image of h under the action of an element w of W. Let $\psi(h)$ be a function on $H \sim R^l$, normalized according to (5.79) but with the integration region extended to the whole R^l. Its physical (invariant) component, which is normalized according to (5.79), is

$$\Phi = N_W^{-1/2}\sum_w \hat{w}\psi(h) = N_W^{-1/2}\sum_w \psi(h_\omega), \tag{5.83}$$

where the summation is carried out over all elements of the Weyl group W and $N_W^{-1/2}$ is a normalization factor. Since W acts linearly in H, in $R^l \sim H$ the Weyl group elements can be represented by $l \times l$ matrices, i.e. with any abstract element w of the Weyl group one can associate an $l \times l$ matrix w_s such that $whw^{-1} = h_{wi}\lambda_i$ where $h_{wi} = (w_s)_{ij}h_j$. The latter relation for the components of h_w and h in an orthonormal basis λ_i, $i = 1, 2, ..., l$, in H is simply written as $h_w = w_s h$ in what follows, until noted otherwise. Moreover, since the Weyl group is generated by reflections in the hyperplanes orthogonal to simple roots, the matrices w_s are orthogonal and $\det w_s = \pm 1$. The normalization factor $N_W^{-1/2}$ is determined by the aforementioned conditions,

$$\int_{K^+} dh\mu|\Phi|^2 = \int_{R^l} dh\mu|\psi|^2 = 1. \tag{5.84}$$

The action of group W on the set of Weyl chambers in R^l is simple transitive. Therefore the number of elements of W can be represented by a symbolic equality, $N_W = V_{R^l}/V_{K^+}$, where V_{R^l} and V_{K^+} are the "volumes" of R^l and K^+, respectively (they are infinite). This explains the normalization factor $N_W^{-1/2}$ in (5.83). It is noteworthy that $N_W = r_1 r_2 \cdots r_l$ [21]. Clearly, Φ is invariant under the Weyl transformations, $\hat{w}\Phi = \Phi$, as W is a group and so is the measure $dh\mu(h)$ because w_s is orthogonal and permutes the roots of X; $w_s\alpha$ is a root if α is a root.

Next, it is proved that the functions (5.80) symmetrized according to rule (5.83), where $\psi = \kappa^{-1}\varphi_E$, and normalized according to (5.84), are regular in R^l. Note that a regular normalized solution of (5.81) may give rise to Φ_E which is singular at the boundaries of the Weyl chambers where $\kappa(h) = 0$, i.e. in the planes orthogonal to all positive roots $(h, \alpha) = h_i\alpha_i = 0$. Moreover, the converse is also true: every regular solution of the Schrödinger equation with the Hamiltonian (5.78) is invariant under Weyl group transformations. The following theorem is invoked [216]. Any polynomial $p(h)$ in H with the property that $p(w_s h) = (\det w_s)p(h)$ can be represented in the form,

$$p(h) = \kappa(h)q(h), \tag{5.85}$$

where $q(h)$ is an invariant polynomial, $q(w_s h) = q(h)$ for any $w \in W$. Note that the theorem also implies that the minimum degree of polynomials with the property $p(w_s h) = (\det w_s)p(h)$ is $(N - l)/2$, which is the degree of $\kappa(h)$.

Since the Weyl group is generated by reflections and $\mu(h) = \kappa^2(h)$ is invariant under W,

$$\kappa(w_s h) = \pm\kappa(h) = (\det w_s)\kappa(h), \quad \det w_s = \pm 1 \tag{5.86}$$

Put $\psi = \kappa^{-1}\varphi$ where φ is a normalized solution of (5.81). Then rule (5.83) yields,

$$\Phi(h) = N_W^{-1/2}\kappa^{-1}(h)\sum_w (\det w_s)\varphi(w_s h) \equiv \kappa^{-1}\tilde{\varphi}(h). \tag{5.87}$$

Since $\Phi(h)$ is W-invariant, $\Phi(w_s h) = \Phi(h)$, therefore $\tilde{\varphi}(w_s h) = (\det w_s)\tilde{\varphi}(h)$. Suppose that potential V is analytic in R^l. Then $\tilde{\varphi}$ can be expanded into a power series. The power series terms can be grouped into homogeneous polynomials $p_n(h)$ of a fixed degree n,

$$\tilde{\varphi}(h) = \sum_{n=0}^{\infty} \tilde{\varphi}_n p_n(h).$$

As $w \in W$ is a homogeneous linear transformation of h, $p_n(w_s h) = (\det w_s)\, p_n(h)$. It follows from (5.85) that $p_n(h) = \kappa(h)q_n(h)$ and $\tilde{\varphi}_n = 0$ if $n < (N - l)/2$. Hence, $\tilde{\varphi}(h) = \kappa(h)\chi(h)$, and

$$\chi(h) = \sum_{n=0}^{\infty} \tilde{\varphi}_n q_n(h) \tag{5.88}$$

is an invariant function. According to (5.86)–(5.88), factor κ^{-1} in (5.87) does not give rise to any pole singularities at the boundaries of the Weyl chambers, and the function $\Phi = \chi(h)$ is regular in R^l.

To prove the converse, consider the eigenvalue problem for the total Hamiltonian,

$$\hat{H}\Psi_E(x) = \left[-\tfrac{1}{2}\Delta + V(x)\right]\Psi_E(x) = E\Psi_E(x)$$

in curvilinear coordinates (5.60), where the Laplace operator is given by (5.77). A solution can always be represented in the form,

$$\Psi_E(x) = \Psi_E(z,h) = \sum_{(n)} \Phi_E^{(n)}(h) Y_{(n)}(z), \tag{5.89}$$

where $Y_{(n)}(z)$ are eigenfunctions of the Casimir operators in the algebra generated by the operators $\hat{\sigma}_a$ and (n) denotes the corresponding set of eigenvalues, $E = E(n)$. The physical subspace is spanned by the functions $\Phi_E^{(0)}(h)$, $Y_{(0)}(z) = \text{const}$. Consider the symmetry transformations of the change of variables (5.60), that is, the set of transformations of $h \in H$ and z under which the left-hand side of (5.60) remains unchanged. This set consists of the group of translations by periods of the compact coset manifold G/G_H (h is not changed) and the Weyl group

$$x \to x, \quad h \to whw^{-1}, \quad S(z) \to \hat{w}S(z) = S(zw). \tag{5.90}$$

Here h and w are viewed as abstract elements of H and W, respectively. The functions (5.89) are invariant under transformations (5.90) as functions of x. Then the functions $\Psi_E^{(0)}(x) \equiv \Phi_E^{(0)}(h)$, which form a basis in the physical Hilbert space, must be invariant under the Weyl group. The regularity of $\Phi_E^{(0)}(h)$ follows immediately from the regularity of $\Psi_E^{(0)}(x)$ (the potential is assumed to be real analytic).

The equality $\Phi_E^{ph}(x) = \Phi_E^{(0)}(h)$ establishes a one-to-one correspondence between the set of analytic gauge-invariant functions in the total configuration space and the set of analytic W-invariant functions in the reduced configuration space. In group theory, this statement follows from the theorem due to Chevalley (see [20, 21]). It asserts that any W-invariant polynomial in H has a unique analytic continuation into X which is invariant under the adjoint action of group G in its Lie algebra X. The conclusion of the theorem also holds for analytic functions in H because the set of all polynomials is dense in the set of analytic functions.

This statement is illustrated with an example of groups from rank 2.

5.4.6 The Chevalley theorem and gauge-invariant wave functions of the harmonic oscillator. Groups of rank 2

The variables $\Phi_{1,2}$ introduced in (5.70) allow one to explicitly construct gauge-invariant wave functions of the harmonic oscillator. To calculate the function $\mu(h)$ in the new variables, Lemma III.3.7 in [216] is invoked. This asserts that

$$\det\left(\frac{\partial p_{r_i}}{\partial h^j}\right) = c'\,\kappa(h), \quad c' = \text{const}, \quad i,j = 1,2,...,\text{rank}\,X.$$

Its application to groups of rank 2 ($r_1 = 2$ and $r_2 \equiv r$) yields $\kappa^2 \sim \Phi_1^{2r}\left(c_2 + c_1\Phi_2 - \Phi_2^2\right)$. The range of Φ_2 is then normalized to the interval $[-1, 1]$

by the change $\Phi_2 \to (\Phi_2 - b)/\sqrt{a}$. The constants a and b were introduced in Section 5.4.4 (see the text between (5.71)) and (5.72). As a result, one finds,

$$\mu(h) = \kappa^2(h) = c\,\Phi_1^{2r}\left(1 - \Phi_2^2\right),$$

where c is a constant. According to the Chevalley theorem, this equation defines the analytic gauge-invariant continuation of the polynomial $\mu(h)$ to the whole configuration space X. It is a polynomial of degree $2r$ constructed out of two independent Casimir polynomials $p_2(x)$ and $p_r(x)$. A gauge-invariant function $\Psi_E(x)$ must be a regular function of $\Phi_{1,2}$. After the substitution $\Psi_E(x) = [\mu(\Phi_1, \Phi_2)]^{-1/2}\varphi_E(\Phi_1, \Phi_2)$ the Schrödinger equation $\hat{H}\Psi_E = E\Psi_E$ in the total configuration space, becomes

$$\left\{-\frac{1}{\Phi_1}\frac{\partial}{\partial\Phi_1}\Phi_1\frac{\partial}{\partial\Phi_1} - \frac{r^2}{\Phi_1^2}\sqrt{1-\Phi_2^2}\,\frac{\partial}{\partial\Phi_2}\sqrt{1-\Phi_2^2}\,\frac{\partial}{\partial\Phi_2} + \Phi_1^2\right\}\varphi_E = 2E\varphi_E. \tag{5.91}$$

This equation is separable and solutions are sought in the form $\varphi_E = F(\Phi_1)f(\Phi_2)$. Equation (5.91) is equivalent to the following two equations:

$$-\left(1 - \Phi_2^2\right)f'' + \Phi_2 f' + cf = 0; \tag{5.92}$$

$$-F'' - \frac{1}{\Phi_1}F' - \left(\frac{cr^2}{\Phi_1^2} - \Phi_1^2 + 2E\right)F = 0, \tag{5.93}$$

where c is the separation variable constant. The regularity of solutions $\Psi_E(x)$ requires the zero boundary condition at the points where μ vanishes, i.e. at $\Phi_1 = 0$ and $\Phi_2 = \pm 1$. Hence, it is demanded that $f(\pm 1) = 0$. Solutions of (5.92) satisfying this condition are given by

$$f_m(\Phi_2) = \sin[(m+1)\cos^{-1}\Phi_2] = \left(1 - \Phi_2^2\right)^{1/2} U_m(\Phi_2),$$

where $U_m(\Phi_2)$ are the Chebyshev polynomials of the second kind [132], $m = 0, 1, 2, \ldots$, and $c = -(m+1)^2$. Equation (5.93) is transformed to the standard form (5.38) with $\alpha = r(m+1) + 1$ and $\beta = -(E - \alpha)/2$ by the substitution $F(\Phi_1) = \Phi_1^{r(m+1)} e^{-\Phi_1^2/2} g(\Phi_1)$, where $g(\Phi_1)$ is a function to be found, with the subsequent change of variables $t = \Phi_1^2$. Thus, the gauge-invariant eigenfunctions are

$$\Psi_{nm} = c_{nm}\Phi_1^{rm} U_m(\Phi_2) L_n^{r(m+1)}\left(\Phi_1^2\right) e^{-\Phi_1^2/2},$$

where c_{mn} are normalization constants, and $L_n^{r(m+1)}$ are Laguerre polynomials. The eigenvalues are given by (5.68) with $r_1 = 2$ and $r_2 = r$. The fact that the ground state energy equals $N/2$ follows here from relation [21] $N = 2r + 2$ for the groups of rank 2.

The eigenfunctions Ψ_{nm} are regular functions of the Casimir polynomials $p_{2,r}$. Indeed, for groups $Sp(4) \sim SO(5)$ and G_2, the factor of the exponential is a polynomial of $p_{2,r}$ since r is an even integer ($r = 4, 6$, respectively), and, therefore, Φ_1^{rm} is a polynomial for any positive integer m. In the case of $SU(3)$, $r = 3$, and for odd m, Φ_1^{3m} is proportional to the non-polynomial factor $[p_2(x)]^{1/2}$. However, as pointed out in Section 5.4.4, the coefficient b which defines the range of Φ_2 in (5.70) vanishes for $SU(3)$. Thanks to this the rescaled variable $\Phi_2 \in [-1, 1]$ used here can be written as $\Phi_2 = \sqrt{6}p_3\Phi_1^{-3}$. Hence, the non-polynomial factor in $\Phi_1^{3m}U_m(\Phi_2)$ is canceled out.

The functions Ψ_{nm} restricted to the Cartan subalgebra H are regular eigenfunctions of the Hamiltonian (5.78) because $\Psi_{nm}(x) = \Psi_{nm}(ShS^{-1}) = \Psi_{nm}(h)$. So, $\Psi_{nm}(h)$ are invariant under the Weyl group transformations. The relation $\Psi_{nm}(h) = \Psi_{nm}(x)$ defines a unique analytic gauge-invariant continuation of the W-invariant functions in H to the total configuration space X in full accord with the Chevalley theorem [20, 21].

5.5 Gauge systems with Grassmann variables

5.5.1 Classical theory. A simple model with an Abelian gauge group

Gauge theories with Fermi fields are just as common as those with Bose fields. Let us turn to studying the dynamics of gauge systems with Grassmann variables. Consider first the simplest system with a minimal number of Grassmann degrees of freedom whose dynamics is nontrivial. The system is defined by the Lagrangian [26, 195, 200],

$$L(\psi, \dot{\psi}, \psi^\dagger, \dot{\psi}^\dagger, y, \dot{y}) = \frac{i}{2}\left[\psi^\dagger D_t\psi - (D_t\psi)^\dagger\psi\right] - V(\psi^\dagger\psi), \tag{5.94}$$

where $D_t\psi = (\partial_t - iy\Gamma)\psi$ is the covariant time derivative, ψ is a column with two components ψ_α being an element of a complex Grassmann algebra, $\psi_\alpha = \theta_1^\alpha + i\theta_2^\alpha$ and θ_j^α are real, $\left(\theta_j^\alpha\right)^* = \theta_j^\alpha$, $\left(\theta_j^\alpha\right)^2 = 0$, $\alpha, j = 1, 2$, Γ is an "electric charge" operator, $\Gamma \equiv \tau_3$ (the Pauli matrix), and $\psi^\dagger$ is the transposed and complex conjugated column ψ, i.e. $\psi^\dagger$ is a row with elements ψ_α^*. The Lagrangian (5.94) is invariant under the gauge transformations

$$\psi \to \exp(i\varepsilon\Gamma)\psi\ , \quad y \to y + \dot{\varepsilon},$$

where ε is an arbitrary function of time. The Hamiltonian formalism for Lagrangians linear in velocity (the first-order formalism) is discussed in Chapter 3 in detail. From the analysis given in Chapter 3 it follows, in particular, that model (5.94) can be viewed as a dynamical system with two primary constraints of the second class (they are not in involution). It is easy to see that, in addition to these, there is another primary constraint of the first class, $\pi_0 = \partial L/\partial\dot{y} = 0$. To go over to the Hamiltonian formalism, the second-class constraints have to be solved so that some of the corresponding canonical variables are eliminated and the Poisson bracket is replaced by the Dirac bracket [148]. The Poisson bracket for Grassmann canonical variables is defined in (1.129) (see also Section 1.11.1). Application of the formalism of Section 3.3.3 to the Lagrangian (5.94) shows that the canonical momenta for $\boldsymbol{\psi}$ are identified with $\boldsymbol{\psi}^\dagger$, the Dirac bracket reads $\left\{\psi_\alpha, \psi_\beta^*\right\}_D = -i\delta_{\alpha\beta}$ and the canonical Hamiltonian is

$$H = V(\boldsymbol{\psi}^\dagger\boldsymbol{\psi}) - y\boldsymbol{\psi}^\dagger\Gamma\boldsymbol{\psi}. \tag{5.95}$$

In what follows, subscript D in the Dirac brackets is omitted, i.e. $\left\{\psi_\alpha, \psi_\beta^*\right\} = -i\delta_{\alpha\beta}$. The system has a secondary constraint,

$$\sigma \equiv \dot{\pi}_0 = \{\pi_0, H\} = \boldsymbol{\psi}^\dagger\Gamma\boldsymbol{\psi} = \psi_1^\dagger\psi_1 - \psi_2^\dagger\psi_2 = 0. \tag{5.96}$$

Function V in (5.95) can always be expanded in the basis of the Grassmann algebra $V = V_0 + \omega\boldsymbol{\psi}^\dagger\boldsymbol{\psi} + \Omega(\boldsymbol{\psi}^\dagger\boldsymbol{\psi})^2$, where V_0, ω, and Ω are real constants. Simple calculations show that $(\boldsymbol{\psi}^\dagger\boldsymbol{\psi})^2 = -\sigma^2$. Setting $V_0 = 0$, the Hamiltonian (5.95) can be written in the form,

$$H = \omega\boldsymbol{\psi}^\dagger\boldsymbol{\psi} - y(t)\boldsymbol{\psi}^\dagger\Gamma\boldsymbol{\psi} - \Omega(\boldsymbol{\psi}^\dagger\Gamma\boldsymbol{\psi})^2. \tag{5.97}$$

Because of (5.96) the term quadratic in constraint has no effect on the Hamiltonian equations of motion. Hence, Ω can be set to zero in (5.97). The Hamiltonian equations of motion have the form,

$$\dot{\boldsymbol{\psi}} = \{\boldsymbol{\psi}, H\} = i(y\Gamma - \omega)\boldsymbol{\psi}, \quad \dot{\boldsymbol{\psi}}^\dagger = -i\boldsymbol{\psi}^\dagger(y\Gamma - \omega).$$

Their integration is elementary,

$$\boldsymbol{\psi}(t) = \exp\left(-i\omega t + i\Gamma\int_0^t y(t')dt'\right)\boldsymbol{\psi}(0). \tag{5.98}$$

Setting $y(t) = -\omega$ (i.e. by choosing a particular gauge), one finds,

$$\psi_1(t) = e^{-2i\omega t}\psi_1(0), \quad \psi_2(t) = \psi_2(0), \tag{5.99}$$

from which it follows that the degree of freedom ψ_2 has a trivial dynamics, while ψ_1 oscillates with double the frequency. As argued above, the latter is evidence that the phase space of the physical degree of freedom is reduced (the notion of the phase space for Grassmann variables is introduced in Section 1.8). Note that here it is impossible to make the non-physical variable zero by choosing a suitable gauge fixing condition. The difference with the bosonic case is that gauge freedom is the multiplication of $\psi_{1,2}$ by complex-valued functions of time with unit absolute value. Such transformations cannot change the dimension of the Grassmann algebra. That is the reason why the non-physical degree of freedom here appears as one having a trivial dynamics (compare (5.99)), or, any preassigned dynamics that is not at all determined by the Hamiltonian.

One could arrive at the conclusion, that the phase space of the physical degree of freedom is reduced, in a different way. Admissible initial conditions must satisfy the constraint (5.96). Hence, by a suitable choice of the gauge (or $y(t)$), one can always achieve $\psi_1(t) = \psi_2(t)$. Indeed, if $y(t) = 0$ in (5.98), then constraint (5.96) demands that $\psi_1(0) = \psi_2(0)$ and, therefore, $\psi_1(t) = \psi_2(t)$. The latter allows for residual gauge transformations, $\psi_1 \to -\psi_1$ $(y(t') \to y(t') + \pi\delta(t-t'))$. The states $\left(\psi_1, \psi_1^\dagger\right)$ and $\left(-\psi_1, -\psi_1^\dagger\right)$ turn out to be gauge equivalent. So the physical phase space, spanned by the pair $\left(\psi_1, \psi_1^\dagger\right)$, is a cone, $\Gamma_{ph}\left(\psi_1, \psi_1^\dagger\right) = \text{cone}(\pi)$. It should be noted that this implies that in expression $\boldsymbol{\psi}(t) = z(t)\boldsymbol{\psi}(0)$, where $z(t)$ is a diagonal complex matrix, the argument of $z(t)$ for the physical degree of freedom is defined only by mod (π) (compare also [217]).

5.5.2 Quantum theory. The simple model

The quantization of model (5.94) can easily be carried out. In accordance with the formalism developed in Chapter 4, the canonical variables ψ_α and $\psi_\alpha^\dagger$ are promoted to operators with the commutation relation induced by the Dirac bracket,

$$\left[\hat{\psi}_\alpha, \hat{\psi}_\beta^\dagger\right]_+ = i\left\{\psi_\alpha, {\psi_\beta}^\dagger\right\} = \delta_{\alpha\beta}. \tag{5.100}$$

As before, constraint $\pi_0 = 0$ can be ignored. The ground state of the system with the Hamiltonian (5.97) (the "vacuum" $|0\rangle$) is defined by the condition $\hat{\psi}_\alpha|0\rangle = 0$. A generic state $|\chi\rangle$ of the Hilbert space has the form,

$$\chi(\hat{\boldsymbol{\psi}}^\dagger)|0\rangle \equiv \left(\chi_0 + \chi_\alpha\hat{\psi}_\alpha^\dagger + \chi_3\hat{\psi}_1^\dagger\hat{\psi}_2^\dagger\right)|0\rangle = |\chi\rangle. \tag{5.101}$$

Its conjugated state reads,

$$\langle\chi| = \langle 0|\left[\chi(\hat{\boldsymbol{\psi}}^\dagger)\right]^\dagger = \langle 0|\left[\chi_0^* + \chi_\alpha^*\hat{\psi}_\alpha + \chi_3^*\hat{\psi}_2\hat{\psi}_1\right], \tag{5.102}$$

so that the scalar product can be written as,

$$\langle\chi|\phi\rangle = \sum_{a=0}^{3}\phi_a^*\chi_a. \tag{5.103}$$

The Hamiltonian of the system is

$$\hat{H} = \omega(\hat{\boldsymbol{\psi}}^\dagger\hat{\boldsymbol{\psi}} - 1). \tag{5.104}$$

Upon quantization, symmetric operator ordering is used, $\boldsymbol{\psi}^\dagger\boldsymbol{\psi} \to (1/2)(\hat{\boldsymbol{\psi}}^\dagger\hat{\boldsymbol{\psi}} - \hat{\boldsymbol{\psi}}\hat{\boldsymbol{\psi}}^\dagger) = \hat{\boldsymbol{\psi}}^\dagger\hat{\boldsymbol{\psi}} - 1$ by analogy with the bosonic case in which $a^\dagger a \to (\hat{a}^\dagger\hat{a} + \hat{a}\hat{a}^\dagger)/2$ upon quantization. The physical states satisfy the condition

$$\hat{\sigma}|\Phi\rangle = \hat{\boldsymbol{\psi}}^\dagger\Gamma\hat{\boldsymbol{\psi}}|\Phi\rangle = 0. \tag{5.105}$$

All solutions of (5.105) are given by linear combinations of the states $|0\rangle$ and $\hat{\psi}_1^\dagger\hat{\psi}_2^\dagger|0\rangle$. Indeed, $\hat{\sigma}\hat{\psi}_\beta^\dagger|0\rangle = (\hat{\boldsymbol{\psi}}^\dagger\Gamma)_\beta|0\rangle \neq 0$ and $\hat{\sigma}\hat{\psi}_1^\dagger\hat{\psi}_2^\dagger|0\rangle = \left(\hat{\psi}_1^\dagger\hat{\psi}_2^\dagger + \hat{\psi}_2^\dagger\hat{\psi}_1^\dagger\right)|0\rangle = 0$. The distance between the energy levels of physical excitations is doubled, $E_2 - E_0 = 2\omega$, in full accordance with the above analysis of classical dynamics (see (5.99)).

5.5.3 A model with an arbitrary compact gauge group. The adjoint representation

As noted, in elementary particle theory dynamical systems with Grassmann variables are just as important as those with commuting (bosonic) variables. This motivates the study of models with Grassmann variables and non-Abelian gauge symmetries. The difficulties here are associated not with the structure of the gauge group itself but rather with the complexity of its representations realized by the dynamical variables. The analysis of elementary representations of non-Abelian groups [218] is rather simple, as the example of group $SO(n)$ indicates, while the problem becomes more involved for the adjoint representation. Its study for fermions may look excessive as fermions usually realize the fundamental representation of gauge groups. However if one recalls that supersymmetric extensions of the Yang–Mills theory contain fermions in the adjoint representation, this is no longer an objection. In addition, gauge systems with fermions in the adjoint representation also turn out to be quite interesting [26, 31, 195].

Consider the Lagrangian (5.94) in which $V = \omega \mathrm{Tr}\, \psi^\dagger \psi$, $\psi = \psi_a \lambda_a$, and $\psi^\dagger = \psi_a^\dagger \lambda_a$, with λ_a being the orthonormal basis of the Lie algebra of the group G just as in Section 5.4, the covariant derivative $\partial_t \psi - iy\Gamma\psi$ is replaced by $\partial_t\psi + [y, \psi]$ where $y = \lambda_a y_a$. Similarly to Section 5.4, put $\mathrm{Tr}\, \psi\psi^\dagger = \psi_a {\psi_a}^\dagger \equiv \psi\psi^\dagger$ to simplify the notation. The Hamiltonian formalism is developed in the same way as in Section 5.5.2; the only difference being that there are $2N$ ($N = \dim G$) first-class constraints (compare Section 5.4),

$$\pi_a = \frac{\partial L}{\partial \dot{y}_a} = 0, \qquad \sigma_a = \psi^\dagger T_a \psi = 0, \quad a = 1, 2, \ldots, N,$$

where T_a are matrices defined by the structure constants, $(T_a)_{bc} = F_{bc}^a$ (see the text after (5.59)). Let us turn to quantum theory. Instead of (5.104), the Hamiltonian assumes the form,

$$\hat{H} = \omega \left(\hat{\psi}^\dagger \hat{\psi} - \frac{N}{2} \right). \tag{5.106}$$

Here $\left[\hat{\psi}_a, \hat{\psi}_b^\dagger\right]_+ = \delta_{ab}$. The physical states are selected by the condition $\hat{\sigma}_a|\Phi\rangle = \psi^\dagger T_a \psi|\Phi\rangle = 0$, similar to (5.105) and hence, can be obtained by the action of invariant polynomials $P(\hat{\psi}^\dagger)$ on the "vacuum" state $|0\rangle$ defined by $\hat{\psi}_a|0\rangle = 0$. Any invariant polynomial is a polynomial of elementary invariant homogeneous polynomials. Since the operators $\hat{\psi}_a{}^\dagger$ are anti-commutative, the elementary polynomials are determined by elementary *totally skew-symmetric* invariant tensors of the adjoint representation, which can be constructed out of the structure constants of the gauge group. A generic homogeneous invariant polynomial of degree k has the form,

$$P_k = \mathrm{Tr}\, (\lambda_{a_1} \ldots \lambda_{a_k}) \hat{\psi}_{a_1}^\dagger \cdots \hat{\psi}_{a_k}^\dagger, \tag{5.107}$$

from which it follows that $P_k = 0$ if k is even because $\mathrm{Tr}\, (\lambda_{a_1} \ldots \lambda_{a_k}) = \mathrm{Tr}\, (\lambda_{a_2} \ldots \lambda_{a_k} \lambda_{a_1})$ while $\hat{\psi}_{a_1}^\dagger \cdots \hat{\psi}_{a_k}^\dagger = -\hat{\psi}_{a_2}^\dagger \cdots \hat{\psi}_{a_k}^\dagger \hat{\psi}_{a_1}^\dagger$. If k is odd, the polynomial (5.107) can be transformed in the following way. Put $\hat{\psi}_a^\dagger \hat{\psi}_b^\dagger \lambda_a \lambda_b = (1/2) F_{ab}^c \lambda_c \hat{\psi}_a^\dagger \hat{\psi}_b^\dagger \equiv \lambda_c B_c = B$. Then

$$P_{2m+1} = \mathrm{Tr}\,(\lambda_{a_1} \dots \lambda_{a_m} \lambda_a) B_{a_1} \dots B_{a_m} \hat{\psi}_a^\dagger. \tag{5.108}$$

The quantities B_a commute with each other (as even elements of the Grassmann algebra). Therefore the trace in (5.108) can be symmetrized in the indices $a_1, \dots, a_m$. In fact, the symmetrization can be carried out in all the indices, i.e. including index a. Indeed, put $\lambda_{a_m} \lambda_a = \lambda_a \lambda_{a_m} + F_{a_m a}^b \lambda_b$ in (5.108). The term $F_{a_m a}^b B_{a_m} \hat{\psi}_a^\dagger = (1/2) F_{a_m a}^b F_{cd}^{a_m} \hat{\psi}_a^\dagger \hat{\psi}_c^\dagger \hat{\psi}_d^\dagger = 0$ vanishes thanks to the Jacobi identity [21],

$$-F_{[ab}^c F_{a']c}^{b'} = F_{ab}^c F_{ca'}^{b'} + F_{ba'}^c F_{ca}^{b'} + F_{a'a}^c F_{cb}^{b'} = 0,$$

where the anti-symmetrization is carried out for the indices in the brackets. Thus, the trace in (5.108) is a symmetric invariant tensor. Every invariant symmetric tensor can be constructed from the elementary symmetric tensors $p_{a_1 \dots a_r}$ that define the Casimir polynomials p_r or degree r (compare (5.66)).

Suppose the trace in (5.108) is a reducible tensor, i.e. it can be written as the tensor product of tensors of smaller rank. Since $r \geq 2$ for all groups, any decomposition of the set of indices $(a_1, \dots, a_m, a)$ into two subsets containing at least two elements a_i, a necessarily generates Casimir polynomials of the form $P_{2r}(\hat{\psi}^\dagger) = p_r(B)$, where $r - 2, \dots, r_l$, $l = \mathrm{rank}\, G$. As established above, all the invariant homogeneous polynomials of even degree vanish identically, $P_{2r}(\hat{\psi}^\dagger) = 0$. There is only one possibility left. The trace $\mathrm{Tr}\,(\lambda_{a_1} \dots \lambda_{a_m} \lambda_a)$ must be an irreducible symmetric tensor of rank $r = m + 1$. Hence, the basis of the physical Hilbert space is generated by polynomials of the form $P_{2m+1}(\hat{\psi}^\dagger) = P_{2r-1}(\hat{\psi}^\dagger)$,

$$\left[\mathrm{Tr}\,(\hat{\psi}^\dagger)^{2r_1 - 1}\right]^{n_1} \cdots \left[\mathrm{Tr}\,(\hat{\psi}^\dagger)^{2r_l - 1}\right]^{n_l} |0\rangle, \tag{5.109}$$

where $n_i = 0, 1$, because odd elements of the Grassmann algebra are nilpotent. According to (5.108) and the definition of B, invariant totally skew-symmetric tensors have the form,

$$p_{a_1 \dots a_{r-1} [a_r} F_{b_1 c_1}^{a_1} \cdots F_{b_{r-1} c_{r-1}]}^{a_r - 1}.$$

The simplest of them (of the smallest rank) coincides with the structure constants F_{bc}^a.

The basis (5.109) corresponds to the following spectrum of the Hamiltonian (5.106):

$$E = \omega \sum_{i=1}^{l} (2r_i - 1) n_i - N\omega/2, \quad n_i = 0, 1.$$

In realistic models (i.e. in field theories), the problem of finding all independent invariant polynomials is more involved because the operators $\hat{\psi}$ may have additional indices associated with other quantum numbers (e.g. spin). They have to be taken into account when constructing invariant polynomials.

5.6 More general mechanical gauge systems with bosonic variables

In the gauge systems studied so far all physical degrees of freedom have a reduced phase space. An increase in the number of degrees of freedom while keeping the number of gauge parameters fixed leads to a new type of model in which some physical variables might seem to have a conventional (plane) phase space, while the phase space of the others is reduced. The actual situation turns out to be quite different, resembling more the case of the adjoint representation studied in the previous sections. A non-standard structure of the physical phase space creates a specific kinematic coupling of physical degrees of freedom that is not generated by the Hamiltonian. Naturally, this has physical consequences for the system dynamics [195, 208].

Another new important characteristic feature of these systems is that the physical configuration space is curved even if the total configuration space is flat. The latter implies, in particular, that the kinetic energy operator in the physical Hamiltonian contains a metric which corresponds to a non-zero Riemann–Christoffel tensor (see (2.35)). Even a simple model of two particles in the plane with the gauge group $SO(2)$, studied in Sections 5.6.1 and 5.6.3, exhibits such a feature. It also occurs in Yang–Mills mechanics [208, 219, 220] (Section 5.6.4) and in gauge field theories [195, 213, 221] (Section 5.8).

Finally, gauge matrix models similar to those studied in this section also appear in the theory of 11-dimensional super-membranes [222, 223], in the dynamics of D-particles [224], and in the matrix theory [225], which describe various dynamical aspects of superstring theory. The analysis presented here is general and applicable to any gauge matrix model.

5.6.1 Two particles in a plane

Consider a system of two particles in a plane. Let the Lagrangian of the system be the sum of two Lagrangians of the form (5.2) [26],

$$L' = \frac{1}{2}[(\partial_t - y_1 T)\mathbf{x}_1]^2 + \frac{1}{2}[(\partial_t - y_2 T)\mathbf{x}_2]^2 - V_1(\mathbf{x}_1) - V_2(\mathbf{x}_2), \tag{5.110}$$

where y_1 and y_2 change independently under gauge transformations. A simple examination of the system shows that the physical phase space of each particle is cone(π), and the energy of the system is the sum of energies of the subsystems. Let us reduce the gauge group $SO(2) \times SO(2)$ of the Lagrangian (5.110) to the subgroup $SO(2)$ by demanding $y_1 \equiv y_2 = y$. The new Lagrangian has the form [208],

$$L = \tfrac{1}{2}[(\partial_t - yT)\mathbf{x}_1]^2 + \tfrac{1}{2}[(\partial_t - yT)\mathbf{x}_2]^2 - V_1(\mathbf{x}_1) - V_2(\mathbf{x}_2). \tag{5.111}$$

It is invariant under infinitesimal gauge transformations,

$$\delta \mathbf{x}_i = \varepsilon T \mathbf{x}_i, \quad \delta y = \dot{\varepsilon}, \quad \delta V_i = 0, \quad i = 1, 2.$$

The standard analysis shows that the Lagrangian (5.111) defines a system with one primary and one secondary constraint,

$$\pi = \frac{\partial L}{\partial \dot{y}} = 0, \qquad \sigma = (\mathbf{p}_1, T\mathbf{x}_1) + (\mathbf{p}_2, T\mathbf{x}_2) = 0, \tag{5.112}$$

and the canonical Hamiltonian

$$H = \tfrac{1}{2}\left(\mathbf{p}_1^2 + \mathbf{p}_2^2\right) + V_1(\mathbf{x}_1) + V_2(\mathbf{x}_2) + y(t)\sigma \equiv H_1 + H_2, \tag{5.113}$$

where each of $H_{1,2}$ is identical to (5.6). The constraint σ generates simultaneous rotations of the particle position vectors $\mathbf{x}_1$ and $\mathbf{x}_2$ and the momenta $\mathbf{p}_1$ and $\mathbf{p}_2$ about the origin. What can be said about the physical phase space of the particles? If the gauge $x_1^{(2)} = 0$ is chosen (here $\mathbf{x}_i = \left(x_i^{(1)}, x_i^{(2)}\right)$), then it seems natural to conclude that $\Gamma\left(x_1^{(1)}, p_1^{(1)}\right) = \mathrm{cone}\,(\pi)$ because the residual discrete gauge group Z_2 identifies the states $\left(x_1^{(1)}, p_1^{(1)}\right)$ and $\left(-x_1^{(1)}, -p_1^{(1)}\right)$, while the phase space of the other physical degrees of freedom have the conventional structure, $\Gamma(\mathbf{x}_2, \mathbf{p}_2) = R^2 \times R^2$. However, instead of $x_1^{(2)}$, the variable $x_2^{(2)}$ could be eliminated with the conclusion that $\Gamma\left(x_2^{(1)}, p_2^{(1)}\right) = \mathrm{cone}(\pi)$. Thus, despite the dynamical independence of the systems described by the Hamiltonians $H_{1,2}$ (note $\{H_1, H_2\} = 0$), questions about the phase space of any of the subsystems separately makes no sense. Only the structure of the phase space of the system as a whole can be discussed.

In order to establish the physical phase-space structure, it is convenient to apply Eq. (5.1). The extended gauge group is the direct product of the group of translations of variable y and the group of simultaneous rotations of vectors $\mathbf{x}_i$ and $\mathbf{p}_i$ about the origin. The total phase space is the direct product of five planes. The canonical variables y and π obviously describe pure non-physical degrees of freedom. Hence, it remains to compute the factor-space $(R^4 \times R^4)|_{\sigma=0}/SO(2)$, where the operator $\hat{\sigma} = \{\sigma, \cdot\}$ is the generator of rotations of the vectors $\mathbf{x}_i$ and $\mathbf{p}_i$. By a suitable gauge transformation, vector $\mathbf{x}_1$ can always be directed along the first coordinate axis, i.e. $x_1^{(2)} = 0$. As a result the physical phase space is contained in the surface of intersection of the gauge fixing and constraint surfaces that is defined by the equations

$$x_1^{(2)} = 0, \qquad p_1^{(2)} = -\frac{1}{x_1^{(1)}}(\mathbf{p}_2, T\mathbf{x}_2). \tag{5.114}$$

The second equation follows from constraint $\sigma = 0$. The surface (5.114) is spanned by six Cartesian variables. But the physical phase space does not coincide with

$R^3 \times R^3$ because the surface (5.114) is mapped onto itself upon gauge transformations which are rotations through angles that are integer multiples of π. Under such transformations

$$\mathbf{x}_{1,2} \to \pm\mathbf{x}_{1,2}, \qquad \mathbf{p}_{1,2} \to \pm\mathbf{p}_{1,2}. \tag{5.115}$$

Therefore the physical phase space is obtained by identifying the points of the surface (5.114) that are related by these discrete gauge transformations,

$$\Gamma_{ph} = (R^4 \times R^4)|_{\sigma=0}/SO(2) = R^3 \times R^3/Z_2, \tag{5.116}$$

where the action of group Z_2 is defined in (5.115). Clearly, the physical phase space cannot be viewed as the direct product of the cone $\times$ cone(π) and two planes because the residual gauge group acts on all canonical variables spanning the surface (5.114). Such a peculiarity of the residual gauge symmetry is of a rather general nature [31, 195, 208]. So the physical phase space is reduced as a whole, which, in turn, implies the existence of a purely kinematic coupling between the physical degrees of freedom not generated by the Hamiltonian.

Thus, the Lagrangian (5.111) provides yet another example of quite unusual properties of theories with a gauge symmetry: The dynamics of two dynamically decoupled systems (no coupling potential) turns out to be related and, moreover, the influence of one subsystem on the other occurs through the degree of freedom y that has no physical significance as a pure gauge parameter (formally, the coupling occurs through the constraint $\sigma = 0$ in (5.112)). This can also be restated differently: after the elimination of all gauge freedom, the kinematic coupling of the physical degrees of freedom occurs through the nontrivial structure of their phase space.

Remark. The conditions $x_1^{(2)} = 0$ and $x_1^{(1)} \geq 0$ seem to completely eliminate the gauge arbitrariness so that the physical configuration space appears to be isomorphic to the half-space R^3 bounded by the plane $x_1^{(1)} = 0$. This is not actually so. In the boundary plane, the residual gauge transformations are $SO(2)$–rotations of the vector $\mathbf{x}_2$. Therefore the boundary plane must be further contracted to a half-line to obtain the physical configuration space.

5.6.2 Quantum theory for the model (5.111). The harmonic oscillator potential

The study of systems with gauge symmetry shows that the structure of the physical phase space affects quantum dynamics. In particular, the distance between the energy levels of a harmonic oscillator with conic phase space is doubled compared to that of the same oscillator but with the conventional plane phase space. If the physical phase space in the model (5.111) were cone(π) $\times R^2 \times R^2$, then for the harmonic oscillator potentials $V_{1,2}$ one would have expected the spectrum to contain the energy levels of two harmonic oscillators with the same frequency and those of a harmonic oscillator with the doubled frequency. But this turns out not to be the case. The distance between the energy levels is doubled for all three oscillators. This provides an independent validation of Eq. (5.116).

Let us now compute the physical spectrum of the model [26, 208]. Put $V_i = \omega_i^2 \mathbf{x}_i^2/2$, where $\omega_1 \neq \omega_2$, in general. Making use of the creation and annihilation operators for the harmonic oscillators, $\hat{\mathbf{a}}_j = (\hat{\mathbf{p}}_j/\sqrt{\omega_j} - i\sqrt{\omega_j}\hat{\mathbf{x}}_j)/\sqrt{2}$, and $\hat{\mathbf{a}}_j^\dagger = (\hat{\mathbf{p}}_j/\sqrt{\omega_j} + i\sqrt{\omega_j}\hat{\mathbf{x}}_j)/\sqrt{2}$, the Hamiltonian (5.113) and the constraint operator $\hat{\sigma}$ in (5.112) can be written in the form,

$$\hat{H} = \sum_{j=1,2} \left[\omega_j \left(\hat{\mathbf{a}}_j^\dagger, \hat{\mathbf{a}}_j\right) - iy\left(\hat{\mathbf{a}}_j^\dagger, T\hat{\mathbf{a}}_j\right) + \omega_j\right], \tag{5.117}$$

$$\hat{\sigma} = -i\left(\hat{\mathbf{a}}_1^\dagger, T\hat{\mathbf{a}}_1\right) - i\left(\hat{\mathbf{a}}_2^\dagger, T\hat{\mathbf{a}}_2\right).$$

Since the ground state $|0\rangle$, i.e. the state annihilated by all $\hat{\mathbf{a}}_j$, is a physical state, $\hat{\sigma}|0\rangle = 0$, all physical states are obtained by action of rotation-invariant polynomials of the creation operators $\hat{\mathbf{a}}_j^\dagger$ on $|0\rangle$. Note that $\hat{\sigma}$ generates simultaneous rotations of the vectors $\hat{\mathbf{a}}_j^\dagger$. Any such polynomial is a polynomial of the four elementary ones,

$$\hat{b}_1 = \left(\hat{\mathbf{a}}_1^\dagger\right)^2, \quad \hat{b}_2 = \left(\hat{\mathbf{a}}_2^\dagger\right)^2, \quad \hat{b}_3 = \left(\hat{\mathbf{a}}_1^\dagger, \hat{\mathbf{a}}_2^\dagger\right), \quad \hat{b}_4 = \varepsilon_{ij}\hat{a}_1^{(i)\dagger}\hat{a}_2^{(j)\dagger}; \tag{5.118}$$

ε_{ij} is the unit skew-symmetric tensor, $\varepsilon_{12} = 1$. Here the following interesting feature of the model should be noted. The operators (5.118) are invariant with respect to the rotation group $SO(2)$. All of them, but $\hat{b}_4$, are also invariant with respect to the larger group $O(2) = SO(2) \times Z_2$. The only nontrivial element of Z_2 corresponds to the mirror reflection of one of the coordinate axes in the plane. The operator $\hat{b}_4$ changes its sign under this transformation. A natural question arises as to whether or not the operator $\hat{b}_4$ is to be included within the set of operators that generate the physical Hilbert space. In other words, what is the gauge group of the model, $SO(2)$ or $O(2)$? From the formal point of view, complete information about the gauge symmetry should be contained in the Lagrangian. The Dirac analysis yields the constraints that are generators of the gauge transformations. However, the generators allow restoration of only the connected component of the group which contains the unit element of the group [20, 21]. Therefore the gauge group should indeed be $SO(2)$. The presence of the discrete group Z_2 is impossible to establish in the framework Hamiltonian formalism for the Lagrangian (5.111). Such a symmetry can only be enforced by an independent additional condition that is neither contained in nor follows from the Lagrangian [208]. Another possibility is to view the system as a subsystem of a larger system with a gauge group that contains $O(2)$ as a subgroup, e.g. $SO(3)$ which has elements corresponding to the mirror reflections of any of the coordinate axes in any particular plane containing the origin in R^3. In either case, the inclusion of discrete groups, such as the group center (by definition, the group center contains elements that commute with all elements of the group in the sense of group multiplication) into the gauge symmetry requires some additional assumptions. This point of view is upheld in what follows, i.e. the operator $\hat{b}_4$ is included within the set of operators generating the physical Hilbert space.

Thanks to the identity $\varepsilon_{ij}\varepsilon_{kl} = \delta_{ik}\delta_{jl} - \delta_{il}\delta_{jk}$, the square of the operator $\hat{b}_4$ can be expressed via the others $\hat{b}_i$ $(i = 1, 2, 3)$. So the physical Hilbert space is the orthogonal sum of two subspaces, $\mathcal{H}_{ph} = \mathcal{H}^0_{ph} \oplus \mathcal{H}^1_{ph}$, whose bases are

$$\hat{b}_1^{n_1}\,\hat{b}_2^{n_2}\,\hat{b}_3^{n_3}\,|0\rangle \in \mathcal{H}^0_{ph}, \qquad \hat{b}_1^{n_1}\,\hat{b}_2^{n_2}\,\hat{b}_3^{n_3}\,\hat{b}_4\,|0\rangle \in \mathcal{H}^1_{ph}, \tag{5.119}$$

where $n_i = 0, 1, \ldots$, and $i = 1, 2, 3$. The physical spectrum of the energy operator (5.117) reads,

$$E = 2n_1\omega_1 + 2n_2\omega_2 + n_3(\omega_1 + \omega_2) + n_4(\omega_1 + \omega_2) + \omega_1 + \omega_2, \quad n_4 = 0, 1. \tag{5.120}$$

It is interesting to note that the physical Hilbert space of the studied model has the graded structure associated with the center group Z_2, similar to supersymmetric theories. Indeed, the spectrum of the system (5.120) contains the spectrum of the supersymmetric harmonic oscillator of the frequency $\omega_1 + \omega_2$.

The structure of the spectrum (5.120) shows that it would be wrong to claim that the phase space of one degree of freedom is the cone cone(π) while the other two physical degrees of freedom have planar phase spaces. The physical phase space is not a direct product of three individual phase spaces for each physical degree of freedom, which causes the aforementioned kinematic coupling of physical degrees of freedom. The kinematic coupling associated with the nontrivial structure of the physical phase space makes it impossible for the system to have a periodic motion with periods $2\pi/\omega_{1,2}$, even though the Hamiltonian seems to allow for such a motion.

The example considered above demonstrates how the very presence of nonphysical degrees of freedom affects the physical ones: the fundamental characteristic of a dynamical system, such as the physical phase space structure, changes. In contrast, the effect of physical degrees of freedom on one another is not so drastic. For instance, if a generic potential $V(\mathbf{x}_1, \mathbf{x}_2)$ invariant under the group $SO(2) \times SO(2)$ is added to the Lagrangian (5.110), the physical phase space of each subsystem remains the same, while the particles exert a potential force on each other.

5.6.3 Quantum theory of the model (5.111). The coordinate representation

It is interesting to investigate how the action of residual gauge symmetry on all physical degrees of freedom emerges in quantum theory. Let x be a matrix whose columns are $\mathbf{x}_{1,2}$ so that the matrix elements read $x_{ij} = x_j^{(i)}$. The matrix p of the canonical momenta is defined in the same fashion. The constraint (5.112) has the form $\sigma = \mathrm{Tr}\ p^T T x = 0$ and the gauge transformation law can be written as,

$$x \to \exp(\omega T)x, \tag{5.121}$$

where the multiplication in the right-hand side is understood as the matrix multiplication. Consider quantum theory in coordinate representation, i.e. when $\hat{p}_{jk} = -i\partial/\partial x_{jk}$. The constraint equation,

$$\hat{\sigma}\Phi_{\rm ph}(x) = 0, \quad \Phi_{\rm ph} \in \mathcal{H}_{\rm ph}, \tag{5.122}$$

can easily be solved in new curvilinear coordinates associated with the gauge transformation law and a suitable gauge condition (see Section 5.3) [208],

$$x = \exp(\theta T)\rho \equiv \Omega\rho, \quad \rho = \begin{pmatrix} \rho_{11} & \rho_{12} \\ 0 & \rho_{22} \end{pmatrix}. \tag{5.123}$$

It follows from (5.121) that the variable θ is translated under the gauge transformations, while the matrix ρ is gauge invariant. Therefore, in the coordinates (5.123) the constraint operator is the generator of translations of θ, $\hat{\sigma} = -i\partial/\partial\theta$. Thus, the solution of (5.122) is $\Phi_{\rm ph}(x) = \Phi(\rho)$.

In order to obtain the Hamiltonian in $\mathcal{H}_{ph}$, one has to calculate the Laplace–Beltrami operator in the coordinates (5.123), which determines the kinetic energy in the total Hamiltonian,

$$\hat{H} = \tfrac{1}{2}{\rm Tr}\,\hat{p}^T\hat{p} + V(x) \tag{5.124}$$

and omit all its terms containing the derivatives with respect to θ on the right. The metric tensor in the new variables (5.123) has the form,

$$\mathrm{Tr}\, dx^T dx = \mathrm{Tr}\,\left[d\rho^T d\rho + d^2\theta\rho^T\rho + (\rho d\rho^T T - T d\rho\rho^T)d\theta\right], \tag{5.125}$$

where $d\rho$ denotes the matrix of the differentials $d\rho_{ij}$. It is convenient to write the metric tensor in (5.125) as a 4×4 matrix,

$$g_{\alpha\beta} = \begin{pmatrix} 1 & 0 & 0 & 0 \\ 0 & 1 & 0 & -\rho_3 \\ 0 & 0 & 1 & \rho_2 \\ 0 & -\rho_3 & \rho_2 & \mathrm{Tr}\,(\rho^T\rho) \end{pmatrix} \equiv \begin{pmatrix} 1 & B \\ B^T & \mathrm{Tr}\,(\rho^T\rho) \end{pmatrix}, \tag{5.126}$$

where $\rho_a = (\rho_{11}, \rho_{12}, \rho_{22})$. The Jacobian is then easy to compute, $\mu \equiv \sqrt{g} = \sqrt{\det g_{\alpha\beta}} = \rho_1$, and the inverse of (5.126) is

$$g^{\alpha\beta} = \begin{pmatrix} \delta_{ab} + \rho_1^{-2}B_aB_b & -B_a\rho_1^{-2} \\ -B_a\rho_1^{-2} & \rho_1 \end{pmatrix}. \tag{5.127}$$

Since the metric tensor is independent of θ, only the 3×3 block $g_{\rm ph}^{ab} = \delta_{ab} + \rho_1^{-2}B_aB_b$ contributes to the physical part of the Laplace–Beltrami operator. As a result, the physical Hamiltonian reads,

$$\hat{H}_{\rm ph} = -\frac{1}{2\mu}\partial_a \circ g_{\rm ph}^{ab}\,\mu\,\partial_b + V(\rho), \tag{5.128}$$

where $\partial_a = \partial/\partial\rho_a$.

The kinematic coupling is straightforward to see from the Hamiltonian (5.128). Even if $V = 0$, the Schrödinger equation is not separable in the variables ρ_a and, therefore, independent excitations are impossible.

Remark. The variables ρ_1 and $\rho_{2,3}$ can be separated in the Schrödinger equation. Set, for instance, $V = (1/2)\rho_1^2\omega_1^2 + (1/2)\omega_2^2\left(\rho_2^2 + \rho_3^2\right)$, as in Section 5.6.2. The ρ_1-oscillator can be excited independently of the other two. Its frequency is doubled. The dynamics of the oscillators described by $\rho_{2,3}$ is not separable. The doubling of the distance between their energy levels is a direct consequence of this kinematic coupling. The fact that the physical variables are associated with curvilinear coordinates plays the crucial role in this conclusion. The Schrödinger equation can be separated in the "collective" variables r, φ defined by $\rho_2 = r\cos\varphi$ and $\rho_3 = r\sin\varphi$. Excitations of the degrees of freedom described by them correspond to simultaneous excitations of the $\rho_{2,3}-$ oscillators. Of course, the physical spectrum coincides with (5.120).

Construction of the quantum theory is completed by determining the scalar product in $\mathcal{H}_{\rm ph}$. It is induced by the scalar product in the total Hilbert space,

$$\int_{R^4} dx = \int_0^{2\pi} d\theta \int_K d^3\rho\mu(\rho). \tag{5.129}$$

The integration region K in (5.129), isomorphic to the physical configuration space, is obtained from the condition that the mapping (5.123) is one-to-one. The range of $\theta \in R$ and $\rho \in R^3$ should be restricted so that the mapping becomes one-to-one. Consider the symmetry transformation of the mapping (5.123), i.e. transformations $\rho \to \rho_s$ and $\theta \to \theta_s$ under which x is not changed, $x(\rho,\theta) = x(\rho_s,\theta_s)$. It is easy to see that these transformations contain the group of translations S_θ:

$$\theta \to \theta + 2\pi n, \quad \rho \to \rho,$$

where n is integer-valued, and the group of reflections Z_2:

$$\theta \to \theta + \pi\ , \quad \rho \to -\rho.$$

The mapping (5.123) becomes a change of variables if the range of the new variables is restricted to the following *modular* domain relative to these transformations:

$$(\rho,\theta) \in \tilde{K} \equiv R(\theta) \times R^3(\rho)/(S_\theta \times Z_2) = [0, 2\pi) \times R^3(\rho)/Z_2.$$

Thus, $K = R^3/Z_2$. In particular, the half-space $\rho_1 \geq 0$, $\rho_{2,3} \in R$ can be taken as region K in R^3.

The group Z_2, that changes the sign of the matrix ρ, coincides with the residual gauge group in the gauge $x_{21} = 0$.

The physical states from $\mathcal{H}_{\rm ph}$ are invaraint under the action of this group. A simple proof of this statement is based on constructing a projection operator onto

$\mathcal{H}_{\rm ph}$. Let $\psi(x)$ be a vector of the total Hilbert space. The projection is defined as the averaging of $\psi(x)$ over the gauge group,

$$\Phi(x) = \frac{1}{2\pi}\int_0^{2\pi} d\omega \psi(e^{\omega T}x) = \frac{1}{2\pi}\int_0^{2\pi} d\omega \psi(e^{\omega T}\rho) = \Phi(\rho), \qquad (5.130)$$

where the second equality is obtained by substituting (5.123) for x, shifting the integration variable, $\omega \to \omega - \theta$, and using the periodicity of the integrand in ω. By construction, the states (5.130) are gauge invariant. Since the integrand is 2π-periodic in ω, the function $\Phi(\rho)$ is symmetric,

$$\Phi(\rho) = \Phi(-\rho). \qquad (5.131)$$

If the functions $\psi(x)$ are chosen to be the eigenfunctions of the total Hamiltonian (5.124), the projected (nonzero) functions $\Phi(\rho)$ are the eigenfunctions of the physical Hamiltonian (5.128). If $\psi(x)$ is regular everywhere, so is $\Phi(\rho)$. Thereby the following important conclusion holds. *The eigenfunctions of the physical Hamiltonian are regular if and only if they are invariant under the residual gauge transformations.* This is the analog of Chevalley's theorem for the model studied. All independent invariants of the transformation $\rho \to -\rho$ are exhausted by the elements of the matrix $\rho^T\rho = x^T x$.

Remark. The matrix ρ in (5.123) may be chosen differently. For example, it can be symmetric $\rho^T = \rho$. In general, it can be any matrix that has one condition imposed on its four components. This condition should not contradict gauge transformation law, meaning that there exists a gauge transformation such that the matrix x can be brought to ρ by it (compare Section 5.3). In this case the residual transformations may not form a group. However, just as in the case of the model discussed in Section 5.3, one can prove that the physical states are invariant under these transformations and, moreover, quantum theories corresponding to different conditions on elements of ρ are unitary equivalent. As in Section 5.3, the proof is straightforward and based on the change of variables associated with the gauge transformation law and compatible with the gauge condition used to constrain the elements of ρ (see also Section 5.7.5).

5.6.4 Yang–Mills quantum mechanics

If in the Yang–Mills theory with the group $SU(2)$ all the vector potentials are assumed to be independent of the spatial coordinates, $A^a_\mu = A^a_\mu(t)$ ($a = 1, 2, 3$ are the isotopic indices), the corresponding dynamical system has a finite number of degrees of freedom and is known as the Yang–Mills mechanics. This model has been intensively studied in the literature (see, e.g. [195, 208, 219, 220, 226]). Here this model is studied as well. In particular, its physical phase space is found and the corresponding quantum theory is developed which takes into account the physical phase-space structure.

The Lagrangian has the form,

$$L = \tfrac{1}{2}\mathrm{Tr}\,(\dot{x} + yx)^T(\dot{x} + yx) - V(x^T x), \tag{5.132}$$

where x is a real 3×3 matrix whose components are the spatially homogeneous Yang–Mills potentials, $x_{ai} = A_i^a(t)$, and y is a 3×3 skew-symmetric matrix, $y_{ab} = -\varepsilon_{abc}A_0^c(t)$. The potential V is obtained from the corresponding Yang–Mills potential when the connection is spatially homogeneous, $A_\mu^a = A_\mu^a(t)$. It can be written as

$$V = \tfrac{1}{4}\left[(\mathrm{Tr}\,(x^T x))^2 - \mathrm{Tr}\,(x^T x)^2\right].$$

However, its explicit form is not essential for the subsequent analysis. The Lagrangian (5.132) is invariant under the gauge transformations,

$$x \to \Omega x\ , \quad y \to \Omega y \Omega^T + \Omega\partial_t\Omega^T\ , \quad \Omega = \Omega(t) \in SO(3). \tag{5.133}$$

Since the Lagrangian is independent of the velocity $\dot{y}$, y is a pure non-physical degree of freedom. In order to select the physical degrees of freedom, note that the gauge transformations (5.133) are simultaneous $SO(3)$ rotations of the columns of matrix x. Therefore by a suitable gauge transformation a generic matrix x can be brought to the upper triangular form. Hence, the system has six physical degrees of freedom. Another way of finding the number of physical degrees of freedom is to construct all independent gauge invariants of matrix x. It is known that all independent $SO(3)$ invariants that can be constructed out of three linearly independent vectors in R^3 are given by their scalar products [153], i.e. any gauge-invariant quantity is a function of six independent components of the symmetric matrix $x^T x$. It is noteworthy that $\det x$ is also an $SO(3)$ invariant, but its square is a polynomial of the components of $x^T x$. For this reason, it is not associated with any independent degree of freedom. Matrix $x^T x$ is invariant under the larger group $O(3)$, while $\det x$ changes its sign under Z_2 transformations of the center of $O(3)$. Just like in the matrix model of Section 5.6.2, the existence of such an invariant leads to the Z_2 graded structure of the physical Hilbert space.

In order to establish the physical phase-space structure and the physical configuration-space structure which determines the integration region in the scalar product for physical states, one should eliminate all non-physical degrees of freedom by using gauge arbitrariness and find out if there exist residual discrete gauge transformations.

This program can be realized as follows. By a suitable gauge transformation the matrix x can be brought to the symmetric form, $x^T = x$. The components of the symmetric matrix describe six physical degrees of freedom of the model. To find the residual gauge transformations that preserve the condition $x^T = x$, the following equation should be solved for the matrix $\hat{s}$:

$$(\hat{s}x)^T = \hat{s}x, \qquad x = x^T, \quad \hat{s} \in SO(3). \tag{5.134}$$

If there is only a trivial solution, then the residual symmetry is absent and the physical phase space is planar. As in Section 5.6.2, the transformations from the center of $O(3)$ are not viewed as gauge transformations.

Put $x = \omega h \omega^T$, where h is a diagonal matrix and $\omega \in SO(3)$. Then it follows from Eq. (5.134) that $\omega_s h = h\omega_s^T$, where $\omega_s = \omega^T \hat{s}\omega \in SO(3)$. Hence, $h = \omega_s^T h \omega_s^T$ or $h = \omega_s h \omega_s$ as $h^T = h$. Multiplying the two latter relations, one arrives at,

$$[h^2, \omega_s] = 0.$$

As h is arbitrary, ω_s must be diagonal. There are only four such matrices in $SO(3)$, $\omega_s = I_\mu$, $\mu = 1, 2, 3, 4$ where $I_1 = \mathrm{diag}(1, -1, -1)$, $I_2 = \mathrm{diag}(-1, 1, -1)$, $I_3 = \mathrm{diag}(-1, -1, 1)$, and $I_4 = 1$ (the unit matrix). Solutions of Eq. (5.134) are then given by $\hat{s}_\mu = \omega(x) I_\mu \omega^T(x)$. The matrices $\hat{s}_\mu$ form a group relative to the matrix multiplication, for example, $\hat{s}_\mu^2 = 1$ (no summation over μ here), $\hat{s}_1\hat{s}_2 = \hat{s}_3$, etc. It should be noted that the group $S = \{s_\mu\}$ depends on the point $x \in R^6$. For generic x, group S has four elements. If any two eigenvalues of x^2 coincide, then ω_s contains continuous rotations forming the group $SO(2)$. If all three eigenvalues of x^2 are the same, ω_s is a generic element of $SO(3)$. However, all such symmetric matrices form a set of zero measure in the set of all symmetric matrices. If K is a modular domain in the set of all symmetric matrices with distinct eigenvalues relative to group S, then the matrices x for which $S(x)$ contain continuous residual gauge transformations belong to the boundary ∂K. Recall that a similar feature also exists in the model with an arbitrary gauge group studied in Section 5.5. Naturally, if one is interested in the geometry of the orbit space R^6/S, the points of the boundary of K related by continuous gauge transformations must be identified. Discussion is postponed to Section 5.8.8 where the orbit space of Yang–Mills theories is analyzed. For the Yang–Mills mechanics, this question is addressed in [226], and various gauge matrix models are studied in this aspect in [227].

To describe K, note that in the decomposition $x = \omega h \omega^T$, the elements of h can always be ordered so that $|h_1| \geq |h_2| \geq |h_3|$, because any two diagonal elements of h can be permuted by a suitable orthogonal transformation. Let $M \subset R^3$ be the open set of all h such that $|h_1| > |h_2| > |h_3|$. The boundaries of M contain the planes $|h_i| = |h_j|$, $i \neq j$, $i, j = 1, 2, 3$. Consider four subsets of M defined by the conditions that $h \in M_{++} \subset M$ if $h_1 > 0$ and $h_2 > 0$, $h \in M_{+-} \subset M$ if $h_1 > 0$ and $h_2 < 0$, and similarly for M_{-+} and M_{--}. It is straightforward to verify that $I_1 M_{+-} = M_{++}$, $I_2 M_{-+} = M_{++}$ and $I_3 M_{--} = M_{++}$. Hence, one can choose $K = \omega M_{++} \omega^T$ for all $\omega \in SO(3)$ and the boundary ∂K is the image of the corresponding orthogonal map of the boundary ∂M_{++}. Thus, the continuous residual gauge transformations can only occur if $x \in \partial K$. Note that, if the gauge group is extended to $O(3)$, then K is formed by positive semi-definite symmetric matrices ρ (i.e. $h_1 > h_2 > h_3 \geq 0$) which is a subset of K. This follows from the polar decomposition of a real matrix $x = \Omega\rho$, $\Omega \in O(3)$.

The structure of the physical phase space can be established as in the model of Section 5.6.1, i.e. by identifying the points of the surface of intersection of $x = x^T$ and $\sigma_a = 0$ in the total phase space which are related by transformations from group S. Here σ_a are the secondary constraints of the model. They are easy to find,

$$\sigma_a = \mathrm{Tr}\,(p^T T_a x) = 0, \tag{5.135}$$

where $p = \partial L/\partial \dot{x}$ is a matrix of the canonical momenta, T_a are 3×3 skew-symmetric matrices, $T_a^{cb} = \varepsilon_{abc}$, the generators of $SO(3)$. The matrix p can always be represented as the sum of symmetric and skew-symmetric matrices, $p_{ij} = \Pi_a \varepsilon_{aij} + (p_x)_{ij}$ where $p_x = p_x^T$. The constraints (5.135) can be rewritten in vector form and solved for the non-physical momenta Π_a,

$$(\mathrm{Tr}\,x - x)\mathbf{\Pi} = -\mathrm{Tr}\,(p_x \mathbf{T} x)\ , \quad \mathbf{\Pi} = -(\mathrm{Tr}\,x - x)^{-1}\mathrm{Tr}\,(p_x \mathbf{T} x),$$

where $(\mathrm{Tr}\,x - x)_{ab} = \delta_{ab}\mathrm{Tr}\,x - x_{ab}$, $\mathbf{\Pi}$ and $\mathbf{T}$ are three-vectors with components Π_a and T_a, respectively. Note that, if $x \in K$ as defined above, then $\det(\mathrm{Tr}\,x - x) = (h_1 + h_2)(h_1 + h_3)(h_2 + h_3) > 0$ where $x = \omega h \omega^T$. To obtain the physical phase space from this surface spanned by the pair of symmetric matrices x and p_x, one should identify the phase-space points $(\hat{s}_\mu x, \hat{s}_\mu p_x)$ for all $\hat{s}_\mu \in S$. The determinant $\det(\mathrm{Tr}\,x - x)$ can change its sign under the S-transformations of x.

The Hamiltonian dynamics of physical degrees of freedom can be cast only in gauge-invariant canonical variables with the phase space as determined above. Consider the change of variables $x = \Omega\rho$ where $\Omega \in SO(3)$ and $\rho^T = \rho \in K$. Under the gauge transformations, Ω undergoes the left group shifts, while ρ remains invariant. It is not hard to verify that the gauge-invariant canonical momentum for ρ is $p_\rho = (1/2)\left(\Omega^T p + p^T \Omega\right) = p_\rho^T$ and $\{\rho_{ij}, (p_\rho)_{kl}\} = \delta_{ik}\delta_{jl}$ where $i \geq j$ and $k \geq l$. Then the momentum matrix can be uniquely represented in the form $p = \Omega(\Pi + p_\rho)$ where Π is a skew-symmetric matrix. The constraints $\sigma_a = 0$ can easily be solved for Π as above if one takes into account that $\Omega^T T_a \Omega = F_{ab}(\Omega) T_b$ with $F_{ab}(\Omega)$ being a non-singular matrix whose components are functions of Ω. The physical Hamiltonian is obtained by substituting the solution into the kinetic energy term, $(1/2)\mathrm{Tr}\,(p^T p) = (1/2)\left(\mathrm{Tr}\,p_\rho^2 + \mathrm{Tr}\,\Pi^2\right)$. Note that the physical kinetic energy contains a metric which turns out to be non-flat, i.e. the orbit space of the model has a nonzero curvature tensor.

The quantum theory that takes into account the correct structure of the physical configuration (or phase) space can be developed similarly by introducing curvilinear coordinates $x = \Omega\rho$ that correspond to the gauge transformation law and the gauge condition chosen (to parameterize the orbit space). Let $\Omega = \Omega(\theta)$, where θ_a are local coordinates on the group manifold $SO(3)$. The constraint equations for the physical states, $\hat{\sigma}_a \Phi_{\mathrm{ph}}(x) = 0$, where $\hat{p}_{kn} = -i\partial/\partial x_{kn}$, are immediately solved in the new variables because by construction the operators $\hat{\sigma}_a$ generate translations of the variables θ_a. Hence, the physical states are independent of θ_a, $\Phi_{\mathrm{ph}}(x) = \Phi(\rho)$. As in the model of Section 5.6.3, the physical

Hilbert space can be obtained by averaging elements of the total Hilbert space over the gauge group $G = SU(2) \sim SO(3)$,

$$\Phi(\rho) = \int_G d\mu_G(\Omega)\psi(\Omega x) = \int_G d\mu_G(\Omega)\psi(\Omega\rho), \tag{5.136}$$

where $d\mu_G$ is the right-invariant measure on G. It follows from (5.136) and the invariance of the measure under group shifts, $d\mu(\Omega\hat{s}_\mu^{-1}) = d\mu(\Omega)$, that the physical states are invariant under group S of residual transformations,

$$\Phi(\hat{s}_\mu\rho) = \Phi(\rho).$$

The quantum Hamiltonian in the physical Hilbert space is found in the same way as (5.128). Let $\delta\theta_a = \omega_{ab}(\theta)d\theta_b$ be 1-forms on the group manifold so that $(\Omega^T d\Omega)_{ab} = \varepsilon_{abc}\delta\theta_c/2 = -T_c^{ab}\delta\theta_c/2$. The metric tensor $ds^2 = \mathrm{Tr}\, dx^T dx$ can be written in the new variables in the same form as in (5.125) and (5.126) where $d\theta$ is replaced by the one-form $\delta\theta$, A is the unit 6×6 matrix corresponding to the term $\mathrm{Tr}\,(d\rho)^2$, D is a symmetric 3×3 matrix, $D_{ab} = (1/4)(\delta_{ab}\mathrm{Tr}\rho^2 - \rho_{ac}\rho_{cb})$, which appears in front of $\delta\theta_a\delta\theta_b$ in ds^2, and, finally, the 6×3 matrix B defines the coefficients at $d\rho_{ab}\delta\theta_c$. A straightforward computation yields, $B_{ab,c} = B_{ba,c} = (\varepsilon_{adc}\rho_{db} + \varepsilon_{bdc}\rho_{da})/4$. The scalar product in the physical Hilbert space is obtained by factorization of the gauge group volume,

$$\int_{R^9} dx = \int_G d\mu_G(\Omega)\int_K d\rho\mu(\rho)\ , \quad \mu = \sqrt{g} = \det(\mathrm{Tr}\rho - \rho), \tag{5.137}$$

where $(\mathrm{Tr}\,\rho - \rho)_{ab} = \delta_{ab}\mathrm{Tr}\,\rho - \rho_{ab}$ and $d\mu_G \sim \prod_a \wedge\delta\theta_a$. Note that $\mu(\rho) > 0$ if $\rho \in K$ according to the above analysis of the structure of K. The physical Hamiltonian has the form (5.128), where $\partial_a \to \partial_{ab} = \partial/\partial\rho_{ab}$ and $g_{ph}^{ab} \to g_{ph}^{ab,a'b'}$,

$$\begin{aligned} g_{ph}^{ab,a'b'} &= \delta^{ab,a'b'} + B_c^{ab}(H^{-1})_{cc'}B_{c'}^{a'b'}, \\ H_{cc'} &= \frac{1}{8}[(\mathrm{Tr}\rho - \rho)^2]_{cc'}. \end{aligned} \tag{5.138}$$

Here $\partial_{ab}\delta^{ab,a'b'}\partial_{a'b'} \equiv \partial_{ab}\partial_{ab}$ by the definition of the symbol $\delta^{ab,a'b'}$. This completes the construction of the quantum theory. The tensor (5.138) also determines the quadratic form of the classical physical kinetic energy constructed above.

In full analogy with the model of Section 5.6.3, one can prove that all regular solutions of the Schrödinger equation are S-invariant by noting that projection (5.136) can be applied to a basis of the total Hilbert space to obtain the corresponding basis of the physical subspace which would contain all regular eigenfunctions of the physical Hamiltonian.

The inverse of (5.138) defines the metrics in the physical configuration space (the orbit space) $ds^2 \equiv g_{ab,a'b'}^{ph} d\rho_{ab}d\rho_{a'b'}$, where $g_{ph}^{ab,cd}g_{cd,a'b'}^{ph} = \delta_{ab}^{a'b'}$. The corresponding Riemann–Christoffel tensor (2.35) does not vanish and therefore the orbit space has a nontrivial curvature (the same applies to the model of Section 5.6.3), which is in contrast to models with a single physical degree of freedom. In Yang–Mills theories, this geometrical feature of the orbit space persists [213, 221, 228–230].

5.7 Systems with Bose and Fermi degrees of freedom

5.7.1 A simple model

The simplest model with both Bose and Fermi degrees of freedom and gauge symmetry is defined by the Lagrangian [26, 212, 213],

$$L = \tfrac{1}{2}\left\{[(\partial_t - yT)\mathbf{x}]^2 - \omega_1^2\mathbf{x}^2\right\} + \psi^\dagger(i\partial_t + y\Gamma)\psi - \omega_2\psi^\dagger\psi, \tag{5.139}$$

where $\Gamma = -iT$. Here the vector ψ with two complex Grassmann components is viewed as a 2×1 matrix and, accordingly, $\psi^\dagger$ is the Hermitian conjugate of the matrix ψ so that the product $\psi^\dagger\psi$ is understood in the sense of matrix multiplication. This Lagrangian is nothing but the sum of the Lagrangians (5.2) and (5.94) with specific (harmonic oscillator) potentials and the replacement of Γ by $-iT$. A straightforward application of the Dirac analysis shows that the Lagrangian (5.139) defines a dynamical system with one primary constraint $p_y = \partial L/\partial\dot{y} = 0$, the Hamiltonian,

$$H = \tfrac{1}{2}\left(\mathbf{p}^2 + \omega_1^2\mathbf{x}^2\right) + \omega_2\psi^\dagger\psi + y(t)\left((\mathbf{p}, T\mathbf{x}) - \psi^\dagger\Gamma\psi\right), \tag{5.140}$$

and the secondary constraint,

$$\sigma = (\mathbf{p}, T\mathbf{x}) - \psi^\dagger\Gamma\psi = 0. \tag{5.141}$$

Since $\{\sigma, H\} = 0$, equality (5.141) is a condition on admissible initial values of the dynamical variables. If $\mathbf{p}(t = t_0) = \mathbf{p}_0$ and $\mathbf{x}(t = t_0) = \mathbf{x}_0$ are viewed as vectors with real-valued components, while $\psi(t = t_0) = \psi_0$ are generators of the Grassmann algebra, the constraint is an even element of the Grassmann algebra. The equality of such elements to zero implies vanishing of all coefficients at the corresponding even basis elements of the Grassmann algebra. In other words, the equality (5.141) seems to be equivalent to two independent conditions,

$$\sigma_1 = (\mathbf{p}, T\mathbf{x}) = 0, \quad \sigma_2 = \psi^\dagger\Gamma\psi = 0. \tag{5.142}$$

The question arises whether this means that in classical theory on supermanifolds there appear additional constraints. The answer to this question is negative. In Section 1.9 it is shown that, when studying the classical dynamics on supermanifolds, the Bose and Fermi degrees of freedom should be viewed as generic even and odd elements of the Grassmann algebra, respectively. In particular, this means that the initial values $\mathbf{x}_0$ and $\mathbf{p}_0$ cannot be viewed as vectors with real-valued components, but rather they must be generic even elements of the Grassmann algebra, otherwise there is a contradiction with the theory of canonical transformations on supermanifolds. A detailed study of the classical Hamiltonian dynamics of the model is given in Section 8.2.3. The constraint (5.141) means that the initial value of the bosonic angular momentum $p_\varphi = (\mathbf{p}, T\mathbf{x}) \equiv \sigma_1$ is an even element of the Grassmann algebra generated by $\psi_0^\dagger$ and ψ_0. Note also that it is easy to find a potential such that, when added to the Hamiltonian (5.140), the quantities $\sigma_{1,2}$ are no longer independent integrals of motion, i.e. $\{\sigma_{1,2}, H\} \neq 0$ while $\{\sigma, H\} = 0$ (see Section 8.2.3 for details). It is then concluded that, although being valid at the initial moment of time $t = t_0$, the equalities (5.142) are violated in due course. If one demands that they should hold for any moment of time regardless of the dynamics, this would lead to additional constraints, $\{\sigma_{1,2}, H\} = \sigma'_{1,2} = 0$ and $\{\sigma'_{1,2}, H\} = \sigma''_{1,2} = 0$, which, in turn, implies an enlargement of the gauge group of the theory (see Section 8.2.3).

The above is in agreement with the definition of the Dirac delta-function of even elements of a Grassmann algebra [231]. In this case, it is the function $\delta(\sigma)$. Its occurrence is unavoidable when formulating the quantum theory of physical degrees of freedom in the framework of the Hamiltonian path integral method (non-physical degrees of freedom are eliminated). Since the Hamiltonian path integral operates only with classical variables, its study may elucidate what happens in classical theory. The Dirac delta-function $\delta(\sigma)$ of an even element σ is defined by the Taylor series [231],

$$\delta(\sigma_1 - \sigma_2) = \delta(\sigma_1) + \sum_{n=1}^{k} \delta^{(n)}(\sigma_1)(-\sigma_2)^n/n!, \tag{5.143}$$

which is terminated at the kth term for which $\sigma_2^{k+1} = 0$ because of the nilpotency of the Grassmann algebra generators. Thus defined the δ-function possesses the characteristic property,

$$\int f(\hat{\xi})\delta(\hat{\xi} - \hat{\xi}_0)d\hat{\xi} = f(\hat{\xi}_0), \tag{5.144}$$

where $\hat{\xi}_0$ is an even element of a Grassmann algebra. A general theory of integration on Grassmann algebras is developed in [231, 232]. It is clear from (5.144) that the occurrence of the δ-function, $\delta(\sigma_1 - \sigma_2)$, where $\sigma_{1,2}$ are classical variables (5.142), does not imply that the measure in the Hamiltonian path integral is reduced so that $\sigma_{1,2}$ vanish independently in time evolution. Thus, in the quantum theory there is only one secondary constraint. As the classical theory is the limiting case of the quantum theory, the former should describe the dynamics with the only secondary constraint. The conclusion becomes evident if one recalls that the gauge arbitrariness is determined by the number of independent generators of gauge transformations. The Hamiltonian (5.140), containing all the information about the dynamics, has only one arbitrary parameter $y(t)$ at the only generator σ. A fixing of $y(t)$, for example, by demanding that $x_2(t) = 0$ exhausts the gauge arbitrariness (modulo the discrete group Z_2 which cannot reduce the dimension of the physical phase space any further).

Let us find the spectrum of the Hamiltonian (5.140). In Fock space representation, the Hamiltonian has the form,

$$\begin{aligned}\hat{H} &= \omega_1(\hat{\mathbf{a}}^\dagger, \hat{\mathbf{a}}) + \omega_2\hat{\psi}^\dagger\hat{\psi} + y(t)\left(i(\hat{\mathbf{a}}^\dagger, T\hat{\mathbf{a}}) - \hat{\psi}^\dagger T\hat{\psi}\right) + (\omega_1 - \omega_2)\\ &\equiv \hat{H}_1 + \hat{H}_2,\end{aligned} \tag{5.145}$$

where $\hat{H}_1$ is identical to (5.43) if $\omega_1 = 1$ and $\hat{H}_2$ coincides with (5.97) if in the latter $\omega = \omega_2$, $\Gamma = -iT$, and $\Omega = 0$ (after quantization). Obviously, $[\hat{H}_1, \hat{H}_2] = 0$. Therefore the bosonic and fermionic subsystems seem dynamically decoupled. The independent gauge-invariant operators that generate the physical Hilbert space from the ground state $|0\rangle$ are

$$\hat{b}_1 = \left(\hat{\mathbf{a}}^\dagger\right)^2, \quad \hat{b}_2 = \hat{\psi}^\dagger\hat{\mathbf{a}}^\dagger, \quad \hat{b}_3 = \hat{\psi}^\dagger T\hat{\psi}^\dagger, \quad \hat{b}_4 = \hat{\psi}^\dagger T\hat{\mathbf{a}}^\dagger \tag{5.146}$$

The operator $\hat{b}_4$ is similar to that in (5.118). It is not invariant under the center of $O(2)$. So its elimination implies that the gauge group is enlarged from $SO(2)$ to $O(2)$. Note that such an extension of the gauge group would also exclude the operator $\hat{b}_3$, while the state $\hat{b}_3|0\rangle$ is the only nontrivial physical state in the model of Section 5.5.2. The latter gives yet another argument to support the claim that the gauge group is indeed $SO(2)$. The states obtained from the vacuum $|0\rangle$ by acting on it with non-negative integer powers of the operators (5.146) form a basis of the physical Hilbert space (analogously to (5.119)). All of them are also eigenvectors of the operator $\hat{H}$. This yields the energy spectrum,

$$E = 2n_1\omega_1 + n_2(\omega_1 + \omega_2) + 2n_3\omega_2 + n_4(\omega_2 + \omega_1) + \omega_1 - \omega_2. \tag{5.147}$$

In contrast to (5.119), here only n_1 attains all non-negative integer values, while the others n_i, $i = 2, 3, 4$, can attain only 0 and 1 because $\hat{b}_i^2 = 0$, and the states $\hat{b}_2\hat{b}_4|0\rangle$ and $\hat{b}_{2,4}\hat{b}_3|0\rangle$ are eliminated from the basis set because of the identities $\hat{b}_2\hat{b}_4 = \hat{b}_1\hat{b}_3$, and $\hat{b}_{2,4}\hat{b}_3 = 0$. Note that, if $\omega_1 = \omega_2$, the ground state energy vanishes (supersymmetry). From the structure of the spectrum (5.147) it follows that as in model (5.111) it is impossible to select a subsystem whose excitations would be independent of the rest of the system and whose spectrum would correspond to either conic or planar phase space. Thus, the physical phase space of the system cannot be represented as the direct product of the conic and planar phase spaces.

5.7.2 The Laplace–Beltrami operator on a supermanifold

In the previous section the quantum constraint equation is solved by constructing all independent gauge-invariant operators without any explicit elimination of non-physical variables by means of a gauge fixing condition. It is instructive to develop a quantum description in the physical configuration space when the non-physical degrees of freedom are eliminated explicitly by finding coordinates canonically conjugated to independent constraints. The goal is usually achieved by going over to curvilinear coordinates in the total configuration space, which are associated with gauge transformation law and an admissible gauge condition. A new feature that emerges here is that the total configuration space is a supermanifold and therefore curvilinear coordinates should be constructed in it in order to solve quantum constraints [31, 195]. Consequently, one should derive the Laplace–Beltrami operator on a supermanifold in order to be able to obtain the physical Hamiltonian operator in coordinate representation. This problem is discussed below.

Consider a quantum mechanical system with Bose and Fermi degrees of freedom. Let its Hamiltonian read,

$$\hat{H} = \tfrac{1}{2}\hat{p}_a^2 + V(\hat{x}, \hat{\psi}^\dagger, \hat{\psi}), \tag{5.148}$$

where $\hat{p}_a = -i\partial/\partial x_a$, $\hat{\psi}_\alpha^\dagger = \psi_\alpha^*$, and $\hat{\psi}_\alpha = \vec{\partial}/\partial\psi_\alpha^*$, i.e. the coordinate representation is used for the bosons, and the holomorphic one for the fermions. The indices $a = 1, 2, \ldots, M$ and $\alpha = 1, 2, \ldots, N$ enumerate bosonic and fermionic degrees of freedom, respectively. State vectors are functions $\varphi(x, \psi^*)$ and the scalar product is defined by,

$$\langle\varphi_1|\varphi_2\rangle = \int_{R^M} d^M x \int \prod_{\alpha=1}^{N} (d\psi_\alpha^* d\psi_\alpha)\, e^{-\psi^*\psi}\, [\varphi_1(x, \psi^*)]^*\, \varphi_2(x, \psi^*), \tag{5.149}$$

where by definition the complex conjugation of a product of the Grassmann variables reverses the ordering in the product, for example, $\left[\psi_{\alpha_1}^*\psi_{\alpha_2}^* \cdots \psi_{\alpha_k}^*\right]^* = \psi_{\alpha_k}\psi_{\alpha_{k-1}} \cdots \psi_{\alpha_1}$.

Let Θ denote a point (x, ψ^*) in the superspace. A change of variables in the superspace is defined by a function $\Theta = \Theta(\theta)$, where $\theta = (y, \xi^*)$ is a point in the new superspace. The subsequent discussion is limited to a special case which is sufficient to study gauge theories. Put

$$x_a = x_a(y)\ ,\quad \psi_\alpha = \Omega_{\alpha\beta}(y)\xi_\beta,\quad \psi_\alpha^* = \xi_\beta^*\Omega^\dagger_{\beta\alpha}(y), \tag{5.150}$$

where $\Omega^\dagger = \Omega^{-1}$ is a unitary matrix. Then the derivative with respect to x_a is written as

$$\frac{\partial}{\partial x^a} = \frac{\partial y_b}{\partial x_a}\frac{\partial}{\partial y_b} + \frac{\partial \xi_\alpha^*}{\partial x_a}\frac{\vec{\partial}}{\partial \xi_\alpha^*}. \tag{5.151}$$

The derivative $\partial\xi_\alpha^*/\partial x_a$ is found by differentiating the equality $\xi^*(\psi^*, x) = \psi^*\Omega(y(x))$ with respect to x_a and substituted into (5.151),

$$\frac{\partial}{\partial x^a} = \frac{\partial y^b}{\partial x^a}(\partial_b + i\hat{\pi}_b); \tag{5.152}$$

$$\hat{\pi}_b = -i\hat{\xi}_\alpha^\dagger(\Omega^\dagger\partial_b\Omega)_{\alpha\beta}\hat{\xi}_\beta\ . \tag{5.153}$$

Here $\partial_b \equiv \partial/\partial y_b$ and the operators $\hat{\xi}_\alpha^\dagger$ and $\hat{\xi}_\alpha$ are taken in the holomorphic representation.

The expression (5.152) can be used to write the Laplace operator $\Delta = -\hat{p}_a^2$ in the Hamiltonian (5.148) in curvilinear supercoordinates (5.150). It is easy to verify that

$$\Delta = \mu^{-1}\left(\partial_a + i\hat{\pi}_a\right) \circ g^{ab}\mu\left(\partial_b + i\hat{\pi}_b\right), \tag{5.154}$$

where $g^{ab} = (\partial y^a/\partial x_c)(\partial y^b/\partial x_c)$ and $\mu = \det J$, $J_b^a = \partial x^a/\partial y^b$, is the Jacobian.

In the new variables the scalar product reads,

$$\langle\varphi_1|\varphi_2\rangle = \int_K d^M y\, \mu(y) \int \prod_{\alpha=1}^{N} \left(d\xi_\alpha^* d\xi_\alpha\right) e^{-\xi^*\xi} [\varphi_1(\theta)]^* \varphi_2(\theta). \tag{5.155}$$

The integration region $K \subset R^M$ is defined by the condition that the mapping $x^a(y):\ R^M \to R^M$ becomes one-to-one when restricted to K. The domain K can be found by studying the set S of symmetry transformations of the mapping (5.150) which are determined by the equation

$$x(y) = x(\hat{s}y)\ , \quad \hat{s} \in S. \tag{5.156}$$

In other words, the points $\hat{s}y$ have the same image under the mapping. Clearly, $K \sim R^M/S$. The subtlety here is that the transformations of the bosonic variables, $y \to \hat{s}y$, necessarily induce transformations of the Grassmann variables $\xi^* \to \hat{s}\xi^*$. The action of $\hat{s}$ on ξ^* can be found from the condition that the variable ψ^* is invariant under the mapping $y \to \hat{s}y$ and $\xi^* \to \hat{s}\xi^*$, which yields $\xi^*\Omega^\dagger(y) = (\hat{s}\xi^*)\Omega^\dagger(\hat{s}y)$ or

$$\hat{s}\xi^* = \xi^*\Omega^\dagger(y)\Omega(\hat{s}y). \tag{5.157}$$

It should be noted that the wave functions in the new curvilinear variables (5.150),

$$\varphi(x, \psi^*) = \varphi(x(y)),\ \psi^*(y, \xi^*) = \tilde{\varphi}(y, \xi^*)$$

are invariant with respect to the S-transformations of the arguments,

$$\tilde{\varphi}(\hat{s}y, \hat{s}\xi^*) = \tilde{\varphi}(y, \xi^*), \tag{5.158}$$

because these transformations do not change x and ψ^* (see (5.156)).

It is convenient to write the Hamiltonian (5.148) in an explicitly Hermitian form,

$$\hat{H} = \tfrac{1}{2}\hat{P}_a g^{ab}\hat{P}_b + V_q + V\ , \tag{5.159}$$

$$V_q = \frac{1}{2\sqrt{\mu}}\left(\partial_a g^{ab}\right)\left(\partial_b\sqrt{\mu}\right) + \frac{1}{2\sqrt{\mu}}\, g^{ab}\left(\partial_a\partial_b\sqrt{\mu}\right)\ , \tag{5.160}$$

where $\hat{P}_a = -i\mu^{-1/2}(\partial_a + i\pi_a) \circ \mu^{1/2}$ is the Hermitian momentum operator, and $V_q \sim \hbar^2$ is an effective quantum potential.

5.7.3 The Chevalley theorem for systems on supermanifolds

The analysis carried out in Section 5.6 for gauge systems with bosonic degrees of freedom shows that the residual discrete gauge symmetry acts on all physical degrees of freedom and, as a consequence, their phase space as a whole is reduced. In quantum theory, it is proved [31, 195] that there exists a one-to-one correspondence between the gauge-invariant functions on the whole configuration space and the functions of physical coordinates which are invariant under the action of the residual discrete gauge transformations.

It is therefore important to find out whether an analogous statement holds for gauge systems with Bose and Fermi degrees of freedom. Consider a dynamical system defined by the Lagrangian

$$L = \tfrac{1}{2}\left[(\partial_t + y)\mathbf{x}\right]^2 + i\psi^\dagger(\partial_t + y)\psi - V(\mathbf{x}, \psi^\dagger, \psi), \tag{5.161}$$

where $\mathbf{x} \in R^3$, ψ is a three-dimensional vector with complex Grassmann components, and y is a 3×3 skew-symmetric matrix (an element of the Lie algebra of $SO(3)$). The Lagrangian is invariant under the gauge transformations that form the group $SO(3)$,

$$\mathbf{x} \to \Omega\mathbf{x}, \quad \psi \to \Omega\psi, \quad \psi^\dagger \to \psi^\dagger\Omega^T, \quad y \to \Omega y \Omega^T + \Omega\partial_t\Omega^T, \tag{5.162}$$

where $\Omega = \Omega(t) \in SO(3)$. The potential V is assumed to be gauge invariant.

The Hamiltonian formalism for this system is obtained in the standard way. As usual, the second-class constraints, associated with the linearity of the Lagrangian in the time derivatives, are treated in the Dirac bracket formalism,

$$\{\psi_a^\dagger, \psi_b\}_D = -i\delta_{ab}\,, \quad a, b = 1, 2, 3.$$

The Hamiltonian has the form,

$$H = \tfrac{1}{2}\mathbf{p}^2 + V(\mathbf{x}, \psi^\dagger, \psi) - (\mathbf{p}, y\mathbf{x}) - i(\psi^\dagger, y\psi), \tag{5.163}$$

because $p_y = \partial L/\partial \dot{y} = 0$, and $\mathbf{p} = \partial L/\partial \dot{\mathbf{x}}$. Put $y_{ab} = y_c\varepsilon_{cab}$, where ε_{abc} is the totally skew-symmetric unit tensor, $\varepsilon_{123} = 1$. Since p_{ya} is canonically conjugated to y_a, the secondary constraints are

$$\sigma_a = \epsilon_{abc}\left(p_b x_c + i\psi_b^\dagger\psi_c\right) = 0. \tag{5.164}$$

They are in involution (first-class constraints),

$$\{\sigma_a, \sigma_b\}_D = -\epsilon_{abc}\sigma_c, \quad \{\sigma_a, H\}_D = \epsilon_{abc}y_b\sigma_c. \tag{5.165}$$

Upon quantization, all the canonical variables become operators whose commutation relations are determined by the rule $\{\ ,\ \}_D \to -i[\ ,\ ]_\pm$ (the commutator and anti-commutator for the bosonic and Grassmann variables, respectively), and the constraint operators are required to annihilate the physical states,

$$\hat{\sigma}_a|\Phi_{ph}\rangle = 0. \tag{5.166}$$

These equations are easy to solve in the Fock space representation, similarly to the analysis carried out in Section 5.7.1. In this case, the set of all independent gauge-invariant operators is,

$$\hat{b}_1^{+} = \hat{a}_c{}^{\dagger}\hat{a}_c{}^{\dagger}\,, \qquad \hat{b}_2^{\dagger} = \epsilon_{abc}\hat{\psi}_a{}^{\dagger}\hat{\psi}_b{}^{\dagger}\hat{a}_c{}^{\dagger}, \tag{5.167}$$

$$\hat{f}_1^{\dagger} = \hat{a}_c{}^{\dagger}\hat{\psi}_c{}^{\dagger}, \qquad \hat{f}_2^{\dagger} = \varepsilon_{abc}\hat{\psi}_a{}^{\dagger}\hat{\psi}_b{}^{\dagger}\hat{\psi}_c{}^{\dagger}, \tag{5.168}$$

where $\hat{a}_c{}^{\dagger} = (\hat{p}_c - i\hat{x}_c)/\sqrt{2}$. The operators $\hat{b}_{1,2}^{\dagger}$ and $\hat{f}_{1,2}^{\dagger}$ generate bosonic and fermionic states, respectively, when they act on the vacuum state $|0\rangle$.

In order to elucidate the role of discrete residual gauge symmetry, the theory should first be formulated in the physical configuration space obtained by eliminating non-physical degrees of freedom. When doing so for theories with Grassmann variables, the following peculiarity should be kept in mind. Components of the vector $\boldsymbol{\psi}$ cannot be turned to zero by any gauge transformation as the number of generators of the Grassmann algebra is preserved under the gauge transformations. So the elimination of non-physical degrees of freedom can only be carried out in the bosonic sector of the theory. Let the quantum theory be described in coordinate and holomorphic representations for bosonic and Grassmann variables, respectively, meaning that the state vectors are functions of $\mathbf{x}$ and $\boldsymbol{\psi}^*$, and the scalar product is defined by Eq. (5.149).

Equations (5.166) are solved in curvilinear coordinates in the superspace spanned by $\mathbf{x}$ and $\boldsymbol{\psi}^*$ [31],

$$\mathbf{x} = \Omega\boldsymbol{\rho}\,, \quad \boldsymbol{\psi}^* = \boldsymbol{\xi}^*\Omega^T, \tag{5.169}$$

$$\Omega = \begin{pmatrix} \sin\theta\cos\varphi & -\sin\varphi & -\cos\theta\cos\varphi \\ \sin\theta\sin\varphi & \cos\varphi & -\cos\theta\sin\varphi \\ \cos\theta & 0 & \sin\theta \end{pmatrix} \in SO(3),$$

where $\boldsymbol{\rho} = (r,0,0)^T$. The substitution of (5.169) into (5.153) yields

$$\hat{\pi}_r = 0\,, \quad \hat{\pi}_\theta = \hat{L}_2\,, \quad \hat{\pi}_\varphi = \sin\theta\,\hat{L}_3 - \cos\theta\,\hat{L}_1,$$

where the operators $\hat{L}_a$ have the form,

$$\hat{L}_a = i\epsilon_{abc}\hat{\xi}_b^{\dagger}\hat{\xi}_c.$$

Direct computation shows that the set of constraint operators $\hat{\sigma}_a$ is a linear non-singular transformation of the set of operators ∂_φ, ∂_θ, and L_1. The coefficients of this linear transformation depend on the angles θ and φ but all of them occur to the left of the operators ∂_φ and ∂_θ. Thus, Eqs. (5.166) are equivalent to the following three equations:

$$\partial_\varphi\Phi_{\rm ph} = \partial_\theta\Phi_{\rm ph} = 0, \qquad \hat{L}_1\Phi_{\rm ph} = 0. \tag{5.170}$$

Hence, the physical wave functions are independent of φ and θ. But this is already evident from the fact that φ and θ are translated under the gauge transformations. The operator $\hat{L}_1$ is the generator of the $SO(2)-$ rotations in the Grassmann plane spanned by $\xi^*_{2,3}$. This gauge group in the Grassmann sector is nothing but a stationary subgroup of the vector $\boldsymbol{\rho}$ (rotations about $\boldsymbol{\rho}$, i.e. in the plane perpendicular to it). The $SO(2)$ rotations leave $\boldsymbol{\rho}$ and ξ^*_1 unchanged, while $\xi_{2,3}$ are transformed by them.

The metric tensor in the coordinates (5.169) is diagonal, $g^{rr} = 1$, $g^{\theta\theta} = r^{-2}$, and $g^{\varphi\varphi} = (r\sin\theta)^{-2}$. The Jacobian is $\mu = r^2$. Then, according to (5.159) and (5.170), the Hamiltonian in the physical Hilbert space reads,

$$\hat{H}_{\rm ph} = -\frac{1}{2r}\partial_r^2 \circ r + \frac{1}{2r^2}\left(\hat{L}_2^2 + \hat{L}_3^2\right) + V(\boldsymbol{\rho}, \hat{\boldsymbol{\xi}}^\dagger, \hat{\boldsymbol{\xi}}). \tag{5.171}$$

As proved in Section 5.5, the gauge symmetry $SO(2) \sim U(1)$ in a pure fermionic sector is equivalent to the Z_2 invariance of all physical states. Therefore the last equation in (5.170) is equivalent to the equation

$$\Phi_{\rm ph}(r, \hat{s}_1\boldsymbol{\xi}^*) = \Phi_{\rm ph}(r, \boldsymbol{\xi}^*), \tag{5.172}$$

where $\hat{s}_1 = \mathrm{diag}(1, -1, -1) \in SO(3)$. Indeed, the only invariant of the group $SO(2)$ generated by $\hat{L}_1$ is $\xi^*_2\xi^*_3$ (apart from the trivial ones, r and ξ^*_1). In turn, it is the only invariant of the group $Z_2 = \{1, \hat{s}_1\}$.

By a suitable gauge transformation (5.162) the vectors $\mathbf{x}$ and $\boldsymbol{\psi}^*$ can be brought to the form $\mathbf{x} = \boldsymbol{\rho}$ and $\boldsymbol{\psi}^* = \boldsymbol{\xi}^*$. There are discrete gauge transformations by which the sign of the physical variables can be changed $r \to \pm r$ without violating the conditions $x_{2,3} = 0$. Under these transformations, the Grassmann vector $\boldsymbol{\psi}^*$ changes as well. It is then natural to expect that in quantum theory the physical states are invariant under this residual discrete gauge transformations which acts simultaneously on both the bosonic and fermionic degrees of freedom. This is indeed so. Consider the transformations of θ, φ, r, and $\boldsymbol{\xi}^*$ under which $\boldsymbol{x}$ and $\boldsymbol{\psi}^*$ in (5.169) do not change. They should include transformations $\boldsymbol{\rho} \to \pm\boldsymbol{\rho}$ and the corresponding transformations of the angles φ and θ such that $\mathbf{x}$ remains unchanged. These transformations form the symmetry group of the spherical coordinates and are exhausted by the following transformations:

$$\begin{aligned}
&\hat{s}_1: \ \theta \to -\theta\ , \ \varphi \to \varphi + \pi\ , \ \ r \to r;\\
&\hat{s}_2: \ \theta \to \theta + \pi, \ \ \varphi \to \varphi\ , \ \ r \to -r;\\
&\hat{s}_3: \ \theta \to -\theta + \pi\ , \ \ \varphi \to \varphi + \pi, \ \ r \to -r.
\end{aligned}$$

In addition, there are also transformations $r \to r$, $\theta \to \theta + 2\pi n$, and $\varphi \to \varphi + 2\pi m$ with n and m being integers. But the latter do not change the matrix Ω in (5.169) and, hence, neither can they change $\boldsymbol{\xi}^*$, contrary to the transformations $\hat{s}$. The action of $\hat{s}$ on $\boldsymbol{\xi}^*$ and $\boldsymbol{\rho}$ can be defined by the $SO(3)$ matrices: $\hat{s}_1 = \mathrm{diag}(1, -1, -1)$, $\hat{s}_2 = \mathrm{diag}(-1, 1, -1)$, and $\hat{s}_3 = \mathrm{diag}(-1, -1, 1)$. For all of them, $\hat{s}^2 = 1$.

The physical states are invariant under the discrete group S, i.e.

$$\Phi(\hat{s}\boldsymbol{\rho}, \hat{s}\boldsymbol{\xi}^*) = \Phi(\boldsymbol{\rho}, \boldsymbol{\xi}^*). \tag{5.173}$$

The proof is elementary. For $\hat{s} = \hat{s}_1$ this relation is already established in (5.172). As for the others, first, note that the eigenfunctions of the total Hamiltonian $\hat{\mathbf{p}}^2/2 + V(\mathbf{x}, \hat{\psi}^\dagger, \hat{\psi})$ are invariant under transformations from S as they do not change $\mathbf{x}$ and ψ^* (compare (5.158)). The eigenfunctions can be expanded into a series over the spherical basis $Y_{lm}(\theta, \varphi)$,

$$\Phi_E(\mathbf{x}, \psi^*) = \sum_{l,m} R^E_{lm}(r, \boldsymbol{\xi}^*) Y_{lm}(\theta, \varphi).$$

From (5.170) it follows that the physical eigenfunctions are $R^E_{00}(r, \boldsymbol{\xi}^*)$. Since the set of $Y_{lm}(\theta, \varphi)$ is complete in the space of square integrable functions on the sphere and $Y_{00} = \text{const} \neq 0$, the functions $R^E_{00}(r, \boldsymbol{\xi}^*) = \Phi(\boldsymbol{\rho}, \boldsymbol{\xi}^*)$ must be invariant under the transformations $\hat{s}_{2,3}$.

An alternative way to prove (5.173) is to use averaging over the gauge group as the projection operator onto the physical subspace of the Hilbert space, analogously to (5.136).

The Chevalley theorem can be extended to gauge systems with both commuting and anti-commuting (Grassmann) variables [31, 195]. Namely, any S-invariant polynomial in the physical configuration space has a unique analytic gauge-invariant extension to the total configuration space. Consider polynomials of $\boldsymbol{\rho}$ and $\boldsymbol{\xi}^*$ invariant under the action of S. It is not hard to verify that they are polynomials of the elementary invariant polynomials (compare (5.167) and (5.168)),

$$\delta_{ab}\rho_a\rho_b, \quad \varepsilon_{abc}\rho_a\xi^*_b\xi^*_c, \quad \delta_{ab}\rho_a\xi^*_b, \quad \varepsilon_{abc}\xi^*_a\xi^*_b\xi^*_c. \tag{5.174}$$

The tensors δ_{ab} and ε_{abc} are invariant tensors of $SO(3)$ and, therefore, the invariants (5.174) are equal to, respectively,

$$\delta_{ab}x_a x_b\,, \quad \varepsilon_{abc}x_a\psi^*_b\psi^*_c, \quad x_a\psi^*_a\,, \quad \varepsilon_{abc}\psi^*_a\psi^*_b\psi^*_c, \tag{5.175}$$

where $\mathbf{x}$ and ψ^* are related to $\boldsymbol{\rho}$ and $\boldsymbol{\xi}^*$ by (5.169). On the other hand, any gauge-invariant polynomial of $\mathbf{x}$ and ψ^* is a polynomial of the elementary ones (5.175). Hence, every polynomial in the physical configuration space that is invariant under the discrete residual gauge transformations has a unique analytic gauge-invariant extension to the total configuration space. This statement also holds for analytic functions because the set of all polynomials is dense in the set of all analytic functions. This establishes a one-to-one correspondence between the set of S-invariant functions in the physical configuration space and the set of all gauge-invariant functions in the total configuration space.

This is a rather general feature that holds for gauge systems with bosonics and fermionic degrees of freedom and with an arbitrary (compact) gauge group (the generalized Chevalley theorem [31]).

5.8 Yang–Mills theories

The definition of physical phase space as the quotient space of the constraint surface relative to the gauge group holds for gauge field theories, i.e. for systems with an infinite number of degrees of freedom. The phase space in a field theory is a functional space, and this gives rise to considerable technical difficulties when calculating the quotient space. One has to specify a functional class to which elements of the phase space, being a pair of functions of the spatial variables, belong. In classical theory it can be a space of smooth functions (e.g. to make the energy functional finite). However, in quantum field theory the corresponding quotient space appears to be insufficient because in the path integral formalism the support of the path integral measure lies in a Sobolev functional class [233], i.e. in the space of distributions, where smooth classical configurations form a zero-measure subset. To circumvent this apparent difficulty, one can, for instance, discretize the space and compactify it into a torus, thus converting a continuum gauge theory into a lattice gauge theory and thereby making the number of degrees of freedom finite. This would make a gauge field model looking more like the mechanical models considered above, where the quotient space can be calculated, and the dynamics in it can be studied. The regularization is then removed by taking an appropriate limit. This is easy to say, but far too difficult to realize for Yang–Mills theories in four-dimensional spacetime (see Sections 5.8.9 and 5.8.10). Here the physical phase space of low-dimensional (topological) Yang-Mills theories and its effects on the quantum theory are studied in detail. Lattice Yang–Mills theories are investigated as well. However further discussion is limited to reviewing possible implications of some relevant results of numerical simulations for the continuum theory because the problem of the continuum limit still needs more rigorous study.

5.8.1 Yang–Mills theory on a cylindrical spacetime

The simplest example of a topological gauge field theory is the Yang–Mills theory on a cylindrical spacetime (space is compactified to a circle S^1) [234]–[243]. Note that in two-dimensional spacetime Yang–Mills theories do not have physical (propagating) degrees of freedom, unless the spacetime has a nontrivial topology [244]–[250]. In the Hamiltonian approach, only space is compactified, thus leading to a cylindrical spacetime. It appears that the physical configuration and phase spaces have an interesting nontrivial structure, and, moreover, in the quantum theory the spectrum of physical excitations is discrete, i.e. the mass gap is a direct consequence of the nontrivial physical phase space geometry [195, 240, 242].

The Lagrangian reads

$$L = -\frac{1}{4}\int_0^{2\pi l} dx(F_{\mu\nu}, F^{\mu\nu}) \equiv -\frac{1}{4}\langle F_{\mu\nu}, F^{\mu\nu}\rangle, \tag{5.176}$$

where $F_{\mu\nu} = \partial_\mu A_\nu - \partial_\nu A_\mu - ig[A_\mu, A_\nu]$, g is a coupling constant, $\mu, \nu = 0, 1$; the Yang–Mills potentials A_μ, being elements of a Lie algebra X of a semi-simple compact group G, are periodic functions of a spatial coordinate, $A_\mu(t, x+2\pi l) = A_\mu(t,x)$, i.e. l is the space radius; the parentheses $(,)$ in the integrand (5.176) stand for the invariant inner product in X defined in Section 8.1.2. Since the vector potential is a periodic function in space, it can be decomposed into a Fourier series. The Fourier components of A_μ are regarded as independent (Cartesian) degrees of freedom in the theory. The Lagrangian is invariant under the gauge transformations,

$$A_\mu \to \Omega A_\mu \Omega^{-1} + ig^{-1}\Omega\partial_\mu\Omega^{-1}, \quad \Omega \in G$$

Consider the Hamiltonian formalism of the theory. The canonical momenta are $E_\mu = \delta L/\delta \dot{A}^\mu = F_{0\mu}$; the overdot denotes the time derivative. The momentum conjugated to A_0 vanishes, $E_0 = 0$, forming the primary constraints. The canonical Hamiltonian has the form

$$H = \langle E_\mu, \dot{A}^\mu\rangle - L = \langle E_1, E_1\rangle/2 - \langle A_0, \sigma\rangle, \tag{5.177}$$

where $\sigma = \nabla(A_1)E_1$ with $\nabla(A_1) = \partial_1 - ig[A_1, \;]$ being the covariant derivative in the adjoint representation. The primary constraints must be satisfied during the time evolution. This yields the secondary constraints

$$\dot{E}_0 = \{E_0, H\} = \partial_1 E_1 - ig[A_1, E_1] = \sigma = 0, \tag{5.178}$$

where the standard symplectic structure,

$$\{A^{a\mu}(x), E^b_\nu(y)\} = \delta^{ab}\delta^\mu_\nu\delta(x-y), \qquad x, y \in S^1, \tag{5.179}$$

has been introduced, and the suffixes a, b refer to the adjoint representation of the Lie algebra. The constraints are in involution,

$$\{\sigma_a(x), \sigma_b(y)\} = if_{ab}{}^c\delta(x-y)\sigma_c(x), \qquad \{\sigma_a, H\} = -f_{ab}{}^c A^b_0\sigma_c, \tag{5.180}$$

with $f_{ab}{}^c$ being the structure constants of X. There are no more constraints in the theory, and all constraints are of the first class.

The primary and secondary (first-class) constraints are independent generators of gauge transformations. As in the mechanical models studied above, the primary constraints $E^a_0 = 0$ generate shifts of the Lagrange multipliers A^a_0, $\delta A^a_0(x) = \{A^a_0, \langle\omega_0, E_0\rangle\} = \omega^a_0(x)$, and leave the phase-space variables E^a_μ and A^a_1 unchanged. Therefore the hyperplane $E^a_0 = 0$ (E_1 and A_1 are fixed) spanned by A^a_0 in the total phase space is the gauge orbit. Thus, A^a_0 and E^a_0 are purely non-physical degrees of freedom and can be discarded in the subsequent analysis.

To simplify the notation, from now on the Lorentz suffix "1" of the field variables is omitted, i.e. instead of E_1 and A_1, E and A are used. The constraints (5.178) generate the following gauge transformations,

$$E \to \Omega E \Omega^{-1} = E^\Omega, \qquad A \to \Omega A \Omega^{-1} + ig^{-1}\Omega\partial\Omega^{-1} = A^\Omega. \tag{5.181}$$

Here and below $\partial_1 \equiv \partial$, while the overdot is used to denote the time derivative ∂_0; $\Omega = \Omega(x)$ takes its values in a semisimple compact group G (X is its Lie algebra). The gauge transformed variables E^Ω and A^Ω must also be periodic functions of x. This results in the periodicity of Ω modulo the center Z_G of G,

$$\Omega(x + 2\pi l) = z\Omega(x), \qquad z \in Z_G. \tag{5.182}$$

Indeed, by definition an element z from the center commutes with any element of X and, therefore, E^Ω and A^Ω are invariant under the shift $x \to x + 2\pi l$. The relation (5.182) is called a twisted boundary condition [252]. The twisted gauge transformations (i.e. satisfying (5.182) with $z \neq e$, e a group unit) form distinct homotopy classes. Therefore they cannot be continuously deformed towards the identity. On the other hand, gauge transformations generated by the constraints (5.178) are homotopically trivial because they are built up by *iterating* the infinitesimal transformations [253]: $\delta E = \{E, \langle\omega,\sigma\rangle\} = ig[E,\omega]$ and $\delta A = \{A, \langle\omega,\sigma\rangle\} = -\nabla(A)\omega$ with ω being an X-valued periodic function of x. This analysis leads to the following conclusion. When determining the structure of the physical phase or configuration space, as the quotient space, only *periodic* (i.e. homotopically trivial) gauge transformations should be used. Such transformations determine a mapping $S^1 \to G$. A collection of all such transformations is called a gauge group and will be denoted $\mathcal{G}$, while an abstract group G is usually called a structure group of the gauge field theory. Similarly to the mechanical matrix models studied in Section 5.6, it is proved below that quantum states annihilated by the operators of the constraints – these are the Dirac physical states – are not invariant under the twisted gauge transformations.

Let $\mathcal{F}$ be a space of all Fourier series with coefficients being elements of the Lie algebra X, i.e. a generic element $f \in \mathcal{F}$ is

$$f(x) = f_0 + \sum_{n=1}^{\infty} \left(f_{s,n} \sin \frac{nx}{l} + f_{c,n} \cos \frac{nx}{l} \right). \tag{5.183}$$

The convergence of the Fourier series is understood in some topology (e.g. $\mathcal{F} \sim C^\infty$ (the space of analytic functions) in the classical theory). Derivatives and integrals of f are understood as those of the Fourier series. The finite-dimensional subspace formed by constant functions is denoted $\mathcal{F}_0$ so that $A = A_0 + \tilde{A}$, where $A_0 \in \mathcal{F}_0$ and $\tilde{A} \in \mathcal{F} \ominus \mathcal{F}_0$. For a generic connection $A(x)$, there exists a *periodic* gauge element $\Omega(x)$ such that the gauge transformed connection A^Ω is homogeneous in space,

$$\partial A^\Omega = 0. \tag{5.184}$$

This means that the Coulomb gauge fixing surface $\partial A = 0$ intersects each gauge orbit at least once. To find $\Omega(x)$, put

$$\omega = -\frac{i}{g}\Omega^{-1}\partial\Omega \in X \tag{5.185}$$

and, hence,

$$\Omega(x) = \mathrm{P}\exp\left[ig\int_x^0 \omega(x')dx'\right], \tag{5.186}$$

where P exp is the path-ordered exponent. Let a sequence $\{x_i\}$, $i = 1, 2, ..., M+1$, be a partition of the interval $[0, x]$ such that $x_1 = x$ and $x_{M+1} = 0$. Put $\Delta x_i = x_{i+1} - x_i$ and $U_i = \exp[igA(x_i^*)\Delta x_i] \in G$ for some $x_i^* \in [x_i, x_{i+1}]$. The path-ordered exponential (5.186) is defined as the limit of $U_M U_{M-1} \cdots U_1$ when the partition is refined, i.e. $M \to \infty$. By definition, it solves Eq. (5.185). After simple algebraic transformations, Eq. (5.184) can be written in the form

$$\nabla(A)\omega = \partial\omega - ig[A, \omega] = -\partial A, \tag{5.187}$$

which has to be solved for the Lie algebra element $\omega(x)$. It is a linear nonhomogeneous differential equation of first order. So its general solution is a sum of a general solution of the corresponding homogeneous equation and a particular solution of the nonhomogeneous equation. Introducing the group element,

$$U_A(x) = \mathrm{P}\exp\left[ig\int_0^x dx' A(x')\right], \tag{5.188}$$

that has simple properties $\partial U_A = igAU_A$ and $\partial U_A^{-1} = -igU_A^{-1}A$, the general solution can be written as

$$\omega(x) = U_A(x)\omega_0 U_A^{-1}(x) - A(x). \tag{5.189}$$

The first term containing an arbitrary *constant* Lie algebra element ω_0 represents a solution of the homogeneous equation, while the second term is obviously a particular solution of the nonhomogeneous equation. The constant ω_0 should be chosen so that the group element (5.186) would satisfy the periodicity condition, which yields

$$\Omega(2\pi l) = \mathrm{P}\exp\left[ig\oint dx\,\omega\right] = e, \tag{5.190}$$

where e is the group unit. This specifies completely function $\omega(x)$, and, hence, $\Omega(x)$ for any generic $A(x)$. So, any configuration $A \in \mathcal{F}$ can be reduced towards a spatially homogeneous configuration by means of a gauge transformation.

The same gauge transformation that transforms A to a homogeneous connection $A_0 \in \mathcal{F}_0$ also transforms the momentum E to $E_0 \in \mathcal{F}_0$, if E lies on the constraint surface. Indeed, substitution of the gauge transformed canonical pair $A^\Omega = A_0 \in \mathcal{F}_0$, E^Ω into the constraint equation $\nabla(A)E = 0$ yields the equation,

$$\nabla(A^\Omega)E^\Omega = \nabla(A_0)E^\Omega = 0. \tag{5.191}$$

The momentum variable is represented as the sum of the homogeneous part E_0 and the nonhomogeneous, $\tilde{E}^\Omega = E^\Omega - E_0$. These two components satisfy equations that are deduced from Eq. (5.191),

$$\sigma_0 \equiv [A_0, E_0] = 0 \ , \tag{5.192}$$
$$\partial \tilde{E}^\Omega - ig[A_0, \tilde{E}^\Omega] = \nabla(A_0)\tilde{E}^\Omega = 0. \tag{5.193}$$

The first equation stems from the $\mathcal{F}_0$-component of the constraint equation (5.191), while the second one is the constraint in the subspace $\mathcal{F} \ominus \mathcal{F}_0$. A general solution of Eq. (5.193) can be written in the form $\tilde{E}^\Omega(x) = U_0(x)\tilde{E}_0^\Omega U_0^{-1}(x)$, where $U_0(x) = \exp[igA_0x]$ and $\partial\tilde{E}_0^\Omega = 0$. For a generic A_0, the solution is *not* periodic in x for all constants $\tilde{E}_0^\Omega \neq 0$. Indeed, for any constant connection $A_0 \in X$, $U_0(x) = \Omega_0 \exp(igax)\Omega_0^{-1}$ where a is an element of the Cartan subalgebra. The periodicity condition yields that the group element $\exp(2\pi igla) = z$ commutes with any $\tilde{E}_0^\Omega \in X$, i.e. z is an element of the group center. Configurations a for which this holds form a lattice in H that is a zero measure set in H (its detailed description will be given in Section 5.8.7). Since $\tilde{E}^\Omega(x)$ must be a periodic function, the constant $\tilde{E}_0^\Omega$ should necessarily vanish. Thus, Eq. (5.193) has only a trivial solution $\tilde{E}^\Omega = 0$, and $E^\Omega = E_0 \in \mathcal{F}_0$.

A useful observation following on from the above analysis is that the operator $\nabla(A_0)$ has no zero modes in the subspace $\mathcal{F} \ominus \mathcal{F}_0$ and, hence, is invertible. The determinant of the operator $\nabla(A_0)$ restricted on $\mathcal{F} \ominus \mathcal{F}_0$ does not vanish (save for configurations from the aforementioned zero measure set). Its explicit form will be found in Section 5.8.3

As a result, a redundant dynamical system is obtained which has $N = \dim X$ degrees of freedom and the constraint (5.192) that generates *homogeneous* gauge transformations of the phase-space variables A_0 and E_0 ($\partial\Omega \equiv 0$). This mechanical system has been studied in Section 5.4. The system is shown to have $r = \text{rank}\ X$ physical degrees of freedom which can be described by Cartan subalgebra components of A_0 and E_0. Since any element of X can be represented in the form $A_0 = \Omega_A a \Omega_A^{-1}$, a an element of the Cartan subalgebra H, $\Omega_A \in G$, configurations A_0 and a belong to the same gauge orbit. Moreover, a spatially homogeneous gauge transformation with $\Omega = \Omega_A^{-1}$ brings the momentum E_0 on the constraint surface (5.192) to the Cartan subalgebra. Indeed, from (5.192), the relation $\left[a, \Omega_A^{-1}E_0\Omega_A\right] = 0$ is derived, and, therefore, $p_a = \Omega_A^{-1}E_0\Omega_A \in H$, by the definition of H. The element a has a stationary group being the Cartan subgroup of G. This means that not all of the constraints (5.192) are independent. There are just $N - r$, $r = \dim H$, independent constraints among (5.192). The continuous gauge arbitrariness is exhausted in the theory. The physical degrees of freedom of the theory are described by the canonical pair a and p_a.

5.8.2 The physical configuration and phase spaces

Analysis of the mechanical gauge model of Section 5.4 ($h \to a$ here) shows that there are discrete gauge transformations of a that do not violate the condition $a \in H$. These transformations form the Weyl group. In the Yang–Mills theory, the Weyl group transformations are homogeneous gauge transformations, i.e. independent of the spatial variable x. Hence, they may not form a set of all residual discrete gauge transformations. If two homogeneous connections from the Cartan subalgebra, a and a_s, belong to the same gauge orbit, then there should exist a gauge group element $\Omega_s(x)$ such that

$$a_s = \Omega_s a \Omega_s^{-1} + \frac{i}{g}\Omega_s \partial \Omega_s^{-1}, \quad \partial a_s = \partial a = 0, \quad a_s, a \in H. \tag{5.194}$$

Clearly, homogeneous solutions ($\partial\Omega_s = 0$) of this equation form the Weyl group and the Cartan subgroup G_H of G. The Cartan subgroup is a stationary group of the connection a (a stabilizer of a). Thus, every connection in the theory is *reducible* [251], i.e. has a nontrivial stabilizer which contains constant gauge transformations from the Cartan subgroup. Transformations from G_H do not lead to any reduction of the configuration space of the physical degrees of freedom.

To find *nonhomogeneous* solutions of Eq. (5.194), the latter is differentiated with respect to x. The result is then transformed to

$$\nabla(a) w_s = 0\,, \qquad w_s = i g^{-1}\Omega_s^{-1}\partial\Omega_s \in \mathcal{F}. \tag{5.195}$$

For a generic a from the Cartan subalgebra the first equation has only a homogeneous solution which is written in the form

$$w_s = a_0 \eta\,, \quad a_0 = (gl)^{-1}\,, \quad \eta \in H. \tag{5.196}$$

Note that Eq. (5.195) can always be transformed into two independent equations by setting $w_s = a_0\eta + \tilde{w}_s$, where $\eta \in H$ and $\tilde{w}_s \in X \setminus H$. Then Eq. (5.195) is equivalent to $\partial\eta = 0$ (so η is constant) and $\nabla(a)\tilde{w}_s = 0$. A general solution of the latter reads

$$\tilde{w}_s(x) = e^{igax} v e^{-igax} = \sum_{\alpha>0}\left(e^{ig(a,\alpha)x} v_\alpha^+ e_\alpha + e^{-ig(a,\alpha)x} v_\alpha^- e_{-\alpha}\right),$$

where $v \in X \setminus H$ is written in the Cartan–Weyl basis and the commutation relation $[a, e_{\pm\alpha}] = \pm(a,\alpha)e_{\pm\alpha}$ has been used (see Section 8). For an arbitrary v, $(a,\alpha)/a_0$ must be an integer for any root α in order for $\tilde{w}_s(x)$ to be periodic. Consider a diagram $D(X)$ in H, which is the union of a finite number of families of equispaced planes in H determined by $(\alpha, a) \in a_0 Z$, where α ranges over the root system and Z stands for the set of all integers. So, if a is not in $D(X)$, then $v=0$ and $\tilde{w}_s(x)=0$. If $a \in D(X)$, then a nontrivial periodic solution $\tilde{w}_s(x) \neq 0$ may exist. Solutions of (5.194) are defined modulo the Weyl group (homogeneous residual gauge transformations). Therefore, one can always impose the condition $\Omega_s(0)=e$ (the group unit). Set $x=0$ in (5.194). Then $a_s = a - w_s(0)$. The condition $a_s \in H$ demands that $w_s(0) \in H$ or $v = 0$, which completes the proof of (5.196).

From the second equation in (5.195) and (5.196) it follows that

$$\Omega_s(x) = \exp\left(iga_0\eta x\right). \tag{5.197}$$

This is still not the whole story because the group element (5.197) must obey the periodicity condition otherwise it does not belong to the gauge group. The periodicity condition yields the restriction on the admissible values of η,

$$\Omega_s(2\pi l) = \exp(2\pi i\eta) = e, \tag{5.198}$$

where e stands for the group unit. The set of elements η obeying this condition is called the unit lattice in the Cartan subalgebra [215, p. 305]. The nonhomogeneous residual gauge transformations do not change the canonical momentum p_a, since $[p_a, \eta] = 0$, and shift the canonical coordinate $a \to a + a_0\eta$, along the unit lattice in the Cartan subalgebra.

Consider then a group T_e of translations in H, $a \to a + a_0\eta$, where η belongs to the unit lattice. The group T_e leaves the diagram $D(X)$ invariant [215, p. 305]. The diagram $D(X)$ is also invariant with respect to the Weyl group transformations. Since W is generated by the reflections in the hyperplanes orthogonal to simple roots, it is sufficient to prove the invariance of $D(X)$ under them. Let $\hat{w}a$ be the image of a under the reflection in the hyperplane orthogonal to a simple root ω. Then for any root α, $(\alpha, \hat{w}a) = a_0 n_\omega$, where $n_\omega = n - 2k_\omega(\omega,\alpha)/(\omega,\omega)$ is an integer because $(a,\omega) = k_\omega a_0$, $k_\omega \in Z$ if $a \in D(X)$. Recall that any root α can be decomposed over the basis formed by simple roots. The coefficients of this decomposition are all either non-negative or non-positive integers. Therefore the number $-2(\omega,\alpha)/(\omega,\omega)$ is a sum of integers since the elements of the Cartan matrix, $C_{\omega\omega'} = -2(\omega,\omega')/(\omega,\omega)$, where ω and ω' are simple roots, are integers. So, $\hat{w}D(X) = D(X)$. Now consider the complement of $D(X)$, $H \setminus D(X)$. This consists of equal open polyhedrons whose walls form the diagram $D(X)$. Each polyhedron is called a cell. A cell inside the positive Weyl chamber K^+ such that its closure contains the origin is called the Weyl cell K_W^+.

Since the Weyl cell plays an important role in the subsequent analysis, let us discuss a few examples before studying the orbit space structure in general. Let $X = su(2)$ (the Lie algebra is denoted by the corresponding small letters of the group notation: $G = SU(2)$). Then $H \sim R$ and the diagram $D(su(2))$ consists of points $na_0\omega/(\omega,\omega)$, $n \in Z$, with ω being the only positive root of $su(2)$, $(\omega,\omega) = 1/2$ (the root normalization with respect to the Cartan–Killing form and their matrix representation is discussed in Section 8.1.2). A cell of $H \setminus D(su(2))$ is an open interval between two neighboring points of $D(su(2))$. In an orthonormal basis of the Cartan subalgebra, $a = \sqrt{2}a_3\omega$, $(a,a) = a_3^2$. Since the Weyl chamber K^+ is isomorphic to the positive half-line, $a_3 > 0$, $a \in H$ belongs to the Weyl cell K_W^+ if a_3 lies in the open interval $(0, \sqrt{2}a_0)$. The translations $a \to a + 2na_0\omega/(\omega,\omega)$, $n \in Z$, form the group T_e, and $W = Z_2$, $\hat{\omega}a = -a$. Thus, the lattice $D(su(2)) \subset H \sim R$ is invariant under translations from T_e and the reflection from the Weyl group W.

The group $G = SU(3)$ has three positive roots, ω_1, ω_2 and $\omega_{12} = \omega_1 + \omega_2$ which have the same norms. The angle between any two neighboring roots equals $\pi/3$.

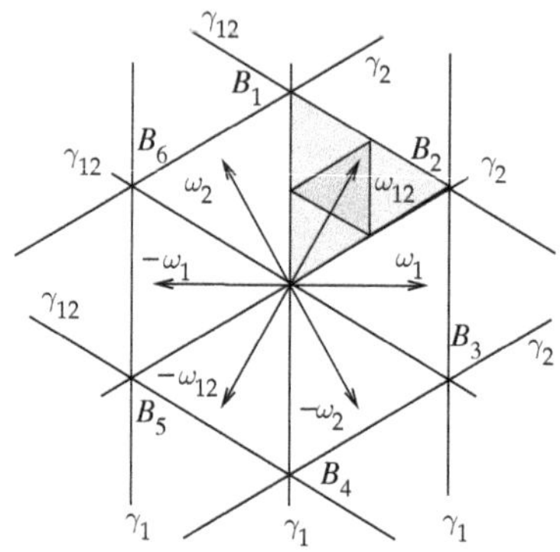

Fig. 5.6. The root pattern of $SU(3)$. The diagram $D(su(3))$ is formed by three families of straight lines perpendicular to the simple roots ω_1, ω_2 and the root $\omega_{12} = \omega_1 + \omega_2$. These families are denoted γ_1, γ_2, and γ_{12}, respectively. The dark grayed equilateral triangle is the Weyl cell of $su(3)$ which is the moduli space of the $su(3)$ connections with respect to homotopically trivial gauge transformations. Had the homotopically nontrivial transformations been included in the gauge group, the moduli space would have been four times smaller and isomorphic to the equilateral triangle whose vertices are the mid-points of the Weyl cell boundaries (see Section 5.8.7 for details). Adapted from [195].

The root pattern of $SU(3)$ is plotted in Fig. 5.6. The diagram $D(su(3))$ consists of three families of equispaced straight lines $(\omega_{1,2,12}, a) = a_0 n_{1,2,12}$, $n_{1,2,12} \in Z$, on the plane $H \sim R^2$. The lines are perpendicular to the roots $\omega_{1,2,12}$, respectively. The complement $H \setminus D(su(3))$ is a set of equilateral open triangles. The Weyl cell K_W^+ is the triangle bounded by lines $(\omega_{1,2}, a) = 0$ (being the boundary of K^+) and $(\omega_{12}, a) = a_0$. It is shown by a light grayed triangle in Fig. 5.6. The group T_e is generated by integral translations through the vectors $2a_0\alpha/(\alpha,\alpha)$, α ranges over $\omega_{1,2,12}$, and $(\alpha,\alpha) = 1/3$ (see Section 8.1.2).

Let W_A denote the group of linear transformations of H generated by the reflections in all the hyperplanes in the diagram $D(X)$. This group is called the affine Weyl group [215, p. 314]. It preserves $D(X)$ and, hence,

$$K_W^+ \sim H/W_A,$$

i.e. the Weyl cell is isomorphic to a quotient of the Cartan subalgebra by the affine Weyl group. Consider a group T_e of translations,

$$a \to a + 2a_0 \sum_{\alpha>0} n_\alpha \alpha/(\alpha,\alpha) \equiv a + a_0 \sum_{\alpha>0} n_\alpha \eta_\alpha, \qquad n_\alpha \in Z.$$

Then W_A is a semidirect product of T_e and W [215, p. 315]. The translation vectors η_α have the property [215, p. 317]

$$\exp(2\pi i \eta_\alpha) = \exp\left[\frac{4\pi i \alpha}{(\alpha,\alpha)}\right] = e. \tag{5.199}$$

A comparison of this relation with (5.198) leads to the conclusion that the residual discrete gauge transformations form the affine Weyl group and the physical configuration space $\mathcal{F}/\mathcal{G} \sim H/W_A$ is isomorphic to the closure of K_W^+.

The total phase space is $\mathcal{F} \times \mathcal{F}$ because it is formed by pairs of elements after $E(x)$, $A(x)$ from $\mathcal{F}$. The quotient with respect to the gauge group reads

$$\Gamma_{ph} = (\mathcal{F} \times \mathcal{F})/\mathcal{G} = (H \times H)/W_A, \tag{5.200}$$

where the action of W_A on $H \times H \sim R^{2r}$ ($r = \operatorname{rank} G$) is diagonal and determined by all possible compositions of the following transformations:

$$\hat{s}_{\alpha,n} p_a = \hat{s}_\alpha p_a = p_a - \frac{2(\alpha, p_a)}{(\alpha,\alpha)}\,\alpha, \tag{5.201}$$

$$\hat{s}_{\alpha,n} a = \hat{s}_\alpha a + \frac{2n_\alpha a_0}{(\alpha,\alpha)}\,\alpha, \tag{5.202}$$

where the element $\hat{s}_{\alpha,n} \in W_A$ acts on a as a reflection in the hyperplane $(\alpha, a) = n_\alpha a_0$, $n_\alpha \in Z$, for any positive root α ($\hat{s}_{\alpha,0} = \hat{s}_\alpha \in W$).

To illustrate the formula (5.200), consider the physical phase space for the simplest case $G = SU(2)$ for which $r = 1$, $W = Z_2$, and $(\omega, \omega) = 1/2$ (see Fig. 5.7). Let p_3 and a_3 be the canonical variables in the standard basis of the phase plane $R^2 \sim H \times H$. The translation group T_e action is then $p_3, a_3 \to p_3, a_3 + 2\sqrt{2}na_0$ and the Weyl reflection is $p_3, a_3 \to -p_3, -a_3$. Computation of the quotient space is illustrated in Fig. 5.7, where $L = \sqrt{2}\,a_0$. The points B and B_1 are related by the gauge transformation from T_e. The strips bounded by the vertical lines $(\gamma\gamma')$ are gauge equivalent through the translations from T_e. The boundary lines $(\gamma\gamma')$ are gauge equivalent to one another, too. So, R^2/T_e is a cylinder. After a cut, this cylinder can be unfolded into the strip $p_3 \in R$ and $a_3 \in (-L, L)$, as shown in Fig. 5.7b. The boundary lines $a_3 = \pm L$ are edges of the cut. They contain the same physical states and later are to be glued back. On the strip one should glue together the points p_3, a_3 and $-p_3, -a_3$ connected by the reflection from the Weyl group (the points B and $-B$ in Fig. 5.7b). This converts the cylinder into a half-cylinder ended by two conic horns at points $p_3 = 0, a_3 = 0, L$. Indeed, the strip can be cut along the line $a_3 = 0$, its right half (the strip $0 < a_3 < L$) can be rotated about the coordinate axis a_3 through the angle π as if the strip were in three-dimensional space (compare a similar procedure for the conic phase space in Section 5.1). The result is shown in Fig. 5.7c. It is important to observe that the half-axis $(L\gamma)$ is gauge equivalent to $(-L\gamma)$ and $(-L\gamma')$ to $(L\gamma')$, while the positive and negative momentum half-axes in Fig. 5.7c are edges of the cut and therefore to be identified too. Next, the strip in Fig. 5.7c is folded along the momentum axis to identify points B and $-B$. Finally, the half-line (γL) is glued back to $(L\gamma)$ and $(p_3 0)$ to $(0 p_3)$ in Fig. 5.7d to obtain the physical phase space (Fig. 5.7e). In the neighborhoods of singular conic points, the physical phase space looks locally like cone(π), studied in Section 5.1, because W_A acts as the

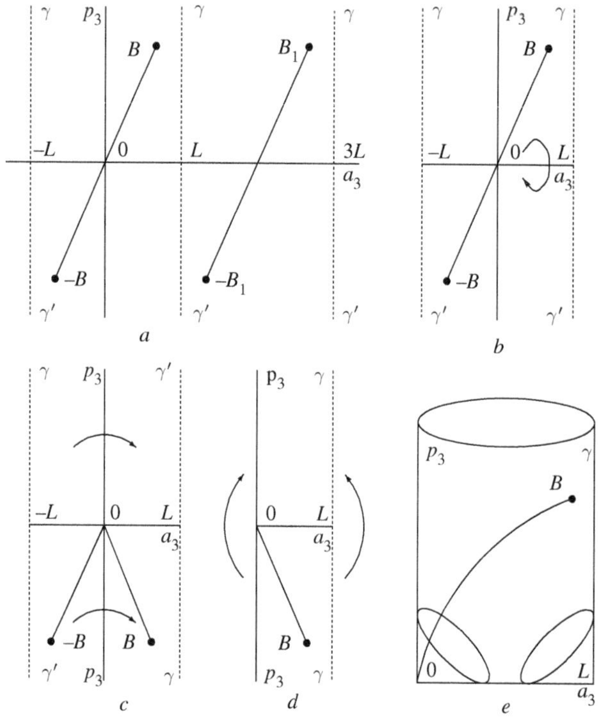

Fig. 5.7. Panel (a): Phase space plane of the physical degrees of freedom spanned by (a_3, p_3) the $SU(2)$ model; the points $B+(2Ln,0)$ and $-B+(2Ln,0)$, $n=0,\pm1,\pm2,...$, are gauge equivalent and must be identified. Here $L=\sqrt{2}a_0$.
Panel (b): The phase-space plane after identification of points (a_3+2Ln, p_3); the edges $(\gamma\gamma')$ of the strip are still to be identified.
Panel (c): The strip shown in Panel (b) is cut along the p_3-axis and the right part is rotated about the a_3-axis through the angle π as if the strip were embedded into three-dimensional space. Then the strip is folded along the p_3-axis. As a result points B and $-B$ are identified. The result is shown in Panel (d).
Panel (d): The edges of the strip must still be identified as indicated by the arrows because they correspond to gauge equivalent states on the edges $(\gamma\gamma')$ and to the cut along the p_3-axis.
Panel (e): The physical phase space of the $SU(2)$ Yang–Mills theory on a cylindrical spacetime. It is a half-cylinder with two conic horns attached to it. It is flat everywhere except the conic singularities where the curvature is infinite. For the sake of presentation the upper part of the cylinder is stretched (the arclength of the cylinder cross-section by a plane perpendicular to the p_3-axis must be $2L$, but it appears larger in the figure). Adapted from [195].

Z_2-reflections (5.201) and (5.202) with $\alpha=\omega$ and $n=0,1$ near $a_3=0,\sqrt{2}a_0$, respectively.

For groups of rank 2, all conic (singular) points of the physical phase space are concentrated on a triangle being the boundary ∂K_W^+ of the Weyl cell (if $G=SU(3)$, ∂K_W^+ is an equilateral triangle with side length $\sqrt{3}a_0$, e.g. the triangle with vertices B_1, B_2, and the origin shown in Fig. 5.6). Define local symplectic coordinates $p_a^\perp$, $a^\perp$ and $p_a^\parallel$, $a^\parallel$ in the neighborhood of a point of ∂K_W^+ (except the triangle vertices) such that $a^\perp$ and $a^\parallel$ vary along lines perpendicular and parallel to ∂K_W^+, respectively. For example, consider the interval (B_1B_2) in Fig. 5.6 which is a part of ∂K_W^+. In the neighborhood of a point of (B_1B_2), the coordinate lines of $a^\parallel$ are parallel to (B_1B_2), while the coordinate lines of $a^\perp$ are perpendicular to it. This local coordinate system is related to the global rectangular coordinate system in H by an orthogonal transformation. The corresponding local momenta $p_a^\parallel$ and $p_a^\perp$ are also obtained by the same orthogonal transformation in the momentum space. The W_A-reflection in the boundary of ∂K_W^+ going through this neighborhood leaves $p_a^\parallel$, $a^\parallel$ invariant, while it changes the sign of the other symplectic pair, $p_a^\perp, a^\perp \to -p_a^\perp, -a^\perp$. Therefore the physical phase space locally coincides with $R^2 \times \text{cone}(\pi)$. At the triangle vertices, two conic singularities going along two triangle edges merge. If those edges are perpendicular, the physical phase space is locally $\text{cone}(\pi)\times\text{cone}(\pi)$, otherwise it is a four-dimensinal cone which cannot be presented as a direct product of two 2D-cones. Its tip is "sharper" than the tip of $\text{cone}(\pi)\times\text{cone}(\pi)$, meaning that the 4D-cone can always be put inside of $\text{cone}(\pi) \times \text{cone}(\pi)$ when the tips of both the cones are placed at the same point. Obviously, a lesser angle between the triangle edges (boundaries of K_W^+) corresponds to a "sharper" 4D-cone (compare Section 5.4).

A generalization of this pattern of singular points in the physical phase space to gauge groups of an arbitrary rank is trivial. The Weyl cell is an r-dimensional polyhedron ($r = \text{rank}\, X$) whose faces are portions of the hyperplanes orthogonal to positive roots. The physical phase space has the most singular local $2r$-dimensional conic structure ($2r$D-cone) at the polyhedron vertices where the r boundary hyperplanes intersect. On the polyhedron edges (the intersection of $r-1$ boundary hyperplanes), it locally appears as $R^2\times 2(r-1)$D-cone. Then at the points of intersection of $r-2$ boundary hyperplanes, it is locally $R^4 \times 2(r-2)$D-cone, etc. Finally, at a generic boundary point in any of the boundary hyperplanes, i.e. in polyhedron faces save for their boundaries, the local phase-space structure looks like $R^{2(r-1)} \times \text{cone}(\pi)$.

Remark. As in the mechanical models studied earlier one can choose various ways to parameterize the physical phase space. When calculating a quotient space, one can, for instance, restrict the values of the canonical momentum $E(x)$. This is equivalent to imposing a gauge condition on the field strength rather than on the connection [214, 254] (quantum theory of topological gauge-field models in the momentum representation is studied in [255–257]). By a gauge rotation E can be brought to the Cartan subalgebra at each point x. So we set $E(x) = E_H(x) \in H$. Put $A(x) = A_H(x) + \bar{A}(x)$ where $A_H(x) \in H$ and $\bar{A}(x) \in X \setminus H$. Then the constraint $\nabla(A)E_H = 0$ is equivalent to two independent constraints: $\partial E_H = 0$ and $[\bar{A}, E_H] = 0$, which are nothing but the projection of the constraint onto H and its orthogonal complement $X\setminus H$, respectively. From the Cartan–Weyl commutation relations it follows that $\bar{A}(x) = 0$. The residual constraints $\partial E_H = 0$ generate the gradient shifts of the corresponding canonical variables: $A_H \to A_H + \partial\omega$, where $\partial\omega$ is a *periodic* function of x. This is the so-called Abelian projection of the Yang–Mills theory [214]. Therefore the physical degrees of freedom can again be described by the pair $E_H(x) = p_a$ and $A_H = a$. Now p_a can be taken into the Weyl chamber by an appropriate Weyl transformation, while a is determined modulo shifts on the periods of the group torus (the shifts along the group unit lattice). Note that one can take $\omega = \eta x$ since $\partial\omega = \eta$ is periodic as a constant function. The necessary restrictions on η follow from the periodicity condition on the corresponding gauge *group* element. This gives another parameterization of the *same* physical phase space such that $p_a \in K^+$ and $a \in H/T_e$, which, obviously, corresponds to another cut of the rD-cone.

5.8.3 Geometry of the gauge orbit space

Before developing quantum theory, the orbit space geometry has to be investigated. In particular, the metric tensor should be computed in order to construct the Hamiltonian in the physical Hilbert space. It is convenient to introduce the following orthogonal decomposition of the space (5.183):

$$\mathcal{F} = \sum_{n=0}^{\infty} \oplus \mathcal{F}_n = \sum_{n=0}^{\infty} \oplus \left(\mathcal{F}_n^H \oplus \bar{\mathcal{F}}_n\right) , \tag{5.203}$$

where $\mathcal{F}_0$ is a space of constant Lie algebra-valued functions (the first term in the series (5.183)), $\mathcal{F}_n, n \neq 0$, is a space of functions with fixed n in the sum (5.183). Each subspace $\mathcal{F}_n$ is finite-dimensional, $\dim \mathcal{F}_0 = \dim X$, $\dim \mathcal{F}_n = 2\dim X, n \neq 0$ (we recall that Lie algebra-valued functions are considered). Functions belonging to $\mathcal{F}_n^H$ take their values in the Cartan subalgebra H, while functions from $\bar{\mathcal{F}}_n$ take their values in (orthogonal) complement of H in X, i.e., in X/H.

From the above analysis of the orbit space of Yang–Mills connections, it follows that a generic connection can always be parameterized as

$$A = \Omega a \Omega^{-1} + ig^{-1}\Omega\partial\Omega^{-1}, \quad \partial a = 0, \quad a \in H, \tag{5.204}$$

where $\Omega \in \mathcal{G}/G_H$, and G_H is the Cartan subgroup (the maximal Abelian subgroup of G) which is isomorphic to the stationary group of the homogeneous connection a. By definition the connection remains invariant under gauge transformations from its stationary group. Equation (5.204) defines a change of variables in the functional space $\mathcal{F}$ if a is restricted to K_W^+. Indeed, the symmetry group of the mapping (5.204) coincides with the affine Weyl group $\Omega \to \Omega\Omega_s^{-1}$, $a \to a_s = \hat{s}a$, $\hat{s} \in W_A$, and a_s is defined in (5.194). So, the points $\Omega\Omega_s^{-1}$ and a_s have the same image A under the mapping (5.204). In the new variables the functional differential $\delta A \in \mathcal{F}$ can be represented in the form,

$$\delta A = \Omega \left(da - ig^{-1}\nabla(a)\delta w\right) \Omega^{-1}, \tag{5.205}$$

where by the definition of the parameterization (5.204), $\delta a = da \in \mathcal{F}_0^H$ and $\delta w(x) = i\Omega^{-1}\delta\Omega \in \mathcal{F} \ominus \mathcal{F}_0^H$. Therefore the metric tensor reads,

$$\begin{aligned}\langle \delta A, \delta A\rangle &= 2\pi l(da, da) - g^{-2}\langle \delta w, \nabla^2(a)\delta w\rangle \\ &\equiv (da, g_{aa} da) + \langle \delta w, g_{ww}\delta w\rangle .\end{aligned} \tag{5.206}$$

The inner product brackets $\langle\cdot\, ,\, \cdot\rangle$ are defined in (5.176). Equality (5.206) results from (5.205) and the relation that $\langle da, \nabla(a)\delta w\rangle = -\langle \nabla(a)da, \delta w\rangle = 0$, which is due to $\partial da = 0$ and $[da, a] = 0$. The operator $\nabla(a)$ acts in the subspace $\mathcal{F} \ominus \mathcal{F}_0^H$. It has no zero mode in this subspace if $a \in K_W^+$ and, hence, is *invertible*. Its determinant is strictly positive on K_W^+. The proof will be given shortly. The metric tensor has the block-diagonal form. The physical block is proportional to the $r \times r$ unit matrix $g_{aa} = 2\pi l$. The nonphysical block is the differential operator, $g_{ww} = -g^{-2}\nabla^2(a)$, and $g_{aw} = g_{wa} = 0$. Thus, the physical metric g_{aa}^{ph} coincides with g_{aa}. That is, the physical configuration space is a flat manifold with (singular) boundaries. As is shown next, the Jacobian of the change of variables (5.204) vanishes at the boundary of K_W^+ (the metric tensor is singular at ∂K_W^+). The physical configuration space has the structure of an orbifold [258].

To obtain the induced volume element on the orbit space, the Jacobian of the change of variables (5.204) has to be calculated,

$$\int_{\mathcal{F}} \prod_{x\in S^1} dA(x)\Phi = \int_{\mathcal{G}/G_H} \prod_{x} dw(x) \int_{K_W^+} da J(a)\Phi \rightarrow \int_{K_W^+} da\kappa^2(a)\Phi, \tag{5.207}$$

$$J^2(a) = \det g_{aa} \det g_{ww} = (2\pi l)^r \det\left[-g^{-2}\nabla^2(a)\right]. \tag{5.208}$$

Here $\Phi = \Phi(A) = \Phi(a)$ is a gauge-invariant functional of A. The induced volume element does *not* coincide with the square root of the determinant of the induced metric on the orbit space. It contains an additional factor, $(\det g_{ww})^{1/2}$, being the volume of the gauge orbit through a generic configuration $A(x) = a$. This volume needs to be renormalized because the integration over $\mathcal{G}/G_H$ produces an infinite factor (proportional to the volume of the gauge group). The renormalization procedure is indicated by the arrow in (5.207). Consider the orthogonal decomposition,

$$\bar{\mathcal{F}}_n = \sum_{\alpha>0} \oplus \mathcal{F}_n^\alpha, \tag{5.209}$$

where $\mathcal{F}_n^\alpha$ contains only functions taking their values in the two-dimensional subspace $X_\alpha \oplus X_{-\alpha}$ of the Lie algebra X spanned by the Cartan–Weyl basis elements $e_{\pm\alpha}$ associated with the positive root α. The subspaces $\mathcal{F}_n^H$, $\mathcal{F}_n^\alpha$ are invariant subspaces of the operator $\nabla(a)$, that is, $\nabla(a)\mathcal{F}_n^H$ is a subspace of $\mathcal{F}_n^H$, and $\nabla(a)\mathcal{F}_n^\alpha$ is a subspace of $\mathcal{F}_n^\alpha$. Hence, the operator $\nabla(a)$ has a block-diagonal form in the decomposition (5.203) and (5.209). Indeed, $\nabla(a) = \partial - ig\mathrm{ad}\, a$, where $\mathrm{ad}\, a = [a,\ ]$ is the adjoint operator acting in X. The operator ∂ is diagonal in the algebra space, and its action does not change periods of functions, i.e. $\mathcal{F}_n^{H,\alpha}$

are its invariant spaces. Obviously, $\mathrm{ad}\, a\mathcal{F}_n^H = 0$ and $\mathrm{ad}\, a\mathcal{F}_n^\alpha = \mathcal{F}_n^\alpha$ if $(\alpha, a) \neq 0$ in accordance with the Cartan–Weyl basis commutation relations (8.12). Therefore an action of the operator $\nabla(a)$ on $\mathcal{F} \ominus \mathcal{F}_0^H$ is given by an infinite-dimensional, block-diagonal matrix. In the real orthogonal basis defined by (8.14), its blocks have the form,

$$\nabla_n^H(a) \equiv \nabla(a)|_{\mathcal{F}_n^H} = \partial|_{\mathcal{F}_n^H} = \left(\otimes \frac{n}{l}\varepsilon\right)^r, \quad n \neq 0, \quad r = \mathrm{rank}\, X, \tag{5.210}$$

$$\nabla_0^\alpha(a) \equiv \nabla(a)|_{\mathcal{F}_0^\alpha} = -ig\, \mathrm{ad}\, a|_{\mathcal{F}_0^\alpha} = g(a, \alpha)\varepsilon, \tag{5.211}$$

$$\nabla_n^\alpha(a) \equiv \nabla(a)|_{\mathcal{F}_n^\alpha} = I_2 \otimes \frac{n}{l}\varepsilon + g(a, \alpha)\varepsilon \otimes I_2, \tag{5.212}$$

where ε is a 2×2 totally antisymmetric matrix, $\varepsilon_{ij} = -\varepsilon_{ji}, \varepsilon_{12} = 1$, and I_2 is the 2×2 unit matrix. In (5.212) the first components in the tensor products correspond to the algebra indices, while the second ones determine the action of $\nabla(a)$ on the functional basis $\sin(xn/l)$, $\cos(nx/l)$. The vertical bars at the operators in Eqs. (5.210)–(5.212) mean restriction of the corresponding operator onto a specified finite-dimensional subspace of $\mathcal{F}$. An explicit matrix form of the restricted operator is easily obtained by applying ∂ to the Fourier basis, and the action of $\mathrm{ad}\, a$, $a \in H$, is computed by means of (8.12). Since $\varepsilon^2 = -1$, the Jacobian can be written in the form,

$$\begin{aligned} J^2(a) &= (2\pi l)^r \prod_{\alpha>0} \det\left(ig^{-1}\nabla_0^\alpha\right)^2 \prod_{n=1}^{\infty} \left[\det\left(ig^{-1}\nabla_n^H\right)^2 \prod_{\alpha>0} \det\left(ig^{-1}\nabla_n^\alpha\right)^2\right] \\ &= (2\pi l)^r \prod_{\alpha>0} (a, \alpha)^4 \prod_{n=1}^{\infty} \left[\left(\frac{n}{gl}\right)^{4r} \prod_{\alpha>0} \left(\frac{n^2}{g^2 l^2} - (a, \alpha)^2\right)^4\right]. \end{aligned} \tag{5.213}$$

Set $J(a) = C(l)\kappa^2(a)$ where the (divergent) factor $C(l)$ is independent of a so that

$$\kappa(a) = \prod_{\alpha>0} \left[\frac{\pi(a, \alpha)}{a_0} \prod_{n=1}^{\infty} \left(1 - \frac{(a, \alpha)^2}{a_0^2 n^2}\right)\right] = \prod_{\alpha>0} \sin \frac{\pi(a, \alpha)}{a_0}, \tag{5.214}$$

$$C(l) = (2\pi l)^{r/2} \left(\frac{a_0}{\pi}\right)^{N_+} \prod_{n=1}^{\infty} \left(n^2 a_0^2\right)^{r+2}, \tag{5.215}$$

where $a_0 = (gl)^{-1}$, the integer $N_+ = (N - r)/2$ is the number of positive roots in X; the last equality in (5.214) results from the product formula [132, p. 37]. The induced volume element is $da\kappa^2(a)$. It vanishes at the boundaries of the Weyl cell (at the boundaries of the physical configuration space in the parameterization considered) since $(a, \alpha)/a_0 \in Z$ for all $a \in \partial K_W^+$. Zeros of function $\kappa(a)$ extended to the whole Cartan subalgebra form the diagram $D(X)$.

5.8.4 Properties of the Jacobian

Here two properties of the function $\kappa(a)$ are established which are important for proving both the gauge invariance of eigenstates of the quantum Yang–Mills

Hamiltonian and the analog of the Chevalley theorem (the existence of a unique analytic gauge-invariant continuation of physical wave functions in K_W^+ to the whole configuration space $\mathcal{F}$).

Let $\hat{s}_\omega \in W$ be the reflection operator in the plane perpendicular to a simple root ω. Any element of W is a product of these generating elements of W. For example, for the group $SU(3)$, the Weyl group has six elements, 1, $\hat{s}_{\omega_1}$, $\hat{s}_{\omega_2}$, $\hat{s}_{\omega_1}\hat{s}_{\omega_2}$, $\hat{s}_{\omega_2}\hat{s}_{\omega_1}$, and $\hat{s}_{\omega_1}\hat{s}_{\omega_2}\hat{s}_{\omega_1} = \hat{s}_{\omega_2}\hat{s}_{\omega_1}\hat{s}_{\omega_2}$. Put $\det \hat{s} = 1$ if $\hat{s} \in W$ has an even number of reflections $\hat{s}_\omega$ and $\det \hat{s} = -1$ if this number is odd. Function $\kappa(a)$ may only change its sign under transformations from the affine Weyl group. The proof is based on $\kappa(a)$ being proportional to the Weyl determinant [259, p. 185],

$$
\begin{aligned}
(2i)^{N_+}\kappa(a) &= \prod_{\alpha>0}\left(e^{i\pi(a,\alpha)/a_0} - e^{-i\pi(a,\alpha)/a_0}\right) \\
&= \sum_{\hat{s}\in W} \det \hat{s} \exp\left[\frac{2\pi i}{a_0}(\hat{s}\rho, a)\right],
\end{aligned} \tag{5.216}
$$

where the element ρ is a half-sum of all positive roots,

$$
\rho = \frac{1}{2}\sum_{\alpha>0}\alpha. \tag{5.217}
$$

From (5.202) and (5.216) it follows that

$$
\begin{aligned}
&(2i)^{N_+}\kappa(\hat{s}_{\beta,n}a) \\
&= \sum_{\hat{s}\in W} \det \hat{s} \exp\left[\frac{2\pi i}{a_0}(\hat{s}\rho, \hat{s}_\beta a)\right] \exp\left[\frac{4\pi i n_\beta}{(\beta,\beta)}(\hat{s}\rho, \beta)\right] \\
&= \det \hat{s}_\beta \sum_{\hat{s}\in W} \det \hat{s} \exp\left[\frac{2\pi i}{a_0}(\hat{s}\rho, a)\right] \exp\left[-\frac{4\pi i n_\beta}{(\beta,\beta)}(\hat{s}\rho, \beta)\right],
\end{aligned} \tag{5.218}
$$

where the sum over the Weyl group has been rearranged by the change $\hat{s} \to \hat{s}_\beta\hat{s}$, and the properties $\hat{s}_\beta^2 = 1$ and $\hat{s}_\beta\beta = -\beta$ are used. Next, the second exponential in (5.218) is proved to be 1 for any β and $\hat{s}$. To this end, note that $(\hat{s}\rho, \beta) = (\rho, \beta')$, where $\beta' = \hat{s}^T\beta$ is also a root that has the same norm as β because the Weyl group preserves the root pattern. Therefore it is sufficient to prove that

$$
n_\rho(\beta) = \frac{2(\rho,\beta)}{(\beta,\beta)}
$$

is an integer. The half-sum of the positive roots has the following properties [215, p. 461],

$$
\frac{2(\omega,\rho)}{(\omega,\omega)} = 1, \tag{5.219}
$$

$$
\hat{s}_\omega\rho = \rho - \omega, \tag{5.220}
$$

for any simple root ω. Since the Weyl group W preserves the root system and the reflection $\hat{s}_\beta$ in the hyperplane $(\beta, a) = 0$ is a composition of reflections $\hat{s}_\omega$, there exists an element $\hat{s} \in W$ and a simple root ω_β such that $\hat{s}\omega_\beta = \beta$. The statement that $n_\rho(\beta)$ is an integer follows from the relation

$$n_\rho(\beta) = \frac{2(\beta, \rho)}{(\beta, \beta)} = \frac{2(\omega_\beta, \hat{s}^T \rho)}{(\omega_\beta, \omega_\beta)} \in Z. \tag{5.221}$$

Indeed, representing $\hat{s}^T$ as a product of the generating elements $\hat{s}_\omega$ and applying (5.219) and (5.220), Eq.(5.221) follows because $2(\omega_\beta, \alpha)/(\omega_\beta, \omega_\beta)$ is an integer for any root α. Recall that a root α can be decomposed into a sum over simple roots with integer-valued coefficients, and the Cartan matrix $2(\omega, \omega')/(\omega, \omega)$ is also integer valued.

Thus, the simple property is established,

$$\kappa(\hat{s}_{\beta,n} a) = (\det \hat{s}_\beta)\, \kappa(a) = -\kappa(a) \tag{5.222}$$

for any root β. Since any element of the affine Weyl group W_A is a composition of the reflections (5.202), the following relation holds:

$$\kappa(\hat{s}a) = (\det \hat{s})\, \kappa(a) = \pm\kappa(a)\,, \qquad \hat{s} \in W_A, \tag{5.223}$$

where by definition $\det \hat{s} = -1$ if $\hat{s}$ contains an odd number of reflections (5.202) and $\det \hat{s} = 1$ for an even number. The Jacobian $\mu = \kappa^2$ is invariant under the affine Weyl group transformations.

The function $\kappa(a)$ is an eigenfunction of the r-dimensional Laplace operator,

$$(\partial_a, \partial_a)\kappa(a) \equiv \Delta_{(r)}\kappa(a) = -\frac{4\pi^2(\rho, \rho)}{a_0^2}\, \kappa(a) = -\frac{\pi^2 N}{6a_0^2}\, \kappa(a), \tag{5.224}$$

where the relation $(\rho, \rho) = N/24$ [21] between the norm of ρ and the dimension N of the Lie algebra has been used. A straightforward calculation of the action of the Laplace operator on $\kappa(a)$ leads to the equality

$$\begin{aligned} \Delta_{(r)}\kappa(a) = &-\frac{4\pi^2}{a_0^2}(\rho, \rho)\, \kappa(a) \\ &+\frac{\pi^2}{a_0^2} \sum_{\alpha \neq \beta > 0} (\alpha, \beta) \left[1 + \cot\frac{\pi(a, \alpha)}{a_0} \cot\frac{\pi(a, \alpha)}{a_0}\right] \kappa(a). \end{aligned} \tag{5.225}$$

The sum over positive roots in (5.225) can be transformed into a sum over the roots $\alpha \neq \beta$ in a plane $P_{\alpha\beta}$ and the sum over all planes $P_{\alpha\beta}$. Each plane contains at least two positive roots. Relation (5.224) follows from

$$\sum_{0<\alpha\neq\beta\in P_{\alpha\beta}} (\alpha,\beta)\,[1+\cot(b,\alpha)\cot(b,\beta)] = 0, \tag{5.226}$$

for any $b \in H$. To prove this, note that the root pattern in each plane coincides with one of the root patterns for Lie algebras of groups of rank 2, $SU(3)$, $Sp(4) \sim SO(5)$ and G_2, because the absolute value of the cosine of the angle between any two roots α and β may take only four values $|\cos\theta_{\alpha\beta}| = 0,\ 1/\sqrt{2},\ 1/2,\ \sqrt{3}/2$, according to the Dynkin diagrams (see Section 8.1). For the groups of rank 2, equality (5.226) can be verified by an explicit calculation. For example, in the case of $SU(3)$, the sum (5.226) is proportional to

$$-\cot b_1 \cot b_2 + \cot b_1 \cot(b_1+b_2) + \cot b_2 \cot(b_1+b_2) + 1 = 0,$$

where $b_{1,2} = (b,\omega_{1,2})$, and ω_1,ω_2 and $\omega_1+\omega_2$ constitute all positive roots of $SU(3)$.

5.8.5 The spectrum of the quantum Yang–Mills theory on a cylinder

Here the spectrum of physical excitations in the 2D Yang–Mills theory is found. It has a mass gap, i.e. there is a finite distance between the energies of the first excited state and the ground state. In fact, the whole spectrum turns out to be discrete, which is a direct consequence of the compactness of the physical configuration space. The problem is solved in the coordinate representation [240]. Following the Dirac quantization method, the canonical variables are promoted to the operators $E(x) \to -i\hbar\delta/\delta A(x)$ and $A(x) \to A(x)$, $A(x) \in \mathcal{F}$, while the physical states are anihilated by the constraint operator. This leads to the quantum field theory in a Schrödinger functional representation [260, 261],

$$\hat{H}\Phi_n[A] = -\frac{\hbar^2}{2}\left\langle \frac{\delta}{\delta A}, \frac{\delta}{\delta A}\right\rangle \Phi_n[A] = E_n\Phi_n[A], \tag{5.227}$$

$$\hat{\sigma}\Phi_n[A] = -i\hbar\nabla(A)\frac{\delta}{\delta A}\Phi_n[A] = 0. \tag{5.228}$$

The states are *functionals* on the space $\mathcal{F}$. To solve the constraint equation and project the Hamiltonian operator onto the physical subspace, the new coordinates (5.204) are used in which the orbit space is parameterized by homogeneous connections a from $K_W^+ \sim \mathcal{F}/\mathcal{G}$, as proved above.

All solutions of the Gauss law (5.228) are given by wave functions independent of Ω. Indeed, put $\delta w = i\Omega^{-1}\delta\Omega \in \mathcal{F} \ominus \mathcal{F}_0^H$. Then the Gauss law can be written in the form,

$$\hat{\sigma}\Phi_n[a,\omega] = -i\hbar g\hat{\Omega}^T\frac{\delta}{\delta w}\Phi_n[a,w] = 0; \tag{5.229}$$

here $\hat{\Omega}$ is a linear transformation in X defined by $(\hat{\Omega}y)_a = (\Omega y\Omega^{-1})_a = (y\hat{\Omega}^T)_a \equiv \hat{\Omega}_{ab}y_b$, $\hat{\Omega}^T\hat{\Omega} = \hat{\Omega}\hat{\Omega}^T = 1$ for any element $y \in X$. Since $\det\hat{\Omega} \neq 0$, the physical states are functionals independent of $w(x)$. To prove (5.229), note first that the following relations can be deduced from (5.205):

$$\delta a \equiv da = \mathcal{P}_0^H \hat{\Omega}^T \delta A \,, \tag{5.230}$$

$$\delta w = -g\nabla^{-1}(a)\left(1-\mathcal{P}_0^H\right)\hat{\Omega}^T \delta A, \tag{5.231}$$

with $\mathcal{P}_0^H$ being a projector on the subspace $\mathcal{F}_0^H$ of spatially homogeneous functions taking their values in the Cartan subalgebra. Now recall that operator $\nabla(a)$ is invertible on $\left(1-\mathcal{P}_0^H\right)\mathcal{F}$. Then a simple computation leads to the desired result,

$$\nabla(A)\frac{\delta}{\delta A} = \nabla(A)\left[\left(\frac{\delta a}{\delta A}, \frac{\partial}{\partial a}\right)_a + \left\langle \frac{\delta w}{\delta A}, \frac{\delta}{\delta w}\right\rangle_w\right] \tag{5.232}$$

$$= \nabla(A)\left[\left(\mathcal{P}_0^H\hat{\Omega}^T\right)^T \frac{\partial}{\partial a} + \left(-g\nabla^{-1}(a)\left(1-\mathcal{P}_0^H\right)\hat{\Omega}^T\right)^T \frac{\delta}{\delta w}\right] \tag{5.233}$$

$$= \hat{\Omega}\nabla(a)\mathcal{P}_0^H\frac{\partial}{\partial a} + g\hat{\Omega}\nabla(a)\left(1-\mathcal{P}_0^H\right)\nabla^{-1}(a)\frac{\delta}{\delta w}$$

$$= g\hat{\Omega}^T\frac{\delta}{\delta w}. \tag{5.234}$$

In (5.232), the subscript of the scalar product brackets denotes variables over whose indices the scalar product is taken, i.e. all indices labeling independent degrees of freedom described by $A(x)$ (the Lie algebra ones and $x \in S^1$) in the scalar products in (5.232) are left free, e.g., if λ_i is an orthonormal basis in X and $w(y) = w_i(y)\lambda_i$, then $\langle \delta w/\delta A, \delta/\delta w\rangle_w \equiv \int dy[\delta w_i(y)/\delta A(x)]\delta/\delta w_i(y)$. Equality (5.233) is obtained by substitution of $\delta a/\delta A(x)$ and $\delta w(y)/\delta A(x)$ which are taken from (5.230) and (5.231), respectively. Finally, Eq. (5.234) follows from the identities $\nabla(a)\mathcal{P}_0^H\partial/\partial a \equiv 0$ and $\nabla(A)\hat{\Omega} = \hat{\Omega}\nabla(a)$.

The norm of the physical states is defined according to the rule (5.207) because they are independent of $w(x)$,

$$\int_{\mathcal{F}} \prod_{x\in S^1} dA(x)\Phi_n^*[A]\Phi_{n'}[A] \to \int_{K_W^+} da\kappa^2(a)\Phi_n^*(a)\Phi_{n'}(a) = \delta_{nn'}, \tag{5.235}$$

where the infinite constant $C(l) = \int_{\mathcal{G}/G_H}\prod_x dw(x)$ is removed by renormalization of the physical states. The renormalization is indicated by the arrow in (5.235). Integration over the non-physical variables w yields an infinite factor, thus making the physical states non-normalizable in the total Hilbert space, even though the gauge orbits are compact. The origin of the divergence is the infinite number of non-physical degrees of freedom. Actually, at each space point x non-physical degrees of freedom contribute a finite factor, proportional to the volume of G, to the norm of a physical state. As one might see from (5.215), one way to get around this difficulty is to make the number of Fourier modes finite, renormalize the physical states, and then remove the regularization. Another possibility is to go over to the corresponding lattice theory in which the spatial variable takes its values in some partition of S^1, thus making the number of degrees of freedom finite. The lattice regularization is then removed after renormalization of the physical states. The lattice Yang–Mills theory will be discussed in detail in Section 5.8.10.

To project the functional Laplace operator in (5.227) onto the physical subspace, one should calculate the Laplace–Beltrami operator in the new functional variables (5.204) and omit all terms containing the variational derivative $\delta/\delta w$. The metric (5.206) is block diagonal. The physical and non-physical parts of the kinetic energy operator are decoupled. Hence, $\langle \delta/\delta A, \delta/\delta A\rangle \sim \kappa^{-2}(\partial_a, \kappa^2\partial_a)$ because the metric on the physical space is flat: $g^{aa} = (2\pi l)^{-1}$. Making use of (5.224), the Schrödinger equation for the physical states can be written in the form,

$$\hat{H}_{ph}\Phi_n(a) = \left[-\frac{\hbar^2}{4\pi l}\frac{1}{\kappa(a)}\Delta_{(r)} \circ \kappa(a) - E_C\right]\Phi_n(a) = E_n\Phi_n(a), \tag{5.236}$$

$$E_C = -\frac{\hbar^2}{4\pi l}\frac{\Delta_{(r)}\kappa}{\kappa} = \frac{\pi\hbar^2}{a_0^2 l}(\rho,\rho) = \frac{\pi\hbar^2 N}{24a_0^2 l}. \tag{5.237}$$

Substitution of $\Phi_n = \kappa^{-1}\phi_n$ into (5.236) shows that ϕ_n is an r-dimensional plane wave, $\phi_n \sim \exp[2\pi i(\gamma_n, a)/a_0]$, where a dimensionless vector $\gamma_n \in H$ is proportional to the momentum in the mechanical model described by (5.236). However, not every vector $\gamma_n \in H$ is admissible because only regular solutions to (5.236) have a physical meaning. The regularity condition requires that the functions $\phi_n(a)$ should vanish wherever $\kappa(a)$ vanishes, i.e. on the hyperplanes orthogonal to positive roots, $(\alpha, a) = n_\alpha a_0$, where n_α is an integer. Since $(\hat{s}\gamma_n, \hat{s}\gamma_n) = (\gamma_n, \gamma_n)$ for any $\hat{s} \in W$, the superposition of the plane waves,

$$\Phi_n(a) \sim [\kappa(a)]^{-1}\sum_{\hat{s}\in W}(\det\hat{s})\exp\left\{\frac{2\pi i}{a_0}(\hat{s}\gamma_n, a)\right\} \equiv [\kappa(a)]^{-1}\phi_n(a) \tag{5.238}$$

is an eigenstate of the physical Hamiltonian with the eigenvalue

$$E_n = \frac{\pi\hbar^2}{a_0^2 l}\left[(\gamma_n,\gamma_n) - (\rho,\rho)\right]. \tag{5.239}$$

The function (5.238) is proved to be regular at the hyperplanes $(a,\alpha) = n_\alpha a_0$, provided the momentum γ_n attains *discrete* values such that the number

$$\frac{2(\gamma_n,\beta)}{(\beta,\beta)} \in Z \tag{5.240}$$

is an integer for any root β.

To prove the regularity condition (5.240) for the functions (5.238), the variable a is decomposed into the sum $a = a^{||} + a^{\perp}$, where $(a^{\perp}, \alpha) = 0$ for a root α. Let $W^{(\alpha)} = W/Z_2^{(\alpha)}$, where $Z_2^{(\alpha)}$ has two elements, the identity transformation and the reflection $\hat{s}_\alpha = \{1, \hat{s}_\alpha\}$ such that $\hat{s}_\alpha\alpha = -\alpha$ and, therefore, $\hat{s}_\alpha a^{\perp} = a^{\perp}$ and

$\hat{s}_\alpha a^{||} = -a^{||}$. Clearly, $\det \hat{s}_\alpha = -1$. By making use of the identities $\det(\hat{s}_\alpha \hat{s}) = -\det \hat{s}$ and

$$(\hat{s}_\alpha \hat{s}\gamma_n, a) = \left(\hat{s}\gamma_n, a^\perp - a^{||}\right) = (\hat{s}\gamma_n, a) - 2\left(\hat{s}\gamma_n, a^{||}\right) , \tag{5.241}$$

the sum in (5.238) can be rewritten as follows:

$$\begin{aligned}\phi_n(a) \sim \sum_{\hat{s}\in W(\alpha)} &\left[\det \hat{s} \exp\left\{\frac{2\pi i}{a_0}(\hat{s}\gamma_n, a)\right\}\right.\\ &\left. + \det(\hat{s}_\alpha \hat{s}) \exp\left\{\frac{2\pi i}{a_0}(\hat{s}_\alpha \hat{s}\gamma_n, a)\right\}\right]\\ = \sum_{\hat{s}\in W(\alpha)} &\det \hat{s} \exp\left\{\frac{2\pi i}{a_0}(\hat{s}\gamma_n, a)\right\}\left[1 - \exp\left\{-\frac{4\pi i}{a_0}(\hat{s}\gamma_n, a^{||})\right\}\right].\end{aligned}$$

In the neighborhood of hyperplane $(a, \alpha) = n_\alpha a_0$ with $n_\alpha \neq 0$, $a^{||} = n_\alpha a_0 \alpha/(\alpha, \alpha) + \epsilon\alpha$ where $\epsilon \to 0$. The sum in the last equation vanishes if the factor in the square brackets vanishes in the limit $\epsilon \to 0$. This yields the condition that $2(\hat{s}\gamma_n, \alpha)/(\alpha, \alpha)$ must be an integer. Since $\hat{s}\alpha = \beta$ is a root, the function (5.238) is regular, provided the momentum γ_n satisfies the condition (5.240).

It should be noted that for some γ_n, the sum over the Weyl group in (5.238) may actually vanish. As is proved in the next section this happens whenever $(\gamma_n, \gamma_n) < (\rho, \rho)$. So the spectrum (5.239) is non-negative. The ground state corresponds to $\gamma_n = \rho$, as follows from the comparison of (5.221) with (5.240).

Thus, the *regularity* condition has a significant effect on the physical spectrum: it turns to be *discrete*, rather than continuous. The structure of the spectrum may already be anticipated from the classical dynamics. After removing the nonphysical degrees of freedom by imposing the Coulomb gauge $\partial A = 0$, the system describes a free particle in R^r. However, the residual gauge symmetry demands confining the particle into the Weyl cell K_W^+. Indeed, the trajectory of a free particle is a straight line which goes through some Weyl cells in H separated by the diagram $D(X)$. The motion in any Weyl cell is gauge equivalent to that in K_W^+. So, the particle keeps bouncing from the boundary ∂K_W^+. It is well known that a particle in a compact region of R^r has a discrete spectrum. There is a difference though with a particle in a box. The reflection from the boundary ∂K_W^+ is not caused by any force (no potential). The momentum has no jump which is easy to see when the motion is viewed in the physical phase space constructed above. The momenta right before the particle hits the boundary and right after that are gauge equivalent and, hence, the phase-space trajectory remains continuous in contrast to the case of a real particle in the well potential.

The latter difference also reveals itself in the fact that the wave functions (5.238) are invariant under the reflections in the planes forming the boundary ∂K_W^+. In fact, they are invariant under the affine Weyl group,

$$\Phi_n(\hat{s}_{\alpha,m} a) = [\kappa(\hat{s}_\alpha a)]^{-1} \phi_n(\hat{s}_\alpha a) = \Phi_n(a) , \quad \hat{s}_{\alpha,m} \in W_A. \tag{5.242}$$

This is a simple consequence of the property (5.222) of the function $\kappa(a)$. The wave function of a particle in the actual K^+_W well potential would not be invariant under such transformation. It would change its sign because of the zero boundary conditions it must satisfy like $\kappa(a)$.

5.8.6 The Chevalley theorem for the 2D Yang–Mills theory

The physical Hamiltonian is self-adjoint and, hence, the functions $\Phi_n(a)$ form a complete set of square integrable functions in K^+_W in the space with the measure κ^2. Any function in H that is regular and invariant under the affine Weyl group can be expanded into a Fourier series over the basis Φ_n. Any such function has a unique analytic gauge-invariant continuation into the total configuration space $\mathcal{F}$. This is the analog of the Chevalley theorem for the physical states in the 2D Yang–Mills theory. Its converse is obviously true. Namely, the reduction of a regular gauge-invariant function in $\mathcal{F}$ onto $\mathcal{F}_0^H$ is certainly regular and invariant under the residual gauge transformations forming the affine Weyl group (as is shown in the preceding section).

It is sufficient to prove that there is a unique analytic gauge-invariant continuation of $\Phi_n(a)$ to $\mathcal{F}$. To do so, the Weyl formula for the characters χ_{Λ_n} of irreducible representations of G [262, p. 909] is invoked together with the relation (5.216) to represent the eigenfunctions in the form,

$$\Phi_n(a) = c_n \frac{\sum\limits_{\hat{s}\in W} (\det \hat{s}) \exp\left\{\frac{2\pi i}{a_0}(\rho + \Lambda_n, \hat{s}a)\right\}}{\sum\limits_{\hat{s}\in W} (\det \hat{s}) \exp\left\{\frac{2\pi i}{a_0}(\rho, \hat{s}a)\right\}} \equiv c_n \chi_{\Lambda_n}\left(e^{2\pi i a/a_0}\right), \tag{5.243}$$

where c_n are normalization constants, $\gamma_n = \rho + \Lambda_n$, and the lattice $\Lambda_n \in K^+$ labels irreducible representations of G. A few remarks about (5.243) are in order. The sum over the Weyl group in (5.238) should vanish for all γ_n such that $(\gamma_n, \gamma_n) < (\rho, \rho)$ because function (5.238) must be regular, which is, in turn, only possible if $(\gamma_n, \gamma_n) \geq (\rho, \rho)$, as one can see from the explicit form (5.216) of function $\kappa(a)$. This condition on the norm of γ_n ensures that spectrum (5.239) is non-negative. If γ_n satisfies (5.240), then so does $\hat{s}\gamma_n$, $\hat{s} \in W$, but the latter does not correspond to an independent Φ_n. So, one can always restrict γ_n to be in the Weyl chamber K^+. According to (5.221), $\gamma_n = \rho \in K^+$ satisfies the regularity condition and the corresponding $\Phi_n = c_n$ as follows from (5.243). It is the ground state ($E_n = 0$). So the excited states correspond to $\gamma_n = \rho + \Lambda_n \in K^+$, where the lattice Λ_n is such that both (5.240) and $(\gamma_n, \gamma_n) \geq (\rho, \rho)$ hold. The lattice with such properties is the lattice that labels irreducible representations of G and the solutions of (5.240) can indeed be written in the form,

$$\gamma_n = \rho + \Lambda_n .$$

The character χ_{Λ_n} has a unique gauge-invariant analytic continuation into $\mathcal{F}$,

$$\chi_{\Lambda_n}\left(\exp\frac{2\pi i a}{a_0}\right) = \mathrm{Tr}\,[\exp 2\pi i g l a]_{\Lambda_n} = \mathrm{Tr}\left[\mathrm{P}\exp ig\oint_{S^1} A dx\right]_{\Lambda_n}, \tag{5.244}$$

where $(e^y)_{\Lambda_n}$ denotes the group element e^y in the irreducible representation Λ_n. The last equality in (5.244) follows from the fact that variable a is related to a generic connection $A(x)$ by a gauge transformation. Formula (5.244) establishes the gauge-invariant analytic continuation of the eigenstates (5.243) to the total configuration space. By construction, the trace of the holonomy elements $\mathrm{P}\exp(ig\oint_{S^1} A dx)$ in all irreducible representations of G are independent gauge-invariant states in the 2D Yang–Mills theory. They are in a one-to-one correspondence with the independent states in the reduced (gauge fixed, $\partial A = 0$) theory that are invariant under the residual discrete gauge transformations.

The wave functions (5.243) are orthogonal with respect to the scalar product (5.235). This follows from the orthogonality of the characters (5.244). For normalization coefficients c_n we obtain,

$$\begin{aligned}\delta_{nn'} &= \int_{K_W^+} da\kappa^2(a)\Phi_n(a)\Phi_n^*(a) \qquad (5.245)\\ &= 2^{2N_+}c_n c_{n'}^* \int_{K_W^+} da \sum_{\hat{s},\hat{s}'\in W} \det(\hat{s}\hat{s}')\exp\left\{\frac{2\pi i}{a_0}(a, \hat{s}\gamma_n - \hat{s}'\gamma_{n'})\right\}.\end{aligned}$$

The integrand in (5.245) is a periodic function on the Cartan subalgebra. Its periods are determined by the geometry of the Weyl cell. Therefore, the integral over the periods vanishes for all $\hat{s} \neq \hat{s}'$, because γ_n and $\gamma_{n'}$ belong to the Weyl chamber and the Weyl group acts simply and transitively on the set of Weyl chambers. Hence, there is no Weyl group element $\hat{s}$ such that $\hat{s}\gamma_n = \gamma_{n'}$ if $\gamma_{n,n'} \in K^+$. For $\hat{s} = \hat{s}'$ the integral differs from zero only for $\gamma_n = \gamma_{n'}$, i.e. when the periodic exponential equals one. Thus,

$$|c_n| = 2^{-N_+}(N_W V_{K_W^+})^{-1/2}, \tag{5.246}$$

where N_W is the number of elements in the Weyl group, $V_{K_W^+}$ is the volume of the Weyl cell.

The energy spectrum (5.239) seems to depend on normalization of the roots in the Lie algebra. Recall, however, that the norms of the roots are fixed by the structure constants in the Cartan–Weyl basis (see Section 8.1.2 for details). If the roots are rescaled by a factor c, which means, in fact, rescaling the structure constants in the Cartan–Weyl basis by a factor c^{-1}, the invariant scalar product $(x, y) = \mathrm{Tr}\,(\mathrm{ad}x, \mathrm{ad}y)$ gets rescaled accordingly, i.e. by c^{-2}. Therefore spectrum (5.239) does not depend on the rescaling factor because factor c^{-2} in the scalar product is canceled against c^2 resulting from rescaling γ_n and ρ by c.

Remark. Non-physical degrees of freedom can be eliminated in the theory by means of the Coulomb gauge prior to quantization. When quantizing the gauge-fixed theory, the ambiguity arises because some boundary conditions at the boundary of the modular domain must be assumed [236]. The choice of boundary conditions is not unique and depends on self-adjoint extensions of the Laplace operator in the Weyl cell. The spectrum also depends on the self-adjoint extension and generally differs from (5.239). The model can also be solved without any explicit parameterization of the gauge orbit space via a gauge fixing. A simple analysis of the lattice version of the model shows that the holonomy group element is the only variable of which all gauge-invariant quantities can depend [234]. So the Schrödinger equation can be written in the holonomy variable and solved [235] (compare also [237]). The results [234, 235, 237] coincide with those presented above thanks to the analog of the Chevalley theorem. The proof of the latter requires specific symmetry conditions for the gauge-fixed physical wave functions which, in turn, stems from the regularity condition at ∂K_W^+. The regularity condition corresponds to a particular self-adjoint extension of the gauge-fixed Hamiltonian. So, in the Dirac approach, no ambiguity arises and the explicit gauge invariance can always be uniquely restored. This is the key difference with the approach based on quantum mechanics on orbifolds [263], where wave functions are, generally, allowed to have singularities at the singular points of the configuration space.

5.8.7 Homotopically nontrivial residual gauge transformations

There are two types of solutions to Eq. (5.194) which determines the residual gauge transformations, homotopically trivial or nontrivial [264]. The latter have been excluded when computing the physical phase space because they cannot be generated by the constraints, that is, two classical states related by homotopically nontrivial transformations are, in fact, two different physical states and, hence, correspond to two different points in the physical phase space. The same is true in quantum theory. Physical wave functions are *not* invariant under homotopically nontrivial residual gauge transformations. The analogy can be made with instanton physics [253, 265] in Yang–Mills theory. An instanton is a classical solution of equations of motion in the Euclidean spacetime which connects two distinct classical minima of the Yang–Mills Hamiltonian related by a homotopically nontrivial gauge transformation. Physical wave functions acquire a phase factor under such a transformation.

To simplify the technicalities, first consider the group $SU(2)$. It has only one positive root ω and, hence, $\rho = \omega/2$. Solutions to (5.240) are given by $\gamma_n = \omega n/2$, where n ranges over *positive* integers. The spectrum and wave functions respectively read,

$$E_n = \frac{\pi\hbar^2}{4a_0^2 l}\,(n^2-1)(\omega,\omega), \quad n = 1,2,...; \tag{5.247}$$

$$\Phi_n = c_n \frac{\sin\left[\pi n(a,\omega)/a_0\right]}{\sin\left[\pi(a,\omega)/a_0\right]}. \tag{5.248}$$

By substituting $n = 2j+1$, $j = 0, 1/2, 1, ...$, into (5.247) E_n is shown to be proportional to eigenvalues of the quadratic Casimir operator of $SU(2)$; $E_n \sim j(j+1)$ where j labels the irreducible representations of $SU(2)$. Define a new variable θ such that $a = a_0\omega\theta/(\omega,\omega)$. When a ranges the Weyl cell K_W^+, the variable θ spans the open interval $(0,1)$. The measure da denotes the Riemann measure in $R^r \sim H$. For the $SU(2)$ case, $da \equiv da_3$, $a = \sqrt{2}\omega a_3$ so that $(a,a) = a_3^2$, $a_3 \in R$, because $(\omega,\omega) = 1/2$ for $SU(2)$. Hence, the normalization coefficients c_n in (5.248) are

$$c_n = \left(\sqrt{2}a_0 \int_0^1 d\theta \sin^2 \pi n\theta\right)^{-1/2} = \left(\frac{a_0}{\sqrt{2}}\right)^{-1/2} . \tag{5.249}$$

The action of homotopically nontrivial elements (5.182) on the argument of wave functions are determined by the shifts $a \to a + i/g\Omega_s\partial\Omega_s^{-1}$, where $\Omega_s = \exp(ix\eta/l)$ and (compare (5.198)),

$$\exp(2\pi i\eta) = z \in Z_G. \tag{5.250}$$

The lattice η is given by integral linear combinations of elements $\alpha/(\alpha,\alpha)$, with α ranging over the root system, because [215]

$$\exp\frac{2\pi i\alpha}{(\alpha,\alpha)} \in Z_G. \tag{5.251}$$

Thus, homotopically nontrivial gauge transformations are generated by the shifts (compare (5.199)),

$$a \to a + \frac{m\alpha a_0}{(\alpha,\alpha)}, \quad m \in Z . \tag{5.252}$$

If they are included into the set of residual gauge transformations, then the modular domain K_W^+ becomes further reduced. In the case of $SU(2)$, the interval K_W^+ becomes twice smaller. The case of $SU(3)$ is shown in Fig. 5.6. The dark triangle inside K_W^+ is the reduced modular domain if homotopically nontrivial transformations are included into the gauge group.

In the matrix representation of $SU(2)$ the only positive root is $\omega = \tau_3/4$ (see Section 8.1.2). Then $\exp(2\pi i\omega/(\omega,\omega)) = \exp(i\pi\tau_3) = -e \in Z_2 = Z_G$ as stated in (5.251). Therefore the wave functions (5.248) are transformed as

$$\Phi_n\left(a + \frac{a_0\omega m}{(\omega,\omega)}\right) = (-1)^{m(n+1)}\Phi_n(a), \tag{5.253}$$

i.e. the physical states acquire a phase factor under homotopically nontrivial gauge transformations.

The analysis can easily be extended to an arbitrary group by using the properties of the root pattern and the Weyl representation of the characters (5.243) under the transformations (5.252). However the use of the gauge-invariant wave functions (5.244) in the total configuration space $\mathcal{F}$ leads to the answer faster. The holonomy group element $U(A) = P\exp(ig\oint Adx)$ transforms as

$U(A^\Omega) = \Omega(0)U(A)\Omega^{-1}(2\pi l)$. A homotopically nontrivial group element $\Omega(x)$ satisfies the twisted periodicity condition (5.182). Therefore,

$$\mathrm{Tr}\left(\mathrm{P}\exp\left[ig\oint A^\Omega dx\right]\right)_{\Lambda_n} = \mathrm{Tr}\left(z\mathrm{P}\exp\left[ig\oint Adx\right]\right)_{\Lambda_n}. \tag{5.254}$$

Thus, the Gauss law (5.228) only provides for the invariance of physical states with respect to gauge transformations which can be continuously deformed towards the identity.

5.8.8 The orbit space as a manifold. Removing reducible connections

The geometry of the orbit space in classical Yang–Mills theories has been studied by Singer [211]. In his analysis, the so-called reducible connections [251], which have a nontrivial stabilizer (or a stationary group, $A^\Omega = A$, $\Omega \neq 1$) in the gauge group, play a special role (a more detailed discussion is given in Section 6.7.3). The reason is that, if there are no reducible connections, then the orbit space is a manifold. This observation can be illustrated with an example of 2D topological Yang–Mills theory. The orbit space is found to be an orbifold with trivial topology (the Weyl cell), and constant gauge transformations are permitted. The latter implies that every constant connection A has a nontrivial stabilizer. Put, for example, $\Omega = \exp(ig\lambda A)$, then $A^\Omega = A$ for any real λ. If the gauge group $\mathcal{G}$ is restricted by excluding constant gauge transformations, then there are no reducible connections, and the orbit space $\mathcal{F}/\mathcal{G}$ coincides with the group manifold G [227, 235]. The group manifold is compact (for compact G) and has a nontrivial topology. One should keep in mind, however, that such a truncation of the gauge group cannot be done in the Lagrangian, and is added to the theory by hand as a supplementary condition on the gauge group. The rationale behind such a restriction comes from boundary conditions usually imposed on gauge group elements at the spatial infinity in four-dimensional Yang–Mills theory whose Lagrangian is given by (5.176) in which $\mu, \nu = 0, 1, 2, 3$, and the integration is carried out over $x \in R^3$. Constant gauge transformations are excluded if one demands that $\Omega(x) \to 1$ as $|x| \to \infty$, i.e. gauge group elements are mappings $\Omega : S^3 \to G$. In other words, the space R^3 is compactified into the three-sphere S^3 by identifying all points at the spatial infinity. This compactification looks rather natural if only densities of physical (gauge-invariant) quantities, like the energy density, are considered because they should vanish at spatial infinity. However, a priori there seems to be no physical reason to eliminate constant gauge transformations under which the Lagrangian is invariant.

To understand the assertion $\mathcal{F}/\mathcal{G} = G$, note first that the elimination of constant gauge transformations implies that Eq. (5.192) has to be excluded from the set of constraints of the theory. The physical degrees of freedom are described by constant connections $A = A_0 \in \mathcal{F}_0$ (compare (5.203)) taking their values in the whole Lie algebra X. The residual gauge transformation satisfies

Eq. (5.195) where a is replaced by A_0. Its solutions are easy to find by means of the decomposition (5.60), $A_0 = \Omega_0 a \Omega_0^{-1}$, where $a \in H$, $\Omega \in G/G_H$, and G_H is the Cartan subgroup of G. Since $\nabla(A_0) = \Omega_0 \nabla(a)\Omega_0^{-1}$, the residual gauge transformations are

$$A_0 \to A_0 + a_0 \sum_{\alpha>0} n_\alpha \Omega_0 \eta_\alpha \Omega_0^{-1}, \qquad n_\alpha \in Z, \tag{5.255}$$

where $\eta_\alpha = 2\alpha/(\alpha,\alpha)$ (compare (5.196)). Group G can be viewed as the set of all adjoint orbits of the Cartan subgroup G_H, i.e. $G = \Omega G_H \Omega^{-1}$ where Ω ranges over G/G_H [21]. Let K_T be an open convex polyhedron in H that contains the zero element $a = 0$ and is bounded by the planes $(a,\alpha) = \pm a_0$, i.e. $a \in K_T$ if for any positive root α, $-a_0 < (a,\alpha) < a_0$. For example, for $G = SU(3)$, K_T is shown in Fig. 5.6. It is the region bounded by the hexagon $B_1B_2...B_6$. For $A_0 = a$ ($\Omega_0 = 1$), the region K_T contains no configurations related by translations (5.255) that form the group T_e. Define the exponential map $X \to \exp(X)$ as $\exp(2\pi i A_0/a_0) \in G$ for any $A_0 \in X$. Since the exponential map $\exp(H)$ is invariant under the action of T_e in H, it can be concluded that $G_H = \exp(K_T) \cup \exp(\partial K_T)$ where ∂K_T is the boundary of K_T. The polyhedron ∂K_T consists of faces which are portions of the hyperplanes $(a,\alpha) = \pm a_0$. Let $\mathcal{P}_\alpha$ denote the face corresponding to a root α. For instance, for $SU(2)$, $\mathcal{P}_{\pm\omega}$ are the points $a = \pm a_0\omega/(\omega,\omega)$. For the group $SU(3)$ (see Fig. 5.6), $\mathcal{P}_{\pm\alpha}$ are parallel sides of the hexagon K_T, e.g. if $\alpha = \pm\omega_1$, then $\mathcal{P}_\alpha = (B_2B_3)$ and $\mathcal{P}_{-\alpha} = (B_6B_5)$. Any $a \in \mathcal{P}_\alpha$ can be decomposed as $a = a_\alpha + a'$ where $a' \in K_T$ and $a_\alpha = a_0\alpha/(\alpha,\alpha)$. The exponential map of a_α, $\alpha > 0$, exhausts all elements of the center Z_G of the group G (see (5.251)). Thus, $G_H/Z_G = \exp(K_T)$ and Z_G is given by the exponential map of special elements $a_\alpha \in \partial K_T$, $\alpha > 0$. Since $A_0 = \Omega a \Omega^{-1}$, the orbit space indeed coincides with G (the exponential map $\exp(X)$ is invariant under the residual gauge transformations (5.255)).

The modular domain $X_G = \Omega K_T \Omega^{-1} \subset X$, with Ω ranging over G, has trivial topology. The nontrivial topology of the orbit space occurs through identifying gauge-equivalent configurations in the boundary of X_G. There are two types of such identifications that should be noted.

First, let $\mathcal{P}_\alpha^X$ be the portion of the boundary ∂X_G obtained from $\mathcal{P}_\alpha$ by the adjoint action of all elements of G on it, so that ∂X_G is the union of $\mathcal{P}_\alpha^X$ over all roots. According to (5.255) connections in $\mathcal{P}_\alpha^X$ are gauge equivalent to those in $\mathcal{P}_{-\alpha}^X$. Thus, the sets $\mathcal{P}_{\pm\alpha}^X$ must be glued together. In particular, $\mathcal{P}_{\pm\alpha}$ may be identified with one another so that the topology of the maximal torus emerges, $(\times S^1)^r \sim G_H$, and then the adjoint action of G is carried out. For example, if $G = SU(3)$, the opposite sides of the hexagon $B_1...B_6$ in Fig. 5.6 are glued together to obtain $G_H \sim S^1 \times S^1$. This can be seen through a simple cut-and-glue procedure. Cut the hexagon along the line (B_4B_1), then glue (B_6B_1) with (B_3B_4), and (B_1B_2) with (B_5B_4) (matching the specified orientation of the segments) to obtain a tube. By gluing the edges of the cut back together and the remaining opposite sides, (B_5B_6) with (B_3B_2), of the hexagon, the torus $S^1 \times S^1$ is obtained (the procedure requires stretching the tube). The other group of rank 2 can be treated similarly. For groups of higher ranks, one should note that, in addition to identifying the opposite faces of K_T, the opposite sides of each face must also be identified. The latter procedure is topologically the same as for K_T of groups whose rank is less by 1.

Second, there are gauge equivalent configurations in $\mathcal{P}_\alpha^X$. For instance, the exponential map of the connection $A_1 = a_0\alpha/(\alpha,\alpha)$ in $\mathcal{P}_\alpha^X$ is the group center element z_α in G_H whose adjoint orbit is trivial, $\Omega z_\alpha \Omega^{-1} = z_\alpha$. Therefore the connections A_1 and $A_2 = \Omega A_1 \Omega^{-1} \in \partial\mathcal{P}_\alpha^X$ are gauge equivalent for any $\Omega \in G$, i.e. they are related by a homotopically trivial gauge transformation. Indeed, the connections $A_{1,2}$ are obtained from the vacuum configuration $A = 0$ by homotopically nontrivial gauge transformations with the group element $\Omega_{1,2}(x) = \exp(igA_{1,2}x)$ (see Section 5.8.7). Hence, they are related by the gauge transformation $\Omega_s(x) = \Omega_1(x)\Omega_2^{-1}(x)$ which is homotopically trivial, $\Omega_s(x+2\pi l) = z_\alpha z_\alpha^{-1}\Omega_s(x) = \Omega_s(x)$. In particular, for $G = SU(2)$ the adjoint orbits of elements in K_T are concentric spheres so that X_G is a ball of radius $a_0/(\omega,\omega)^{1/2}$ (ω is the only simple root of $SU(2)$). All connections in the boundary sphere $(A, A) = a_0^2/(\omega,\omega)$ are gauge equivalent to a single one $A = a_0\omega/(\omega,\omega)$. The group manifold $SU(2)$ is obtained by contracting the boundary sphere to a single point which corresponds to the only nontrivial element of the group center Z_2.

So, if constant gauge transformations are excluded, then the quantum theory of physical degrees of freedom in the $2D$ Yang–Mills theory describes a quantum particle on a group manifold.

5.8.9 The orbit space in 4D Yang–Mills theories

In 4D Yang–Mills theory ($\mu,\nu = 0,1,2,3$ in (5.176) and integration with respect to x is extended to R^3), the total configuration space is an affine space of all connections $A_i(x)$ in R^3. For classical (smooth) connections, the orbit space can be analyzed by means of the Morse theory [319–321]. A detailed description of this procedure is postponed to Section 6.7.3 where the path integral formalism on the orbit space is discussed. In particular, it is shown in Section 6.7.3 that the orbit space can be parameterized by transverse connections, i.e. satisfying the Coulomb gauge $\partial_i A_i = 0$. This gauge is incomplete because there are residual gauge transformations (Gribov copies [209]). The Gribov horizon is the set of transverse connections for which the lowest eigenvalue of the Faddeev–Popov operator $\Delta_{FP} = -\partial_i \nabla_i(A)$ vanishes. If constant gauge transformations are excluded from the gauge group, then the boundary of the modular domain K has common points with the Gribov horizon. It is a convex subset, and its interior contains the zero configuration $A_i = 0$. As any such subset of an affine space, K has a trivial topology. A nontrivial topology of the orbit space emerges through identifying gauge equivalent configurations within the boundary of K. It can be shown that the orbit space contains non-contractable spheres of any dimension [211]. All these features are naturally present in the studied 2D Yang–Mills theory.

Indeed, when constant gauge transformations are excluded, then all constant connections $A \in X$, $\partial A = 0$, can be used to parameterize the orbit space (see Section 5.8.8). The Faddeev–Popov determinant $\det(-\partial\nabla(A)) \sim \det(-i\nabla(A))$ (here $-\partial\nabla(A) = (-i\partial)(-i\nabla(A))$ and $\det(-i\partial)$ is factored out) can be calculated by the same method as in Section 5.8.3. For any A there is a group element Ω and a Cartan subalgebra element a such that $A = \Omega a \Omega^{-1}$. Since Ω is spatially homogeneous, $\det(-i\nabla(A)) = \det(-i\nabla(a))$, but the functional space in which the operator $-i\nabla(a)$ acts must be reduced by eliminating the subspace of constant functions because constant gauge transformations are not allowed. This amounts to discarding the block (5.211) in $-i\nabla(a)$. Removing the factor corresponding to this block in (5.213) (the factor $\prod_{\alpha>0}\det(ig^{-1}\nabla_0^\alpha)$), one infers that

$$\det\Delta_{FP}(A) = \det\Delta_{FP}(a) \sim \prod_{\alpha>0}\frac{\sin^2(\pi(a,\alpha)/a_0)}{(\pi(a,\alpha)/a_0)^2}. \tag{5.256}$$

Since the right-hand side of Eq. (5.256) is invariant under the Weyl group, it has a unique invariant analytic extension into the whole algebra X according to the Chevalley theorem (see Sections 5.4.5 and 5.4.6). It is then straightforward to verify that this extension (or $\det\Delta_{FP}(A)$) is proportional to the Haar measure [215] on the group manifold (an explicit form of the Haar measure, which is suitable for this algebraic exercise, is given in Section 8.1.1). The determinant $\det\Delta_{FP}(a)$ appears regular on hyperplanes through the origin $a = 0$ and orthogonal to positive roots. So the Gribov horizon consists of such constant connections A which, when transformed to the Cartan subalgebra by the adjoint action of the group, satisfy the condition $(a,\alpha) = \pm a_0$, $A = \Omega a\Omega^{-1}$, and $a \in H$. Thus, the modular domain is $X_G \subset X$, introduced in Section 5.8.8. It has also been shown in Section 5.8.8 that the boundary ∂X_G contains gauge equivalent configurations by whose identification the topology of the group manifold is obtained. The orbit space is isomorphic to the group manifold. In Section 6.7.3 the path integral formalism on the orbit space is developed by which it can be shown that the metric on the orbit space indeed coincides with that of the group manifold. Thus, the physical degrees of freedom in the theory describe a dynamics on the group manifold.

Remark. The eigenfunctions of the Faddeev–Popov operator are determined by the eigenvectors of the blocks (5.212) because $-i\partial$ and $-i\nabla(a)$ commute. So, in each subspace $\mathcal{F}_n^\alpha$, the eigenvalue λ of Δ_{FP} is given by $\lambda = l^{-2}n(n \pm (a,\alpha)/a_0)$, where $a_0 = (gl)^{-1}$ and $n = 1, 2, \ldots$. The lowest eigenvalue vanishes when $(a,\alpha) = \pm a_0$, which defines the Gribov horizon. The Faddeev–Popov operator is strictly positive in the interior of the modular domain ($|(a,\alpha)| < a_0$ for any root α), which also holds in the 4D case.

The configuration space in quantum field theory is much larger than the space of analytic functions (see, e.g. [233]). Smooth classical functions form a subset of zero measure in Sobolev functional spaces used to describe configurations of quantum fields [233]. The Sobolev space $\mathcal{S}_k^p$, where $1 \leq p < \infty, k = 0, 1, 2, ...$, consists of fields all of whose derivatives up to and including order k have integrable pth power. The smaller the indices p and k the larger the space of the fields. The result of Singer on the absence of a global continuous gauge fixing for smooth classical field configurations can be extended to the Sobolev space [226, 273], provided

$$p(k+1) > n, \tag{5.257}$$

where n is the dimension of the base (spacetime) manifold. Since the gauge transformation law of the connection involves the derivatives of the group elements, the latter must have one derivative more than the connections, i.e. they must be from the Sobolev space (of the group valued functions) $\mathcal{S}_{k+1}^p$. Only under condition (5.257) does the gauge group possess the structure of a Lie group and acts smoothly on the space of connections $\mathcal{S}_k^p$ [274]. The condition (5.257) is crucial for continuity of gauge transformations as functions of a point of the base manifold. For instance [226], the function $|x|^{-\epsilon}$ is singular at the origin ($\epsilon > 0$), but the pth power of its kth derivative is integrable if $(n-1)-(\epsilon+k)p > -1$, which holds for any $\epsilon < 1$, provided $p(k+1) > n$. If $p(k+1) = n$ ($p \neq 1$), then there may exist a singularity $(-\ln|x|)^{1-1/p-\epsilon}$. Thus, the necessity of the condition (5.257) for continuity is clear. The condition (5.257) ensures also the existence of a local gauge fixing and the structure of the principal fiber bundle in the configuration space [275].

If $p(k+1) < n$, the very notion of gauge fixing becomes meaningless [226]. Ordinary conditions like the Coulomb gauge will not be any gauge fixing even locally. Indeed, consider the transformation $A_i(x) \to A_i^\lambda(x) = \lambda A_i(\lambda x)$ of connection in R^n. The Coulomb gauge condition $\partial_i A_i = 0$ implies that $\partial_i A_i^\lambda = 0$. Next, consider a transverse connection A and its Gribov copy. For sufficiently large λ, their λ-transformations become arbitrary close to the zero field in the sense of the topology of $\mathcal{S}_k^p$, while they remain transverse. This follows from the scaling property of the distance between the zero connection and A_i^λ

$$\begin{aligned} \left\|\partial_{j_1}\cdots\partial_{j_k}A^\lambda\right\|_p &\equiv \left\{\int d^n x\,\left(\partial_{j_1}\cdots\partial_{j_k}A^\lambda, \partial_{j_1}\cdots\partial_{j_k}A^\lambda\right)^{p/2}\right\}^{1/p} \\ &= \lambda^{k+1-n/p}\left\|\partial_{j_1}\cdots\partial_{j_k}A\right\|_p . \end{aligned} \tag{5.258}$$

The right-hand side of Eq. (5.258) tends to zero as $\lambda \to \infty$ if $p(k+1) < n$. Thus, any neighborhood of the zero field in the set of transverse connections contains Gribov copies. The noncompactness of the base is not important here because both A and its copy can be taken near the vacuum configuration.

In the Sobolev space of connections satisfying (5.257) there is an improved version of the theorem of Singer which is due to Soloviev [226]. It asserts that the gauge orbit fiber bundle in non-Abelian field theory does not admit reduction to a finite-dimensional Lie group. In other words, there is no gauge condition that would fix the gauge arbitrariness globally modulo some finite *subgroup* of the gauge group. Observe that in all models discussed so far, one can always find a gauge condition that removes the gauge arbitrariness completely up to a discrete *subgroup* of the gauge group. In contrast, in the Yang–Mills theory the residual gauge symmetry in any gauge would not form a finite subgroup of the gauge group. Soloviev's result gives the most precise characterization of the problem of eliminating non-physical degrees of freedom in the Yang–Mills theory.

This problem is essential for the path integral formalism in Yang–Mills theory because a class of fields on which the functional integral measure has support depends on the field model in question (compare the Minlos–Backner theorem [276]). The property of continuity discussed above is decisive for Singer's analysis, while quantum fields in four dimensions would typically have the singularity "$|x|^{-1}$" almost everywhere [233]. Note that the topological 2D Yang–Mills theory has no propagating degrees of freedom and the analysis carried out in the previous sections holds for the quantum theory as well. In 4D spacetime, the theory should be regularized first in some gauge-invariant way in order to develop a quantum theory on the orbit space. A lattice regularization seems natural for this purpose.

5.8.10 Kogut–Susskind lattice gauge theory

A finite spatial lattice regularization turns the Yang–Mills theory into a mechanical system with a gauge symmetry and a finite number of degrees of freedom (the time is not discretized). Non-physical degrees of freedom in this theory can be eliminated by means of a gauge fixing so that residual discrete gauge transformation form a group (in contrast to the continuum case (compare Soloviev's theorem stated in the preceding section)). This group acts in the phase space of physical degrees of freedom. The latter allows us to calculate the physical phase space as the quotient space (5.1). However, the continuum limit remains an open question.

Let points of the three-dimensional cubic lattice be designated by three-vectors with integer components, denoted x or y. The lattice is finite and periodic in every spatial direction (three-torus). The total configuration space is formed by the link variables $u_{xy} = u_{yx}^{-1} \in G$, where $y = x + ae_k$, e_k is the unit vector in the direction of the kth coordinate axis, and a is the grid spacing. For brevity, $x + ae_k \equiv x + k$ in what follows. If $A_k(x)$ is the (Lie algebra-valued) vector potential at the site x, then

$$u_{xy}(t) \equiv u_{x,k}(t) = e^{igaA_k(x,t)}.$$

The dynamics is defined by the Lagrangian,

$$L = \tfrac{a}{2} \sum_{x,k} \left(ig^{-1}\dot{u}_{x,k}u_{x,k}^{-1} - A_x + u_{x,k}A_{x+k}u_{x,k}^{-1} \right)^2 - V(u), \tag{5.259}$$

where $A_x \in X$ is the time component of the Yang–Mills vector potential A_μ at site x, $A_0(x,t) \equiv A_x(t)$, $V(u) = (2/ag^2)\sum_p \mathrm{Re}\chi(1-u_p)$, u_p is the product of the link variables around the plaquette p, χ is the character of the group G in the adjoint representation (in matrix representation, $\chi(1-u_p) = \mathrm{Tr}\,(1-u_p)$), and the square in the first term of (5.259) is understood in the sense of the Cartan–Killing scalar product in X. In the formal continuum limit, $a \to 0$, the first term in (5.259) yields $(1/2)\int F_{0k}^2 dx$ (the color electric field energy) and the potential $V(u)$ produces the color magnetic field energy, $(1/4)\int F_{jk}^2 dx$. The Lagrangian is invariant under the gauge transformations

$$u_{x,k} \to \Omega_x u_{x,k}\Omega_{x+k}^{-1}, \quad A_x \to \Omega_x A_x \Omega_x^{-1} + ig^{-1}\dot{\Omega}_x\Omega_x^{-1},$$

where $\Omega_x \in G$. The gauge group is $\mathcal{G} = \prod_x(\times G)$.

The Hamiltonian formalism is straightforward to deduce. The canonical momenta are defined by $\pi_x = \partial L/\dot{A}_x = 0$ (the primary constraint) and $\pi_{x,k} = \partial L/\partial \dot{u}_{x,k}$. It is more convenient to introduce non-canonical momentum variables which take their values in the Lie algebra,

$$E_{x,k} = iu_{x,k}\pi_{x,k} = \textstyle\sum_b \lambda_b E_{x,k}^b,$$

where $\{\lambda_b\}$ is the orthogonal basis set in X (associated with the Cartan–Weyl basis (see Section 8.1)) so that $[\lambda_a, \lambda_b] = iF_{ab}^c\lambda_c$. In contrast to the canonical momenta $\pi_{x,k}$, the variables $E_{x,k}$ are transformed locally at each site under the gauge transformations

$$E_{x,k} \to \Omega_x E_{x,k}\Omega_x^{-1}. \tag{5.260}$$

The Poisson bracket of the real components $E_{x,k}^a$ is

$$\{E_{x,k}^a, E_{x',k'}^b\} = \delta_{xx'}\delta_{kk'}F_c^{ab}E_{x,k}^c. \tag{5.261}$$

In the continuum limit, $E_{x,k}/a^2$ is finite and coincides with the color electric field which is canonically conjugated to the kth spatial Yang–Mills connection. This is easy to see by multiplying both sides of (5.261) by a^{-4} and using the scaling rule for the Dirac δ-function, $\delta_{xx'} \sim a^3\delta(x-x')$, in three-dimensional space. The Poisson bracket for the color electric fields $E_{x,k}/a^2$ vanishes as $O(a)$ in the continuum limit ($a \to 0$), and the Hamiltonian formalism of the Yang–Mills theory is recovered. The Hamiltonian and the secondary constraint (the Gauss law) have the form,

$$H = \tfrac{g^2}{2a}\sum_{x,k}\left(E_{x,k}^b\right)^2 + V(u), \tag{5.262}$$

$$\sigma_x = \sum_k \left(E_{x,k} - u_{x-k,k}^{-1}E_{x-k,k}u_{x-k,k}\right) = 0. \tag{5.263}$$

The kinetic energy in the Hamiltonian (5.262) is nothing but the sum of kinetic energies of rotators (free particles on the group manifold) for each x and k

according to (5.261). The free motion of each rotator is determined by the values of r independent Casimir operators $d^{(j)}_{b_1 b_2 \ldots b_j} E^{b_1}_{x,k} E^{b_2}_{x,k} \cdots E^{b_j}_{x,k}$, where $d_{b_1 b_2 \ldots b_j}$ are irreducible invariant symmetric tensors of G and j ($j = r_j$) ranges over their degrees listed in Table 8.1 of Section 8.1. Note that $d^{(2)}_{b_1 b_2} = \delta_{b_1 b_2}$ is an invariant tensor for every G. The corresponding Casimir operator is the kinetic energy of the rotator. By construction, the Poisson bracket of Casimir operators among each other vanishes. So they are integrals of motion of a free rotator. Constraint (5.263) imposes no restriction on allowed values of the Casimir operators (they are gauge invariant) of the rotators. The potential $V(u)$ describes a coupling of the rotators so that the Casimir operators are no longer the integrals of motion. The same picture holds in quantum theory when $E^b_{x,k}$ are promoted to operators with the commutation relation according to the canonical quantization rule, $[\cdot\,,\cdot] = i\hbar\{\cdot\,,\cdot\}$ (see (5.261)).

In a non-invariant approach, one can, for instance, bring the momentum variables $E_{x,1}$ to the Cartan subalgebra H, $E_{x,1} = E^H_{x,1}$, by a suitable gauge transformation (5.260). This condition exhausts all the non-Abelian gauge arbitrariness modulo the discrete Weyl group acting at each site x independently. However, in the continuum limit, the condition $E_{x,1} = E^H_{x,1}$ is not invariant under the spatial rotations. On the other hand, the $SO(3)$ group is a global symmetry group of the continuum theory, and it might be desired to preserve it when eliminating the non-physical degrees of freedom. This can be achieved by means of the following (polar) decomposition:

$$E_{x,k} = \sum_{n=1}^{3} U_{x,kn}\mathcal{E}_{x,n}, \quad (\mathcal{E}_{x,n}, \mathcal{E}_{x,n'}) = 0, \quad \text{if} \quad n \neq n',$$

where $(\cdot\,,\cdot)$ denotes the Cartan–Killing form in the Lie algebra. The 3×3 matrix U_x is orthogonal, $UU^T = U^T U = 1$. This representation follows from the observation that the 3×3 symmetric matrix $(E_{x,k}, E_{x,k'})$ can be brought to a diagonal matrix $(\mathcal{E}_{x,n}, \mathcal{E}_{x,n'})$ by an orthogonal transformation. Under a spatial rotation $E_{x,k} \to \Omega_{lk} E_{x,k}$ described by an orthogonal matrix Ω_{lk}, the matrices $U_{x,kn}$ are changed as $\Omega_{lk} U_{x,kn}$. Hence, the momentum variables $\mathcal{E}_{x,n}$ are invariant under the spatial rotations and obey the gauge transformation law (5.260). So any gauge condition imposed on $\mathcal{E}_{x,n}$ is invariant under the spatial rotations in the continuum limit. In particular, $\mathcal{E}_{x,1} = \mathcal{E}^H_{x,1}$. Although the continuous non-Abelian gauge transformation have been exhausted, there are discrete non-Abelian gauge transformations that form the Weyl group at each lattice site x. In order to calculate the action of the Weyl group on all physical variables $\mathcal{E}_{x,i}$, the Weyl group elements, being reflections in the plane orthogonal to the roots, should be realized as the adjoint action of a group element on a root α,

$$w_\alpha h w_\alpha^{-1} = h - 2(h,\alpha)\alpha/(\alpha,\alpha)$$

for any $h \in H$, where $w_\alpha \in G/G_H$. Making use of the real Cartan–Weyl basis (8.14), a solution of this equation is

$$w_\alpha = \exp\left[i\pi(\alpha,\alpha)^{-1}s_\alpha\right]. \tag{5.264}$$

Every other solution is obtained from it by the adjoint action of the Cartan subgroup G_H, because the decomposition (5.60) is unique if $h \in K^+$ and the Weyl group acts simply transitively on the set of Weyl chambers. Thus, modulo the Abelian gauge transformations, the residual non-Abelian gauge transformations in the gauge $\mathcal{E}_{x,1} = \mathcal{E}^H_{x,1}$ are generated by group elements (5.264) acting independently at each lattice site, i.e. they form the group $\prod_x(\times W)$.

The continuous Abelian gauge freedom that still remains is the Cartan subgroup of the gauge group, $\mathcal{G}_H = \prod_x(\times G_H)$. The momentum variable $\mathcal{E}_{x,k}$ can be expanded in the real Cartan–Weyl basis (8.14),

$$\mathcal{E}_{x,k} = \mathcal{E}^H_{x,k} + \sum_{\alpha>0}\left(\mathcal{E}^{c\alpha}_{x,k}c_\alpha + \mathcal{E}^{s\alpha}_{x,k}s_\alpha\right).$$

Under residual Abelian transformations (5.260) with $\Omega_x = e^{ih_x}$, where $h_x \in H$, the Cartan component $\mathcal{E}^H_{x,k}$ is invariant, while the non-Cartan components associated with each positive root α are rotated,

$$\begin{aligned}
\mathcal{E}^{c\alpha}_{x,k} &\to \cos(\alpha,h_x)\mathcal{E}^{c\alpha}_{x,k} + \sin(\alpha,h_x)\mathcal{E}^{s\alpha}_{x,k},\\
\mathcal{E}^{s\alpha}_{x,k} &\to \cos(\alpha,h_x)\mathcal{E}^{s\alpha}_{x,k} - \sin(\alpha,h_x)\mathcal{E}^{c\alpha}_{x,k}.
\end{aligned}$$

This follows from the commutation relations (8.12) and the definition (8.14). By suitable choice of h_x, the components $\mathcal{E}^{s\omega}_{x,2}$, where ω is a simple root, can be nullified. Indeed, put $\tan\varphi^\omega_x \equiv -\mathcal{E}^{c\omega}_{x,2}/\mathcal{E}^{s\omega}_{x,2}$. Then the sought-for h_x satisfies a system of $r = \dim H$ linear equations,

$$(\omega, h_x) = \varphi^\omega_x.$$

The gauge variable can be expanded in the basis of simple roots, $h_x = \sum_\omega 2h^\omega_x\omega/(\omega,\omega)$. The coefficients satisfy the equation $\sum_{\omega'} C_{\omega\omega'}h^{\omega'}_x = \varphi^\omega_x$, where $C_{\omega\omega'} = 2(\omega,\omega')/(\omega,\omega)$ is the Cartan matrix which is not singular and, hence, h^ω_x are uniquely defined. Thus, the residual Abelian gauge freedom can be eliminated by the condition $\mathcal{E}^{s\omega}_{x,2} = 0$. There are residual discrete Abelian gauge transformations that form the group $(\times Z_2)^r$ at each site x independently. They can change the sign of the physical variables $\mathcal{E}^{c\omega}_{x,2}$ independently for each simple root and each site x. Given a simple root ω', consider $h_x = h_x(\omega')$ that satisfy the equations

$$(\omega', h_x) = \pi, \quad (\omega, h_x) = 0, \quad \omega \neq \omega',$$

where ω ranges over all simple roots save for ω'. For each ω' the system has a unique solution according the above analysis. Under the gauge transformation (5.260) with $\Omega_x = \exp(ih_x)$, the condition $\mathcal{E}^{s\omega}_{x,2} = 0$ is preserved, while $\mathcal{E}^{c\omega}_{x,2}$ changes its sign if $\omega = \omega'$, otherwise the sign is not changed.

The continuous gauge freedom is exhausted. The residual discrete gauge transformations form the group $\mathcal{S} = \prod_x W \times (\times Z_2)^r$. The Hamiltonian dynamics in the physical phase space can be formulated in the non-invariant approach, i.e. by solving the constraints for the non-physical coordinates in $u_{x,k}$ canonically conjugated to the eliminated momenta. When studying the dynamics the physical phase space structure must be taken into account. With the above choice of physical canonical variables, the phase space points related by transformations from $\mathcal{S}$ must be identified to obtain the physical phase space.

5.9 Simple effects of the physical phase space structure in quantum theory

In the preceding sections, it is shown that classical trajectories depend on phase space structure. Therefore it is natural to expect that the quasiclassical description of quantum dynamics is affected when the phase space structure changes.

In the quasiclassical approach, the spectrum of a quantum system is usually determined by means of the WKB quantization method for periodic classical solutions [277]. Another WKB problem of calculating transition amplitudes is usually solved by the stationary phase method for the path integral [277]. The analysis of this problem is postponed to Chapter 6 where the path integral formalism that takes into account the true structure of the physical phase space is developed.

Here other simpler methods are employed to illustrate the effects of the nontrivial phase space structure on the WKB approximation, quantization of wormhole sizes in quantum cosmology with gauge fields [278], and the pole structure of quantum Green's functions [212].

5.9.1 *The Wentzel–Kramers–Brillouin (WKB) method*

Consider a mechanical system with just one degree of freedom. Suppose that the dynamics is such that there are periodic solutions of the classical equations of motion (the corresponding phase space trajectories are closed curves). The period $T = T(E)$ is a function of the system energy E. The Bohr–Sommerfeld quantization rule asserts that the energy levels are determined by the equation

$$W(E) \equiv \oint pdq = \int_0^T p\dot{q}dt = 2\pi\hbar\left(n + \tfrac{1}{2}\right), \qquad n = 0, 1, \ldots, \tag{5.265}$$

where the integral is taken along the classical trajectory. The quantization rule (5.265) generally determines the spectrum in the quasiclassical approximation up to higher orders of $\hbar$ [279] (e.g. it does not account for tunneling effects). So this is not a full quantum theory. Nonetheless it turns out to be sufficient to qualitatively distinguish between the effects of a non-planar phase space structure and potential forces on quantum theory. Yet, it is explicitly independent of the choice of canonical variables in the phase space because the line integral $\oint pdq$ is invariant under canonical transformations. For a harmonic oscillator with a frequency ω, the WKB spectrum determined by (5.265) is exact. Indeed, the classical trajectory of the system with energy E is given by $q(t) = \left(\sqrt{2E}/\omega\right)\sin(\omega t)$ and $p(t) = \sqrt{2E}\cos(\omega t)$, where the period is $T = 2\pi/\omega$. Therefore $W(E) = 2\pi E/\omega$, and the spectrum reads $E_n = \hbar\omega(n + 1/2)$.

What changes when the phase space is $\Gamma = \text{cone}(\pi)$? If the dynamics is such that every periodic trajectory is invariant under the transformations $(q, p) \to (-q, -p)$ (e.g. the potential is an even function, $V(-q) = V(q)$ and has a minimum at $q = 0$), then the physical period T^c is twice less than that in the system with the standard phase space $\Gamma = R^2$ and the same energy E,

$$T^c(E) = \tfrac{1}{2}T(E). \tag{5.266}$$

The reason is simple (see Section 5.2). The conic phase space is obtained from the planar one by identifying the points (q, p) and $(-q, -p)$ and the time spent by the system in the region $q < 0$ is the same as that in the region $q > 0$. However the motions in these two regions are physically indistinguishable and, hence, (5.266) holds. Thus, for conic phase space, the quasiclassical quantization rule (5.265) should be modified as,

$$W^c(E) = \int_0^{T/2} p\dot{q}dt = \tfrac{1}{2}W(E) = 2\pi\hbar\left(n + \tfrac{1}{2}\right), \qquad n = 0, 1, \ldots, . \tag{5.267}$$

It follows from (5.267) and (5.265) that the physical energy levels for the system with the phase space $\Gamma = \text{cone}(\pi)$ correspond to only even values of n in (5.265). In particular, for the harmonic oscillator, one finds $E_n^c = 2\hbar\omega(n + 1/2)$. The distance between the energy levels is doubled as though the physical frequency of the oscillator were 2ω. Note that the frequency as the parameter of the Hamiltonian is not changed, and the conclusion holds for any choice of canonical variables. The entire effect is therefore due to the conic structure of the phase space. The ground state energy appears to be $E_0^c = \hbar\omega$ no matter from which gauge system the harmonic oscillator with the conic phase space cone(π) originates, while in the full quantum theory it depends on the gauge group, as shown in Section 5.2 (compare (5.40)). This is the consequence of the aforementioned deficiency of the quasiclassical quantization rule (5.265).

The conic phase space has a singularity at the origin. One may expect that the conic singularity of the phase space does not affect the motion localized in phase space regions which do not contain the origin. Such motion would be indistinguishable from the motion in the planar phase space. The simplest example of this kind is the harmonic oscillator whose equilibrium is not located at the origin [183]. Consider the gauge model studied in Section 5.1 where the potential has the form,

$$V = \frac{\omega^2}{2}\left(|\mathbf{x}| - r_0\right)^2 . \tag{5.268}$$

The motion is easy to analyze in the local gauge-invariant variables (p_r, r), when the cone is cut along the momentum axis. As long as the energy does not exceed a critical value $E_0 = \omega^2 r_0^2/2$, i.e. the oscillator cannot reach the origin $r = 0$, the

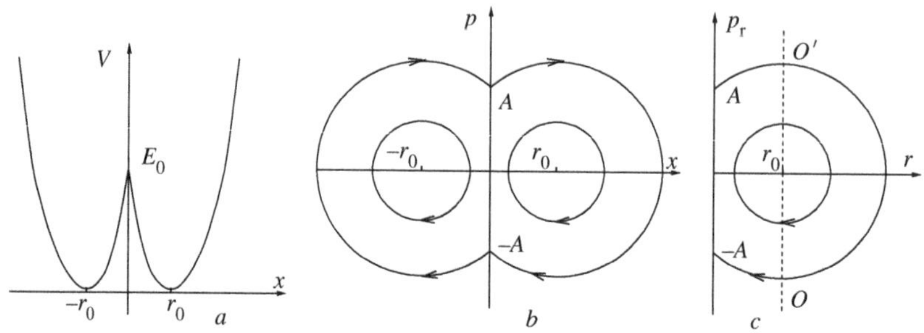

Fig. 5.8. *a*. Oscillator double-well potential.
b. Phase space trajectories in the flat phase space. For $E < E_0 = \omega^2 r_0^2/2$ there are two periodic trajectories associated with two minima of the double-well potential.
c. The same motion in the conic phase space. It is obtained from the corresponding motion in the flat phase space by identifying the points (p, x) with $(-p, -x)$. The local coordinates p_r and r are related to the parameterization of the cone when the cut is made along the momentum axis (the states A and $-A$ are physically identical). Adopted from [195].

period of classical trajectory remains $2\pi/\omega$. The WKB method yields the spectrum of the ordinary harmonic oscillator. However the gauge system differs from the corresponding system with the phase space being a full plane. As shown in Fig. 5.8 (panel b), the latter system has two periodic trajectories with the energy $E < E_0$ associated with two minima of the oscillator double-well potential. Therefore in quantum theory the low energy levels must be doubly degenerate. Due to the tunneling effect the degeneracy is removed. Instead of one degenerate level with $E < E_0$ there must be two close levels (meaning that $(E_0 - E)/E_0 \ll 1$). In contrast, there is no doubling of classical trajectories in the conic phase space (see Fig. 5.8 (panel c)), and no splitting of the energy levels should be expected. These qualitative arguments are made rigorous in the framework of the instanton calculus developed in Chapter 6.

When the energy is greater than E_0, the particle can go over the potential barrier. In the planar phase space there is only one trajectory with fixed energy E exceeding E_0. From the symmetry arguments it is also clear that this trajectory is mapped onto itself upon the reflection $(p, x) \to (-p, -x)$. Identifying these points of the planar phase space, the corresponding trajectory in the conic phase space with $E > E_0$ is obtained. It is also continuous and periodic. In Fig. 5.8 (panel c) the semi-axes $p_r < 0$ and $p_r > 0$ on the line $r = 0$ are identified in accordance with the chosen parameterization of the cone.

Suppose that the initial state of the gauge system is at the phase space point O in Fig. 5.8 (panel c), i.e. $r(0) = r_0$. Let t_A be the time when the system approaches the state $-A$. As soon as the system passes state $-A$, it emerges in the state A as if the momentum had suffered a sudden jump. But states A and $-A$ lie on the cut of the cone and, hence, correspond to the same state of the system. No jump of the physical momentum occurs at $t = t_A$ when the trajectory is viewed on the physical (conic) phase space. From the symmetry arguments it follows that the system returns to the initial state in the time

$$T_c = \frac{\pi}{\omega} + 2t_A. \tag{5.269}$$

It takes $t = 2t_A$ to go from state O to $-A$ and then from A to O'. From state O' the system reaches the initial state O in half of the period of the harmonic oscillator, π/ω. The time t_A depends on the energy of the system. The phase space trajectory of the oscillator with initial state at point O traveling toward the point $-A$ (as shown in Fig. 5.8 (panel c)) is given by the equations $r(t) = \sqrt{2E_0}/\omega - \left(\sqrt{2E}/\omega\right)\sin(\omega t)$ and $p_r(t) = \dot{r}(t) = -\sqrt{2E}\cos(\omega t)$, where E is the oscillator energy. If $E < E_0$, then $r(t) > 0$ for any t. When $E = E_0$, the oscillator reaches the origin at time $\pi/(2\omega)$, but still $r(t) \geq 0$ for any t. When $E > E_0$, the time t_A is found by solving the equation $r(t_A) = 0$ and is given by,

$$t_A = \frac{1}{\omega}\sin^{-1}\sqrt{\frac{E_0}{E}} \leq \frac{\pi}{2\omega}, \qquad E \geq E_0 .$$

The quasiclassical quantization rule yields the equation for energy levels,

$$\begin{aligned} W_c(E) &= W(E) - 2E\int_{t_A}^{\frac{\pi}{\omega}-t_A} \cos^2\omega t\, dt \\ &= W(E)\left(\frac{1}{2} + \frac{\omega t_A}{\pi} + \frac{1}{2\pi}\sin 2\omega t_A\right) = 2\pi\hbar\left(n + \frac{1}{2}\right). \end{aligned} \tag{5.270}$$

Here $W(E) = 2\pi E/\omega$ is the Bohr–Sommerfeld functional for the harmonic oscillator of frequency ω. The function $W_c(E)$ for the conic phase space is obtained by subtracting a contribution of the portion of the ordinary oscillator trajectory between the states $-A$ and A for negative values of the canonical coordinate, i.e. for $t \in [t_A, \pi/\omega - t_A]$.

Two limiting cases might be of interest. First, when the energy is sufficiently large, $E \gg E_0$, the time $2t_A$ is much smaller than the half-period π/ω. The classical oscillator returns to the initial state in $T_c \approx \pi/\omega$. So the actual frequency of oscillations is doubled (approximately equal to 2ω). This effect is reflected in the WKB quantization, $W_c(E) \sim \frac{1}{2}W(E)$, leading to a doubling of the distance between the energy levels. In this case typical fluctuations have the amplitude much larger than the distance from $r = r_0$ to the singular point of the phase space at $r = 0$, and the potential barrier at $r = 0$ has no significant effect. The second limiting case concerns energies small compared to E_0. The classical system never reaches the singular point, and the phase space structure has no effect on the dynamics; the latter is fully driven by the potential force so that the oscillator actual frequency coincides with ω. The WKB spectrum turns into the harmonic oscillator equidistant spectrum with spacing $\hbar\omega$. As soon as the energy attains the critical value E_0 the distance between energy levels starts growing, tending to its asymptotic value $\Delta E = 2\hbar\omega$. However, as noted, the quantization rule (5.265) (or (5.267)) does not take into account tunneling. The quantum system may penetrate into classically forbidden domains. The wave functions of the states with $E < E_0$ do not vanish under the potential barrier. So even for $E < E_0$ there are fluctuations that can reach the conic singularity of the phase space. As a result a small shift of the oscillator energy levels for $E \ll E_0$ occurs. The shift can be calculated by means of the instanton technique. It is easy to see that there should exist an instanton solution that starts at the classical vacuum $r = r_0$, goes to the origin, and then returns back to the initial state. Thus, even though in some regimes the classical dynamics may not be sensitive to phase space structure, in quantum theory the influence of the phase space geometry on dynamics may be well exposed. The effects of the phase space structure on instanton amplitudes and energy shifts due to tunneling are studied in Section 6.8.

The lessons to be learned from this simple qualitative consideration are that both the potential force and the phase space geometry affect the behavior of the classical gauge system. In some regimes the dynamics may be strongly affected by the phase-space structure. There might also exist dynamical regimes where the potential force mostly determines the evolution of the gauge system, and the phase-space structure has little influence on it. In contrast, in quantum mechanics, the dynamics appears to be more sensitive to the physical phase-space structure than in the case of the classical one because the system wave function penetrates into classically forbidden regions.

5.9.2 A quantum soliton with conic phase space

In quantum theory, solitons have internal excitations whose spectrum can be determined by the WKB method. If the field whose classical excitations admit solitons has a non-planar phase space, then the spectrum of internal excitations of the quantum soliton may be affected. To illustrate this, consider a simple Abelian gauge field theory in the $1+1$ spacetime whose Lagrangian density is

$$\mathcal{L}(x,t) = -\tfrac{1}{4}F_{\mu\nu}^2 + |(\partial_\mu + igA_\mu)\phi|^2 + \frac{m^4}{\lambda}\left[\cos\left(\frac{\sqrt{\lambda}|\phi|}{m}\right) - 1\right], \qquad (5.271)$$

where $F_{\mu\nu} = \partial_\mu A_\nu - \partial_\nu A_\mu$ is the electromagnetic field strength, $\mu,\nu = 0,1$, ϕ is a complex scalar field, and m, λ, and g are constants. The Lagrangian (5.271) is invariant under the gauge transformations $A_\mu \to A_\mu - \partial_\mu\omega$ and $\phi \to \exp(ig\omega)\phi$. The time component of the vector potential A_0 is the Lagrange multiplier in the theory (a pure non-physical degree of freedom).

The canonical momentum for A_1 is $E = F_{01}$, and $p = \partial_0\phi^* - igA_0\phi^*$ is canonically conjugated to ϕ. Then it is easy to verify that the constraint generating the gauge transformations of A_1 and ϕ is,

$$\partial_1 E + ig(p^*\phi - p\phi^*) = 0.$$

In the dynamical sector of the theory which has no excitation of the electromagnetic field $F_{\mu\nu} = 0$, the scalar field can have just one degree of freedom (at each spatial point) with the conic phase space, cone(π), which is similar to the mechanical model of Section 5.1.

In this sector the physical motion described by real solutions of the Hamiltonian equations of motions, $A_\mu = 0$, $\phi = \rho/\sqrt{2}$, where $\rho(x,t)$ is a real function and $p = \partial_t\phi$ is also real so that the constraint is trivially satisfied. The classical field equations for ρ coincide with those in the sine-Gordon model and, therefore, have soliton solutions [277]. In particular, consider the breather solution,

$$\rho = \rho_{br}(x,t) = \frac{4m}{\sqrt{\lambda}} \frac{\sin\left[mvt(1+v^2)^{-1/2}\right]}{v\cosh\left[mx(1+v^2)^{-1/2}\right]},$$

where v is a parameter of the solution. The breather is a periodic solution with period $T = 2\pi(1+v^2)^{1/2}/(mv)$. In quantum theory it is interpreted as a bound state of two solitons (orbiting about each other).

The energy levels of internal (rotational) excitations of the quantum breather in the ordinary sine-Gordon theory (no gauge symmetry) are determined by the WKB method [277] which goes as follows. First, the action of the breather solution is computed as a function of the period,

$$S_{br}(T) = \int_0^T dt \int_{-\infty}^{\infty} dx\, \mathcal{L}(x,t)\Big|_{breather} .$$

Then the period $T = T(E)$ as a function of the energy is found from the equation $E = -dS_{br}/dT$. Finally, the quantum energy levels are obtained from the Bohr–Sommerfeld quantization rule,

$$W(E) = S_{br}(T(E)) + ET(E) = 2\pi n, \quad n = 1, 2, \ldots . \tag{5.272}$$

Here and below $\hbar = 1$. In quantum field theory this rule differs slightly from that in quantum mechanics (5.265). In the right-hand side of Eq. (5.272) there is no contribution for vacuum oscillations. The point is that the number of degrees of freedom in field theory is infinite. Therefore the vacuum energy diverges and a renormalization procedure is required. The renormalized vacuum energy is included in the so-called renormalized action, i.e. $S_{br}(T) \to S_{br}^{ren}(T)$ in (5.272) (to renormalize the vacuum energy, necessary counter terms are introduced into the classical action). The details of this procedure are not relevant for what follows. For the breather in the sine-Gordon theory, they can be found in [277] where it is shown that solutions to Eq. (5.272) are,

$$E_n = 16m\gamma^{-1}\sin(n\gamma/16), \quad n = 1, 2, \ldots, < 8\pi\gamma^{-1},$$

where $\gamma = (\lambda/m^2)[1 - \lambda/(8\pi m^2)]$. For $n > 8\pi\gamma^{-1}$ the breather decays into a soliton and anti-soliton because its period becomes infinite, $T(E_n) \to \infty$ when n approaches the critical value $8\pi\gamma^{-1}$ from below. The appearance of the upper bound for n is the effect of renormalization (the breather's interaction with quantum fluctuations changes its physical properties, just like an electron acquires an effective mass in a solid due to its coupling with phonons) [277].

In the phase space (ρ, p_ρ) where $p_\rho = \partial_0\rho$, the trajectory of the breather solution at a fixed spatial point x is an ellipse centered at the origin. In the gauge model (5.271), the physical phase space of the scalar field is the cone cone(π), thanks to the constraint (just like in the mechanical model). So points (ρ, p_ρ) and $(-\rho, -p_\rho)$ correspond to the same physical state. The physical period of breather oscillations become twice less, $T^c = T/2$. All quantities related to the conic phase space carry the index "c". Accordingly, when computing the action S^c_{br} of the breather with the conic phase space as a function of its period T^c, the integration over time is restricted to the interval $[0, T^c]$. Since $\mathcal{L}$ is an even functional of ρ,

$$S^c_{br} = \tfrac{1}{2} S_{br}(2T^c).$$

Period T^c as a function of the energy E^c is defined similarly, $E^c = -dS^c_{br}(T^c)/dT^c$, which yields $T^c(E^c) = T(E^c)/2$. Finally, the Bohr–Sommerfeld quantization rule $W^c(E^c) = S^c_{br}(T^c) + E^c T^c(E^c) = 2\pi n$ leads to $(1/2)W(E^c) = 2\pi n$ and, hence,

$$E^c_n = E_{2n},$$

where E_{2n} are the energy levels of a quantum breather with the planar phase space. Thus, some of the internal excitations of the quantum breather are excluded due to the phase space reduction.

5.9.3 The phase space of Yang–Mills fields in cosmology

Consider the Einstein–Yang–Mills theory. The theory is complicated for a general analysis, but one can introduce a set of simplifying assumptions and study closed cosmologies with $R \times S^3$ topology. These are known as *minisuperspace* cosmological models [280] (the prefix "super" does not refer to supersymmetry here). They are used to study wormhole dynamics [281]. Wormholes are Riemannian manifolds which have two or more asymptotically Euclidean regions (a suitable analogy might be a surface that consists of two parallel planes with holes connected by a tube). They are believed to play an important role in quantum gravity [280, 282, 283]. It is known, however, that there is no wormhole solution of the Einstein equations in a vacuum [284–286]. The presence of matter changes the situation [284]. The case when only gauge fields are present is studied here. Gauge fields on a homogeneous space are described by the $SO(4)$-invariant Ansatz [281, 287, 288]. The reduced system contains only a finite number of degrees of freedom of gravitational and gauge fields. The effect of a non-standard physical phase space structure of gauge fields on the dynamics is studied here.

In the minisuperspace approach to the Einstein–Yang–Mills system, the prototype of a four-dimensional wormhole may be described by the $SO(4)$ symmetric metric. The most general form of such a metric, i.e. a metric which is spatially homogeneous and isotropic in the spacetime of $R \times S^3$ topology, is given by the Friedmann–Robertson–Walker Ansatz [289]–[291],

$$g_{\mu\nu}dx^\mu dx^\nu = \frac{2G_g}{3\pi}\left[-N^2(t)dt^2 + \rho^2(t)\theta^i\theta^i\right], \tag{5.273}$$

where $N(t)$ and $\rho(t)$ are arbitrary nonvanishing functions of time, G_g is the gravitational constant and θ^i are the left-invariant one-forms ($i = 1, 2, 3$) on the three-sphere S^3 satisfying the condition $d\theta^i = -\varepsilon_{ijk}\theta^j \wedge \theta^k$ (compare (8.6)). The Ansatz for gauge fields in the metric (5.273) is proposed in [281] for the group $SU(2)$ and generalized to an arbitrary group in [287, 288]. The degrees of freedom of gauge fields with the $SO(n)$ group, $n > 4$, are described by a scalar $z(t) \in R$, a vector $\mathbf{x}(t) \in R^l, l = n - 3$, and a real antisymmetric $l \times l$ matrix $y = y^a T_a$, with T_a being generators of $SO(l)$. The Einstein–Yang–Mills action for the remaining degrees of freedom reads,

$$S = \frac{1}{2}\int dt\, \frac{N}{\rho}\left\{-\left(\frac{\rho}{N}\dot{\rho}\right)^2 + \left(\frac{\rho}{N}\dot{z}\right)^2 + \left(\frac{\rho}{N}D_t\mathbf{x}\right)^2 - 2V\right\}, \tag{5.274}$$

where $D_t\mathbf{x} = \dot{\mathbf{x}} + y\mathbf{x}$ is the covariant derivative of $\mathbf{x}$. The potential has the form,

$$V = \frac{\alpha_g}{3\pi}\left[\left(z^2 + \mathbf{x}^2 - \frac{3\pi}{2\alpha_g}\right)^2 + 4z^2\mathbf{x}^2\right] - \frac{1}{2}\rho^2 + \frac{\lambda}{2}\rho^4, \tag{5.275}$$

with $\alpha_g = g^2/(4\pi)$ being the Yang–Mills coupling constant, $\lambda = 2G_g\Lambda/(9\pi)$, and Λ is the cosmological constant.

The action is invariant under the same gauge transformations of $\mathbf{x}$ and y as in the model (5.31) (the variable z is invariant) and also time reparameterizations,

$$t \to t(t'), \quad N(t) \to N(t')\frac{dt'}{dt}. \tag{5.276}$$

Therefore the analysis of the phase-space structure given in Section 5.1 applies here. The gauge fields in the action (5.274) have two physical degrees of freedom. As z is gauge invariant, it has a planar phase space, while the other physical degree of freedom $|\mathbf{x}|$ has a conic phase space, cone(π). A quantum theory can be developed similarly to Section 5.2. It is natural to expect that the semiclassical approximation, where wormhole solutions play the role of a stationary point, is changed due to the conic phase-space structure. However here only the classical effects caused by the conic structure of the physical phase space on the wormhole dynamics are studied, in particular, on wormhole size quantization [281]. Wormhole size quantization is first observed in [281] for Yang–Mills fields with the group $SU(2)$. In this case $SU(2) \sim SO(3)$, i.e. $l = 0$ in the minisuperspace model. So, the physical phase space is a plane. Gauge groups of higher rank are needed to see the effect of the conic physical phase space.

The wormholes are solutions to the equations of motion in the Euclidean spacetime ($t \to -i\tau, y \to iy$) for the action (5.274) with a particular behavior for $\rho(\tau)$: $\rho^2(\tau) \sim \tau^2$ as $\tau \to \pm\infty$. The simplest example of a wormhole is known as the Tolman wormhole [284]. This is a closed radiation-dominated universe, and,

$$\rho^2(\tau) = 4b^2 + \tau^2. \tag{5.277}$$

The positive constant b is identified as the wormhole radius (or size). The idea is to find solutions of the minisuperspace Einstein–Yang–Mills system which have an asymptotic behavior as (5.277). It turns out that such solutions exist, provided the constant b is quantized [281],

$$b = b_n \sim \Lambda^{-1/2} \exp\left(-\pi n/\sqrt{2}\right). \tag{5.278}$$

In the gauge sector the solutions are determined modulo gauge transformations associated by various choices of the Lagrange multiplier $y(\tau)$. Hence, the gauge can be fixed so that $x_i(\tau) = \delta_{i1}x(\tau)$. The time reparameterization gauge freedom is fixed by going over to the conformal time $d\eta = d\tau/\rho(\tau)$. Use of the conformal time has the advantage that the equations of motion for ρ and gauge fields are decoupled. They are furnished by the stationary action principle for the gauge-fixed action (2.574) ($y = 0$ and $x_i = \delta_{i1}x$),

$$\frac{d^2x}{d\eta^2} = \frac{\partial V}{\partial x}, \quad \frac{d^2z}{d\eta^2} = \frac{\partial V}{\partial z}. \tag{5.279}$$

On any line $x = az$ the potential (5.275) has the form of a double-well. Therefore the equations of motion (5.279) should have periodic solutions oscillating around the local minimum $x = z = 0$ of the Euclidean potential $-V$. For every periodic solution in the gauge sector, one can find a periodic solution for ρ [287]. The solution $\rho(\eta)$ is interpreted as a wormhole connecting two points in the *same* space. Therefore the gauge fields should be the same at both sides of the wormhole. Since $z(\eta)$ and $x(\eta)$ are periodic (with periods $T_{z,x}$), the period T_ρ (the time τ between two ρ-maxima) should be an integer multiple of their periods [281],

$$T_\rho = n_1 T_z = n_2 T_x, \tag{5.280}$$

where n_1 and n_2 are positive integers. Relation (5.280) leads to quantization of the wormhole size [281, 287]. For the gauge group $SU(2)$, the integer n determines the wormhole size quantization (5.278). For the group $SO(n)$, $n > 4$, the wormhole size depends on both the integers n_1 and n_2 [287].

The phase space of the x-degree of freedom is a cone unfoldable into a half-plane. Since $x(\eta)$ oscillates around the origin $x = 0$, the corresponding phase-space trajectory winds about the phase-space origin. Therefore the physical degree of freedom x needs half the time to return to the initial state, that is, $T_x^c = \frac{1}{2}T_x$, thus leading to modification of the wormhole size quantization rule,

$$T_\rho = n_1 T_z = n_2' T_x^c = \frac{n_2'}{2} T_x, \tag{5.281}$$

i.e. n_2 in (5.280) may take half-integer values. If the theory contains fields realizing different representations of the gauge group, the periods of their physical oscillations would be determined by degrees of the independent Casimir operators for a given representation [278]. Modification of the wormhole size quantization would have an effect on quantum tunneling in quantum gravity involving wormholes. The minisuperspace quantum theory with gauge fields and fermions is discussed in [292].

5.9.4 Quantum Green's functions and the phase-space structure

Consider a quantum gauge theory in physical variables (i.e. when all non-physical degrees of freedom have been eliminated). As shown in Section 5.3, the fact that the physical phase-space structure is non-standard is accounted for in quantum theory by demanding that the physical states be invariant under a certain set of discrete transformations, and their scalar product is appropriately modified (the integration region and integration measure are changed). Therefore the physical phase-space structure affects quantum mechanical amplitudes. The most interesting of the latter are Green's functions, i.e. the quantities $\langle T(\hat{q}(t_1)\hat{q}(t_2)\cdots\hat{q}(t_n))\rangle_0$ where $\hat{q}(t) = e^{iHt}\hat{q}e^{-iHt}$ is the Heisenberg operator of the coordinate in the physical configuration space, and T denotes the time ordering. The expectation value is calculated relative to the ground state. The pole structure of the Fourier transform of Green's functions contains information about the spectrum of the system.

In quantum mechanics, the quantity $\langle T(\hat{q}_a(t)\hat{q}_{a'}(0)\rangle_0$, where $\hat{q}_a(t) = \hat{U}_t^+ \hat{q}_a(0)\hat{U}_t$, $\hat{U}_t = \exp(-i\hat{H}t)$, $\hat{q}_a(0) \equiv \hat{q}_a$, is analogous to the propagator in quantum field theory. Consider, for instance, the system of Bose and Fermi oscillators defined by the Hamiltonian operator

$$\hat{H} = \hat{b}^\dagger\hat{b} + \hat{f}^\dagger\hat{f},$$

where $[\hat{b}, \hat{b}^\dagger] = [\hat{f}, \hat{f}^\dagger]_+ = 1$. Then the position operator for the Bose oscillator has the form $\hat{q} = (\hat{b}^\dagger + \hat{b})/\sqrt{2}$. Set $\hat{b}|0\rangle = \hat{f}|0\rangle \equiv 0$. Then,

$$D_b(t) = \langle T(\hat{q}(t)\hat{q})\rangle_0 = \tfrac{1}{2}\theta(t)e^{-it} + \tfrac{1}{2}\theta(-t)e^{it},$$

$$D_f(t) = \langle T(\hat{f}(t)\hat{f}^+)\rangle_0 = \theta(t)e^{-it}.$$

The quantities $D_{b,f}(t)$ are the Green's functions for the classical equations of motion of the Bose and Fermi oscillators,

$$\left(-\partial_t^2 - 1\right) D_b(t) = \left(i\partial_t - 1\right) D_f(t) = i\delta(t). \tag{5.282}$$

Consider the Fourier transforms of the functions $D_{b,f}(t)$,

$$\tilde{D}(\omega) = \int_{-\infty}^{\infty} dt e^{i\omega t} D(t).$$

It follows from (5.282) that

$$\tilde{D}_b(\omega) = i(\omega^2 - 1 + i\varepsilon)^{-1}, \qquad \tilde{D}_f(\omega) = i(\omega - 1 + i\varepsilon)^{-1},$$

where $\varepsilon \to 0^+$. The poles of the Fourier transforms of the Green's functions correspond to the first (lowest) excitations of the oscillators.

Consider now the gauge model with the Lagrangian (5.139). In the coordinate representation for the Bose degrees of freedom and the holomorphic representation for the Fermi ones (see Section 5.7.2), the curvilinear (polar) coordinates (compare (1.174) and (5.150)) in the superspace spanned by $(\mathbf{x}, \boldsymbol{\psi}^*)$

$$\mathbf{x} = \exp(\varphi T)\boldsymbol{\rho}, \qquad \boldsymbol{\psi} = \exp(i\varphi\Gamma)\boldsymbol{\xi}, \tag{5.283}$$

where $\boldsymbol{\rho} = (r, 0)^T$, can be used to solve the constraint equation in quantum theory,

$$\hat{\sigma}\Phi_{\rm ph}(\mathbf{x}, \boldsymbol{\psi}^*) = \left(\hat{\mathbf{p}}T\hat{\mathbf{x}} - \hat{\boldsymbol{\psi}}^\dagger\Gamma\hat{\boldsymbol{\psi}}\right)\Phi_{\rm ph}(\mathbf{x}, \boldsymbol{\psi}^*) = 0.$$

The variables r and $\boldsymbol{\xi}^*$ describe the physical degrees of freedom, while φ is a pure non-physical variable (it is translated under the gauge transformations, $\varphi \to \varphi + \omega$). By making use of relations (5.152) and (5.153), one infers $\hat{\pi}_r = 0$, $\hat{\pi}_\varphi = \hat{\boldsymbol{\xi}}^\dagger\Gamma\hat{\boldsymbol{\xi}}$, and $\hat{\sigma} = -i\partial_\varphi$. Therefore the physical states are independent of φ,

$$\Phi_{\rm ph}(\mathbf{x}, \boldsymbol{\psi}^*) = \Phi(r, \boldsymbol{\xi}^*) \in \mathcal{H}_{\rm ph}. \tag{5.284}$$

The functions (5.284) are normalized according to the condition (5.155) which, in this case, is written in the form,

$$\langle\Phi_1|\Phi_2\rangle = \int_0^\infty dr r \int \prod_{a=1}^{2} (d\xi_a^* d\xi_a) \exp(-\boldsymbol{\xi}^*\boldsymbol{\xi}) \left[\Phi_1(r, \boldsymbol{\xi}^*)\right]^* \Phi_2(r, \boldsymbol{\xi}^*) \tag{5.285}$$

(the factor $\int_0^{2\pi} d\varphi$ is included into the norm of the physical states). The Hamiltonian in the physical subspace $\mathcal{H}_{\rm ph}$ is obtained from (5.148) and (5.154) by omitting the terms containing the derivative ∂_φ. As a result, the Schrödinger equation has the form,

$$\left(-\frac{1}{2}\partial_r^2 - \frac{1}{2r}\partial_r + \frac{1}{2r^2}\hat{\pi}_\varphi^2 + \frac{1}{2}r^2 + \hat{\boldsymbol{\xi}}^\dagger\hat{\boldsymbol{\xi}} - 1\right)\Phi_E = E\Phi_E; \tag{5.286}$$

here the fermion operators are ordered as $\boldsymbol{\psi}^*\boldsymbol{\psi} \to \frac{1}{2}(\hat{\boldsymbol{\psi}}^\dagger\hat{\boldsymbol{\psi}} - \hat{\boldsymbol{\psi}}\hat{\boldsymbol{\psi}}^\dagger) = \hat{\boldsymbol{\psi}}^\dagger\hat{\boldsymbol{\psi}} - 1 = \hat{\boldsymbol{\xi}}^\dagger\hat{\boldsymbol{\xi}} - 1$, and, for the sake of simplicity, the oscillator frequencies are set to one, $\omega_1 = \omega_2 = 1$.

To solve (5.286), the Hilbert space $\mathcal{H}_{\rm ph}$ is split into four orthogonal subspaces $\mathcal{H}_{\rm ph}^{(\mu)}$ ($\mu = 0, 1, 2, 3$) whose elements are $\Phi_E^{(0)}(r)$, $\Phi_E^{(a)} = \xi_a^* F_E^{(a)}(r)$ (here $a = 1, 2$), and $\Phi_E^{(3)} = \xi_1^*\xi_2^* F_E^{(3)}(r)$, where $F_E^{(a)}$ and $F_E^{(3)}$ are wave functions of the physical bosonic degree of freedom. Since $\hat{\pi}_\varphi 1 = \hat{\pi}_\varphi \xi_1^*\xi_2^* = 0$ and $\hat{\pi}_\varphi\boldsymbol{\xi}^* = \Gamma\boldsymbol{\xi}^*$, Eq. (5.286) can be solved in each of the subspaces independently. The operators $\hat{\pi}_\varphi{}^2$ and $\hat{\boldsymbol{\xi}}^\dagger\hat{\boldsymbol{\xi}}$ are c-numbers in each $\mathcal{H}_{\rm ph}^{(\mu)}$. Therefore Eq. (5.286) is equivalent to four equations of the type (5.37). The normalized eigenfunctions are,

$$\Phi_n^{(0)} = \frac{\sqrt{2}}{n!} L_n(r^2)\, e^{-r^2/2}, \qquad E_n^{(0)} = 2n; \tag{5.287}$$

$$\Phi_n^{(a)} = \frac{\sqrt{2}}{n!\sqrt{n+1}} r\xi_a^* L_n^1(r^2)\, e^{-r^2/2}, \qquad E_n^{(a)} = 2n+2; \tag{5.288}$$

$$\Phi_n^{(3)} = \xi_1^*\xi_2^*\Phi_n^{(0)}(r^2)\,, \qquad E_n^{(3)} = 2n+2, \tag{5.289}$$

where $n = 0, 1, 2, \ldots$. The vacuum state is not degenerate, while all excitations are four-fold degenerate and for each E_n there are two Bose (even in Grassmann variables) and two Fermi states. The system is supersymmetric [32].

In accordance with the generalized Chevalley theorem (see Section 5.7.3), the Z_2 invariance of the functions (5.287)–(5.289) ($r \to \pm r$, $\boldsymbol{\xi}^* \to \pm\boldsymbol{\xi}^*$) allows for their unique gauge-invariant analytic continuation to the whole configuration space spanned by $(\mathbf{x}, \boldsymbol{\psi}^*)$. This is done by the following change of arguments in the functions (5.287)–(5.289): $r^2 = \boldsymbol{x}^2$, $\xi_1^*\xi_2^* = \psi_1^*\psi_2^*$, $r\xi_1^* = z\psi_1^*$, and $r\xi_2^* = z^*\psi_2^*$, where $z = x_1 + ix_2$. Thus obtained wave functions satisfy the Schrödinger equation in the total configuration space and are explicitly gauge invariant.

Let us turn to the analysis of the Green's functions [212, 213],

$$D_b(t) = \langle T(\hat{r}(t)\hat{r}(0))\rangle_0\,, \quad D_f^{ab}(t) = \left\langle T\left(\hat{\xi}_a(t)\hat{\xi}_b^\dagger(0)\right)\right\rangle_0. \tag{5.290}$$

Calculation of the Green's functions (5.290) is illustrated with the following obvious chain of equalities:

$$D_b(t) = \theta(t)\langle \hat{r}e^{-it\hat{H}_{\rm ph}}\hat{r}\rangle_0 + \theta(-t)\langle \hat{r}e^{it\hat{H}_{\rm ph}}\hat{r}\rangle_0, \tag{5.291}$$

$$\begin{aligned}\langle \hat{r}e^{-it\hat{H}_{\rm ph}}\hat{r}\rangle_0 &= \sum_{\mu=0}^{3}\sum_{n=0}^{\infty}\left\langle 0|\hat{r}|\Phi_n^{(\mu)}\right\rangle\left\langle \Phi_n^{(\mu)}|e^{-itH_{\rm ph}}\hat{r}|0\right\rangle\\ &= \sum_{n=0}^{\infty}\left|\left\langle 0|\hat{r}|\Phi_n^{(0)}\right\rangle\right|^2 e^{-i2nt};\end{aligned} \tag{5.292}$$

similarly for $D_f^{ab}(t)$. The amplitude $\left\langle 0|\hat{r}|\Phi_n^{(0)}\right\rangle$ is calculated by means of the scalar product (5.285). Carrying out the Fourier transform of Green's functions (5.291), one infers,

$$\tilde{D}_b(\omega) = \sum_{n=0}^{\infty}\frac{\Gamma^2(n-1/2)}{4n!^2}\,\frac{in}{\omega^2 - 4n^2 + i\varepsilon}; \tag{5.293}$$

$$\tilde{D}_f^{ab}(\omega) = \delta^{ab}\sum_{n=0}^{\infty}\frac{\Gamma^2(n+1/2)}{4n!^2(n+1)}\,\frac{i}{\omega - 2n - 2 + i\varepsilon}. \tag{5.294}$$

According to the De Morgan theorem (see [293]), the series (5.293) and (5.294) converge absolutely and define analytic functions of ω with simple poles in the complex plane.

In contrast to the case of an oscillator with planar phase space, the quantum Green's functions (5.293) and (5.294) contain poles corresponding to all eigenvalues (5.287)–(5.289) of the Hamiltonian. Moreover, despite the linearity of the theory, the functions $D_b(t)$ and $D_f^{ab}(t)$ no longer satisfy equations (5.282). Thus, modification of the physical phase structure leads to drastic changes of the analytic properties of the quantum Green's functions.

Let us analyze the reason for such an exotic pole structure of the Green's functions to occur for oscillators with a non-standard phase space. From the mathematical point of view, this is related to a reduction of the integration region in the scalar product (5.285). The scalar product in the quantum theory of a conventional harmonic oscillator contains the integration over the whole real line. This leads to only one term $\langle 0|\hat{x}|n\rangle \sim \delta_{n1}$ differing from zero in the series (5.292). Reduction of the integration region to the positive semi-axis automatically yields a nonzero result for all terms in the series (5.292).

As the eigenfunctions $\Phi_n^{(0)}$ are even, the integration region can be extended to the whole real line, $\int_0^\infty dr r = (1/2)\int_{-\infty}^\infty dr|r|$, while preserving the orthonormality of the eigenstates $\Phi_n^{(0)}$, and thereby going over to conventional configuration space (and a planar phase space). However, the states $\hat{r}\Phi_E$ and $\hat{\xi}_a\Phi_E$ are not Z_2 invariant. If they are analytically continued into the non-physical region, then the Green's functions would vanish, $D_b(t) = D_f^{ab}(t) = 0$. Therefore these states must be continued into the non-physical region in a Z_2-invariant way in order to preserve correct values for all the amplitudes.

Thus, when going over to the theory on the entire real line, $r \in R$, operators $\hat{r}$ and $\hat{\xi}_a$ should be replaced by their Z_2-invariant extension to the non-physical region $r < 0$,

$$\hat{r}_Q \equiv r_Q(\hat{r}) = \theta(\hat{r})\hat{r} - \theta(-\hat{r})\hat{r} = \hat{r}\varepsilon(\hat{r}), \tag{5.295}$$

$$\hat{\xi}_Q^\dagger \equiv \boldsymbol{\xi}_Q^*(\hat{r}, \hat{\boldsymbol{\xi}}^\dagger) = \theta(\hat{r})\hat{\boldsymbol{\xi}}^\dagger - \theta(-\hat{r})\hat{\boldsymbol{\xi}}^\dagger = \hat{\boldsymbol{\xi}}^\dagger\varepsilon(\hat{r}), \tag{5.296}$$

where $\varepsilon(r)$ is the sign function. The set of Z_2-invariant polynomials (polynomials of r^2 and $r\xi_a^*$) is dense in the space of continuous Z_2-invariant function. Therefore quantities (5.295) and (5.296) can be represented as a series over the gauge-invariant polynomials that are polynomials of $r^2 = \mathbf{x}^2$ and $r\xi_a^* = z\psi_a^*$. Hence, the excitations described by variables r_Q and $\boldsymbol{\xi}_Q^*$ are not elementary in the total configuration space. Precisely this fact is reflected in the pole structure of the Green's functions (5.293) and (5.294) (see also [206, 294] and the discussion in Section 5.2.1)

Thus, it can be concluded that *the dynamics of an elementary degree of freedom (either bosonic or fermionic) with a reduced phase space is equivalent to the dynamics of a composite object in standard (planar) phase space.*

Another important observation resulting from the analysis of Green's functions is as follows. The gauge-invariant approach discussed in Sections 5.2.1 and 5.6.2 shows that physical excitations are described by (composite) variables that are quadratic in the initial dynamical variables. On the other hand, by eliminating the non-physical degrees of freedom by means of the gauge condition $x_2 = 0$, the theory is obtained in which physical excitations seem to be generated by operators linear in physical variables. A resolution of this paradox is that *in the latter theory the physical degrees of freedom have a reduced phase space and, in essence, their dynamics is nothing but that of composite objects* like (5.295) and (5.296). The simplest excitations in the theory are described by Z_2-invariant operators quadratic in the physical variables, r^2 and $r\boldsymbol{\xi}^*$. Their Green's functions contain only the poles corresponding to the lowest excitation of the oscillator [195, 212, 294].

Thus, the reduced (or gauge fixed) theory "remembers" about its "gauge" origin.

5.9.5 Green's functions and the choice of physical variables

The physical phase (or configuration) space is defined in a coordinate-independent way. When studying the dynamics of the physical degrees of freedom, local canonical variables are introduced in the physical phase (or configuration) space. Their choice is quite arbitrary if no peculiarities of the dynamics are known to make a particular choice preferable. It is therefore of interest to investigate how quantum Green's functions depend on the choice of physical variables that are used to parameterize the physical phase or configuration space [212].

Let the dynamics of the physical Bose degree of freedom in model (5.145) be described by the variable $u \in R$ that spans a curve $\mathbf{x} = \mathbf{f}(u)$. In order to develop a correct quantum mechanical description, the curvilinear coordinates are introduced,

$$\mathbf{x} = \exp(\varphi T)\mathbf{f}(u), \quad \boldsymbol{\psi} = \exp(i\varphi\Gamma)\boldsymbol{\zeta}. \tag{5.297}$$

They represent a generalization of the curvilinear coordinates (5.283) to the superspace spanned by pairs $(\mathbf{x}, \boldsymbol{\psi})$. The Bose sector of this change of variables is described in detail in Section 5.3. In the variables (5.297), the constraint operator has a simple form, $\hat{\sigma} = -i\partial_\varphi$. Therefore the physical states are functions of u and $\boldsymbol{\zeta}^*$. The metric tensor in the new variables is no longer diagonal in contrast to (5.283). The method for calculating the Laplace–Beltrami operator proposed in Section 5.7.2 can be applied to calculate the physical Hamiltonian of the oscillators. It has a rather complicated form,

$$\begin{aligned}\hat{H}_{ph} &= \frac{1}{2}\hat{P}_u \circ g(u)\hat{P}_u + \frac{1}{2}\hat{P}_u \circ \frac{(\mathbf{f}, T\mathbf{f}')}{\mu^2}\hat{\pi}_\varphi + \frac{1}{2}\hat{\pi}_\varphi \frac{(\mathbf{f}, T\mathbf{f}')}{\mu^2}\hat{P}_u \\ &\quad + \frac{\mathbf{f}'^2}{\mu^2}\hat{\pi}_\varphi^2 + V_q + \frac{1}{2}\mathbf{f}^2 + (\hat{\boldsymbol{\zeta}}^\dagger, \hat{\boldsymbol{\zeta}}) - 1,\end{aligned} \tag{5.298}$$

where $\hat{P}_u = -i\mu^{-1/2}\partial_u \circ \mu^{1/2}$ is the Hermitian momentum operator, $\hat{\pi}_\varphi = -i\partial_\varphi$, $\mu(u) = (\mathbf{f}', \mathbf{f})$, $\mathbf{f}' = \partial_u \mathbf{f}$, $g(u) = \mathbf{f}^2/\mu^2$, and the quantum potential is given by,

$$V_q = \frac{1}{2}\mu^{-1/2}\partial_u(g(u)\partial_u\mu^{1/2}) .$$

The scalar product in the theory is

$$\langle\Phi_1|\Phi_2\rangle = \int_K du\mu(u) \int \prod_{a=1}^{2} (d\zeta_a^* d\zeta_a) \exp[-(\boldsymbol{\zeta}^*, \boldsymbol{\zeta})] \left[\Phi_1(u, \boldsymbol{\zeta}^*)\right]^* \Phi_2(u, \boldsymbol{\zeta}^*); \tag{5.299}$$

the integration region K is described in detail in Section 5.3. Thus, the quantum theory in the gauge $\mathbf{x} = \mathbf{f}(u)$ is constructed and quantum Green's functions, $D_b(t) = \langle T(\hat{u}(t)\hat{u}(0))\rangle_0$ and $D_f^{ab}(t) = \left\langle T\left(\hat{\zeta}_a(t)\hat{\zeta}_b^\dagger(0)\right)\right\rangle_0$, can now be analyzed.

To solve the Schrödinger equation with the Hamiltonian (5.298), it is convenient to use the theorem about the unitary equivalence of all quantum theories corresponding to different choices of $\mathbf{f}$ (different gauges), proved in Section 5.3. According to this theorem, there exists a change of variables that transforms the Hamiltonian (5.298) into that in the Schrödinger equation (5.286) and the scalar product (5.299) into (5.285). This change of variable is a generalization of (5.47) to the superspace case and defined by,

$$\mathbf{f}(u) = \exp(\theta(u)T)\boldsymbol{\rho}, \quad \boldsymbol{\zeta}^* = \exp(i\theta(u)\Gamma)\boldsymbol{\xi}^*, \quad u \in K, \tag{5.300}$$

where $\exp(i\theta(u)) = z_f/r$ and $z_f = f_1 + if_2$. Indeed, the isomorphism of quantum theories in the coordinates (5.283) and (5.300) can be established by substituting the relations

$$\partial_u = \frac{\partial r}{\partial u}\partial_r + \frac{\partial \boldsymbol{\zeta}^*}{\partial u}\frac{\partial}{\partial \boldsymbol{\zeta}^*} = \frac{\mu}{r}\partial_r + i\theta'(u)\hat{\pi}_\varphi,$$
$$\theta'(u) = \partial_u\theta(u) = -\frac{(\mathbf{f}, T\mathbf{f}')}{r^2},$$
$$\hat{\pi}_\varphi = (\hat{\boldsymbol{\zeta}}^\dagger, \Gamma\hat{\boldsymbol{\zeta}}) = (\hat{\boldsymbol{\xi}}^\dagger, \Gamma\hat{\boldsymbol{\xi}})$$

into (5.298) and (5.299) and making use of the integration measure transformation rule $\int_K du\mu = \int_0^\infty drr$ proved in Section 5.3. Thus, the eigenfunctions and eigenvalues of the Hamiltonian (5.298) are given in (5.287)–(5.289) in which the following changes should be made: $r^2 \to \mathbf{f}^2(u)$, $r\xi_{1,2}^* \to z_f\zeta_{1,2}^*$, and $\xi_1^*\xi_2^* \to \zeta_1^*\zeta_2^*$ according to (5.300).

Thus obtained wave functions are invariant under the set of discrete transformations S (see Section 5.3),

$$\Phi_n^{(\mu)}(\hat{s}u, \hat{s}\boldsymbol{\zeta}^*) = \Phi_n^{(\mu)}(u, \boldsymbol{\zeta}^*) , \quad u \in K,$$
$$\hat{s}u = u_s(u) , \quad \hat{s}\boldsymbol{\zeta}^* = \exp\left[i\theta_s(u)\Gamma\right]\boldsymbol{\zeta}^* , \quad u \in K, \tag{5.301}$$

where the functions $u_s(u)$ and $\theta_s(u)$ are defined in (5.48) and (5.49). The transformations (5.301) along with the translations $\varphi \to \varphi + \theta_s(u)$ form the set of symmetry transfomations of the change of variables (5.297). This set is not a group (see Section 5.3.2).

Consider the Bose Green's function,

$$D_b(t) = \theta(t) \sum_{n=0}^{\infty} \left| \left\langle 0|\hat{u}|\Phi_n^{(0)} \right\rangle \right|^2 e^{-2itn} + (t \to -t). \tag{5.302}$$

For the amplitudes in the series (5.302) the following equality holds:

$$\begin{aligned} \left\langle 0|\hat{u}|\Phi_n^{(0)} \right\rangle &= \int_K d u \mu(u) \left[\Phi_0^{(0)}(\mathbf{f}^{\,2}(u)) \right]^* u \, \Phi_n^{(0)}(\mathbf{f}^{\,2}(u)) \\ &= \int_0^{\infty} drr \left[\Phi_0^{(0)}(r^2) \right]^* u(r) \, \Phi_n^{(0)}(r^2) = \left\langle 0|u(\hat{r})|\Phi_n^{(0)} \right\rangle . \end{aligned} \tag{5.303}$$

Relations (5.302) and (5.303) show that an explicit form of the Green's functions does depend on the choice of physical variables (local coordinates in the physical phase or configuration space), but their pole structure remains invariant relative to it. Thus, only the poles of quantum Green's functions have a physical meaning.

The variables $\hat{u}$ and $\hat{\zeta}^\dagger$ have a non-standard phase space and therefore cannot describe elementary excitations. For example, the states created by their action on the vacuum state are not S-invariant. Hence, when extending the integration region in the scalar product (5.299) to the whole axis (i.e. when going over to the corresponding theory with the standard (planar) phase space), $\int_K d u \mu = \int_{-\infty}^{\infty} du |\mu(u)| N_u^{-1}$, where N_u is the number of elements $\hat{s}$ for a given value of u, the operators $\hat{u}$ and $\hat{\zeta}^\dagger$ turn into the S-invariant composite objects, respectively,

$$u_Q(\hat{u}) = \sum_S \Theta_{K_s}(\hat{u}) u_s^{-1}(\hat{u}); \tag{5.304}$$

$$\zeta_Q^*(\hat{u}, \hat{\zeta}^\dagger) = \sum_S \Theta_{K_s}(\hat{u}) \exp\left[i\Gamma \theta_s^{-1}(\hat{u}) \right] \hat{\zeta}^\dagger, \tag{5.305}$$

where $K_s = \hat{s}K$ and $\Theta_K(u)$ is the characteristic function of the region K, i.e. $\Theta_K(u) = 1$ if $u \in K$ and vanishes if $u \notin K$. The functions u_s^{-1} are the inverse to u_s $\left(u_s^{-1} : K_s \to K\right)$ (see Section 5.3); the function $\theta^{-1}(u)$ is defined on the set K_s by the condition,

$$\theta_s^{-1}(u_s(u)) = -\theta_s(u), \quad u \in K. \tag{5.306}$$

Functions $u_Q(u)$ and $\zeta_Q^*(u, \zeta^*)$ define an S-invariant extension of the variables u and ζ^* into the non-physical region $R \setminus K$. A proof is elementary. It is easy to verify that

$$u_Q(u) = u, \quad \zeta_Q^*(u, \zeta^*) = \zeta^*, \qquad u \in K.$$

If $u \in K$, then only one term in the sums (5.304) and (5.305), with $\hat{s}$ being the identity transformation, differs from zero. By definition, $\theta_s(u) = 0$ if $\hat{s} = 1$. According to the analysis given in Section 5.3, the S-invariance means that

$$\begin{aligned} u_Q(u_{s'}(u)) &= u_Q(u) = u \ , \quad u \in K \\ \zeta_Q^*(u_{s'}(u), \exp[i\theta_{s'}(u)\Gamma]\zeta^*) &= \zeta_Q^*(u, \zeta^*) = \zeta^* \ , \quad u \in K \end{aligned} \tag{5.307}$$

for any $\hat{s}' \in S$. It is not hard to see that these equalities do hold. Now the only nonzero contribution to the sums (5.304) and (5.305) comes from the term with $s = s'$, and cancelation of the phase factors in (5.307) follows from the definition (5.306) of the function θ_s^{-1}.

Thus, *Green's functions of bosonic and fermionic degrees of freedom with a non-standard (reduced) phase space correspond to Green's functions of composite objects made of corresponding degrees of freedom with conventional (planar) phase space.*

This important statement is also proved in the framework of the path integral formalism (see Section 6.8.2). According to this, physical bosonic and fermionic degrees of freedom with a reduced phase space should be interpreted as composite objects. More to the point, due to the theorem, proved above, about a one-to-one correspondence between S-invariant and gauge-invariant states, the excitations of degrees of freedom with a reduced phase space correspond to some gauge-invariant excitations. This brings the discussion back to the conclusion of Section 5.8.2 which asserts that the physical phase space structure is some kind of memory of the reduced (gauge-fixed) system about its gauge origin. This observation is relevant for confinement in gauge theories (see Chapter 7).

Remark. Under certain circumstances perturbative Green's functions may not be sensitive to a non-planar structure of the phase space. A simple example is the double-well potential discussed at the end of Section 5.9.1. The potential has a minimum at $r = r_0$, so the *perturbative* Green's functions of the operator $\hat{\rho}$ that describe small fluctuations around minima of the classical potential, $\hat{\rho} = \hat{r} - r_0$, are not significantly affected by the conic singularity of the phase space. Indeed, $\hat{\rho}_Q = r_Q(\hat{r}) - r_0 \approx \hat{r} - r_0$ (see (5.295)) as long as $(\langle r \rangle - r_0)/r_0 \ll 1$ for the states close to the perturbative (oscillator) ground state (compare also the Bohr–Sommerfeld quantization of the system discussed in Section 5.9.1). If constant gauge transformations are eliminated in the 2D Yang–Mills theory studied in Section 5.8, then the orbit space is a group manifold and small fluctuations around the classical state $A_\mu = 0$ are not sensitive to its topology. A non-perturbative analysis of Green's functions can be done in the $(1 + 1)$ QCD on cylindrical spacetime (the 2D Yang–Mills theory with fermions in the fundamental representation). The residual gauge transformations from the affine Weyl group would lead to a specific anomaly because the Dirac sea (the fermionic vacuum) is not invariant under them [239].

In 4D Yang–Mills theory, the non-perturbative propagator in the Coulomb gauge can be calculated numerically in the framework of lattice gauge theory [295, 296]. It is found that in the continuum limit,[4] the Fourier transform of

[4] The definition of the continuum limit in Monte-Carlo simulations of quantum field theory on a lattice can be found, e.g. in [297, 298]. It requires a specific scaling behavior of gauge invariant correlation functions calculated by the Monte-Carlo method as the lattice spacing decreases. The formal continuum limit is never achieved in simulations. The stated results should be understood in this sense.

the propagator behaves as $D_{\mu\nu}(k^2) \sim [\delta_{\mu\nu} - k_\mu k_\nu/k^2]D(k^2)$ where $D(k^2) \sim (k^2)^\alpha$, $k^2 = k_\mu^2$, k_μ is the 4-momentum, and $\alpha \approx 1.5$. Recall that for the perturbative propagator $D(k^2) = (k^2)^{-1}$. So, the non-perturbative propagator cannot be fitted as a sum of single particle poles with positive residues, which is certainly an unacceptable feature of the propagator of a physical particle because it violates the Källen–Lehmann representation [299]. However it still remains unclear which effect is the most relevant for such a behavior of the nonperturbative gluon propagator: that of the configuration space reduction (the Gribov horizon in the Coulomb gauge [209]), or the effects of a strong self-interaction. For instance, the influence of the Gribov copies in the Coulomb gauge on the gauge-fixed correlation functions in lattice QCD has been studied in [300]. It has been observed that the residual gauge symmetry does not appear to be relevant. However, the authors of [300] have also noted that the effect may become important on bigger lattices. Yet, invoking a special technique of solving Schwinger–Dayson equations [301], in which the residual gauge symmetry and the existence of the Gribov horizon is ignored, modification of the propagator is shown to be only due to the strong self-interaction of gluon fields. Thus, a rigorous conclusion about the effects of the residual gauge transformations on Green's functions in a gauge-fixed Yang–Mills theory is yet to be reached.

6
Path integrals in gauge theories

6.1 Preliminary remarks

It is clear that the above studied reduction of the physical phase space affects the Hamiltonian path integral formalism for quantum gauge systems. It is therefore important to find out how the path integral formalism is to be modified because this formalism represents a natural tool for studying quantum theory beyond the perturbation theory approach. Before turning to quantum gauge field theories, it is instructive to begin with simple mechanical models in which the physical phase space is not standard.

The basic object of the path integral formalism studied in this chapter is the evolution operator kernel ($\hbar = 1$),

$$\langle q, t | q' \rangle = \langle q | \exp(-it\hat{H}) | q' \rangle = U_t(q, q'), \tag{6.1}$$

where $\hat{H}$ is the Hamiltonian of a system. The amplitude (6.1) is the probability amplitude that a system which was in point q' of the configuration space at the initial moment of time will be found in point q in time t. In Chapter 2 it is shown that this amplitude can be determined by the Hamiltonian path integral,

$$U_t(q, q') = \int \prod_{\tau=0}^{t} \left(\frac{dp(\tau) dq(\tau)}{(2\pi)^N} \right) \exp\left[i \int_0^t d\tau (p\dot{q} - H(p, q)) \right], \tag{6.2}$$

where N is the number of degrees of freedom, $H(p, q)$ is the Hamiltonian, and the integration in carried out in the infinite limits for each canonical variable (the phase space of the system is the standard one, $R^N(q) \times R^N(p)$).

A direct extension of (6.1) and (6.2) to systems with gauge symmetry is not possible because the procedure does not take into account constraints which must be satisfied in the system evolution (see Chapter 4). In other words, not all configurations in phase space should give contributions to the integral (6.2), but only those that fulfill the constraints $\varphi_a(p, q) = 0$. A naive way to solve the problem could be that factor $\prod_a \prod_{\tau=0}^{t} \delta(\varphi_a(p(\tau), q(\tau)))$ is inserted into the integration measure in (6.2), which would automatically reduce the integration region to the surface of constraints in the phase space. However, the action of the system is invariant under gauge transformations generated by the constraints φ_a (see Chapter 3). Therefore the integral (6.2) would diverge. Indeed, the gauge arbitrariness implies that the action is independent of some of variables (their number is obviously equal to the number of independent constraints). So, while the integration with respect to momenta eliminates the delta-functions of constraints the integral with respect to the coordinates is still divergent. This problem was first studied in [171].

There are also other, more subtle, reasons for the insufficiency of such a direct approach.

The path integral formalism is equivalent to the operator formalism (see Chapter 2). The correct operator formalism has been established for gauge theories in Chapter 5. So it is a natural starting point to develop the corresponding path integral formalism. A self-sufficient path integral formalism for gauge theories (with no reference to the operator formalism) is developed in Section 6.7, where it is also proved that the operator formalism of Chapter 5 follows from it.

6.2 Hamiltonian path integral for gauge systems with conic phase space

Before developing a general method of constructing the Hamiltonian path integral for gauge systems, consider a simple gauge system in which the physical phase space is the cone unfoldable into a half-plane. The operator formalism is developed in Chapter 5. The problem is to obtain an equivalent path integral formalism.

6.2.1 Harmonic oscillator with discrete gauge group Z_2

To elucidate the key issue of the problem, consider the simplest quantum system, the one-dimensional harmonic oscillator, in which the discrete gauge group Z_2 is postulated, i.e. it is viewed as the characteristic feature of the physical phase space in the Hamiltonian formalism. The problem can easily be solved [26, 302]. Let x denote the position of an oscillating particle. The transition amplitude (6.1) for the conventional harmonic oscillator (no gauge symmetry) can be found using spectral decomposition [63],

$$
\begin{aligned}
U_t(x,x') &= \sum_E \langle x|E\rangle\langle E|x'\rangle e^{-itE} \\
&= \sum_{n=0}^{\infty} c_n^2 H_n(x)H_n(x')e^{-(x^2+x'^2)/2}e^{-iE_nt}, \qquad (6.3)
\end{aligned}
$$

where $E=E_n=n+1/2$ are eigenvalues of the Hamiltonian, $\langle x|E\rangle = c_n H_n(x)e^{-x^2/2}$ are the corresponding eigenfunctions with c_n being the normalization constants and $H_n(x)$ being the Hermite polynomials. The coordinates x and x' range over the whole real line. As is shown in Chapter 5, the condition that physical states are invariant relative to the discrete gauge group Z_2 changes the spectrum of the Hamiltonian because only even eigenfunctions belong to the physical Hilbert space. Thus, for a harmonic oscillator with conic phase space, the orthonormal basis in the physical Hilbert space consists of the functions

$$\psi_{2n}(x) = \tilde{c}_{2n} H_{2n}(x) e^{-x^2/2}\ , \quad \tilde{c}_{2n} = \sqrt{2} c_{2n}\ , \quad n = 0, 1, \ldots . \tag{6.4}$$

The physical wave functions are normalized over the half-axis $x \in [0, \infty)$ (the physical configuration space). This explains the change of normalization constants, $c_{2n} \to \tilde{c}_{2n}$. Let $U_t^c(x, x')$ denote the transition amplitude for the system with conic phase space. Then it follows from (6.3) that

$$\begin{aligned} U_t^c(x, x') &= \sum_{n=0}^{\infty} \tilde{c}_{2n}^2 H_{2n}(x) H_{2n}(x') \exp\left[-\frac{1}{2}(x^2 + x'^2) - iE_{2n} t\right] \\ &= U_t(x, x') + U_t(x, -x'); \end{aligned} \tag{6.5}$$

variables x and x' in (6.5) range over the positive half-axis. The time evolution of any state is determined by,

$$\psi_t(x) = \int_0^{\infty} dx' U_t^c(x, x') \psi_0(x') = \int_{-\infty}^{\infty} dx' U_t(x, x') \Psi_0(x'), \tag{6.6}$$

where

$$\Psi_0(x) = \int_0^{\infty} dx' Q(x, x') \psi_0(x')\ , \quad Q(x, x') = \delta(x - x') + \delta(x + x'), \tag{6.7}$$

and $\psi_t(x) = \psi(x, t)$ is the state wave function at time t. Obviously, the function ψ_t is even and its norm on the whole real axis coincides with that of Ψ_0. It is easy to see from (6.7) that Ψ_0 is the even extension of ψ_0 into the non-physical region $x < 0$: $\Psi_0(x) = \psi_0(x)$ if $x \geq 0$, and $\Psi_0(x) = \psi_0(-x)$ if $x < 0$. Thus, function Ψ_0 is defined on the whole real line, and the operator $\hat{Q}$ with the kernel (6.7) extends ψ_0 into the non-physical region $x < 0$. According to (6.6) and (6.7), the following relation for the transition amplitudes holds:

$$U_t^c(x, x') = \int_{-\infty}^{\infty} dx'' U_t(x, x'') Q(x'', x'). \tag{6.8}$$

From relation (6.8) it is clear how the Hamiltonian path integral of the harmonic oscillator

$$U_t(x, x') = \int \prod_{\tau=0}^{t} \left(\frac{dp(\tau) dx(\tau)}{2\pi} \right) \exp\left\{ i \int_0^t d\tau \left[p\dot{x} - \tfrac{1}{2}(p^2 + x^2) \right] \right\}, \tag{6.9}$$

where $x = x(t)$ and $x(0) = x'$, is modified when the planar phase space is changed to a conic one.

A few remarks are in order.

1. Equations (6.8) and (6.9) give a complete solution of the problem. It is useful to be convinced that the solution is self-consistent. The path integral representation of the kernel of the evolution operator $\hat{U}_t = \exp(-it\hat{H})$ is based on the semigroup property $\hat{U}_{t+t'} = \hat{U}_t \hat{U}_{t'}$ [63]. For the operator $\hat{U}_t$ defined in (6.3) this

is obviously true, while the validity of the semigroup property for $\hat{U}_t^c$ should be verified. The following relation is to be proved:

$$U_{t+t'}^c(x,x') = \int_0^\infty dx'' U_t^c(x,x'') U_{t'}^c(x'',x'). \tag{6.10}$$

In the framework of the operator formalism, this can be reformulated as the self-consistency of two conditions,

$$\hat{U}_t^c = \hat{U}_t \hat{Q} \quad \text{and} \quad \hat{U}_{t+t'}^c = \hat{U}_t^c \hat{U}_{t'}^c \,.$$

Combining them, one infers $\hat{U}_{t+t'}\hat{Q} = \hat{U}_t\hat{Q}\hat{U}_{t'}\hat{Q} = \hat{U}_t\hat{U}_{t'}\hat{Q}$, i.e. the following relation must hold:

$$\hat{Q}\hat{U}_t\hat{Q} = \hat{U}_t\hat{Q}\,. \tag{6.11}$$

The above operator equalities require some clarification. The point is that the second argument of the kernel $Q(x,x')$ ranges over the physical region $x' > 0$, while the first one takes values in the whole configuration space $x \in R$. Since the operator $\hat{Q}$ makes the even extension of functions in the physical region into the whole real line, it acts on an even function as the unit operator. Therefore Eq. (6.11) means that the kernel of $\hat{U}_t\hat{Q}$ is even in the first argument. Furthermore, since the the kernel $Q(x,x')$ is even in the first argument, a sufficient condition on $\hat{U}_t$ for (6.11) to hold is $U_t(-x,x') = U_t(x,-x')$, which is obviously true for the kernel (6.3). Actually, the self-consistency of the path integral formalism is guaranteed by a weaker condition (see Section 2.2),

$$U_\epsilon(-x,x') = U_\epsilon(x,-x') + O(\epsilon^2), \tag{6.12}$$

which is used in what follows.

2. From Eqs. (6.3), (6.5), and (6.8) it is clear that the operator $\hat{Q}$ eliminates the contribution of the non-physical states, being odd functions, to the physical kernel $U_t^c(x,x')$. This is the meaning of $\hat{Q}$.

3. It is also instructive to compare the role of $\hat{Q}$ in (6.8) with that of a similar operator that emerges in the path integral formulation of quantum mechanics of a particle on the semi-axis with the zero boundary condition at the origin (a totally reflective barrier at the origin) which is studied in Chapter 2. It is found there that the kernel of $\hat{Q}$ has the form $\tilde{Q}(x,x') = \delta(x-x') - \delta(x+x')$. The kernel $\tilde{Q}$ ensures that the transition amplitude satisfies the zero boundary condition. The meaning of its second term is that, in addition to direct trajectories from x' to x in the physical region, trajectories reflected from the origin give contributions to the path integral. The difference in the sign of the second delta-function in Q and $\tilde{Q}$ corresponds to different physics in the two systems. In the problem considered here the motion in the region $x < 0$ is not distinguishable from that in $x > 0$ in the opposite direction (because of gauge symmetry). So, when the particle passes the origin $x = 0$ there is no actual reflection and, hence, the phase of the wave function

is not changed, while for a quantum particle in a half-line [96, 124] the reflection is real and the phase is changed by π. This can be interpreted in the following way. For the infinitesimal transition amplitude U_ϵ, the main contribution to the sum over paths comes from a straight line connecting initial and final points. In the case of the totally reflective barrier, the sum of contributions of the straight and reflected trajectories to the infinitesimal transition amplitude U_ϵ equals the sum of contributions of two straight trajectories from x' and $-x'$ to x, but the contribution of the second trajectory is taken with the opposite sign to uphold the zero boundary condition at the origin. Similarly, in the Z_2 gauge model, the main contribution to the infinitesimal transition amplitude comes from the two straight trajectories from x' and $-x'$ to x, but with the same sign. This guarantees that the physical amplitude is even and, hence, invariant under the Z_2 gauge group.

4. Despite the reduction of the phase space, the measure in the Hamiltonian path integral does not acquire the factor 1/2 as one might naively expect. The reason is that the integration region in the scalar product is changed since the physical variable ranges only over the physical configuration space being the half-axis, not the whole real line. This is exactly the reason for the renormalization of the wave functions, $c_n \to \tilde{c}_n = \sqrt{2}c_n$, in (6.4) and (6.5).

In conclusion, the following is noteworthy. When going over to polar coordinates, a similar extension operator $\hat{Q}$ occurs in the path integral for a particle in the plane [120] (compare (2.197)),

$$Q(r, \phi, r', \phi') = \delta(r - r') \sum_{n=-\infty}^{\infty} \delta(\theta - \theta' + 2\pi n) + \delta(r + r') \sum_{n=-\infty}^{\infty} \delta(\theta - \theta' + \pi + 2\pi n).$$

Is it consistent with Eq. (6.7) for model (5.2) (a particle in a plane with gauge group $SO(2)$)? It is easy to see that, indeed, it is. As the physical states are independent of the angular variable θ thanks to the constraint condition (5.25), the integration with respect to θ' in

$$\int_0^{2\pi} d\theta' \int_0^{\infty} dr' r' Q(r, \theta, r', \theta') \psi_0(r')$$

eliminates all the delta-functions in the sum. The resulting kernel is identical (up to the difference in notation) to the kernel (6.7). In fact, this is an alternative derivation of (6.6) and (6.7), which emphasizes that the obtained results are a direct consequence of the constraint equations imposed on physical states in the Dirac quantization scheme [148]. Peculiarities associated with the physical degree of freedom in the model (5.2) being described by the curvilinear coordinate r are discussed below in Section 6.2.3.

6.2.2 Harmonic oscillator with gauge group $SO(3)$

In the previous section the simplest model is studied where the discrete gauge group Z_2 is postulated. Consider now the path integral formalism for a simple model [26, 302] in which the gauge symmetry is defined by the gauge-invariant Lagrangian (5.31), where T_a are generators of $SO(3)$, $a = 1, 2, 3$, and $\mathbf{x}$ and $\mathbf{y}$ are vectors in R^3. The constraints $\sigma_a = (\mathbf{p}, T_a\mathbf{x})$ are the angular momentum components (see Section 5.2). So the operator corresponding to the classical Hamiltonian (5.32) reads [75] (for the reader's convenience, all calculations are carried out in detail without referring to Section 5.2),

$$\hat{H} = \frac{1}{2}\left(-\partial_r^2 - \frac{2}{r}\partial_r + \frac{\hat{\sigma}^2}{r^2} + r^2\right) + (\mathbf{y}, \hat{\boldsymbol{\sigma}}) = \frac{1}{2}\left(\hat{p}_r^2 + \frac{\hat{\sigma}^2}{r^2} + r^2\right) + (\mathbf{y}, \hat{\boldsymbol{\sigma}}), \quad (6.13)$$

where $\hat{p}_r = -ir^{-1}\partial_r \circ r$. As the physical states are subject to the condition $\hat{\sigma}_a\Phi = 0$, the eigenvalue problem is reduced to the equation,

$$\frac{1}{2}\left(\hat{p}_r^2 + r^2\right)\Phi = E\Phi. \quad (6.14)$$

Physical solutions are regular at the origin $r = 0$. Functions that behave as $\sim 1/r$ near $r = 0$, though being normalizable (recall that the scalar product involves integration with respect to r with the weight r^2), satisfy a non-homogeneous Schrödinger equation with the delta-function $\delta(\mathbf{x})$ in the right-hand side [65] (this fact is easily established by noting that $\Delta(1/r) \sim \delta(\mathbf{x})$ for the Laplace operator Δ in R^3). Put $\psi = r\Phi$. Function ψ satisfies the Schrödinger equation for the conventional one-dimensional harmonic oscillator whose eigenfunctions are $\psi_n = c_n H_n(r)\exp(-r^2/2)$. The regularity condition for Φ demands that the basis in the physical Hilbert space is the set of solutions with odd n, i.e.

$$\Phi_{2k+1} = \tilde{c}_{2k+1}\frac{H_{2k+1}(r)}{r}e^{-r^2/2}, \quad k = 0, 1, \ldots. \quad (6.15)$$

The energy spectrum is given by $E_k = 2k + 3/2$, in accordance with (5.68). Evoking Eq. (6.3), the physical transition amplitude is found to be [302],

$$U_t^c(r, r') = \frac{1}{rr'}\left[U_t(r, r') - U_t(r, -r')\right] = \int_{-\infty}^{\infty}\frac{dr''}{rr''}U_t(r, r'')Q(r'', r'), \quad (6.16)$$

where $U_t(r, r')$ is the transition amplitude for a one-dimensional harmonic oscillator defined in (6.3), and kernel Q has the same form as in (6.7). The final result is similar to (6.6) and (6.7),

$$\psi_t(r) = \int_0^{\infty} dr' r'^2 U_t^c(r, r')\psi_0(r') = \int_{-\infty}^{\infty} dr'\,\frac{r'}{r}\,U_t(r, r')\Psi_0(r'),$$
$$\Psi_0(r) = \int_0^{\infty} dr' Q(r, r')\psi_0(r'). \quad (6.17)$$

The difference with (6.6) and (6.7) is only in the integration measure. The physical configuration space retains information about the whole configuration space, which includes physical and non-physical degrees of freedom, and the gauge group. The non-physical variables do not disappear tracelessly (a consequence of being described by curvilinear coordinates in the total configuration space).

6.2.3 The model with gauge group $SO(n)$

The peculiarity of the model with gauge group $SO(3)$ is that the physical wave functions coincide with those of the corresponding one-dimensional quantum system (up to the measure factor). For this very reason it appears possible to relate the Hamiltonian path integral representation of the transition amplitude for the one-dimensional system with a planar phase space to that for the same system with a conic phase space, in a simple way as given in (6.16).

For a generic orthogonal group $n \neq 3$ such a simple relation between the wave functions in the theories with the planar and conic phase spaces disappears. A mathematical reason for this is that for $n \neq 3$ the quantum mechanical correction to the classical potential, $V_q = (n-1)(n-3)/(8r^2)$, does not vanish. Consider the problem of developing the Hamiltonian path integral formalism in this case [197]. The method of analytic continuation of the unit operator kernel $\langle r|r'\rangle$ into the non-physical region $r < 0$ is proposed in [120, 124]. As is shown in Section 5.2, the physical wave functions $\Phi_E(r)$ must be even. Therefore the following resolution of unity holds:

$$\langle r|r'\rangle = \sum_E \Phi_E(r)\Phi_E^*(r') = (rr')^{(1-n)/2}\delta(r-r'), \quad r, r' \geq 0, \tag{6.18}$$

where the summation is carried out over the spectum of $\hat{H}_{\text{ph}}$. Because of $\Phi_E(-r) = \Phi_E(r)$, the unit operator kernel should have the property, $\langle -r|r'\rangle = \langle r|r'\rangle$ for $r, r' \geq 0$. Hence,

$$\begin{aligned}\langle r|r'\rangle &= (rr')^{(1-n)/2}\,\delta(r-r') + (-rr')^{(1-n)/2}\delta(r+r') \\ &= \int_{-\infty}^{\infty} \frac{dr''}{(rr'')^{(n-1)/2}}\,\delta(r-r'')Q(r''-r') \\ &= \int_{-\infty}^{\infty} \frac{dr''}{(rr'')^{(n-1)/2}} \int_{-\infty}^{\infty} \frac{dp}{2\pi}\, e^{ip(r-r'')}Q(r''-r'),\end{aligned} \tag{6.19}$$

where $r \in R$, $r' \geq 0$, and the kernel Q is defined in (6.7). For the infinitesimal evolution operator kernel,

$$U_\varepsilon^c(r, r') = \left[1 - i\varepsilon\hat{H}_{\text{ph}}(r)\right]\langle r|r'\rangle + O(\varepsilon^2) \tag{6.20}$$

it follows that

$$U_\varepsilon^c(r, r') = \int_{-\infty}^{\infty} \frac{dr''}{(rr'')^{(n-1)/2}} U_\varepsilon^{\text{eff}}(r, r'')Q(r'', r') + O(\varepsilon^2); \tag{6.21}$$

$$U_\varepsilon^{\rm eff}(r,r'') = \int_{-\infty}^{\infty} \frac{dp}{2\pi} \exp i\Big[p(r-r'') - \varepsilon H^{\rm eff}(p,r)\Big]; \tag{6.22}$$

$$H^{\rm eff}(p,r) = \frac{1}{2}p^2 + V(r^2) + V_q\ ; \qquad V_q = \frac{(n-1)(n-3)}{8r^2}\ . \tag{6.23}$$

Equation (6.21) can be written in the operator form, $\hat{U}_\varepsilon^c = \hat{U}_\varepsilon^{\rm eff}\hat{Q}$. In order to prove this for a finite time interval, the convolution rule,

$$\hat{U}_{2\varepsilon}^c = \hat{U}_\varepsilon^c \hat{U}_\varepsilon^c = \hat{U}_\varepsilon^{\rm eff}\hat{Q}\hat{U}_\varepsilon^{\rm eff}\hat{Q} = \hat{U}_{2\varepsilon}^{\rm eff}\hat{Q} \tag{6.24}$$

for two operators $\hat{U}_\varepsilon^c$ should be verified where,

$$U_{2\varepsilon}^{\rm eff}(r,r') = \int_{-\infty}^{\infty} dr'' U_\varepsilon^{\rm eff}(r,r'')U_\varepsilon^{\rm eff}(r'',r'). \tag{6.25}$$

It is easy to see that

$$U_\varepsilon^c(-r,r') = U_\varepsilon^c(r,r')\ . \tag{6.26}$$

Indeed, $\hat{H}_{\rm ph}(-r) = \hat{H}_{\rm ph}(r)$ and $\langle -r|r'\rangle = \langle r|r'\rangle$ therefore Eq. (6.26) follows from (6.20). By the definition of the scalar product in $\mathcal{H}_{\rm ph}$, the convolution (6.24) can be written as

$$\begin{aligned} U_{2\varepsilon}^c(r,r') &= \int_0^\infty dr_1 r_1^{n-1}\Big[(rr_1)^{(1-n)/2}U_\varepsilon^{\rm eff}(r,r_1) + (r_1\to -r_1)\Big]\,U_\varepsilon^c(r_1,r') \\ &= \int_0^\infty dr_1\left[\left(\frac{r_1}{r}\right)^{(n-1)/2} U_\varepsilon^{\rm eff}(r,r_1) + (r_1\to -r_1)\right]U_\varepsilon^c(r_1,r') \\ &= \int_{-\infty}^\infty dr''\left(\frac{r''}{r}\right)^{(n-1)/2} U_\varepsilon^{\rm eff}(r,r'')\int_0^\infty dr_1 Q(r'',r_1)U_\varepsilon^c(r_1,r') \\ &= \int_{-\infty}^{\infty} dr''\left(\frac{r''}{r}\right)^{(n-1)/2} U_\varepsilon^{\rm eff}(r,r'')U_\varepsilon^c(|r''|,r'). \end{aligned} \tag{6.27}$$

For positive r'' the absolute value in the argument of the kernel $U_\varepsilon^c(|r''|,r')$ can be omitted. For negative r'' the following relation holds:

$$\begin{aligned} U_\varepsilon^c(|r''|,r') &= (|r''|r')^{(1-n)/2}\,U_\varepsilon^{\rm eff}(|r''|,r') + (-|r''|r')^{(1-n)/2}\,U_\varepsilon^{\rm eff}(|r''|,-r') \\ &= (-r''r')^{(1-n)/2}\,U_\varepsilon^{\rm eff}(-r'',r') + (r''r')^{(1-n)/2}\,U_\varepsilon^{\rm eff}(-r'',-r'). \end{aligned} \tag{6.28}$$

Changing the integration variable $p \to -p$ in (6.22) and taking into account that the effective Hamiltonian (6.23) is an even function in both p and r, it is straightforward to establish that

$$U_\varepsilon^{\rm eff}(-r'',-r') = U_\varepsilon^{\rm eff}(r'',r'). \tag{6.29}$$

The substitution of (6.29) into (6.28) shows that the absolute value of the first argument of the kernel U_ε^c in (6.27) may be omitted.[1] As a result substitution of (6.21) into (6.27) yields the desired equality,

$$U_{2\varepsilon}^c(r,r') = \int_{-\infty}^{\infty} \frac{dr''}{(rr'')^{(n-1)/2}}\, U_{2\varepsilon}^{\text{eff}}(r,r'')Q(r'',r'), \tag{6.30}$$

where the kernel $U_{2\varepsilon}^{\text{eff}}(r,r'')$ is given in (6.25).

The proof of (6.24) or (6.30) is based on the relation (6.29). Therefore representation (6.24) would also be valid for a finite time interval if Eq. (6.29) were shown to hold not only for an infinitesimal ε, but also for any time $t > 0$,

$$U_t^{\text{eff}}(-r,-r') = U_t^{\text{eff}}(r,r'). \tag{6.31}$$

The latter follows from (6.29) and the representation,

$$U_t^{\text{eff}}(r,r') = \int_{-\infty}^{\infty} \prod_{i=1}^{N} dr_i U_\varepsilon^{\text{eff}}(r,r_1)U_\varepsilon^{\text{eff}}(r_1,r_2)\cdots U_\varepsilon^{\text{eff}}(r_N,r'), \tag{6.32}$$

where $t = (N+1)\varepsilon$.

Thus, the transition amplitude, which takes into account the true structure of the physical phase space of the system and that the physical degree of freedom is described by curvilinear coordinates, is given by the following Hamiltonian path integral:

$$U_t^c(r,r') = \int_{-\infty}^{\infty} \frac{dr''}{(rr'')^{(n-1)/2}}\, U_t^{\text{eff}}(r,r'')Q(r'',r'), \tag{6.33}$$

$$U_t^{\text{eff}}(r,r'') = \int_{-\infty}^{\infty} \prod_{\tau=0}^{t} \left(\frac{dp(\tau)dr(\tau)}{2\pi}\right) e^{iS^{\text{eff}}}, \tag{6.34}$$

$$S^{\text{eff}} = \int_0^t d\tau \Big[p\dot{r} - H^{\text{eff}}(p,r)\Big],$$

where $r(0) = r''$ and $r(t) = r$. These equations show that in the case $n \neq 3$ ($n > 1$), the Hamiltonian path integral does not coincide with that for the one-dimensional mechanical system with conic phase space because the "effective" classical Hamiltonian (6.23) contains a quantum potential V_q ($\sim \hbar^2$). The integration measure is different as well. By construction of the Hamiltonian path integral, potential V_q takes into account the operator ordering in the Hermitian quantum Hamiltonian and, thereby, guarantees unitarity of the evolution operator (6.33). The Hamiltonian path integral for an arbitrary parameterization of the orbit space (arbitrary gauge condition) is considered in Section 6.5.

The above derivation of the Hamiltonian path integral relies explicitly on the operator formalism (compare (6.20)) which is obtained by the Dirac quantization

[1] Equations (6.28) and (6.29) offer an alternative proof of (6.26) based on the explicit integral form (6.21) of the kernel (6.20).

method. In Section 6.7 an alternative method of deriving the path integral, proposed in [195, 303], is given which uses only the gauge transformation law of the classical Lagrangian, the gauge fixing condition, and equivalence rules discussed in Chapter 2. The physical phase space structure of the corresponding classical theory as well as the Dirac physical Hamiltonian operator (with the correct operator ordering) can be deduced from this modified path integral formalism. Hence, the modified path integral is shown to be fully self-contained and independent of the Dirac operator formalism, while being equivalent to the latter. In particular, when applied to topological Yang–Mills theories or Kogut–Susskind lattice gauge theory, it resolves the Gribov obstruction arising in the conventional Faddeev–Popov approach to constructing the path integral formalism in gauge theories in a way that is consistent with an explicitly gauge-invariant operator formalism.

6.3 Models with more complicated structures of the physical phase space

6.3.1 Model with an arbitrary gauge group. Adjoint representation

It is instructive to study the path integral formalism for a model with an arbitrary gauge group where dynamical variables realize the adjoint representation [26, 197, 304]. The model is discussed in Section 5.4 where the corresponding operator formalism is developed. After solving the constraint equations for physical states, the quantum dynamics of a particle in an l-dimensional space with an unusual normalization of state vectors is obtained. To investigate the evolution operator kernel, consider a quantum system with the Hamiltonian (5.78) and the scalar product (5.79) in which the integration is extended to the whole R^l. It is straightforward to construct the needed kernel, e.g. by means of the Feynmann–Kac spectral decomposition formula (defined in (6.3)). What is to be changed in it when the gauge invariance with respect to the Weyl group is required, i.e. when the integration domain in (5.79) is to be reduced to the Weyl chamber K^+? Even though the Hamiltonian (5.78) is expressed only in physical variables, not all solutions of the equation $\hat{H}_{\rm ph}\psi_E = E\psi_E$ are physical. In order to select physical solutions, the eigenfunctions of the Hamiltonian (5.78) must be symmetrized with respect to the Weyl group according to (5.83). This observation leads to the sought-for procedure for obtaining the physical evolution operator kernel U_t^c. In the spectral decomposition of the latter where all eigenfunctions ψ_E of the operator (5.78) are included (the analog of (6.3)), one should select the terms that contain only the physical eigenfunctions Φ_E (the analog of (6.5)). Here index E symbolizes a complete set of quantum numbers that label the physical eigenstates. Making use of (5.83), the physical evolution operator kernel is obtained,

$$U_t^c(h,h') = \sum_E \Phi_E(h,t)\Phi_E^*(h',t)$$

$$= N_W^{-1} \sum_{w,w' \in W} U_t\left(h_w, h'_{w'}\right), \qquad h,h' \in K^+, \tag{6.35}$$

where $U_t(h,h')$ is the evolution operator kernel in the system with the Hamiltonian (5.78) and the normalization in the whole R^l. It is straightforward to show that the following relation holds:

$$U_\varepsilon(h,h') = (\kappa\kappa')^{-1} \int \frac{dp}{(2\pi)^l} \exp\left\{ i \left[p(h-h') - \varepsilon \left(\frac{p^2}{2} + V(h) \right) \right] \right\}, \tag{6.36}$$

as $\varepsilon \to 0$, where $\kappa' \equiv \kappa(h')$ (to simplify notation). Note that the eigenfunctions of the Hamiltonian $\hat{H}' = -(1/2)\partial_i^2 + V$ are $\kappa\psi_E$ according to (5.78). This explains factor $(\kappa\kappa')^{-1}$ in (6.36). It is convenient to use notation introduced right after Eq. (5.83) in which the action of the Weyl group element w on h is written as an orthogonal transformation $h_w = w_s h$. The kernel (6.36) is then shown to have the property

$$U_\varepsilon(w_s h, h') = U_\varepsilon\left(h, w_s^{-1} h'\right) + O(\varepsilon^2), \tag{6.37}$$

for any (regular) potential $V(h)$ and any element of the Weyl group. If in addition V is invariant under the Weyl group, $V(w_s h) = V(h)$, the term $O(\varepsilon^2)$ is absent (compare (6.12)). This can be proved by changing the integration variable $p \to w_s p$ in (6.36). As w_s is an orthogonal matrix, dp and p^2 are invariant under this transformation, and the statement follows. Setting $t = \varepsilon$ in (6.35) and making use of (6.37), the infinitesimal evolution operator kernel can be written as

$$U_\varepsilon^c(h,h') = \sum_{w \in W} U_\varepsilon\left(h, h'_w\right) = \int dh'' U_\varepsilon(h,h'') Q(h'',h'), \tag{6.38}$$

$$Q(h,h') = \sum_{w \in W} \delta\left(h - h'_w\right). \tag{6.39}$$

Unless otherwise specified, the integration is carried out over the whole R^l.

Formula (6.38) is a generalization of (6.16) to the case of an arbitrary group. Similarly to (6.16), the significance of kernel Q is twofold. First, it eliminates the contribution of non-physical states to the evolution operator kernel. On the other hand, it defines an extension of the initial wave function $\Phi_0(h)$ to the whole space R^l, analogously to (6.17). Indeed, from (6.39) and the relation $\delta(h_w) = \delta(h)$, which holds because h_ω and h are related by an orthogonal transformation, it follows that

$$Q(h_w, h') = Q(h,h'). \tag{6.40}$$

The evolved state is then given by,

$$\Phi_\varepsilon(h) = \int_{K^+} dh'\mu(h')U^c_\varepsilon(h,h')\Phi_0(h')$$
$$= \int dh'\mu(h')U_\varepsilon(h,h')\Phi_0^Q(h'), \tag{6.41}$$
$$\Phi_0^Q(h) = \int_{K^+} dh' Q(h,h')\Phi_0(h'), \tag{6.42}$$

where (6.38) and the invariance of the measure $\mu(h) = \kappa^2(h)$ relative to the Weyl group have been used. Equation (6.41) shows that extension Φ^Q of the initial state Φ_0 to R^l is invariant under the residual discrete gauge group (the Weyl group).

Because of (6.37) and (6.39), the equation analogous to Eq. (6.11) also holds for this model. Hence, a repeated action of $\hat{U}^c_\varepsilon$ on the initial state yields operator $\hat{U}_t\hat{Q}$ in which the kernel of $\hat{U}_t$ is given by the path integral,

$$U_t(h,h') = (\kappa\kappa')^{-1}\int \prod_{\tau=0}^{t} \frac{dp(\tau)dq(\tau)}{(2\pi)^l} \exp\left\{ i\int_0^t d\tau\left[p\dot{q} - \frac{p^2}{2} - V(q)\right]\right\}, \tag{6.43}$$

where $h = q(t)$, $h' = q(0)$, and $\kappa' = \kappa(h')$. Relations (6.37)–(6.43) together with $\hat{U}^c_t = \hat{U}_t\hat{Q}$ and the scalar product definition (5.79) solve the problem. In the above formulas the integration with respect to h is carried out with the weight $g^{1/2} = \mu$ (compare (5.77)) save for the integral (6.42) (or (6.38)). It is easy to achieve a total uniformity by redefining $Q \to Q_\kappa = (\kappa\kappa')^{-1}Q$,

$$\int_{K^+} dh' Q(h,h')\Phi_0(h') = \int_{K^+} dh'\mu(h')Q_\kappa(h,h')\Phi_0(h') \ .$$

An explicit form of kernel Q is determined by the transformation properties of physical states relative to the Weyl group (cf. (6.7) and (6.16)).

After integrating over the momentum variables in (6.43), the Lagrangian path integral is obtained,

$$U^c_t(h,h') = \sum_{w\in W}\left[\kappa(h)\kappa(h'_w)\right]^{-1} \int_{h(0)=h'_w}^{h(t)=h} \mathcal{D}h\ e^{i\int_0^t d\tau[\dot{h}^2/2 - V(h)]}, \tag{6.44}$$

where $\mathcal{D}h$ symbolizes the formal path integral measure. The integral in the exponential is just the gauge-fixed action of the model (5.57) ($y = 0$ and $x = h \in H$). In contrast to the conventional path integral obtained by eliminating non-physical degrees of freedom in the classical theory, with a subsequent formal path integral quantization of the gauge-fixed theory, the path integral (6.44) contains a sum over residual gauge transformations associated with the gauge chosen and the measure factor. It is worth noting that κ vanishes at the Weyl chamber boundary. Due to the Weyl invariance of the action, the sum over the Weyl group can be interpreted as contributions of trajectories reflected from the boundary of the Weyl chamber (recall the analysis of a harmonic oscillator in Fig. 5.5 and discussion in Section 5.4.3). That is, a trajectory outgoing from h'_w and ending at $h \in K^+$ has the same action as the trajectory outgoing from $h' \in K^+$, reflecting from the boundary ∂K^+ (maybe not once) and ending at $h \in K^+$. Even if the Hamiltonian has no dynamical coupling between the physical degrees of freedom, the transition amplitude cannot be factorized over the degrees of freedom because of the kinematic coupling through the boundary. This is in full contrast with the conventional gauge-fixed path integral. An example of group $SU(3)$ is plotted in Fig 6.1.

It should be stressed again that the reflections are not caused by any force action (no infinite potential well as in the case of a particle in a box or in a sector (compare Fig. 2.2)). The physical state of the system is not changed at the very moment of reflection. Thanks to the factor kk' in the denominator of Eq. (6.44), the amplitude does not vanish when either the initial or final point lies on the boundary of the Weyl chamber, that is, the system can reach the boundary with nonzero probability. This is in contrast to the infinite well case (zero boundary conditions at ∂K^+). The occurrence of reflected trajectories in the path integral measure is the price that must be paid when cutting the hyperconic physical phase space to unfold it into a part of the Euclidean space R^{2l} spanned by the canonical variables h and p_h and, thereby, to establish the relation between

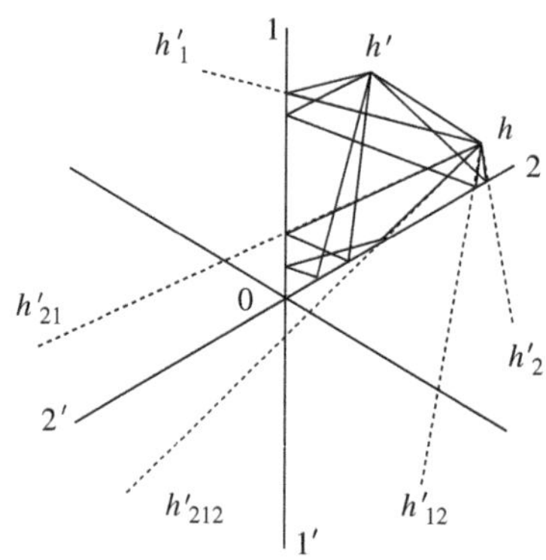

Fig. 6.1. The case of $SU(3)$. The modular domain is the Weyl chamber being sector 102 with angle $\pi/3$. In addition to a trajectory connecting the initial configuration h' and final configuration h, there are five trajectories connecting them and containing reflections from the Weyl chamber boundary. All these trajectories contribute to the transition amplitude even if there is no dynamical coupling between the physical degrees of freedom, $V = 0$. Points $h'_{1,2,12,21,212}$ are the Weyl images of h' obtained by all compositions of the mirror reflections with respect to the lines $101'$ and $202'$. For instance, h'_{12} is obtained by the reflection in $101'$ and then in $202'$, etc. Such an interaction with the boundary induces the kinematic coupling of the physical degrees of freedom. The transition amplitude cannot be factorized, even though the Hamiltonian would have *no* interaction between the physical degrees of freedom. Adapted From [195].

the the path integral measure on the hypercone and the conventional Liouville phase-space measure $dp_h dh$ in the R^{2l} phase space. The Weyl symmetry of the probability amplitude guarantees that the physical state of the system does not change when passing through the cut, which means that the system does not feel the discontinuity of the phase-space trajectory associated with a particular choice of canonical variables on the hypercone.

A final remark is that the path integral (6.44) is invariant relative to the Weyl transformations. Therefore it has a *unique, gauge-invariant* analytic continuation into the total configuration space in accordance with the theorem of Chevalley proved in Section 5.4.6. It is a function of the independent Casimir polynomials $P_{r_i}(x)$ and $P_{r_i}(x')$, $i = 1, 2, \ldots l = \operatorname{rank} X$. The result can also be anticipated from the Feynman–Kac spectral decomposition of the transition amplitude over gauge-invariant eigenstates of the total Hamiltonian. Thus, the path integral (6.44) (or (6.43)) does not depend on any particular parameterization of the gauge orbit space and represents a genuine gauge-invariant transition amplitude.

6.3.2 Topological Yang–Mills theory

Consider the Yang–Mills theory on a cylindrical spacetime introduced in Section 5.8. The orbit space and physical phase space have been analyzed there. In particular, the orbit space is spanned by configurations from the Weyl cell when the Coulomb gauge is used. The residual discrete gauge group is not finite compared to the mechanical models studied so far. A generic configuration in the gauge fixing surface has countably many copies. How is the path integral representation to be modified in order to resolve the problem of copies?

The simplest case of group $SU(2)$ is considered first. The physical phase space is the half-cylinder with two conic horns attached, as shown in Fig. 5.7. Consider canonical variables such that the variable $\theta = (a, \omega)/a_0$ introduced in Section 5.8.7 is used to span the physical configuration space. The Weyl cell is the open interval $\theta \in (0, 1)$ and $\kappa(\theta) = \sin(\pi\theta)$. The affine Weyl transformations are,

$$\theta \to \theta_{s,n} = s\theta + 2n \ , \quad s = \pm 1,$$

where n ranges over all integers. The interval $[0, 1]$ is the quotient of the real line by the affine Weyl group. A transition amplitude is a solution to the Schrödinger equation ($\hbar = 1$),

$$\left[-\frac{1}{2b\sin(\pi\theta)} \frac{\partial^2}{\partial\theta^2} \circ \sin(\pi\theta) - E_C \right] U_t^{\text{ph}}(\theta, \theta') = i\partial_t U_t^{\text{ph}}(\theta, \theta'), \tag{6.45}$$

where E_C is given in (5.237). All the parameters of the kinetic energy in the Hamiltonian (5.236) are included into the constant $b = 4\pi l a_0^2 = 4\pi/(lg^2)$. The solution must be regular at the boundaries $\theta = 0, 1$ and satisfy the initial condition,

$$U^{\rm ph}_{t=0}(\theta,\theta') = \langle\theta|\theta'\rangle = [\sin(\pi\theta)\sin(\pi\theta')]^{-1}\,\delta(\theta-\theta'), \tag{6.46}$$

where $\theta, \theta' \in (0,1)$. From (6.45) and (6.46) it is clear that the solution is symmetric in θ and θ'. So the substitution $U^{\rm ph}_t = [\sin(\pi\theta)\sin(\pi\theta')]^{-1}U_t$ transforms the problem to that for a free particle on a line, which is easily solved. The analysis of Section 5.8.5 shows that the regularity condition can be satisfied if the solution is invariant under the affine Weyl group W_A. Symmetrization of the free-particle solution U_t with the measure factor $[\sin(\pi\theta)\sin(\pi\theta')]^{-1}$ relative to W_A gives the sought-for solution of the problem [195, 198],

$$U^{\rm ph}_t(\theta,\theta') = (2\pi i t b)^{-1/2}\sum_{s=\pm 1}\sum_{n=-\infty}^{\infty}\frac{\exp\left\{\frac{i(\theta-\theta'_{s,n})^2}{2tb}+iE_C t\right\}}{\sin(\pi\theta)\sin\left(\pi\theta'_{s,n}\right)} \tag{6.47}$$

$$= \frac{e^{iE_C t}}{(2\pi i t b)^{1/2}}\sum_{n=-\infty}^{\infty}\frac{\exp\left\{\frac{i(\theta-\theta'+2n)^2}{2bt}\right\}-\exp\left\{\frac{i(\theta+\theta'+2n)^2}{2bt}\right\}}{\sin(\pi\theta)\,\sin(\pi\theta')}. \tag{6.48}$$

The sum in (6.47) is extended over the residual gauge transformations (the affine Weyl group), or, in other words, over the copies of the initial configuration θ' in the gauge fixing surface. The regularity of the transition amplitude at $\theta = n$ or $\theta' = n$ is easy to verify. The numerator and the denominator in the sum in (6.48) and vanish if either θ or θ' attains an integer value, but the ratio remains finite because the zeros are simple. The exponential in (6.47) is just the evolution operator kernel of a free particle on a line. It can be written as the path integral with the standard measure which involves no restriction of the integration region to the modular domain (the Weyl cell). The action of a free particle coincides with the 2D Yang–Mills action (5.176) in the Coulomb gauge $\partial_1 A_1 = 0$ and $A_1 \in H$. Indeed, put $A_0 = 0$ and $A_1 = a_0\omega\theta/(\omega,\omega)$ in (5.176) where ω is the only simple root of $SU(2)$, the Lagrangian (5.176) becomes $L = \dot\theta^2/(2b)$ (compare the exponential in (6.47)). The sum over the copies of the initial configuration θ' in the gauge fixing surface can be interpreted as contributions of the trajectories that reflect from the boundary of the Weyl cell (the Gribov horizon) multiple times before they reach the final point θ (compare Fig. 6.1). The amplitude does not vanish if the initial or final point is on the horizon.

A generalization to an arbitrary compact group is straightforward [195, 198],

$$U^{\rm ph}_t(a,a') = \sum_{\hat w\in W_A}[\kappa(a)\kappa(\hat w a')]^{-1}\int_{a(0)=\hat w a'}^{a(t)=a}\mathcal{D}a\; e^{i\int_0^t d\tau[\pi l\dot a^2+E_C]} \tag{6.49}$$

$$= \sum_{\hat w\in W_A}[\kappa(a)\kappa(\hat w a')]^{-1}\,(it/l)^{-r/2}e^{i\pi l(a-\hat w a')^2/t}e^{itE_C}, \tag{6.50}$$

where the action of elements $\hat w$ of the affine Weyl group W_A on any $a \in H$ is defined in (5.202) (any $\hat w$ can be written as a product of the operators $\hat s_{\alpha,n}$). The first term in the exponential in (6.49) coincides with the 2D Yang–Mills action (5.176) in the physical variables (the gauge-fixed action), i.e. if $A_0 = 0$ and $A_1 = a \in H$ in (5.176), then $L = \pi l \dot{a}^2$. The second term is the quantum correction to the classical action defined in (5.237). The function $\kappa(a)$ and its transformation properties under the action of the affine Weyl group transformations are defined in (5.214) and (5.222), respectively. The path integral for a free particle in $r = \dim H$ dimensions has the standard measure. The transition amplitude obviously satisfies the Schrödinger evolution equation. One can also verify the validity of the representation (6.49) by direct summation of the spectral representation of the transition amplitude because the explicit form of the eigenstates is known (see (5.243)), although the method of extending the unit operator kernel into the non-physical region (the complement of the Weyl cell K_W^+) leads to the result faster.

Consider a spectral decomposition of the unit operator kernel,

$$\langle a|a'\rangle = \sum_E \Phi_E(a)\Phi_E^*(a') = [\kappa(a)\kappa(a')]^{-1}\delta(a-a'), \quad a, a' \in K_W^+. \tag{6.51}$$

The eigenfunctions $\Phi_E(a)$ are the gauge-invariant eigenfunctions (5.243), (5.244) reduced on the gauge-fixing surface. Therefore the kernel (6.51) is, in fact, a genuine unit operator kernel on the gauge orbits space, which does not depend on any particular parameterization of the latter. Clearly, $\Phi_E(a)$ are invariant under the residual gauge transformations, under affine Weyl transformations, which can be used to extend the unit operator kernel to the non-physical region $a \in H$, i.e. to the whole covering space H of the modular domain K_W^+. The following property should hold: $\langle a|\hat{w}a'\rangle = \langle a|a'\rangle$, because $\Phi_E(\hat{w}a) = \Phi_E(a)$. Therefore, by taking into account the transformation property $\kappa(\hat{w}a) = \pm\kappa(a)$ (see (5.222)), one infers,

$$\begin{aligned}\langle a|a'\rangle &= \sum_{\hat{w}\in W_A} [\kappa(a)\kappa(\hat{w}a')]^{-1}\delta(a-\hat{w}a') \\ &= \int_H \frac{da''}{\kappa(a)\kappa(a'')}\,\delta(a-a'')Q(a'',a') \\ &= \int_H \frac{da''}{\kappa(a)\kappa(a'')} \int_H \frac{dp}{(2\pi)^r}\, e^{ip(a-a'')}Q(a'',a'),\end{aligned} \tag{6.52}$$

where $a \in H$ and $a' \in K_W^+$, $pa \equiv (p,a)$; the kernel of the operator $\hat{Q}$ is defined as before,

$$Q(a,a') = \sum_{\hat{w}\in W_A} \delta(a-\hat{w}a'). \tag{6.53}$$

The infinitesimal transition amplitude is given by

$$U_\varepsilon^{\mathrm{ph}}(a,a') = \left(1 - i\varepsilon\hat{H}_{\mathrm{ph}}(a)\right)\langle a|a'\rangle + O(\varepsilon^2), \tag{6.54}$$

where the physical Hamiltonian is obtained from the Dirac quantization method (see (5.236)). Applying the physical Hamiltonian in (6.54) to the Fourier transform of the unit operator kernel (6.52), the following representation is derived:

$$U_\varepsilon^{\rm ph}(a,a') = \int_H \frac{da''}{\kappa(a)\kappa(a'')} U_\varepsilon(a,a'')Q(a'',a') + O(\epsilon^2), \tag{6.55}$$

$$U_\varepsilon(a,a'') = \int_H \frac{dp}{(2\pi)^r} \exp\left\{ip(a-a'') - i\varepsilon\left(\frac{p^2}{4\pi l} - E_C\right)\right\} \tag{6.56}$$

$$= (i\varepsilon/l)^{-r/2} \exp\left\{\frac{i\pi l(a-a'')^2}{\varepsilon} + i\varepsilon E_C\right\}. \tag{6.57}$$

Function $p^2/(4\pi l) = H_{\rm ph}$ in (6.56) is the classical gauge-fixed Hamiltonian, the addition E_C is a quantum correction to it resulting from operator ordering (see (5.237)).

Convolution $(\hat U_\varepsilon \hat Q)^N$ can be calculated in the same fashion as it has been in Section 6.3.1. The only difference is that the integration in the scalar product is carried out over the Weyl cell which is compact. As a result of the invariance of amplitude $U_\epsilon^{\rm ph}(a,a')$ relative to the affine Weyl transformations, all the operators $\hat Q$ in the convolution can be pulled over to the right with the result that integration over the Weyl cell is replaced by integration over the whole Cartan subalgebra in the convolution $(\hat U_\varepsilon)^N$ (thanks to the sum over the affine Weyl group generated by $\hat Q$). Thus, formula (6.49) is recovered again.

The amplitude (6.49) has a unique analytic continuation into the original functional configuration space $\mathcal{F}$, which results from spectral decomposition of the transition amplitude and representation (5.244) for the eigenfunctions. It is a function of characters of two holonomy group elements for the initial and final configurations of the vector potential. Therefore the probability amplitude does not depend on any particular parameterization of the gauge orbit space, which has been used to compute the corresponding path integral. It is a genuine *coordinate-free* transition amplitude on the gauge orbit space $\mathcal{F}/\mathcal{G}$. It should be stressed that, in contrast to the recipe proposed in [209] to resolve the gauge-fixing ambiguity problem, the integration domain in the path integral (6.49) is *not* restricted to the fundamental modular domain (the Weyl cell K_W^+).

Replacing time t by the imaginary $t \to -i\beta$, one can calculate the partition function,

$$Z(\beta) = {\rm Tr}\, e^{-\beta\hat H} = \int_{K_W^+} da\kappa^2(a)U_\beta^{\rm ph}(a,a) = \sum_{\Lambda_n} e^{-\beta E_n}, \tag{6.58}$$

where E_n coincides with the spectrum of the system (5.239) and the sum is extended over the irreducible representations Λ_n. This shows that the constructed path integral reproduces the gauge-invariant spectrum of the 2D Yang–Mills theory. The proof can be done by direct evaluation of the integral in (6.58) (by using the explicit form of $U_\beta^{\rm ph}(a,a)$ (see (6.50)), the property $\kappa(\hat w a) = \pm\kappa(a)$ (to get rid of the measure κ^2 in (6.58)), and the symmetry properties of the remaining integrand under the affine Weyl transformations (5.202)). This approach is rather technical. An alternative way of proving the last equality in (6.58) is to note that the transition amplitude (6.50) satisfies the time-dependent Schrödinger equation with Hamiltonian (5.236) and, hence, can be expanded into the spectral sum over its eigenfunctions which are the characters of the irreducible representations as proved in (5.243). After substitution of the spectral decomposition into (6.58), the desired conclusion follows immediately from the orthogonality of the characters. The result (6.58) coincides with the calculations in [234, 246] of the partition function of the 2D Yang–Mills theory on a finite periodic 2D lattice where no gauge fixing is needed since the path integral is just a finite multiple integral.

6.3.3 Hamiltonian path integral for gauge matrix models

The models with non-standard physical phase space discussed in Sections 6.2.2 and 6.3.1 possess a remarkable property that the symmetrization of the transition amplitude in the conventional phase space with respect to the residual discrete gauge group leads to the correct Hamiltonian path integral. Unfortunately, this property does not hold for more complicated gauge systems. A simple example discussed in Section 6.2.3 shows that operator ordering must be taken into account in the Hamiltonian path integral, which results in modification of the classical action (the latter acquires additional terms of order $\sim \hbar^2$). In this section the issues related to the curvature of the physical configuration space are studied in the framework of the Hamiltonian path integral formalism. Discussion is limited to the case when the system has several physical degrees of freedom and the residual gauge transformations form a group. As an example, consider the gauge matrix model analyzed in Section 5.6.1. It describes two particles in a plane with gauge group $SO(2)$. A generalization to more complicated similar matrix models introduced in Section 5.6, like, e.g. Yang–Mills quantum mechanics, is straightforward.

The operator formalism for the model is developed in Section 5.6.3. Its peculiarity is that the metric of the physical configuration space (the tensor g^{ab}_{ph} in (5.128)) is non-Euclidean. Introducing the Hermitian momentum operators $\hat{P}_a = -i\mu^{-1/2}\partial_a \circ \mu^{1/2}$, the physical Hamiltonian (5.128) can be written in the form,

$$\hat{H}_{\text{ph}} = \tfrac{1}{2}\hat{P}_a g^{ab}_{\text{ph}} \hat{P}_b + V_q(\rho) + V(\rho), \tag{6.59}$$

where

$$V_q = \frac{1}{2\sqrt{\mu}}\left(\partial_a g^{ab}_{\text{ph}}\right)\partial_b\sqrt{\mu} + \frac{1}{2\sqrt{\mu}} g^{ab}_{\text{ph}}\partial_a\partial_b\sqrt{\mu} = -\frac{1}{8\rho_1^2}. \tag{6.60}$$

Here the explicit form of the Jacobian $\mu = \rho_1$ and metric tensor (see (5.127)) has been used in local coordinates ρ_a. The presence of the metric depending on local coordinates in (6.59) is a distinct feature of the model studied.

To develop the Hamiltonian path integral formalism, the method of continuation of the unit operator kernel proposed in Section 6.2.3 is applied. An alternative method that does not rely on the operator formalism is given in Section 6.7. The unit operator kernel can be extended into the non-physical region by means of the Z_2-invariance of the physical states (5.131). As a result,

$$\langle\rho|\rho'\rangle = \left(\rho_1\rho_1'\right)^{-1/2}\delta(\rho-\rho') + \left(-\rho_1\rho_1'\right)^{-1/2}\delta(\rho+\rho'), \tag{6.61}$$

where the δ-function of the triangular matrix ρ is understood as the product of the δ-functions of its three independent elements, $\delta(\rho) = \delta(\rho_1)\delta(\rho_2)\delta(\rho_3)$, $\rho_a = (\rho_{11}, \rho_{12}, \rho_{22})$. A proof of (6.61) goes along the lines of deriving (6.19) from (6.18). As a result the following integral representation of the kernel (6.61) is established,

$$\langle\rho|\rho'\rangle = \int_{-\infty}^{\infty}\frac{d^3\rho''}{\left(\rho_1\rho_1''\right)^{1/2}}\int_{-\infty}^{\infty}\frac{d^3p}{(2\pi)^3}e^{i\,\mathrm{Tr}[p^T(\rho-\rho'')]}\,Q(\rho'',\rho), \tag{6.62}$$

where p is an upper-triangular matrix, and

$$Q(\rho,\rho') = \delta(\rho-\rho') + \delta(\rho+\rho'). \tag{6.63}$$

Applying the infinitesimal evolution operator $\hat{U}_\varepsilon^{\mathrm{ph}} = \exp(-i\varepsilon\hat{H}_{\mathrm{ph}}) = 1 - i\varepsilon\hat{H}_{\mathrm{ph}}(\rho) + O(\varepsilon^2)$ to the kernel (6.62), one infers, up to terms $O(\varepsilon^2)$, that

$$U_\varepsilon^{\mathrm{ph}}(\rho,\rho') = \int_{-\infty}^{\infty}\frac{d^3\rho''}{\left(\rho_1\rho_1''\right)^{1/2}}U_\varepsilon^{\mathrm{eff}}(\rho,\rho'')\,Q(\rho'',\rho'), \tag{6.64}$$

$$U_\varepsilon^{\mathrm{eff}}(\rho,\rho'') = \int_{-\infty}^{\infty}\frac{d^3p}{(2\pi)^3}\exp i\left\{\mathrm{Tr}[p^T(\rho-\rho'')] - \varepsilon H^{\mathrm{eff}}(p,\rho)\right\}, \tag{6.65}$$

$$H^{\mathrm{eff}} = \frac{1}{2}g_{\mathrm{ph}}^{ab}(\rho)p_a p_b - \frac{i}{2\rho_1^2}(p_2\rho_2 + p_3\rho_3) - \frac{1}{8\rho_1^2} + V\ , \tag{6.66}$$

where $p_a = (p_{11}, p_{12}, p_{22})$. The operator ordering in $\hat{H}_{\mathrm{ph}}$ is fixed uniquely by the second and third terms in the effective Hamiltonian (6.66) ($\sim\hbar$ and $\hbar^2$, respectively), i.e. they uniquely determine the "inverse" transition from the "classical" Hamiltonian H^{eff} in the path integral to the quantum Hermitian operator (6.59) in the Schrödinger equation (for details see Chapter 2).

Representation (6.30) holds for the convolution of two kernels (6.64) if in the former r'', r', r, and dr'' are replaced by ρ'', ρ', ρ, and $d^3\rho''$, respectively, and also set $r'' = \rho_1''$, $r = \rho_1$, and $n = 2$ in the Jacobian factor of $U_\varepsilon^{\mathrm{eff}}$, while quantity $|\rho''|$ is understood as the matrix $\theta(\rho_1)\rho - \theta(-\rho_1)\rho$ if the half-space $\rho_1 \geq 0$, $\rho_{2,3} \in R$ is chosen as the modular domain K in (5.129). With all these changes, a proof of representation (6.30) for the kernel (6.64)–(6.66) is similar to that given in Section 6.2. It is based on the Z_2-invariance of the effective Hamiltonian (6.66), $H^{\mathrm{eff}}(-p,-\rho) = H^{\mathrm{eff}}(p,\rho)$. As a result, after repeating steps (6.31)–(6.33), one concludes that the representation (6.64) is also valid for any finite $t > 0$, where

$$U_t^{\text{eff}}(\rho, \rho'') = \int_{-\infty}^{\infty} \prod_{\tau=0}^{t} \left(\frac{d^3p d^3\rho}{(2\pi)^3} \right) \exp\left\{ i \int_0^t d\tau \left[\text{Tr}(p^T \dot{\rho}) - H^{\text{eff}}(p, \rho) \right] \right\}, \quad (6.67)$$

with $\rho(0) = \rho''$ and $\rho(t) = \rho$ being the boundary conditions for the Hamiltonian path integral.

To conclude, consider the question about the gauge invariance of the Hamiltonian path integral constructed above. As is established in Section 5.6.3, any Z_2-invariant function of the matrix ρ has a unique gauge-invariant extension into the total configuration space (the space of all 2×2 matrices x). In particular, this extension for the kernel (6.61) is not difficult to find (see Section 8.5.2) [208],

$$\langle \rho | \rho' \rangle = \langle x | x' \rangle_{ph} = |\det x \det x'|^{1/2} \delta(x^T x - x'^T x') \left(1 + \frac{\det x}{\det x'} \right), \quad (6.68)$$

where the matrix δ-function is understood as the product of the δ-functions of each of the independent elements of the symmetric matrix $(x^T x - x'^T x')_{ij}$ (for instance, $ij = 11, 12, 22$). To verify equality (6.68), it is sufficient to represent x in the form $\exp(\theta T)\rho$ and make use of the rule of changing the argument of a multi-dimensional δ-function.

The kernel $U_t^{\text{ph}}(\rho, \rho')$ is also Z_2-invariant in both the arguments. Indeed, for the second argument ρ', this follows from the Z_2-invariance of the kernel Q. To prove the equality $U_t^{\text{ph}}(-\rho, \rho') = U_t^{\text{ph}}(\rho, \rho')$, the representation (6.32), in which $r, r' \to \rho, \rho'$, for the kernel (6.67) should be invoked, then the integration variables are changed $\rho'' \to -\rho''$ in (6.64). The result follows from the Z_2-invariance of the kernel (6.63). Hence, there exists a unique gauge-invariant extension of the kernel $U_t^{\text{ph}}(\rho, \rho')$ to the whole configuration space such that

$$U_t^{\text{ph}}(\rho, \rho') = U_t^{\text{ph}}(x, x').$$

Therefore the Hamiltonian path integral developed above does not depend on any particular parameterization of the orbit space and is a function only of gauge-invariant quantities $x^T x$, $x'^T x'$, $\det x$, and $\det x'$ which determine the initial and final states of the system.

Thus, the unitarity of the evolution operator and its explicit gauge invariance in the Hamiltonian path integral formalism is provided by taking into account both the true structure of the physical phase space (the operator Q) and the operator ordering (effective quantum corrections to the classical (gauge-fixed) action). In Sections 6.5 and 6.6, this statement is shown to be valid for all gauge systems.

6.4 Models with Grassmann variables

Let us find out how the path integral formalism is modified in gauge models with Grassmann variables [26, 203].

Matrix elements of the evolution operator read,

$$\langle\phi|e^{-i\hat{H}t}|\xi\rangle = \int d\theta d\theta' \langle\phi|\theta\rangle\langle\theta|e^{-i\hat{H}t}|\theta'\rangle\langle\theta'|\xi\rangle, \tag{6.69}$$

for any two states $|\phi\rangle$ and $|\xi\rangle$. Here the representation introduced in Section 8.2 is used in which the fermion creation and annihilation operators are realized as multiplication and differentiation, $\langle\theta|\hat{\psi}_\alpha|\xi\rangle = \partial/\partial\theta_\alpha\langle\theta|\xi\rangle$ and $\langle\theta|\hat{\psi}_\alpha^\dagger|\xi\rangle = \theta_\alpha\langle\theta|\xi\rangle$, $\left[\hat{\psi}_\alpha, \hat{\psi}_\beta^\dagger\right]_+ = \delta_{\alpha\beta}$, and a set of Grassmann variables θ_α labels states in this representation. For the evolution operator kernel, the spectral decomposition holds,

$$U_t(\theta,\theta') = \langle\theta|e^{-i\hat{H}t}|\theta'\rangle = \sum_E \langle\theta|E\rangle e^{-iEt}\langle E|\theta'\rangle, \tag{6.70}$$

where the sum is taken over the spectrum of $\hat{H}$. As an example, the spectral decomposition is applied to compute the evolution operator kernel $U_t^c(\boldsymbol{\theta},\boldsymbol{\theta}')$ for the model in Section 5.5.2 ($\boldsymbol{\theta} = (\theta_1,\theta_2)$). The spectrum of the Hamiltonian (5.104) under constraint condition (5.105) contains only two points, $E = 0, 2\omega$ (for convenience, the energy is defined relative to the ground state energy, $E \to E - E_0$). Equation (6.70) then yields,

$$U_t^c(\boldsymbol{\theta},\boldsymbol{\theta}') = \langle\boldsymbol{\theta}|0\rangle\langle 0|\boldsymbol{\theta}'\rangle + e^{-2i\omega t}\langle\boldsymbol{\theta}|2\omega\rangle\langle 2\omega|\boldsymbol{\theta}'\rangle. \tag{6.71}$$

The basis wave functions in the model have the form $\langle\boldsymbol{\theta}|0\rangle = 1$, $\langle\boldsymbol{\theta}|\omega\rangle_1 = \theta_1$, $\langle\boldsymbol{\theta}|\omega\rangle_2 = \theta_2$, and $\langle\boldsymbol{\theta}|2\omega\rangle = \theta_1\theta_2$. Their conjugated states are

$$\langle 0|\boldsymbol{\theta}\rangle = \theta_1\theta_2, \quad \langle\omega|\boldsymbol{\theta}\rangle_1 = -\theta_2, \quad \langle\omega|\boldsymbol{\theta}\rangle_2 = \theta_1, \quad \langle 2\omega|\boldsymbol{\theta}\rangle = 1,$$

according to the conjugation rule given in Section 8.2. As a result, the following representation of the physical evolution operator kernel is obtained:

$$U_t^c(\boldsymbol{\theta},\boldsymbol{\theta}') = \theta_1'\theta_2' + e^{-i2\omega t}\theta_1\theta_2. \tag{6.72}$$

For the corresponding model without gauge symmetry, i.e. without condition (5.105), the spectral decomposition of the evolution operator kernel would contain additional terms,

$$e^{-i\omega t}\left(\langle\boldsymbol{\theta}|\omega\rangle_1\langle\omega|\boldsymbol{\theta}'\rangle_1 + \langle\boldsymbol{\theta}|\omega\rangle_2\langle\omega|\boldsymbol{\theta}'\rangle_2\right),$$

so that

$$U_t(\boldsymbol{\theta},\boldsymbol{\theta}') = \theta_1'\theta_2' + e^{-i\omega t}\left(\theta_2\theta_1' - \theta_1\theta_2'\right) + e^{-2i\omega t}\theta_1\theta_2. \tag{6.73}$$

It turns out that the kernels U_t and U_t^c are related in the same way as in the case of the bosonic gauge system discussed in Section 6.1,

$$U_t^c(\boldsymbol{\theta}, \boldsymbol{\theta}') = \int d\theta_2'' d\theta_1'' U_t(\boldsymbol{\theta}, \boldsymbol{\theta}'') Q(\boldsymbol{\theta}'', \boldsymbol{\theta}'), \tag{6.74}$$

$$Q(\boldsymbol{\theta}, \boldsymbol{\theta}') = \tfrac{1}{2}\left[\delta(\boldsymbol{\theta} - \boldsymbol{\theta}') + \delta(\boldsymbol{\theta} + \boldsymbol{\theta}')\right], \tag{6.75}$$

where $\delta(\boldsymbol{\theta}) = \theta_1\theta_2$, that is, $Q(\boldsymbol{\theta}, \boldsymbol{\theta}') = \theta_1\theta_2 + \theta_1'\theta_2'$ (δ-functions of Grassmann variables have been studied in [231, 232, 305–307]). The validity of (6.74) is proved by direct calculation of Grassmann integrals. For instance,

$$\int d\theta_2'' d\theta_1'' e^{-i\omega t} \left(\theta_2\theta_1'' - \theta_1\theta_2''\right)\left(\theta_1''\theta_2'' + \theta_1'\theta_2'\right) = 0.$$

Note factor 1/2 in (6.75) which does not occur in the analogous expression (6.7) for bosonic gauge systems. Its absence in the latter is associated with reduction of the integration domain to the half-axis. For the Grassmann variables the very notion of a "half-axis" does not even exist, which explains factor 1/2 in (6.75).

It is not difficult to verify that kernel (6.74) has the required semigroup property,

$$\int d^2\theta'' U_t^c(\boldsymbol{\theta}, \boldsymbol{\theta}'') U_{t'}^c(\boldsymbol{\theta}'', \boldsymbol{\theta}) = U_{t+t'}^c(\boldsymbol{\theta}, \boldsymbol{\theta}'). \tag{6.76}$$

Equations (6.73)–(6.75) solve the problem because the path integral representation for U_t can be obtained in a standard way, thanks to (6.76).

In the holomorphic representation (see Section 8.2) the Hamiltonian path integral is constructed as follows. The physical states in this representation are $\langle \boldsymbol{\psi}^* | 0\rangle = 1$ and $\langle \boldsymbol{\psi}^* | 2\omega\rangle = \psi_1^*\psi_2^*$. They satisfy the constraint equation (5.105). The same arguments that led to (6.73)–(6.75) yield

$$U_t^c(\boldsymbol{\psi}^*, \boldsymbol{\psi}'') = \int \prod_{\alpha=1}^{2} \left(d\tilde{\psi}_\alpha^* d\tilde{\psi}_\alpha\right) \exp\left(-\tilde{\boldsymbol{\psi}}^*\tilde{\boldsymbol{\psi}}\right) U_t(\boldsymbol{\psi}^*, \tilde{\boldsymbol{\psi}}) Q(\tilde{\boldsymbol{\psi}}^*, \boldsymbol{\psi}''), \tag{6.77}$$

$$Q(\boldsymbol{\psi}^*, \boldsymbol{\psi}') = \tfrac{1}{2}\left[\exp\left(\boldsymbol{\psi}^*\boldsymbol{\psi}'\right) + \exp\left(-\boldsymbol{\psi}^*\boldsymbol{\psi}'\right)\right] = \cosh\left(\boldsymbol{\psi}^*\boldsymbol{\psi}'\right). \tag{6.78}$$

Here $U_t(\boldsymbol{\psi}^*, \boldsymbol{\psi}')$ is the evolution operator kernel for the corresponding system without gauge symmetry for which there is a standard path integral representation,

$$U_t(\boldsymbol{\psi}^*, \boldsymbol{\psi}') = \int \prod_{\tau=0}^{t} \left[d\boldsymbol{\psi}^*(\tau) d\boldsymbol{\psi}(\tau)\right] \exp\left(\tfrac{1}{2}\phi + iS[\boldsymbol{\psi}^*, \boldsymbol{\psi}']\right) \tag{6.79}$$

$$S[\boldsymbol{\psi}^*, \boldsymbol{\psi}'] = \int_0^t d\tau \left[\tfrac{1}{2i}(\dot{\boldsymbol{\psi}}^*\boldsymbol{\psi} - \boldsymbol{\psi}^*\dot{\boldsymbol{\psi}}) - V(\boldsymbol{\psi}^*\boldsymbol{\psi})\right],$$

where $\phi = \boldsymbol{\psi}^*(t)\boldsymbol{\psi}(t) + \boldsymbol{\psi}^*(0)\boldsymbol{\psi}(0)$. It should be pointed out that variables $\boldsymbol{\psi}(t)$ and $\boldsymbol{\psi}^*(0)$ are integration variables in (6.79), while $\boldsymbol{\psi}^*(t)$ and $\boldsymbol{\psi}(0)$ are fixed by the boundary condition for the path integral, $\boldsymbol{\psi}^*(t) = \boldsymbol{\psi}^*$ and $\boldsymbol{\psi}(0) = \boldsymbol{\psi}'$.

It is noteworthy that the operator $\hat{Q}$ with kernel (6.78) in the relation $\hat{U}_t^c = \hat{U}_t\hat{Q}$ (see (6.77)) acts as the projection operator onto the physical subspace of the total Hilbert space determined by the Dirac condition (5.105). The projection operator $\hat{Q}$ can be defined as the averaging over the gauge group, i.e. for any state $|\phi\rangle$,

$$\langle \psi^* |\hat{Q}|\phi\rangle = (2\pi)^{-1} \int_0^{2\pi} d\varphi \langle e^{-i\varphi\Gamma}\psi^*|\phi\rangle.$$

Here $\Gamma = \tau_3$, the Pauli matrix (see (5.94)). Consequently, the relation between the evolution operator in the physical subspace and that in the total Hilbert space is given by,

$$\hat{U}_t^c = \hat{Q}\hat{U}_t\hat{Q} = \hat{U}_t\hat{Q}. \tag{6.80}$$

The latter equality follows from $\hat{Q}^2 = \hat{Q}$ and the commutativity of the Hamiltonian (5.104) with the projection operator. Thus, the relation (6.77) may be expressed in an alternative form,

$$U_t^c(\psi^*, \psi') = (2\pi)^{-1} \int_0^{2\pi} d\varphi U_t(\psi^*, e^{i\varphi\Gamma}\psi'). \tag{6.81}$$

The limit $t \to 0$ in (6.81) shows that the kernel (6.78) is obtained by group averaging of the unit operator kernel in the total Hilbert space. Observe that representation (6.80) is rather general and not specific to any particular gauge model, provided projection $\hat{Q}$ is suitably defined [207, 308, 309]. The evolution operator kernel is automatically gauge invariant and no explicit elimination of non-physical degrees of freedom is required. Use of the projection operator method to develop the path integral formalism for general gauge systems is discussed in Sections 6.5.2 and 6.7. As an example, kernel Q in the holomorphic representation for bosonic gauge systems is calculated in Section 8.2, and relation (6.80) is illustrated there as well.

6.5 Hamiltonian path integral in an arbitrary gauge

In Section 5.3 it is shown that residual discrete gauge transformations acting in the phase space of the physical degrees of freedom may have a rather complicated structure with an unsuitable choice of gauge condition (or local canonical variables in the physical phase space). Moreover, they do not always form a group. In realistic gauge theories, the gauge condition choice is often motivated by, e.g. the Lorentz covariance of the gauge fixed theory rather than by the geometrical structure of the orbit space. This causes the well-known problem of gauge fixing ambiguities [209, 211].

Here the Hamiltonian path integral formalism is developed for an arbitrary choice of physical variables (local coordinates in the orbit space). The discussion is limited to not too complicated, but sufficiently representative, gauge fixing conditions with an example of the model studied in Section 5.3. Then the procedure is generalized to mechanical gauge systems with an arbitrary gauge group. Yang–Mills theories are studied in Sections 6.3.2 and 6.7 (see also [195, 198, 213, 221]).

6.5.1 The gauge group SO(2)

The unit operator kernel $\langle u|u'\rangle_{\rm ph}$ in the theory specified by the Schrödinger equation (5.53) and the scalar product (5.54) is defined in the modular domain, $u, u' \in K$. Thanks to the S-invariance of the physical states (see (5.56)), the kernel can be extended into the non-physical region $u \in R$ in full analogy with the extension method used in Section 6.2.3,

$$\langle u|u'\rangle_{\rm ph} = \sum_E \Phi_E(u)\Phi_E^*(u') = \int_{-\infty}^{\infty} \frac{du''}{(\mu\mu'')^{1/2}}\, \delta(u-u'')Q(u'',u'), \tag{6.82}$$

$$Q(u,u') = \sum_S \delta\left(u - u_s'\right), \tag{6.83}$$

where $u \in R$, $u' \in K$, $\mu = \mu(u)$, $\mu'' = \mu(u'')$, and the functions $u_s' \equiv u_s(u')$ are defined in Section 5.3.2. Kernel (6.82) is S-invariant, meaning that for any $u, u' \in K$ the following equality holds:

$$\langle u_s(u)|u'\rangle_{\rm ph} = \langle u|u'\rangle_{\rm ph} = [\mu(u)\mu(u')]^{-1/2}\delta(u-u'), \quad \forall s \in S. \tag{6.84}$$

This is a natural consequence of the spectral resolution of unity (6.82) and the S-invariance of the physical wave functions (5.56). It is instructive though to give an alternative proof based only on the integral representation also given in (6.82).

The functions u_s are defined implicitly as solutions of Eq. (5.49), $r^2(u_s) = r^2(u)$ where $u \in K$. By differentiating this equality the transformation law for the measure μ is found,

$$\mu(u_s) = \frac{1}{2}\frac{d}{du_s} r^2(u_s) = \frac{1}{2}\left[\frac{du_s(u)}{du}\right]^{-1} \frac{d}{du} r^2(u) = \left[\frac{du_s(u)}{du}\right]^{-1} \mu(u). \tag{6.85}$$

If argument u is replaced by $u_s(u)$ in (6.82), then only one term contributes to the sum over S in the right-hand side of (6.82) (see (6.83)) so that

$$\langle u_s(u)|u'\rangle_{\rm ph} = [\mu(u_s(u))\mu(u_s(u'))]^{-1/2}\delta(u_s(u) - u_s(u')), \quad u, u' \in K. \tag{6.86}$$

The equation $u_s(u) = u_s(u')$ has only one solution, $u = u'$. Therefore, after changing the argument of the δ-function by using rule (6.85), equality (6.84) follows. The proof is completed.

The Hamiltonian path integral can be obtained by a conventional procedure based on the operator formalism developed in Section 5.3.3. The infinitesimal evolution operator kernel has the form,

$$U_\varepsilon^{\rm ph}(u,u') = \langle u|e^{-i\hat{H}\varepsilon}|u'\rangle_{\rm ph} = \left[1 - i\varepsilon\hat{H}(u)\right]\langle u|u'\rangle_{\rm ph} + O(\varepsilon^2), \tag{6.87}$$

where $\varepsilon \to 0$ and $\hat{H}(u)$ is given in (5.53). The integral representation (6.82) of $\langle u|u'\rangle_{\rm ph}$, in which

$$\delta(u-u'') = (2\pi)^{-1}\int dp \exp[ip(u-u'')],$$

is then substituted into (6.87). When calculating the action of $\hat{H}(u)$, operator ordering should be taken into account [96, 208]. In order to do so, it is convenient to write the Hamiltonian in terms of the Hermitian momentum operator, $\hat{P}_u = -i\mu^{-1/2}\partial_u \circ \mu^{1/2}$,

$$\hat{H}_u = \frac{1}{2}\hat{P}_u g(u)\hat{P}_u + V_q + V, \qquad V_q = \frac{1}{2\mu^{1/2}}\partial_u\left(g\partial_u\mu^{1/2}\right), \tag{6.88}$$

where $g = r^2(u)/\mu^2(u)$. Computing all the derivatives as required in (6.87), the following integral representation for the infinitesimal evolution operator kernel is derived up to terms of order $O(\varepsilon^2)$ [208, 213]:

$$U_\varepsilon^{\rm ph}(u,u') = \int_{-\infty}^{\infty} \frac{du''}{[\mu(u)\mu(u'')]^{1/2}} U_\varepsilon^{\rm eff}(u,u'')Q(u'',u'), \tag{6.89}$$

$$U_\varepsilon^{\rm eff}(u,u'') = \int_{-\infty}^{\infty} \frac{dp}{2\pi} \exp i\left[p(u-u'') - \varepsilon H^{\rm eff}(u,p)\right], \tag{6.90}$$

$$H^{\rm eff} = \tfrac{1}{2}gp^2 + \tfrac{i}{2}p\partial_u g + V_q + V. \tag{6.91}$$

In Eqs. (6.89)–(6.91), $u \in R$ and $u' \in K$ because the functions $u_s(u')$ in the kernel Q are defined in the region $K = \bigcup_\alpha K_\alpha$ (see Section 5.3.2).

Next the convolution rule for two kernels (6.89) should be found. By the scalar product definition,

$$U_{2\varepsilon}^{\rm ph}(u,u') = \int_K du_1\mu(u_1)\sum_S \left[\mu(u)\mu(u_s(u_1))\right]^{-1/2} U_\varepsilon^{\rm eff}(u,u_s(u_1))U_\varepsilon^{\rm ph}(u_1,u'). \tag{6.92}$$

This expression is further transformed by means of the measure property (6.85) to obtain

$$U_{2\varepsilon}^{\rm ph}(u,u') = \int_K du_1 \sum_S \left|\frac{du_s(u_1)}{du_1}\right| \left[\frac{\mu(u_s(u_1))}{\mu(u)}\right]^{1/2} U_\varepsilon^{\rm eff}(u,u_s(u_1))U_\varepsilon^{\rm ph}(u_1,u'). \tag{6.93}$$

In each term of the sum over S a change of the integration variable is then carried out, $u_1 \to u_s(u_1)$. When doing so, the orientation of the integration region may change if $du_s(u)/du$ is negative in K for some particular u_s, in which case the new upper integration limit becomes less than the lower one. To avoid any confusion in what follows, it is set,

$$\int_{K_s} du \geq 0, \qquad \forall s \in S. \tag{6.94}$$

By definition (6.94) integration over the whole axis can be written as

$$\int_{-\infty}^{\infty} du = \sum_S \int_{K_s} du, \tag{6.95}$$

while the rule for changing the integration variable reads,

$$\int_K du \left| \frac{du_s}{du} \right| = \int_{K_s} du. \tag{6.96}$$

The absolute value of the intergrand in the left-hand side of (6.96) provides for preservation of the positive orientation of the integral (see (6.94)) under the change of variables $u \to u_s(u)$. Applying the rule (6.96), one infers,

$$U_{2\varepsilon}^{\rm ph}(u,u') = \sum_S \int_{K_s} du_1 \left[\frac{\mu(u_1)}{\mu(u)}\right]^{1/2} U_\varepsilon^{\rm eff}(u,u_1) U_\varepsilon^{\rm ph}\left(u_s^{-1}(u_1),u'\right). \tag{6.97}$$

For each fixed $s \in S$, the integration variable ranges over K_s. Therefore function $u_s^{-1}(u_1)$ in (6.97) is well defined. Recall that $u_s^{-1}: K_s \to K$ is the mapping inverse to $u_s :\ K \to K_s$ which is one-to-one.

Kernel $U_\varepsilon^{\rm ph}(u,u')$ has the property,

$$U_\varepsilon^{\rm ph}\left(u_s^{-1}(u),u'\right) = U_\varepsilon^{\rm ph}(u,u')\ , \quad u \in K_s\ , \quad u' \in K. \tag{6.98}$$

It follows from representation (6.87) and two equalities,

$$\hat{H}(u_s^{-1}(u)) = \hat{H}(u), \tag{6.99}$$

$$\left\langle u_s^{-1}(u)|u'\right\rangle_{\rm ph} = \langle u|u'\rangle_{\rm ph}, \qquad u \in K_s, \quad u' \in K. \tag{6.100}$$

To prove the S-invariance of the Hamiltonian (6.99), the measure property (6.85) and the chain rule $d/du = (du_s(u)/du)d/du_s$ should be used. Relation (6.100) means that the equality

$$\left\langle u_s^{-1}(u)|u'\right\rangle_{\rm ph} = \langle u|u'\rangle_{\rm ph} = [\mu(u)\mu\ u_s(u')]^{-1/2}\,\delta(u - u_s(u')) \tag{6.101}$$

should hold for any $s \in S$, $u \in K_s$, and $u' \in K$. Its proof is analogous to the proof of the S-invariance of the kernel (6.82) (see (6.84)–(6.86)).

Note that relation (6.98) is, generally speaking, valid only up to terms $O(\varepsilon^2)$ if the kernel $U_\varepsilon^{\rm ph}(u,u')$ is represented by integrals (6.89) and (6.90). Indeed, the exact equality holds only for kernel (6.87) which differs from (6.89) by terms $O(\varepsilon^2)$.

Substitution of (6.98) into (6.97) together with rule (6.95) yields,

$$U_{2\varepsilon}^{\rm ph}(u,u') = \int_{-\infty}^{\infty} du_1 \left[\frac{\mu(u_1)}{\mu(u)}\right]^{1/2} U_\varepsilon^{\rm eff}(u,u_1) U_\varepsilon^{\rm ph}(u_1,u') \tag{6.102}$$

$$= \int_{-\infty}^{\infty} \frac{du''}{[\mu(u)\mu(u'')]^{1/2}} U_{2\varepsilon}^{\rm eff}(u,u'')Q(u'',u'), \tag{6.103}$$

$$U_{2\varepsilon}^{\text{eff}}(u,u'') = \int_{-\infty}^{\infty} du_1 U_\varepsilon^{\text{eff}}(u,u_1) U_\varepsilon^{\text{eff}}(u_1,u''), \tag{6.104}$$

where representation (6.89) has been used.

Set $t = \varepsilon N$ where $\varepsilon \to 0$ and $N \to \infty$, while t remains fixed. The physical transition amplitude for a finite time t is given by the convolution of N kernels (6.89),

$$U_t^{\text{ph}}(u,u') = \int_K du_1 \mu(u_1) \cdots du_{N-1} \mu(u_{N-1}) \prod_{i=0}^{N-1} U_\varepsilon^{\text{ph}}(u_i, u_{i+1}), \tag{6.105}$$

where $u_0 \equiv u$ and $u_N \equiv u'$. Simple repetition of the above calculations for the convolution of two kernels $U_\varepsilon^{\text{ph}}$ on the very left in this expression gives

$$\begin{aligned} U_t^{\text{ph}}(u,u') = &\int_K \prod_{i=2}^{N-1} (du_i \mu(u_i)) \sum_S [\mu(u)\mu\ u_s(u_2)]^{-1/2}\, U_{2\varepsilon}^{\text{eff}}(u, u_s(u_2)) \\ &\times U_\varepsilon^{\text{ph}}(u_2,u_3) \cdots U_\varepsilon^{\text{ph}}(u_{N-1},u'). \end{aligned}$$

The calculation is then carried out further iteratively in the same fashion by adding one kernel $U_\varepsilon^{\text{ph}}$ to the kernel $U_{k\varepsilon}^{\text{eff}}$ at each step, $k = 2, 3, ..., N-1$. The result reads,

$$U_t^{\text{ph}}(u,u') = \int_{-\infty}^{\infty} \frac{du''}{[\mu(u)\mu(u'')]^{1/2}} U_t^{\text{eff}}(u,u'') Q(u'',u'), \tag{6.106}$$

$$\begin{aligned} U_t^{\text{eff}}(u,u'') &= \lim_{N\to\infty} \int_{-\infty}^{\infty} du_1 \cdots du_{N-1} \prod_{i=0}^{N-1} U_\varepsilon^{\text{eff}}(u_i,u_{i+1}) \\ &= \int_{-\infty}^{\infty} \prod_{\tau=0}^{t} \left[\frac{dp(\tau)du(\tau)}{2\pi}\right] \exp\left\{ i\int_0^t d\tau\, [p\dot{u} - H^{\text{eff}}(p,u)]\right\}, \end{aligned} \tag{6.107}$$

where $u(t) = u \equiv u_0$ and $u(0) = u'' \equiv u_N$ are the initial conditions for the path integral.

Thus, the problem of developing the Hamiltonian path integral formalism in an arbitrary gauge is solved in (6.106) and (6.107). The path integral (6.106) takes into account both the fact that the physical variables are described by curvilinear coordinates and the structure of the physical phase and configuration spaces. The former determines the effective Hamiltonian $H^{\text{eff}}(p,u)$ from the classical gauge-fixed Hamiltonian (compare H^* in (Eq. 8.71)) and the measure μ, while the latter specifies the operator Q that symmetrizes over the residual discrete gauge symmetry. It follows from (6.106) and (6.107) that the problem of "copies", i.e. the gauge equivalent configurations on the gauge fixing surface (spanned by $u \in R$) in the total configuration space, is solved by symmetrization of the kernel $U_t^{\text{eff}}(u,u'')$, which is given by the Hamiltonian path integral with the standard measure, with respect to the residual gauge transformations S acting on the point u'' or u as proposed in [213], rather than by a reduction of the integration domain in the path integral. The same conclusion has been reached for the 2D Yang–Mills in Section 6.3.2 (compare (6.49) and (6.106)). It has also been noted there that method used in the derivation of (6.49) for solving the gauge fixing ambiguity problem in the path integral formalism fully agrees with an explicitly gauge-invariant lattice approach [234, 235] (where no gauge fixing is required). In Section 6.7, an alternative proof of this result for general gauge systems is offered that is solely based on the gauge-invariant path integral formalism and does not use the Dirac operator formalism [195, 303, 310], i.e. without constructing the physical Hamiltonian operator, like in Eq. (6.87), and extension of the unit operator kernel, similar to (6.82).

6.5.2 Systems with an arbitrary gauge group

Reduction of the phase space of physical variables is taken into account in the Hamiltonian path integral by operator $\hat{Q}$ which symmetrizes the evolution operator kernel with respect to the residual discrete gauge group. An explicit form of $\hat{Q}$ depends on the choice of physical variables (local coordinates on the gauge orbit space) as argued in Section 6.5.1. On the other hand, with an example of a simple model from Section 5.3.3, it is shown that the amplitudes (scalar products of states) are independent of this choice because the physical state vectors are functions of gauge-invariant combinations of dynamical variables of the theory.

Here it is demonstrated that operator $\hat{Q}$ is, in fact, independent of system dynamics and fully determined by the gauge group, its representation realized by degrees of freedom of the system, and the choice of physical variables. Making use of this property, the path integral formalism for an arbitrary gauge can be developed for a rather general class of gauge systems which is in agreement with the Dirac operator method [31, 208].

Consider a quantum theory defined by the Schrödinger equation,

$$\left[-\frac{1}{2}\left\langle\frac{\partial}{\partial x},\frac{\partial}{\partial x}\right\rangle+V(x)\right]\psi_E=E\psi_E \tag{6.108}$$

and the normalization condition

$$\int d^N x\psi_E^*(x)\psi_{E'}(x)=\delta_{EE'}. \tag{6.109}$$

Here variable x realizes a linear representation of a compact group G, and

$$\langle x,y\rangle=\sum_{i=1}^{N}x_iy_i \tag{6.110}$$

is an invariant scalar product in the representation space with x_i and y_i being real independent components of the elements x and y, respectively, i.e. the integration in (6.109) is carried out over R^N. The potential V in (6.108) is assumed to be a G-invariant function.

A quantum theory with a gauge symmetry is obtained from (6.108) and (6.109) by demanding that the physical solutions of (6.108) are also annihilated by generators of transformations from G, $\hat{\sigma}_a \Phi_E = 0$ [148]. The states Φ_E form a basis in the physical Hilbert space $\mathcal{H}_{\rm ph}$. Since $\hat{\sigma}_a$ generate linear transformations from G, for any state $\exp(\omega_a \hat{\sigma}_a)\psi(x) = \psi(T(\omega)x)$, where $T(\omega)$ is an element of G in the representation realized by x. Hence, physical states are gauge invariant,

$$\Phi(T(\omega)x) = \Phi(x). \tag{6.111}$$

Relations (6.108)–(6.111) define a quantum gauge theory in the total configuration space.

Consider averaging over group G of a wave function,

$$\hat{P}_G \psi(x) = \frac{1}{V_G} \int_G dg(\omega) \psi(T(\omega)x), \tag{6.112}$$

where $V_G = \int dg(\omega)$ is the group volume, $dg(\omega)$ is an invariant measure on G. The operator $\hat{P}_G$ is nothing but an orthogonal projector of the total Hilbert space $\mathcal{H}$ onto $\mathcal{H}_{\rm ph}$. The unit operator kernel in $\mathcal{H}$ has the spectral decomposition,

$$\langle x|x'\rangle = \sum_E \Phi_E(x)\Phi_E^*(x') + \sum_{\tilde{E}} \psi_{\tilde{E}}(x)\psi_{\tilde{E}}^*(x') = \delta^N(x - x'), \tag{6.113}$$

where the states Φ_E and $\psi_{\tilde{E}}$ form orthonormal bases in $\mathcal{H}_{\rm ph}$ and the non-physical subspace $\mathcal{H} \ominus \mathcal{H}_{\rm ph}$, respectively. The latter is always possible because the constraint operators generate a Lie algebra of G. The first term in (6.113) is the unit operator kernel in $\mathcal{H}_{\rm ph}$. The action of the projector $\hat{P}_G$ on the unit operator (6.113) yields the unit operator in $\mathcal{H}_{\rm ph}$,

$$\langle x|x'\rangle_{\rm ph} = \langle x|\hat{P}_G|x'\rangle = V_G^{-1} \int_G dg(\omega)\delta^N(x - T(\omega)x'). \tag{6.114}$$

Spectral decomposition of the kernel (6.114) is given by the first term in the right-hand side of the first equality in (6.113). Its gauge invariance is evident.

Representation (6.114) allows for an explicitly gauge-invariant representation of the physical evolution operator kernel without any particular choice of parameterization of the orbit space (physical variables). The evolution operator kernel in the theory (6.108) and (6.109) is given by the standard Hamiltonian path integral,

$$U_t(x, x') = \int \prod_{\tau=0}^{t} \left[\frac{d^N p(\tau) d^N x(\tau)}{(2\pi)^N}\right] \exp(iS[p, x]), \tag{6.115}$$

$$S[p, x] = \int_0^t d\tau \left[\langle p, \dot{x}\rangle - \frac{1}{2}\langle p, p\rangle - V(x)\right],$$

where $x(0) = x'$ and $x(t) = x$. On the other hand, the evolution operator kernel can be expanded into the spectral sum,

$$U_t(x, x') = \sum_E \Phi_E(x)\Phi^*(x')e^{-iEt} + \sum_{\tilde{E}} \psi_{\tilde{E}}(x)\psi^*_{\tilde{E}}(x')e^{-i\tilde{E}t}, \tag{6.116}$$

in which the first term is the physical evolution operator kernel $U_t^{ph}(x, x')$ (the transition amplitude on the orbit space). Therefore the physical evolution operator $\hat{U}_t^{\mathrm{ph}}$ is the product $\hat{P}_G\hat{U}_t\hat{P}_G = \hat{U}_t\hat{P}_G$ whose kernel reads,

$$U_t^{\mathrm{ph}}(x, x') = \int d^N x'' U_t(x, x'')\langle x''|\hat{P}_G|x'\rangle. \tag{6.117}$$

Relations similar to (6.114)–(6.117) can be obtained in the coherent state representation (see Section 8.5). An explicit form of the physical projection operator kernel (6.114) is also given there for the models from Sections 5.2, 5.4, and 5.6 (see also [31, 195, 208, 308, 311, 312]).

Although in practical terms calculation of the gauge group averaging might be a tedious task, representation (6.114) shows that the kernel $\langle x|x'\rangle_{\mathrm{ph}}$ has a universal structure, i.e. it depends only on group G and its representation used in the system, not on peculiarities of the dynamics (potential V). It is therefore of significance to establish a relation between the operator $\hat{Q}$ and (6.114). In order to do so, one has to introduce physical variables (local coordinates in the orbit space) and go over a description of the theory in the physical configuration space. Let the dimension of the orbit space be M. Then the number of independent constraints is $N - M$. Define curvilinear coordinates θ and u by,

$$x = T(\theta)\tilde{x}(u). \tag{6.118}$$

Here $x = \tilde{x}(u)$, $u \in R^M$, are parametric equations of a gauge fixing surface $\chi(x) = 0$, i.e. $\chi(\tilde{x}(u)) \equiv 0$ for all $u \in R^M$. The gauge condition should be chosen so that for every x there exists a gauge transformation $T(\omega)$ such that $T(\omega)x$ lies in the gauge fixing surface. In other words, every gauge orbit must intersect the gauge fixing surface at least once. With such a choice, for every $x \in R^N$ there is at least one pair (θ, u) for which (6.118) holds. The Jacobian of the change of variables is factorized, $d^N x = \mu(u)d^M u d\mu_G(\theta)$, where $d\mu_G(\theta)$ is the measure on the group G manifold. This can be established by writing the volume N-form, $dx_1 \wedge dx_2 \wedge \cdots \wedge dx_N$, in the new variables (6.118). Function $\mu(u)$ may vanish only on a zero measure set (otherwise (6.118) is not a change of variables).

In the new variables, the physical wave functions are independent of θ, $\tilde{\Phi}(x(u, \theta)) = \bar{\Phi}(u, \theta) = \Phi(u)$ because $\hat{\sigma}_a$ generate translations of θ.

The basis $\Phi_E(u)$ in $\mathcal{H}_{\mathrm{ph}}$ is formed by solutions of (6.108) in the curvilinear coordinates (6.118),

$$\hat{H}_{\rm ph}(u)\Phi_E(u) = E\Phi_E(u), \tag{6.119}$$

$$\hat{H}_{\rm ph}(u) = \tfrac{1}{2}\hat{P}_\alpha g^{\alpha\beta}_{\rm ph}(u)\hat{P}_\beta + V_q(u) + V(\tilde{x}(u)), \tag{6.120}$$

where $\hat{P}_\alpha = -i\mu^{-1/2}\partial/\partial u_\alpha \circ \mu^{1/2}$ is the Hermitian momentum operator, $\alpha,\ \beta = 1, 2, \ldots M$, $g^{\alpha\beta}_{\rm ph}$ is the metric tensor in the physical configuration space. It coincides with the uu-block of the inverse Euclidean metric tensor in the coordinates (6.118) (just as $g^{\alpha\beta}_{\rm ph}$ introduced in Sections 5.6.3 and 5.6.4). The quantum potential is given by

$$V_q = \frac{1}{2\sqrt{\mu}}\left(\partial_\alpha g^{\alpha\beta}_{\rm ph}\right)\partial_\beta\sqrt{\mu} + \frac{1}{2\sqrt{\mu}} g^{\alpha\beta}_{\rm ph}\partial_\alpha\partial_\beta\sqrt{\mu}, \tag{6.121}$$

where $\partial_\alpha \equiv \partial/\partial u_\alpha$. The first two terms in the physical Hamiltonian (6.120) are the "radial" part of the Laplace operator $-\Delta/2 = -1/2\langle\partial/\partial x, \partial/\partial x\rangle$ (see (6.108)) in the coordinates (6.118) (i.e. the Laplace operator in which all terms with derivatives with respect to the non-physical variables θ_a are omitted). By construction, for every $x \in R^N$ there might exist more than one pair (u, θ) for which (6.118) holds. Let $\tilde{K}$ be a subset of pairs $(u, \theta) \in R^N$ for which the mapping (6.118) is one-to-one. Let the range of u be K when (u, θ) ranges over $\tilde{K}$. Then the physical states $\Phi = \Phi(u)$ are normalized with the measure induced by the change of variables,

$$\int d^N x = \int_{G/H} d\mu_G(\theta) \int_K d^M u\mu(u) = V_G V_H^{-1} \int_K d^M u\mu(u), \tag{6.122}$$

where V_G is the group manifold volume and V_H is the volume of a stationary subgroup H of a generic configuration on the gauge fixing surface (if the representation of G realized by x is such that H is trivial, then $G/H = G$ and $V_H = 1$). In fact, $V_G V_H^{-1}\mu(u)$ is the volume of the gauge orbit through $x = \tilde{x}(u)$ [26, 313]. This follows from an obvious decomposition $\tilde{K} = G/H \times K$ where $\theta \in G/H$ and $u \in K$.

In order to determine K, consider transformations of the pair (u, θ) under which x in (6.118) is not changed (the symmetry transformations of the mapping (6.118)). They include shifts S_θ of θ by integer multiples of periods of the compact manifold G/H while u is fixed, and a set S of transformations when both elements of the pair change, $\theta \to \theta'_s(\theta, u)$ (modulo S_θ) and $u \to u_s(u)$. By definition, the functions $u_s(u)$ and $\theta'_s(u, \theta)$ satisfy the condition $x = T(\theta)\tilde{x}(u) = T(\theta'_s)\tilde{x}(u_s)$. The index s labels the transformations from S. By construction, for every fixed u the points $x_s = \tilde{x}(u_s)$ are the points of intersection of the gauge fixing surface $\chi(x) = 0$ with the gauge orbit through $x = \tilde{x}(u)$. Indeed, the gauge orbit through $x = \tilde{x}(u)$ is $T(\theta_s)\tilde{x}(u)$ where θ_s is a parameter along the orbit. Its intersection with $\chi(x) = 0$ is determined by the equation, $\chi(T(\theta_s)\tilde{x}(u)) = 0$ for θ_s where $\chi(\tilde{x}(u)) = 0$ for any u. If this equation has a nontrivial solution, $\theta_s = \theta_s(u)$, then there is a nontrivial transformation from S. The function $u_s = u_s(u)$ is then found from $T(\theta_s)\tilde{x}(u) = \tilde{x}(u_s)$. The functions $\theta'_s(u,\theta)$ are then determined by the condition $T(\theta'_s) = T(\theta)T^{-1}(\theta_s(u))$. Thus, the group elements $T(\theta_s)$ define residual discrete gauge transformations in the gauge $\chi(x) = 0$. The region K is therefore the set of u in R^M such that every gauge orbit intersects that part of the gauge fixing surface $x = \tilde{x}(u)$, $u \in K \subset R^M$, only once. Thus, $\tilde{K} = R^N/(S_\theta \times S) = (R^{N-M}/S_\theta)\bigcup(R^M/S) = (G/H)\bigcup K$, i.e. K is the fundamental modular domain in the space of parameters $u \in R^M$, that span the gauge fixing surface, with respect to the set of transformations $S:\ u \to u_s(u)$. These transformations relate values of the parameters u that correspond to gauge-equivalent configurations in the gauge fixing surface.

Study of a simple example in Section 5.3.2 shows that the set S may depend on a point, i.e. for different u there might exist different numbers of elements of S (mappings $u_s:\ K \to R^M$). In this case, as before, it is assumed that R^M can be split into subsets R^M_α such that for $u \in R^M_\alpha$ the set of symmetry transformations $S = S_\alpha$ has a fixed number of elements, and $S = \prod_\alpha S_\alpha$, where each S_α acts in its own subset R^M_α. The modular domain is then split into subsets $K_\alpha = R^M_\alpha/S_\alpha$, $K = \bigcup K_\alpha$, so that $u_s(u) = u_{s\alpha}(u) : K_\alpha \to R^M_\alpha$ are smooth functions of residual gauge transformations $T(\theta_s)$. The residual gauge transformations do not form a group in general. This happens when the product of two group elements $T(\theta_s)T(\theta_{s'})$ does not generate another copy of $x = \tilde{x}(u)$, i.e. $T(\theta_s)T(\theta_{s'})\tilde{x}(u)$ is not in the gauge fixing surface (see Section 5.3.2). By choosing a particular subset of R^M_α that is isomorphic to K_α, an explicit form of functions u_s (or a representation of S_α) is fixed as one-to-one mappings, $u_s:\ K_\alpha \to K_{\alpha s}$ where $R^M_\alpha = \bigcup_S K_{\alpha s}$. For developing the quantum theory and, in particular, the Hamiltonian path integral formalism, it is sufficient to know only the mappings $u_s:\ K \to K_s$. According to the procedure described above, to find an explicit form of $u_s(u)$, the equation $\chi(T(\theta_s)\tilde{x}(u)) = 0$, $u \in K$, has to be solved first for $\theta_s = \theta_s(u)$. For smooth χ and $\tilde{x}$, K (or K_α) is the region in which $\theta_s(u)$ is smooth (by the implicit function theorem). The functions $u_s = u_s(u)$ are then defined implicitly by the equation $T(\theta_s(u))\tilde{x}(u) = \tilde{x}(u_s)$.

The quantum theory is similar to that developed in Section 5.3.3. The scalar product in $\mathcal{H}_{\rm ph}$ is given by (5.54) where $du \to d^M u$ and the factor $V_G V_H^{-1}$ is included into the norm of physical states. Let an orthonormal basis in the total Hilbert space be $\psi_E(x)$. The basis in $\mathcal{H}_{\rm ph}$ can be determined as an induced one by the projection (6.112),

$$\Phi_E(u) = \hat{P}_G\psi_E(x) = \hat{P}_G\psi_E(T(\theta)\tilde{x}(u)) = \hat{P}_G\psi_E(\tilde{x}(u)). \tag{6.123}$$

The last equality in (6.123) follows from the invariance of the measure $dg(\omega)$ in (6.112) under group shifts. Thanks to the invariance of the measure $dg(\omega)$ and the definition $\tilde{x}(u_s) = T(\theta_s)\tilde{x}(u)$, the physical states are S-invariant, $\Phi_E(u_s) = \Phi_E(u)$. The latter allows us to extend the unit operator kernel $\langle u|u'\rangle_{\rm ph}$ in $\mathcal{H}_{\rm ph}$ to the non-physical region $u \in R^M$, similarly to (6.82) (if du'' and

$\delta(u - u'')$ are replaced by $d^M u''$ and $\delta^M(u - u'')$, respectively). The kernel of operator $\hat{Q}$ also has the corresponding form (6.83). If the Jacobian $\mu(u) = \kappa^2(u)$ with $\kappa(u)$ being real analytic in R^M, then $(\mu\mu'')^{1/2} \to \kappa\kappa''$ in (6.82). Examples of such theories are given in Sections 5.4 and 5.8 (see also Section 6.3).

The method to develop the Hamiltonian path integral formalism proposed in Section 6.5.1 is readily generalized to the general case discussed here. The transition amplitude for the Hamiltonian (6.120) has the form,

$$U_t^{\text{ph}}(u, u') = \int_{R^M} \frac{d^M u''}{(\mu(u)\mu(u''))^{1/2}} U_t^{\text{eff}}(u, u'') Q(u'', u'), \tag{6.124}$$

$$Q(u, u') = \sum_S \delta^M(u - u_s(u')), \tag{6.125}$$

$$U_t^{\text{eff}}(u, u'') = \int_{R^M} \prod_{\tau=0}^{t} \left[\frac{d^M p(\tau) d^M u(\tau)}{(2\pi)^M}\right] \exp\left(iS[p, u]\right), \tag{6.126}$$

$$S[p, u] = \int_0^t d\tau \left[p_\alpha \dot{u}_\alpha - H^{\text{eff}}(p, u)\right],$$
$$H^{\text{eff}}(p, u) = \tfrac{1}{2} g_{\text{ph}}^{\alpha\beta} p_\alpha p_\beta + \tfrac{i}{2} \partial_\alpha g_{\text{ph}}^{\alpha\beta} p_\beta + V_q(u) + V(\tilde{x}(u)). \tag{6.127}$$

The functional integral (6.126) is computed with the initial conditions $u(0) = u''$ and $u(t) = u$. By definition, the initial and final configurations, u'' and u, are in the physical region K. The second and third terms in the effective Hamiltonian (6.127) take into account the operator ordering, providing thereby the unitarity of the evolution operator $\hat{U}_t^{\text{ph}}$. As seen, an addition of more physical degrees of freedom does not lead to any significant modification of the structure of the Hamiltonian path integral. Equations (6.124)–(6.127) define the Hamiltonian path integral formalism for a generic bosonic quantum mechanical gauge system with an arbitrary compact gauge group, including lattice Yang–Mills theories (the Kogut–Susskind Hamiltonian formulation given in Section 5.8.10) and the Yang–Mills quantum mechanics studied in Section 5.6.4.

Kernel (6.125) of operator $\hat{Q}$ can also be found by means of the projection method (6.114). Substituting $x = T(\theta)\tilde{x}(u)$ and $x' = T(\theta')\tilde{x}(u')$ into the latter, one infers,

$$\langle u|u'\rangle_{\text{ph}} = V_H^{-1} \int_G dg(\omega) \delta^N(\tilde{x}(u) - T(\omega)\tilde{x}(u')). \tag{6.128}$$

The dependence on θ and θ' has been eliminated by using the invariance of the measure $dg(\omega)$ under group shifts. The kernel (6.128) is the unit operator kernel in the physical configuration space relative to the measure $\int_K d^M u \mu(u)$. Therefore it is obtained from kernel (6.114) by multiplying the latter by the volume of the non-physical space $V_G V_H^{-1}$, which explains the factor V_H^{-1} in (6.128). Taking the limit $t \to 0$ in (6.124), one finds $\langle u|u'\rangle_{\rm ph} = |\mu(u)|^{-1}Q(u,u')$. Comparison of this relation with (6.128) yields a universal expression for the kernel of $\hat{Q}$ which holds for any compact gauge group and any (within reasonable limits) gauge condition,

$$Q(u,u') = \frac{|\mu(u)|}{V_H}\int_G dg(\omega)\delta^N\left(\tilde{x}(u) - T(\omega)\tilde{x}(u')\right), \tag{6.129}$$

where $u \in R^M$ and $u' \in K$. Integration over the group manifold automatically reproduces the sum (6.83), i.e. determines the set S of symmetry transformations.

Equation (6.129) shows that operator $\hat{Q}$ does not depend on the dynamics of the gauge system and is fully determined by the gauge group, its representation, and the choice of physical variables (local coordinates on the orbit space).

What is the practical lesson of the above studies? The first thing to note is that with an inappropriate, awkward choice of gauge condition, the description of a quantum gauge system in the framework of the Hamiltonian path integral formalism might turn out to be an extremely difficult problem, because of the complexity of residual gauge transformations. The second remark is that Lorentz-invariant gauge conditions, like, e.g. the Fermi gauge, are known to be inadequate for non-perturbative treatments of Yang–Mills fields because of their ambiguities, i.e. precisely because of the complexity of the residual gauge transformations (see Section 5.8.9 and [209–211]). Therefore the problem of a correct and constructive non-perturbative description of these theories by explicitly Lorentz-invariant path integrals seems rather difficult (see also [212, 213, 221]).

6.6 Hamiltonian path integrals for gauge systems with bosons and fermions

In Section 5.7 gauge systems with bosonic and fermionic degrees of freedom have been studied. It is shown there that after elimination of all non-physical degrees of freedom, there might still exist discrete gauge transformations that cannot further reduce the number of physical degrees of freedom, but rather they reduce the phase space of the latter. Reduction of the physical phase space has dynamical consequences (the spectrum of the Hamiltonian and Green's functions are changed). The residual discrete gauge transformations act on all physical degrees of freedom, bosonic and fermionic, in accordance with their gauge transformation law. The physical states are invariant under such transformations. A natural question arises: How is this residual symmetry taken into account in the path integral formalism for gauge systems with bosonic and fermionic degrees of freedom? To begin with, a simple gauge model is considered. Then a general procedure to develop the path integral formalism for gauge systems with fermions and bosons is given in the case of a compact gauge group.

6.6.1 A model with the gauge group SO(2)

Consider a quantum theory for the model defined by the Lagrangian (5.139) in an arbitrary gauge. The Hamiltonian operator and scalar product are given by (5.298) and (5.299), respectively. The harmonic oscillator potential is not relevant for what follows and may be replaced by an arbitrary regular (gauge-invariant) function on the superspace. As is shown in Section 5.9.5, physical transition amplitudes $\langle \Phi_1 | \Phi_2 \rangle$ are independent of the choice of physical variables (the choice of the functions $f_i(u)$). Important features of the quantum theory are a nontrivial integration measure $\int_K du\mu$ in the scalar product and the S-invariance of state vectors. These have to be taken into account in the Hamiltonian path integral formalism in order to achieve equivalence of the latter to the operator description.

To derive the Hamiltonian path integral, the method of extending the unit operator kernel to the non-physical domain is applied as proposed in Section 6.5. A novel feature here to be taken care of is that the residual discrete gauge transformations act simultaneously on the Bose and Fermi variables of the theory. By the S-invariance of the physical eigenfunctions of the Hamiltonian (5.298) and the scalar product (5.299), the spectral resolution of unity yields,

$$\langle q|q'\rangle_{\text{ph}} = \sum_E \Phi_E(q)\Phi_E^*(q') = \sum_S \left[\mu(u)\mu(\hat{s}u')\right]^{1/2} \delta(q, \hat{s}q'), \tag{6.130}$$

where q denotes a point in the superspace, $q = (u, \boldsymbol{\zeta}^*)$, the operators $\hat{s} \in S$ are defined in (5.301), and

$$\delta(q, q') = \delta(u - u') \exp\left(\boldsymbol{\zeta}^* \boldsymbol{\zeta}'\right) \tag{6.131}$$

is the δ-function on the superspace. Its characteristic property,

$$\int_{-\infty}^{\infty} dq' \delta(q, q')\Phi(q') \equiv \int_{-\infty}^{\infty} du' \int \prod_{a=1}^{2} (d\zeta_a'^* d\zeta_a') \exp\left(-\boldsymbol{\zeta}'^* \boldsymbol{\zeta}'\right) \delta(q, q')\Phi(q') = \Phi(q) \tag{6.132}$$

is easily verified. In (6.130) $u' \in K$ and $u \in R = \cup_s K_s$. The kernel (6.130) can also be written in the form,

$$\langle q|q'\rangle_{\text{ph}} = \int_{-\infty}^{\infty} dq'' \left[\mu(u)\mu(u'')\right]^{-1/2} \delta(q, q'') Q(q'', q'), \tag{6.133}$$

$$Q(q, q') = \sum_S \delta(q, \hat{s}q'). \tag{6.134}$$

The distribution (6.130) is S-invariant,

$$\langle \hat{s}q|q'\rangle_{\text{ph}} = \langle q|q'\rangle_{\text{ph}}, \tag{6.135}$$

for any $\hat{s} \in S$ and $u, u' \in K$. The condition $u \in K$ is necessary in order for $\hat{s}u$ to be well defined (see (5.301)). To prove (6.135), note that when $u, u' \in K$ in the left-hand side of Eq. (6.134), only one term contributes to the sum (6.130),

$$\begin{aligned}\langle \hat{s}q|q'\rangle_{\rm ph} &= \left[\mu(u_s(u))\mu(u_s(u'))\right]^{-1/2}\delta(u_s(u)-u_s(u'))\\ &\quad\times\exp\left(e^{i\theta_s(u)\Gamma}\zeta^* e^{-i\theta_s(u')\Gamma}\zeta'\right)\\ &= \left[\mu(u)\mu(u')\right]^{-1/2}\exp\left(\zeta^*\zeta\right) = \langle q|q'\rangle_{\rm ph}, \quad u,u'\in K, \end{aligned}\qquad (6.136)$$

where relations (6.85) and (6.86) have been used.

Consider the infinitesimal physical evolution operator kernel,

$$U_\varepsilon^{\rm ph}(q,q') = \langle q|\exp\left\{-i\varepsilon\hat{H}_{\rm ph}\right\}|q'\rangle_{\rm ph} = \left[1-i\varepsilon\hat{H}_{\rm ph}(q)\right]\langle q|q'\rangle_{\rm ph} + O(\varepsilon^2). \quad (6.137)$$

The δ-function of the bosonic variables in (6.131) is replaced by its Fourier integral representation. Then (6.133) is substituted into (6.137). The differentiation with respect to $\boldsymbol{\zeta}$ and u, which is involved in the action of $\hat{H}_{\rm ph}(q)$ in (6.137), yields the following expression for the infinitesimal evolution operator kernel up to terms $O(\varepsilon^2)$:

$$U_\varepsilon^{\rm ph}(q,q') = \int_{-\infty}^{\infty}\frac{dq''}{\left[\mu(u)\mu(u'')\right]^{1/2}}U_\varepsilon^{\rm eff}(q,q'')Q(q'',q'), \qquad (6.138)$$

$$U_\varepsilon^{\rm eff} = \int_{-\infty}^{\infty}\frac{dp}{2\pi}\exp\left(\zeta^*\zeta''\right)\exp\left\{iS^{\rm eff}(p,q^*,q'')\right\}, \qquad (6.139)$$

$$S^{\rm eff} = p(u-u'') - \varepsilon H^{\rm eff}(p,q^*,q''), \qquad (6.140)$$

$$H^{\rm eff} = \tfrac{1}{2}p^2 + p\,\frac{(\mathbf{f},T\mathbf{f}')}{\mu^2}\,\pi_\varphi + \tilde{V}_q + \frac{\mathbf{f}'^2}{\mu^2}\pi_\varphi^2 + \tilde{V}. \qquad (6.141)$$

In the effective Hamiltonian (6.141) the functions $\mathbf{f}$ and μ are taken at point u, $\pi_\varphi = \boldsymbol{\zeta}^*\Gamma\boldsymbol{\zeta}$, and the potential $\tilde{V}$ is obtained from the operator $\hat{V} = V(\hat{\mathbf{f}}(u),\hat{\boldsymbol{\zeta}}^\dagger\hat{\boldsymbol{\zeta}})$ by pulling all the operators $\hat{\boldsymbol{\zeta}}$ over to the very right (using the necessary commutation relations) with subsequent replacement of $\hat{\boldsymbol{\zeta}}$ by $\boldsymbol{\zeta}''$ and $\hat{\boldsymbol{\zeta}}^\dagger$ by $\boldsymbol{\zeta}^*$ (this is the result of the action of the operator $\hat{\boldsymbol{\zeta}} = \partial/\partial\boldsymbol{\zeta}^*$ on $\exp(\boldsymbol{\zeta}^*\boldsymbol{\zeta}'')$ in the kernel (6.133)). The potential $\tilde{V}_q$ is related to the operator ordering in the kinetic energy operator,

$$\tilde{V}_q = V_q(u) - \frac{i}{2}\left[p\partial_u g(u) + \partial_u\left(\frac{(\mathbf{f},T\mathbf{f}')}{\mu}\right)\pi_\varphi\right] + V_F, \qquad (6.142)$$

$$V_F = \frac{\mathbf{f}'^2(u)}{\mu^2(u)}\,\hat{\pi}_\varphi\pi_\varphi = \frac{\mathbf{f}'^2(u)}{\mu^2(u)}\,\boldsymbol{\zeta}^*\boldsymbol{\zeta}''. \qquad (6.143)$$

Here the expression $\hat{\pi}_\varphi\pi_\varphi$ denotes the action of the operator $\hat{\pi}_\varphi = \hat{\boldsymbol{\zeta}}^\dagger\Gamma\hat{\boldsymbol{\zeta}}$ on the function $\pi_\varphi = \boldsymbol{\zeta}^*\Gamma\boldsymbol{\zeta}$ introduced in (6.141). If the dependence on Planck's constant is restored, then $V_q \sim \hbar^2$ (see (5.160)), while all other terms in (6.142) and V_F in (6.143) are proportional to $\hbar$.

In order to obtain the evolution operator kernel for a finite time interval, it is necessary to carry out the iterative convolution (6.105) taking into account the Grassmann degrees of freedom. The first step is to calculate the kernel of the operator $\hat{U}_{2\varepsilon}^{\rm ph} = \hat{U}_\varepsilon^{\rm ph}\hat{U}_\varepsilon^{\rm ph}$, i.e. to repeat algebraic transformations (6.92)–(6.103) including the contribution of Grassmann variables. Equation (6.93) holds for kernel (6.138) if $\int_K du_1$ is replaced by $\int_K dq_1$ (measure dq_1 is defined in (6.132)) because kernels (6.138) and (6.89) have the same structure. In the integral with respect to Grassmann variables in the equation analogous to (6.93) one should change the integration variables to $\exp(i\theta_s(u_1)\zeta_1^*$ and $\exp(-i\theta_s(u_1))\zeta_1$. The integration measure dq_1 remains the same (a change of variables in integrals on supermanifolds is discussed in Sections 5.7.2 and 8.2.1). The change of the bosonic integration variable $u_1 \to u_s(u_1)$ is carried out analogously to (6.94)–(6.96). There is a small additional subtlety, though. The argument of the function $\theta_s(u_1)$ should be properly transformed according to definition (5.306) of the inverse transformation $\hat{s}^{-1}$ on the superspace. All functions $\theta_s(u_1)$ are first replaced by $-\theta_s^{-1}(u_s(u_1))$ and then the new integration variable $u_s(u_1)$ is introduced. In doing so, after the change of variables the integrand in (6.97) contains the kernel $U_\varepsilon^{\text{ph}}(\hat{s}^{-1}q_1, q')$, where $\hat{s}^{-1}q_1 = \left(u_s(u_1), \exp\left[i\theta_s^{-1}(u_1)\Gamma\right]\zeta_1^*\right)$ and $u_1 \in K_s$. Hence, the integral $\int_{K_s} dq_1$ is correctly defined.

Relation (6.98) is also generalized to the case studied here, i.e.

$$U_\varepsilon^{\text{ph}}(\hat{s}^{-1}q, q') = U_\varepsilon^{\text{ph}}(q, q'), \quad u \in K_s, \quad u' \in K. \tag{6.144}$$

A proof of (6.144) goes along the same lines as (6.99)–(6.101) where, in addition, the definition (5.306) of the inverse function $\theta_s^{-1}(u)$ should be taken into account. Therefore, with the replacement of du_1 by dq_1 and du'' by dq'' the formulas (6.102)–(6.103) hold for the model with bosonic and fermionic degrees of freedom.

Repeating the calculations (6.105), the Hamiltonian path integral is obtained for the model studied with an arbitrary choice of physical variables (arbitrary gauge) and with taking into account the physical phase space structure,

$$U_t^{\text{ph}}(q, q') = \int_{-\infty}^{\infty} \frac{dq''}{[\mu(u)\mu(u'')]^{1/2}} U_t^{\text{eff}}(q, q'')Q(q'', q'), \tag{6.145}$$

$$U_t^{\text{eff}}(q, q'') = \int_{-\infty}^{\infty} \prod_{\tau=0}^{t} \left[\frac{dp(\tau)du(\tau)}{2\pi} \prod_{a=1}^{2} d\zeta_a^*(\tau)d\zeta_a(\tau)\right] e^{\phi + iS^{\text{eff}}}, \tag{6.146}$$

$$S^{\text{eff}} = \int_0^t d\tau \left[p\dot{u} + \tfrac{1}{2i}\left(\zeta^*\dot{\zeta} - \dot{\zeta}^*\zeta\right) - H^{\text{eff}}\right], \tag{6.147}$$

where the phase $\phi = 1/2[\zeta^*(t)\zeta(t) + \zeta^*(0)\zeta(0)]$ accounts for the standard initial conditions in Grassmann path integrals in the holomorphic representation, $\zeta^*(t) = \zeta^*$ and $\zeta(0) = \zeta''$. The initial conditions in the bosonic sector are $u(t) = u$ and $u(0) = u''$. The effective Hamiltonian in the action (6.147) is given in (6.141).

Equations (6.145)–(6.147) together with the rule of acting the evolution operator $\hat{U}_t^{\text{ph}}$ on an initial wave function $\Phi_0(q)$,

$$\hat{U}_t^{\rm ph}\Phi_0(q) = \Phi_t(q) = \int_K dq'\mu(u')U_t^{\rm ph}(q,q')\Phi_0(q'), \tag{6.148}$$

solve the problem of the Hamiltonian path integral in an arbitrary gauge for the gauge system in question.

6.6.2 Hamiltonian path integral in curvilinear supercoordinates and general gauge systems with bosons and fermions

There is a close relation between the path integral in curvilinear coordinates and the path integral in gauge theories. To investigate this relation, consider the Hamiltonian path integral formalism for the quantum theory defined by the Hamiltonian operator (5.159) and the scalar product (5.155) [31]. The system has M bosonic degrees of freedom and N fermionic ones.

The problem here amounts to changing variables according to (5.150) in the standard path integral for the theory with Hamiltonian (5.148) and scalar product (5.149). It is solved by the method of extending the unit operator kernel to the non-physical region $R^M \setminus K$ of the new bosonic variables. Recall that the mapping $q = (y, \xi^*) \to (x, \psi^*)$ is one-to-one (i.e. a change of variables) if $y \in K \subset R^M$. By definition,

$$\langle q|q'\rangle = \delta^M(x(y) - x(y'))\exp\left[\psi^*(y,\xi^*)\psi(y',\xi')\right] \tag{6.149}$$

$$= \sum_S \left[\mu(y)\mu(\hat{s}y')\right]^{-1/2}\delta^M(y - \hat{s}y')\exp[\xi^*\hat{s}\xi'] \tag{6.150}$$

$$= \int_{R^M} \frac{dq''}{[\mu(y)\mu(y'')]^{1/2}}\,\delta(q,q'')Q(q'',q'), \tag{6.151}$$

$$Q(q,q') = \sum_S \delta(q,\hat{s}q'), \quad y \in R^M\,, \quad y' \in K, \tag{6.152}$$

where $dq = d^M y \prod_1^N [d\xi_\alpha^* d\xi_\alpha]\exp(-\xi^*\xi)$ is the integration measure on the superspace and R^M indicates the integration region of the bosonic variables. The delta-function $\delta(q,q'')$ in (6.151) has the same form as the right-hand side of (6.149) if x, x', ψ^*, and ψ' are replaced by y, y', ξ^*, and ξ, respectively. Equation (6.150) follows from the identity $|\mu(y)| = [\mu^2(y)]^{1/2}$ and the rule of changing the argument of the delta-function of several variables under the assumption that $y \in R^M$ and $y' \in K$. The transformations $\hat{s} \in S$ are defined by (5.156) and (5.157). In general, they do not form a group, i.e. the composition of $\hat{s}$ and $\hat{s}'$ is not defined for $\hat{s}, \hat{s}' \in S$.

The unit operator kernel (6.151) has the same structure as (6.130). Therefore the kernel of the infinitesimal evolution operator $\hat{U}_\varepsilon = 1 - i\varepsilon\hat{H} + O(\varepsilon^2)$ is obtained analogously to (6.138)–(6.141) with the effective Hamiltonian being

$$H^{\rm eff}(q,q'') = \tfrac{1}{2}P_a g^{ab}P_b + \tilde{V}_q + \tilde{V}, \tag{6.153}$$

$$\tilde{V}_q = V_q(y) + \tfrac{i}{2}\partial_a g^{ab}P_b - \tfrac{1}{2}g^{ab}\xi_\alpha^*(\partial_a\Omega^\dagger\partial_b\Omega)_{\alpha\beta}\xi''_\beta. \tag{6.154}$$

Here $P_a = p_a + \pi_a$, $\pi_a = -i\xi^*_\alpha(\Omega^\dagger \partial_a \Omega)_{\alpha\beta}\xi''_\beta$. The dependence on ξ'' occurs through the action of the operators $\hat{\xi}$ on the kernel $\delta(q, q'')$ in (6.151). The function $\tilde{V}$ is derived from the operator $V(\hat{x}, \hat{\psi}^\dagger, \hat{\psi})$ by the substitution (5.150), pulling over all the operators $\hat{\xi}$ to the very right, and subsequent replacement of $\hat{y}$, $\hat{\xi}^\dagger$, and $\hat{\xi}$ by y, ξ^*, and ξ'', respectively. The effective quantum correction (6.154) to the classical potential is associated with the specific operator ordering resulting from the change of variables in the quantum Hamiltonian.

Operator $\hat{H}$ and the kernel (6.149) are S-invariant because they depend only on x, $\psi^\dagger$, and ψ which do not change under the transformations (5.156) and (5.157) (by the definition of S). Therefore convolution of the infinitesimal evolution operator kernels according to the scalar product of the theory yields the following Hamiltonian path integral in curvilinear supercoordiantes (compare (6.145)–(6.147)) [31],

$$U_t(q, q') = \int_{R^M} \frac{dq''}{[\mu(y)\mu(y'')]^{1/2}}\, U_t^{\text{eff}}(q, q'')Q(q'', q'); \tag{6.155}$$

$$U_t^{\text{eff}} = \int_{R^M} \prod_{\tau=0}^{t} \left[\frac{d^M p(\tau) d^M y(\tau)}{(2\pi)^M} \prod_{\alpha=1}^{N} d\xi^*_\alpha(\tau) d\xi_\alpha(\tau)\right] e^{\phi + iS^{\text{eff}}}, \tag{6.156}$$

$$\phi = \tfrac{1}{2}\left[\xi^*_\alpha(t)\xi_\alpha(t) + \xi^a_\alpha(0)\xi_\alpha(0)\right],$$

$$S^{\text{eff}} = \int_0^t d\tau \left[p_a \dot{y}_a - \tfrac{1}{2i}\left(\xi^*_\alpha \dot{\xi}_\alpha - \dot{\xi}^*_\alpha \xi_\alpha\right) - H^{\text{eff}}\right]. \tag{6.157}$$

The boundary conditions for the path integral (6.156) read $\xi^*(t) = \xi^*$, $\xi(0) = \xi''$, $y(t) = y$, and $y(0) = y''$. As an example, consider the quantum theory of a two-dimensional supersymmetric harmonic oscillator, $N = M = 2$, $\hat{V} = \hat{\mathbf{x}}^2/2 + \hat{\boldsymbol{\psi}}^\dagger \hat{\boldsymbol{\psi}}$. In the superspace spanned by $(\mathbf{x}, \boldsymbol{\psi})$, new coordinates are introduced,

$$x_1 = r\cos\varphi, \quad x_2 = r\sin\varphi, \quad \boldsymbol{\psi} = \exp(i\varphi)\boldsymbol{\xi}. \tag{6.158}$$

This is the simplest generalization of the polar coordinates to the case of a superspace. The Hamiltonian path integral in the polar coordinates for a bosonic system is derived in Chapter 2. Its generalization to the coordinates (6.158) is given by (6.155)–(6.157). The Hamiltonian of this system in the new variables (6.158) is obtained from the general expression (5.159) where $\mu = r$, $g^{ab} = \text{diag}(1, r^{-2})$, $\hat{P}_1 \equiv \hat{P}_r = -ir^{-1/2}\partial_r \circ r^{1/2}$, $\hat{P}_2 \equiv \hat{P}_\varphi = -i\partial_\varphi + \hat{\pi}_\varphi$, $\hat{\pi}_\varphi = \hat{\xi}^\dagger_\alpha \hat{\xi}_\alpha \equiv \hat{\boldsymbol{\xi}}^\dagger \hat{\boldsymbol{\xi}}$, $V_q = -1/(8r^2)$. The effective Hamiltonian is calculated according to (6.153) and (6.154) where $P_1 = p_r, P_2 = p\hat{\varphi} + \boldsymbol{\xi}^*\boldsymbol{\xi}, \Omega = e^{i\varphi}, \partial_1 = \partial_r$, and $\partial_2 = \partial_\varphi$. The bosonic integration variables in (6.156) are p_r, p_φ, r, and φ. Operator $\hat{Q}$ in (6.155) is determined by the symmetry of the mapping (6.158). It is easy to verify that the symmetry transformations form a group in this particular case, which contains two subgroups, $S = S_t \times S_r$, where S_t and S_r are translations and reflections defined, respectively, by

$$S_t\colon\ \varphi \to \varphi + 2\pi n\,, \quad r \to r, \quad \boldsymbol{\xi}^* \to \boldsymbol{\xi}^*, \tag{6.159}$$

$$S_{\hat{r}}:\ \varphi \to \varphi + \pi\ ,\quad r \to -r\ ,\quad \boldsymbol{\xi}^* \to -\boldsymbol{\xi}^*, \tag{6.160}$$

where n ranges over all integers. Accordingly the kernel of operator $\hat{Q}$ has the form,

$$Q(q,q') = Q_\varphi(\varphi,\varphi')\delta(r-r')\exp(\boldsymbol{\xi}^*\boldsymbol{\xi}') \\ + Q_\varphi(\varphi,\varphi'+\pi)\delta(r+r')\exp(-\boldsymbol{\xi}^*\boldsymbol{\xi}'), \tag{6.161}$$

$$Q_\varphi(\varphi,\varphi') = \sum_{n=-\infty}^{\infty} \delta(\varphi-\varphi'+2\pi n)\ , \tag{6.162}$$

where $r \in R$, $r' \geq 0$, $\varphi \in R$, and $\varphi' \in [0,2\pi)$.

Equations (6.155)–(6.157) solve the problem of the Hamiltonian path integral formalism for the theory (5.148) in the curvilinear coordinates (5.150). Now suppose that this theory has a global symmetry,

$$x \to T_\Omega x\ ,\quad \psi \to \Omega\psi,\quad \psi^* \to \psi^*\Omega^\dagger, \tag{6.163}$$

where $M \times M$ matrices T_Ω realize a linear representation of a compact group G, and, for simplicity, it is assumed that $N \times N$ matrices Ω realize a unitary representation of G. The global symmetry can be turned into the corresponding gauge symmetry by demanding that physical states satisfy additional conditions, the quantum equations of constraints, in the same way as in Section 6.5.2 for pure bosonic systems, $\hat{\sigma}_a\Phi = 0$. The operators $\hat{\sigma}_a$ generate the gauge transformations (6.163) and, by definition, $[\hat{H},\hat{\sigma}_a] \approx 0$, i.e. the Hamiltonian is weakly gauge invariant. Note that such operators $\hat{\sigma}_a$ exist because, due to the global symmetry of the theory (5.148) there are Noether integrals of motion. Modulo a possible operator ordering problem, these integrals of motion become the constraint operators upon the canonical quantization.

Equations (6.155)–(6.157) can be used to develop the Hamiltonian path integral formalism for this gauge theory in an arbitrary gauge. Let $\chi(x) = 0$ be a gauge condition which is assumed to be admissible, i.e. for any configuration x there is a gauge transformation (6.163) such that the gauge transformed configuration $\tilde{x}$ lies in the gauge fixing surface, $\chi(\tilde{x}) \equiv 0$. Let $x = \tilde{x}(u)$ be parametric equations of the gauge fixing surface. Define new coordinates in the superspace by

$$x = T_\Omega(\omega)\tilde{x}(u),\quad \psi^* = \xi^*\Omega^\dagger(\omega),\quad y = (\omega,u) \in R^M, \tag{6.164}$$

where ω is a set of local coordinates parameterizing the group manifold G. By construction, variables u are physical, while ω are non-physical because they are translated upon the gauge transformations. The operators $\hat{\sigma}_a$ generate these translations and, hence, in the new variables, solutions of the constraint equations $\hat{\sigma}_a\Phi(\omega,u,\xi^*) = 0$ are given by functions independent of ω. For the sake of simplicity, the discussion is limited to the case when a generic configuration x has no nontrivial stationary subgroup in G.

Discussion of the symmetries of mapping (6.164) and the modular domain $\tilde{K}$ such that the mapping becomes one-to-one when y is restricted to $\tilde{K} \subset R^M$

follows pretty much that of the pure bosonic case given in the text between Eqs. (6.122) and (6.123). The only difference is that the symmetry transformations also act on the Grassmann variables. Ignoring the subtleties related to possible partitioning of $\tilde{K}$ and the related partitioning of the symmetry transformations, one can say that the latter contain the group of translations on ω by integer multiples of the periods of the group manifold G (the analog of (6.159)) and transformations of the form $u \to \hat{s}u$, $\xi^* \to \hat{s}\xi^*$, and $\omega \to \hat{s}\omega$ which can be realized by functions of u. In order to determine these functions, for a fixed u one should find all the solutions $\omega_s = \omega_s(u)$ of the equation $\chi(\tilde{x}_s) = 0$, where $\tilde{x}_s = T_\Omega(\omega_s)\tilde{x}(u)$. The configurations $\tilde{x}_s$ are intersections of the gauge fixing surface and the gauge orbit though $x = \tilde{x}(u)$. For a *generic* element of the gauge fixing surface (a generic u), the transformations $T_\Omega(\omega_s)$ are discrete, otherwise the condition $\chi(x) = 0$ does not eliminate the continuous gauge freedom the very existence of which implies that there are non-physical degrees of freedom among the variables u. As $\tilde{x}_s$, fixed now by $\omega_s = \omega_s(u)$, lies in the gauge fixing surface, there should exist $u_s = u_s(u)$ such that $\tilde{x}_s(u) = \tilde{x}(u_s)$, i.e. $u_s(u)$ are implicitly defined by the equation, $T_\Omega(\omega_s(u))\tilde{x}(u) = \tilde{x}(u_s)$. As a result, the configurations $\hat{s}u = u_s(u)$, $\hat{s}\omega = \omega'_s(u,\omega)$, where ω'_s are defined by $T_\Omega(\omega'_s) = T_\Omega(\omega)T_\Omega^{-1}(\omega_s)$, $\omega_s = \omega_s(u)$, have the same image x under the mapping (6.164). Transformations $\hat{s}\xi^*$ are induced by $\hat{s}\omega$ so that ψ^* remains unchanged; they are defined by the equality $\xi^*\Omega^\dagger(\omega) = (\hat{s}\xi^*)\Omega^\dagger(\hat{s}\omega)$. The implicit function theorem establishes the existence of functions $u_s(u)$ and $\omega'_s(\omega, u)$ in a neighborhood of $(u,\omega) \in R^N$. In general, these functions cannot be analytically extended to the whole R^N as illustrated in Section 6.5.2 with an example of a gauge bosonic system. Consider a restriction of the mapping (6.164) to a region $(u,\omega) \in \tilde{K} \subset R^M$ such that the restricted mapping is one-to-one. Then the symmetry transformations are uniquely defined by functions $u_s(u)$ and $\omega'_s(\omega, u)$ constructed by the above procedure for $(u,\omega) \in \tilde{K}$. Let K be the range of u in $\tilde{K}$, i.e. K is the domain of the functions $u_s(u)$.

The unit operator kernel (6.149) in the total configuration space in the curvilinear coordinates (6.164) can then be written in the form (6.151). The structure of the kernel of $\hat{Q}$ is analogous to (6.161),

$$Q(q,q') = \sum_S Q_G(\omega, \hat{s}\omega')\delta(u - \hat{s}u')\exp(\xi^*\hat{s}\xi'), \tag{6.165}$$

$$Q_G(\omega,\omega') = \sum_{(n)} \delta(\omega - \omega' + \lambda_{(n)}), \tag{6.166}$$

where $\hat{s}\xi' \equiv (\hat{s}\xi'^*)^*$, and $\lambda_{(n)}$ are elements of a lattice formed by integer multiples of the group manifold periods in the space of parameters $\omega \in R^L$, $L = \dim G$ (compare (6.162)), i.e. $T_\Omega(\omega + \lambda_{(n)}) = T_\Omega(\omega)$. The unit operator kernel in $\mathcal{H}_{\text{ph}}$ can be obtained by group averaging the kernel (6.151), similarly to the pure bosonic case (6.128), because group averaging (6.112) is an orthogonal projection onto the physical subspace $\mathcal{H}_{\text{ph}}$ of the Hilbert space of the system. Since group transformations in the new variables $y = (\omega, u)$ mean shifts of ω, the group averaging of the kernel (6.149) implies $\int dg(\varphi)\,\langle q|q'_\varphi\rangle$, where $q'_\varphi = (y'_\varphi, \xi')$ and $y'_\varphi = (\omega'+\varphi, u')$ (the pair (u, ξ) spans the physical configuration superspace), and $dg(\varphi)$ is the invariant measure on G. Due to the translation invariance of measure dg (the transformation $\varphi \to \varphi + \omega'$ means a shift on the group manifold), the group averaging of (6.151) is simply equivalent to the integration with respect to ω' in $Q(q, q')$ given by (6.165). The following equalities are proved to hold:

$$\int_G dg(\omega')Q_G(\omega, \hat{s}\omega') = \int_G dg(\omega')Q_G(\omega, \omega') = \int_G dg(\omega')\delta(\omega - \omega'). \qquad (6.167)$$

The first equality (6.167) results from $dg(\omega) = dg(\hat{s}\omega)$ which is a consequence of the invariance of the measure under group shifts and the definition $T_\Omega(\hat{s}\omega) = T_\Omega(\omega)T_\Omega^{-1}(\omega_s)$, where ω_s is independent of ω, while the second one is a consequence of the invariance of the measure $dg(\omega + \lambda_{(n)}) = dg(\omega)$ under translations of ω by periods of the group manifold. Therefore the integrals with respect to ω'' (contained in the integral with respect to q'' which includes $\int dy'' = \int d\omega'' du''$ in (6.151)) and ω' (the averaging variables) are factorized in (6.151) and equal one. Note that the Jacobian $\mu(y) \equiv J(\omega, u)$ in (6.151) is factorized similarly to the pure bosonic case (6.118): $d^M x = \mu(y) d^M y = J(\omega, u) d^L\omega d^{M-L}u = dg(\omega)\mu(u) d^{M-L}u$ (see the end of the paragraph below Eq. (6.118)). This explains that the integrals with respect to ω'' and ω' yield just the unity factor. So, the result of group averaging may be written in exactly the same form as (6.151) if the set of variables q is now assumed to contain only u and the Grassmann variables, $\mu(y) \to \mu(u)$ (as defined above), and the kernel $Q(q, q')$ now has the form,

$$Q(q, q') = \sum_{\hat{s}\in S} \delta(u - \hat{s}u') \exp(\xi^* \hat{s}\xi'),$$

where $u \in R^{M-L}$, and $u' \in K$. In particular, the averaging of the kernel (6.161) over the group $SO(2)$ yields the kernel of the operator $\hat{Q}$ in the gauge model studied in Section 6.6.1 if $f_a = \delta_{a1}u \equiv \delta_{a1}r$.

To complete construction of the Hamiltonian path integral formalism, the effective Hamiltonian analogous to (6.153) should be derived. With this purpose, consider the Schrödinger equation $\hat{H}\psi = E\psi$ with the Hamiltonian (5.159) in the coordinates (6.164). Both sides of this equation can be averaged over the group manifold G. The resulting equation reads $\hat{H}_{\rm ph}\Phi = E\Phi$ where the function $\Phi = \Phi(u, \xi^*)$ is S-invariant and belongs to $\mathcal{H}_{\rm ph}$. The operator $\hat{H}_{\rm ph}$ is obtained from $\hat{H}$ by removing terms containing the derivative with respect to ω (thanks to the gauge invariance, the potential V is independent of ω and S-invariant). The right-hand side of the Schrödinger equation in $\mathcal{H}_{\rm ph}$ is S-invariant, and so should the operator $\hat{H}_{\rm ph}$ be,

$$\hat{H}_{\rm ph}(\hat{s}u, \hat{s}\xi^*) = \hat{H}_{\rm ph}(u, \xi^*).$$

The resulting quantum theory has an S-invariant Hamiltonian and unit operator kernel. Therefore the corresponding path integral has the form (6.155)–(6.157), where $y \in R^M$ is to be replaced by $u \in R^{M-L}$ and in H^{eff} the tensor g^{ab}_{ph} should be used in place of g^{ab} (g^{ab}_{ph} is the uu-block of the matrix g^{ab}). In particular, the Hamiltonian path integral for the model studied in Section 6.6.1 in the gauge $x_2 = 0$ ($f_a = \delta_{a1}u \equiv \delta_{a1}r$) can also be derived by this method from the Hamiltonian path integral in curvilinear coordinates (6.158).

The above analysis establishes a relation between Hamiltonian path integrals in curvilinear supercoordinates and the path integral formalism for gauge systems with bosonic and fermionic degrees of freedom in an arbitrary gauge.

6.7 The Kato–Trotter product formula for gauge theories

The path integral quantization is regarded as an independent quantization method from which the corresponding operator formalism can be derived. So far the other way around has been explored. Namely, the infinitesimal evolution operator kernel is obtained from the physical Hamiltonian operator (in which all the non-physical degrees of freedom are eliminated). The form of this operator is essential to derive quantum corrections to the classical action in the path integral which takes into account operator ordering in the physical Hamiltonian. It is therefore of interest to put forward the following question. Is it possible to develop a self-sufficient path integral quantization of gauge systems which is equivalent to the Dirac operator formalism and does not require an explicit form of the physical Hamiltonian operator? The answer is affirmative [303]. The idea is to combine the Kato–Trotter product formula for the evolution operator in the total Hilbert space and the projection on the physical (Dirac) subspace. In such an approach no gauge fixing is needed in the path integral formalism [207, 308–310]. The gauge-invariant path integral can then be reduced onto any gauge fixing surface. In doing so, the quantum corrections as well as the physical phase space structure (the S-symmetry) are correctly reproduced depending on the choice of the gauge condition (or parameterization of the orbit space).

6.7.1 The projection formalism and Kato–Trotter formula

Consider a bosonic gauge system with a compact gauge group G. It is assumed that the total configuration space is R^N, x are Cartesian coordinates in it and operators T_a realize a linear representation of G in R^N (similarly to a general gauge model discussed in Section 6.5.2). Let $\hat{\sigma}_a$ be generators of the gauge transformations (the operators of constraints). Consider the projection operator,

$$\hat{P}_G = \int_G dg(\omega) e^{i\omega_a \hat{\sigma}_a}, \tag{6.168}$$

where the measure is normalized on unity, $\int_G dg = 1$, e.g. it can be the Haar measure of group G. The operators of constraints are assumed to be Hermitian. So, $\hat{P}_G = \hat{P}_G^\dagger = \hat{P}_G^2$. In the coordinate representation, the Dirac physical states are obtained by applying the projection operator (6.168) to all states in the total Hilbert space according to (6.112). The projection formalism and subsequent application to path integrals can be generalized to the case of non-compact gauge groups [182]. One should take a sequence of rescaled projection operators $c_\delta \hat{P}_G^\delta$ where c_δ is a scaling factor depending on a real parameter δ, and $\hat{P}_G^\delta$ projects on the subspace $\mathcal{H}_{\text{ph}}^\delta$: $\langle \Phi_\delta | \sum_a \hat{\sigma}_a^2 | \Phi_\delta \rangle \leq \delta$, $|\Phi_\delta\rangle \in \mathcal{H}_{\text{ph}}^\delta$, where $\langle \Phi_\delta | \Phi_\delta \rangle = 1$, i.e. the expectation value of the master constraint operator, which is, by definition, $\hat{\sigma} = \sum_a \hat{\sigma}_a^2$, does not exceed $\delta > 0$ in the subspace $\mathcal{H}_{\text{ph}}^\delta$. The Dirac subspace corresponds to the limit $\delta \to 0$. The apparent problem with this limit in the case of non-compact gauge groups is evident – the Dirac states would have an infinite norm in the total Hilbert space. For example, if the gauge group is the group of translations along a particular coordinate, then the Dirac states are ill-defined because they are given by functions independent on this coordinate, and, hence, their norm diverges. Of course, in this particular case, this divergence can be eliminated by omitting the integration with respect to the non-physical variable. However, for a generic non-compact gauge group, the "renormalization" procedure of the Dirac states becomes much more subtle because (non-compact) gauge orbits might have a complicated geometry. In the projection formalism, it is proved that for a given set of constraints the scaling factor c_δ can be chosen so that the limit Hilbert space $\mathcal{H}_{\text{ph}}^\delta$, $\delta \to 0^+$, is not empty and can be identified with the Dirac physical space. The mathematical details are not given here and can be found in [182]. For a general treatment of this procedure, the coherent state formalism appears quite helpful [182, 309]. An explicit form of the projection operator kernel in the coherent state representation for some gauge models can be found in [26, 207, 309, 311] and Section 8.5.

Suppose the total Hamiltonian of the system has the form $\hat{H} = \hat{H}_0 + \hat{V}$ where $\hat{V}$ is a potential and $\hat{H}_0$ is a free Hamiltonian (e.g. the kinetic energy operator). The Kato–Trotter product formula asserts that for any two self-adjoint operators $\hat{H}_0$ and $\hat{V}$ in a separable Hilbert space such that $\hat{H} = \hat{H}_0 + \hat{V}$ is self-adjoint on the intersection of the domains of the operators $\hat{H}_0$ and $\hat{V}$, the following relation holds [91]–[95] (see also Section 2.1.2):

$$\hat{U}_t = e^{it(\hat{H}_0 + \hat{V})} = \lim_{n\to\infty} \left(e^{it\hat{H}_0/n} e^{it\hat{V}/n} \right)^n . \tag{6.169}$$

Therefore the evolution operator kernel $U_t(x, x') = \langle x|\hat{U}_t|x'\rangle$ can be written as a convolution of n corresponding infinitesimal evolution operator kernels $U_\varepsilon(x, x') = \langle x| \exp(i\varepsilon \hat{H}_0) \exp(i\varepsilon \hat{V})|x'\rangle$, where $\varepsilon = t/n$. The convolution is calculated according to the scalar product of the theory which defines the resolution of unity, $1 = \int dx|x\rangle\langle x|$. The path integral representation of the transition amplitude is defined as the limit (6.169) of this convolution. Suppose that the kinetic energy operator has the standard form $\hat{H}_0 = -(\partial, \partial)/2$, where $\partial = \partial/\partial x$ and $(\cdot, \cdot)$ denotes an invariant scalar product, and $\hat{V}$ contains no derivatives in the coordinate representation, then,

$$U_\varepsilon(x, x') = (2\pi\varepsilon)^{-N/2} \exp\left[\frac{i(x - x')^2}{2\varepsilon}\right] \exp\left[i\varepsilon V(x')\right]. \tag{6.170}$$

To avoid excessive technicalities, it is assumed that $[\hat{V}, \hat{\sigma}_a] = [\hat{H}, \hat{\sigma}_a] = 0$, i.e. the total Hamiltonian and potential are gauge invariant and, hence, so is the kinetic energy. The physical Hamiltonian is obtained by the projection,

$$\hat{H}_{\text{ph}} = \hat{P}_G \hat{H} \hat{P}_G = \hat{P}_G \hat{H}_0 \hat{P}_G + \hat{P}_G \hat{V} \hat{P}_G = \hat{P}_G \hat{H}_0 + \hat{P}_G \hat{V}.$$

Therefore the physical evolution operator is given by the Kato–Trotter product formula (6.169),

$$\hat{U}_t^{\text{ph}} = e^{it\hat{H}_{\text{ph}}} = \lim_{n\to\infty} \left(e^{it\hat{H}_0/n} \hat{P}_G e^{it\hat{V}/n}\right)^n, \tag{6.171}$$

where the simplifying assumption of the exact gauge invariance of $\hat{H}_0$ and $\hat{V}$ has been used, e.g. $\exp(it\hat{P}_G\hat{V}\hat{P}_G) = \hat{P}_G \exp(it\hat{V})$ because $\hat{P}_G^2 = \hat{P}_G$. Inserting the resolution of unity, $1 = \int dx|x\rangle\langle x|$ between the operators in the product (6.171) and using that the projection operator is Hermitian, it is easy to see that the physical transition amplitude is given by the convolution of n infinitesimal amplitudes,

$$U_\varepsilon^{\text{ph}}(x, x') = (2\pi i\varepsilon)^{-N/2} \int_G dg(\omega) \exp\left[\frac{i(x - T(\omega)x')^2}{2\varepsilon}\right] \exp\left[i\varepsilon V(x')\right]. \tag{6.172}$$

As the quadratic form is invariant under the group action, the quantity $(x - T(\omega)x')^2$ in (6.172) is nothing but the squared distance between two points, one of which is on the gauge orbit through x while the other lies in the gauge orbit through x'. The path integral representation of the physical transition amplitude is then defined as the limit (6.171) for the convolution of the infinitesimal physical transition amplitudes given by (6.172). Due to the invariance of measure $dg(\omega)$ under group shifts, the amplitude $U_t^{\text{ph}}(x, x')$ is a gauge-invariant function of both its arguments and, hence, is a genuine transition amplitude on the orbit space R^N/G. There is no gauge-fixing procedure involved.

From the mathematical point of view, the limiting procedure leads to a certain measure for the averaging *functions* $\omega = \omega(t)$ in the continuum limit. By making use of the classical probability theory of Kolmogorov, one can show that this measure is a countably additive *probability* measure for $\omega(t)$ such that *any* set of values of $\omega(t)$ at *any* set of distinct times is equally likely [310, 314].

At every finite n, the convolution of n kernels (6.172) is well defined and, according to the Kato–Trotter product formula, the limit exists. In the formal continuum limit of the convolution of kernels (6.170), one usually assumes that $(x - x')^2/\varepsilon \approx \varepsilon\dot{x}^2$ so that the exponential in (6.170) tends to the classical action $S(x,x') = \int_0^\varepsilon d\tau[\dot{x}^2/2 - V(x)]$ in the time interval $\tau \in [0,\varepsilon]$ for trajectories such that $x(0) = x'$ and $x(\varepsilon) = x$. In this way the transition amplitude can be interpreted as a sum over paths weighed with the exponential of the classical action. The situation changes for the convolution of physical kernels (6.172). Formally, one can still obtain the gauge-invariant action in the exponential in (6.172). Set $\omega = \varepsilon y$ and $T(\omega) = T(\varepsilon y) \approx 1 + \varepsilon y$ where $y = y_a T_a$ with T_a being generators of gauge transformations of x. The action of linear operators T_a on x is determined by the representation of G realized by x. Then $(x - T(\omega)x')^2/\varepsilon \approx \varepsilon(D_t x)^2$ where $D_t x = \dot{x} - yx$ is the covariant derivative (the gauge transformation law for y that ensures the invariance of $(D_t x)^2$ is obvious). Therefore the exponential in (6.172) can be formally written as the classical gauge-invariant action, $S(x,x',y) = \int_0^\varepsilon d\tau[(D_\tau x)^2/2 - V(x)]$, where y is an arbitrary function, $x(0) = x'$, and $x(\varepsilon) = x$. The problem is to interpret the product of the measures $dg(\varepsilon y)$ in the continuum limit ($\varepsilon \to 0$). Naively, this seems to be proportional to the formal measure $\prod_{\tau,a} dy_a(\tau)$. However, the path integral would not exist because it diverges and, hence, must be regularized. The gauge-invariant transition amplitude defined by the Kato–Trotter product formula (6.171) gives a specific prescription for how this regularization must be done, namely, in the same way as in lattice gauge theories. Then the averaging measure is well defined in the continuum limit as the Kolmogorov probability measure. Recall that in numerical simulations of lattice gauge theories, no gauge fixing is required in the path integral. Note well that in the original theory no lattice regularization is assumed, while the "lattice" action in the exponential (6.172) naturally occurs through the elimination of contributions of non-physical states into the transition amplitude by means of the projection formalism (the operator $\hat{P}_G$ in (6.171)).

Although the above procedure accurately defines the path integral representation of the transition amplitude on the gauge orbit space, its calculation involves the group averaging integral. The next step is to develop a technique to compute the group averaging integral. The goal is to derive the gauge fixed path integral (6.106)–(6.107) directly from the Kato–Trotter product formula and, thus, to eliminate any use of the operator formalism in quantizing the gauge theory. Two essential elements are to be reproduced, the quantum corrections to the gauge fixed classical action and the symmetrization with respect to residual discrete gauge transformations associated with a possible incompleteness of the gauge condition chosen to parameterize the orbit space.

To introduce local coordinates on the orbit space, define new curvilinear coordinates (6.118) associated with the gauge transformation law and a particular choice of the (admissible) gauge condition, $x = T(\theta)f(u)$ where parameters θ on the group manifold are chosen so that, when $\theta = 0$ and $T(0) = 1$, the relation $x = f(u)$, $u \in R^M$, is the parametric equation of the gauge fixing surface in the total configuration space. The variables u serve as local coordinates on the orbit space. As before, the range of u is restricted to the fundamental modular domain

K so that the orbit space is isomorphic to K, $R^N/G \sim K$, and the discrete maps, $\hat{s}u = u_s(u)$, $u \in K$, define gauge equivalent points on the gauge fixing surface, i.e. $x = f(u)$ and $x_s = f(u_s)$ lie on the same gauge orbit. Due to the gauge invariance of the amplitude (6.172), $U_\varepsilon^{\rm ph}(x,x') = U_\varepsilon^{\rm ph}(f(u),f(u')) \equiv U_\varepsilon^{\rm ph}(u,u')$. Note that the transition amplitude as a function of u and u' depends on the choice of f. To avoid cumbersome notations as well as to match the notations with those used in Section 6.5, this dependence is not explicitly indicated in what follows. The potential is gauge invariant, $V(x) = V(f(u))$ and therefore its presence has no effect on the group averaging in (6.172). The phase factor in (6.172) containing the potential can be omitted and easily restored after the group averaging has been carried out. So, the infinitesimal transition amplitude (6.172) can then be written in the form,

$$U_\varepsilon^{\rm ph}(u,u') = (2\pi i\varepsilon)^{-N/2}\int_G dg(\omega)\exp\left\{\frac{i\,[f(u)-T(\omega)f(u')]^2}{2\varepsilon}\right\}, \qquad (6.173)$$

where the invariance of the averaging measure with respect to group shifts has been used. Since the operator with the kernel (6.173) acts on physical wave functions according to (6.148), $u' \in K$ in (6.173), while $u \in R^M$ so that $U_\varepsilon^{\rm ph}(\hat{s}u,u') = U_\varepsilon^{\rm ph}(u,u')$ for any $\hat{s} \in S$ and $u \in K$, thanks to the gauge invariance of the amplitude.

In the continuum limit $\varepsilon \to 0$, the group average integral in (6.173) can be computed in the stationary phase approximation. To avoid excessive technicalities associated with a generic gauge theory in a generic gauge, only the key elements of the procedure are outlined below. However, its applications to the gauge models studied above are given in detail in Section 6.7.2. Two important observations are in order.

First, due to the S-invariance of the amplitude (6.173), more than one stationary point exists. To see this, note that when $u - u' = \Delta$ is small, there should exist a stationary point $\omega = \omega_c(u,u')$ so that $T(\omega_c)$ is close to group unity. This allows us to solve the equation,

$$\partial_\omega(f(u), T(\omega)f(u')) = 0, \qquad (6.174)$$

for the stationary point by expanding $T(\omega)$ up to quadratic terms in ω. When deriving (6.174) from (6.173), the group transformations are assumed to preserve the norm: $(T(\omega)f, T(\omega)f) = (f,f)$. Let $\omega = \lambda_j\omega^j$, where λ_j form a basis of the Lie algebra in the representation realized by x. By construction λ_j are skew-symmetric linear transformations with respect to the inner product (because $T(\omega)$ are orthogonal). Put $T(\omega) = e^\omega = 1 + \lambda_j\omega^j + (1/2)\Lambda_{jk}\omega^j\omega^k + O(\omega^3)$, where $\Lambda_{jk} = (\lambda_j\lambda_k + \lambda_j\lambda_k)/2$. Then the stationary point $\omega^k = \omega_c^k$ satisfies the linearized Eq. (6.174),

$$(f(u), \Lambda_{jk}f(u'))\omega^k = -(f(u),\lambda_j f(u')) = (f(u),\lambda_j f'(u)\Delta) + O(\Delta^2), \qquad (6.175)$$

where $f'(u)$ denotes the matrix of derivatives of the components of f with respect to components of u. Having found the stationary point ω_c when

$u, u' \in K$, the other stationary points must satisfy Eq. (6.174) or (6.175) where $f(u') \to T(\omega_s)f(u') = f(\hat{s}u')$, $\hat{s} \in S$ (the set of residual gauge transformations in the gauge $x = f(u)$). Indeed, the kernel (6.172) is a gauge-invariant function of both the arguments. Suppose $u' \in K$. Then it follows that $U_\varepsilon(x, x') = U_\varepsilon(f(u), f(u')) = U_\varepsilon(f(u), f(\hat{s}u'))$ for any $\hat{s} \in S$. Therefore, if the equation (6.174) has a solution, then it should also have a solution when $f(u') \to f(\hat{s}u')$. When the squared distance is expanded in powers of Δ, the quadratic term

$$(f(u) - T(\omega_c)f(u'))^2 = \Delta^\alpha g^{\rm ph}_{\alpha\beta}\Delta^\beta + O(\Delta^3) \tag{6.176}$$

determines the metric $g^{\rm ph}_{\alpha\beta}$ on the orbit space in local coordinates u^α. Its inverse appears in the physical kinetic energy operator in (6.120). The relation (6.176) is illustrated with examples of the $SO(2)$ gauge matrix model in Section 6.7.2.4 and 4D Yang–Mills theory in Section 6.7.3.

Second, in the continuum limit not only the leading term in the stationary phase approximation must be computed, but also the two next-to-leading corrections. This means that the group element $T(\omega)$ is to be decomposed up to order $(\Delta\omega)^4$, $\Delta\omega = \omega - \omega_c$, near every stationary point ω_c because the terms $(\Delta\omega)^4/\varepsilon$ produce contributions of order ϵ. This is proved by rescaling the integration variable $\Delta\omega \to \sqrt{\varepsilon}\Delta\omega$ in the stationary phase approximation, which is similar to the procedure discussed in Section 2.8 to recover all relevant extra terms in the path integral quantization method. The group average measure should also be decomposed up to the necessary order to control the relevant ε-terms. All together these terms would yield *quantum corrections* to the classical action that are associated with operator ordering in the kinetic energy operator on the orbit space. Note well that the averaging procedure gives a *unique* ordering so that the integral is invariant under general coordinate transformations on the orbit space, i.e. it is independent of any specific parameterization of the orbit space, and, hence, gauge invariant. Thus, the transition amplitude can be written in the form,

$$\begin{aligned} U^{\rm ph}_\varepsilon(u, u') &= (2\pi i\varepsilon)^{-M/2}\sum_{\hat{s}\in S} D^{-1/2}(u, \hat{s}u') \\ &\quad\times\left\{\exp\left[i\left(f(u) - F(u, \hat{s}u')\right)^2/2\varepsilon - i\varepsilon\bar{V}_q(u, \hat{s}u')\right] + O(\varepsilon^2)\right\} \\ &\equiv \sum_{\hat{s}\in S} D^{-1/2}(u, \hat{s}u')\tilde{U}_\varepsilon(u, \hat{s}u'), \end{aligned} \tag{6.177}$$

where $F(u, u') = T(\omega_c(u, u'))f(u')$, and $D(u, u') = \sqrt{\det \mathcal{D}}$ and $\mathcal{D}$ is the matrix of second derivatives with respect to ω of the exponential in (6.173) taken at the stationary point $\omega = \omega_c$, i.e. $\mathcal{D} = \partial^2_\omega(f(u), T(\omega)f(u'))|_{\omega_c}$. It is the standard determinant arising from multiple Gaussian integration in the stationary phase approximation (see Section 8.3.1). The function $\bar{V}_q(u, u')$ denotes all contributions (from the measure and the phase) of the relevant next-to-leading order terms. Thus, the symmetrization with respect to residual gauge transformations arises as a consequence of the existence of multiple stationary points in the stationary phase approximation of the group averaging integral.

Convolution of the infinitesimal kernels (6.177) yields the path integral (6.124)–(6.127), and the quantum correction to the classical action has the form (6.121). The second part of this statement is technically the most tedious because determinant D also contributes to the quantum corrections to the classical action in the continuum limit. A proof of the first part requires only examination of the leading terms in the limit $\varepsilon \to 0$. When $\varepsilon \to 0$, the leading contribution comes from the first term in the exponential of (6.177) which is proportional to the quadratic form (6.176), so that in the limit the right-hand side of (6.177) is proportional to $D^{-1/2}(u', u')[\det g^{ph}(u')]^{-1/2}\delta^M(\Delta)$. Here it is assumed that $u, u' \in K \subset R^M$. The extension to $u \in R^M$ requires only the replacement of u' by $\hat{s}u'$. On the other hand, taking the limit in (6.173) and then carrying out the group average, one should obtain the physical unit operator kernel, $|\mu(u')|^{-1}\delta^M(\Delta)$. Hence,

$$D(u' + \Delta, u') = \mu^2(u')/\det g^{\rm ph}(u') + O(\Delta). \tag{6.178}$$

Consider the convolution of two infinitesimal evolution operator kernels,

$$U_{2\varepsilon}^{\rm ph}(u, u') = \int_K du_1 \mu(u_1) U_\varepsilon^{\rm ph}(u, u_1) U_\varepsilon^{\rm ph}(u_1, u'),$$

in which for kernel $U_\varepsilon^{\rm ph}(u, u_1)$ representation (6.177) is substituted. The integration region can be extended to the whole covering space R^M by noting that $U_\varepsilon^{\rm ph}(u_1, u') = U_\varepsilon^{\rm ph}(\hat{s}u_1, u')$ and $du_s\mu(u_s) = du\mu(u)$. Hence, changing the integration variable $u_1 \to \hat{s}u_1$ in each term of the sum over S, one infers from (6.177) that

$$U_{2\varepsilon}^{\rm ph}(u, u') = \int du_1 |\mu(u_1)| D^{-1/2}(u, u_1) \tilde{U}_\varepsilon(u, u_1) U_\varepsilon^{\rm ph}(u_1, u'). \tag{6.179}$$

The absolute value of the density μ accounts for a possible sign change of μ outside K, when extending the integration over K to the whole space R^M using the sum over the residual transformations S in (6.177). The procedure can be repeated from left to right for any number of infinitesimal kernels, thus removing the restriction of the integration domain and the sum over S for all intermediate configurations (integration variables) in the convolution. Only the sum over S for the initial configuration u' remains in the convolution. Thus, combining relations (6.177), (6.178), and (6.179), a formal continuum limit can be written in the path integral form [303],

$$U_t^{\rm ph}(u, u') = \sum_{\hat{s} \in S} [\mu(u)\mu(\hat{s}u')]^{-1/2} \int_{u(0)=\hat{s}u'}^{u(t)=u} \mathcal{D}u \sqrt{\det g^{\rm ph}}\, e^{iS_{\rm eff}[u]}, \tag{6.180}$$

$$S_{\rm eff} = \int_0^t d\tau \left[(\dot{u}, g^{\rm ph}\dot{u})/2 - V_q(u) - V(f(u))\right], \tag{6.181}$$

where, for brevity, $\dot{u}^\alpha g^{\rm ph}_{\alpha\beta}\dot{u}^\beta \equiv (\dot{u}, g^{\rm ph}\dot{u})$. The local density $\prod_{\tau=0}^t \sqrt{\det g^{ph}}$ can also be understood as the result of the integration over the momenta in the corresponding Hamiltonian path integral where the kinetic energy is $p_\alpha g^{\alpha\beta}_{\rm ph} p_\beta/2$, with p_α being a canonical momentum for u^α and $g^{\alpha\beta}_{\rm ph}$ is the inverse of the matrix $g^{\rm ph}_{\alpha\beta}$. Note that relation (6.178) explains the cancelation of the absolute value of μ in all intermediate points in the convolution (6.179) because the terms $O(\Delta)$ and $O(\Delta^2)$ only contribute to V_q in the continuum limit (see examples given in Section 6.7.2).

To calculate the operator ordering corrections $V_q(u)$ in the continuum limit, the procedure of Section 2.8 can be employed. The determinant $D(u, u')$ has to be decomposed up to order Δ^2, while the exponential in (6.177) up to order Δ^4 because the path integral measure has support on paths for which $\Delta^2 \sim \epsilon$ and $\Delta^4 \sim \epsilon^2$. The terms Δ^{2n} are then converted into terms ϵ^n by the equivalence rules (2.47),

$$\Delta^{\alpha_1} \cdots \Delta^{\alpha_{2k}} \to g_{\rm ph}^{-1/2} (i\epsilon\hbar)^k g_{\rm ph}^{\alpha_1 \cdots \alpha_{2k}} .$$

In this way all the quantum corrections to the classical action are recovered and the base point in the path integral definition is also fixed (see Section 2.5). Following the formalism of Section 2.8, the Schrödinger equation for the physical amplitude (6.177) can be derived. The corresponding Hamiltonian operator has the form ($\hbar = 1$) [303],

$$\hat{H}_{\rm ph} = -\frac{1}{2\mu}\, \partial_\alpha \circ \mu\, g^{\alpha\beta}_{\rm ph}\, \partial_\beta + V(f(u)). \tag{6.182}$$

It can easily be transformed to the explicitly Hermitian form (6.120) by introducing the Hermitian momenta $\hat{P}_\alpha$. The spectrum of the Hamiltonian operator (6.182) is independent of any particular parameterization of the orbit space (compare Section 5.3.3). It should be emphasized that the kinetic energy operator does *not* generally coincide with the Laplace–Beltrami operator on the orbit space because $\left[\det g^{\rm ph}_{\alpha\beta}\right]^{1/2} \neq \mu$ (recall that measure μ depends on the action of the gauge group in the total configuration space, e.g. in the $SO(n)$ models $g^{\rm ph} \sim 1$, while $\mu \sim r^{n-1}$ (see Section 6.2.3)). This completes the development of a self-consistent path integral quantization of gauge theories that takes into account both the true structure of the physical configuration (or phase) space and the operator ordering corrections to the effective gauge fixed action, which are both important for the gauge invariance of the path integral.

Remark. It is instructive to compare the above formalism with the gauge fixed path integral obtained by means of the substitution of the Faddeev–Popov identity [171, 315],

$$1 = \det \Delta_{FP}(x) \int_G dg(\omega) \delta^{N-M} \left(\chi(T(\omega)x)\right) \tag{6.183}$$

into the divergent path integral measure in the total configuration space (this measure corresponds to the formal continuum limit discussed after Eq. (6.172)). Here $\chi(x) = 0$ is the gauge condition and Δ_{FP} is the matrix (Faddeev–Popov operator) $\partial\chi(T(\omega)x)/\partial\omega|_{\omega=0}$, and $\det \Delta_{FP}$ is the Faddeev–Popov determinant (shortcomings of this procedure are assessed in Section 8.4.1). Suppose $x = f(u)$ is the parametric equation of the gauge fixing surface, $\chi(f(u)) \equiv 0$ for any u. Then a simple calculation shows that $\Delta_{FP} = \mu$. The procedure formally leads to the path integral (6.180) without the symmetrization with respect to S, and the classical action (6.181) does not contain the quantum correction term $V_q(u)$. Note that the group averaging in (6.183) may have contributions from several (gauge equivalent) points in the gauge fixing surface [316]. However, proper inclusion of these contributions into the gauge-fixed path integral remains obscure, when dealing with the formal (divergent) path integral measure. The path integral measure is characterized by correlators of neighboring points on paths. Because of the locality of the identity (6.183), such information cannot be obtained from it, which, in particular leads to the loss of the quantum correction term V_q. In contrast, the procedure based on (6.173) is free of these shortcomings. Without the quantum correction term, the amplitude (6.180) does not satisfy the time-dependent Schrödinger equation with Hamiltonian (6.182) and, hence, would generally be gauge dependent (see Section 5.3.3). A further discussion of the quantum theory resulting from the Faddeev–Popov gauge-fixed path integral is given in Section 8.4.

6.7.2 The projection formalism. Examples

Let us illustrate the main features of the path integral quantization method based on the projection formalism and the Kato–Trotter formula. To begin, consider the simplest example of the $SO(n)$ model introduced in Section 5.2. The oscillator potential is not relevant for the following discussion as explained in the text above Eq. (6.173). For pedagogical reasons, the model is investigated in two ways. First, the averaging integral is calculated *exactly* for the infinitesimal transition amplitude (6.173) and then the result is used to obtain the path integral. Second, the path integral is obtained by means of the general procedure presented in the previous section, i.e. by the stationary phase approximation of the gauge group average integral. The latter approach is more powerful and general since it does not require performing the averaging integral exactly.

6.7.2.1 The $SO(n)$ model

The quantum potential reads $V_q = (n-3)(n-1)/(8r^2)$ if the unitary gauge, $x_1 = r, x_i = 0, i \neq 1$, is used to parameterize the orbit space, and the residual discrete gauge group is $S = Z_2$. The Hamiltonian path integral is given by (6.33) and (6.34). The task is to recover this without any use of the operator formalism. Assuming a spherical coordinate system (as the one associated with the chosen gauge and the gauge transformation law), the group average integral in (6.173) for the infinitesimal amplitude reads,

$$U_\varepsilon^{ph}(r,r') = \frac{\Omega_{n-1}e^{i(r^2+r'^2)/2\varepsilon}}{\Omega_n(2\pi i\epsilon)^{n/2}}\int_0^\pi d\theta\,\sin^{n-2}\theta\,\exp\left\{-\frac{irr'}{\epsilon}\cos\theta\right\} \tag{6.184}$$

$$= \frac{\Omega_{n-1}}{\Omega_n(\pi i)^{n/2}}\frac{\Gamma(\nu+1/2)\Gamma(1/2)}{2\varepsilon(rr')^{\nu/2}}\,J_\nu\left(\frac{rr'}{\varepsilon}\right)e^{i(r^2+r'^2)/2\varepsilon}, \tag{6.185}$$

where Ω_n is the volume of the n-sphere of unit radius, θ is the angle between $\mathbf{x}$ and $\mathbf{x}'$, J_ν is the Bessel function where $\nu = n/2-1$. The factor Ω_n^{-1} is inserted to normalize the averaging measure on unity. For $n=3$ the Bessel function in question is a trigonometric function and the representation (6.16) follows from (6.185) for a finite time, $\varepsilon \to t$. If $n \neq 3$, the asymptotic expansion of the Bessel function in (6.185) for a large argument should be used [132] (as $\varepsilon \to 0$),

$$J_\nu(z) \approx \sqrt{\frac{2}{\pi z}}\left\{\cos z_\nu - \sin z_\nu\,\frac{\Gamma(\nu+3/2)}{2z\Gamma(\nu-1/2)}\right\},$$

where $z_\nu = z - \pi\nu/2 - \pi/4$ and $z = rr'/\varepsilon$. Up to terms of order $O(\varepsilon^2)$,

$$U_\varepsilon^{ph}(r,r') = (2\pi i\varepsilon)^{-1/2}\int_{-\infty}^{\infty}\frac{dr''}{(rr'')^{(n-1)/2}}\,e^{i(r-r'')^2/(2\varepsilon)-i\varepsilon V_q(r)}Q(r'',r')$$

$$Q(r'',r') = \delta(r''-r') + \delta(r''+r'), \tag{6.186}$$

where $V_q = (n-1)(n-3)/(8r^2)$ is the quantum potential. Observe that there are no absolute value bars in the denominator of the integrand. The phase is determined by the phases of the two exponentials $e^{\pm iz_\nu}$ in the asymptote of the Bessel function. The result of Section 6.2.3 is recovered with no reference to the operator formalism. The projection method automatically reproduces the density r^{n-1} of the scalar product measure in the factor at the exponential. The quantum potential barrier near the singularity $r=0$ affects the phase with which trajectories reflected from the physical configuration space boundary contribute to the sum over paths.

In the stationary phase approximation of the average integral (6.184), control over the corrections of order ε in the exponential is more delicate. The stationary points are $\theta = 0$ and $\theta = \pi$. In the vicinity of $\theta = 0$, in the exponential $\cos\theta \approx 1-\theta^2/2+\theta^4/24$, while in the measure $\sin\theta \approx \theta-\theta^3/6$. The cubic and quartic terms give the contribution of order ε. This can be immediately inferred by rescaling the integration variable $\theta \to \theta/\sqrt{\varepsilon}$. Keeping only the ε- and r-dependencies and the phase factors, while omitting all other numerical factors, the contribution of the first stationary point $\theta = 0$ to the averaging integral can be written in the form,

$$\frac{e^{i(r-r')^2/2\varepsilon}}{(i\varepsilon)^{n/2}}\int_0^\infty d\theta\varepsilon^{n/2-1}\left(\theta-\frac{\varepsilon\theta^3}{6}\right)^{n-2}\left(1-\frac{irr'}{24}\varepsilon\theta^4\right)e^{-irr'\theta^2/2}.$$

The contributions of the averaging measure and group element $T(\omega)$ in (6.173) to the next-to-leading order of the stationary phase approximation are given, respectively, by the θ^3- and θ^4-terms in parentheses. All of the quantum corrections are determined by these. Indeed, computation of this integral yields

$$\frac{e^{i(r-r')^2/2\varepsilon}}{(2\pi i\varepsilon)^{1/2}(rr')^{(n-1)/2}}\left(1-\frac{i\varepsilon(n-1)(n-3)}{8rr'}\right)+O(\varepsilon^2), \tag{6.187}$$

where all the omitted numerical factors have now been restored. The expression in parentheses is nothing but the exponential of the quantum potential up to terms of order $\varepsilon^{3/2}$. Similarly, the contribution of the second stationary point $\theta = \pi$ is computed. The result has the form (6.187) where $r' \to -r'$ because $\cos\pi = -1$. By adding the contributions of both the stationary points, the result (6.186) is recovered again.

The above exercise illustrates the significance of the next-to-leading order corrections to the stationary phase approximation in the average integral in (6.173).

6.7.2.2 Yang–Mills theory in (0+1) spacetime

Consider the model from Section 5.4. The path integral representation of the corresponding quantum theory has been obtained in Section 6.3.1. The distance in the exponential (6.173), where $f(u) \to h \in H$, is taken with respect to the Cartan–Killing form defined in Section 8.1.1: $(h - \exp(i\mathrm{ad}\omega)h')^2$. Here components of $\omega \in X$ are assumed to be real in the real Cartan–Weyl basis (8.14) (this explains i in the exponential) and $\mathrm{ad}\omega(h') \equiv [\omega, h']$. Equation (6.174) for stationary points is $\partial_\omega(h, e^{i\mathrm{ad}\omega}h') = 0$ where ω lies in $X\backslash H$, i.e. $\mathrm{ad}\omega$ is a linear combination of the operators $\mathrm{ad}e_{\pm\alpha}$ (for all positive roots α). Making use of this fact and that $\mathrm{ad}h(h') = 0$ for any $h, h' \in H$, the equation for stationary points can be written in the form

$$i\left(h, \mathrm{ad}\, e_{\pm\alpha}(e^{i\mathrm{ad}\omega_c}h')\right) = 0, \tag{6.188}$$

Equation (6.188) has a trivial solution $\omega_c = 0$ because $\mathrm{ad}e_{\pm\alpha}(h') = [e_{\pm\alpha}, h']$ is orthogonal to any element of the Cartan subalgebra H. All nontrivial solutions are exhausted by the elements of the Weyl group, $\exp(i\mathrm{ad}\,\omega_c) = \hat{w} \in W$. The explicit form of ω_c is given in (5.264). The averaging measure is the Haar measure [215] (see the end of Section 8.1.1),

$$dg(\omega) = d\omega \det\left\{(i\mathrm{ad}\,\omega)^{-1}\left[e^{i\mathrm{ad}\,\omega} - 1\right]\right\}. \tag{6.189}$$

The determinant has to be decomposed up to second order in $\mathrm{ad}\,\omega$, while the exponential in the distance formula, up to fourth order, similarly to the case of the $SO(n)$ model studied above (this follows from rescaling the integration variable $\omega \to \omega/\sqrt{\varepsilon}$). The second variation of the distance at the stationary point is obtained from the decomposition,

$$\begin{aligned}(h - e^{i\mathrm{ad}\,\omega}h')^2 &= \left(h - h' - i\mathrm{ad}\,\omega(h') + \frac{1}{2}(\mathrm{ad}\,\omega)^2 h'\right)^2 + O(\omega^3)\\ &= (h-h')^2 - (\omega, \mathrm{ad}\,h\,\mathrm{ad}\,h'(\omega)) + O(\omega^3).\end{aligned} \tag{6.190}$$

Therefore, the determinant in (6.177) reads,

$$\begin{aligned}D^{1/2}(h,h') &= [\det(-\mathrm{ad}\,h\,\mathrm{ad}\,h')]^{1/2}\\ &= [\det(i\mathrm{ad}\,h)\det(i\mathrm{ad}\,h')]^{1/2} = \kappa(h)\kappa(h').\end{aligned} \tag{6.191}$$

For other stationary points, the configuration h' has to be replaced by $h'_w = wh'w^{-1} \equiv \hat{w}h'$, $w \in W$ (the linear operator $\hat{w}$ is orthogonal with respect to the Cartan–Killing scalar product). There is a subtlety here that has to be addressed. Indeed, $\det(i\mathrm{ad}\,h') = \kappa^2(h') = \kappa^2(\hat{w}h')$, while function $\kappa(h')$ can change its sign under the Weyl transformations (compare (5.86)). The question is: How is the square root defined in (6.191)? If the absolute value bars in the right-hand side of Eq. (6.191) were put as seems to be formally correct, the corresponding transition amplitude would not have coincided with (6.38). That is, the *phase* with which the trajectories reflected from the boundary of the modular domain K^+ (the Weyl chamber) contribute to the sum over paths would have been incorrect. A more elaborate and accurate treatment of the group average integral leads to the correct answer. Note that, if $\hat{D}$ is a strictly positive operator, then

$$\int d\omega \exp[-(\omega, \hat{D}\omega)] \sim D^{-1/2}, \quad D = \det\hat{D}. \tag{6.192}$$

If $\hat{D} \to \pm i\hat{D}$, the integral (6.192) is obtained by an *analytic* continuation of the left-hand side of (6.192). In the present case $\hat{D}\omega = i\mathrm{ad}\,h\,(\mathrm{ad}\,h'(\omega)) = i[h,[h',\omega]]$, $\omega \in X \ominus H$, because the distance (6.190) is multiplied by i in the exponential (6.173). Making use of the orthogonal Cartan–Weyl basis (8.15), the quadratic form can be written as

$$(\omega, \hat{D}\omega) = i\sum_{\alpha>0}(h,\alpha)(h',\alpha)\left[(\omega_c^\alpha)^2 + (\omega_s^\alpha)^2\right]. \tag{6.193}$$

The operator $-i\hat{D} = \mathrm{ad}\,h\mathrm{ad}\,h'$ is strictly positive if $h, h' \in K^+$ because $(h,\alpha) > 0$ for any positive root α. Recall that a positive root α is a linear combination of simple roots with non-negative integer coefficients, and, by definition, a scalar product of $h \in K^+$ and any simple root is strictly positive. Replacement of h or h' by $\hat{w}h$ or $\hat{w}h'$, respectively, induces permutations and reflections of the roots in the product $\prod_{\alpha>0}(h,\alpha)(h',\alpha)$ which emerges after integration over $\omega^\alpha_{c,s}$ since $(\hat{w}h,\alpha) = (h,\hat{w}^{-1}\alpha)$ and the Weyl group preserves the root pattern ($\hat{w}^{-1}\alpha$ is also a root for any $w \in W$). Permutations do not change the product. A reflection in the hyperplane orthogonal to a positive root α, $\hat{w}\alpha = -\alpha$, changes the sign of an odd number of factors in it and may make some permutations among other positive roots, too. Indeed, for any two positive roots, β and γ, distinct from α, the reflection can only occur pairwise: $\hat{w}\beta = -\gamma$ and $\hat{w}\gamma = -\beta$ because

$\hat{w}^2 = 1$. By the analytic continuation of (6.192), each of the integrals over $\omega^\alpha_{\beta,c}$ (α fixed) contributes the phase factor $\exp(-i\pi/2)$ when h' is replaced by $\hat{w}h'$ and $\det\hat{w} = -1$ (reflection), thus making together the phase $\exp(-i\pi) = -1 = \det\hat{w}$, while the pairwise reflections give rise to the total phase $[\exp(-i\pi/2)]^{4k} = 1$, $k = 0, 1, \ldots$ (k is the number of such pairs of roots β and γ changed under the action of $\hat{\omega}$). Therefore, an analytic continuation of (6.191) assumes the form,

$$D^{1/2}(h, \hat{w}h') = \det\hat{w}\,(\det(-\mathrm{ad}\,h\,\mathrm{ad}\,\hat{w}h'))^{1/2} = \det\hat{w}\;|\kappa(h)\kappa(\hat{w}h')|$$
$$= \kappa(h)\kappa(\hat{w}h'), \quad \det\hat{\omega} = \pm 1. \tag{6.194}$$

No absolute value of κ occurs, and the result (6.38) follows immediately. This is the power and the beauty of the projection formalism. A change of the probability amplitude phase, when the system hits the boundary of the physical configuration space, is uniquely determined whatever parameterization of the orbit space is used. In contrast, in the reduced phase-space quantization (i.e. when non-physical degrees of freedom are eliminated prior to quantization) the phase change is not unique and depends on a self-adjoint extension of the kinetic energy operator in the modular domain. Needless to say, the very construction of a self-adjoint extension may be an extremely hard technical problem, given the fact that the modular domain depends on the gauge choice.

6.7.2.3 The mass gap in the 2D Yang–Mills theory

The number of stationary points in the averaging integral (6.173) can be infinite. This indicates that the physical configuration space may be *compact* in certain directions. Feynman conjectured [317] that a compactification of the configuration space in certain directions due to gauge symmetry might be responsible for the mass gap in the spectrum of (2+1) Yang–Mills theory (a finite gap between the ground state energy and the first excited state energy). For Yang–Mills theory in (1+1) spacetime, the statement is proved by the operator formalism in Section 5.8.5. Here the same result is established in the framework of the path integral formalism developed, i.e. without solving the Schrödinger equation [303]. The gauge group averaging integral is now a *functional* integral over the gauge group $\mathcal{G}/G_H$. A rigorous definition of the normalized averaging measure can be given via a lattice regularization of the theory (see Section 5.8.10). To achieve the goal, it is sufficient to calculate the leading order of the stationary phase approximation for the averaging integral, in which case no lattice regularization is actually needed. The key observation is that the sum over an infinite number of stationary points has a similar effect on the spectrum of *free* motion (there is no magnetic field in the 2D Yang–Mills theory) as the sum over the winding numbers in the free particle transition amplitude discussed in Section 2.7; the spectrum becomes discrete.

Let us turn to the details. The quadratic form in the exponential in (6.172) assumes the form $\langle A - A'^{\Omega}\rangle^2$ for any two configurations $A(x)$ and $A'(x)$ (notations are introduced in Section 5.8). It is the distance between two configurations $A(x)$ and $A'^{\Omega}(x)$ introduced by Feynman [317]. The scalar product has the form $\langle \, , \rangle = \int_0^{2\pi l} dx(\, ,)$. The orbit space can be parameterized by constant connections $A(x) = a$ taking their values in the Cartan subalgebra H. An infinitesimal gauge transformation of a has the form $\delta a = \nabla(a)\omega$ where $\omega(x) \in \mathcal{F} \ominus \mathcal{F}_0^H$ (here $\Omega \equiv e^{ig\omega}$, compare Section 5.8.3). Thus, if $\omega(x) = 0$ is a stationary configuration, $\Omega(0) = e$, then all other stationary configurations $\omega_c(x)$ in the functional averaging integral (6.173) are given by the residual gauge transformations of the gauge fixed potential $A(x) = a$, i.e. $\Omega(\omega_c)$ generate transformations from the affine Weyl group. To find function $D(a, a')$, the distance function is decomposed up to second order in the vicinity of each stationary point,

$$\begin{aligned}\left\langle a - a'^{\Omega}\right\rangle^2 &= \left\langle a - a' - \nabla(a')\omega + \frac{ig}{2}[\nabla(a')\omega, \omega]\right\rangle^2 + O(\omega^3) \\ &= \left\langle a - a'\right\rangle^2 - \langle \omega, \nabla(a)\nabla(a')\omega\rangle + O(\omega^3).\end{aligned}$$

Here the identity $\nabla(a')\omega = \nabla(a)\omega - ig[a' - a, \omega]$ has been used for one of the covariant derivatives $\nabla(a')\omega$ in the term $\langle\nabla(a')\omega, \nabla(a')\omega\rangle$. The Gaussian functional integration over ω yields

$$D^{1/2}(a, a') = [\det(-\nabla(a)\nabla(a'))]^{1/2} \sim \kappa(a)\kappa(a') \,, \qquad (6.195)$$

where $\kappa^2(a) \sim \det[i\nabla(a)]$ is given by (5.214). One should be careful when taking the square root in (6.195) for other stationary points, i.e. when $a' \to \hat{s}a'$, where $\hat{s}$ is from the affine Weyl group (see Section 5.8.2), or when a or a' is outside the modular domain which is the Weyl cell. By making use of the representation (5.210)–(5.212) and the analyticity arguments similar to those given above to prove (6.194), it is not hard to be convinced that the absolute value bars must be omitted when taking the square root in (6.195). Formula (6.192) is applied to the operator $-\nabla(a)\nabla(a')$ which is strictly positive in $\mathcal{F} \ominus \mathcal{F}_0^H$ if a and a' are in the modular domain K_W^+ (the Weyl cell), as shown in Section 5.8.3.

Thus, the infinitesimal transition amplitude (6.57) is recovered modulo the constant term in the exponential proportional to E_C (see (5.237)). The latter is the correction of order $O(\varepsilon)$ and must come from the next-to-leading in the stationary phase approximation, in the same way as the effective quantum potential in the example of the $SO(N)$ model (note that $E_C \sim \hbar^2$). Its accurate calculation requires lattice regularization of the averaging measure (the Haar measure at each lattice site in the Kogut–Susskind formulation (see Section 8.1)) as there are contributions of order $O(\varepsilon)$ coming from the measure similar to the case of the $SO(N)$ model (compare the derivation of (6.187)). The technicalities are omitted here. Finally, convolution of the kernels (6.57) with the subsequent continuum limit yields the path integral representation of the transition amplitude. It is obtained without any use of the operator formalism. Moreover, the spectrum of the physical excitations in the theory can immediately be inferred from the pole structure of the resolvent,

$$\mathrm{Tr}\,\hat{R}(E) = \mathrm{Tr}\left(E - i\hat{H}\right)^{-1} = \int_0^\infty dt e^{-Et}\,\mathrm{Tr}\,\hat{U}_t^{\mathrm{ph}},$$

$$\mathrm{Tr}\,\hat{U}_t^{\mathrm{ph}} = \int_{K_W^+} da\kappa^2(a)U_t^{\mathrm{ph}}(a,a),$$

in the complex plane of E. The latter integral can be transformed to the sum of Gaussian integrals via the Fourier transform of the amplitude $U_t^{\mathrm{ph}}(u,u)$ (see (6.55), (6.56), and the discussion of (6.58)) so that the integral over the compact region K_W^+ can be computed using the symmetry properties of the integrand under the affine Weyl transformations. By definition of the resolvent $\hat{R}(E)$, the Laplace transform of the trace $\mathrm{Tr}\,\hat{U}_t^{\mathrm{ph}}$ has pure imaginary poles $E = -iE_n$ on the complex E-plane where E_n are eigenvalues of the Hamiltonian $\hat{H}$. Of course, the latter are found to coincide with the spectrum of the physical Hamiltonian (5.239). It is important to emphasize that modification of the path integral due to the existence of countably many stationary points in the group average of the infinitesimal transition amplitude as well as the next-to-leading order corrections (combined with the equivalence rules) are essential to recovering the correct spectrum of the Yang–Mills theory. This would be impossible, had the conventional Faddeev–Popov recipe been used. Neither would the recipe be successful based on formal restriction of the integration domain in the path integral suggested in [209] as the resulting path integral is not even Gaussian.

6.7.2.4 The SO(2) matrix model

To elucidate how the curvature of the orbit space is accounted for in the projection path integral formalism, consider a simple gauge matrix model studied by the operator formalism in Section 5.6.3. The orbit space in this model is curved. Let the orbit space be parameterized by triangular matrices ρ, $\rho_{21} \equiv 0$ (see Eq. (5.123); the gauge x_{21} or $\theta = 0$). The residual gauge transformations form the group $S = Z_2$: $\rho \to \pm\rho$. The modular domain is a positive half-space $\rho_{11} > 0$. According to the analysis of Section 5.6.3, $\mu(\rho) = \rho_{11}$. The averaging measure in (6.173) reads $(2\pi)^{-1}d\omega$ and the integration is extended over the interval $[0, 2\pi]$. The quadratic form in the exponential in (6.173) has the form,

$$\left(\rho - e^{\omega T}\rho'\right)^2 = (\rho,\rho) + (\rho',\rho') - 2(\rho,\rho')\cos\omega - 2(\rho,T\rho')\sin\omega. \tag{6.196}$$

Here $(\rho\,,\,\rho') \equiv \mathrm{Tr}\,(\rho\,,\,\rho')$ to simplify the notation. A distinguished feature of this model, when compared to the models considered above, is that the stationary point is a function of ρ and ρ'. Taking the derivative of (6.196) with respect to ω and setting it to zero, the stationary points are found,

$$\omega_c = \tan^{-1}\frac{(\rho,T\rho')}{(\rho,\rho')}\,, \quad \omega_c^s = \omega_c + \pi.$$

The second stationary point ω_c^s is associated with the transformation $\rho \to -\rho$.

A geometrical meaning of the transformation $\rho' \to \exp(\omega_c T)\rho'$ is transparent. The distance $[(\rho - \rho')^2]^{1/2}$ between two points on the gauge fixing plane is greater than the distance between the two gauge orbits through $x = \rho$ and $x' = \rho'$ (which is defined as the distance between two sets in a metric space, $D(A, B) = [\inf(x - x')^2]^{1/2}$, where the infinitum is taken over $x \in A$ and $x' \in B$ for two sets A and B). This happens when the angle at which the gauge fixing plane $x_{21} = 0$ intersects gauge orbits is not $\pi/2$. The distance between $x = \rho$ and $x'_c = \exp(\omega_c T)\rho'$ (the configuration $x' = \rho'$ shifted along the orbit) is the distance between the gauge orbits through x and x'. Through this position-dependent stationary point, the metric on the orbit space emerges in the projection path integral formalism. To find its explicit form, the stationary point $\omega = \omega_c(\rho, \rho')$ is substituted into (6.196). The result is decomposed in powers of Δ where $\rho' = \rho - \Delta$. The quadratic term (the leading term) determines the metric. Indeed,

$$(\Delta, g^{\mathrm{ph}}(\rho)\Delta) = (\Delta, \Delta) + (\Delta, T\rho)(T\rho, \Delta)/(\rho, \rho), \tag{6.197}$$

which coincides with the inverse of g^{ab}_{ph} in (5.128).

In the stationary phase approximation the cosine and sine in (6.196) should be decomposed up to fourth order in the vicinity of each stationary point to account for quantum corrections (technical details are standard and omitted here). The infinitesimal transition amplitude on the orbit space is

$$U^{\mathrm{ph}}_{\varepsilon}(\rho, \rho') = D^{-1/2}(\rho, \rho')\tilde{U}_{\varepsilon}(\rho, \rho') + D^{-1/2}(\rho, -\rho')\tilde{U}_{\varepsilon}(\rho, -\rho'), \tag{6.198}$$

$$\tilde{U}_{\varepsilon}(\rho, \rho') = (2\pi i\varepsilon)^{-3/2} e^{iS_{\varepsilon}(\rho, \rho')}, \tag{6.199}$$

$$S_{\varepsilon}(\rho, \rho') = \frac{1}{2\varepsilon}\left[(\rho, \rho) + (\rho', \rho') - 2D(\rho, \rho')\right] - \frac{\varepsilon}{8D(\rho, \rho')} - \varepsilon V(\rho), \tag{6.200}$$

where $-2D(\rho, \rho')$ is given by the two last terms in Eq. (6.196) at the stationary point $\omega = \omega_c$. To calculate the contribution of factor $D^{-1/2}$ in (6.198) in the continuum limit, function $D(\rho, \rho')$ must be expanded up to order Δ^2 with subsequent application of the equivalence rules. The details of this standard procedure are omitted. Here only the question of taking the square root of D is addressed, because it is crucial for the correct symmetrization of the transition amplitude with respect to the residual group $S = Z_2$. In the leading order,

$$D(\rho, \rho') = \mu(\rho)\mu(\rho') \det{}^{-1} g^{\mathrm{ph}}(\rho') + O(\Delta^2), \tag{6.201}$$

where the explicit form of metric (6.197) has been used and $\mu = \rho_{11}$ to compute $\det g^{\mathrm{ph}} = \mu^2/(\rho, \rho)$. As in the 2D Yang–Mills theory, an analytic continuation of the Gaussian integral (6.192) must be applied to obtain $D^{-1/2}$ outside the modular domain $\rho_{11} > 0$. The phase of $D^{-1/2}$ is determined only by the sign of the Faddeev–Popov determinant μ at points ρ and ρ' because $\det g^{\mathrm{ph}} > 0$. In the continuum limit, only the leading term in (6.201) determines the phase of $D^{-1/2}(\rho, \rho')$. Therefore the phase is also invariant under permutations of ρ and

ρ' in (6.201) because the terms $O(\Delta^2)$ arising from $\det g^{\rm ph}$ do not affect it in the continuum limit.

Consider the convolution of $N+1$ kernels (6.198)

$$\int_K \prod_{k=1}^{N} (d\rho_k \mu(\rho_k)) U_\varepsilon^{\rm ph}(\rho_{N+1}, \rho_N) \cdots U_\varepsilon^{\rm ph}(\rho_1, \rho_0)$$

in which the invariance of the kernels $U_\varepsilon^{\rm ph}$ relative to the Z_2-symmetry in both its argument is used to extend the integration region to whole space and, at the same time, to convert the product of N kernels $U_\varepsilon^{\rm ph}$ into the corresponding product of kernels $\tilde{U}_\varepsilon$ similarly to the derivation of (6.106). The resulting integrand also contains the density (see (6.178)):

$$\frac{|\mu_N| \cdots |\mu_1|\, |\mu_0|}{[D_{N+1,N} \cdots D_{2,1} D_{1,0}]^{1/2}} = \frac{\prod_{k=0}^{N} \det^{1/2} g_k^{\rm ph}}{[\mu_{N+1}\mu_0]^{1/2}} + O(\epsilon), \tag{6.202}$$

with N being the number of integrations in the convolution; $\mu_k = \mu(\rho_k)$, $D_{k,k-1} = D(\rho_k, \rho_{k-1})$, $g_k^{\rm ph} = g^{\rm ph}(\rho_k)$, $k = 0, 1, ..., N+1$, and ρ_0, ρ_{N+1} are initial and final configurations, respectively. All the terms $O(\Delta^2)$ are assumed to have been converted into corresponding terms $O(\varepsilon)$ by means of the equivalence rules. In the numerator of the right-hand side of (6.202), the density at the initial state $\det^{1/2} g_0^{\rm ph}$ can be replaced by the density at the final state $\det^{1/2} g_{N+1}^{\rm ph}$. The choice depends on the base point (pre-point or post-point) in the definition of the path integral on a curved space (see Section 2.5). In other words, the infinitesimal action (6.200) in amplitude (6.199) can be decomposed in powers of Δ either at point ρ (post-point) or at point ρ' (pre-point). Both representations differ in terms of order ε. Here the pre-point decomposition for D and S_ε has been chosen. The base point can be changed by the rules formulated in Section 2.5. To go over to the Hamiltonian path integral, the Fourier transform in Δ for each kernel (6.199) in the convolution has to be carried out. The $N+1$ factors in the numerator of (6.202) are canceled against the same factors resulting from the Fourier integrals over the momentum variables. In the continuum limit, the procedure yields the conventional local Liouville measure of the Hamiltonian path integral. The number of momentum integrals should be exactly $N+1$ because it exceeds by one the number of integrals over configurations (see Section 2.2.1).

6.7.3 The Morse theory, orbit space geometry, and the projection operator formalism in Yang–Mills theory

The way the curvature of the orbit space emerges in the projection operator formalism is universal and not specific to the above gauge matrix model. The non-flat metric on the orbit space of 4D Yang–Mills theory [230] can be deduced in exactly the same way. This can be established by means of the Morse theory [318]. Let $\mathbf{A}$ be the spatial Yang–Mills connection. The time component A_0 is a pure non-physical variable (the Lagrange multiplier) and may be excluded so that the total configuration space is an affine space of all connections $\mathbf{A}$. In Eq. (6.172), $(x - T(\omega)x')^2 \to \langle \mathbf{A} - \mathbf{A}'^{\Omega}\rangle^2$, where, similarly to (5.176), the brackets $\langle \cdot, \cdot \rangle = \int d^3x\, (\cdot, \cdot)$ include the Cartan–Killing scalar product in the Lie algebra and the integration over the space, $\mathbf{A}'^{\Omega}$ is the gauge transform of $\mathbf{A}'$ with a gauge group element $\Omega = \Omega(\omega)$ (see Section 5.8.1 for notation), and the potential energy $V(x) \to V(\mathbf{A}) = (1/4)\langle F_{ij}\rangle^2$. Here and in what follows, the notation $\langle B\rangle^2 \equiv \langle B, B\rangle$ is used for brevity. The Yang–Mills potential energy is not relevant for what follows (as argued in Section 6.7.1).

The first problem is to introduce parameterization of the orbit space in order to go over to the representation (6.173). The goal is achieved by means of the Morse theory of critical points of smooth functions on Banach manifolds as was suggested in [319] and further developed in [320, 321]. Given a connection $\mathbf{A}$, the Morse function is defined by,

$$M_A(\Omega) = \left\langle \mathbf{A}^{\Omega}\right\rangle^2 = \left\langle \Omega\mathbf{A}\Omega^{-1} + ig^{-1}\Omega\partial\Omega^{-1}\right\rangle^2. \tag{6.203}$$

The minima of the Morse functional (6.203) relative to Ω carry information about the topology of the gauge orbit through $\mathbf{A}$. Taking $\Omega = e^{ig\omega}$ and expanding the Morse functional around the critical point $\omega = 0$, one finds,

$$M_A(\omega) = \langle \mathbf{A}\rangle^2 + 2\langle \omega, \partial_j A_j\rangle - \langle \omega, (\partial_i \nabla_i)\omega\rangle + O(w^3), \tag{6.204}$$

where $\nabla_j\omega = \partial_j\omega - ig[A_j, \omega]$ is the covariant derivative of ω. From (6.204) it follows that the Morse function attains its relative minima if the connection satisfies the Coulomb gauge $\partial_j A_j = 0$ and the Faddeev–Popov operator $\Delta_{FP} = -\,\partial_j\nabla_j$ is a symmetric, positive operator. The positivity of this operator ensures that connection $\mathbf{A}$ has the property that $\Omega = 1$ is a *minimum* of M_A. The *Gribov horizon* is defined by the condition that the *lowest* eigenvalue of the operator Δ_{FP} vanishes. The configurations on the Gribov horizon are *degenerate* critical points of the Morse function. A Gribov region K_G is defined as the set of all minima of the Morse functional. K_G has the property that each gauge orbit intersects it at least once, and it is convex and bounded [319, 322].

It may happen that two *relative* minima inside the region K_G are related by a gauge transformation, i.e. they are on the same gauge orbit. To obtain the modular domain K which contains only one representative of each gauge orbit, one has only to take the absolute minima of the Morse functional. Let $\mathbf{A}$ and the gauge transform of it $\mathbf{A}^{\Omega}$ both be from the region K_G. Then it is straightforward to show [264, 266] that

$$\langle \mathbf{A}^{\Omega}\rangle^2 - \langle \mathbf{A}\rangle^2 = \left\langle \Omega^{-1}, \partial_i \nabla_i^f(\mathbf{A})\Omega\right\rangle, \tag{6.205}$$

where $\boldsymbol{\nabla}^f\Omega = \boldsymbol{\partial}\Omega - ig\mathbf{A}\Omega$ is the covariant derivative in the fundamental representation. Since the operator Δ_{FP} is positive in K_G, the absolute minima of the Morse function can be defined in terms of the absolute minima over the gauge group of the right-hand side of Eq. (6.205). A configuration $\mathbf{A}$ from the region K_G belongs to the modular domain K if the minimum of the functional (6.205) over the gauge group vanishes. This condition simply selects the absolute minima of the Morse function out of its relative minima. Since the operator Δ_{FP} is linear in $\mathbf{A}$, all configurations of the line segment $s\mathbf{A}_{(1)} + (1-s)\mathbf{A}_{(2)}$, where $s \in [0,1]$ and $\mathbf{A}_{(1,2)} \in K$, also belong to K. That is, the modular domain is convex.

In a similar way the modular domain can be described for a background gauge $\nabla_i(\bar{\mathbf{A}})A_i = 0$ ($\bar{\mathbf{A}}$ is a background (fixed) connection) [322]. In this case the Morse functional is

$$M_A(\Omega) = \langle \mathbf{A}^\Omega - \bar{\mathbf{A}} \rangle^2 \,, \tag{6.206}$$

and now the Faddeev–Popov operator has the form, $\Delta_{FP} = -\nabla_j(\bar{\mathbf{A}})\nabla_j(\mathbf{A})$.

The main properties of the modular domain for *classical* (smooth) configurations are as follows [266]. First, its boundary has common points with the Gribov horizon. Second, the modular domain K has a trivial topology as any convex subset in an affine space, but its boundary contains gauge-equivalent configurations. A nontrivial topology of the gauge orbit space occurs through their identification. In fact, the orbit space contains non-contractable spheres of any dimension [211]. Third, in the interior of the modular domain there are configurations that are related by homotopically nontrivial gauge transformations. This feature has been illustrated in Section 5.8.7. Any point on the Gribov horizon has a *finite* distance from the origin of the field space and one can derive a uniform bound for it [209, 323].

Although the above procedure to determine the modular domain applies to general background connections, some properties of K_G and K may depend on the choice of the background connection. In particular, reducible and irreducible background configurations have to be distinguished [270, 324]. A connection $\mathbf{A}$ is said to be reducible if it has a nontrivial stationary group $\mathcal{G}_A$ (the stabilizer) such that $\mathbf{A}^\Omega = \mathbf{A}$ for all $\Omega \in \mathcal{G}_A$. If $\mathcal{G}_A$ coincides with the center Z_G of the structure group G (the gauge group is denoted $\mathcal{G}$ which is a collection of maps $R^3 \to G$, see Section 5.8.1), then the connection is irreducible. From the identity $\mathbf{A}^\Omega = \mathbf{A} + ig^{-1}\Omega\boldsymbol{\nabla}(\mathbf{A})\Omega^{-1}$ it follows that $\Omega \in \mathcal{G}_A$ if $\boldsymbol{\nabla}(\mathbf{A})\Omega = 0$. Any stabilizer $\mathcal{G}_A$ is isomorphic to a subgroup of G [325]. This can be understood as follows. Recall that for any $\mathbf{A}$, $\mathcal{G}_A$ is isomorphic to the centralizer of the holonomy group of $\mathbf{A}$ relative to the structure group G [251]. By definition, the centralizer G'_c of a subgroup G' of G consists of all elements of G which commute with all elements of G'. Clearly, G'_c is a subgroup of G. On the other hand, the holonomy group is a Lie subgroup of G [251], i.e. its centralizer is a subgroup of G.

The orbit space has the structure of a so-called stratified variety which can be regarded as the disjoint sum of strata that are smooth manifolds [267–269]. Each stratum of the variety consists of orbits of connections whose stabilizers are conjugate subgroups of $\mathcal{G}$. In other words, the stabilizers of the connections of a fixed stratum are isomorphic to one another. A stratum that consists of orbits of all *irreducible* connections is called a main stratum. The set of orbits of *reducible* connections (singular strata) is a closed subset in the orbit space which is *nowhere dense*. Accordingly, the main stratum is dense in the orbit space, and any singular strata can be approximated arbitrarily well by irreducible connections [270]. If all reducible connections are excluded from the total configuration space, then the orbit space is a manifold (compare Section 5.8.8).

The Morse functional (6.206) can also be viewed as the distance between $\mathbf{A}^{\Omega}$ and $\bar{\mathbf{A}}$. Let $\bar{\mathbf{A}}$ be an irreducible connection. Any two connections $\mathbf{A}_{1,2}$, $\mathbf{A}_1 \neq \mathbf{A}_2$, on the gauge fixing surface $\nabla_i(\bar{\mathbf{A}})A_i = 0$ that are sufficiently close to $\bar{\mathbf{A}}$ belong to distinct gauge orbits. For reducible backgrounds, the gauge fixing surface does not possess such a property. The Morse functional (6.206) has a degeneracy for reducible backgrounds. Indeed, if $\bar{\Omega} \in \mathcal{G}_{\bar{A}}$, i.e. $\bar{\Omega}$ satisfies $\bar{\mathbf{A}} = \bar{\mathbf{A}}^{\bar{\Omega}}$, then one has

$$M_A(\Omega\bar{\Omega}) = \left\langle \mathbf{A}^{\Omega\bar{\Omega}} - \bar{\mathbf{A}} \right\rangle^2 = \left\langle \mathbf{A}^{\Omega\bar{\Omega}} - \bar{\mathbf{A}}^{\bar{\Omega}} \right\rangle^2 = M_A(\Omega) \,, \tag{6.207}$$

because the Morse functional is invariant under simultaneous gauge transformations of $\mathbf{A}^{\Omega}$ and $\bar{\mathbf{A}}$. It is also not hard to see that, if $M_A(\Omega\bar{\Omega}) = M_A(\Omega)$ holds true for any $\mathbf{A}$, then $\bar{\Omega}$ should be an element of the stabilizer $\mathcal{G}_{\bar{A}}$.

In the case of a reducible background, the Faddeev–Popov operator always has zero modes. Let $\bar{\mathbf{A}}^{\Omega} = \bar{\mathbf{A}}$, i.e. $\Omega \in \mathcal{G}_{\bar{A}}$. As $\mathcal{G}_{\bar{A}}$ is isomorphic to a Lie group (a subgroup of G), there exists a family $\Omega_\lambda \in \mathcal{G}_{\bar{A}}$ such that all elements are connected to the group unity, $\Omega_{\lambda=0} = 1$. The Lie algebra-valued function $\psi(\mathbf{x}) = \partial_\lambda \Omega_\lambda(\mathbf{x})|_{\lambda=0}$ is covariantly constant, $\nabla_j(\bar{\mathbf{A}})\psi = 0$, and therefore it is a zero mode of the Faddeev–Popov operator, $\Delta_{FP}\psi = -\nabla_j(\bar{\mathbf{A}})\nabla_j(\mathbf{A})\psi = -\nabla_j(\mathbf{A})\nabla_j(\bar{\mathbf{A}})\psi = 0$, thanks to the symmetry of Δ_{FP}. In particular, for $\bar{\mathbf{A}} = 0$ one gets $\mathcal{G}_{\bar{A}=0} \sim G$, i.e., $\mathcal{G}_{\bar{A}=0}$ is a group of constant gauge transformations. So, by removing constant gauge transformations from the gauge group, a systematic degeneracy of the Faddeev–Popov operator for the Coulomb gauge is removed. Next, observe that, because of the degeneracy (6.207), the collection of all absolute minima of the Morse functional can no longer serve as the fundamental modular domain K if $\mathcal{G}_{\bar{A}}$ is nontrivial. There are gauge equivalent configurations inside the set of the absolute minima. The identification in the interior of K precisely amounts to dividing out the stabilizer $\mathcal{G}_{\bar{A}}$ [270], i.e. the orbit space is $K/\mathcal{G}_{\bar{A}}$. This has been illustrated with an example of 2D Yang–Mills theory in Section 5.8.8.

Fianlly, it should be mentioned that stratification of the orbit space of the classical $SU(2)$ Yang–Mills theory is studied in detail in [270]. It is noteworthy that classical trajectories in the Hamiltonian formalism are always contained in one fixed stratum (in one smooth manifold) [271]. A principal bundle is known to have isomorphism classes characterized by the instanton number which can be any integer. Connections with different instanton numbers satisfy different asymptotic conditions at infinity. If one allows asymptotic conditions associated with all instanton numbers, then the fundamental modular domain will be the disjoint sum of modular domains for every instanton number [272].

As explained in Section 5.8.9, the above analysis of the modular domain that is used to parameterize the gauge orbit space in Yang–Mills theory may not be directly applied to the quantum case without a proper regularization of the theory, e.g. lattice regularization. The above analysis of the modular domain (the Coulomb gauge) has been extended to lattice Yang–Mills theory and to Yang–Mills theory with space compactified to a small 3-torus [266, 326–329].

Once the modular domain K and parameterization of the orbit space is established, the average integral in (6.173) can be calculated. The orbit space is known to have a non-flat metric tensor [230]. The next objective is to show that the average integral in (6.173) reproduces the correct metric on the orbit space in Yang–Mills theory. Suppose $\mathbf{A}$ and $\mathbf{A}'$ are transverse connections from the modular domain K in the quadratic form $\langle \mathbf{A} - \mathbf{A}'^{\Omega}\rangle^2$ that appears in the group average integral (6.173). This quadratic form coincides with the Morse functional (6.206). Its minimum is attained at $\Omega = \Omega_c$ such that

$$\nabla_i(\mathbf{A})A_i'^{\Omega_c} = 0. \tag{6.208}$$

Indeed, if $\mathbf{A}'$ satisfied the background gauge $\nabla_i(\mathbf{A})A_i' = 0$, then the minimun would have been attained at $\Omega = 1$ as the above analysis shows. However, $\mathbf{A}'$ is transverse $\partial_i A_i' = 0$, so it needs to be gauge transformed to satisfy the background gauge where $\mathbf{A}$ serves as the background connection. Let $\mathbf{A} - \mathbf{A}' = \delta\mathbf{A}$ be small (the formal continuum limit in the path integral). Then there is a particular $\Omega_c = \exp(-ig\omega_c)$ which is close to group unity. Linearizing the exponential in (6.208) one infers the analog of Eq. (6.175),

$$\nabla_i(\mathbf{A})\left(A_i' - \nabla_i(\mathbf{A}')\right)\omega_c = 0,$$

whose solution is

$$\omega_c = (\boldsymbol{\nabla}(\mathbf{A}), \boldsymbol{\nabla}(\mathbf{A}'))^{-1}\nabla_i(\mathbf{A})A_i' = -\Delta^{-1}(\mathbf{A})\nabla_i(\mathbf{A})\delta A_i + O(\delta\mathbf{A}^2), \tag{6.209}$$

where $\Delta(\mathbf{A}) = (\boldsymbol{\nabla}(\mathbf{A}), \boldsymbol{\nabla}(\mathbf{A}))$; the latter equality in (6.209) follows from the relation $\nabla_i(\mathbf{A})A_i' = -\nabla_i(\mathbf{A})\delta A_i$ (recall that $\nabla_i(\mathbf{A})A_i = \partial_i A_i = 0$). The expansion of $\Omega(\omega_c + \omega)$ is powers of ω in the quadratic form $\langle \mathbf{A} - \mathbf{A}'^{\Omega}\rangle^2$ contains the leading term ($\omega \equiv 0$), and higher-order terms ω^n, $n = 2, 3, ...,$. The linear is absent because $\omega = 0$ is the minimum. The quadratic term determines the leading contribution of the stationary phase approximation (the determinant $D^{-1/2}(\mathbf{A}, \mathbf{A}')$ in (6.177)), while the terms of ω^3 and ω^4 are needed to calculate quantum corrections to the classical action in the physical variables. The metric on the orbit space is defined by the zero-order term (see (6.176)). At the stationary point, the quadratic form in the leading order in $\delta\mathbf{A}$ reads,

$$\left\langle A_j - A_j'^{\Omega_c}\right\rangle^2 \approx \langle \delta A_j + \nabla_j(\mathbf{A}')\omega_c\rangle^2 = \left\langle \delta A_j, g_{jk}^{\mathrm{ph}}\delta A_k\right\rangle, \tag{6.210}$$

$$g_{jk}^{\mathrm{ph}}(\mathbf{A}) = P_{jk} - P_{jn}\nabla_n(\mathbf{A})\Delta^{-1}(\mathbf{A})\nabla_m(\mathbf{A})P_{mk}, \tag{6.211}$$

where $P_{jk} = \delta_{jk} - \partial_j(\partial^2)^{-1}\partial_k$ is the projector on transverse vector fields. In contrast to the mechanical case, here the metric tensor g_{jk}^{ph} is an operator in the functional space of connections. The first equality in (6.210) is obtained by linearizing Ω_c in ω_c (all higher-order terms are neglected). Then the explicit form

(6.209) of ω_c has been used, and the expression in the inner product bracket is linearized in $\delta\mathbf{A}$ to obtain $\delta A_j - \nabla_j\Delta^{-1}\nabla_k\delta A_k \equiv \Pi_{jk}\delta A_k$. The operator Π is the projector on covariantly transverse fields, $\Pi^2 = \Pi$, and Π is Hermitian relative to the inner product $\langle\, ,\, \rangle$. So, the metric becomes $g^{\rm ph} = P\Pi P$; the projector P is needed because Π acts only on the transverse fields, $\delta A_j = P_{jk}\delta A_k$, and, therefore, $\langle \delta A_j, B_j\rangle = \langle \delta A_j, P_{jk}B_k\rangle$ for any vector field B_j. The metric (6.211) is written in an explicitly Hermitian form. If one agrees, that $g^{\rm ph}$ acts only in the subspace of transverse fields, then $P_{jk} \to \delta_{jk}$ and $P_{mk} \to \delta_{mk}$ in (6.211), i.e. $g^{\rm ph} = 1 - P(1-\Pi)$. The latter form is convenient to calculate the inverse $(g^{\rm ph})^{-1} \equiv g_{\rm ph}$ by means of the geometric series summation,

$$g_{\rm ph} = [1-P(1-\Pi)]^{-1} = \sum_{n=0}^{\infty}(P(1-\Pi))^n\,, \qquad g_{\rm ph}^{jk} = \delta_{jk} + P_{ji}\nabla_i D^{-1}\nabla_k, \quad (6.212)$$

where $D = \Delta_{FP}(\partial^2)^{-1}\Delta_{FP}$ and $\Delta_{FP} = -\partial_j\nabla_j$. The inverse metric (6.212) appears in the kinetic energy of the physical Hamiltonian of the gauge-fixed Yang–Mills theory. It can be shown by further calculations [221] that the transition amplitude for the Yang–Mills theory (defined by the formal path integral (6.180) and (6.181)) satisfies the functional Schrödinger equation with the Hamiltonian (6.182) where the metric is given by (6.212), $\partial/\partial u \to \delta/\delta\mathbf{A}$, the contraction of indices is replaced by $\langle\, ,\, \rangle$, and $\mu = \det\Delta_{FP}(\mathbf{A})$.

Due to locality of the kinetic energy, the quantum corrections would contain a non-physical infinite factor $\hbar^2[\delta^3(0)]^2$, which typically results from the operator ordering in the kinetic energy operator of any local field theory that contains a non-Euclidean metric in the field space, and also be singular at the Gribov horizon. Thus, the Schrödinger equation in field theory requires regularization of the local product of operators involved [101, 175]. Various approaches have been developed to understand the effect of singular points where $\det\Delta_{FP} = 0$ (the Gribov horizon) [333]–[336], [266, 326], [330]–[332] by means of some reduction of degrees of freedom to make the operator formalism well defined. However the presence of a strong self-interaction of Yang–Mills fields makes it hard to prove that the effects associated with the horizon are of any significance as compared to those due to self-interaction. The same conclusion applies to the path integral approach (see the end of Section 5.9.5).

The above procedure can be rigorously formulated for the Kogut–Susskind lattice gauge theory (see Section 5.8.10). The structure of the gauge-fixed transition amplitude remains as in (6.177), i.e. there is a symmetrization over the residual gauge transformations rather than a restriction of the integration domain to the modular domain. Note that in the 2D case, this structure of the path integral was crucial to recover the correct spectrum of the Hamiltonian from the partition function (compare (6.58) and (6.49)). The symmetrization also changes the path integral representation of (gauge-fixed) Green's functions in the lattice gauge theory, as explained in Section 6.8.2.

Remark. If in the 2D Yang–Mills theory the constant gauge transformations were forbidden, then the projection formalism would have led to the path integral that coincides with that for a particle on a group manifold (there are no reducible connections and the orbit space is the group manifold (Section 5.8.8)). The metric (6.211) equals the metric tensor on the group manifold. Summation over the residual gauge transformations in (6.180) is the summation over translations by the periods of the group manifold (the group T_e introduced in Section 5.8.2).

6.8 Simple consequences of the modification of the path integral for gauge systems

The form of operator $\hat{Q}$ in the path integral representation of quantum gauge theories is determined by the structure of the physical phase space. As has been already pointed out, this modification of the path integral formalism is usually ignored (see also Section 8.4). The path integral formalism naturally allows us to determine a semi-classical transition amplitude. Let us investigate a relation between the physical phase-space structure and the semi-classical description of gauge systems (see also [200, 204, 304]).

6.8.1 Instantons in gauge quantum mechanics

Consider the simplest example of a gauge theory where the found modification of the path integral affects the instanton calculus in gauge quantum mechanics. Instantons are used in quantum theory to calculate tunneling effects [265, 277, 313, 337]. Consider a one-dimensional quantum system with a periodic potential [277]. The ground state in the vicinity of each potential minimum is degenerate. The degeneracy is removed due to tunneling effects, and the ground states turn into a zone. It turns out that knowledge of the solutions of Euclidean equations of motion (the equations of motion in imaginary time $t \to -i\tau$) allows us to calculate approximately the energy levels in the zone and to find the corresponding wave functions (the θ-vacua).

Consider the $SU(2)$ model from Section 5.4 with periodic potential $V(x) = 1 - \cos[(x,x)^{1/2}]$. Since the cosine is an even function, the potential is a regular function of the only independent Casimir polynomial $p_2(x) = (x,x)$. The analogous one-dimensional model has been well studied (see, e.g. [277] and references therein). The only physical degree of freedom has the phase space being the cone cone(π).

Consider the Euclidean version of the theory. In the Lagrangian (5.57) the following replacements are to be made: $t \to -i\tau$ and $y \to iy$ (τ is called the Euclidean time). Recall that y is analogous to the time component of the Yang–Mills potential which requires factor i in the Euclidean formulation [277]. The Lagrangian assumes the form $L \to L_E = (D_\tau x)^2/2 + V(x)$. The dynamics of the only physical degree of freedom is described by the element of the Cartan subalgebra $x = h\lambda_1 \in H$ (λ_1 is the only basis element of $H \sim R$, $(\lambda_1, \lambda_1) = 1$). Solutions of the equations of motion,

$$\frac{d}{d\tau}\frac{\partial L_E}{\partial \dot{x}} = \frac{\partial L_E}{\partial x}, \qquad \frac{\partial L_E}{\partial \dot{y}} = 0,$$

where the overdot denotes the Euclidean time derivative ∂_τ, depend on the arbitrary functions $y = y(\tau)$ whose variations generate the gauge transformations of the classical solutions $x(\tau)$ (see also Section 3.3). After removing the gauge arbitrariness by imposing condition $y = 0$, the equations of motion are equivalent to,

$$\ddot{h} = \sin h. \tag{6.213}$$

The instanton solution of Eq. (6.213) has the form [277],

$$h(\tau) = h_{\text{inst}}(\tau) = 4\tan^{-1}\exp(\tau - \tau_c) + 2\pi m, \quad \tau_c = \text{const}, \quad m = 0, \pm 1, \cdots \tag{6.214}$$

It connects the local minima of the potential: $x^2_{\text{inst}} \to (2\pi m)^2$ as $\tau \to -\infty$, and $x^2_{\text{inst}} \to [2\pi(m+1)]^2$ as $\tau \to \infty$, where $x_{\text{inst}}(\tau) = h_{\text{inst}}(\tau)\lambda_1$ in the chosen gauge.

Equation (6.213) coincides with the Euler–Lagrange equation for the the analogous one-dimensional model $\tilde{L}_E = \dot{h}^2/2 + 1 - \cos h$, $h \in R$, i.e. with the Euclidean phase space R^2. For this model the wave function of the θ-vacuum is calculated as follows [277]. First, one finds the amplitude $U_\tau(2\pi m, 2\pi m')$ in the stationary phase approximation of the corresponding path integral. The instanton solution serves as the stationary point. In the limit $\tau \to \infty$, the main contribution comes from the states of the lowest zone (the contributions of higher levels are exponentially suppressed),

$$\begin{aligned} U_\tau(2\pi m, 2\pi m') &= \langle 2\pi m | e^{-\tau\hat{H}} | 2\pi m'\rangle \\ &\approx \int_0^{2\pi} d\theta \langle 2\pi m|\theta\rangle\langle\theta|2\pi m'\rangle e^{-\tau E_\theta}, \end{aligned} \tag{6.215}$$

as $\tau \to \infty$, where θ parameterizes the energy levels E_θ in the lowest zone. The amplitude $\langle 2\pi m|\theta\rangle$ is extracted from the path integral in the stationary phase approximation for the instanton solution (6.214). The details can be found in [277] where it is shown that, in the asymptote $\tau \to \infty$,

$$U_\tau(2\pi m, 2\pi m') \approx \int_0^{2\pi} \frac{d\theta}{2\pi^{3/2}} e^{-i(m-m')\theta} e^{-\tau E_\theta}, \tag{6.216}$$

$$E_\theta = \frac{1}{2} - e^{-S_0} S_0 K \cos\theta;$$

here S_0 is the action computed for the instanton solution, K is a constant independent of θ (the instanton determinant [277]). The amplitude $\langle 2\pi m|\theta\rangle \sim \exp(-im\theta)$ follows from the comparison of (6.215) and (6.216). It determines the value of the vacuum wave function $\langle h|\theta\rangle$ in the local minima of the potential, $h = 2\pi m$. Therefore the wave function $\langle h|\theta\rangle$ can be approximated by the superposition

$$\langle h|\theta\rangle \approx C \sum_{m=-\infty}^{\infty} e^{-im\theta}\langle h|2\pi m\rangle, \quad C = \text{const.}, \tag{6.217}$$

where $\langle h|2\pi m\rangle \sim \exp[-(h-2\pi m)^2/2]$ is the ground state wave function in the oscillator approximation in the vicinity of each potential minimum.

To find how the above calculations are modified in the case when the physical degree of freedom has a conic phase space, one has to take the amplitude $U^c_\tau(2\pi m, 2\pi m')$ instead of $U_\tau(2\pi m, 2\pi m')$ in (6.215). Since the algebra $su(2)$ is isomorphic to $so(3)$, the relation between these amplitudes is given by (6.16) (where $r \to h$ and $t \to -i\tau$). The result reads [204],

$$U^c_\tau(2\pi m, 2\pi m') \approx \int_0^{2\pi} \frac{d\theta}{\pi^{3/2}} \frac{\sin(m\theta)\sin m'\theta}{(2\pi)^2 mm'} e^{-\tau E_\theta}. \tag{6.218}$$

Therefore the change of the phase space structure does not affect distribution of the energy levels in the lowest zone. However, it does affect the amplitudes $\langle 2\pi m|\theta\rangle$, thus leading to a modification of the wave function of the θ-vacuum,

$$\langle h|\theta\rangle^c \approx \text{const}\cdot \sum_{m=-\infty}^{\infty} \frac{\sin m\theta}{2\pi m}\langle h|2\pi m\rangle. \tag{6.219}$$

By the obvious relation $\langle -h|2\pi m\rangle = \langle h|-2\pi m\rangle$, function (6.219) is even, $\langle -h|\theta\rangle^c = \langle h|\theta\rangle^c$, i.e. it is invariant under the residual Weyl transformations and normalized according to (5.79), while the function (6.217) does not have definite parity.

That the energy level distribution in the lowest zone is not sensitive to the conic phase space structure holds, in general, only for the continuum spectrum. The analogy with the case of a free particle is suitable here. The change of the phase space structure from the plane to the cone has no effect on the spectrum. In contrast, a discrete spectrum, e.g. the harmonic oscillator, is sensitive to the change of phase space structure. A similar observation exists for instantons as well. Consider, for example, the double-well potential $V = (x^2 - v^2)^2$ and the gauge group $SU(2)$. The corresponding one-dimensional system has been well studied [265] where it was shown that the lowest zone contains only two levels because the classical ground state is doubly degenerate. In the gauge model, the classical ground state is also doubly degenerate. In the gauge $x = h\lambda_1 \in H$ and $h = \pm v$ are classical minima of the potential. Therefore the lowest zone would also seem to contain two levels. The lower level has an odd wave function, while the upper one has an even wave function. Such an unusual parity property (typically one expects the lowest level to have an even wave function) is a consequence of the fact that wave functions of the corresponding one-dimensional system have to be multiplied by the odd density factor $\kappa(h) \sim h^{-1}$ to obtain the wave functions of the gauge system (compare Section 5.4). Reduction of the phase space from plane to cone implies that the odd functions are to be excluded (those that are not invariant under the residual gauge transformations Z_2).

Analysis of more complicated gauge systems would not add essentially new features. Given a classical solution, one should evaluate the gauge-fixed path integral in the stationary phase approximation, multiply it by the density $\mu^{-1/2}$ at initial and final configurations, and symmetrize the result relative to the residual gauge transformations [200].

6.8.2 Greens functions for gauge systems with a non-standard physical phase space and the path integral method

In Section 5.9.4 by means of the Dirac operator formalism for gauge theories it is shown that the evolution of boson and fermion degrees of freedom with a reduced physical phase space should be viewed as the evolution of composite objects. In particular, for the model discussed in Section 6.6.1, the evolution of elementary excitations of the physical degrees of freedom $\hat{u}\Phi_0$ and $\hat{\zeta}_a^\dagger\Phi_0$, with Φ_0 being the ground state, looks like the evolution of the composite states $\hat{u}_Q\Phi_0$ and $\hat{\zeta}_Q^\dagger\Phi_0$ (see (5.304) and (5.305)). Let us show how this interpretation emerges in the path integral method.

Consider the path integral (6.145). It determines the time evolution of any state according to the rule (6.148). The integral in the right-hand side of (6.148) can be transformed in the following way:

$$
\begin{aligned}
\Phi_t(q) &= \int\limits_K dq'\mu(u')\sum_S(\mu(u)\mu(\hat{s}u'))^{-1/2}U_t^{\text{eff}}(q,\hat{s}q')\Phi_0(q') \\
&= \sum_S\int\limits_{K_s} dq'\left(\frac{\mu(u')}{\mu(u)}\right)^{1/2} U_t^{\text{eff}}(q,q')\Phi_0(\hat{s}^{-1}q'). \qquad (6.220)
\end{aligned}
$$

The technicalities here are the same as in calculating the kernel of $\hat{U}_{2\varepsilon}^{\text{ph}} = \hat{U}_\varepsilon^{\text{ph}}\hat{U}_\varepsilon^{\text{ph}}$ in Section 6.6.1 (see also the text after (6.93)).

The factor $(\mu(u')/\mu(u))^{1/2}$ in (6.220) can be eliminated by changing terms linear in momenta in the effective action (6.140) because

$$
\left(\frac{\mu(u')}{\mu(u)}\right)^{1/2} = \exp\left[-\frac{1}{2}\int\limits_0^t d\tau\frac{\mu'(u)}{\mu(u)}\dot{u}\right], \qquad (6.221)
$$

where $u(t) = u$ and $u(0) = u'$ by the definition of the Hamiltonian path integral. Then

$$
\begin{aligned}
\Phi_t(q) &= \sum_S\int\limits_{K_s} dq'\tilde{U}_t^{\text{eff}}(q,q')\Phi_0(\hat{s}^{-1}q') \\
&= \int\limits_{-\infty}^{\infty} dq''\tilde{U}_t^{\text{eff}}(q,q'')\Phi_0^Q(q''), \qquad (6.222)
\end{aligned}
$$

where $\tilde{U}_t^{\text{eff}}$ is given by the Hamiltonian path integral (6.146) with the modified action that takes into account factor (6.221), and

$$\Phi_0^Q(q'') = \sum_S \int_{K_s} dq' \delta(q'', q') \Phi_0(\hat{s}^{-1} q') \tag{6.223}$$

is the S-invariant extension of the function Φ_0 into the non-physical domain. The S-invariance means that the following equality holds:

$$\Phi_0^Q(\hat{s}q) = \Phi_0^Q(q), \quad u \in K, \quad \forall \hat{s} \in S. \tag{6.224}$$

Let us prove that function (6.223) satisfies (6.224). Evidently, $\Phi_0^Q(q'') = \Phi_0(q'')$ if $u'' \in K$ in (6.223) for the only term with $\hat{s} = 1$ gives a nontrivial contribution to the sum (6.223). Accordingly, when $q'' = \hat{s}q$, $q \in K$, only one term in the sum differs from zero, i.e. when $u \in K$,

$$\Phi_0^Q(\hat{s}q) = \int_{K_s} dq' \delta(\hat{s}q, q') \Phi_0(\hat{s}^{-1} q') = \Phi_0(\hat{s}^{-1}\hat{s}q) = \Phi_0(q) = \Phi_0^Q(q),$$

by the definition of the function $\delta(q, q')$. The proof is complete.

Equation (6.222) relates the description of the time evolution of the same systems with the standard and reduced phase spaces. Namely, the non-standard structure of the phase (configuration) space is taken into account by replacing the initial state Φ_0 by its S-invariant extension Φ_0^Q to the non-physical region of the configuration space. It follows then that the state $\hat{q}\Phi_0(q)$ with $\Phi_0(q)$ being the vacuum state and $\hat{q} = \left(\hat{u}, \hat{\zeta}_a^+\right)$ evolves as the state $q_Q(\hat{q})\Phi_0(q)$ in which the composite operators $q_Q = \left(u_Q, \zeta_Q^*\right)$ are defined in (5.304) and (5.305). Indeed, thanks to the S-invariance of the physical vacuum ($\Phi_0 \in \mathcal{H}_{\rm ph}$), when applied to the function $\hat{q}\Phi_0(q) = q\Phi_0(q)$, Eq. (6.223) yields,

$$(q\Phi_0)^Q(q) = \sum_S \int_{K_s} dq' \delta(q, q') \hat{s}^{-1} q' \Phi_0(q') = q_Q(q)\Phi_0(q).$$

The functions $q_Q(q) = \left(u_Q(u), \zeta_Q^*(u, \zeta^*)\right)$ coincide with (5.304) and (5.305). Therefore the evolution of elementary excitations with the reduced phase space is equivalent to the evolution of some composite excitations in the corresponding theory with standard phase space.

The latter statement is rather general and valid for any gauge theory with a non-standard physical phase space structure. To prove this, it is sufficient to repeat the above calculations (6.220)–(6.222) for the model with an arbitrary gauge group discussed in Sections 6.5.2, 6.6.2, and 6.7. In the case of Yang–Mills theory the path integral representation first needs to be regularized by, e.g. using the Kogut–Susskind lattice gauge theory in a particular gauge to parameterize the physical phase (configuration) space (see Section 5.8.10)). The transition amplitude would have the same structure $\hat{U}_t^{\rm ph} = \hat{U}_t^{\rm eff}\hat{Q}$ where the extension operator $\hat{Q}$ depends on the set of residual discrete gauge transformations S in the parameterization (gauge) chosen and $\hat{U}_t^{\rm eff}$ is given by the conventional gauge-fixed path integral (with possible quantum corrections in the effective action (see Section 6.7). Hence, Eq. (6.222) holds, and this brings us again to the conclusion reached in Section 5.9.2 that the evolution of elementary excitations $\hat{q}\Phi_0$ is described by the composite excitations $q^Q(\hat{q})\Phi_0$ (excitations non-invariant under S are excluded), i.e. the phase space structure is a kind of memory of the reduced system about its gauge origin. Thus, thanks to the universal structure of the path integral, the conclusion has a universal nature and applies to all gauge theories including Yang–Mills theories (see also [212, 213, 221]).

7
Confinement

7.1 Introduction

7.1.1 A brief historical overview

While the prehistory of the question begins with the discovery of strong interactions (the nucleon model of atomic nuclei was developed in 1932), the origin of the confinement story may, perhaps, be referred to the year of 1961. By the early 1960s quite a number of strongly interacting particles had been discovered (π, K; $\bar{K}$; N, Λ, Σ, Ξ, Δ, Σ^*, Ξ^*), and the problem of finding some kind of systematic order in this zoo of "elementary building blocks" of matter became rather urgent. M. Gell-Mann and J. Neeman (see [338–340]) proposed a classification of these particles by the group $SU(3)$ (in modern terminology, the flavor group $SU(3)_f$). The classification scheme was known as the "eightfold way": mesons (π, K, η) and barions (N, Λ, Σ, Ξ) were organized into octet representations of $SU(3)_f$. However, the observed particles realized only particular representations of $SU(3)_f$ (the octet and decuplet (Δ, Σ^*, Ξ^*, Ω^-) ones); the role of other representations was not clear. As is well known, all representations of the group $SU(3)$ can be constructed from the two simplest triplets, ξ and ξ^*, being the so-called fundamental representation and its dual [218]. M. Gell-Mann and G. Zweig (see [341, 342]) assumed that all strongly interacting particles (hadrons) are built of elementary objects realizing the fundamental representation of $SU(3)_f$ ("quarks" [341] or "aces" [342]). The question, whether the quarks are real particles or pure mathematical, auxiliary objects, remained open. Acknowledgment of their existence created new problems:

(1) the quarks, while being fermions, were described by symmetric wave functions in all quantum numbers known at the time;
(2) there were no experimental observations of diquarks and other possible exotic bound states;
(3) isolated quarks were not observed either (confinement).

The first problem was solved by adding a new quantum number to quarks [343–345] (the color, in the modern terminology) which had three values; it was assumed thereby that there are three kinds of quarks, and their wave function in a nucleon is anti-symmetric in color.

Interest in quarks increased after the discovery of the scaling invariance and the parton model of hadrons [346, 347], when it became clear that the latter were built of point-like objects. It was natural to identify them with quarks, but then the question about the nature of quark interactions became of paramount significance. The hypothesis was made that strong interactions are described by a gauge theory [348]–[350] (see also [351], [352]); in [350] the model was proposed that is now known as quantum chromodynamics (QCD). The modern theory of strong interactions emerged as a result of the extension of global color symmetry [343]–[345] to the local one (i.e. to a gauge symmetry (see [353] for a historical overview)). Thus, by 1973 quantum chromodynamics had been formulated as a theory of strong interactions. The question about quark statistics was solved automatically, while the problem of confinement (as well as that of diquarks) remained open. This problem was immediately attacked by theoreticians. Two approaches emerged: a physical one based on common sense, experience, and physical intuition; and a "mathematical physics" approach in which the starting point was the QCD Lagrangian as the foundation of the theory. Among works devoted to the first approach, the "bag" model [354] and the model of a vacuum condensate of monopoles [355, 356] should be noted. Among other works, the second program was formulated in [357]–[359]. So far there is no general agreement on the theory of confinement. Some researchers insist that instantons play the principal role [360], others tend to disagree [361]. Strong quantum fluctuations might also be dominant for the confinement mechanism [362]. The program of studying quark confinement from first principles was developed by K. Wilson (see [363]) in the most comprehensive and systematic way. His pioneering work is remarkable in several aspects. First, he formulated lattice gauge theories; second, the "Wilson loops" were introduced; third, the "Wilson criterion" was formulated; and, finally, the area law was obtained for lattice electrodynamics in the strong coupling limit (for massive charges). Non-Abelian theories were studied numerically in Wilson's approach.

In addition to the aforementioned works devoted to confinement, studies related to the Schwinger model (see [364]; the electrodynamics of massless fermions in (1+1) spacetime) should be noted. The interest in this model is associated with that in one-dimensional electrodynamics: the confinement takes place automatically (the Coulomb potential is linear with distance). In the massless Schwinger model, the electric charge is totally screened (i.e. a complete "bleaching" always occurs) [365–367] (see also a review [206]).

7.1.2 General remarks about the confinement problem

The confinement problem is formulated in many ways. At the beginning, it was reduced [348] to the question of why physical quarks are not observed. Another question is also appropriate [368]: why do quarks occur only in colorless combinations? Now, perhaps, one could highlight three levels in understanding of the problem of quark confinement [294]:

(1) a formal proof of the absence of colored objects in the physical sector of the theory;
(2) clarifying the nature and character of forces confining colored objects;
(3) developing a complete theory of hadrons (i.e. the theory that would allow one to calculate, analytically or numerically, the structure of mesons and nucleons).

In each of the two last points, an affirmative solution of the first is assumed.

In the framework of the Dirac quantization scheme (see [148]), solution of the first problem is trivial [370, 371] because only gauge-invariant operators and gauge-invariant states are admissible in the physical sector of the theory. Quarks and gluons, described by *local* fields, carry color charges, i.e. they change under gauge transformations, and, therefore, they are excluded from normal physical degrees of freedom. This is a formal answer to this "naive" question of why isolated quarks are not seen (see also Section 7.4.4). There is a physical subtlety here. Such a formal answer does not address the issue of a physical scale at which isolated quarks are not observed. The statement that only gauge-invariant objects may exist in nature does not preclude the existence of gauge-invariant combinations of quarks and gluons of a large size, e.g. of the size of the Universe; and the very notion of an "isolated" quark becomes obscure. In other words, the dynamics plays a far more significant role in the problem than a formal kinematics associated with gauge invariance.

The second question deals with gauge field dynamics and needs much more serious investigation. Clarification of the nature of quark confinement requires clarifying subtle features of gauge field theory. Having agreed that only gauge-invariant objects have a physical meaning, the problem of classifying such objects and their dynamical properties arises. Only in this way, can one hope to reveal the physical nature of the confinement. The key point is in the following. It is safe to assume that the forces holding quarks together are characterized by a linearly rising potential at large distances. Modern quantum field theories (including the gauge ones) are causal (local) in their nature; the principle of local action is in their very foundation (no "actio ad distantia"). This means that the interaction of physical objects occurs through the fields surrounding them. What are the forms in which such fields may appear or be excited? It is known that in electrodynamics charges at rest excite static fields around themselves that provide for their Coulomb interaction. The "physical electron" (i.e. an electron together with its own electrical field) is described by a non-local gauge invariant operator [65]. In non-Abelian gauge theories, it is also not so difficult to construct gauge-invariant objects. For example, a "string" (the path ordered exponential of gauge potentials) with a quark and antiquark at its ends is one such object (see Section 7.2 for further discussion). In this case, it is not hard to see that the gluon field is excited only along the integration contour of the P-exponential between the quarks, and the energy of this system is given by a function linearly rising with contour length in the case of massive quarks. At first glance, this might look like a plausible answer to the question about the nature of confining forces in chromodynamics. But, as a point of fact, the same object can also be constructed in electrodynamics [372], i.e. such arguments are insufficient. The problem lies in studying the stability of field configurations (surrounding the charges) that are allowed by the gauge invariance. If the dynamics only permits stable configurations that are spread like the Coulomb field of two opposite charges, then the potential should be expected to decrease with distance. This is essentially a dynamical question. In Section 7.4.2, it is shown that string-like excitations are unstable in electrodynamics. Thus, in non-Abelian theories, one should answer the question whether or not string-like excitations of gauge fields between color charges are dynamically stable (at least, when the charges are viewed classically) and do not become spread in space in due course of time evolution similarly to the case of electrodynamics. It is interesting to note that asymptotic freedom leads to a reasonable conjecture that the QCD vacuum is a *dia-electric* medium (its dielectric constant is less than one, in contrast to ordinary dielectric media where it exceeds one), and in such a medium the electric field of a dipole occupies a finite volume, unlike the dipole Coulomb field in an ordinary dielectric [373, 374]. In reality, the situation may be far more complicated – see Section 7.4.5, but by now these subtleties may be put aside. From the kinematic point of view, for quarks, realizing the fundamental representation of the group $SU(3)_c$, only linear excitations of fields are possible (i.e. excitations localized on one-dimensional sets in space). This can be demonstrated from very basic principles (as one says, first principles) of gauge theories. This chapter is devoted to the analysis of gauge-invariant objects and their properties.

As far as the third issue is concerned, this is an extremely tough problem. One should only hope for development of a working scheme for approximate calculations. However with the advent of a lattice QCD [297, 363, 375] and the ever increasing computational power of modern computers, the task may be solved by *ab initio* numerical simulations where a particular nature of confinement is not relevant as the confinement appears as a feature of a numerical solution of the whole theory. The lattice QCD has made tremendous progress in recent years [376]–[378]; masses of many hadrons (including the proton and neutron) and mesons are calculated with a few per cent accuracy.

7.1.3 Peculiarities of gauge field theories

Among the experts there is an unspoken agreement that confinement takes place only in gauge theories. Why? What is so special about gauge theories that is not present in non-gauge theories? The answer is straightforward: the former have first-class constraints [148], while the latter do not. This observation, although rather trivial at first glance, is the most distinctive feature of gauge theories; it is exactly this feature that distinguishes gauge theories from the others and, to a large extent, depreciates the only well-developed calculation technique, the perturbation theory method. In contrast to equations of motion, constraints do not contain time derivatives; they are conditions imposed on the canonical variables at each moment of time, i.e. on instantaneous field configurations. In other words, constraints select the physically admissible field states. They are absolute – constraints cannot be violated because otherwise the gauge invariance would have been broken, which implies renouncing the original (gauge-invariant) Lagrangian (see also [206]).

Thus, in gauge theories it is desirable to first figure out the physically admissible configurations of gauge fields (i.e. "solve the constraints") and then to study their time evolution. However, it would be too hasty to take the recommendation to "solve the constraints" literally. This is not easy in classical theory and potentially leads to misconceptions in quantum theories (first-class constraints cannot be understood as operator equalities, see [26, 148]; and also Chapter 4). Fortunately, a way around this exists. Recall that first-class constraints are generators of gauge transformations [148], therefore they should vanish on the physical states and commute with physical operators, that is, the physical operators and states must be gauge invariant. The construction and classification of gauge-invariant field states means, in fact, solving the constraints.

What are the consequences of taking into account the constraints in gauge theories? In order to answer this question, let us turn first to the simplest gauge field theory, electodynamics. It is well known that the "physical" electron corresponds to the non-local gauge-invariant operator $\hat{\Psi} = \exp(-ie\Delta^{-1}\boldsymbol{\partial}\cdot\hat{\mathbf{A}})\hat{\psi}$, where $\Delta = \boldsymbol{\partial}\cdot\boldsymbol{\partial}$, [65, 164] that describes a charged particle with its own static Coulomb field (the exponentials correspond to the Coulomb field [65]). The use of perturbation theory obscures this important fact: in the leading order $\hat{\Psi}_0 = \hat{\psi}$, whereas the field $\hat{\psi}$ is local and changes under gauge transformations. It is the latter fields that appear in perturbative calculations, and, because of their locality, there arises an illusion that gauge theories are essentially the same as non-gauge theories since in both of them one works with local fields.

What are the consequences of the existence of static fields surrounding the charges? Apparently, these extended fields have a certain energy – this is an electromagnetic mass of a charged particle. It is also clear that, if the field energy of two charges is calculated relative to the sum of their own energies, one obtains the interaction energy which is a function of the distance between the charges, namely, the Coulomb potential. Thus, in electrodynamics, the static interaction of charges is, in fact, a consequence of the gauge invariance. It is this observation that plays the central role in all gauge theories, the differences are just in technical details. It turns out that, in the framework of the fiber bundle theory [379]–[382] applied to non-Abelian gauge theories, the extended fields accompanying colored objects do not fill the whole space, like the Coulomb field in electrodynamics; they are localized on lines originating from the colored source and whose number depends on the dimension of the representation of the gauge group realized by the source. If the classical source realizes the fundamental representation, then its accompanying field is localized on just one such line. The field of two charges of opposite signs is localized on a contour connecting the (point-like) charges (usually such objects are identified with strings). It is not difficult to verify that, for instance, in gluodynamics, the static interaction of massive quarks is determined by a linearly rising potential in the strong coupling limit, provided *the string connecting the quarks is stable.* Quark confinement is said to take place because the quarks cannot move away from one another as the energy of their accompanying field increases with the distance between them. Generally speaking, this is the origin of the phenomenon called confinement (or "imprisonment") of quarks, if a theory, defined by a local Lagrangian, is meant.

The arguments outlined above comprise the "ideological" foundation for the subsequent analysis. Studies of the problem give convincing reasons why the most direct way to the goal of elucidating the logical structure of the analysis should be related to the formulation of gauge field theories as a fiber bundle theory [379]–[382]; the vector gauge fields are identified with connections in the principal fiber bundle, while matter ("charged") fields form associated bundles. In this formulation of the problem, construction and analysis of gauge invariants are reduced to a standard, purely mathematical problem discussed in textbooks on differential geometry. In doing so, the dynamical properties of the theory are separated from the geometrical ones. The dynamics is introduced by postulating a specific Lagrangian that defines the law according to which the objects in question evolve with time. A quantum dynamical description is obtained by applying an appropriate quantization procedure. This plan is adopted through out this chapter.

7.2 Kinematics. Gauge fields and fiber bundle theory

7.2.1 Gauge fields as connections in the fiber bundle theory

There are a number of textbooks on differential geometry where the fiber bundle theory and its relation to Yang–Mills theories with matter fields can be found. Here the basic concepts of this geometrical interpretation of gauge theories are used to study the aforementioned problem of constructing gauge invariants [385] (see also [206, 383]). Consider a principal fiber bundle with the base being the Minkowski space and fibers being a simple compact Lie group G. For the sake of definitiveness, G is assumed to be $SU(n)$. Let a field $\psi(x) \equiv \psi_x$ realize a unitary representation of G, i.e.

$$\psi_x^{'} = U_g(x)\psi_x, \quad \psi_x^{'*} = \psi_x^* U_g^\dagger(x), \quad g \in G, \quad U_g^\dagger U_g = 1, \tag{7.1}$$

where ψ^* is the complex conjugate of ψ. The invariants bilinear in ψ and ψ^* are sought. The problem for fields ψ_x and ψ_x^* taken at the same base point is simple;

it is reduced to constructing a trivial representation out of nontrivial ones (it is solved by listing all irreducible invariant tensors for the given representations of the group G). A more important task is to find invariants that contain the matter fields at different base points, for instance, $\psi_{x'}$ and ψ^*_x where $x \neq x'$. The problem makes sense only if the parallel transport is defined. By definition (according to the fiber bundle theory), the parallel transport of ψ from a point x' to an infinitesimally close one $x = x' + dx$ is given by

$$\tilde{\psi}_x = (1 + A_\mu dx^\mu)\psi_{x'} \equiv P_{xx'}\psi_{x'}. \tag{7.2}$$

The connection A_μ is a vector field whose components take their values in the Lie algebra of the group G (in the representation realized by ψ). A bilinear invariant of fields $\psi_{x'}$ and ϕ_x realizing the same representation of G is

$$\phi^*_x \tilde{\psi}_x = \text{inv}. \tag{7.3}$$

From (7.1)–(7.3) one infers:

(1) the expression for the covariant derivative,

$$D_\mu = \partial_\mu - A_\mu \tag{7.4}$$

as a consequence of (7.2): $\psi_x - \tilde{\psi}_x \approx (\partial_\mu - A_\mu)\psi_x dx^\mu)$;

(2) the transformation law for A_μ which follows from the invariance of the form (7.3),

$$\phi^*_x U^\dagger_g(x) \left(1 + A'_\mu dx^\mu\right) U_g(x')\psi_{x'} = \phi^*_x \left(1 + A_\mu dx^\mu\right) \psi_{x'}, \tag{7.5}$$

and, hence,

$$A'_\mu(x) = U_g(x) A_\mu(x) U^\dagger_g(x) - U_g(x)\partial_\mu U^\dagger_g(x). \tag{7.6}$$

Equations (7.4) and (7.6) allow one to identify the connection A_μ in the fiber bundle space with the vector gauge field.

A parallel transport of ψ along a contour $c(x, x')$ of finite length connecting points x' and x is given by,

$$\tilde{\psi}_x = P[c(x, x')]\psi_{x'}, \tag{7.7}$$

where, according to (7.2), the parallel transport operator

$$P[c(x, x')] = P\exp\left(\int_{c(x,x')} A_\mu dx^\mu\right) \equiv P_{xx'}, \tag{7.8}$$

is obtained by successive applications of the elementary operation (7.2) along the contour $c(x, x')$; it is nothing but the path-ordered exponential (or P- exponential). It is assumed that point x' corresponds to the smallest value of an ordering parameter along the contour. According to (7.2) and (7.5), $P_{xx'}$ is transformed by the rule,

$$P'_{xx'} = U_g(x) P_{xx'} U^+_g(x'), \tag{7.9}$$

i.e. in contrast to ψ_x, the operator $P_{xx'}$ is a *bilocal* tensor. From this geometrical analysis it is clear that the fields ψ, ϕ, P-exponential, and invariant irreducible tensors (similar to the fully skew-symmetric unit tensor $\epsilon_{\alpha\beta\ldots}$) are the only objects from which non-local invariants can be built. The simplest one reads,

$$\phi^*_x P_{xx'}\psi_{x'} \tag{7.10}$$

according to (7.7) and (7.9). It is the bilocal tensor $P_{xx'}$ that is associated with the external field of color charges, analogous to the Coulomb field in electrodynamics.

It should be emphasized that, first, the above analysis is solely based on the geometrical interpretation of gauge fields (the dynamics has not yet been introduced). Second, the equations obtained also hold in the case when G is an Abelian group (electrodynamics).

Remark. When constructing invariants, the connection in (7.8) can be shifted by any structure group tensor C_μ that transforms as $C_\mu \to U_g C_\mu U_g^\dagger$, i.e. $A_\mu \to A_\mu^C = A_\mu + C_\mu$. Indeed, the new "connection" A_μ^C would have the same transformation law (7.6). The local tensor C_μ can be built from other fields present in the theory. For example, if $G = SU(n)$, C_μ can depend on the tensor product, $C_\mu = \gamma_\mu\psi \otimes \bar{\psi}$ where the spinor ψ realizes the fundamental representation of $SU(n)$ and the tensor product is taken with respect to the color indices, while the spinor indices are contracted with the Dirac matrix γ_μ. If no other (matter) fields are present, then according to the reconstruction theorem [230, 379], C_μ can only be built from $F_{\mu\nu} \sim [D_\mu, D_\nu]$ and its covariant derivatives, e.g. $D_\nu F_{\nu\mu}$; here A_μ is a generic connection, not assumed to satisfy the classical Yang–Mills equations of motion.

7.2.2 Invariants in gauge theories with the groups $U(1)$, $SU(2)$, and $SU(3)$

Before studying the dynamics, let us investigate invariants in gauge theories with the simplest unitary groups.

Electrodynamics. To begin with, consider the Abelian gauge group $U(1)$. All its complex representations are one-dimensional. Let the vector field A_μ and all charged fields ψ be transformed according to the rules,

$$A_\mu \to A_\mu + \partial_\mu\Lambda, \quad \psi_x \to \exp(ig\Lambda)\psi_x. \tag{7.11}$$

The constant g defines, in fact, a representation. In dynamics, it is identified with the coupling constant (with the electric charge of the field ψ). Taking into account the gauge transformation law of charged fields, it is important to know objects B linear in A_μ whose gauge transformations are reduced to a shift $B \to B + \Lambda$. Once all such B's are known, the desired gauge invariants are easy to construct. Here is the list [384]:

$$B_1 = \int_{-\infty}^{x} A_\mu dx^\mu, \quad dx_\mu^2 < 0, \tag{7.12}$$

$$B_2 = \Delta_2^{-1}(\nabla_\perp \mathbf{A}_\perp), \quad \Delta_2 = (\nabla_\perp, \nabla_\perp) = \partial_1^2 + \partial_2^2, \tag{7.13}$$

$$B_3 = \Delta^{-1}(\nabla, \mathbf{A}), \quad \Delta = \partial_1^2 + \partial_2^2 + \partial_3^2, \tag{7.14}$$

$$B_4 = -\Box^{-1}\partial_\mu A_\mu, \quad \Box = -\partial_\mu^2 = -\partial_0^2 + \Delta. \tag{7.15}$$

With the help of B_k, the gauge-invariant combinations of the fields are found:[1]

$$\Psi_k(x) = \exp\Big(-igB_k(x)\Big)\psi(x) \equiv P_k(x)\psi(x), \quad k = 1, \dots, 4. \tag{7.16}$$

All of them are non-local. The field A_μ is localized on the integration contour in B_1, in the (x_1, x_2) plane for B_2, in the whole space for B_3, and in the whole spacetime for B_4. At first glance, this contradicts the statement of the previous section that the only fundamental structure in gauge theories is P-exponential. The paradox is easily resolved: all new non-local structures can be viewed as composite, constructed out of P-exponentials. Before proving this proposition, it should be noted that the field Ψ_4 is a special case because it is "non-local" also in time, and its evolution with time is not dynamical as it is determined by the operator $\Box^{-1}$ in (7.15). The invariants of interest should describe only static fields (the constraints do not contain time derivatives). But in this case the invariant (7.15) turns into (7.14). Indeed,

$$\Box^{-1}\partial_\mu A_\mu = \int d^4x' \Box^{-1}_{xx'}\partial_\mu A_\mu(x') = (2\pi)^{-4}\int d^4k k^{-2} e^{ikx} ik_\mu a_\mu(k),$$

and for a static field $a_\mu(k) = 2\pi\delta(k_0)\tilde{a}_\mu(\mathbf{k})$, the configuration Ψ_4 becomes Ψ_3.

Let us turn to Ψ_2 and Ψ_3. It is shown [384] that P_3 in (7.16) can be composed out of linear structures $P_1(x) = \exp(-igB_1(x))$. Put

$$P_3(N, x) = \prod_{i,j}^{N} \exp\left(-ig\int_{-\infty}^{x} A_\mu(y_{ij})dy_{ij}^\mu\right). \tag{7.17}$$

The expression (7.17) is the product of N exponentials with integration contours being rays coming out from point x. The rays are enumerated by indices i and j which determine the position of an infinitesimal solid angle that contains the (ij) ray: $(4\pi/N)_{ij} = \sin\theta_j\Delta\theta_j\Delta\phi_i$ (the area element of the unit sphere per each ray piercing through the sphere). Assuming the product $Ng = e$ to be fixed as $N \to \infty$, one obtains,

[1] The combinations (7.16) are invariant under local gauge transformations with arbitrary functions Λ_x vanishing at the spatial infinity ($\Lambda_x \to 0$ as $|\mathbf{x}| \to \infty$). This group can be extended by global transformations (allowing $\Lambda_x \to$ const as $|\mathbf{x}| \to \infty$). Such an extension of the gauge group is admissible and useful: it allows one to prove the superselection rules for the electric charges [370]; moreover, it follows that the total electric charge of the closed universe equals zero [370].

$$\lim_{N\to\infty} P_3(N,x) = \lim_{N\to\infty} \exp\left[\frac{ie}{4\pi}\sum_{i,j}\sin\theta_j\Delta\theta_j\Delta\phi_i \int_0^\infty dr A_r(r,\theta_j,\phi_i)\right]$$
$$= \exp\left[\frac{ie}{4\pi}\int_0^{2\pi} d\phi \int_0^\pi d\theta \sin\theta \int_0^\infty dr A_r\right] \equiv e^{ieI_3}, \qquad (7.18)$$

where $A_\mu dy^\mu_{ij} \equiv A_r(r,\theta_j,\phi_i)dr$. The integral in the exponential (7.18) can be written in the form,

$$I_3 = \frac{1}{4\pi}\int d^3y r^{-2} A_r = -\frac{1}{4\pi}\int d^3y(\partial_r r^{-1})A_r$$
$$= \frac{1}{4\pi}\int d^3y r^{-1}\left[r^{-2}\partial_r(r^2A_r)\right] = -B_3. \qquad (7.19)$$

The latter equality in (7.19) is deduced by means of the relation $-(1/4\pi)|\mathbf{x}-\mathbf{y}|^{-1} \equiv \Delta^{-1}(\mathbf{x},\mathbf{y})$, the expression of the divergence in the spherical coordinates: $(\nabla,\mathbf{A}) = r^{-2}\partial_r(r^2A_r) + (r\sin\theta)^{-1}\left[\partial_\theta(\sin\theta A_\theta) + \partial_\phi A_\phi\right]$, and the fact that $A_\theta = A_\phi = 0$. It follows from (7.17) and (7.19) that the factor $P_3(x)$, describing the Coulomb field of a charged particle, can be viewed as composite, a combination of linear exponentials ("strings"). The exponential $P_2(x)$ in (7.16) has a similar representation [384].

Thus, the existence of the invariants (7.16) for $k = 2,3,4$ in the Abelian theory does not contradict the statement of the previous section that all invariants can be built of bilocal tensors $P_{xx'}$ and local charged fields ψ.

Remark. In [369, 386], the hypothesis was analyzed that the number N is large, but finite for elementary particles ($N \sim 10^{40}$). Some experimental consequences of this hypothesis were also discussed therein.

The group $SU(3)$. In the case $G = SU(3)$ the following new features emerge. First, only one string can be attached to the field ψ realizing the fundamental representation of the group because this representation is characterized by just one index taking three values. Second, the strings can branch. Recall that in this case there are invariant totally skew-symmetric unit tensors,

$$\epsilon^{\alpha\beta\gamma}, \quad \epsilon_{\alpha\beta\gamma}, \quad \alpha,\beta,\gamma = 1,2,3, \qquad (7.20)$$

appearing in the unimodularity condition of the matrices $U_g(x)$: $\det U_g(x) = 1$. Because of this fact, three strings in the fundamental representation can be attached to one another at a point, i.e. besides obvious invariants (see (7.10)),

$$\bar\psi^\beta_x P^\alpha_\beta[c(x,x')]\psi_{\alpha x'}, \quad \mathrm{Tr}\, P[c(x,x)], \qquad (7.21)$$

the following invariant structures can be constructed with these tensors:

$$\Theta(x,x') = \epsilon^{\alpha\beta\gamma}(P_{xx'})^{\alpha'}_\alpha (P'_{xx'})^{\beta'}_\beta (P''_{xx'})^{\gamma'}_\gamma \epsilon_{\alpha'\beta'\gamma'}, \qquad (7.22)$$
$$B_{x_1x_2x_3}(x) = \epsilon_{\alpha'\beta'\gamma'}(P_{xx_1})^{\alpha'}_\alpha (P_{xx_2})^{\beta'}_\beta (P_{xx_3})^{\gamma'}_\gamma \bar\psi^\alpha_{x_1}\bar\psi^\beta_{x_2}\bar\psi^\gamma_{x_3}. \qquad (7.23)$$

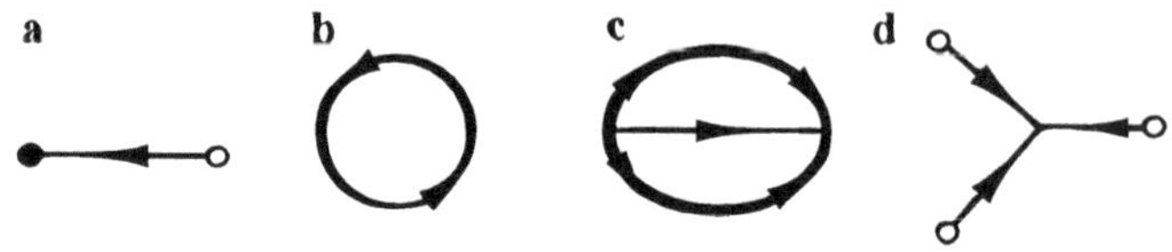

Fig. 7.1. The simple invariants.

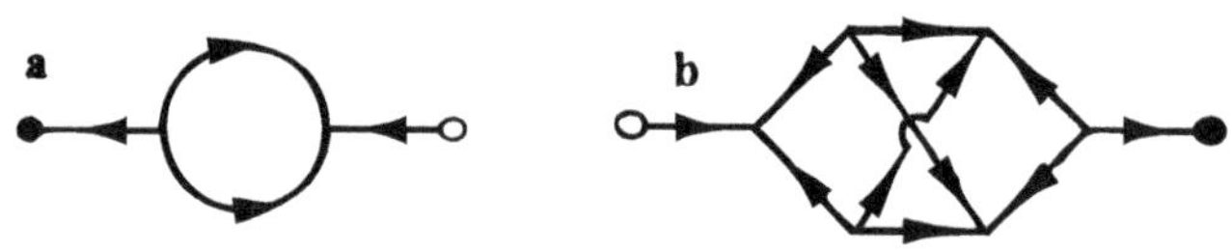

Fig. 7.2. String branching (mesons).

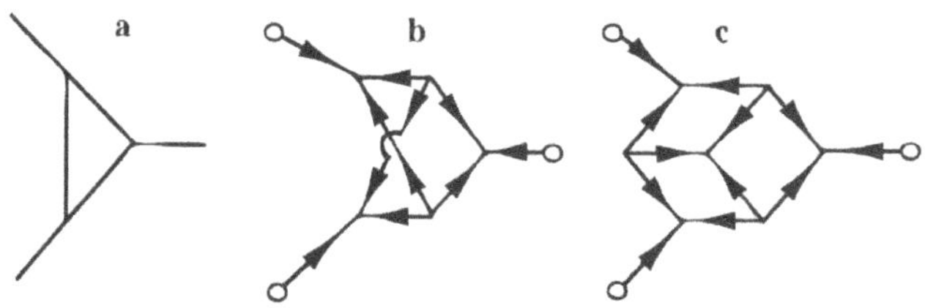

Fig. 7.3. String branching (barions).

In Eq. (7.22), the P-exponentials have different integration contours. It is clear from the expressions (7.21)–(7.23) that invariant structures can be conveniently represented by diagrams [384] in which strings are depicted as oriented line segments. For instance, the bilocal tensor $(P_{xx'})^{\alpha}_{\beta}$ is pictured as a line segment from x' to x oriented by an arrow directed from the upper index to the lower ones. The diagrams for the structures (7.21)–(7.23) are shown in Figs. 7.1 and 7.2. Filled circles correspond to the quark field creation operator, the unfilled one to the quark field annihilation operator. Note that quarks can be connected to strings of nontrivial topology thanks to string branching (see Fig. 7.2).

The diagrams introduced have little in common with Feynman diagrams. The former describe instantaneous string configurations ("irreducible solutions of the constraints"), while the latter are associated with the probability amplitudes of events. The peculiarity of string diagrams is that a diagram constructed with non-oriented line segments may be meaningless. For example, the diagram of the topology, as depicted in Panel **a** of Fig. 7.3 is impossible because each vertex may only have all three lines oriented either inward or outward.

The group $SU(2)$. The case of $G = SU(2)$ differs substantially from the previous one. For the group $SU(2)$ the fundamental representation and the one conjugated to it are unitary equivalent [387]: the invariant totally skew-symmetric unit tensors of this group $\epsilon^{\alpha\beta}$ and $\epsilon_{\alpha\beta}$ $(\alpha, \beta = 1, 2)$ convert spinors with upper indices into spinors with lower indices and vice versa. This drastically changes admissible string structures. First, the triple vertices disappear, i.e. strings can no longer branch; second, the string orientation in diagrams becomes unnecessary because the representations with upper and lower indices are unitary equivalent. Thus, the non-Abelian theories with the groups $SU(2)$ and $SU(n)$, $n > 2$, are different in structures (topologies) of static fields. The former is more similar to electrodynamics in which string branching is also absent. Perhaps, this is related to the fact that, in electroweak interaction theory, the gauge fields of exactly these groups (i.e. $U(1)$ and $SU(2)$) are mixed.

Remark. Here only the simplest invariants of gauge groups have been considered. Simplicity is determined by:

(1) the rank of invariants with respect to Lorentz transformations (for example, (7.21) and (7.22) are scalars, while (7.23) is a spin-tensor; it is not hard to construct invariants that are Lorentz tensors of an arbitrary rank);
(2) the topology (the number of vertices in the diagrams like those in Figs. 7.1–7.3);
(3) the number of local tensors (matter fields) involved in the invariant, i.e. the number of strings ending at charges;
(4) the reducibility; a gauge tensor is said to be *irreducible* if it is not a tensor product of two other (nontrivial) tensors [384].

Finally, after turning on the dynamics, it would make sense to additionally classify the invariants by their energies [385].

7.2.3 Gauge-invariant vector fields in the Higgs phase

In gauge field theories with scalar (Higgs) fields realizing a nontrivial representation of the gauge group, there are also invariants *local* in spacetime and *linear* in connection. In a dynamical description, such invariants are useful for identifying the physical degrees of freedom in gauge theories with "spontaneous symmetry breaking" [388, 389]. Constraints in a gauge theory generate gauge transformations and should hold regardless of any peculiarities of the scalar field dynamics, i.e. the gauge symmetry cannot be broken [390]. On the other hand, if the dynamics is such that the scalar field develops a nonzero vacuum expectation value, then the phenomenon known as "spontaneous gauge symmetry breaking" takes place; the gauge group is reduced to its (possibly trivial) subgroup. How can this apparent paradox be resolved? Simple examples of the $U(1)$ and $SU(2)$ are considered first. Then the electroweak interactions are studied [200, 391].

7.2.3.1 Abelian case

Let φ be a complex scalar field and A_μ be the connection. The gauge transformation law reads,

$$A_\mu \to A_\mu + \partial_\mu \omega, \qquad \varphi \to e^{i\omega}\varphi. \tag{7.24}$$

The covariant derivative is defined by $D_\mu = \partial_\mu - iA_\mu$ so that, under gauge transformations (7.24), $D_\mu\varphi \to e^{i\omega}D_\mu\varphi$. Consider the polar decomposition of the scalar field,

$$\varphi = e^{i\phi}\rho, \quad \rho = |\varphi|. \tag{7.25}$$

The variable ϕ describes the phase degree of freedom of the scalar field. It is translated under the gauge transformation (7.24): $\phi \to \phi + \omega$. The following gauge-invariant combination can be made:

$$W_\mu = ie^{-i\phi}D_\mu e^{i\phi} = A_\mu - \partial_\mu\phi. \tag{7.26}$$

The gauge-invariant vector field W_μ can be expressed in terms of the original variables of the theory. Consider the following gauge-invariant combination $\varphi^* D_\mu\varphi = \rho\partial_\mu\rho - iW_\mu\rho^2$, where the polar decomposition (7.25) has been used. The vector field W_μ appears only in the imaginary part of this combination. Hence,

$$W_\mu = -\frac{i\left((D_\mu\varphi)^*\varphi - \varphi^* D_\mu\varphi\right)}{2\varphi^*\varphi}. \tag{7.27}$$

The gauge $\varphi = \rho$ (i.e, $\phi = 0$) is called a *unitary* gauge. The gauge-invariant vector field W_μ coincides with the vector field A_μ in this gauge and, therefore, can be used to describe the physical degrees of freedom of the gauge fields in the Higgs phase (the vacuum expectation value of ρ does not vanish, $\langle\rho\rangle = v \neq 0$).

In the Hamiltonian formalism, the variable A_0 is non-physical. The physical degrees of freedom of the theory are the transverse connection, $\partial_j A_j^\perp \equiv 0$, $\mathbf{A}^\perp = \mathbf{A} - \boldsymbol{\partial}\Delta^{-1}(\partial_j A_j)$, $\rho = (\varphi^*\varphi)^{1/2}$, and the linear combination $\theta = \Delta^{-1}\partial_j A_j - \phi$. They are explicitly gauge invariant. If the dynamics is such that the non-local variable θ describes the longitudinal component of the vector field, $\mathbf{W} = \mathbf{A}^\perp + \boldsymbol{\partial}\theta$, then the scalar field has only one physical degree of freedom ρ. This happens in the Higgs phase. Otherwise the variable θ can be put into the phase of the scalar field $\varphi_{ph} = \exp(i\theta)\rho$. This invariant field describes charged particles with their own Coulomb field (see Section 7.4). So, the constraint is fulfilled. The gauge symmetry is not broken. The normal and Higgs phases of the theory are distinguished by peculiarities of the dynamics of the variable θ.

7.2.3.2 $SU(2)$ gauge theory

The idea of transferring physical degrees of freedom between the gauge and scalar fields can be extended to the non-Abelian case. However the details would depend on the gauge group and the scalar field representation. Let the scalar field φ realize the fundamental representation of $SU(2)$ (it has two complex components). The gauge transformation law reads,

$$A_\mu \to \Omega A_\mu \Omega^\dagger + i\Omega\partial_\mu\Omega^\dagger, \qquad \varphi \to \Omega\varphi, \tag{7.28}$$

where $\Omega \in SU(2)$. The covariant derivative is $D_\mu = \partial_\mu - iA_\mu$ so that, under gauge transformations (7.28): $D_\mu\varphi \to \Omega D_\mu\varphi$. Put $\rho = (\varphi^\dagger\varphi)^{1/2}$ and consider the polar decomposition of the scalar field (the analog of (7.25)),

$$\varphi = U\mathbf{e}\,\rho, \quad U = \frac{1}{\rho}\begin{pmatrix} \varphi_2^* & \varphi_1 \\ -\varphi_1^* & \varphi_2 \end{pmatrix}, \quad U^\dagger U = 1, \quad \mathbf{e} = \begin{pmatrix} 0 \\ 1 \end{pmatrix}, \tag{7.29}$$

where $U \in SU(2)$ and the star* denotes the complex conjugation. Any complex two-vector φ can always be written in the form (7.29). The first column of U is the vector $\tilde\varphi$ dual to φ, $\tilde\varphi_\alpha = \varepsilon_{\alpha\beta}\varphi^*_\beta$, where $\varepsilon_{\alpha\beta}$ is a totally skew-symmetric unit tensor, $\varepsilon_{12} = 1$, $\alpha, \beta = 1, 2$. Since $\varepsilon_{\alpha\beta}$ is an invariant tensor of $SU(2)$, i.e. $\varepsilon_{\alpha'\beta'}\Omega_{\alpha'\alpha}\Omega_{\beta'\beta} = \varepsilon_{\alpha'\beta'}\Omega^*_{\alpha'\alpha}\Omega^*_{\beta'\beta} = \varepsilon_{\alpha\beta}$ for any $\Omega \in SU(2)$, the gauge transformation law of $\tilde\varphi$ is: $\tilde\varphi \to \Omega\,\tilde\varphi$. Under the gauge transformations (7.28) the group element U is shifted: $U \to \rho^{-1}\left(\Omega\tilde\varphi \;\; \Omega\varphi\right) = \Omega U$. The following gauge-invariant combination can be made (the analog of (7.26)):

$$W_\mu = iU^\dagger D_\mu U = U^\dagger A_\mu U + iU^\dagger \partial_\mu U. \tag{7.30}$$

Note that W_μ has the same quantum numbers as A_μ, i.e. it is an element of the $SU(2)$ algebra. Its independent components are obtained, e.g. in the basis of the Pauli matrices τ_a: $W_\mu = \tau_a W_\mu^a$. The fields W_μ^a can be expressed in an explicitly gauge-invariant form via the the original variables of the theory. For what follows it is convenient to use the complex vector fields $W_\mu^+ = W_\mu^1 - iW_\mu^2$ and $W_\mu^- = (W_\mu^+)^*$.

Consider the following gauge-invariant combinations:

$$\varphi^\dagger D_\mu\varphi = \rho\partial_\mu\rho - iW_\mu^a\,\mathbf{e}^\dagger\tau_a\mathbf{e}\,\rho^2 = \rho\partial_\mu\rho + iW_\mu^3\rho^2,$$
$$\tilde\varphi^\dagger D_\mu\varphi = \rho\partial_\mu\rho\,\tilde{\mathbf{e}}^\dagger\mathbf{e} - iW_\mu^a\,\tilde{\mathbf{e}}^\dagger\tau_a\mathbf{e}\,\rho^2 = -iW_\mu^+\rho^2,$$

where $\tilde{\mathbf{e}}_\alpha = \varepsilon_{\alpha\beta}\mathbf{e}_\beta$; the first equalities follow from (7.29) and (7.30), and the last ones are obtained by means of the relations $\mathbf{e}^\dagger\tau_a\mathbf{e} = -\delta_{3a}$, $\tilde{\mathbf{e}}^\dagger\mathbf{e} = 0$, and $\tilde{\mathbf{e}}^\dagger\tau_a\mathbf{e} = 1, -i, 0$, for $a = 1, 2, 3$, respectively. Hence, the analog of (7.27) reads,

$$W_\mu^3 = \frac{i\left((D_\mu\varphi)^\dagger\varphi - \varphi^\dagger D_\mu\varphi\right)}{2\varphi^\dagger\varphi}, \quad W_\mu^+ = \frac{i\tilde\varphi^\dagger D_\mu\varphi}{\varphi^\dagger\varphi}, \quad W_\mu^- = \left(W_\mu^+\right)^*. \tag{7.31}$$

The gauge-invariant vector fields W_μ^3 and $W_\mu^\pm$ can also be represented in a uniform way. Define $\overleftrightarrow{D}_\mu = \overrightarrow{D}_\mu - (\overleftarrow{D}_\mu)^\dagger = \overleftrightarrow{\partial}_\mu - 2iA_\mu$, i.e. $\phi^\dagger \overleftrightarrow{D}_\mu \varphi = \phi^\dagger\partial_\mu\varphi - \partial_\mu\phi^\dagger\varphi - 2i\phi^\dagger A_\mu\varphi$ for any vectors ϕ and φ. Then all the invariants (7.31) can be written in the same fashion,

$$W_\mu^3 = \frac{i\varphi^\dagger \overleftrightarrow{D}_\mu \varphi}{2\varphi^\dagger\varphi}, \qquad W_\mu^+ = -\frac{i\tilde\varphi^\dagger \overleftrightarrow{D}_\mu \varphi}{2\varphi^\dagger\varphi}, \qquad W_\mu^- = \left(W_\mu^+\right)^*.$$

The invariants (7.31) are meaningful, provided the dynamics is such that $\rho \neq 0$ (i.e. in the Higgs phase).

Remark. What is the analog of the variable θ? This is not a simple question. Formally, following the analogy with the Abelian case, the connection can be written in the form $\mathbf{A} = U_\chi \mathbf{B} U_\chi^\dagger + iU_\chi \boldsymbol{\partial} U_\chi^\dagger$ (the time component A_0 is non-physical) where the components of $\mathbf{B}$ are not independent and subject to some conditions $\chi_a(\mathbf{B}) = 0$, $a = 1, 2, 3$ so that the number of independent variables in $\mathbf{A}$ and in the pair U_χ and $\mathbf{B}$ is the same (in the Abelian case, $\mathbf{B} = \mathbf{A}^\perp$, that is, $\chi(\mathbf{B}) = \partial_j B_j = 0$ and $U_\chi = \exp(i\Delta^{-1}\partial_j A_j)$). The analog of $e^{i\theta}$ would be the matrix $U_\theta \equiv U_\chi^\dagger U$. However, the statement that U_θ is a combination of the longitudinal component of the gauge field and the phase of the Higgs field is not true in the non-Abelian case. Even if $\partial_j B_j^a = 0$, the group element U_χ depends on all components of $\mathbf{A}$ (such as that $U_\chi(\mathbf{A})$ minimizes the Morse functional (6.203)). More to the point, the invariant scalar field $\phi_{ph} = \rho U_\theta \mathbf{e}$ describes scalar particles with their own external field, as is shown in Section 7.4 (note that U_χ depends on $\mathbf{A}$, and, therefore, $U_\theta = U_\theta(\mathbf{A})$). The form of the particle's external field cannot be inferred solely from the gauge invariance. The dynamics plays a dominant role here. In fact, the Coulomb field in the Abelian case is distinguished by the dynamics as the only stable one (see Section 7.4.2) and so is the choice $\partial_j B_j = 0$. In the non-Abelian case, the external field appears in the form of a string-like excitation that provides for the confinement (see Section 7.4.4). An explicit form of $U_\chi(\mathbf{A})$ that describes such a *stable* excitation is not known (to some extent, its discovery is equivalent to solving the confinement problem). The choice of χ is extensively debated in studies of the confinement mechanism in lattice gauge theories. These studies show [392, 393] that a possible choice of χ is the so-called maximal Abelian gauge (a particular case of Abelian projections suggested earlier by 't Hooft [214]): $\left(\partial_j \pm iB_j^3\right) B_j^\pm = 0$, $\partial_j B_j^3 = 0$. Numerical simulations also show [393, 394] that in the confinement phase, the field B_j^3 contains monopole-like configurations [395] that are responsible for the formation of a stable string-like excitation between non-Abelian charges [393]. Furthermore, an effective theory of these configurations can be developed [396, 397]. An interesting conjecture is proposed [396]–[399] that this effective theory supports string-like excitations of gauge fields in the form of knot solitons [400, 401] (which might be natural candidates for glueballs, thus leading to a solution of the mass gap problem in Yang–Mills theory because the solitons have a nonzero mass). It appears, however, that perturbative studies [402] of the Wilsonian effective action [397] are insufficient to prove the conjecture, and further studies are needed (see [403] for non-perturbative lattice simulations).

7.2.3.3 Electroweak theory

The electroweak model has the gauge group $G = SU(2) \times U(1)$. The covariant derivative is written in the form $D_\mu = \partial_\mu - ig\tau_a A_\mu^a/2 - ig'YA_\mu^0/2 \equiv D_\mu^\tau - ig'YA_\mu^0/2$, where τ_a are the Pauli matrices ($a = 1, 2, 3$) being the generators of the subgroup $SU(2)$, and Y is the hypercharge (the subgroup $U(1)$ generator). If φ is the Higgs field and ψ_L is the left fermion doublet, then $Y\varphi = \varphi$ and

$Y\psi_L = -\psi_L$, while for the right fermions, $Y\psi_R = -2\psi_R$ [299]. The fields ψ_L and φ also realize the fundamental representation of the subgroup $SU(2)$ (in the simplest case scenario), while ψ_R is an $SU(2)$ singlet. For one fermion generation the Lagrangian has the form,

$$
\begin{aligned}
L = & -\frac{1}{4}\left(F^a_{\mu\nu}\right)^2 - \frac{1}{4}\left(F^0_{\mu\nu}\right)^2 + (D_\mu\varphi)^\dagger D_\mu\varphi \\
& + \left[i\bar{\psi}_L\gamma_\mu D_\mu\psi_L + i\bar{\psi}_R\gamma_\mu D^Y_\mu\psi_R + \lambda(\bar{\psi}_L\varphi)\psi_R + h.c.\right] - V, \qquad (7.32)\\
V = & \frac{\lambda'}{4}\left(\varphi^\dagger\varphi - v^2\right)^2,
\end{aligned}
$$

where $D^Y_\mu = \partial_\mu - ig'YA^0_\mu/2$ is the $U(1)$ covariant derivative, $F^a_{\mu\nu}$ and $F^0_{\mu\nu}$ are the non-Abelian and Abelian field strength tensors, respectively, λ is the Yukawa coupling constant, V is the Higgs field potential with a coupling constant λ'.

Given a Higgs field configuration, consider the following unitary matrix:

$$
U_Y = \frac{e^{i\alpha\frac{Y}{2}}}{\sqrt{\varphi^\dagger\varphi}}\begin{pmatrix} e^{i\frac{\alpha}{2}}\varphi_2 & -e^{i\frac{\alpha}{2}}\varphi_1 \\ e^{-i\frac{\alpha}{2}}\varphi_1^* & e^{-i\frac{\alpha}{2}}\varphi_2^* \end{pmatrix}, \quad \varphi = \begin{pmatrix}\varphi_1\\ \varphi_2\end{pmatrix}, \qquad (7.33)
$$

where α is an arbitrary function in spacetime. The action of U_Y on the fields in the theory depends on the value of the hypercharge Y realized by them ($Y = 1, -1, -2$ for φ, ψ_L, and ψ_R, respectively). Since, ψ_R realizes a trivial representation of $SU(2)$, the action of U_2 on ψ_R yields, by definition, a phase factor $e^{i\alpha Y/2} = e^{i\alpha}$ as $Y = 2$. Put

$$
U_1\varphi \equiv \begin{pmatrix}0\\ \rho\end{pmatrix}, \quad \chi_L = U_{-1}\psi_L \equiv \begin{pmatrix}\nu_L\\ f_L\end{pmatrix}, \qquad (7.34)
$$

$$
f_R \equiv U_{-2}\psi_R, \quad U_{-2} \equiv e^{-i\alpha}, \qquad (7.35)
$$

$$
g\frac{\tau_a}{2}B^a_\mu + g'\frac{Y}{2}B^0_\mu \equiv U_Y\left(g\frac{\tau_a}{2}A^a_\mu + g'\frac{Y}{2}A^0_\mu - \frac{1}{i}\partial_\mu\right)U_Y^\dagger. \qquad (7.36)
$$

It follows from the first relation in (7.34) that the unitary matrix U_Y describes the phase degrees of freedom of the Higgs field as $\rho = (\varphi^\dagger\varphi)^{1/2}$. Let $\tilde{\varphi}$ denote the dual of φ (as in Section 7.2.3.2). The new local variables defined in (7.34)–(7.36) can be expressed via the original variables of the Lagrangian (7.32),

$$
\rho = \left(\varphi^\dagger\varphi\right)^{1/2}, \ \nu_L = \frac{\tilde{\varphi}^\dagger\psi_L}{\rho}, \quad f_L = \frac{\varphi^\dagger\psi_L}{\rho}, \quad f_R = e^{-i\alpha}\psi_R, \qquad (7.37)
$$

$$
B^0_\mu = A^0_\mu + \frac{1}{g'}\partial_\mu\alpha, \quad B^3_\mu = \frac{1}{ig}\frac{\varphi^\dagger \overleftrightarrow{D^\tau_\mu}\varphi}{\rho^2} + \frac{1}{g}\partial_\mu\alpha, \qquad (7.38)
$$

$$
\sqrt{2}W^+_\mu = B^1_\mu - iB^2_\mu = -\frac{1}{ig}\frac{\tilde{\varphi}^\dagger \overleftrightarrow{D^\tau_\mu}\varphi}{\rho^2}e^{i\alpha}, \quad W^-_\mu = \left(W^+_\mu\right)^*, \qquad (7.39)
$$

where $\overleftrightarrow{D}^{\tau}_{\mu}=\overrightarrow{D}^{\tau}_{\mu}-\left(\overleftarrow{D}^{\tau}_{\mu}\right)^{\dagger}=\overleftrightarrow{\partial}_{\mu}-ig\tau_a A^a_{\mu}$ (as in Section 7.2.3.2). It is straightforward to verify that the variables (7.37)–(7.39) are invariant under gauge transformations from the $SU(2)$ subgroup of the gauge group. More to the point, the Lagrangian (7.32) is independent of the function α when expressed via the new variables. A simple analysis of the transformation law of the new variables (7.37)–(7.39) under variations of α and Lorentz transformations shows that the residual gauge group can be identified with the electromagnetic $U(1)$ group generated by the electric charge, whereas the $SU(2)$-invariant fields ν_L, f_L, and $W^{\pm}_{\mu}$ are identified with the neutrino, corresponding lepton, and charged vector bosons, respectively. The neutral vector boson and photon are given by the $SU(2)$-invariant combinations,

$$Z^0_{\mu}=B^3_{\mu}\cos\theta_W-B^0_{\mu}\sin\theta_W=\frac{\varphi^{\dagger}\overleftrightarrow{D}^{\tau}_{\mu}\varphi}{ig\varphi^{\dagger}\varphi}\cos\theta_W-A^0_{\mu}\sin\theta_W, \tag{7.40}$$

$$A_{\mu}=B^3_{\mu}\sin\theta_W+B^0_{\mu}\cos\theta_W=\frac{\varphi^{\dagger}\overleftrightarrow{D}^{\tau}_{\mu}\varphi}{ig\varphi^{\dagger}\varphi}\sin\theta_W+A^0_{\mu}\cos\theta_W+\frac{1}{e}\partial_{\mu}\alpha, \tag{7.41}$$

where $\tan\theta_W=g'/g$ defines the Weinberg angle and $e=gg'/\sqrt{g^2+g'^2}$ is the electric charge associated with the photon field A_{μ}.

The Lagrangian (7.32) written in the new $SU(2)$-invariant variables coincides with the celebrated Glashow–Weinberg–Salam Lagrangian of the electroweak theory with the "spontaneously broken" gauge symmetry ($SU(2)_L\times U(1)_Y\to U(1)_e$). This is easy to verify by noting that the transformation of the new variables formally coincides with a gauge transformation generated by the matrix (7.33). The change of variables (7.34)–(7.36) is singular for $\rho=0$. Therefore the Lagrangian in the new variables is meaningful if the dynamics is such that the field ρ cannot vanish. This is possible when its vacuum expectation value does not vanish $\langle\rho\rangle_0\neq 0$, i.e. the so called "spontaneous symmetry breaking" takes place. However, the new field variables are already $SU(2)$ gauge invariant and so are the system states associated with their excitations. Therefore, no breaking of the gauge symmetry actually occurs. The constraints are fulfilled (the gauge symmetry is not broken)[2] – the observed physical degrees of freedom are described by gauge-invariant combinations and their dynamics looks as though the gauge symmetry $G=SU(2)_L\times U(1)_Y$ were broken to $U(1)_e$. In other words, one cannot refer to the spontaneous breaking of a *gauge* symmetry as to a phenomenon fully analogous to the spontaneous breaking of a *global* symmetry (the latter does imply the actual breaking of the global symmetry; the system states are not invariant under the global symmetry transformations). It is interesting

[2] The constraints written in the new variables coincide with the Gauss law corresponding to the gauge group $U(1)_e$; the constraints associated with the $SU(2)$ subgroup of $SU(2)_L\times U(1)_Y$ have been solved.

to point out that the physical degrees of freedom of the theory are described by the *composite* objects (7.37)–(7.41).

7.3 Dynamics. Quantization

7.3.1 Introducing the dynamics

With the exception of the processes shown in Figs. 7.2 and 7.3, the analysis has so far only been carried out from a purely kinematic point of view: the basic elements of the theory have been treated as elements of the fiber bundle theory [379–381]. The invariant structures found become subject to dynamics (evolution with time) after postulating the Lagrangian. If the Lagrangian is invariant under gauge transformations from the group G, then the above analysis retains its significance: the strings may now evolve and transform, but no other structures than invariants of the type studied above would emerge in due course.

The standard gauge-invariant Lagrangian ($G = SU(n)$) has the form,

$$L = -\frac{1}{4}\mathrm{Tr}\, F_{\mu\nu}^2 + \sum_f \bar{\psi}_f(i\hat{D} - m_f)\psi_f. \tag{7.42}$$

Here $\hat{D}_\mu = D_\mu\gamma_\mu$, $D_\mu = \partial_\mu - igA_\mu$ (the vector field has been rescaled $A_\mu \to igA_\mu$ and the coupling constant g has been introduced), $A_\mu \equiv A_\mu^a\lambda_a$, whereas γ_μ and λ_a are the Dirac matrices and matrices analogous to Gell–Mann ones; $\mathrm{Tr}\,\lambda_a\lambda_b = \delta_{ab}$, $a, b = 1, \ldots, n^2 - 1$; $F_{\mu\nu} = ig^{-1}(D_\mu D_\nu - D_\nu D_\mu)$, and the index f enumerates different spinor fields ψ. In the case of quantum chromodynamics (QCD), $n = 3$, $G = SU(3)$, and the index f enumerates quark flavors. So, the dynamical law is now set. The equations of motion are defined by the Lagrangian (7.42). According to them the invariant structures determined by the constraints evolve with time. The Hamiltonian formalism is introduced by means of the standard procedure described in Chapter 3. The Hamiltonian is given by

$$H = \int d^3y \left[\frac{1}{2}\left(\mathbf{E}_a^2 + \mathbf{H}_a^2\right) + \sum_f \bar{\psi}_f(\gamma_0\hat{\mathbf{D}} + \gamma_0 m_f)\psi_f\right], \tag{7.43}$$

where $\hat{\mathbf{D}} = \boldsymbol{\gamma}\cdot(\boldsymbol{\partial} - ig\mathbf{A})$, and $\mathbf{E}_a$ and $\mathbf{H}_a$ are the color electric and magnetic fields, respectively. The classical Poisson brackets are defined in Chapter 1 (here the corresponding components of the vectors $\mathbf{A}_a$ and $\mathbf{E}_a$ are canonically conjugated variables).

7.3.2 Quantization

Quantization is carried out in accordance with the Dirac procedure (see Chapter 4). The following canonical commutation relations are used in what follows ($x_0 = y_0$):

$$\left[A_k^a(x), E_b^l(y)\right] = i\delta_b^a\delta_k^l\delta(\mathbf{x}-\mathbf{y}), \quad \left[\psi_\rho(x), \psi_\lambda^\dagger(y)\right]_+ = \delta_{\rho\lambda}\delta(\mathbf{x}-\mathbf{y}); \qquad (7.44)$$

here the indices k, l $(= 1, 2, 3)$ and a, b enumerate, respectively, vector components and components of the adjoint representation, whereas the combined index $\rho(\equiv i, \alpha)$ includes color indices $i(= 1, 2, ..., n)$ as well as the spinor ones $\alpha(= 1, 2, 3, 4)$. A stumbling point of the Dirac quantization procedure is sometimes the operator ordering problem in the Hamiltonian and constraints. In the framework of the present approach, this problem might only occur in invariant structures. The entire classical analysis might be rendered significantly devalued if classical invariant structures are not preserved upon quantization. Fortunately, this is not the case. In the Abelian theory, only mutually commuting fields A_μ occur in exponentiated line integrals for space-like contours, there is no necessity to order operators in them. In non-Abelian theories, the ordering of fields (elements of the Lie algebra) along the integration contour must be introduced in the P-exponential already in the classical theory. This ordering is carried over to the quantum theory, retaining its meaning even when field operators do not commute, as in the case of time-like integration contours.

So, the results of the geometrical analysis are fully preserved after introducing the dynamics and subsequent quantization. However, an additional new and quite important feature emerges – this is the dynamics of string branching, which is the next topic to be discussed.

7.3.3 The problem of string branching

The gauge group $G = SU(3)$ is considered in [384]. String branching (the existence of structures like those depicted in Figs. 7.1–7.3) has been discussed so far as an abstract mathematical possibility. Postulating the Lagrangian brings a new peculiarity into the analysis because the question arises: can the string topology change during the time evolution? Can, for instance, the configuration shown in Panel **a** of Fig. 7.2 turn into the configuration shown in Panel **b** of the same figure? Apparently, this may happen only if the following transition is dynamically possible:

$$\left(1 + igA_\mu dx^\mu\right)^\gamma_{\gamma'} \to \epsilon^{\alpha\beta\gamma}\left(1 - igA_\mu dx^\mu\right)^{\alpha'}_\alpha \left(1 - igA_\mu dx^\mu\right)^{\beta'}_\beta \epsilon_{\alpha'\beta'\gamma'}.$$

A necessary condition for this transition to exist is the presence of the invariant tensor ϵ-or the d-vertices in the Lagrangian since

$$\epsilon^{\alpha\beta\gamma} A_\alpha^{\alpha'} A_\beta^{\beta'} A_\gamma^{\gamma'} \epsilon_{\alpha'\beta'\gamma'} = Tr\left(A[A, A]_+\right) = \sqrt{2}\, d^{abc} A^a A^b A^c, \qquad (7.45)$$

where $A \sim A_\mu dx^\mu$, $\mathrm{Tr}A = 0$, and $d^{abc} = (1/\sqrt{2})\mathrm{Tr}(\lambda^a[\lambda^b, \lambda^c]_+)$; the matrices λ differ from the standard ones by a factor $\sqrt{2}$, which explains the factor $(1/\sqrt{2})$ in

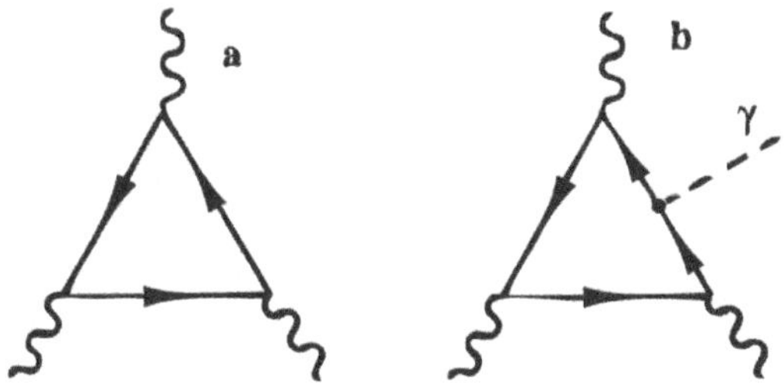

Fig. 7.4. Illustration of the string branching problem in the standard model (explanations are given in the text).

the definition of tensor d. However, the Lagrangian (7.42) has neither the d- nor ϵ-tensors. The first term in the sum (7.42) contains only the invariant tensor f^{abc} (the group structure constants), whereas the second term has just the matrices λ_a. The tensor d is impossible to build from the tensor f^{abc}, $(\mathrm{Tr}\,(F^a[F^b, F^c]_+) = 0$ where $(F^a)_{bc} = f^{abc}$ [387]. Hence, in pure gluodynamics $(L = -\mathrm{Tr}\,F^2_{\mu\nu}/4)$, strings in the fundamental representation cannot branch out, i.e. the string topology is preserved in this case [384]. They also cannot break because, by assumption, there are no particles realizing the fundamental representation, while open strings are physically meaningless (they are not gauge invariant). Vertices like (7.45) can only be expected in an effective Lagrangian of chromodynamics with spinors. However, the triangular Feynman diagram shown in Panel **a** of Fig. 7.4 generates only the f-vertex because there is also a contribution from the same diagram but with the reversed arrows so that the summed contribution of all the λ-matrices involved yields the factor $\mathrm{Tr}\,(\lambda^a[\lambda^b, \lambda^c]_-) \sim f^{abc}$. Triangular diagrams in higher orders of perturbation theory change nothing. The desired vertex may only emerge from diagrams with even numbers of γ-vertices in the fermion loop, i.e. when other vector fields interacting with fermions are present. The diagram shown in Panel **b** of Fig. 7.4 gives such an example where the dashed line represents a photon (a vector boson). This implies that every vertex in diagrams shown in Figs. 7.1–7.3 should come with the factor $g^3 e = (4\pi\alpha_s)^2(\alpha/\alpha_s)^{1/2}$. The latter is true only in the local limit. In the general case, the factor has to contain a function of momenta (a "form factor").

7.4 External fields of charges and static forces. Confinement

7.4.1 Electrodynamics

It has been established that the gauge invariance (the presence of constraints) leads to the conclusion that charges should be accompanied by their own external fields. Following Dirac (see [65]), here the form of these fields is studied in electrodynamics. Let Φ be an eigenvector of the electric field operator: $\hat{E}^j\Phi = E^j_\Phi\Phi$. Consider the state $\Psi_3^+(x)\Phi$ (see (7.14) and (7.16)). One has,

$$\hat{E}^j(x)\Psi_3^\dagger(y)\Phi = -i\frac{\delta}{\delta A_j(x)}\left\{\exp\left(ig\Delta^{-1}\boldsymbol{\partial}\cdot\mathbf{A}(y)\right)\psi^\dagger(y)\Phi\right\}$$
$$= -\left(\frac{g}{4\pi}\frac{(x-y)^j}{|\mathbf{x}-\mathbf{y}|^3} - E_\Phi^j\right)\Psi_3^+(y)\Phi. \tag{7.46}$$

It is clear that an additional electric field created by the operator $\Psi_3^\dagger$ is identical to the Coulomb field of a point charge. A similar calculation for the state created by the operator $\Psi_1^\dagger$ gives,

$$\hat{E}^j(x)\Psi_1^\dagger(y)\Phi = \left\{g\int_{-\infty}^{0} d\sigma\frac{dz^j}{d\sigma}\delta\left(\mathbf{z}(\sigma)-\mathbf{x}\right) + E_\Phi^j\right\}\Psi_1^\dagger(y)\Phi, \tag{7.47}$$

where the vector function $\mathbf{z}(\sigma)$ defines the integration contour, $\mathbf{z}(0) = \mathbf{y}$. It follows from (7.47) that, first, the electric field created by $\Psi_1^\dagger$ differs from zero only along the integration contour, second, only its component tangent to the contour does not vanish, and third, the field magnitude is infinite ($\sim \delta^{(2)}(0)$).

Let us calculate the energy of a pair of static sources. The interaction energy is defined as the total energy of the system from which the own field energy of each individual source is subtracted. Owing to this definition, there is no need to calculate the contributions of the fermion (matter) fields to the total energy. Thus, according to (7.16) the energy of the following states is calculated:

$$P_3(x)P_3(x')|0\rangle, \quad P_{xx'}|0\rangle = e^{ig\int A_j dx^j}|0\rangle. \tag{7.48}$$

Here $|0\rangle$ denotes the physical vacuum; the integration contour in $P_{xx'}$ is chosen to be a straight line segment connecting points $\mathbf{x}$ and $\mathbf{x}'$. The commutation relations,

$$\left[E^j(y), P_{xx'}\right]_- = P_{xx'}g\int_{x'}^{x} dz^j\delta(\mathbf{z}-\mathbf{y}), \tag{7.49}$$

$$\left[E^j(y), P_{3x}\right]_- = P_{3x}g\int d^3z(\partial^j\Delta^{-1})_{xz}\delta(\mathbf{z}-\mathbf{y}). \tag{7.50}$$

turn out to be useful. In the case of the Coulomb sources, the expectation value of the quantum electromagnetic field Hamiltonian, $H_0 = \int d^3y[\mathbf{E}^2 + \mathbf{H}^2]/2$, is

$$\left\langle P_{3x'}^\dagger P_{3x}^\dagger \hat{H}_0 P_{3x}P_{3x'}\right\rangle_0$$
$$= C_0 + \frac{g^2}{2}\int d^3z_1 d^3z_2\left[(\partial^j\Delta^{-1})_{xz_1} + (\partial^j\Delta^{-1})_{x'z_1}\right]\delta(\mathbf{z}_1-\mathbf{z}_2)$$
$$\times\left[(\partial^j\Delta^{-1})_{xz_2} + (\partial^j\Delta^{-1})_{x'z_2}\right], \tag{7.51}$$

where $C_0 \equiv \langle H_0\rangle_0$. The second term in (7.51) comes from the term $\mathbf{E}^2/2$ in the Hamiltonian H_0. Subtracting the own Coulomb energies of the sources (the terms $g^2\int d^3z[(\boldsymbol{\partial}\Delta^{-1})_{xz}]^2/2$), the interaction energy is found,

$$V_{xx'} = g^2\int d^3z(\partial^j\Delta^{-1})_{xz}(\partial^j\Delta^{-1})_{x'z} = \frac{g^2}{4\pi}\frac{1}{|\mathbf{x}-\mathbf{x}'|}. \tag{7.52}$$

In the case of the field localized on the contour (a straight string between the sources), one obtains [369, 386],

$$\left\langle P^{\dagger}_{xx'}\hat{H}_0 P_{xx'}\right\rangle_0 = C_0 + \frac{g^2}{2}\int_{x'}^{x} dz_1^j \int_{x'}^{x} dz_2^j \delta(\mathbf{z}_1-\mathbf{z}_2) = C_0 + \frac{g^2}{2}\delta^{(2)}(0)|\mathbf{x}-\mathbf{x}'|. \quad (7.53)$$

So, the interaction energy of two static sources coincides with the Coulomb energy for the first state in (7.48), while for the second one it is proportional to the distance between the sources according to (7.53). A further discussion regarding the meaning of the factor $\delta^{(2)}(0)$ can be found in Section 7.4.3. Note that the terms corresponding to the own energies of the sources are absent in (7.53). If the state $P_{xx'}|0\rangle$ were stationary, its energy could have been identified with the potential energy of the static force field. As is shown in Section 7.4.2, only the Coulomb external field is stable in electrodynamics, while any other external field excitation compatible with constraints would eventually decay into it by emitting radiation [404].

7.4.2 "String" evolution and static forces in electrodynamics

Equation (7.53) shows that the gauge-invariant kinematics allows for string-like excitations of external fields of static sources, and their energy grows with the distance between the sources. Nevertheless, no confining potential is observed in electrodynamics; the static interaction of the sources is described by the Coulomb law. This apparent "discrepancy" leads to the conclusion that there is a *dynamical* principle that distinguishes the states generated by the invariant operator $\Psi_3^{\dagger}$ (the charge with its own Coulomb field) among all other possible invariant excitations. It appears that these states are the only ones that are dynamically stable. All other gauge-invariant excitations would decay by emitting radiation. In the case of localized string-like excitations, the radiation fields carry away most of the string energy, leaving behind only the part that is exactly equal to the Coulomb energy. For this very reason there are no confining forces in electrodynamics.

For a better understanding of this assertion, it is instructive to begin by studying the dynamics of coherent excitations of a free massless scalar field. In what follows, spacetime points are denoted by $x = (\mathbf{x}, x_0)$ and the scalar product $px \equiv p_\mu x^\mu$ is defined with respect to the Minkowski metrics. The Lagrangian has the form,

$$L = \frac{1}{2}\left(\partial^2_\mu \varphi\right)^2.$$

The quantization is carried out in the standard way (no constraints). The quantum Hamiltonian reads,

$$\hat{H} = \frac{1}{2}\int d\mathbf{x}\left[\hat{\pi}^2 + (\boldsymbol{\partial}\hat{\varphi})^2\right] = \int d\mu(\mathbf{p})\omega_p \hat{a}^{\dagger}(\mathbf{p})\hat{a}(\mathbf{p}) + \mathcal{E}_0, \quad (7.54)$$

where $\mathcal{E}_0$ is the energy of vacuum fluctuations,

$$\begin{aligned}\hat{\varphi}(\mathbf{x}) &= \int d\mu(\mathbf{p})\, \left[\hat{a}^\dagger(\mathbf{p})e^{-ipx} + \hat{a}(\mathbf{p})e^{ipx}\right], \\ [\hat{\varphi}(x), \hat{\pi}(y)]\Big|_{x_0=y_0} &= i\delta(\mathbf{x}-\mathbf{y}), \\ \left[\hat{a}(\mathbf{p}), \hat{a}^\dagger(\mathbf{p}')\right] &= (2\pi)^3 2\omega_p \delta(\mathbf{p}-\mathbf{p}'),\end{aligned} \tag{7.55}$$

$\omega_p = |\mathbf{p}|$, and $d\mu(\mathbf{p}) = d\mathbf{p}/[(2\pi)^3\omega_p]$. It is not difficult to obtain the Heisenberg canonical field operators that are solutions of the Heisenberg equations of motion. Put $\mathrm{ad}\,\hat{H} = [\hat{H}, \cdot]$ (the adjoint action of the operator $\hat{H}$). The following chain equalities can be inferred from the canonical commutation relations (7.55):

$$\begin{aligned}\hat{\varphi}(\mathbf{x},t) = e^{it\hat{H}}\hat{\varphi}(\mathbf{x},0)e^{-it\hat{H}} &= \sum_{n=0}^{\infty} \frac{(it)^n}{n!}(\mathrm{ad}\,\hat{H})^n \hat{\varphi}(\mathbf{x},0) \\ &= \cos\left(t\sqrt{-\Delta}\right)\hat{\varphi}(\mathbf{x},0) + \frac{\sin\left(t\sqrt{-\Delta}\right)}{\sqrt{-\Delta}}\hat{\pi}(\mathbf{x},0).\end{aligned} \tag{7.56}$$

To obtain Eq. (7.56), the sum has been split into the sums over even and odd n with the subsequent use of the relations $(\mathrm{ad}\,\hat{H})^2\hat{\varphi} = -\Delta\hat{\varphi}$ and $(\mathrm{ad}\,\hat{H})^2\hat{\pi} = -\Delta\hat{\pi}$. It is convenient to introduce the kernel $D(\mathbf{x}-\mathbf{y}, t)$ defined by

$$\frac{\sin\left(t\sqrt{-\Delta}\right)}{\sqrt{-\Delta}} f(\mathbf{x}) = \int d\mathbf{y}\, D(\mathbf{x}-\mathbf{y}, t) f(\mathbf{y}). \tag{7.57}$$

Making use of the Fourier transform in (7.57), an explicit form of $D(\mathbf{x}, t)$ can be found,

$$\begin{aligned}D(\mathbf{x},t) = (2\pi)^{-3}\int d\mathbf{k}\, e^{i\mathbf{k}\cdot\mathbf{x}} \frac{\sin(t|\mathbf{k}|)}{|\mathbf{k}|} &= \frac{1}{4\pi|\mathbf{x}|}\left(\delta(t-|\mathbf{x}|) - \delta(t+|\mathbf{x}|)\right) \\ &= (2\pi)^{-1}\varepsilon(t)\delta(t^2-\mathbf{x}^2) \equiv (2\pi)^{-1}\varepsilon(x_0)\delta(x^2),\end{aligned} \tag{7.58}$$

where $\varepsilon(t)$ is the sign function. The Fourier integral is easy to evaluate in the spherical coordinates. The kernel (7.58) is known as the Pauli–Jordan permutation function, $D(\mathbf{x}-\mathbf{y}, x_0-y_0) = i[\hat{\varphi}(x), \hat{\varphi}(y)]$. The Heisenberg field canonical operators may now be written in the form,

$$\hat{\varphi}(\mathbf{x},t) = \partial_t \int d\mathbf{y} D(\mathbf{x}-\mathbf{y},t)\hat{\varphi}(\mathbf{y},0) + \int d\mathbf{y} D(\mathbf{x}-\mathbf{y},t)\hat{\pi}(\mathbf{y},0), \tag{7.59}$$

$$\hat{\pi}(\mathbf{x},t) = \partial_t\hat{\varphi}(\mathbf{x},t). \tag{7.60}$$

Consider a coherent excitation,

$$|J\rangle = \exp\left[i\int d\mathbf{x}\hat{\varphi}(\mathbf{x},0)J(\mathbf{x})\right]|0\rangle \equiv \hat{P}_J|0\rangle, \tag{7.61}$$

where $|0\rangle$ is the ground state defined by $\hat{a}|0\rangle = 0$ ($\hat{H}|0\rangle = \mathcal{E}_0|0\rangle$), and $J(\mathbf{x})$ is a real-valued function. A characteristic property of the coherent state (7.61) is

$$\varphi_J(\mathbf{x},0) = \langle J|\hat{\varphi}(\mathbf{x},0)|J\rangle = 0, \quad \pi_J(\mathbf{x},0) = \langle J|\hat{\pi}(\mathbf{x},0)|J\rangle = J(\mathbf{x}), \tag{7.62}$$

i.e. the field is excited in the region where $J(\mathbf{x}) \neq 0$, and $J(\mathbf{x})$ coincide with the expectation value of the canonical momentum in this state. Recall that a free field can be viewed as a system of coupled harmonic oscillators. Then, function $J(\mathbf{x})$ determines a displacement of the oscillation center in the phase space $(\varphi(\mathbf{x},0), \pi(\mathbf{x},0))$. The state (7.61) is created at the initial time moment $t=0$. It evolves into the state $|J,t\rangle = \exp(-it\hat{H})|J\rangle$ in time $t>0$. Therefore the expectation values (7.62) change in due course. A simple calculation, based on the solutions (7.59) and (7.60) of the Heisenberg equations and the initial conditions (7.62), shows that

$$
\begin{aligned}
\varphi_J(\mathbf{x},t) &= \langle J,t|\hat{\varphi}(\mathbf{x},0)|J,t\rangle = \langle J|\hat{\varphi}(\mathbf{x},t)|J\rangle \\
&= \int d\mathbf{y} D(\mathbf{x}-\mathbf{y},t)J(\mathbf{y}), \quad t>0, \qquad (7.63)\\
\pi_J(\mathbf{x},t) &= \langle J,t|\hat{\pi}(\mathbf{x},0)|J,t\rangle = \langle J|\hat{\pi}(\mathbf{x},t)|J\rangle = \partial_t\varphi_J(\mathbf{x},t). \qquad (7.64)
\end{aligned}
$$

The expectation values (7.63) and (7.64) solve the initial value problem for the wave equation,

$$
\left(\partial_t^2 - \Delta\right)\varphi_J(\mathbf{x},t) = 0, \quad \partial_t\varphi_J(\mathbf{x},t)|_{t=0} = J(\mathbf{x}), \quad \varphi_J(\mathbf{x},t)\big|_{t=0} = 0.
$$

An important observation following from the above analysis is that, even if the field is initially excited in a bounded region, in due course its excitations can go away to spatial infinity, and the field excitation energy in the region where the initial perturbation is created, $J(\mathbf{x}) \neq 0$, is depleted (carried away by field excitations propagating to spatial infinity).

Now the time evolution of the string-like excitation in electrodynamics described by the second state in (7.48) can easily be understood. Let $\mathbf{x} = \mathbf{x}(s)$ be parametric equations of the contour connecting the points $\mathbf{x}_1 = \mathbf{x}(0)$ and $\mathbf{x}_2 = \mathbf{x}(1)$. Put

$$
\mathbf{J}(\mathbf{x}) = e\int_0^1 \delta\left(\mathbf{x}-\mathbf{x}(s)\right)\frac{d\mathbf{x}(s)}{ds}\,ds. \qquad (7.65)
$$

Then

$$
-\boldsymbol{\partial}\cdot\mathbf{J}(\mathbf{x}) = e\int_0^1 \frac{d}{ds}\delta\left(\mathbf{x}-\mathbf{x}(s)\right)ds = e\delta(\mathbf{x}-\mathbf{x}_2) - e\delta(\mathbf{x}-\mathbf{x}_1) \equiv J_0(\mathbf{x}). \quad (7.66)
$$

The vector potential is decomposed into the sum of the transverse and conservative (longitudinal) components,

$$
\mathbf{A} = \mathbf{A}_\perp + \boldsymbol{\partial}\Delta^{-1}(\boldsymbol{\partial}\cdot\mathbf{A}), \quad \boldsymbol{\partial}\cdot\mathbf{A}_\perp = 0. \qquad (7.67)
$$

In fact, this equation is the definition of the transverse potential $\mathbf{A}_\perp$. With the help of (7.65)–(7.67) the coherent state describing a string-like excitation of the electric field can be written in the form,

$$
\begin{aligned}
|\mathbf{J}\rangle \equiv P_{x_1x_2}|0\rangle &= \exp\left(i\int d\mathbf{x}\,\mathbf{J}\cdot\hat{\mathbf{A}}\right)|0\rangle \\
&= \exp\left(i\int d\mathbf{x}\,\mathbf{J}\cdot\hat{\mathbf{A}}_\perp\right)\exp\left[i\int d\mathbf{x}\,\left(\Delta^{-1}J_0\right)\boldsymbol{\partial}\cdot\hat{\mathbf{A}}\right]|0\rangle \\
&= \exp\left(i\int d\mathbf{x}\,\mathbf{J}\cdot\hat{\mathbf{A}}_\perp\right)P_3(x_2)P_3^\dagger(x_1)|0\rangle. \qquad (7.68)
\end{aligned}
$$

The electric field operator $\hat{\mathbf{E}}(\mathbf{x},0)$ is the canonical momentum conjugated to $\hat{\mathbf{A}}(\mathbf{x},0)$, and the magnetic field operator is $\hat{\mathbf{B}}(\mathbf{x},0)=\boldsymbol{\partial}\times\hat{\mathbf{A}}(\mathbf{x},0)$. Consider the Heisenberg field operators $\hat{\mathbf{E}}(\mathbf{x},t)=e^{it\hat{H}}\hat{\mathbf{E}}(\mathbf{x},0)e^{-it\hat{H}}$ and $\hat{\mathbf{B}}(\mathbf{x},t)=e^{it\hat{H}}$ $\hat{\mathbf{B}}(\mathbf{x},0)e^{-it\hat{H}}$, where $\hat{H}=(1/2)\int d\mathbf{x}(\hat{\mathbf{E}}^2+\hat{\mathbf{B}}^2)$ is the electromagnetic Hamiltonian. Due to the linearity of Maxwell's theory, the Heisenberg equations coincide with Maxwell's equations for a free electromagnetic field. This implies, in particular, that only the transverse component of the electric field operator evolves,

$$
\hat{\mathbf{E}}(\mathbf{x},t)=\hat{\mathbf{E}}_\perp(\mathbf{x},t)+\boldsymbol{\partial}\Delta^{-1}(\boldsymbol{\partial}\cdot\hat{\mathbf{E}})(\mathbf{x},0)
$$

because the Hamiltonian contains only $\hat{\mathbf{A}}_\perp$: $\hat{\mathbf{B}}=\boldsymbol{\partial}\times\hat{\mathbf{A}}=\boldsymbol{\partial}\times\hat{\mathbf{A}}_\perp$; the transverse component $\hat{\mathbf{E}}_\perp(\mathbf{x},t)$ satisfies the wave equation.

Just like the case of the free scalar massless field, consider the Heisenberg field expectation values,

$$
\mathbf{E}_J(\mathbf{x},t)=\langle\mathbf{J}|\hat{\mathbf{E}}(\mathbf{x},t)|\mathbf{J}\rangle,\quad \mathbf{B}_J(\mathbf{x},t)=\langle\mathbf{J}|\hat{\mathbf{B}}(\mathbf{x},t)|\mathbf{J}\rangle.
$$

They describe the expectation values of the electromagnetic fields in the evolving state $|\mathbf{J},t\rangle=e^{-it\hat{H}}|\mathbf{J}\rangle$ (cf. (7.63) and (7.64)). They satisfy Maxwell's equations with the initial conditions,

$$
\mathbf{E}_J(\mathbf{x},0)=\mathbf{J}(\mathbf{x}),\quad \mathbf{B}_J(\mathbf{x},0)=\mathbf{0}. \qquad (7.69)
$$

The solution of this initial value problem reads [404, 405],

$$
\begin{aligned}
\mathbf{E}_J(\mathbf{x},t)&=\partial_t\int d\mathbf{y}D(\mathbf{x}-\mathbf{y},t)\mathbf{J}(\mathbf{y}) \\
&\quad+\boldsymbol{\partial}\int d\mathbf{y}\theta(t-|\mathbf{x}-\mathbf{y}|)\left(\Delta^{-1}J_0\right)(\mathbf{y}), \qquad (7.70)\\
\mathbf{B}_J(\mathbf{x},t)&=\boldsymbol{\partial}\times\int d\mathbf{y}D(\mathbf{x}-\mathbf{y},t)\mathbf{J}(\mathbf{y}), \qquad (7.71)
\end{aligned}
$$

where θ is the Heaviside step function. The evolution of the coherent string-like excitation is depicted in Fig. 7.5 for the case of the contour being a straight line segment. The initial electric field is shown in the upper left panel of the figure. The first term in (7.70) and the magnetic field (7.71) describe the radiation field. The region occupied by them is shadowed by dots and can be constructed by the Huygens principle. Each point of the initial line segment expands into a sphere whose radius increases with the speed of light (in Fig. 7.5 the speed of light is set to unity, and the radii of the spheres coincide with the time t). The

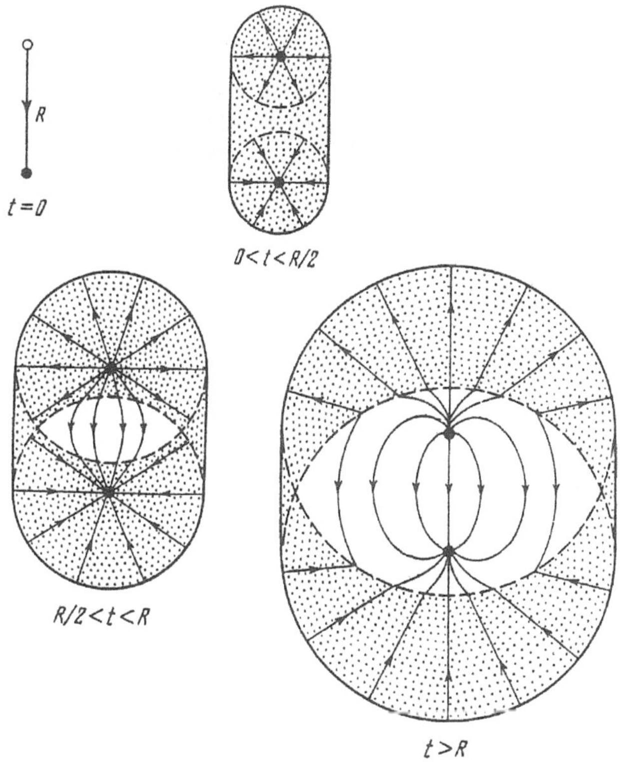

Fig. 7.5. The decay of a string-like field excitation in electrodynamics (see explanations in the text). Adapted from [404] and [405].

dotted region is filled out by these spheres expanding from every point of the line segment. The second term in (7.70) describes the creation of the Coulomb field around two charges. Its force lines are shown by solid lines with arrows. Due to the Heaviside function, the Coulomb field of each source is confined in a sphere centered at the source; the fronts of these two spherical regions expand at the speed of light and do not intersect each other for $t < R/2$ where R is the distance between the sources (the speed of light is one in the units used here). This stage is shown in the upper right panel of the figure. When $t > R/2$, a bubble free of the radiation field emerges; the radii of Huygens fronts become large enough so that no front passes through this region. It is filled by the Coulomb field of two sources (the lower left panel of the figure). Note that the combined Coulomb field is formed only in this radiation-free bubble. For a sufficiently large time $t \gg R$, the Huygens fronts would form a nearly spherical layer of width R and radius $t \gg R$, expanding away from the sources. The region bounded by this layer is filled with the Coulomb field of two static opposite charges (see the lower right panel of the figure). Thus, the energy that was initially localized on the contour radiates away; only the Coulomb interaction energy is left behind.

An alternative way to see this is to study the electromagnetic energy in a small region V (that contains the contour) as a function of time,

$$\mathcal{E}_V(t) = \langle \mathbf{J}, t| \frac{1}{2} \int_V d\mathbf{x} \left[\hat{\mathbf{E}}^2 + \hat{\mathbf{B}}^2 \right] |\mathbf{J}, t\rangle = \frac{1}{2} \int_V d\mathbf{x} \left[\mathbf{E}_J^2(\mathbf{x}, t) + \mathbf{B}_J^2(\mathbf{x}, t) \right],$$

where the energy of vacuum fluctuations is omitted. According to the solutions (7.70) and (7.71) the electromagnetic fields in the region V are

$$\mathbf{E}_J(\mathbf{x}, t) = \boldsymbol{\partial} \left(\Delta^{-1} J_0 \right)(\mathbf{x}), \quad \mathbf{B}_J(\mathbf{x}, t) = \mathbf{0}, \quad \mathbf{x} \in V,$$

when $t \geq t_V \sim R_V$, where R_V is the diameter of V (defined as the largest distance possible for any two points in V). Thus, even if $\mathcal{E}_V(0)$ is given by (7.53), then by the time moment t_V, only the Coulomb part of the energy remains in the region V,

$$\mathcal{E}_V(t) = \frac{1}{2} \int_V d\mathbf{x} \left[\boldsymbol{\partial} \left(\Delta^{-1} J_0 \right)(\mathbf{x}) \right]^2, \quad t > t_V. \tag{7.72}$$

This study proves that string-like excitations of the field between two charges, although admissible by gauge invariance, are dynamically unstable. After radiating the excess of energy, they turn into the Coulomb field whose energy determines the interaction of static sources in electrodynamics.

Remark 1. The external field excitation described by the invariant Ψ_2 in (7.16) can be studied in a similar way [404]. It is also unstable and decays into the Coulomb field and radiation. Since the initial electric field is localized in a plane, the radiation-free bubble of the Coulomb field never forms in a finite time; the function $\mathbf{J}$ has an unbounded support (the plane) and therefore the Huygens fronts always intersect any bounded region containing the sources. However, asymptotically (as $t \to \infty$), the energy $\mathcal{E}_V(t)$ converges to (7.72).

Remark 2. A closed string-like excitation is obtained when periodic boundary conditions $\mathbf{x}(0) = \mathbf{x}(1)$ (or $\mathbf{x}_1 = \mathbf{x}_2$) are imposed on the contour in (7.65) and, hence, $J_0 = 0$ in (7.66). Such a coherent excitation of the electric field disperses completely; all its energy is radiated away because $J_0 = 0$ in (7.72) (see also [372, 404, 405]).

7.4.3 Non-Abelian theories

States with specific values of the chromoelectric field E_a^j do not belong to the physical subspace because $\left[E_a^j(\mathbf{x}), D_i^{bc} E_c^i(\mathbf{y}) \right] = -igf^{abc} E_c^j \delta(\mathbf{x} - \mathbf{y}) \neq 0$, where $D_i^{ab} E_b^i = \delta^{ab}(\partial_i E_b^i + g f_{bc}^d A_i^c E_d^i)$ is the covariant derivative in the adjoint representation; therefore there is no analogy for (7.47) in non-Abelian theories. Let us calculate the energy of two static sources in the fundamental representation (heavy quarks, $m \to \infty$, $G = SU(n)$), i.e. the expectation value of the Hamiltonian (7.43) in the state $|\mathcal{M}_{xx'}\rangle = \mathcal{M}_{xx'}|0\rangle$ where

$$\mathcal{M}_{xx'} = \bar{\psi}_x P_{xx'} \psi_{x'}, \tag{7.73}$$

and $|0\rangle$ is the physical vacuum; the integration contour in $P_{xx'}$ is a straight line segment. It is proved that in the limit $m \to \infty$ the energy of the gluon field excitation is proportional to the distance between the quarks. When calculating the energy, the following facts are to be taken into consideration.

First, let $\psi = \psi^{(+)} + \psi^{(-)}$, $\bar{\psi} = \bar{\psi}^{(+)} + \bar{\psi}^{(-)}$, where $\psi^{(+)}$ ($\bar{\psi}^{(+)}$) contains the quark (antiquark) creation operator. Generally speaking, $\psi^{(+)}|0\rangle \neq 0$, but as $m \to \infty$ it is obvious that

$$\psi^{(+)}|0\rangle \to 0, \quad \bar{\psi}^{(+)}|0\rangle \to 0, \tag{7.74}$$

because the quark degrees of freedom are "frozen" (cannot be excited) in this limit.

Second, define the combined indices $\rho = (i, \alpha)$ and $\lambda = (j, \beta)$ and consider the simultaneous anti-commutation relations,

$$S_{\rho\lambda} = \left[\psi_\rho^{(+)}(x), \bar{\psi}_\lambda^{(-)}(y)\right]_+ = \int \frac{d^3p}{(2\pi)^3 2E_p} e^{-ip(x-y)} u_\alpha^\sigma(p) \bar{u}_\beta^\sigma(p) \delta_{ij} \tag{7.75}$$

and a similar expression for $\tilde{S}_{\rho\lambda} = \left[\psi_\rho^{(-)}(x), \bar{\psi}_\lambda^{(+)}(y)\right]_+$ which is obtained from (7.75) by the replacements $u(p) \to v(p)$ and $p \to -p$ in the exponential. In Eq. (7.75) the standard notations and normalization are used: $u^\sigma(p)\bar{u}^\sigma(p) = (\hat{p} + m)$, $v^\sigma(p)\bar{v}^\sigma(p) = (\hat{p} - m)$, $E_p = (\mathbf{p}^2 + m^2)^{1/2}$, and σ defines the fermion polarization. For heavy quarks, one has,

$$S_{\rho\lambda} \to \frac{1}{2}(\gamma_0 + 1)_{\alpha\beta} \delta_{ij} \delta(\mathbf{x} - \mathbf{y}), \quad \tilde{S}_{\rho\lambda} \to \frac{1}{2}(\gamma_0 - 1)_{\alpha\beta} \delta_{ij} \delta(\mathbf{x} - \mathbf{y}). \tag{7.76}$$

Third, the normalization of the state $|\mathcal{M}\rangle$ created by the operator (7.73) is: $\langle\mathcal{M}|\mathcal{M}\rangle \equiv C = 2\mathrm{Tr}\, 1[\delta^{(3)}(0)]^2$, $\mathrm{Tr}\, 1 = n$. The difference,

$$V_{xx'} = \left(\langle\mathcal{M}_{xx'}|\hat{H}|\mathcal{M}_{xx'}\rangle - \langle\mathcal{M}_{xx}|\hat{H}|\mathcal{M}_{xx}\rangle\right) C^{-1} \tag{7.77}$$

is calculated where $\hat{H}$ is the Hamiltonian (7.43). The technicalities are similar to those in electrodynamics only difference being that Eq. (7.77) contains fermion fields and the ordinary exponential is replaced by the path-ordered one,

$$P_{xx'} = P \exp\left(ig \int_0^1 A_\mu^a(z(\tau)) \lambda_a(\tau) \dot{z}^\mu(\tau) d\tau\right). \tag{7.78}$$

Here $\dot{z}^\mu = dz^\mu/d\tau$, $z(0) = x'$, and $z(1) = x$; following Feynman (see [71]) the matrices λ_a depending on the ordering parameter are introduced – the symbol P determines their ordering along the contour in (7.78). The commutation relation analogous to (7.49) reads,

$$\left[E_a^k(y), P_{xx'}\right] = gP\left\{P_{xx'}\int_0^1 \lambda_a(\tau)\dot{z}^k(\tau)\delta\left(\mathbf{z}(\tau)-\mathbf{y}\right)d\tau\right\}. \tag{7.79}$$

Note that the ordering symbol P is related to all the matrices in the curly brackets, including the matrices in the operator $P_{xx'}$. Taking into account the identity $P_{xx'}P^\dagger_{xx'} = 1$, Eqs. (7.74), (7.76), and (7.79), and that the simultaneous commutators of the field operators A^a_μ vanish, the difference (7.77) is transformed to the vacuum expectation value,

$$\begin{aligned} V_{xx'} = \frac{g^2}{2\mathrm{Tr}\,1}\Bigg\langle \mathrm{Tr}\int d\mathbf{y}\Bigg\{\tilde{P}\left[P^\dagger_{xx'}\int_0^1 \lambda_a(\sigma)\dot{z}^k(\sigma)\delta\left(\mathbf{z}(\sigma)-\mathbf{y}\right)d\sigma\right] \\ \times P\left[P_{xx'}\int_0^1 \lambda_a(\tau)\dot{z}^k(\tau)\delta\left(\mathbf{z}(\tau)-\mathbf{y}\right)d\tau\right]\Bigg\}\Bigg\rangle_0; \end{aligned} \tag{7.80}$$

the symbol $\tilde{P}$ stands for the ordering reverse to the ordering prescribed by the symbol P, for example, $P^\dagger_{xx'} = \tilde{P}\exp(-ig\int A_\mu dx^\mu)$. It follows immediately from (7.80) that the sought-for expectation value of the gluon field energy is

$$V_{xx'} = \frac{g^2}{2\mathrm{Tr}\,1}\mathrm{Tr}\,\left(\lambda_a^2\right)\,\delta^{(2)}(0)\,|\mathbf{x}-\mathbf{x}'|. \tag{7.81}$$

So, it is indeed the parallel transport operator (7.8) that describes the external field, created by static color charges, whose energy rises linearly with the distance between the sources. As in the case of electrodynamics, if the state $|\mathcal{M}\rangle$ were proved to be dynamically stable, the energy (7.81) could have been identified with the linearly rising static potential. This kind of potential was postulated to explain the linearity of Regge trajectories, spectra of quarkonia, and confinement.

7.4.4 Discussion

It is often acceptable to identify the P-exponential (7.8) with a string because the corresponding field excitation is one-dimensional and its energy depends linearly on the distance between the sources. It should be emphasized that such an identification would have indeed been possible if it were proved that the string-like excitation remains a linearly extended object during its evolution. As has been demonstrated, string-like excitations are unstable in electrodynamics. A similar investigation for non-Abelian theories still remains an unsolved and challenging problem in theoretical and mathematical physics, although there is little doubt that a stable field excitation between two static charges is localized in the vicinity of a straight line connecting the charges. This picture is supported by Monte Carlo simulations of lattice gauge theories which do provide compelling evidence that the static energy can be well approximated by a linearly rising potential [406, 407]. However, the actual mechanism of the string stability have not yet been established.

Lattice gauge theories provide a natural short-distance regularization of the theory, but they might also have dynamical effects that do not exist in the continuous theories. Such differences exist even for the Abelian theory. In particular, it is straightforward to verify that in the strong coupling limit, string-like excitations become stable in lattice electrodynamics, while the exact solution given in Section 7.4.2 predicts exactly the converse because the analysis does not depend on the coupling constant. Indeed, in the Kogut–Susskind formulation of lattice gauge theories (see Section 5.8.10), a string-like excitation is described by the state

$$|\mathcal{M}_{xx'}\rangle = \prod_{(y,k)\in C} \hat{u}_{y,k}|0\rangle, \qquad \hat{u}_{yk} = e^{iga\hat{A}_k(y)}$$

where the product is taken over the link variables along the contour C connecting the static sources localized at points $y = x$ and $y = x'$. In the formal strong coupling limit ($g \to \infty$) in the Hamiltonian (5.262), the electric field energy dominates over the magnetic energy because the potential V is a bounded function of the link variables (V is defined in the text after Eq. (5.259)), in contrast to the continuum case. The interaction between neighboring field oscillators disappears in this limit and the string-like excitation becomes an eigenvector of the Hamiltonian with the eigenvalue being $g^2L(C)/(2a^2)$ ($L(C)$ is the length of the contour C). Thus, the strong coupling limit of the lattice gauge theory is not appropriate to solve the confinement problem. One of the challenging problems here is to find a quantum formulation of the theory that would be equally applicable in the strong and weak coupling limits. The latter is necessary to control the renormalization effects due to the self-interaction of field oscillators in the non-Abelian case [408, 409].

Let us formulate in brief the assumptions which have been used to obtain Eq. (7.81): (1) the static sources (large quark mass limit, $m \to \infty$); (2) the "static string" (no dynamics). The peculiarity of the problem is that the position of the sources as well as that of the string is fixed in the space. In reality, the string can change its configuration. Can this be taken into account in some kind of "parametric" way, like the condition $m \to \infty$ for the quark mass? Apparently, the answer is affirmative. In the case of a curved contour of length L, Eq. (7.81) would contain the factor g^2L in place of $g^2|\mathbf{x} - \mathbf{x}'|$. It is then clear that in the limit $g^2 \to \infty$ even a small deviation of the string from the straight line segment would cost a lot of energy and, hence, string fluctuations (vibrations) would be suppressed.

What happens at finite m and g? The case of infinite fermion masses corresponds to pure gluodynamics. The consequence of fermion dynamics is the possibility of breaking strings in the fundamental representation – now the quarks can be created as they must be attached to the string end points. This leads to a new parameter in the theory, the critical length r_c at which the string breaks. At distances $r > r_c$ the potential becomes meaningless. Furthermore, according to the analysis of Section 7.3, strings in the fundamental representation of $SU(n)$

($n > 2$) can branch in this case (if an additional (specific) interaction is present). This dynamical feature affects the potential even for shorter distances $r < r_c$. It should also be noted that the potential (7.81) contains the infinite factor $\delta^2(0)$. Simplifying the matter, it can be replaced by const./πr_s^2, where r_s is the string radius; alternatively, it can be included into the string tension σ: $V = \sigma r^2$ with r being the distance between the sources. In reality, this parameter should be derived from the renormalization of a quantum theory whose formulation is applicable beyond the conventional perturbation theory. Apparently, the three parameters (r_c, m, σ) are related: $r_c \sim m/\sigma$, i.e. only two of them are independent. If concrete values of these parameters are chosen for ad hoc reasons, not set (or found) by solving the actual problem from first principles, then one could obtain models which may substantially differ from one another in their physical properties even though they stem from the very same theory (the same Lagrangian). This feature leads to many obliques in which the confinement may reveal itself.

7.4.5 Obliques of confinement

The phenomenon of confinement, when viewed broadly, may have appearances with different physical properties. Here is the list of possible obliques of confinement.

Weak confinement. In its simplest form, the notion of confinement is reduced to the statement that only colorless (charge-free) states form the space of physical states. Weak confinement takes place in any gauge theory. This is a rather trivial statement; it merely follows from the Gauss law (the constraints of the theory). There are a few subtleties worth noting. If string-like excitations are dynamically stable, then every charged object would be accompanied by such an excitation. A finite length string should have charges at its ends as required by the Gauss law so that the total charge of this state vanishes. Infinite strings are physically inadmissible because they have infinite energy at a finite string tension. Note also that, if the string is extended to spatial infinity, then such an object would not be invariant under global gauge transformations [370, 371]. The case when the static interaction decreases with the distance is a lime less obvious (e.g. string-like excitations are unstable in electrodynamics; the static interaction decreases with the distance and is described by the Coulomb law). If the field strength vanishes at spatial infinity, then there is a dilemma as to whether or not the gauge group should be extended to include global gauge transformations. If they are included into the gauge group, then, integrating the Gauss law over a ball of radius tending to infinity, one infers that the total charge of the closed universe must vanish [370] (the total charge is the generator of global gauge transformations).

So, the confinement, in a broad understanding of this term, is not necessarily related to a rising interaction potential. It is merely equivalent to the statement that the force lines ("strings") cannot be extended to spatial infinity; they originate from and end on charges. Recent numerical simulations of the gluon field structure in hadrons and mesons can be found in [410, 411].

Strong confinement. This is a standard, commonly accepted, understanding of confinement when the static interaction of massive particles is described by a (linearly) rising potential, i.e. when oppositely (color) charged particles are connected by a (stable) string which cannot break.

Confinement in the physics of hadrons. In the real world, strings can break at a certain length r_c. As a result, a quark–antiquark pair is created at the breaking point. The question about observation of an isolated quark is related to the actual value of the parameter r_c (the question of why one cannot observe physical quarks). In order to be registered by modern devices, a charged particle should travel an atomic size distance r_a. If $r_c < r_a$, then an isolated quark cannot be detected by the Wilson chamber or other similar device. Such an observation opportunity would only appear if $r_c > r_a$. As a point of fact, the confinement still takes place in both the cases! Hence, the aforementioned question about the imprisonment of quarks makes sense only for specific observation methods. Experiments on the deep inelastic scattering of leptons on hadrons (i.e. the use of a "lepton microscope") already allows one not only to register isolated quarks but also to determine their quantum numbers (spin, isospin, electric charge). Thus, the imprisonment of quarks does not imply that they become unobservable particles – in many aspects they retain the "rights" of regular particles,[3] but exist in a rather specific environment, though, and because of that special devices are required to study them.

Screening (decoloration). If $r_c \to 0$, then string-like field configurations do not occur. A finite length string collapses, decaying into fragments of vanishing length (a phenomenon analogous to the screening of an electric charge in conductors – the string breaking means the creation of a pair). This phenomenon is well studied in the Schwinger model [364] where massless fermions can be created in any numbers [365–367]. Strings in the adjoint representation are expected to behave similarly.

Other cases. Let r_h be a parameter characterizing a hadron size. The ratio r_c/r_h defines two limiting cases. When $r_c \gg r_h$, and especially when $r_c \gg r_a$, the quarks may be separated by macroscopic distances while being confined in a bound state. Of course, such confined particles could be registered by ordinary experimental devices (like the Wilson chamber) if they have an electric charge – the confinement is no obstacle here). A similar situation may exist for a sufficiently small string tension, $\sigma r_a^2 \ll 1$ (*a soft confinement*). In contrast, when $r_a \gg r_c \geq r_h$, confined particles cannot be observed as isolated particles. This takes place at a large string tension, $\sigma r_h^2 \sim 1$ (*a hard confinement*). The latter is exactly the case in hadron physics.

[3] Although a certain deprivation of the rights does occur – they cannot be separated.

The electroweak interaction model is a special case as it has two types of invariants, string-like excitations and those obtained in Section 7.2.3 (see also *Remark* at the end of Section 7.2.3.2). The quantum dynamics here requires special investigation as the physical vacuum plays a key role (the Higgs phenomenon takes place). The invariant variables of Section 7.2.3 seem appropriate for a perturbation theory analysis when $\langle\rho\rangle_0 \neq 0$ (the residual $U(1)_e$ gauge symmetry leads to stable invariant structures describing charged particles accompanied by their Coulomb field). Then why are the other types of gauge-invariant excitations not dynamically favorable? This question is ultimately related to the question: what happens when fluctuations of ρ become large enough to reach the value $\rho = 0$? This question is addressed in extensive numerical lattice studies of phase transitions between the Higgs and non-Higgs phases (see, e.g. [412, 413]).

8
Supplementary material

8.1 A brief survey of the group theory

8.1.1 Basic definitions

An abstract set G is said to be a *group* if [20]:

(1) for any two elements g_1 and g_2 of G an associative product is defined $g_1 g_2 \in G$ (in general, the product is not commutative);
(2) there is an element $e \in G$, called the identity element, such that $eg = ge = g$ for any $g \in G$;
(3) for any $g \in G$ there is the inverse element g^{-1} such that $g^{-1}g = gg^{-1} = e$.

A group G is called *parametric* if elements of G can be parameterized by a set of (real) parameters:

$$g = g(t_1, t_2, ..., t_n), \tag{8.1}$$

i.e. $g(t_1, t_2, ..., t_n)$ defines a map from a subset of R^n onto G. The parameterization can be local, i.e. different coordinate charts can be used in different subsets of G. The group multiplication law $g_3 = g_1 g_2$ can now be expressed in the analytic form,

$$\mathbf{t}(g_3) = \mathbf{F}(\mathbf{t}(g_1), \mathbf{t}(g_2)),$$

where the vector function $\mathbf{t}$ has components $t_i(g)$, $i = 1, 2, ..., n$. A parametric group G is called a *Lie group* if the function $\mathbf{F}$ that defines the multiplication law is analytic in both the arguments, and the same also holds for the group inversion operation.

A *representation* of a group G on a vector space E is a continuous map,

$$g \to T_g \tag{8.2}$$

of G into the group of all linear transformations in E which preserves the group structure: $T_{g_1 g_2} = T_{g_1} T_{g_2}$, from which it follows that $T_e = 1$ and $T_{g^{-1}} = T_g^{-1}$. The mapping is assumed to be single valued. A representation $\hat{T}_g$ is called *dual* or *contragredient* to T_g if there exists a non-singular bilinear form,

$$(x, \hat{x}), \qquad x \in E, \qquad \hat{x} \in \hat{E} \tag{8.3}$$

invariant under the simultaneous action of T_g and $\hat{T}_g$,

$$(T_g x, \hat{T}_g \hat{x}) = (x, \hat{x}). \tag{8.4}$$

Non-singularity means that $\forall x \in E$ there exists $\hat{x} \in \hat{E}$ such that $(x, \hat{x}) \neq 0$, and also the converse: $\forall \hat{x} \in \hat{E}$ there exists $x \in E$ such that $(x, \hat{x}) \neq 0$. With a suitable basis choice, the form (8.3) can be diagonalized (i.e. the space E and its dual $\hat{E}$ become identical) and the condition (8.4) is equivalent to

$$T_g^T = T_g^{-1},$$

where T_g^T is the transpose of T_g.

A *Lie algebra* X is a vector space with a binary operation $[\cdot, \cdot] : X \times X \to X$ (the Lie bracket, or commutator) that is skew-symmetric: $[x_1, x_2] = -[x_2, x_1]$; bilinear: $[ax_1 + bx_2, x_3] = a[x_1, x_3] + b[x_2, x_3]$, where a and b are real or complex (or, generally, can be elements of any other field); and satisfies the Jacobi identity,

$$[x_1[x_2, x_3]] + [x_2[x_3, x_1]] + [x_3[x_1, x_2]] = 0.$$

Let $\{e_i\} \in X$, $i = 1, 2, ..., n = \dim X$ be a basis in X. Because of the bilinearity, the multiplication law is fully determined by the basis commutation relations,

$$[e_i, e_j] = f_{ijk} e_k. \tag{8.5}$$

The quantities f_{ijk} are called *the structure constants* of the Lie algebra X.

If a Lie algebra X is defined over the field of real or complex numbers, then any element of X can be expanded over the basis: $x = x^i e_i$ where the expansion coefficients x^i are real or complex. A Lie algebra is isomorphic to a real or complex Euclidean space (if x^i are real or complex, respectively).

A relation between Lie groups and Lie algebras is established in the Lie theorem: any Lie group G defines an associated real Lie algebra X_G. In the case of real matrix groups, it can be formulated via the exponential map, or the matrix exponent. The Lie algebra X_G consists of those matrices x for which $\exp(tx) \in G$ for all real numbers t [20]. In general, given a Lie group, a Lie algebra can be viewed as the tangent space to the group identity element.

Consider a one-form for a Lie group G,

$$\omega = g^{-1} dg, \tag{8.6}$$

where the differential dg can be defined by means of a parameterization (8.1) (local coordinates on the group manifold). If G is a matrix group, then dg is a matrix composed of differentials of elements of the matrix g. The form (8.6) is invariant under the left shifts $g \to hg$, $h \in G$. Similarly, the form

$$\tilde{\omega} = dg g^{-1}$$

is invariant under the right shifts $g \to gh$. The form (8.6) can be used to define a left-invariant metric on G,

$$ds^2 = \mathrm{Tr}\, \omega^2$$

(G is assumed to be a matrix group). It is easy to verify that

$$\mathrm{Tr}\, \omega^2 = \mathrm{Tr}\, \left(g^{-1} dg g^{-1} dg\right) = \mathrm{Tr}\, \left(dg g^{-1} dg g^{-1}\right) = \mathrm{Tr}\, \tilde{\omega}^2, \tag{8.7}$$

i.e. the left- and right-invariant metrics coincide.

The invariant metric on G can be used to define an invariant measure on G (the *Haar measure*) as the volume of an n-dimensional parallelepiped with adjacent sides being independent infinitesimal displacements in the matrix $\omega = \omega(g)$. This measure has the following properties [20]:

$$d\mu_{gg_0} = d\mu_{g_0 g} = d\mu_{g^{-1}} = d\mu_g.$$

The integral over the group G is defined by

$$\int_G d\mu_g.$$

The following theorem holds [216]: *Let dx be the Riemann measure on a Lie algebra X of a group G ($X \sim R^n$ and dx stands for the Riemann measure on the corresponding Euclidean space), and let $d\mu_g$ have the form dx in a neighborhood of the group identity, $(d\mu_g)_e = dx$. Then for any function $f(g)$ with compact support in the region* $\exp X \subset G$, *the following relation holds:*

$$\int_G f(g) d\mu_g = \int_X f(e^x) \det\left(\frac{1 - e^{-\operatorname{ad} x}}{\operatorname{ad} x}\right) dx.$$

The linear operator $\operatorname{ad} x$ acts in X by the rule $\operatorname{ad} x y = [x, y]$ for any $x, y \in X$ and is determined by the structure constants of the Lie algebra.

The action of the linear operator $\operatorname{ad} x$ on $y = y_j e_j$ can be represented as a square $n \times n$ matrix acting on the column whose entries are the components y_j. Indeed, using the commutation relation (8.5), one has $[x, y] = [e_i, e_j] x_i y_j = e_k (x_i f_{ijk} y_j)$, i.e. $y_j \to x_i f_{ikj} y_k \equiv (\operatorname{ad} x)_{jk} y_k$. For any two elements x and y of X, the quantity

$$(x, y) = \operatorname{Tr}(\operatorname{ad} x \operatorname{ad} y), \tag{8.8}$$

is called the *Cartan–Killing form* [21], where $\operatorname{ad} x$ and $\operatorname{ad} y$ are understood as the matrices $(\operatorname{ad} x)_{jk}$ and $(\operatorname{ad} y)_{jk}$, respectively.

8.1.2 A brief survey of Lie algebras

Consider a subset H of all mutually commutative elements of X. Obviously, H is a subalgebra of X (a subspace of X closed under the Lie bracket is called a *Lie subalgebra*); it is called the *Cartan subalgebra.* The dimension of the Cartan subalgebra $l = \dim H$ is called the *rank* of a Lie algebra.

A subalgebra I of a Lie algebra X is called an *ideal* in X if $[I, X] \subset I$. A Lie algebra is said to be *simple* if the commutator in it is not identically zero and all its ideals are the trivial one (zero) and the algebra itself. A direct sum of simple Lie algebras is called a *semisimple Lie algebra.* A Lie algebra in which the commutator is identically zero is called an Abelian Lie algebra. A Lie algebra is

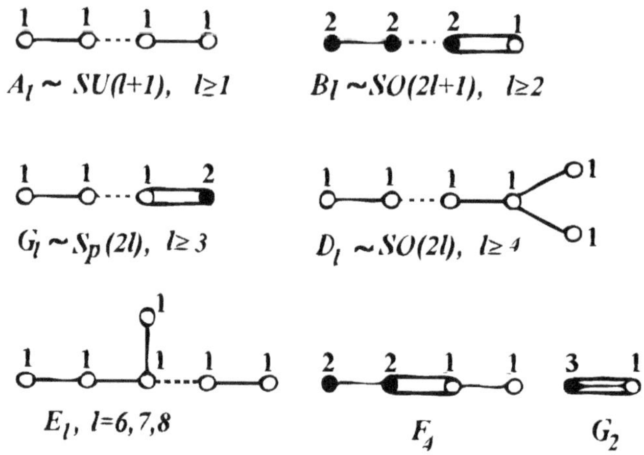

Fig. 8.1. The Dynkin diagrams.

said to be *reductive* if it is the direct sum of a semisimple and an Abelian Lie algebra. All simple Lie algebras are classified by means of the Dynkin diagrams. The theorem due to Dynkin asserts that *any simple Lie algebra is characterized by one of the diagrams shown in Fig. 8.1 where the number of circles equals the rank l of the algebra* [21].

Any simple Lie algebra can be reconstructed from the Dynkin diagram in the following way. A vector $\boldsymbol{\omega} \in R^l$ is assigned to each circle in the diagram. The length of $\boldsymbol{\omega}$ is given by the Euclidean norm $|\boldsymbol{\omega}| = (\boldsymbol{\omega}, \boldsymbol{\omega})^{1/2}$. The number over the circle indicates the square of a relative Euclidean norm of $\boldsymbol{\omega}$, i.e. $(\boldsymbol{\omega}, \boldsymbol{\omega})/(\boldsymbol{\omega}_0, \boldsymbol{\omega}_0)$ where $\boldsymbol{\omega}_0$ is a fixed vector. The links between any pair of circles denote the angles between the corresponding vectors:

○ • 90°, ○—• 120°, ○═• 135°, •≡○ 150°.

Thus, each Dynkin diagram defines a linearly independent set of l vectors in R^l. The span of these vectors ($\sim R^l$) is called the *root space* of the Lie algebra, whereas the $\boldsymbol{\omega}$ vectors are *simple roots* in it. Simple roots determine all roots of a Lie algebra. Any root is either positive or negative. A root $\boldsymbol{\alpha}$ is said to be positive if $\boldsymbol{\alpha} = \sum_{i=1}^{l} n_i \boldsymbol{\omega}_i$ where the decomposition coefficients n_i are non-negative integers. For every positive root $\boldsymbol{\alpha} > 0$ there is a negative root $-\boldsymbol{\alpha}$ ($n_i \to -n_i$) (and vice versa). The integers n_i are determined by the relation

$$-\frac{2(\boldsymbol{\alpha}, \boldsymbol{\beta})}{(\boldsymbol{\alpha}, \boldsymbol{\alpha})} = p + q, \tag{8.9}$$

which holds for any roots $\boldsymbol{\alpha}$ and $\boldsymbol{\beta}$ and where $p = \min n$ and $q = \max n$ in the series of roots $\boldsymbol{\gamma}_n = \boldsymbol{\beta} + n\boldsymbol{\alpha}$, i.e. $\boldsymbol{\gamma}_n$ is a root only for $n = p, p+1, ..., q-1, q$.

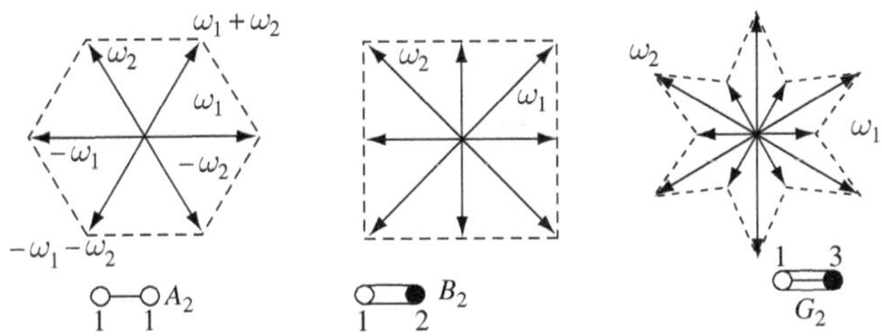

Fig. 8.2. The root systems for groups of rank 2.

Note that, if $\boldsymbol{\alpha}$ and $\boldsymbol{\beta}$ are simple roots, then $p = 0$, for $\boldsymbol{\alpha} - \boldsymbol{\beta}$ cannot be a root (for any root, all the n_i's are either non-negative or non-positive). When reconstructing the root system, one should keep in mind the fact that the multiplicity of any root is always one.

Example: A_2 $\quad {}^1 \circ\!\!-\!\!\circ\, {}^1$ (the algebra $SU(3)$). It follows from (8.9) that $-2(\boldsymbol{\omega}_1, \boldsymbol{\omega}_2) = q = 1$, therefore, $\boldsymbol{\omega}_1 + \boldsymbol{\omega}_2$ is a root. A simple analysis of the relation (8.9) shows that there are no other positive roots. The diagrams ${}^1\circ\!\!=\!\!\bullet{}^2$ and ${}^1\circ\!\!\equiv\!\!\bullet{}^3$ can be examined in a similar way. The root patterns of these algebras are shown in Fig. 8.2.

Equation (8.9) states that $4(\boldsymbol{\alpha}, \boldsymbol{\beta})^2/(\boldsymbol{\alpha}, \boldsymbol{\alpha})(\boldsymbol{\beta}, \boldsymbol{\beta}) = 4\cos^2\theta_{\alpha\beta}$ is an integer ($\theta_{\alpha\beta}$ is the angle between the roots $\boldsymbol{\alpha}$ and $\boldsymbol{\beta}$). Hence, the admissible angles between the roots obey the conditions,

$$\cos^2\theta_{\alpha\beta} = 0,\ \frac{1}{4},\ \frac{1}{2},\ \frac{3}{4}. \tag{8.10}$$

This implies, in particular, that the mutual orientation of all roots in the plane through any two roots $\boldsymbol{\alpha}$ and $\boldsymbol{\beta}$ coincides with one of the cases depicted in Fig. 8.2 (of course, there is also the case when $\boldsymbol{\alpha}$ is orthogonal to $\boldsymbol{\beta}$ and the plane contains only $\pm\boldsymbol{\alpha}$ and $\pm\boldsymbol{\beta}$). If the exceptional algebra G_2 is excluded from consideration, then all other root systems can be reconstructed from the root patterns of A_2 and B_2 (by finding all new roots in planes through any pair of known roots).

Once the root system is reconstructed, the whole algebra can be restored by means of the Cartan–Weyl basis. The root space is isomorphic to the Cartan subalgebra H in X (the maximal commutative subalgebra in X), i.e. with every simple root vector $\boldsymbol{\omega}_i \in R^l$, an element $\omega_i \in X$, also called a simple root of X, is associated so that all simple roots (as elements of X) commute among each other: $[\omega_i, \omega_j] = 0$. Any element of H is a linear combination of simple roots ω_i. Any simple algebra X can be decomposed into the orthogonal sum,

$$X = H \oplus \sum_{\alpha>0} \oplus\, (X_\alpha \oplus X_{-\alpha})\,,$$

where the sum is taken over all positive roots $\alpha > 0$, and $\dim X_{\pm\alpha} = 1$. Let a basis element of X_α be e_α and that of $X_{-\alpha}$ be $e_{-\alpha}$. Then any element x of X can be expanded in this basis,

$$x = \sum_{\omega} x_\omega \omega + \sum_{\alpha>0} \left(x_\alpha^+ e_\alpha + x_\alpha^- e_{-\alpha}\right), \tag{8.11}$$

where ω ranges over all simple roots. The basis elements satisfy the following commutation relations (the *Cartan–Weyl basis*):

$$[e_\alpha, e_{-\alpha}] = \alpha, \quad [h, e_\alpha] = (h, \alpha)e_\alpha, \quad [e_\alpha, e_\beta] = N_{\alpha,\beta} e_{\alpha+\beta}, \tag{8.12}$$

where (h, α) is the Cartan–Killing form (8.8), $h \in H$, and the numbers $N_{\alpha,\beta} = 0$ if $\alpha+\beta$ is not a root, otherwise $N_{\alpha,\beta}^2 = \frac{1}{2}q(1-p)(\alpha, \alpha)$, $p = \min(n)$ and $q = \max(n)$ in the root series $\beta + n\alpha$; $N_{\alpha,\beta} = N_{-\alpha,-\beta}$. The correct signs of the structure constants $N_{\alpha,\beta}$ are restored from the Jacobi identity for all the basis elements. Thus, any simple Lie algebra can be reconstructed from its Dynkin diagram. The procedure is illustrated with examples of $SU(2)$ and $SU(3)$ below.

The commutation relations (8.12) imply a definite choice of the norms of the elements $e_{\pm\alpha}$ relative to the Cartan–Killing form (8.8) [215, p. 167],

$$(e_{\pm\alpha}, e_{\pm\alpha}) = 0, \qquad (e_\alpha, e_{-\alpha}) = 1. \tag{8.13}$$

These relations can be proved by computing the action of $\operatorname{ad} e_{\pm\alpha}$ on (8.11) and then expressing the result as a linear transformation of components $x_\alpha^\pm$ and x_ω to find the matrix elements of $\operatorname{ad} e_{\pm\alpha}$ in the basis (8.12). Norms of simple roots are also fixed in (8.12). Consider, for instance, the Lie algebra of $SU(2)$. There is just one positive root ω. Let its squared norm be $\gamma = (\omega, \omega)$. The Cartan–Weyl basis reads $[e_\omega, e_{-\omega}] = \omega$ and $[\omega, e_{\pm\omega}] = \pm\gamma e_{\pm\omega}$. The number γ can be calculated in this basis. By definition $\gamma = (\omega, \omega) = \operatorname{Tr}(\operatorname{ad}\omega)^2$. In the Cartan–Weyl basis, the operator $\operatorname{ad}\omega$ is a 3×3 diagonal matrix with $0, \pm\gamma$ being its diagonal elements as follows from the basis commutation relations and the definition of the operator $\operatorname{ad}\omega$. Thus, $\operatorname{Tr}(\operatorname{ad}\omega)^2 = 2\gamma^2 = \gamma$, i.e. $\gamma = 1/2$.

The algebra of $SU(3)$ has two equal-norm simple roots $\boldsymbol{\omega}_1$ and $\boldsymbol{\omega}_2$ with the angle between them equal to $2\pi/3$. For the corresponding Cartan subalgebra elements, put $(\omega_1, \omega_1) = (\omega_2, \omega_2) = \gamma$ and $(\omega_1, \omega_2) = -\gamma/2$. The root system is given by six elements $\pm\omega_1, \pm\omega_2$ and $\pm(\omega_1 + \omega_2) \equiv \pm\omega_{12}$. It is readily seen that $(\omega_{12}, \omega_{12}) = \gamma$ and $(\omega_1, \omega_{12}) = (\omega_2, \omega_{12}) = \gamma/2$. All the roots have the same norm and the angle between two neighbor roots is equal to $\pi/3$. Having obtained the root pattern, the number γ can be calculated. The Cartan–Weyl basis consists of eight elements, $\omega_{1,2}$, $e_{\pm 1}$, $e_{\pm 2}$ and $e_{\pm 12}$ where simplified notations $e_{\pm\omega_1} \equiv e_{\pm 1}$, etc., are introduced. Similarly, $N_{12,-1} \equiv N_{\alpha,\beta}$ where $\alpha = \omega_{12}$ and $\beta = -\omega_1$, and so on. The operators $\operatorname{ad}\omega_{1,2}$ are 8×8 diagonal matrices as follows from (8.12) and $[\omega_1, \omega_2] = 0$. Hence, $\gamma = \operatorname{Tr}(\operatorname{ad}\omega_{1,2})^2 = 3\gamma^2$ and, therefore, $\gamma = 1/3$. Once root norms are established, the structure constants $N_{\alpha,\beta}$ can be found. For $SU(3)$, it follows from the root pattern and the definition of $N_{\alpha,\beta}$ (see the text below Eq. (8.12)) that $N_{1,2}^2 = N_{12,-1}^2 = N_{12,-2}^2 = 1/6$ and all others vanish (notice that $N_{\alpha,\beta} = -N_{-\alpha,-\beta}$ and $N_{\alpha,\beta} = -N_{\beta,\alpha}$). The latter determines the structure constants up to a sign. The transformation $e_\alpha \to -e_\alpha$, $N_{\alpha,\beta} \to -N_{\alpha,\beta}$ leaves the Cartan–Weyl commutation relations (8.12) unchanged. Therefore, only relative signs of the structure constants must be fixed. Fulfilling the Jacobi identity for elements e_{-1}, e_1, e_2 and e_{-2}, e_1, e_2 results in $N_{1,2} = -N_{12,-1}$ and $N_{1,2} = N_{12,-2}$, respectively. Now one can set $N_{1,2} = N_{12,-2} = -N_{12,-1} = 1/\sqrt{6}$, which completes determination of the structure constants for $SU(3)$.

It follows from (8.13), the basis (8.12) is not orthogonal with respect to the Cartan–Killing form. An orthonormal real basis in X is given by the set of elements c_α, s_α, and h_ω where

$$c_\alpha = (e_\alpha + e_{-\alpha})/\sqrt{2}\,, \qquad s_\alpha = i(e_\alpha - e_{-\alpha})/\sqrt{2}, \tag{8.14}$$

and h_ω are elements of an orthonormal basis in H (obtained by a suitable orthogonalization of the basis of simple roots) [215, p. 181]. The orthonormal basis commutation relations are

$$[h, s_\alpha] = i(h,\alpha)c_\alpha\,, \qquad [h, c_\alpha] = -i(h,\alpha)s_\alpha\,, \quad [s_\alpha, c_\alpha] = i\alpha, \tag{8.15}$$

where $h \in H$. The commutation relations amongst s_α and c_β ($\alpha \neq \beta$) are readily obtained from the third relation in (8.12). The structure constants in the real orthonormal basis are imaginary, while in the basis (8.12) they are real.

According to the theorem of Ado [414, 415], there is a matrix representation for any X. One can prove that $(x, y) = C_l \mathrm{Tr}\,(x_m y_m)$, where x_m and y_m are matrix representations of the abstract algebra elements x and y, respectively (in what follows the index m will be omitted), and C_l is a number that depends on X. For classical Lie algebras, the numbers C_l are listed in [215, pp. 187–190]. For example, $C_l = 2(l+1)$ for $SU(l+1)$. This relation is illustrated with examples of $SU(2)$ and $SU(3)$. Let the bases in matrix representations of the corresponding Lie algebras be realized by the Pauli and Gell–Mann matrices, respectively. The Pauli matrices (see, e.g. [416, p.17]) are 2×2 traceless, Hermitian matrices that satisfy the commutation relation $[\tau_a, \tau_b] = 2i\varepsilon_{abc}\tau_c$, and are normalized by the condition $\mathrm{Tr}\,(\tau_a\tau_b) = 2\delta_{ab}$, $a, b, c = 1, 2, 3$, where ε_{abc} is a totally skew-symmetric unit tensor, $\varepsilon_{123} = 1$. Hence, their Cartan–Killing norm is defined by $(\tau_a, \tau_b) = -4\varepsilon_{ab'c'}\varepsilon_{bc'b'} = 8\delta_{ab} = 4\mathrm{Tr}\,(\tau_a\tau_b)$ in full accordance with $C_l = 2(l+1)$, $l = 1$. In particular, one can set $\omega = \tau_3/4$, $s_\omega = \varphi\tau_1$, and $c_\omega = \varphi\tau_2$ where $1/\varphi = 2\sqrt{2}$. A similar analysis of the structure constants in the basis of the Gell–Mann matrices λ_a in $SU(3)$ [416, p.18] yields $\omega_1 = \lambda_3/6$, $s_1 = \varphi\lambda_1$, $c_1 = \varphi\lambda_2$, $\omega_2 = \left(\sqrt{3}\lambda_8 - \lambda_3\right)/12$, $s_2 = \varphi\lambda_6$, $c_2 = \varphi\lambda_7$, $\omega_{12} = \left(\sqrt{3}\lambda_8 + \lambda_3\right)/12$, $s_{12} = \varphi\lambda_5$, and $c_{12} = -\varphi\lambda_4$ where $1/\varphi = 2\sqrt{3}$. This choice is not unique. Actually, the identification of non-diagonal generators λ_a, $a \neq 3, 8$, with the elements (8.14) depends on a representation of the simple roots $\omega_{1,2}$ by the diagonal matrices $\lambda_{3,8}$. One could choose $\omega_1 = \lambda_3/6$ and $\omega_2 = -\left(\sqrt{3}\lambda_8 + \lambda_3\right)/12$, which would lead to another matrix realization of the elements (8.14).

For any $z \in X$, the adjoint action of the group element $U = e^z \in G$ on any $x \in X$ is defined by $\mathrm{ad}\,U(x) = \exp(\mathrm{ad}\,z)x$. In a matrix representation it has a more familiar form, UxU^{-1}. The Cartan–Killing form is invariant under the adjoint action of the group,

$$(\operatorname{ad} U(x), \operatorname{ad} U(y)) = (x, y). \tag{8.16}$$

In a real Cartan–Weyl basis, the adjoint action of the group is an orthogonal transformation. In a matrix representation this is a simple statement, $\operatorname{Tr}(UxU^{-1}UyU^{-1}) = \operatorname{Tr}(xy)$. A great advantage of the Cartan–Weyl basis is that it allows one to carry out computations *without* dealing with peculiarities of any particular representation of a Lie algebra.

Any Lie group contains a discrete finite subgroup W, called the *Weyl group*, which is generated by reflections in all the hyperplanes orthogonal to simple roots. Group W is isomorphic to a group of permutations of the roots, i.e. the group of linear transformations in H that preserves the root system. The fundamental domain in H with respect to W is called the *Weyl chamber*: $K^+ = H/W$. It is customary to choose K^+ to be the intersection of all positive half-spaces bounded by the hyperplanes orthogonal to simple roots (the "positivity" here is defined relative to the root orientation); the Weyl chamber is an open concave cone in H [129]. In other words, $(h, \omega) = h_i \omega_i > 0$ for any $h \in K^+$ and any simple root ω (here h_i and ω_i are components of h and ω, respectively, in an orthonormal basis in H).

Consider polynomials in H invariant under the action of the Weyl group. The Chevalley restriction theorem asserts [21] that any polynomial in H invariant under the action of the Weyl group can be uniquely extended to the whole X and the extended polynomial $p(x)$ is an invariant of the adjoint action of G on X, i.e. $p(x) = p(\Omega x \Omega^{-1})$, $\Omega \in G$. The following theorem classifies all invariant polynomials in X.

Theorem [20, 216]. *Any polynomial in the Lie algebra X of a simple group G, that is invariant under the adjoint action of G on X, is a polynomial of the generating invariant polynomials* $\operatorname{Tr} x^{r_i}$, $i = 1, 2, \ldots l = \operatorname{rank} G$, *where r_i are the degrees of independent Casimir operators in X* (see Table 8.1).

It follows from this theorem that any polynomial in H invariant under the action of the Weyl group is a polynomial of the generating independent polynomials $\operatorname{Tr} h^{r_i}$.

Table 8.1. *Degrees of Casimir operators*

Group	r_i $i = 1, 2, \ldots, l$	$N = \dim G$
$A_l \sim SU(l+1)$	$i+1$	$l(l+2)$
$B_l \sim SO(2l+1)$	$2i$	$l(2l+1)$
$C_l \sim Sp(2l)$	$2i$	$l(2l+1)$
$D_l \sim SO(2l)$	$2i,\ l\ (i \neq l)$	$l(2l-1)$
G_2	$2,\ 6$	14
F_4	$2, 6, 8, 12$	52
E_6	$2, 5, 6, 8, 9, 12$	78
E_7	$2, 6, 8, 10, 12, 14, 18$	133
E_8	$2, 8, 12, 14, 18, 20, 24, 30$	248

Invariant homogeneous polynomials in X define invariant symmetric tensors. For example, $\mathrm{Tr}\, x^n \equiv d_{i_1\cdots i_n} x^{i_1} \cdots x^{i_n}$,

$$d_{i_1\cdots i_n} = \mathrm{sym}\,\mathrm{Tr}\left(\lambda_{\{i_1} \cdots \lambda_{i_n\}}\right),$$

where the symmetrization, denoted by sym, is carried out for all indices in the curly brackets, $x = x^i\lambda_i$, and the set $\{\lambda_i\}$ is an orthonormal basis in a matrix representation of X. Clearly, any invariant symmetric tensor can be expanded into a tensor product over irreducible invariant symmetric tensors determined by the basis (independent) symmetric polynomials $\mathrm{Tr}\, x^{r_i}$, $i = 1, 2, \ldots, l = \mathrm{rank}\, X$. A similar statement holds for tensors in H invariant under the action of the Weyl group.

8.1.3 Properties of the measure density $\kappa^2(h)$

In Section 5.4 the measure density is introduced $\mu = \kappa^2(h)$. Its explicit form can be calculated using the Cartan–Weyl basis (8.12) [196, 200, 216]. The Cartan–Weyl basis is not orthogonal relative to the Cartan–Killing form. In particular, it follows from (8.11) and (8.13) that

$$(x, x) = (h, h) + \sum_{\alpha>0} x_\alpha^+ x_\alpha^-, \tag{8.17}$$

where $h = \sum_\omega x_\omega \omega \in H$ and $x - h \in X \ominus H$. In order to calculate the determinant of the matrix $\omega(h)$ in (5.76), the orthonormal basis (8.14) in $X \ominus H$ should be used,

$$x - h = \sum_{\alpha>0} \left(x_\alpha^c c_\alpha + x_\alpha^s s_\alpha\right).$$

Then the second term in (8.17) becomes

$$\sum_{\alpha>0} \left[\left(x_\alpha^c\right)^2 + \left(x_\alpha^s\right)^2\right].$$

Now the matrix elements $\omega_{\alpha\beta}(h)$ are easy to find. By definition (5.76), $[h, \lambda_\alpha] = -i\omega_{\alpha\beta}(h)\lambda_\beta$, where λ_α denote the elements of an orthonormal basis in $X \ominus H$. Put $\{\lambda_{l+1}, \lambda_{l+2}, \ldots, \ldots \lambda_N\} = \{c_{\alpha_1}, s_{\alpha_1}, c_{\alpha_2}, s_{\alpha_2}, \ldots\}$ (assuming that the positive roots are ordered in some way). It follows from (8.15) that the matrix $\omega_{\alpha\beta}(h) = i([h, \lambda_\alpha], \lambda_\beta)$ is block-diagonal, and each block is a 2×2 real skew-symmetric matrix $(h, \alpha)\varepsilon_{jk}$, where $\varepsilon_{jk} = -\varepsilon_{jk}$, $j, k = 1, 2$, and $\varepsilon_{12} = 1$, for every positive root α. Thus, up to a sign, one has

$$\kappa(h) = \left(\det \omega(h)\right)^{1/2} = \prod_{\alpha>0} (\alpha, h), \tag{8.18}$$

i.e. $\kappa(h)$ is a polynomial of h of degree $(N - l)/2$. Note that $N - l$ is always even (see Table 8.1). By the construction (5.75) and (5.76), the quantity $V_G V_H^{-1} \mu(h)$

is the volume of a gauge orbit through h [313]. Here V_G is the volume of the group manifold and V_H is the volume of the stationary group of h; $V_H = (2\pi)^l$ because the Cartan subgroup is isomorphic to $[\times U(1)]^l$.

It is not hard to infer from (8.18) that $\kappa(h)$ is a harmonic function, $\partial_i^2\kappa(h) = 0$, and, as a consequence, the quantum potential V_q vanishes,

$$V_q = \frac{1}{2\kappa}\partial_i^2\kappa = \frac{1}{2}\sum_{\alpha\neq\beta>0}\frac{(\alpha,\beta)}{(\alpha,h)(\beta,h)} = \frac{1}{2}\sum_P\sum_{\alpha\neq\beta\in P}\frac{(\alpha,\beta)}{(h,\alpha)(h,\beta)} = 0,$$

where the sum over all positive roots $\alpha \neq \beta > 0$ is decomposed into the sum over all $\alpha \neq \beta > 0$ in one plane P and the sum over all such planes. The sum over the roots in each plane P vanishes. This is verified by direct calculation, for the mutual orientation of the roots in P is known and is determined by (8.10), i.e. it is sufficient to verify the result for groups of rank $l = 2$: $SU(3)$, $Sp(4) \sim SO(5)$, and G_2. The same conclusion can also be established by geometric means (see [216], Theorem P.5.33).

8.2 Grassmann variables

8.2.1 Calculus on Grassmann algebras

Consider a set of n independent real elements θ_j, $j = 1, 2, ..., n$, of an algebra that anticommute with each other but commute with ordinary (complex) numbers,

$$\theta_i\theta_j = -\theta_j\theta_i, \quad \theta_i x = x\theta_i, \tag{8.19}$$

for all i, j, and any real or complex x. Setting $i = j$ in (8.19), one infers that $\theta_i^2 = 0$ (no summation over i). Consider 2^n homogeneous polynomials of θ_i,

$$\Theta^{(k)}_{j_1 j_2\cdots j_k} = \theta_{j_1}\theta_{j_2}\cdots\theta_{j_k}, \quad j_1 < j_2 < \cdots < j_k, \quad k = 0, 1, \ldots, n, \tag{8.20}$$

by definition $\Theta^{(0)} = 1$. Homogeneous polynomials of higher degrees identically vanish because of (8.19). The span of the basis set (8.20) over the real (or complex) numbers is called the *Grassmann algebra* G_n, whereas the elements θ_i are called *generating elements* of G_n or simply *generators*. Any element of G_n has the form

$$f(\theta) = \sum_{k=0}^{n} f^{(k)}_{j_1 j_2\cdots j_k}\Theta^{(k)}_{j_1 j_2\cdots j_k}, \tag{8.21}$$

where tensors $f^{(k)}$ are skew-symmetric in all indices with elements being real (or complex) numbers.

The functions

$$f_{\text{even}}(\theta) = \sum_{k=0}^{[n/2]} f^{(2k)}_{j_1 j_2 \cdots j_{2k}} \Theta^{(2k)}_{j_1 j_2 \cdots j_{2k}},$$

$$f_{\text{odd}}(\theta) = \sum_{k=0}^{[n/2]} f^{(2k+1)}_{j_1 j_2 \cdots j_{2k+1}} \Theta^{(2k+1)}_{j_1 j_2 \cdots j_{2k+1}}$$

are said to be *even* and *odd* elements of a Grassmann algebra, respectively, where $[n/2]$ denotes the integer part of a number $n/2$. Even elements commute with any element of G_n, while the odd ones anticommute among each other. Therefore, a *parity* ε can be defined in G_n. The parity of odd elements is set to be 1, while for the even ones it is 0: $\varepsilon(f_{\text{odd}}) = 1$ and $\varepsilon(f_{\text{even}}) = 0$. Because of the anticommutativity of odd elements, the square of any such element vanishes: $f^2_{\text{odd}}(\theta) = 0$.

The *left derivative* $\vec{\partial}/\partial\theta_j$ of the basis element $\Theta^{(k)}$ is the following element of G_n:

$$\frac{\vec{\partial}}{\partial\theta_j}\Theta^{(k)}_{j_1 j_2 \cdots j_k} = \delta_{jj_1}(\theta_{j_2}\theta_{j_3}\cdots\theta_{j_k}) - \delta_{jj_2}(\theta_{j_1}\theta_{j_3}\cdots\theta_{j_k}) + \cdots + (-1)^{k-1}\delta_{jj_k}(\theta_{j_1}\theta_{j_2}\cdots\theta_{j_{k-1}}). \quad (8.22)$$

In particular,

$$\frac{\vec{\partial}}{\partial\theta_j}1 = 0, \quad \frac{\vec{\partial}\theta_i}{\partial\theta_j} = \delta_{ij}, \quad \frac{\vec{\partial}}{\partial\theta_j}\theta_i\theta_k = \delta_{ij}\theta_k - \delta_{jk}\theta_i. \quad (8.23)$$

The element

$$\Theta^{(k)}_{j_1 j_2 \cdots j_k}\frac{\overleftarrow{\partial}}{\partial\theta_j} = (-1)^{k-1}\frac{\vec{\partial}}{\partial\theta_j}\Theta^{(k)}_{j_1 j_2 \cdots j_k}, \quad (8.24)$$

is called the *right derivative* $\overleftarrow{\partial}/\partial\theta_j$ of the basis element $\Theta^{(k)}$. The left and right derivatives of odd elements of G_n coincide, while they differ by the sign for even elements. By definition, the operation $\partial/\partial\theta_j$ (left or right) is linear and therefore the derivatives of a generic element of G_n is fully determined by (8.22) and (8.24). It is not difficult to verify that

$$\frac{\partial}{\partial\theta_j}\frac{\partial}{\partial\theta_i} = -\frac{\partial}{\partial\theta_i}\frac{\partial}{\partial\theta_j}.$$

Consider a set of generating elements $\{d\theta_j\}$, $j = 1, 2, \ldots, n$, such that

$$[d\theta_i, d\theta_j]_+ = [d\theta_i, \theta_j]_+ = [\theta_i, \theta_j]_+ = 0, \quad (8.25)$$

where $[\cdot,\cdot]_+$ denotes the anticommutator. By definition, put [305]

$$\int \theta_j d\theta_i = \delta_{ij}, \qquad \int d\theta_i = 0. \quad (8.26)$$

The rules (8.26) together with (8.25) define the *integral* of any element of the Grassmann algebra; multiple integrals are understood as iterated integrals (see also [417]). For example,

$$\int (\theta_{j_1} \cdots \theta_{j_n}) d\theta_1 \cdots d\theta_n = (-1)^P,$$

where $(-1)^P$ is the sign of the permutation (P is odd or even)

$$\begin{pmatrix} j_1 & j_2 & \cdots & j_n \\ n & n-1 & \cdots & 1 \end{pmatrix}.$$

Similarly to conventional calculus, there is an integration by parts in the Grassmann integral calculus,

$$\int f_1(\theta) \frac{\vec{\partial}}{\partial \theta_j} f_2(\theta) d\theta = \int f_1(\theta) \frac{\overleftarrow{\partial}}{\partial \theta_j} f_2(\theta) d\theta;$$
$$\int f_1(\theta) \left(f_2(\theta) \frac{\overleftarrow{\partial}}{\partial \theta_j} \right) d\theta = (-1)^{n+1} \int \left(\frac{\vec{\partial}}{\partial \theta_j} f_1(\theta) \right) f_2(\theta) d\theta,$$

where $d\theta = d\theta_1 d\theta_2 \cdots d\theta_n$ for brevity of notation. To prove these relations, it is sufficient to verify them for basis elements in place of $f_{1,2}$. In [231, 232] a general theory of integral calculus on Grassmann algebras was developed.

Consider a mixed algebra with several commuting and anticommuting generators. The former are denoted by $x_{i'}$, $i' = 1, 2, ..., n'$. A function $f(x, \theta)$ has the form (8.21) where the expansion coefficients $f^{(k)}$ are real (or complex) valued functions of $x_{i'}$. By definition, $[x_{i'}, \theta_i] = 0$; the differentials $dx_{i'}$ and derivatives $\partial / \partial x_{i'}$ also commute with all θ_i, $d\theta_i$, and the derivatives with respect to θ_i. The integral of $f(x, \theta)$ with respect to the Grassmann variables is defined by the rules (8.26), and the integration with respect to commuting variables is standard.

A *change of variables* is a parity preserving mapping of one set of generating elements to another one [27, 418],

$$x_{i'} = x_{i'}(y, \vartheta), \quad \theta_i = \theta_i(y, \vartheta). \tag{8.27}$$

The block matrix

$$R = \begin{pmatrix} A_{i'j'} & B_{i'i} \\ C_{ii'} & D_{ij} \end{pmatrix} = \begin{pmatrix} \frac{\partial x_{i'}}{\partial y_{j'}} & \frac{x_{i'} \overleftarrow{\partial}}{\partial \vartheta_i} \\ \frac{\partial \theta_i}{\partial y i'} & \frac{\vec{\partial} \theta_i}{\partial \vartheta_j} \end{pmatrix}$$

is associated with the change of variables (8.27). Apparently, elements of the matrices A and D are even, while elements of the matrices B and C are odd. Put

$$J = \det R \equiv \det(A - BD^{-1}C) \det D. \tag{8.28}$$

The quantity (8.28) is called the *superdeterminant* of the matrix R. The following relation holds:

$$\int f\Big(x(y,\vartheta),\theta(y,\vartheta)\Big)J(y,\vartheta)dyd\vartheta = \int f(x,\theta)dxd\theta,$$

where $d\vartheta, d\theta$ and dy, dx are integration measures for the Grassmann and ordinary (commuting) variables, respectively. A proof of this relation can be found in [418].

8.2.2 Quantum description of systems with Grassmann variables

Besides the Dirac quantum formalism given in Section 5.5.2, it is useful to have a specific representation of the algebra (5.100) which is required for the path integral formalism. A natural and convenient choice of such a representation is the following (the holomorphic representation). Let ξ_α and ξ_α^*, $\alpha = 1,2,\dots n$, be generators of a complex Grassmann algebra, $\xi_\alpha = \theta_1^\alpha + i\theta_2^\alpha$, $\xi_\alpha^* = \theta_1^\alpha - i\theta_2^\alpha$, where $(\theta_i^\alpha)^* = \theta_i^\alpha$ $(i = 1,2)$ are real generators. Any state $|\chi\rangle$ is represented by the Grassmann algebra element

$$\langle\xi|\chi\rangle = \sum_{k=0}^{n}\sum_{\{\alpha\}} \chi_{\alpha_1\cdots\alpha_k}\xi_{\alpha_1}^*\cdots\xi_{\alpha_k}^* = \chi(\xi^*),$$

and the conjugated state $\langle\chi|$ corresponds to the element

$$\langle\chi|\xi\rangle = \sum_{k=0}^{n}\sum_{\{\alpha\}} \chi_{\alpha_1\cdots\alpha_k}^*\xi_{\alpha_k}\cdots\xi_{\alpha_1} = \overline{\langle\xi|\chi\rangle} \equiv \chi^*(\xi). \tag{8.29}$$

The operator $\hat{\xi}_\alpha^\dagger$ is the multiplication operator in this representation, $\hat{\xi}_\alpha^\dagger\chi(\xi^*) = \xi_\alpha^*\chi(\xi^*)$, and the action of the operator $\hat{\xi}_\alpha$ is realized as the differentiation, $\hat{\xi}_\alpha\chi(\xi^*) = (\partial/\partial\xi_\alpha^*)\,\chi(\xi^*)$. The scalar product is defined in the following way:

$$\langle\varphi|\chi\rangle = \int \exp\left[-\sum_\alpha \xi_\alpha^*\xi_\alpha\right]\varphi^*(\xi)\chi(\xi^*)\prod_\alpha(d\xi_\alpha^* d\xi_\alpha) = \varphi_0^*\chi_0 + \varphi_\alpha^*\chi_\alpha + \cdots, \tag{8.30}$$

where the integration over a complex Grassmann algebra is defined analogously to (8.26),

$$\int d\xi = \int d\xi^* = 0, \quad \int \xi d\xi = \int \xi^* d\xi^* = 1. \tag{8.31}$$

The above construction provides a representation of the abstract algebraic relations (5.100)–(5.103).

If the integration with respect to ξ_α is carried out in Eq. (8.30), then the scalar product is written in the form,

$$\langle\varphi|\chi\rangle = \int \bar{\varphi}(\xi^*)\chi(\xi^*)\prod_\alpha d\xi_\alpha^*, \tag{8.32}$$

where [26, 419]

$$\bar{\varphi}(\xi^*) = P \int \exp\left[-\sum_\alpha \xi^*_\alpha \xi_\alpha\right] \varphi^*(\xi) \prod_\alpha d\xi_\alpha, \qquad (8.33)$$

and P is the parity of the permutation,

$$\prod_\alpha (d\xi^*_\alpha d\xi_\alpha) = P \prod_\alpha d\xi_\alpha \prod_\beta d\xi^*_\beta.$$

It is straightforward to verify that the multiplication and differentiation operations, ξ^*_α and $\partial/\partial\xi^*_\alpha$, are the Hermitian conjugates of one another with respect to the scalar product (8.30) (or (8.32) and (8.33)). The operation (8.33) is known as the Martin conjugation [305]. For example, if $n = 2$, one obtains

$$\begin{aligned}
\langle\xi|\chi\rangle &= \chi(\xi^*) = \chi_0 + \chi_\alpha\xi^*_\alpha + \chi_3\xi^*_1\xi^*_2, \\
\bar{\varphi}(\xi^*) &= \varphi^*_0\xi^*_1\xi^*_2 - \varphi^*_1\xi^*_2 + \varphi^*_2\xi^*_1 + \varphi^*_3 \equiv \langle\varphi|\xi\rangle, \\
\langle\varphi|\chi\rangle &= \int \bar{\varphi}(\xi^*)\chi(\xi^*) d\xi^*_2 d\xi^*_1 = \sum_{\alpha=0}^{3} \varphi^*_\alpha\chi_\alpha.
\end{aligned}$$

The variable ξ^* in equations like (8.32) can be viewed as a real Grassmann variable θ, for the formalism (8.32) uses only ξ^*, and the complex conjugation converts the amplitude (8.32) into

$$\begin{aligned}
\langle\varphi|\chi\rangle^* = \langle\chi|\varphi\rangle &= \int \bar{\varphi}^*(\xi)\chi^*(\xi) \prod_\alpha d\xi_\alpha = \int \bar{\varphi}^*(\xi^*)\chi^*(\xi^*) \prod_\alpha d\xi^*_\alpha \\
&= \int \bar{\chi}(\xi^*)\varphi(\xi^*) \prod_\alpha d\xi^*_\alpha.
\end{aligned}$$

The third equality follows from (8.31) (the properties of the algebras generated by ξ and ξ^* are identical), and the last equality is a consequence of (8.32) and (8.33). The variables should be distinguished in an algebra with involution, i.e. with generating elements $\{\xi, \xi^*\}$.

Let $O(\psi^*, \psi)$ be the kernel of an operator in the holomorphic representation. By definition, the action of the operator $\hat{O}$ on a vector $\chi(\psi^*)$ is given by the equality

$$\begin{aligned}
\hat{O}\chi(\psi^*) &= \int \prod_\alpha (d\psi'^*_\alpha d\psi'_\alpha) \exp\left[-\sum_\alpha \psi'^*_\alpha\psi'_\alpha\right] O(\psi^*, \psi')\chi(\psi'^*) \\
&= \int \prod_\alpha d\psi'^*_\alpha \bar{O}(\psi^*, \psi'^*)\chi(\psi'^*), \\
\bar{O}(\psi^*, \psi'^*) &= P \int \prod_\alpha d\psi'_\alpha \exp\left[-\sum_\alpha \psi'^*_\alpha\psi'_\alpha\right] O(\psi^*, \psi'); \qquad (8.34)
\end{aligned}$$

Eq. (8.34) defines the Martin conjugate of the kernel of the operator $\hat{O}$. It is easy to be convinced that the Martin conjugate of the kernel (6.77), defined according to (8.34), coincides with the kernel (6.74) because,

$$U_t(\boldsymbol{\theta},\boldsymbol{\theta}') = P\int \exp\left[-(\boldsymbol{\psi}'^*,\boldsymbol{\psi}')\right] U_t(\boldsymbol{\psi}^*,\boldsymbol{\psi}')d\psi_1' d\psi_2'\Big|_{\psi'^*\equiv\theta',\psi^*\equiv\theta}.$$

8.2.3 Solving classical equations of motion for mixed systems with constraints

As pointed out in Section 5.7, there are subtleties in interpreting the classical equations of constraints in systems with commuting and anticommuting variables. In fact, the issue is much broader and concerns also the interpretation of solutions of the equations of motion. With an example of the model (5.139) it will be shown here that the initial values of bosonic variables should be understood as even elements of the Grassmann algebra, not as conventional real (or complex) numbers. The Hamiltonian equations of motion are solved in two sets of variables related by canonical transformations under the assumption that the initial values of bosonic variables are ordinary numbers in both cases. The solutions obtained appear to be qualitatively distinct: in one set of canonical variables the bosonic and fermionic degrees of freedom become mixed in due course, while in the other set they evolve independently. The invariance of the Hamiltonian dynamics under canonical transformations can be restored, if the initial values of bosonic (commuting) variables are allowed to be generic even elements of the Grassmann algebra.

Let us turn to details. When solving the Hamiltonian equations of motion, put $\omega_1 = 0$ and $\omega_2 \equiv \omega$ in the Hamiltonian (5.140) (just for the sake of simplicity). Since the Poisson bracket of the constraint with the Hamiltonian vanishes, the constraint imposes a restriction on admissible initial conditions of the dynamical variables, i.e.

$$(\mathbf{p}_0, T\mathbf{x}_0) - \boldsymbol{\psi}_0^\dagger\Gamma\boldsymbol{\psi}_0 = 0, \tag{8.35}$$

where the subscript 0 refers to the canonical variables at initial time $t = 0$. Suppose that the vectors $\mathbf{p}_0$ and $\mathbf{x}_0$ have components being ordinary real numbers. Since the Grassmann algebra is a vector space with the basis (8.20), Eq. (8.35) implies two independent equations: $\sigma_B = (\mathbf{p}_0, T\mathbf{x}_0) = 0$ and $\sigma_F = \boldsymbol{\psi}_0^\dagger\Gamma\boldsymbol{\psi}_0 = 0$. Consider the Hamiltonian equations of motion $\dot{\theta} = \{\theta, H\}$, where $\theta(t)$ is any of the canonical variables $\mathbf{p}$, $\mathbf{x}$, $\boldsymbol{\psi}^\dagger$, or $\boldsymbol{\psi}$ with the initial value $\theta(0) = \theta_0$. The solution that obeys the conditions $\sigma_{B,F} = 0$ has the form,

$$\mathbf{x}(t) = (1 + t/t_0)\Omega_T(t)\mathbf{x}_0, \quad \mathbf{p}(t) = \Omega_T(t)\mathbf{x}_0\lambda_0 \tag{8.36}$$

$$\boldsymbol{\psi}(t) = \Omega_\Gamma(t)\exp(-i\omega t)\boldsymbol{\psi}_0, \quad \boldsymbol{\psi}^\dagger(t) = \Omega_\Gamma^\dagger(t)\exp(i\omega t)\boldsymbol{\psi}_0^\dagger, \tag{8.37}$$

where $\Omega_T(t) = \exp\left[-\int_0^t d\tau y(\tau) T\right]$, $\Omega_\Gamma(t) = \exp\left[i\int_0^t d\tau y(\tau)\Gamma\right]$, $t_0 = \text{const.}$, and $\lambda_0 = \text{const.}$ ($\mathbf{p}_0 = \lambda_0 \mathbf{x}_0$ is the solution of $\sigma_B = 0$). The generating elements $\boldsymbol{\psi}_0^\dagger$ and $\boldsymbol{\psi}_0$ of the Grassmann algebra are not independent because of the condition $\sigma_F = 0$. The general solution of the equation $\sigma_F = 0$ reads $\psi_{0\alpha}^\dagger = c_{\alpha\beta}\psi_{0\beta}$, where $c_{\alpha\beta} = c_{\beta\alpha}$ and the matrices c and Γ commute. The solutions (8.36) and (8.37) show that the bosonic and fermionic degrees of freedom evolve independently.

On the other hand, consider the following canonical transformation on the super phase space (see Section 1.10),

$$\mathbf{x} = \exp(T\varphi)\boldsymbol{\rho}, \quad \boldsymbol{\psi} = \exp(i\Gamma\varphi)\boldsymbol{\xi}, \tag{8.38}$$

$$p_r = (\mathbf{x}, \mathbf{p})/r, \quad p_\varphi = (\mathbf{p}, T\mathbf{x}) - \boldsymbol{\psi}^\dagger\Gamma\boldsymbol{\psi}, \quad \boldsymbol{\psi}^\dagger = \exp(-i\Gamma\varphi)\boldsymbol{\xi}^\dagger, \tag{8.39}$$

where $\boldsymbol{\rho} = (r, 0)$, and $r^2 = \mathbf{x}^2$. Since $\mathbf{p}^2 = p_r^2 + (\mathbf{p}, T\mathbf{x})^2/r^2$, the Hamiltonian in the new variables has the form,

$$H = \frac{1}{2}p_r^2 + \frac{\left(p_\varphi + \boldsymbol{\xi}^\dagger\Gamma\boldsymbol{\xi}\right)^2}{2r^2} + \omega\boldsymbol{\xi}^\dagger\boldsymbol{\xi} - y p_\varphi, \tag{8.40}$$

whereas the constraint (5.141) is equivalent to vanishing the momentum canonically conjugated to the angular variable φ: $p_\varphi = 0$. Therefore only the evolution of the variable φ depends on an arbitrary function $y(t)$ (φ is a non-physical variable),

$$\dot{\varphi} = \{\varphi, H\}, \quad \varphi(t) = \varphi_0 + \int_0^t d\tau \left[\frac{\sigma_F}{r^2(\tau)} - y(\tau)\right],$$

where, after computing the Poisson bracket, the constraint $p_\varphi = 0$ has been used to integrate the equation along with the fact that $\sigma_F = \boldsymbol{\xi}^\dagger\Gamma\boldsymbol{\xi}$ is an integral of motion: $\{\sigma_F, H\} = 0$. The Hamiltonian equations of motion for the remaining variables are

$$\dot{r} = p_r, \quad \dot{p}_r = \frac{\sigma_F^2}{r^3} \tag{8.41}$$

$$\dot{\boldsymbol{\xi}} = -i\left(\omega + \frac{\sigma_F}{r^2}\Gamma\right)\boldsymbol{\xi}, \quad \dot{\boldsymbol{\xi}}^\dagger = i\left(\omega + \frac{\sigma_F}{r^2}\Gamma\right)\boldsymbol{\xi}^\dagger,$$

subject to the initial conditions $\boldsymbol{\xi}^\dagger(0) = \boldsymbol{\xi}_0^\dagger$, $\boldsymbol{\xi}(0) = \boldsymbol{\xi}_0$ ($\boldsymbol{\xi}_0^\dagger$ and $\boldsymbol{\xi}_0$ are the generating elements of the Grassmann algebra), $r(0) = r_0$, and $p_r(0) = p_{r0}$. The quantity σ_F is nilpotent, $\sigma_F^3 = 0$. Hence, the functions $r(t)$ and $p_r(t)$ can be represented in the following form:

$$r(t) = R_0(t) + R_1(t)\sigma_F + R_2(t)\sigma_F^2 \tag{8.42}$$

$$p_r(t) = P_0(t) + P_1(t)\sigma_F + P_2(t)\sigma_F^2, \tag{8.43}$$

where R_ν and P_ν are ordinary real valued functions of time, $\nu = 0, 1, 2$. Equations for these are obtained by substituting (8.42) and (8.43) into (8.41),

$$\dot{R}_\nu = P_\nu, \tag{8.44}$$

$$\dot{P}_{0,1} = 0, \quad \dot{P}_2 = R_0^{-3}. \tag{8.45}$$

If once again it is assumed that the initial values of the bosonic variables r_0 and p_{r0} are ordinary (real) numbers, then $R_{1,2}(0) = 0$, $P_{1,2}(0) = 0$, $R_0(0) = r_0$, and $P_0(0) = p_{r0}$ are the initial conditions for Eqs. (8.44) and (8.45). The solution of this initial value problem reads,

$$r(t) = r_0 + p_{r0}t + \frac{\sigma_F^2}{2r_0^2 p_{r0}^2}\left(p_{r0}t - r_0 + \frac{r_0^2}{r_0 + p_{r0}t}\right), \tag{8.46}$$

$$p_r(t) = p_{r0} + \frac{\sigma_F^2}{2r_0^2 p_{r0}}\left(1 - \frac{r_0^2}{(r_0 + p_{r0}t)^2}\right), \tag{8.47}$$

$$\boldsymbol{\xi}(t) = \exp\left[-i\omega t - i\frac{\sigma_F}{r_0 p_{r0}}\left(1 - \frac{r_0}{r_0 + p_{r0}t}\right)\Gamma\right]\boldsymbol{\xi}_0,$$

i.e. the bosonic and fermionic variables become mixed in due course.

It is easy to see that substitution of the solutions (8.36) and (8.37) into (8.38) and (8.39) does not give the functions (8.46) and (8.47). Therefore the trajectories defined by (8.36) and (8.46), (8.47) are different. The reason is that these trajectories originate from distinct points of the super phase space, i.e. they correspond to different initial values of the canonical variables. Indeed, by inverting the relations (8.38) and (8.39) one infers that

$$\mathbf{x}(0) = \mathbf{x}_0 \equiv \exp(T\varphi_0)\boldsymbol{\rho}_0, \tag{8.48}$$

$$\mathbf{p}(0) = \mathbf{p}_0 \equiv \mathbf{x}_0\,\frac{p_{r0}}{r_0} + T\mathbf{x}_0\,\frac{\sigma_F}{r_0^2}. \tag{8.49}$$

The initial values (8.48) and (8.49) satisfy the constraint (8.35), while the components of $\mathbf{p}_0$ are even elements of the Grassmann algebra. The solution (8.36) is a particular case of the solution (8.46) and (8.47) which corresponds to an *additional* condition imposed on the initial values of the phase-space variables, $\sigma_F = 0$. The latter restricts the set of solutions of the constraint (8.35). This restriction does not correspond to the gauge symmetry of the model. It is a direct consequence of the assumption that initial values for the bosonic variables of the model should be understood as ordinary numbers rather than generic even elements of the Grassmann algebra. The example studied shows that such an assumption contradicts the theory of canonical transformations. If, in a particular set of canonical variables, the initial values of bosonic variables are set to be real numbers, then these initial values may well become generic even elements of the Grassmann algebra when another set of canonical variables is used. A priori there is no preference in the choice of canonical variables in the Hamiltonian mechanics. Therefore the initial values of bosonic variables in mixed systems should be treated as generic elements of a Grassmann algebra. As a consequence of this statement, one concludes that Eq. (5.141) (or (8.35)) cannot be interpreted as two independent equations $\sigma_{B,F} = 0$. In particular, the equality

$$\mathbf{p} = \lambda(t)\mathbf{x} + \frac{\sigma_F}{\mathbf{x}^2} T\mathbf{x}$$

determines a general solution of the constraint equation (5.141); here $\lambda(t)$ is function of time to be determined by the equations of motion. The solutions in Cartesian and "polar" coordinates, (8.38) and (8.39), coincide if $\lambda(t) = p_r(t)/r(t)$ where the functions $r(t)$ and $p_r(t)$ are given in (8.46) and (8.47), respectively.

8.3 Gaussian integrals, the Poisson summation formula, kernel Q_n, and Van Fleck determinant

8.3.1 Multi-dimensional Gaussian integrals

Integration variables in multi-dimensional Gaussian integrals occurring in practical applications may be real or complex, commuting or anticommuting (Grassmann). The following relations hold:

Real variables

$$\int [d\xi] \exp\left[-\frac{1}{2}\xi A\xi + \nu\xi\right] = c(\det A)^{-\sigma/2} \exp\left[\frac{\sigma}{2}\nu A^{-1}\nu\right], \tag{8.50}$$

where $\sigma = \pm 1$, the upper and lower signs refer to commuting (bosonic) and anticommuting (fermionic) variables, respectively; in Eq. (8.50) $[d\xi] = \prod_i d\xi_i$ (or $\prod_t d\xi(t)$ when a functional integral is considered) and the matrix (or the operator) A is symmetric for $\sigma = +1$ and skew-symmetric for $\sigma = -1$, and $c = (2\pi)^{(1-\sigma)N/4}$ (which is easily verified).

Complex variables

$$\int [d\xi^*][d\xi] \exp\left[-\xi^* A\xi + \nu^*\xi + \xi^*\nu\right] = c(\det A)^{-\sigma} \exp\left[\nu^* A^{-1}\nu\right].$$

Here A is Hermitian and anti-Hermitian in the bosonic and fermionic cases, respectively, and $c = \pi^{(1+\sigma)N/2}$. The cases of complex and real variables are related by the following transformations:

$$\xi^* A\xi = \frac{1}{2}\begin{pmatrix} \xi \\ \xi^* \end{pmatrix}\begin{pmatrix} 0 & \sigma A^T \\ A & 0 \end{pmatrix}\begin{pmatrix} \xi \\ \xi^* \end{pmatrix} \equiv \frac{1}{2}\Phi A\sigma\Phi, \qquad \Phi \equiv \begin{pmatrix} \xi \\ \xi^* \end{pmatrix};$$

$$\nu^*\xi + \xi^*\nu = \begin{pmatrix} \xi \\ \xi^* \end{pmatrix}\begin{pmatrix} 0 & \sigma \\ 1 & 0 \end{pmatrix}\begin{pmatrix} \nu \\ \nu^* \end{pmatrix} \equiv \Phi\Sigma\Theta, \qquad \Theta \equiv \begin{pmatrix} \nu \\ \nu^* \end{pmatrix}.$$

A general case (real fermionic and bosonic variables)

$$\int [d\xi] \exp\left[-\frac{1}{2}\xi Q\xi\right] = c\det Q_\psi (\det Q)^{-1/2}, \qquad \det Q = \det Q_\psi \det(B + bQ_\psi^{-1}b),$$

where

$$\xi \equiv \begin{pmatrix} \psi \\ \phi \end{pmatrix}, \quad Q = \begin{pmatrix} Q_\psi & b \\ -b & B \end{pmatrix},$$

ψ and ϕ denote the fermionic and bosonic components, respectively, and

$$[d\xi] = [d\psi][d\phi], \quad \xi Q \xi = \psi Q_\psi \psi + \psi b \phi - \phi b \psi + \phi B \phi,$$

whereas $\psi^2 = b^2 = 0$ and $\psi b = -b\psi$.

8.3.2 Derivation of Eq. (2.131)

The derivation of Eqs. (2.131) and (2.205) is based on the Poisson summation formula [141],

$$\sum_{m=-\infty}^{\infty} \delta(p - m\hbar) = \frac{1}{\hbar} \sum_{m=-\infty}^{\infty} \exp\left(i2\pi \frac{p}{\hbar} m\right). \tag{8.51}$$

The following chain of equalities that leads to Eq. (2.131) is easy to verify,

$$\begin{aligned}\sum_{m=-\infty}^{\infty} f(m\hbar) e^{im(\phi-\phi'')} &= \sum_{m=-\infty}^{\infty} \int_{-\infty}^{\infty} dp f(p) e^{ip(\phi-\phi'')/\hbar} \delta(p - m\hbar) \\ &= \sum_{m=-\infty}^{\infty} \int_{-\infty}^{\infty} \frac{dp}{\hbar} f(p) e^{ip(\phi-\phi'')/\hbar} e^{i2\pi pm/\hbar} \\ &= \sum_{m=-\infty}^{\infty} \iint_{-\infty}^{\infty} \frac{dp d\phi'}{\hbar} f(p) e^{ip(\phi-\phi')/\hbar} \delta(\phi' - \phi'' + 2\pi m).\end{aligned}$$

The second equality is a consequence of (8.51); the third is an identity. The first equality would also have been obvious if it did not assume restoration of an analytic function from its values at the integer points $p/\hbar = m$ (when reading this equality from right to left, it seems trivial). According to the theorem due to Carlson (see [420]), a function $f(\zeta)$ regular for $\mathrm{Re}\,\zeta > 0$ is uniquely determined by its values at the points $\zeta = m$, $m = 0, \pm 1, \pm 2, ...$, provided $|f(\zeta)| \le \exp(k|\zeta|)$, $k < \pi$. At first glance, the theorem is not applicable, for instance, to Eq. (2.203) because, according to (2.204), one should deal with functions of the form $\exp(im^2/2z)$ for which the conditions of the theorem are not fulfilled. However only the asymptote $\exp(im^2/2z) \approx 1 + im^2/2z$, $z \sim 1/\varepsilon$, as $\epsilon \to 0$, is required for the path integral derivation, i.e. only power functions are used.

8.3.3 Another derivation of Eq. (2.236) for the kernel Q_n

Up to a factor, the kernel Q_n in (2.236) is the n-dimensional δ-function in spherical coordinates. Here this is proved for the dimensions $n = 2$ and $n = 3$.

Two-dimensional space. Making use of the polar coordinates, one has,

$$\delta^{(2)}(\mathbf{x}-\mathbf{x}') = \delta(r\cos\phi - r'\cos\phi')\delta(r\sin\phi - r'\sin\phi'). \tag{8.52}$$

The first δ-function has support at $r' = r\cos\phi/\cos\phi'$. Substituting r' into the argument of the second δ-function and pulling the factor $r\cos\phi$ out, (i.e. using the well-known formula $\delta(ax) = |a|^{-1}\delta(x)$), Eq. (8.52) is rewritten in the form,

$$\delta^{(2)}(\mathbf{x}-\mathbf{x}') = |r\cos\phi|^{-1}\delta(r\cos\phi - r'\cos\phi')\delta(\tan\phi - \tan\phi'). \tag{8.53}$$

Solutions to the equation $\tan\phi = \tan\phi'$ are $\phi = \phi' + \pi n$, $n = 0, \pm 1, \ldots$. With the help of the formula

$$\delta(f(x)) = \sum_n \frac{\delta(x - x_n)}{|f'(x_n)|}, \quad f(x_n) = 0, \tag{8.54}$$

the right-hand side of Eq. (8.53) can be transformed to

$$\frac{1}{|r|}\delta\left(r - r'\frac{\cos\phi'}{\cos\phi}\right)\sum_{n=-\infty}^{\infty}\delta(\phi - \phi' + \pi n). \tag{8.55}$$

The sum in (8.55) is split into the sums over even and odd integers,

$$\sum_{n=-\infty}^{\infty}\left[\delta(\phi - \phi' + 2\pi n) + \delta(\phi - \phi' + 2\pi n + \pi)\right] = Q(\phi-\phi') + Q^{\pi}(\phi-\phi'). \tag{8.56}$$

It follows then from $\cos(\phi + \pi n) = (-1)^n\cos\phi$ that

$$\begin{aligned}\delta^{(2)}(\mathbf{x}-\mathbf{x}') &= |r|^{-1}\left[\delta(r - r')Q(\phi - \phi') + \delta(r + r')Q^{\pi}(\phi - \phi')\right] \\ &= \frac{1}{|\sqrt{g}|}Q_2(r,\phi;r',\phi'),\end{aligned}$$

where $Q_2(r,\phi;r',\phi')$ coincides with (2.197); the factor at Q_2 can also be represented as $|(gg')^{-1/4}|$.

In the case of the n-dimensional δ-function, the procedure remains essentially the same. Therefore only key points are highlighted for $n = 3$. One begins with the analog of (8.52),

$$\begin{aligned}\delta^{(3)}(\mathbf{x}-\mathbf{x}') &= \delta(r\sin\theta\cos\phi - r'\sin\theta'\cos\phi') \\ &\quad\times\delta(r\sin\theta\sin\phi - r'\sin\theta'\sin\phi')\delta(r\cos\phi - r'\cos\phi'). \end{aligned}\tag{8.57}$$

The support of the first δ-function is determined by the equality $r'\sin\theta' = r\sin\theta\cos\phi/\cos\phi'$ which is substituted into the argument of the second δ-function so that the latter becomes,

$$\frac{1}{|r\sin\theta\cos\phi|}\delta(\tan\phi - \tan\phi'),$$

in which, according to (8.53)–(8.56),

$$\delta(\tan\phi - \tan\phi') = \cos^2\phi\left[Q(\phi - \phi') + Q^{\pi}(\phi - \phi')\right]. \tag{8.58}$$

Next, the support of the last δ-function in (8.57), $r' = r\cos\theta/\cos\theta'$, is substituted into the argument of the first one, and the factors $r\cos\theta$ and $\cos\theta$ are then pulled out from the first and third δ-functions. The result reads,

$$\delta^{(3)}(\mathbf{x}-\mathbf{x}') = \frac{1}{|r^2\sin\theta|\cos^2\theta}\delta\left(r - r'\frac{\cos\theta'}{\cos\theta}\right)$$
$$\times\,\delta\left(\tan\theta - \tan\theta'\frac{\cos\phi'}{\cos\phi}\right)\left[Q(\phi-\phi') + Q^{\pi}(\phi-\phi')\right].$$

Making use of Eq. (8.58) once again, one infers that

$$\delta^{(3)}(\mathbf{x}-\mathbf{x}') = \frac{1}{|r^2\sin\theta|}\left[\delta(r-r')Q_2(\theta,\phi;\theta',\phi') + \delta(r+r')Q_2^{\pi}(\theta,\phi;\theta',\phi')\right]$$
$$= \left|g^{-1/2}\right| Q_3(r,\theta,\phi;r',\theta',\phi').$$

Derivation of the corresponding formulas (2.234)–(2.236) for an arbitrary n is now straightforward.

8.3.4 The Van Fleck determinant

Here an alternative derivation of the pre-exponential in Eq. (2.106) is given following the ideas of [115]. Consider an ensemble of particles with a uniform constant density in the phase space. The number dN of particles in an infinitesimal phase space volume is proportional to the volume,

$$dN = C\,dp'dq', \quad C = \text{const}.$$

The number of particles in volume dq at time t that were in volume dq' at time $t' < t$ is given by

$$dN = C\det\left\|\frac{\partial p'}{\partial q}\right\| dqdq', \tag{8.59}$$

where $p_i' = -\partial S(q,q')/\partial q'^i$ according to (1.50) and (2.91), for the transition $q', p' \to q, p$ is a canonical transformation with the generating function being the classical action $S(q,q') = -F_1(q',q)$ in accordance with the Hamilton–Jacoby theory (Section 1.10.1).

In the quantum mechanical description of the system, the space is partitioned into cells $\Delta q'$, and it is assumed that only one particle occupies each cell so that the wave function of a particle is localized over one cell and the cell volume is small enough. Then one has,

$$\psi_t(q) = \int U_{t-t'}(q,q')\psi_{t'}(q')\sqrt{g'}dq' \approx U_{t-t'}(q,q')\psi_{t'}(q')\sqrt{g'}\Delta q'. \tag{8.60}$$

Taking into account the normalization condition

$$\int |\psi_{t'}(q')|^2\sqrt{g'}dq' \approx |\psi_{t'}(q')|^2\sqrt{g'}\Delta q' = 1,$$

and relation (8.60), one obtains,

$$dN = |\psi_t(q)|^2\sqrt{g}\Delta q = |U_{t-t'}(q,q')|^2|\psi_{t'}(q')|^2|\sqrt{g'}\Delta q'|^2\sqrt{g}\Delta q$$
$$= |U_{t-t'}(q,q')|^2\sqrt{g'}\Delta q'\sqrt{g}\Delta q. \tag{8.61}$$

Comparison of (8.59) with (8.61) yields,

$$(gg')^{1/2}|U_{t-t'}(q,q')|^2 = |C|^2 \det \| -\partial^2 S/\partial q^j \partial q'^i\| \equiv |C|^2 D,$$

where C is a constant, i.e.

$$|U_{t-t'}(q,q')| = |C|(gg')^{-1/4}D^{1/2}.$$

Constant C is determined by the initial condition, $U_{t-t'}(q,q') \to \delta(q,q')$ (being the unit operator kernel) as $t \to t'$ [142] and equals $(2\pi i\hbar)^{-n/2}$.

8.4 Elimination of gauge arbitrariness and residual gauge transformations

8.4.1 Recipes to eliminate gauge arbitrariness

In order to remove the divergence of the path integral measure in (6.2) as well as to account for the constraints, the following procedure was proposed to develop the path integral formalism for gauge systems [171].

Let all the constraints φ_α, $\alpha = 1,2,...,M$ be of the first class. They are assumed to be independent. Consider M supplementary conditions,

$$\chi_\alpha(p,q) = 0, \tag{8.62}$$

such that

$$\det|\{\chi_\alpha, \varphi_\beta\}| \neq 0, \tag{8.63}$$
$$\{\chi_\alpha, \chi_\beta\} = 0. \tag{8.64}$$

Then the Hamiltonian mechanics of the original constrained system is equivalent to the conventional unconstrained Hamiltonian mechanics with $N - M$ degrees of freedom and the physical phase space defined by the equations $\varphi_\alpha = \chi_\alpha = 0$.

Canonical variables of the *reduced* system can be obtained in the following way. Thanks to the conditions (8.64), the new canonical variables can be chosen so that χ_α coincide with the first M generalized coordinates,

$$q = (\chi_\alpha, q^*), \tag{8.65}$$

where q^* denotes the remaining set of the generalized coordinates. Let

$$p = (p_\alpha, p^*) \tag{8.66}$$

be the corresponding canonical momenta. Because of the invariance of the Poisson brackets with respect to canonical transformations (8.65) and (8.66), condition (8.63) in the new canonical variables becomes

$$\det\left|\frac{\partial\varphi_\alpha}{\partial p_\beta}\right| \neq 0.$$

Therefore the constraints $\varphi_\alpha(p,q) = 0$ can be solved for p_α. As a result the reduced phase space is determined by the equations

$$\chi_\alpha \equiv q_\alpha, \qquad p_\alpha = p_\alpha(p^*, q^*), \tag{8.67}$$

where q^* and p^* are the canonically conjugated variables. The Hamiltonian of the reduced system has the form,

$$H^* = H^*(p^*, q^*) = H(p,q)\Big|_{\varphi=0,\chi=0},$$

where $H(p,q)$ is the Hamiltonian of the original constrained system.

The equivalence of the reduced dynamics to the original means the following. Consider the Hamiltonian equations of motion of the original system,

$$\dot{p} = \{p, H_T\} = -\frac{\partial H}{\partial q} - y_\alpha\frac{\partial\varphi_\alpha}{\partial q}, \tag{8.68}$$

$$\dot{q} = \{q, H_T\} = \frac{\partial H}{\partial p} + y_\alpha\frac{\partial\varphi_\alpha}{\partial p}, \tag{8.69}$$

where $y_\alpha = y_\alpha(t)$ are arbitrary functions of time and

$$H_T = H + y_\alpha\varphi_\alpha \tag{8.70}$$

is the generalized Hamiltonian of the constrained system (see Chapter 3). Solutions to the equations of motion (8.68) and (8.69) compatible with the equations of constraints $\varphi_\alpha = 0$ contain arbitrary functions y_α whose variations determine gauge transformations of $p_\alpha(t)$ and $q_\alpha(t)$. The supplementary conditions $\chi_\alpha = 0$ eliminate this arbitrariness by fixing y_α and defining thereby a gauge condition (see Section 3). Indeed, since they must hold at any moment of time, then $\dot{\chi}_\alpha = \{\chi_\alpha, H_T\} = 0$ or $\{\chi_\alpha, H\} = -y_\beta\{\chi_\beta, \varphi_\alpha\}$ which can be solved for y_α because the matrix $\{\varphi_\beta, \chi_\alpha\}$ is not singular (compare (8.63)). Substitution of y_α (as the functions of canonical variables), relations (8.67), and $\chi_\alpha = 0$ into (8.68) and (8.69) yields the equations of motion for the variables p^* and q^*. They coincide with the Hamiltonian equations of motion generated by the Hamiltonian H^* [171],

$$\dot{q}^* = \{q^*, H^*\} = \frac{\partial H^*}{\partial p^*}; \quad \dot{p}^* = \{p^*, H^*\} = -\frac{\partial H^*}{\partial q^*}.$$

It is then proposed [171] that the quantum mechanical transition amplitude between two points, q'^* and q^*, of the $(N-M)$-dimensional physical configuration space be determined by the following Hamiltonian path integral:

$$U_t(q^*, q'^*) = \int \prod_{\tau=0}^{t}\left(\frac{dp^*(\tau)dq^*(\tau)}{(2\pi)^{N-M}}\right)\exp\left\{i\int_0^t d\tau\Big[p^*\dot{q}^* - H^*(p^*, q^*)\Big]\right\} \tag{8.71}$$

with the boundary conditions $q^*(0) = q'^*$ and $q^*(t) = q^*$.

In practice, constraints are not always easy to solve. Therefore a representation of the path integral in the original canonical variables is desired. The amplitude (8.71) can be transformed to the following form:

$$U_t(q^*, q'^*) = \int \prod_{\tau=0}^{t} \left[\frac{dpdq}{(2\pi)^N} \det \|\{\varphi_\alpha, \chi_\beta\}\| \prod_\alpha \delta(\chi_\alpha) \frac{dy_\alpha}{2\pi} \right] e^{iS_T}, \tag{8.72}$$

$$S_T = \int_0^t d\tau \Big[p\dot{q} - H_T(p, q, y) \Big]. \tag{8.73}$$

Indeed, after integration with respect to y_α, the measure of the path integral contains the factor

$$\prod_{\tau=0}^{t} \left[\det \|\{\varphi_\alpha, \chi_\beta\}\| \prod_\alpha \delta(\varphi_\alpha)\delta(\chi_\alpha) \right] . \tag{8.74}$$

When written in new canonical variables, this factor becomes

$$\prod_{\tau=0}^{t} \left[\det \|\frac{\partial \varphi_\alpha}{\partial p_\beta}\| \prod_\alpha \delta(\varphi_\alpha)\delta(q_\alpha) \right] = \prod_{\tau=0}^{t} \left[\prod_\alpha \delta(q_\alpha)\delta\Big(p_\alpha - p_\alpha(q^*, p^*)\Big) \right] . \tag{8.75}$$

As a result of the integration with respect to p_α and q_α, the path integral (8.72) turns into (8.71).

Remark. Suppose that the gauge transformations of canonical variables are linear (the constraints generating them are bilinear in canonical coordinates and momenta (compare (3.66)). Gauge conditions are imposed on the canonical coordinates only, $\chi_\alpha = \chi_\alpha(q)$. The identity (6.183) in which $\Delta_{FP} = \det \|\{\chi_\alpha, \varphi_\beta\}\|$ can be inserted into the formal path integral measure $\prod_t (dpdqdy)$ for the theory defined by the action (8.73). A change of variables $q \to T(\omega)q$, where $T(\omega)q$ is a gauge transformation of q, and similarly for p is then carried out. Since a gauge transformation is a canonical transformation, the measure $dpdq$ is invariant, while the action S_T is modified only by a possible linear transform of the Lagrange multipliers y (compare Section 3.3.1). By a suitable change of variables in the integral over y, this modification of S_T can be eliminated. The resulting integral coincides with (8.72) modulo the (divergent) factor $\int \prod_t dg(\omega)$. So, the formal procedure of inserting the identity (6.183) often used to "divide out" the divergence of the path integral in gauge theories is equivalent to the postulate (8.71).

The transition amplitude (8.71) corresponds to the quantum theory which is obtained by the canonical quantization of the classical theory derived from the original gauge theory by elimination of all non-physical degrees of freedom. Physical degrees of freedom are not generally described by Cartesian coordinates, while the operations of the canonical quantization and general coordinate transformations do not commute. For this reason the quantum theory obtained by canonical quantization after the elimination of non-physical degrees of freedom might differ from that defined by the Dirac quantization procedure when non-physical degrees of freedom are eliminated after quantization (see Section 4.1). Multiple examples illustrating this statement are studied in Chapter 5. The procedure of eliminating non-physical degrees of freedom (8.71)–(8.75) is not generally acceptable for the following reasons. First, it does not take into account a possible nontrivial structure of the physical phase space. Second, without additional modifications, i.e. as it stands, the procedure is not gauge invariant. Let us elucidate this in more detail.

The path integral (8.71) is well defined if the integration with respect to p^* and q^* is carried out in the infinite limits, i.e. a priori the physical phase space is assumed to be Euclidean, $\Gamma(p^*, q^*) = R^{N-M}(p^*) \times R^{N-M}(q^*)$. However in Chapter 5 it is shown that such an assumption is not justified at all – the physical phase space often has a nontrivial geometrical structure in gauge theories. Therefore the problem of an appropriate modification of the path integral to take into account the physical phase space structure arises. It should be noted that this problem cannot be solved by a mere reduction of the integration domain in (8.71) (for example, by reducing the plane to a half-plane in the case of the conic phase space). In particular, the transition amplitude for a free particle or harmonic oscillator with conic phase space can easily be found by the operator formalism. However the Gaussian integral (8.71) over the conic phase space (actually, over the half-plane $q^* \geq 0$) cannot be evaluated analytically at all even for the free particle or harmonic oscillator system.

Lastly, one should prove that the amplitude (8.71) is independent of the choice of supplementary conditions (gauge) χ_α. In the classical theory, two reduced theories corresponding to two different χ_α's are related by a canonical transformation. The gauge invariance demands that the quantum theories defined by the path integral (8.71) for two different χ_α's should be unitary equivalent, which is not generally the case, for the operation of canonical transformations and that of the canonical quantization does not commute (see Section 2.6.4). Analysis of the operator formalism given in Section 5.3 shows that the quantum theories corresponding to two different choices of physical canonical variables are unitary equivalent if and only if the curvilinear coordinates associated with both the gauge transformation law and the gauge condition chosen are used to eliminate non-physical degrees of freedom in the original theory quantized by the Dirac method. In the quantum theory defined by the path integral (8.71), this requirement is not fulfilled because, as already pointed out above, the non-physical degrees of freedom are removed prior to quantization. Hence, the amplitude (8.71) may generally depend on the choice of gauge condition, leading to all undesired consequences (the evolution operator defined by the kernel (8.71) may not even be unitary and so on).

Example. The above general statements can be illustrated with an example of the model from Section 5.3. With the help of the canonical transformation (5.50) non-physical degrees of freedom can be eliminated so that the Hamiltonian dynamics of the physical variables $(p_u, u) \equiv (p^*, q^*)$ is determined by the

Hamiltonian $H^*(p_u, u) = H_{ph}(p_u, u)$, which is defined by Eq. (5.52). According to the conventional definition of the path integral (see Chapter 2), Eq. (8.71) means the convolution of $N+1$ infinitesimal evolution operator kernels, or, in the operator form, $\hat{U}_t = \hat{U}_\varepsilon \cdots \hat{U}_\varepsilon$, where $\varepsilon = t/(N+1)$, $N \to \infty$ and $\varepsilon \to 0$, whereas $t = (N+1)\varepsilon$ remains fixed. By the path integral definition, the kernel of the operator $\hat{U}_{2\varepsilon} = \hat{U}_\varepsilon \hat{U}_\varepsilon$ has the form,

$$\begin{aligned} U_{2\varepsilon}(u, u') &= \int_{-\infty}^{\infty} du'' U_\varepsilon(u, u'') U_\varepsilon(u'', u'), \\ U_\varepsilon(u, u') &= \int_{-\infty}^{\infty} \frac{dp_u}{2\pi} \exp\left[ip_u(u - u') - i\varepsilon H^*(p_u, u)\right] \\ &= \left\{1 - i\varepsilon\left[-g(u)\partial_u^2 + V(u)\right]\right\} \int_{-\infty}^{\infty} \frac{dp_u}{2\pi}\, e^{ip_u(u-u')} + O(\varepsilon^2), \end{aligned} \tag{8.76}$$

where $g(u) = r^2/\mu^2$ (see (5.52)). On the other hand, by the definition of the evolution operator kernel,

$$U_\varepsilon(u, u') = \langle u| \exp\left(-i\varepsilon \hat{H}^*\right) |u'\rangle = [1 - i\varepsilon H^*(\hat{p}_u, \hat{u})] \langle u|u'\rangle + O(\varepsilon^2), \tag{8.77}$$

where $\langle u|u'\rangle = (2\pi)^{-1} \int_{-\infty}^{\infty} dp_u \exp[ip_u(u - u')] = \delta(u - u')$ is the unit operator kernel. By comparing (8.76) and (8.77), the quantum Hamiltonian $\hat{H}^*$ is found that corresponds to the path integral (8.71) (this can always be done because the path integral uniquely defines the quantum Hamiltonian and vice versa (see Chapter 2)),

$$\hat{H}^* = H^*(\hat{p}_u, \hat{u}) = \frac{1}{2}\, g(\hat{u})\hat{p}_u^2 + V(\hat{u}), \qquad \hat{p}_u = -i\partial_u. \tag{8.78}$$

It is easy to see that the Hamiltonian (8.78) does not coincide with the Hamilton operator in the Schrödinger equation (5.53), which leads to a gauge-dependent quantum theory as is shown in Section 5.3.3. More to the point, the operator (8.78) is not Hermitian $(\hat{H}^*)^\dagger \neq \hat{H}^*$ because of the non-commutativity of $g(u)$ and $\hat{p}^2$. Therefore the evolution operator $\hat{U}_t = \exp(-it\hat{H}^*)$ is not unitary.

If supplementary conditions $\chi_\alpha = 0$ do not completely eliminate the gauge arbitrariness, then the gauge is said to be *ambiguous*. The physical phase space structure and gauge fixing ambiguities appear to be related in the path integral formalism. In Section 5.3 it is shown that after eliminating all non-physical variables by means of fixing a gauge there remains a gauge arbitrariness associated with discrete gauge transformations which depends on the gauge condition chosen. The configurations related by these residual gauge transformations correspond to the same physical state of the system and, therefore, all but one such configurations should be removed from the path integral (the sum over physical configurations) in order to avoid a "double counting".

In [209] it is proposed that the gauge fixing ambiguity problem can be resolved by restricting the integration domain in the path integral to the region defined by the condition $\det \Delta_{FP} > 0$ where $(\Delta_{FP})_{\alpha\beta} = \{\chi_\alpha, \varphi_\beta\}$ is the Faddeev–Popov (operator) matrix. As demonstrated with multiple examples in Chapter 6, a formal restriction of the integration domain to remove gauge equivalent configurations on the gauge fixing surface defined by $\chi_\alpha = 0$ does not make any sense in the path integral. Besides the path integral formalism, the condition $\det \Delta_{FP} > 0$ cannot even be considered as a general method to solve the gauge fixing ambiguity problem that is suitable for any gauge.

Consider the model studied in Section 5.3 and set the gauge condition $x_2 = 0$ if $x_1 < 0$ and $x_2 = f(x_1) = x_1(x_1 - a)$ if $x_1 \geq 0$. In this case $\Delta_{FP} = \{\chi, \sigma\} = \{x_2 - f(x_1), (\mathbf{p}, T\mathbf{x})\} = x_1 + f'f = x_1\left(2x_1^2 - 3ax_1 + a^2 + 1\right)$. The equation $\Delta_{FP} = 0$ defines the points at which the gauge orbit and gauge fixing curve $x_2 = f(x_1)$ are tangent (their tangent vectors are parallel); its solutions are $x_1^{(1,2)} = [3a \pm (a^2 - 8)^{1/2}]/4$, and $x_1^{(3)} = 0$. Real non-trivial solutions exist only if $a^2 \geq 8$. In Fig. 5.3 they are the points at which curve 3 is tangent to the circles $S(r_1)$ and $S(r_0)$. The function $\Delta_{FP} = 2x_1\left(x_1 - x_1^{(1)}\right)\left(x_1 - x_1^{(2)}\right)$ is negative in the open interval $\left(x_1^{(1)}, x_1^{(2)}\right)$. The removal of this interval, which corresponds to the motion between points B and B' in Fig. 5.3, does not solve the gauge fixing ambiguity because the interval $r \in (r_0, r_1)$ would be counted twice along the remaining portion of the gauge fixing curve: $(AB) \sim (r_0, r_1)$ and $(B'C) \sim (r_0, r_1)$. For curve 2 the same fact is illustrated by a symbolic contour of the radial variable $r = |\mathbf{x}|$: for the segments with the arrows between points r_1 and r_2, $\Delta_{FP} > 0$, but they correspond to gauge equivalent states as one can see in Fig. 5.3.

The determinant $\det \Delta_{FP}$ cannot generally be used to determine whether a particular gauge condition is admissible or not (i.e. the gauge fixing surface intersects every gauge orbit). For the straight line DD' in Fig. 5.3, $x_2 = x_1 + b$, $b > 0$, one easily finds that $\Delta_{FP} = 2x_1 + b > 0$ if $x_1 > b/2$, while this is an inadmissible gauge condition. More to the point, in this particular model, the case $\det \Delta_{FP} = \text{const} \neq 0$ corresponds to an inadmissible gauge. Indeed, put $x_2 = f(x_1)$, then the equation $\Delta_{FP} = x_1 + f'f = c > 0$ can easily be integrated with the result $(x_1 - c)^2 + f^2(x_1) = R^2$, with R being the integration constant. So the gauge fixing curve is a semi-circle ($x_2 \geq 0$ or $x_2 \leq 0$) centered at $x_1 = c$ and $x_2 = 0$ with radius R, which is clearly an inadmissible gauge condition (for example, for $0 < c < R$, the physical states corresponding to gauge orbits with $r > c + R$ are excluded). For a general gauge $\chi(x_1, x_2) = 0$, the equation $\Delta_{FP} = \{\chi, \sigma\} = c > 0$ has no solution for single-valued $\chi(x_1, x_2)$ defined in the *entire* plane. Indeed, in polar coordinates, one has $\Delta_{FP} = \partial\chi/\partial\theta = c$ and, hence, $\chi = c\theta + \theta_0$, but the function $\theta = \tan^{-1}(x_2/x_1)$ is not a single-valued function in the plane.

In Section 6.7.3 it is shown that the Coulomb gauge $\partial_i A_i = 0$ can be used to define local affine coordinates on the orbit space in the classical Yang–Mills theory. The Morse theory provides a more accurate description of the modular domain than the condition $\det \Delta_{FP} > 0$ [209]. It turns out that there are gauge equivalent configurations in the boundary of the modular domain (see also Section 5.8.8). A nontrivial topology of the orbit space emerges after their identification. So a formal reduction of the integration region in the path integral (8.71) to the modular domain would imply, besides the aforementioned shortcomings, that the topology of the orbit space is ignored. Yet, in 4D spacetime the very notion of gauge fixing may even become meaningless unless the theory is regularized (see Section 5.8.9)). For a lattice Yang–Mills theory there is an analog of the Coulomb gauge for which the properties of the Faddeev–Popov operator Δ_{FP} can be used in combination with the lattice formulation of the Morse theory to resolve the gauge fixing ambiguity problem. Still, reduction of the integration domain to the modular domain would lead to an incorrect answer. In Section 6.3.2 with an example of the Yang–Mills theory on the cylindrical spacetime, a path integral on the orbit space (consistent with the Dirac quantization method) in the Coulomb gauge is given. It does not involve any reduction of the integration domain and yields the correct spectrum of the theory which can also be obtained in the framework of the axiomatic quantum field theory [243].

Thus, the conclusion has to be made that the postulate (8.71), amended by a possible reduction of the integration domain, cannot be accepted as general; it requires a modification if there is an operator ordering problem for physical variables and the phase space of the latter is non-standard.

Remark 1. A correct path integral formalism on the orbit space developed in Chapter 6 requires an explicit form of all residual gauge transformations (compare (6.177)). This condition is hard to fulfill for gauges like, e.g. the Coulomb gauge. Even in lattice simulations usually only configurations from the modular domain are determined and used to calculate gauge-fixed Green's functions [295, 296, 300] (compare Sections 5.9.4 and 6.8.2). Unitary gauges described in Section 5.8.10 have an advantage in this regard as the residual gauge transformations are explicitly known.

Remark 2. The Faddeev–Popov gauge-fixed path integral has a symmetry known as the BRST symmetry [422–424]. Since the discovery of the BRST symmetry, the operator version of the BRST quantization has been developed (see, e.g., [425]) which is widely used and equivalent to the Dirac method. A natural question arises: how do the aforementioned shortcomings of the Faddeev-Popov path integral quantization reveal themselves in the BRST path integral? It should be noted first that the operator BRST scheme does not use any gauge fixing to determine the physical Hilbert space. The gauge fixing appears when the BRST evolution operator and its path integral representation are constructed. The BRST Hamiltonian contains the so-called gauge fixing fermion that depends on the gauge fixing condition [425]. It is shown [426] that, if gauge orbits have nontrivial topology, the gauge fixing fermion that satisfies all the conditions required by the BRST scheme may not exist. In particular, for the model from Section 5.3, it can be proved that no single valued gauge fixing fermion exists and the

Batalin-Vilkovisky theorem [427] about the independence of physical amplitudes from the gauge fixing fermion fails [426] (see also [428, 429]). Thus, the BRST quantization does not offer any "cure" to the Faddeev-Popov method. In particular, the Gribov problem cannot be circumvented in the BRST scheme without a modification of the latter (see, however, a scheme based on a supersymmetry different from the BRST symmetry [430–432] and a similar approach in [433]).

8.4.2 An example of the set of residual gauge transformations S

In (5.47) put $f_1 = -a$, $f_2 = -\alpha(2a+u)$ if $u < -a$ and $f_1 = u$, $f_2 = \alpha u$, if $u > -a$; here $a > 0$ and $\alpha > 0$. The functions $u_s(u)$ are nontrivial solutions of the equation

$$f_1^2(u) + f_2^2(u) = f_2^2(u_s) + f_1^2(u_s) \tag{8.79}$$

(see (5.49) and Fig 5.3). The curve $\mathbf{x} = \mathbf{f}(u)$ is tangent to concentric circles (gauge orbits) with the radii $r = a$ and $r = a(1+\alpha^2)^{1/2}$. In the interval $R_1 = (-a\gamma, \gamma a)$, where $\gamma = (1+\alpha^2)^{-1/2} < 1$, it is a stright line segment that lies in the disk $r < a$. Therefore for $u \in R_1$, Eq. (8.79) has the only nontrivial solution $u_s = -u$, hence, $S_1 = Z_2$ and $K_1 = R_1/S_1 \sim (0, \gamma a)$ (see Section 6.3.2). The curve $\mathbf{x} = \mathbf{f}(u)$ intersects any circle of the radius $r \in (a, a/\gamma)$ four times, i.e. Eq. (8.79) has three nontrivial solutions. The region R_2 is the union of two intervals $R_2 = (-3a, -a\gamma) \cup (a\gamma, a)$. Put $K_2 = R_2/S_2 = (a\gamma, a)$ and, hence, $u \in K_2$ and $u_s \in R_2$ in (8.79). The three nontrivial solutions are,

$$u_{s_1}(u) = -u; \quad u_{s_1} : (a\gamma, a) \to (-a, -a\gamma); \tag{8.80}$$
$$u_{s_2}(u) = -2a + \frac{1}{\alpha\gamma}\sqrt{a^2\gamma^2 - u^2}; \quad u_{s_2} : (a\gamma, a) \to (-a, -2a); \tag{8.81}$$
$$u_{s_3}(u) = -2a - \frac{1}{\alpha\gamma}\sqrt{a^2\gamma^2 - u^2}; \quad u_{s_3} : (a\gamma, a) \to (-2a, -3a). \tag{8.82}$$

Equations (8.80)–(8.82) show that the functions $u_s(u)$ do not have a unique analytic continuation to the whole region R_2. Therefore their composition (and, hence, the composition of elements $\hat{s} \in S$) is not defined. As noted in Section 5.3, the explicit form of the functions u_s depends on the choice of fundamental domain. For example, put $K_2 = (-2a, a)$, then the corresponding mappings u_s read,

$$u_{s_1}(u) = -4a - u; \quad u_{s_1} : (-2a, -a) \to (-3a, -2a); \tag{8.83}$$
$$u_{s_2}(u) = -\gamma\sqrt{a^2 + \alpha^2(2a+u)^2}; \quad u_{s_2} : (-2a, -a) \to (-a, -\gamma a); \tag{8.84}$$
$$u_{s_3}(u) = \gamma\sqrt{a^2 + \alpha^2(2a+u)^2}; \quad u_{s_3} : (-2a, -a) \to (\gamma a, a). \tag{8.85}$$

Any circle of the radius $r > a\gamma$ intersects the curve $\mathbf{x} = \mathbf{f}(u)$ twice. Therefore, $R_3 = (-\infty, -3a) \cup (a, \infty)$. Choosing $K_3 = (a, \infty)$, one obtains the only nontrivial solution of (8.79),

$$u_s(u) = -\gamma\sqrt{a^2 + \alpha^2(2a+u)^2}, \quad u_s:\ (a,\infty) \to (-\infty, -3a). \tag{8.86}$$

For every function u_s there exists a function $\theta_s(u)$ such that

$$\mathbf{f}(u) = e^{T\theta_s(u)}\mathbf{f}(u_s(u)) \tag{8.87}$$

(see (5.48)). The latter equality means that the points $\mathbf{f}(u)$ and $\mathbf{f}(u_s)$ belong to the same gauge orbit and, hence, can be transformed to one another by a suitable gauge transformation $e^{T\theta_s(u)} \in SO(2)$. Let $K_1 = (0, \gamma a)$. Then it is obvious that $\theta_s(u) = \pi$; the group element $e^{T\theta_s(u)}$ defines the rotation through angle π. If the mappings u_s are chosen as in (8.80)–(8.82) in the region R_2, then

$$\theta_{s_1}(u) = \pi, \tag{8.88}$$
$$\theta_{s_2}(u) = 3\pi/2 - \sin^{-1}(a\gamma/u) - \tan^{-1}\alpha, \tag{8.89}$$
$$\theta_{s_3}(u) = \pi/2 + \sin^{-1}(a\gamma/u) - \tan^{-1}\alpha. \tag{8.90}$$

Finally, for the mapping (8.86) the solution of (8.87) is

$$\theta_s(u) = 3\pi/2 + \sin^{-1}(a\gamma/u) - \tan^{-1}\alpha.$$

Note that a different choice of the fundamental domain K_2 in R_2 leads to modification of the functions (8.88)–(8.90). Their explicit form for the case $K_2 = (-2a, -a)$ can be found by substituting (8.83)–(8.85) into (8.87).

8.5 Gauge-invariant representations of the unit operator kernel

8.5.1 The coherent state representation

Quantum theory is often formulated in the coherent state representation [421] in which state vectors are functions of the complex variables $a_j = (x_j + ip_j)/\sqrt{2}$ j enumerates the degrees of freedom, and the operators $\hat{a}_j = (\hat{x}_j + i\hat{p}_j)/\sqrt{2}$ and $\hat{a}_j^\dagger$ are, respectively, the operators of differentiation $\partial/\partial a_j$ and multiplication by a_j. The kernel $P_G(a, a^*)$ of the projection operator on the physical subspace can also be found in this representation. The Hamiltonian path integral for gauge systems can be constructed without any explicit elimination of non-physical degrees of freedom [207],

$$U_t^{\text{ph}}(a, a^*) = \int d\mu(b, b^*) U_t(a, b^*) P_G(b, a^*), \tag{8.91}$$

where $d\mu(a, a^*) = dada^* e^{-aa^*}$ is the scalar product measure in the coherent state representation, and $U_t(a, a^*)$ is the evolution operator kernel of the unconstrained system in the coherent state representation for which there is a standard path integral representation. The projector P_G eliminates contributions of gauge non-invariant eigenvectors of the Hamiltonian to the evolution operator. A straightforward way to find P_G is to use its spectral decomposition. For example, the states (5.46) $\langle \mathbf{a}|\Phi_k\rangle = \Phi_k(\mathbf{a})$ ($\langle \mathbf{a}|\Phi_0\rangle = 1$) form an orthonormal basis in the physical subspace; hence [207],

$$P_G(\mathbf{a}, \mathbf{a}^*) = \sum_{k=0}^{\infty} \Phi_k(\mathbf{a})\Phi_k^*(\mathbf{a}) = \Gamma\left(\frac{n}{2}\right) \xi^{1-n/2} I_{n/2-1}(2\xi), \tag{8.92}$$

where $\xi = (\mathbf{a}^2\mathbf{a}^{*2})^{1/2}/2$, and $I_\nu(x) = J_\nu(ix)$ is the modified Bessel function. Equation (8.92) shows that the kernel $P_G(\mathbf{a}, \mathbf{a}^*)$ is an analytic function of the invariant $\mathbf{a}^2\mathbf{a}^{*2}$.

For models with an arbitrary gauge group discussed in Section 6.5.2, the gauge transformation law is the same for x_j and a_j. Therefore $\Phi(a) \in \mathcal{H}_{\rm ph}$ are functions of gauge invariant polynomials of a_j from which it follows that the representation (6.112) for the projection operator is also valid for the coherent state representation if x is replaced by a. A peculiarity of the coherent state representation is that the unit operator kernel in the total Hilbert space is $\exp\langle a, a^*\rangle$ where $\langle\cdot,\cdot\rangle$ denotes the invariant scalar product in the representation realized by x or a. Thus, for a generic gauge system with a compact gauge group, the projection operator kernel in the coherent state representation reads [207],

$$P_G(a, a^*) = \frac{1}{V_G}\int_G dg(\omega) \exp\langle a^*, T(\omega)a\rangle, \tag{8.93}$$

where $T(\omega)a$ is a gauge transform of a and $\int_G dg(\omega) = V_G$. Clearly, the kernel (8.93) is universal for a given representation and gauge group; it does not depend on the system dynamics.

In some cases integration over the group manifold can be carried out explicitly in (8.93) or in (6.114) (see Section 8.1). For example, for the matrix model from Section 5.6.1, one has $\langle a^*, T(\omega)a\rangle = \sum_i (\mathbf{a}_i, \exp(\omega T)\mathbf{a}_i^*)$ where $i = 1, 2$, $V_G = 2\pi$, and

$$\int_G dg(\omega) = \int_0^{2\pi} d\omega\ ;$$

then Eq. (8.93) leads to

$$P_G(a, a^*) = \frac{1}{2}\Big(I_0(\xi_+) + I_0(\xi_-)\Big)\,,$$

where $\xi_\pm = \sum_i [(\mathbf{a}_i^*, \mathbf{a}_i) \pm (\mathbf{a}_i, T\mathbf{a}_i^*)]$, and I_0 is the modified Bessel function. The projection operator formalism in the coherent state representation can also be extended to the path integral formalism for gauge systems with non-compact gauge groups [182].

8.5.2 The coordinate representation

The gauge-free path integral (8.91) can also be obtained in the coordinate representation (see (6.117)). As an example, the derivation of (6.68) from (6.114) is given. Let x be the 2×2 matrix whose columns are the vectors $\mathbf{x}_i$ as defined in the model from Section 5.6.1. Then

$$P_G(x,x') = \frac{1}{2\pi}\int_0^{2\pi} d\omega \delta^4(x - e^{\omega T}x');$$

here the δ-function of a matrix is understood as the product of δ-functions of its matrix elements. After the change of variables $x = e^{\theta T}\rho$, $x' = e^{\theta' T}\rho'$, where ρ and ρ' are lower-triangular matrices, the integral is easily calculated [208],

$$P_G(x,x') = P_G(\rho,\rho') = \frac{1}{\pi}|\det\rho|\delta^3(\rho^T\rho - \rho'^T\rho')\left(1 + \frac{\det\rho}{\det\rho'}\right) \tag{8.94}$$

$$= \frac{1}{\pi}|\det x \det x'|^{1/2}\delta^3(x^Tx - x'^Tx')\left(1 + \frac{\det x}{\det x'}\right). \tag{8.95}$$

Equality (8.94) allows one to find the kernel $Q(\rho,\rho')$. Taking into account the different normalization of $\langle\rho|\rho'\rangle_{\rm ph}$ (see (6.128)) and the kernel P_G, one infers that

$$Q(\rho,\rho') = \delta^3(\rho - \rho') + \delta^3(\rho + \rho'),$$

in full accordance with Section 5.6. The residual gauge group $S = Z_2$ acts on all the components of ρ simultaneously. Note that the representation (8.95) can be used to calculate the kernel of $\hat{Q}$ for any choice of the physical variables (any parameterization of the orbit space), i.e. instead of $\rho_{12} = 0$ any other admissible gauge condition on the matrix ρ can be imposed. It is sufficient to put $x = \exp(T\theta)\rho$ (similarly for x') in (8.95) where ρ is subject to the chosen condition in order to find $\langle\rho|\rho'\rangle_{\rm ph}$ from which $Q(\rho,\rho')$ immediately follows.

For an arbitrary gauge group (the model from Section 5.4) the kernel $P_G(x,x')$, $x,x'\in X$ (the Lie algebra), is determined by the Casimir polynomials. A proof of this is based on an analytic extension of the unit operator kernel in the physical subspace,

$$\langle h|h'\rangle_{\rm ph} = \frac{1}{\kappa^2(h)}\sum_{w\in W}\delta(h - wh')\,, \quad h\in H\,, \quad h'\in K^+\,, \tag{8.96}$$

to the total configuration space X, where W is the Weyl group. It should also be noted that the kernel (8.96) can be obtained by direct calculation of the average integral in (6.128); in so doing, the theorem [216, Thm. II.5.35] and the fact that the stationary group of a generic $x\in X$ is the Cartan group (i.e. that $V_H = (2\pi)^l$ ($l = \text{rank } G$)) are helpful. The following relation holds [216, Eq. III.3.7]:

$$\det\left|\frac{\partial p_{r_i}(h)}{\partial h_j}\right| = c\kappa(h), \qquad c = \text{const.},$$

where $p_{r_i} = \text{Tr } h^{r_i}$, r_i are the degrees of the independent Casimir operators (see Table in Section 8.1), and $\kappa(h)$ is defined in (8.18). Using the rule of changing the argument of a multi-dimensional δ-function, Eq. (8.96) can be transformed to the following form [308]:

$$\langle h|h'\rangle_{\mathrm{ph}} = c(\kappa^2(h)\kappa^2(h'))^{1/4}\prod_{i=1}^{l}\delta\Big(p_{r_i}(h) - p_{r_i}(h')\Big). \tag{8.97}$$

Here the invariance of the measure $\kappa^2(h)$ with respect to the Weyl groups (see Section 8.1.3) and the definition of $|\kappa(h)| = (\kappa^2(h))^{1/2}$ have been used. The polynomial $\kappa^2(h)$ can be expressed via the Casimir polynomials p_{r_i} as any Weyl-invariant polynomial in H [20]. Because of the equality $p_{r_i}(x) = p_{r_i}(S(z)hS^{-1}(z)) = p_{r_i}(h)$ (see Section 8.1.2), the elements $h, h' \in H$ can be replaced by $x, x' \in X$, respectively, in (8.97). Finally, the kernel $P_G(x, x')$ is obtained from (8.97) by multiplying the latter by $V_H V_G^{-1} = (2\pi)^l V_G^{-1}$. Equation (8.97) shows that the kernel P_G (and, hence, Q) is fully determined by characteristic properties of the gauge group alone, the Casimir polynomials, and does not depend on the dynamics.

References

[1] H. Goldstein, *Classical Mechanics* (Addison-Wesley Pub. Co., Inc., Reading, MA, 1959).
[2] L. D. Landau, E. M. Lifshitz, *The Course of Theoretical Physics*, Vol. 1, Mechanics, (Pergamon Press, Oxford, 1969).
[3] V. I. Arnold, *Mathematical Methods of Classical Mechanics*, (Springer-Verlag, Berlin, 1989).
[4] B. M. Barbashov, V. V. Nesterenko, *Introduction to the Relativistic String Theory* (World Scientific Publishing Co, Singapore, 1990).
[5] M. V. Ostrogradsky, Les equations differentielles, *Mem. Ac. St. Petersburg.*, **4**, 385 (1850).
[6] V. I. Arnol'd, V. V. Kozlov, A. I. Neishtadt, *Mathematical aspects of classical and selestian mechanics*, In Encyclopedia of Mathematical Science, Vol. 3, Dynamical systems (Springer-Verlag, Berlin, 1988)
[7] J. A. Schouten, *Ricci-Calculus.* An *Introduction to Tensor Analysis and its Geometrical Applications*, (Springer-Verlag, Berlin, 1954).
[8] H. Poincaré,
Les Méthodes Nouvelle de la Méchanique Céleste, Vol. 1–3 (Gauthier-Villars, Paris, 1892); Engl. Transl. (Dover Publications, NY, 1957).
[9] L. A. Pars, *A Treatise on Analytical Dynamics* (Heinemann, London, 1965).
[10] H. Poincaré, *Compt. rend.* **123**, 530 (1896).
[11] E. Witten, *Nucl. Phys. B* **223**, 422 (1983).
[12] H. von Helmholtz, *Journ. f. d. reine u. angew. Math.* **100**, 137 (1887).
[13] J. Douglas, *Trans. Am. Math. Soc.* **50**, 71128 (1941).
[14] E. Noether, "Invariante Variationsprobleme". Nachr. d. Knig. Gesellsch. d. Wiss. zu Gttingen, Math-phys., 235257 (1918); Engl. Translation in: arXiv:physics/0503066v1 [physics.hist-ph].
[15] C. Lanczos, *The Variational Principles of Mechanics* (Dover Publications, New York, 1970).
[16] S. I. Anisimov, B. S. Lukyanchuk, *Phys. Uspekhi* **172**, 301 (2002).
[17] I. B. Gornushkin, S. V. Shabanov, N. Omenetto, J. D. Winefordner, *J. Appl. Physics* **100**, 073304 (2006).
[18] N. P. Baxter, S. V. Shabanov, *J. Math. Phys.* **49**, 093101 (2008).
[19] J. Harnad, P. Winternitz, G. Sabidussi, eds. *Integrable Systems: From Classical to Quantum.* (American Mathematical Society, 2000).
[20] D. P. Zhelobenko, *Lectures in the Lie Group Theory*, (JINR, Dubna, 1965) (in Russian), p.344.
[21] D. P. Zhelobenko, *Compact Lie Groups and their Representations*, Translations of Mathematical Monographs, Vol. 40 (AMS, Providence, 1973).
[22] L. V. Prokhorov, E. A. Sazonov, *Vestnik Leningr. Univ., Series 4*, Issue 2, 12 (1980).
[23] L. V. Prokhorov, E. A. Sazonov, *Vestnik Leningr. Univ., Series 4*, Issue 1, 21 (1981).
[24] R. Finkelstein, M. Villasante, *Phys. Rev. D* **33**, 1666 (1986).
[25] J. Govaerts, *Int. J. Mod. Phys.* A **5**, 3625 (1990).
[26] L. V. Prokhorov, S. V. Shabanov, *Sov. Phys. Uspekhi*, **34(2)**, 108 (1991).
[27] F. A. Berezin, In: *Introduction to Superanalysis*, ed. A.A. Kirillov, (D. Reidel Publ. Co., Dordrecht, Holland, 1987).
[28] B. DeWitt, *Supermanifolds* (Cambridge Univ. Press, Cambridge, 1984).

[29] I. A. Batalin, E. S. Fradkin, *Nucl. Phys. B* **326**, 701 (1989).
[30] F. A. Berezin, *Izv. Akad. Nauk SSSR Math.* (USSR - Izvestija Ser. Mat.) **38**, 1112 (1974).
[31] S. V. Shabanov, *J. Phys. A: Math. Gen.* **24**, 1199 (1991).
[32] L. E. Gendenshtein, I. V. Krive, *Soviet Phys. Uspekhi* **28**, 645 (1985).
[33] F. Cooper, A. Khare, U. Sukhatme, *Supersymmetry in Quantum Mechanics* (World Scientific Publishing Co, Singapore, 2001).
[34] S. V. Shabanov, *The proceedings of the XXVI international symposium on elementary particle physics* (Wendisch-Reiz, Germany, 1992), DESY preprint, DESY 93-013, 1993, p.276.
[35] S. V. Shabanov, *Mod. Phys. Lett.* A **10**, 941 (1995).
[36] I. Ya. Aref'eva, I. V. Volovich, *Quantum Group Particles and Non-Archimedian Geometry*, CERN preprint, CERN-TH.6137/91, 1991.
[37] L. V. Prokhorov, *Phys. Part. Nucl.* **39**, 810 (2008).
[38] H. Weyl, *The Theory of Groups and Quantum Mechanics* (Dover Publications, New York, 1950).
[39] J. Schwinger, *Proc. Natl. Acad. of Sci.* **46**, 385 (1960).
[40] H. Rampacher, H. Stumpf, F. Wagner, *Fotrschr. Phys.* **13**, 385 (1965).
[41] M. Yamamura, *Prog. Theor. Phys.* **62**, 681 (1979).
[42] I. Saavedra, C. Utreras, *Phys. Lett. B* **98**, 74 (1981).
[43] V. G. Drinfeld, *Dokl. Akad. Nauk SSSR* **238**, 1960 (1985).
[44] M. Jimbo, *Lett. Math. Phys.* **10**, 63 (1985).
[45] L. D. Faddeev, N. Yu. Reshitikhin, L. A. Takhtajan, *Advanced Series in Mathematical Physics*, Vol. **9** (Eds. C. N. Yang, M. L.Ge; (World Scientific Publishing Co, Singapore, 1989).
[46] A. J. Macfarlane, *J. Phys. A: Math. Gen.*, **22**, 4581 (1989).
[47] G. Brodimas, A. Janussis, R. Mignani, *J. Phys. A: Math. Gen.*, **25**, L329 (1992).
[48] S. V. Shabanov, *Phys. Lett. B* **293**, 117 (1992).
[49] S. V. Shabanov, *J. Phys. A: Math. Gen.*, **25**, L1245 (1992).
[50] S. V. Shabanov, *J. Phys. A: Math. Gen.*, **26**, 2563 (1992).
[51] Y. Nambu, *Phys. Rev. D* **7**, 2405 (1973).
[52] T. Curtright, C. Zachos, *Phys. Rev. D* **68**, 085001 (2003).
[53] V. I. Arnold, *Ordinary Differential Equations* (The MIT Press, Cambridge, 1973).
[54] L. A. Takhtajan, *Commun. Math. Phys.* **160**, 295 (1994).
[55] S. Shnider, S. Sternberg, *Quantum Groups. From Coalgebras to Drinfel'd Algebras. A Guided Tour*, Graduate Texts in Mathematical Physics, II (International Press, Cambridge, MA, 1993).
[56] L. V. Prokhorov, A. S. Ushakov, *Vestnik SpBGU*, Ser. 4, Issue 1, 29 (2010).
[57] L. V. Prokhorov, A. S. Ushakov, *Doklady Mathematics*, **78**, 925 (2008).
[58] S. L. Adler, A. Bassi, *Science* **325**, 275 (2009).
[59] L. V. Prokhorov, *Phys. Part. Nucl.* **38**, 364 (2007).
[60] N. Wiener, *J. Math. Phys. Sci.* **2**, 131 (1923).
[61] N. Wiener, *Nonlinear Problems in Random Theory* (Cambridge, Massachusetts, The MIT Press, 1958).
[62] R. P. Feynman, *Rev. Mod. Phys.* **20**, 367 (1948).
[63] R. P. Feynman, A. R. Hibbs, *Quantum Mechanics and Path Integrals* (McGraw-Hill, New York, 1965).
[64] P. A. M. Dirac, *Sov. Phys.* **3**, 64 (1933).
[65] P. A. M. Dirac, *The Principles of Quantum Mechanics* (Clarendon Press, Oxford, 1958).
[66] W. Pauli, *Pauli Lectures on Physics*, Vol. 6, (Cambridge, Massachusetts, The MIT Press, 1973).
[67] W. Janke, H. Kleinert, *Let. Nouvo Cim.* **25**, 297 (1979).
[68] H. Kleinert, *Path Integrals in Quantum Mechanics, Statistics, and Polymer Physics* (WSPC, Singapore, 1990).
[69] B. DeWitt, *Rev. Mod. Phys.* **29**, 377 (1957).

[70] A. G. Nuramatov, L. V. Prokhorov, *Vestnik Lening. Univ., Ser.* 4: *Phys. Chem.*, Issue 1, 66 (1981).
[71] R. P. Feynman, *Phys. Rev.* **84**, 108 (1951).
[72] W. Tobocman, *Nouvo Cim.* **3**, 1213 (1956).
[73] H. Davies, Proc. *Cambridge Phil. Soc.* **59**, 147 (1963).
[74] C. Garrod, *Rev. Mod. Phys.* **38**, 483 (1966).
[75] V. A. Fock, *Foundations of Quantum Mechanics*, (Nauka, Moscow, 1976) (in Russian).
[76] I. M. Gelfand, A. M. Yaglom, *Uspekhi Mat. Nauk* (Russian Math. Survey), **11**, 77 (1956).
[77] I. M. Gelfand, R. A. Minlos, A. M. Yaglom, *Path Integrals.* In: *Proceedings of the 3rd National Mathematical Meeting*, Vol. 3, p. 521 (USSR Academy of Science, Nauka, Moscow, 1958).
[78] Yu. L. Daletsky, *Uspekhi Mat. Nauk* (Russian Math. Survey) **17**, 3 (1962).
[79] I. M. Kovalchik, *Uspekhi Mat. Nauk* (Russian Math. Survey) **18**, 97 (1963).
[80] I. M. Kovalchik, L. A. Yanovich, *Generalized Wiener Integral and its Applications*, (Nauka i Tekhnika, Minsk, 1989) (in Russian).
[81] L. Streit, *Acta Phys. Austrica. Suppl.* **II**, 2 (1965).
[82] F. A. Berezin, *The Method of Second Quantization* (Academic Press, New York, 1966).
[83] F. A. Berezin, *Sov. Phys. Uspekhi* **132**, 497 (1980).
[84] S. A. Alberverio, R. J. Høegh-Krohn, *Mathematical Theory of Feynman Path Integrals*, (Springer-Verlag, Berlin, 1976).
[85] C. DeWitt-Morette, A. Macheshwari, B. Nelson, *Phys. Rept.* **50**, 255 (1979).
[86] C. DeWitt-Morette, *Commun. Math. Phys.* **28**, 47 (1972).
[87] B. Simon, *Functional Integration and Quantum Physics* (Academic Press, New York, 1979).
[88] R. H. Cameron, *J. Math. and Phys.* **39**, 126 (1960).
[89] E. Nelson, *J. Math. Phys.* **5**, 332 (1964).
[90] M. A. Efgrafov, *Doklady Akad. Nauk USSR* **191**, 979 (1970).
[91] O. G. Smolyanov, E. T. Shavgulidze, *Path Integrals* (Moscow State Univ., Moscow, 1990).
[92] G. Röpsdorff, *Path Integral Approach to Quantum Physics. An Introduction* (Springer-Verlag, Berlin, 1994).
[93] L. S. Schulmann, *Techniques and Applications of Path Integration* (Wiley, New York, 1981).
[94] T. Kato, H. F. Trotter, *Pacific Math. J.* **8**, 887 (1958).
[95] H. Trotter, *Proc. Amer. Math.* **10**, 545 (1959).
[96] L. V. Prokhorov, *Phys. Part. Nucl.* **13**, 1094 (1982).
[97] L. V. Prokhorov, In: *Proceedings of the 3rd International Seminar on Problems in High Energy Physics and Quantum Field Theory*, Vol. 1, p. 103 (IHEP, Protvino, 1981).
[98] L. V. Prokhorov, *Vestnik Leningr. Univ.*, Ser. 4: Phys. Chem., Issue 1, 106 (1982).
[99] Yu. P. Malyshev, L. V. Prokhorov, *Vestnik Leningr. Univ.*, Ser. 4: Phys. Chem., Issue 1, 11 (1984).
[100] Yu. P. Malyshev, L. V. Prokhorov, *Quantization by the Hamiltonian Path Integral Method*, VINITI manuscript 7676-84 (VINITI, Moscow, 1984).
[101] Yu. P. Malyshev, L. V. Prokhorov, *Sov. J. Nucl. Phys.* [Yad. Fiz.] **48**, 890 (1988).
[102] L. V. Prokhorov, *Vestnik Leningr. Univ.*, Ser. 4: Phys. Chem., Issue 2, 67 (1983).
[103] S. Edwards, Y. V. Gulyaev, *Proc. Roy. Soc. Ser. A* **279**, 229 (1964).
[104] A. Arthurus, *Proc. Roy. Soc. Ser. A* **313**, 445 (1969).
[105] F. A. Berezin, *Theor. Math. Phys.* **6**, 194 (1971).
[106] I. W. Mayes, J. S. Dowker, *Proc. Roy. Soc. Ser. A* **327**, 131 (1972).
[107] I. W. Mayes, J. S. Dowker, *J. Math. Phys.* **14**, 434 (1973).
[108] D. McLaughlin, L. Schulman, *J. Math. Phys.* **12**, 2520 (1971).
[109] F. Langouche, D. Roekaerts, E. Tirapegui, *Phys. Rev. D* **20**, 419 (1979).
[110] A. N. Vassiliev, *Functional Methods in Quantum Field Theory and Statistics* (Leningr. State Univ. Press, Leningrad, 1976) (in Russian).
[111] M. Moshinsky, T. Seligman, *Ann. Phys.* **114**, 243 (1978).

[112] M. Moshinsky, T. Seligman, *Ann. Phys.* **120**, 402 (1978).
[113] J. Deenen, M. Moshinsky, T. Seligman, *Ann. Phys.* **127**, 458 (1980).
[114] J. Gervais, A. Jevicki, *Nucl. Phys. B* **110**, 93 (1976).
[115] J. Van Vleck, *Proc. Nat. Acad. Sci.* **14**, 178 (1928).
[116] M. Berry, K. Mount, *Repts. Progr. Phys.* **35**, 315 (1972).
[117] L. V. Prokhorov, *Sov. J. Nucl. Phys.* **35**, 498 (1982).
[118] F. Santos, L. Oliveira, T. Kodama, *Nouvo Cim. B* **58**, 251 (1980).
[119] H. Fukutaka, T. Kashiwa, *Ann. Phys.* **185**, 301 (1988).
[120] L. V. Prokhorov, *Sov. J. Nucl. Phys.* **39**, 496 (1984).
[121] M. S. Marinov, M. V. Terentiev, *Sov. J. Nucl. Phys.* **28**, 1418 (1978).
[122] M. S. Marinov, M. V. Terentiev, *Fortschr. Phys.* **27**, 511 (1979).
[123] T. A. Miura, *J. Math. Phys.* **31**, 1189 (1990).
[124] L. V. Prokhorov, *Vestnik Leningr. Univ.*, Ser. 4: Phys. Chem. Issue 1, 14 (1983).
[125] H. S. M. Coxeter, *Ann. of Math.* **35**, 588 (1934).
[126] J. E. Humphreys, *Reflection Groups and Coxeter Groups*, Cambridge Studies in Advanced Mathematics, **29** (1990).
[127] M. W. Davis, *The Geometry and Topology of Coxeter Groups*, (Princeton Univ. Press, Princeton, 2008).
[128] E. B. Vinberg, Absence of crystallographic groups of reflections in Lobachevski spaces of large dimension, *Trudy Moskov. Mat. Obshch.*, Vol. 47 (1984)
[129] O. Loos, *Symmetric Spaces. I. General Theory* (Benjamin, New York/Amsterdam, 1969).
[130] I. Chetouani, L. Guechi, T. F. Hamman, *Nuovo Cim.* **101**, 547 (1988).
[131] R. E. Crandall, *J. Phys. A: Math. Gen.* **16**, 513 (1983).
[132] I. S. Gradshteyn, I. M. Ryzhyk, *Tables of Integrals, Series, and Products* (Academic Press, NY, 1965).
[133] B. Gaveau, L. S. Schulmann, *J. Phys. A: Math. Gen.* **19**, 1833 (1986).
[134] C. Grosche, *Annalen Phys.* **2**, 557 (1993).
[135] S. V. Shabanov, In Proceedings of Yamada Conference on Mathematical Physics. (Eds A. Arima, *et al.*) (World Scientific, Singapore, 1995), p.423.
[136] P. Choquard, *Helv. Phys. Acta* **28**, 89 (1955).
[137] N. Bleistein, R. Handelsman, *Asymptotic Expansions of Integrals* (Dover, New York, 1975).
[138] V. I. Smirnov, *A Course in Higher Mathematics*, Vol. 5 (Fizmat GIZ, Moscow, 1959).
[139] M. Reed, B. Simon, *Methods of Modern Mathematical Physics* (Academic Press, New York, 1972).
[140] M. Abramowitz, I. A. Stegun, *Handbook of Mathematical Functions* (National Bureau of Standards, Washington DC, 1972).
[141] M. Morse, H. Feshbach, *Methods of Theoretical Physics* (McGraw-Hill, New York, 1953).
[142] C. Morette, *Phys. Rev.* **81**, 848 (1951).
[143] H. Kleinert, S. V. Shabanov, *Phys. Lett. A* **232**, 327 (1997).
[144] R. C. T. da Costa, *Phys. Rev. A* **23**, 1982 (1981).
[145] B. M. Smirnov, *Physics of Fractal Clusters* (Nauka, Moscow, 1991) (in Russian)
[146] H. Takayasu, *Fractals in the Physical Sciences* (Manchester Univ. Press, New York, 1990).
[147] V. G. Nevzglyadov, *Theoretical Mechanics* (Fiz. Mat. GIZ, Moscow, 1959) (in Russian).
[148] P. A. M. Dirac, *Lectures on Quantum Mechanics* (Yeshiva Univ. Press., Massachusetts, 1964).
[149] V. V. Nesterenko, A. M. Chervyakov, *Singular Lagrangians*, JINR lectures for young scientists, Issue 36 (JINR, Dubna, 1986).
[150] P. N. Pyatov, A. V. Razumov, *Int. J. Mod. Phys.* A **4**, 3211 (1989).
[151] P. N. Pyatov, *The Lagrangian Formalism for Constraint Systems.* I. *Lagrangian and Hamiltonian Constraints*, Preprint IHEP 90-35; II. Gauge theories, Preprint IHEP 90-148 (IHEP, Protvino, 1990).
[152] T. Maskawa, H. Nakajima, *Prog. Theor. Phys.* **56**, 1295 (1976).

[153] H. Weyl, *The Classical Groups, Their Invariants and Representations* (Princeton University Press, Princeton, 1939).
[154] L. D. Faddeev, S. L. Shatashvili, *Phys. Lett. B* **167**, 225 (1986).
[155] I. A. Batalin, E. S. Fradkin, *Phys. Lett. B* **180**, 157 (1987); Errata **236**, 528 (1990).
[156] I. A. Batalin, E. S. Fradkin, *Nucl. Phys. B* **279**, 514 (1987).
[157] I. A. Batalin, E. S. Fradkin, T. E. Fradkina *Nucl. Phys. B* **332**, 723 (1990).
[158] I. A. Batalin, E. S. Fradkin, T. E. Fradkina *Nucl. Phys. B* **314**, 158 (1989).
[159] I. A. Batalin, S. L. Lyakhovich, I. V. Tyutin, *Mod. Phys. Lett. A* **7**, 1931 (1992).
[160] M. V. Karasev, V. P. Maslov, *Nonlinear Poisson Brackets, Geometry and Quantization*, Translations of Math. Monographs, vol. 119, (AMS, Providence, RI, 1993).
[161] E. Fermi, *Rend. Acad. Lincei.* **9**, 881 (1929).
[162] E. Fermi, *Rev. Mod. Phys.* **4**, 87 (1932).
[163] W. Heisenberg, W. Pauli, *Z. Phys.*, **59**, 168 (1930).
[164] L. V. Prokhorov, *Sov. Phys. Uspekhi* **154**, 299 (1988).
[165] P. A. M. Dirac, *Can. J. Math.* **2**, 129 (1950).
[166] P. A. M. Dirac, *Can. J. Math.* **3**, 1 (1951).
[167] P. A. M. Dirac, *Proc. Roy. Soc. A* **246**, 326 (1958).
[168] P. G. Bergmann, *Phys. Rev.* **75**, 680 (1949).
[169] J. L. Anderson, P. G. Bergmann, *Phys. Rev.* **83**, 1018 (1951).
[170] P. G. Bergmann, *Rev. Mod. Phys.* **33**, 510 (1961).
[171] L. D. Faddeev, *Theor. Math. Phys.* (USSR) **1**, 3 (1969).
[172] E. S. Fradkin, In: *Proc. of the 10th Winter School of Theoretical Physics in Karpacs*, Issue 207, 93 (1973).
[173] I. A. Batalin, E. S. Fradkin, *Riv. Nuovo Cim.* **9**, 1 (1986).
[174] R. Jackiw, *Rev. Mod. Phys.* **52**, 661 (1980).
[175] N. H. Christ, T. D. Lee, *Phys. Rev. D* **22**, 939 (1980).
[176] T. D. Lee, *Particle Physics and Introduction to Field Theory* (Harwood Acad. Publ. New York, 1981).
[177] M. Henneaux, *Phys. Rept.* **126**, 1 (1985).
[178] A. J. Hanson, T. Regge, C. Teitelboim, *Constrained Hamiltonian Systems* (Acad. naz. Lincei, Roma, 1976).
[179] D. M. Gitman, I. V. Tyutin, *Quantization of Fields with Constraints* (Springer-Verlag, Berlin, 1990).
[180] N. P. Konopleva, V. N. Popov, *Gauge Fields* (Atomizdat, Moscow, 1972) (in Russian).
[181] V. N. Popov, *Path Integrals in Quantum Field Theory and Statistical Physics* (Atomizdat, Moscow, 1976) (in Russian).
[182] J. R. Klauder, Quantization of Constrained Systems, *Lect. Notes Phys.* **572**, 143 (2001).
[183] L. V. Prokhorov, Sov. *J. Nucl. Phys.* **35**, 229 (1982).
[184] A. Ashtekar, G. T. Horowitz, *Phys. Rev. D* **26**, 3342 (1982).
[185] C. J. Isham, *Quantum Gravity*, Imperial College Preprint TP/85-86/39 (London, 1986).
[186] R. Jackiw, *Topics in Planar Physics*, MIT Report No. CTP 1824 (1989).
[187] L. V. Prokhorov, *Vestnik Leningr. Univ.* Ser. 4: Phys. Chem. Issue 4, 174 (2009).
[188] L. V. Prokhorov, *Vestnik Leningr. Univ.* Ser. 4: Phys. Chem. Issue 3, 3 (1988).
[189] L. V. Prokhorov, A. G. Nuramatov, *Vestnik Leningr. Univ.* Ser. 4: Phys. Chem. Issue 3, 86 (1991).
[190] D. Lüst, S. Theisen, *Lectures in String Theory.* Lecture Notes in Physics, Vol. 346 (Springer-Verlag, Berlin, 1989).
[191] L. H. Ryder, *Quantum Field Theory* (Cambridge Univ. Press, Cambridge, 1985).
[192] V. Fock, *Zs. f. Phys.* **38**, 242 (1926).
[193] H. Kragh, *Am. J. Phys.* **52**, 1024 (1984).
[194] L. V. Prokhorov, *Phys. Part. Nucl.* **31**, 47 (2000).
[195] S. V. Shabanov, *Phys. Rept.* **326**, 1 (2000).
[196] L. V. Prokhorov, S. V. Shabanov, *Phys. Lett. B* **216**, 341 (1989).
[197] S. V. Shabanov, *Theor. Math. Phys.* (USSR) **78**, 292 (1989).
[198] S. V. Shabanov, *Phys. Lett. B* **318**, 323 (1993).

[199] C. R. Ordónez, M. Rubin, D. Zwanziger, *Phys. Rev. D* **40**, 4056 (1989).
[200] S. V. Shabanov, *The Phase Space Structure in Gauge Theories.* Lectures for young scientists, Issue 54, P2-89-533 (JINR, Dubna, 1989).
[201] J. R. Klauder, , *Int. Jour. Mod. Phys. D* **12**, 1769 (2003).
[202] Yu. P. Malyshev, L. V. Prokhorov, *Vestnik Leningr. Univ.* Ser. 4: Phys. Chem. Issue 3, 99 (1986).
[203] L. V. Prokhorov, S. V. Shabanov, *Vestnik Leningr. Univ.* Ser. 4: Phys. Chem. Issue 1, 3 (1990).
[204] L. V. Prokhorov, S. V. Shabanov, In: *Topological Phases in Quantum Theory*, pp.354–370, (WSPC, Singapore, 1989).
[205] L. V. Prokhorov, S. V. Shabanov, *Phase Space of Yang–Mills Fields*, JINR preprint E2-90-207 (JINR, Dubna, 1990).
[206] L. V. Prokhorov, *Phys. Part. Nucl.* **25**, 559 (1994).
[207] S. V. Shabanov, *Path Integral in Holomorphic Representation Without Gauge Fixation*, Preprint E2-89-687 (JINR, Dubna, 1989); In Proc. Intern. Seminar "Path Integrals: Theory and Applications", Eds. V. S. Yarunin, M. A. Smondyrev, p. 133 (JINR, Dubna, 1996).
[208] S. V. Shabanov, *Int. J. Mod. Phys. A* **78**, 163 (1991).
[209] V. N. Gribov, *Nucl. Phys. B* **139**, 1 (1978).
[210] M. A. Soloviev, *Theor. Math. Phys.* (USSR) **60**, 7 (1989).
[211] I. M. Singer, *Commun. Math. Phys.* **60**, 7 (1978).
[212] S. V. Shabanov, *Mod. Phys. Lett. A* **6**, 909 (1991).
[213] S. V. Shabanov, *Lectures on Quantization of Gauge Theories by the Path Integral Method*, In: *Proceedings of the 4th Hellenic School on High Energy Physics*, National Technical University of Athens, Vol. 2, p. 272, Eds. E. N. Gazis, G. Koutsoumbas, N. D. Tracas, G. Zoupanos (Athens, 1994); IFM preprint 16/92 (Lisbon, IFM, 1992).
[214] G. 't Hooft, *Nucl. Phys. B* **190** [FS3], 455 (1981).
[215] S. Helgason, *Differential Geometry, Lie Groups, and Symmetric Spaces* (Academic Press, New York, 1978).
[216] S. Helgason, *Groups and Geometric Analysis* (Academic Press, New York, 1984).
[217] S. Weinberg, *Prog. Theor. Phys. Suppl.* **86**, 43 (1986).
[218] A. Salam, *Semisimple Groups and Classification of Elementary Particles*, Vol. 1, p.51 (JINR, Dubna, 1965).
[219] S. G. Matinyan, G. K. Savvidy, N. Ter-Arutyunyan-Savvidy, *Sov. Phys. JETP* **53**, 421 (1981).
[220] G. K. Savvidy, *Phys. Lett. B* **159**, 325 (1985).
[221] S. V. Shabanov, *Phys. Lett. B* **255**, 398 (1991).
[222] B. de Witt, J. Hoppe, H. Nicolai, *Nucl. Phys. B* **305**, 545 (1988).
[223] B. de Witt, J. Łüscher, H. Nicolai, *Nucl. Phys. B* **320**, 135 (1989).
[224] E. Witten, *Nucl. Phys. B* **443**, 85 (1995).
[225] T. Banks, W. Fischer, S.H. Shenker, L. Susskind, *Phys. Rev. D* **55**, 5112 (1997).
[226] M. A. Soloviev, *Theor. Math. Phys.* (USSR) **73**, 3 (1987).
[227] T. Pause, T. Heinzl, *Nucl. Phys. B* **524**, 695 (1998).
[228] O. Babelon, C.-M. Viallet, *Phys. Lett. B* **85**, 246 (1979).
[229] O. Babelon, C.-M. Viallet, *Commun. Math. Phys.* **81**, 515 (1981).
[230] M. Daniel, C.-M. Viallet, *Rev. Mod. Phys. B* **52**, 175 (1980).
[231] L. V. Prokhorov, *Theor. Math. Phys.* (USSR) **47**, 210 (1981).
[232] L. V. Prokhorov, *Vestnik Leningr. Univ.* Ser. 4: Phys. Chem., Issue 3, 66 (1982).
[233] J. R. Klauder, *Beyond Conventional Quantization* (Cambridge Univ. Press, Cambridge 1999).
[234] A. A. Migdal, *Sov. Phys. JETP* **42**, 413 (1976).
[235] S. G. Rajeev, *Phys. Lett. B* **212**, 203 (1988).
[236] J. E. Hetrick, Y. Hosotani, *Phys. Lett. B* **230**, 88 (1989).
[237] M. Mickelsson, *Phys. Lett. B* **242**, 217 (1990).
[238] E. Langmann, G. W. Semenoff, *Phys. Lett. B* **296**, 117 (1992).

[239] E. Langmann, G. W. Semenoff, *Phys. Lett. B* **303**, 303 (1993).
[240] S. V. Shabanov, *Phys. Lett. B* **318**, 323 (1993).
[241] J. E. Hetrick, *Int. J. Mod. Phys.* A **9**, 3153 (1994).
[242] S. V. Shabanov, *Commun. Theor. Phys.* **4**, 1 (1995).
[243] A. Ashtekar, J. Lewandowski, D. Marolf, J. Mourao, T. Thiemann, *J. Math. Phys.* **38**, 5453 (1997).
[244] H. G. Dosch, V.F. Muller, *Fortschr. Phys.* **27**, 647 (1979).
[245] V. Kazakov, I. Kostov, *Nucl. Phys. B* **179**, 283 (1981).
[246] B. Rusakov, *Mod. Phys. Lett. A* **5**, 693 (1990).
[247] D. Fine, *Commun. Math. Phys.* **134**, 273 (1990).
[248] E. Witten, *Commun. Math. Phys.* **141**, 153 (1991).
[249] E. Witten, *J. Geom.Phys.* **9**, 303 (1992).
[250] M. Blau, G. Thompson, *J. Mod. Phys. A* **7**, 3781 (1991).
[251] S. Donaldson, P. Kronheimer, *The Geometry of Four Manifolds* (Oxford Univ. Press, Oxford, 1990).
[252] G. 't Hooft, *Nucl. Phys. B* **153**, 141 (1979).
[253] R. Jackiw, *Rev. Mod. Phys.* **52**, 661 (1980)
[254] A. G. Izergin, V. E. Korepin, M. A. Semenov-Tyan-Shanskii, L. D. Faddeev, *Theor. Math. Phys.* (USSR) **38**, 1 (1979).
[255] D. Amati, E. Elitzur, E. Rabinovici, *Nucl. Phys. B* **418**, 45 (1994).
[256] D. Cangemi, R. Jackiw, *Phys. Rev. D* **50**, 3913 (1994).
[257] A. Yu. Alekseev, P. Schaller, T. Strobl, *Phys. Rev. D* **52**, 7146 (1995).
[258] M. Nakahara, *Geometry, Topology, and Physics* (IOP, Bristol, 1993).
[259] N. Burbaki, *Group et Algebres de Lie*, ch.4,5 et 6 (Masson, Paris, 1981).
[260] H. G. Loos, *Phys. Rev.* ***188***, 2342 (1969).
[261] H. G. Loos, *J. Math. Phys.* **11**, 3258 (1970).
[262] *Encyclopedic Dictionary of Mathematics*, Vol. II, ed. by Kiyosi Ito (MIT Press, Cambridge, 1987).
[263] C. Emmrich and H. Römer, *Commun. Math. Phys.* **129**, 69 (1990).
[264] P. van Baal, *Nucl. Phys. B* **369**, 259 (1992).
[265] A. I. Vainshtein, V. I. Zakharov, V. A. Novikov, M. A. Shifman, *Sov. Phys. Uspekhi* **25**, 195 (1982).
[266] P. van Baal, In: *Confinement, Duality, and Nonperturbative Aspects of QCD*, Ed. by P. van Baal, NATO ASI Series, Vol. B368 (Plenum Press, New York, 1998).
[267] I. M. Singer, *Physica Scripta* **24**, 817 (1981).
[268] W. Kondracki, P. Sadowski, *J. Geom. Phys.* **3**, 421 (1983).
[269] A. Heil, A. Kersch, N. Papadopoulos, B. Reifenhäuser, F. Scheck, *J. Geom. Phys.* **7**, 489 (1990).
[270] J. Fuchs, M.G. Schmidt, C. Schweigert, *Nucl. Phys. B* **426**, 107 (1994).
[271] J. M. Arms, J. E. Marsden and V. Moncrief, *Commun. Math. Phys.*, **78**, 455 (1981).
[272] M. Asorey, P. K. Mitter, *Ann. Inst. Poincaré A* **45**, 61 (1986).
[273] M. S. Narasimhan, T. R. Ramadas, *Commun. Math. Phys.* **67**, 121 (1979).
[274] K. K. Uhlenbeck, *Commun. Math. Phys.* **83**, 31 (1982).
[275] M. A. Soloviev, *Kratk. Soobshch. Fiz.* **3**, 29 (1985).
[276] I. M. Gelfand, G. E. Shilov, *Generalized Functions*, Vol. 4 (Academic Press, New York, 1964).
[277] R. Rajaraman, *An Introduction to Solitons and Instantons in Quantum Field Theory* (Elsevier, New York, 1982).
[278] S. V. Shabanov, *Phys. Lett. B* **272**, 11 (1991).
[279] V. P. Maslov, M. V. Fedoriuk, *Semi-Classical Approximation in Quantum Mechanics* (D. Reidel Publ. Comp., Dortrecht, Holland, 1981).
[280] S. W. Hawking, *Phys. Lett B* **195** (1987); *Phys. Rev. D* **37**, 904 (1988).
[281] Y. Verbin, A. Davidson, *Phys. Lett. B* **299**, 364 (1989).
[282] G. V. Lavrelashvili, V. A. Rubakov, P. G. Tinyakov, *JETP Lett.* **46**, 167 (1987).
[283] S. Coleman, *Nucl. Phys. B* **307**, 867, (1988); *Nucl. Phys. B* **310**, 643 (1988).

[284] G. W. Gibbons, C. N. Pope, *Commun. Math. Phys.* **66**, 627 (1979).
[285] E. Witten, *Commun. Math. Phys.* **80**, 381 (1981).
[286] P. Schoen, S. T. Yau, *Phys. Rev. Lett.* **42**, 547 (1979).
[287] O. Bertolami, J. M. Mourao, R. F. Picken, I. P. Volobuev, *Int. J. Mod. Phys. A* **6**, 4149 (1991).
[288] O. Bertolami, J. M. Mourao, *Class. Quant. Grav.* **8**, 1271 (1991).
[289] B. DeWitt, *Phys. Rev. D* **160**, 1113 (1967).
[290] C. Isham, J. E. Nelson, *Phys. Rev. D* **10**, 3226 (1974).
[291] W. E. Blyth, C. Isham, *Phys. Rev. D* **11**, 768 (1975).
[292] S. V. Shabanov, In: *The Proceedings of the First Iberian Meeting on Classical and Quantum Gravity*, ed. M. C. Bento *et al*, (World Scientific, Singapore, 1993); p.322.
[293] E. T. Whitteker, G. N. Watson, *A Course of Modern Analysis* (Oxford University Press, Oxford, 1927); Vol.1, Sec. 2.37.
[294] L. V. Prokhorov, *The Weyl Group and Confinement*, Preprint OCIP-89-03 (The Ottawa University, Ottawa, 1989).
[295] P. Marenzoni, G. Martinelli, N. Stella, M. Testa, *Phys. Lett. B* **318**, 511 (1993).
[296] C. Bernard, C. Parrinello, A. Soni, *Phys.Rev. D* **49**, 1585 (1994).
[297] M. Creutz, *Quarks, Gluons and Lattices* (Cambridge University Press, Cambridge, 1983).
[298] I. Montvay and G. Münster, *Quantum Fields on a Lattice* (Cambridge University Press, Cambridge, 1997).
[299] S. Weinberg, *The Quantum Theory of Fields*, Vol II, (Cambridge University Press, Cambridge, 1996).
[300] M. L. Paciello, C. Parrinello, S. Petrarca, B. Taglienti, A. Vladikas, *Phys. Lett. B* **289**, 405 (1992).
[301] M. Stingl, *Phys. Rev. D* **34**, 3863 (1986).
[302] L. V. Prokhorov, S. V. Shabanov, *Vestnik Leningr. Univ.*, Series 4: Phys. Chem., Issue 1, p. 68 (1988).
[303] S. V. Shabanov, J. R. Klauder, *Phys. Lett. B* **456**, 38 (1999).
[304] L. V. Prokhorov, S. V. Shabanov, *Vestnik Leningr. Univ.*, Series 4: Phys. Chem., Issue 2, p. 8 (1988).
[305] J. L. Martin, *Proc. Roy. Soc. A* **251**, 536 (1959).
[306] J. L. Martin, *Proc. Roy. Soc. A* **251**, 543 (1959).
[307] V. I. Ogievetski, L. Mezinchesku, *Soviet Phys. Uspekhi* **117**, 637 (1975).
[308] S. V. Shabanov, *The Role of Gauge Invariance in the Path Integral Construction*, JINR preprint E2-89-688 (JINR, Dubna, 1989).
[309] J. R. Klauder, *Ann. Phys.* (NY) **254**, 419 (1997).
[310] J. R. Klauder, S. V. Shabanov, *Phys. Lett. B* **398**, 116 (1997).
[311] J. Govaerts, J. R. Klauder, *Ann. Phys.* **274**, 251 (1999).
[312] A. Kempf, J. R. Klauder, *J. Phys. A: Math. Gen.* **34**, 1019 (2001).
[313] A. S. Schwarz, *Quantum Field Theory and Topology* (Springer-Verlag, Berlin, 1993).
[314] J. R. Klauder, S. V. Shabanov, In: *Mathematical Methods of Quantum Physics* eds. C. C. Bernido *et al.*, (Gordon and Breach Science Publishers, Amsterdam, 1999), pp. 117–130.
[315] L. D. Faddeev, V. N. Popov, *Phys. Lett. B* **25**, 30 (1967).
[316] G. 't Hooft, *Nucl. Phys. B* **33**, 173 (1971).
[317] R. Feynman, *Nucl. Phys. B* **188**, 479 (1981).
[318] Y. Matsumoto, *An introduction to Morse Theory,* Translations of Mathematical Monographs, Vol. 208 (Amer. Soc., Providence, 2002)
[319] M. A. Semenov-Tyan-Shanskii, V. A. Franke, *Zapiski Nauch. Sem. Len. Otdel. Mat. Inst. im. V. A. Steklov* AN SSSR, **120**, 159 (1982); Translation: Plenum Press, NY, 1986; p. 999.
[320] G. Dell'Antonio, D. Zwanziger, In *Probabilistic Methods in Quantum Field Theory and Quantum Gravity*, ed. P. H. Damgraad *et al.* (Plenum Press, New York, 1990).
[321] G. Dell'Antonio, D. Zwanziger, *Commun. Math. Phys.* **138**, 291 (1991).
[322] D. Zwanziger, *Nucl. Phys. B* **209**, 336 (1982).

[323] G. Dell'Antonio, D. Zwanziger, *Nucl. Phys. B* **378**, 333 (1989).
[324] J. Fuchs, *The Singularity Structure of the Yang-Mills Configuration Space*, ArXiv preprint, hep-th/9506005, 1995.
[325] W. Kondracki, J. S. Rogulski, *Dissertationes Mathematicae*, Warsaw, **CCL**, 1 (1986).
[326] M. Lüscher, *Nucl. Phys. B* **219**, 233 (1983).
[327] J. Koller, P. van Baal, *Ann. Phys.* **174**, 288 (1987).
[328] J. Koller, P. van Baal, *Nucl. Phys. B* **302**, 1 (1988).
[329] P. van Baal, B. van den Heuvel, *Nucl. Phys. B* **417**, 215 (1994).
[330] D. Zwanziger, *Nucl. Phys. B* **345**, 461 (1990).
[331] D. Zwanziger, *Nucl. Phys. B* **412**, 657 (1994).
[332] D. Zwanziger, *Nucl. Phys. B* **518**, 237 (1998).
[333] R. E. Cutkosky, *Phys. Rev. Lett.* **51**, 538 (1983).
[334] R. E. Cutkosky, *Phys. Rev. D* **30**, 447 (1984).
[335] R. E. Cutkosky, *J. Math. Phys.* **25**, 939 (1984).
[336] R. E. Cutkosky, K. C. Wang, *Phys. Rev D* **36**, 3825 (1987).
[337] S. R. Coleman, In *Aspects of Symmetry* (Cambridge University Press, Cambridge, 1985), pp. 265–350.
[338] M. Gell-Mann, *Phys. Rev.* **125**, 1067 (1962); Caltech Synchrotron Report CTSL-20 (1961).
[339] Y. Ne'eman, *Nucl. Phys.* 26, 222 (1961).
[340] M. Gell-Mann, Y. Ne'eman, *The Eightfold Way* (W.A. Benjamin, New York, 1964).
[341] M. Gell-Mann, *Phys. Lett.* **8**, 214 (1964).
[342] G. Zweig, CERN Preprints 8182/TH-401, 8419/TH-412 (CERN, Geneva, 1964)
[343] M. Han, Y. Nambu, *Phys. Rev.* **139**, 1006 (1965).
[344] N. N. Bogolyubov, B. V. Struminsky, A. N. Tavkhelidze, *On the Composite Models in Elementary Particle Theory*, JINR preprint (Dubna, JINR, 1965).
[345] Y. Miyamoto, *Prog. Theor. Phys. Suppl.* (Special issue), 187 (1965).
[346] R. Feynman, *Phys. Rev. Lett.* **23**, 1415 (1969).
[347] J. Bjorken, E. Pasckos, *Phys. Rev.* **185**, 1975 (1969).
[348] S. Weinberg, *Phys. Rev. Lett.* **31**, 494 (1973).
[349] J. Pati, A. Salam, *Phys. Rev. D* **8**, 1240 (1973).
[350] H. Fritzsch, M. Gell-Mann, H. Leutwyller, *Phys. Lett. B* **47**, 365 (1973).
[351] L. V. Prokhorov, *JETP Lett.* **16**, 561 (1972).
[352] L. V. Prokhorov, *The Unified Model of Weak, Electromagnetic, and Strong Interactions*, Preprint (Publ. House Len. St. Univ., Leningrad, 1972).
[353] L. O'Raifeartaigh, N. Straumann, *Rev. Mod. Phys.* **72**, 1 (2000).
[354] A. Chodos, R. L. Jaffe, K. Johnson, *Phys. Rev. D* **9**, 3471 (1974); ibid. **10**, 2599 (1974).
[355] S. Mandelstam, *Phys. Rept.* **23**, 245 (1976).
[356] G. t'Hooft, In *High Energy Physics*, ed. A. Zichichi (Editrice Composizitori, Bolonga, 1976).
[357] D. Amati, M. Testa, *Phys. Lett. B* **48**, 227 (1974).
[358] A. M. Polyakov, *Nucl. Phys. B* **120**, 429 (1977).
[359] G. t'Hooft, *Nucl. Phys. B* **138**, 1 (1978).
[360] E. Shuryak, *Phys. Rept.* **264**, 357 (1996).
[361] I. Horvath, N. Isgur, J. McCune, H. B. Thacker, *Phys. Rev. D* **65**, 014502 (2002).
[362] E. Witten, *Nucl. Phys. B* **149**, 285 (1979).
[363] K. Wilson, *Phys. Rev. D* **10**, 2445 (1974).
[364] J. Schwinger, *Phys. Rev.* **128**, 2425 (1962).
[365] J. Lowenstein, J. Swieca, *Ann. Phys.* **68**, 172 (1971).
[366] A. Casher, J. Kogut, L. Susskind, *Phys. Rev. D* **10**, 732 (1974).
[367] G. S. Danilov, I. T. Dyatlov, V. V. Petrov, *Nucl. Phys. B* **174**, 68 (1980).
[368] M. B. Voloshin, K. A. Ter-Martirosyan, *The Theory of Gauge Interactions of Elementary Particles* (Energoatomizdat, Moscow, 1984) (in Russian).
[369] L. V. Prokhorov, *The String Model of Electric Charge*, Preprint OCIP-89-04 (Ottawa, 1989).
[370] L. V. Prokhorov, *Lett. Math. Phys.* **19**, 245 (1990).

[371] T. Kugo, I. Ogima, *Suppl. Prog. Theor. Phys.* **66**, 1 (1979).
[372] D. V. Fursaev, L. V. Prokhorov, S. V. Shabanov, *Mod. Phys. Lett.* A **7**, 3441 (1992).
[373] J. B. Kogut, L. Susskind, *Phys. Rev. D* **9**, 3501 (1974).
[374] C. Quigg, *Gauge Theories of the Strong, Weak, and Electromagnetic Interactions* (Perseus Westview Press, Boulder, CO, 1997), Sec. 8.8.
[375] M. Creutz, *Phys. Rev. D* **21**, 2308 (1980).
[376] C. E. DeTar, S. Gottlieb, *Phys. Today* **57**, 45 (2004).
[377] A. S. Kronfeld, *Science* **322**, 1198 (2008).
[378] F. Wilczek, *Nature* **456**, 449 (2008).
[379] S. Kobayashi, K. Nomidzu, *Foundations of Differential Geometry*, Vol. I and II, (John Wiley & Sons, New York, 1963).
[380] B. A. Dubrovin, S. P. Novikov, A. T. Fomenko, *Modern Geometry – Methods and Applications, Parts I and 2*, (Springer-Verlag, Berlin, 1991).
[381] A. Lichnerowizc, *Geometry of Groups of Transformations* (Springer-Verlag, Berlin, 1977).
[382] B. Schutz, *Geometrical Methods of Mathematical Physics* (Cambridge University Press, Cambridge, 1980).
[383] L. V. Prokhorov, *Vestnik St. Peter. Univ.*, Series 4: Phys. Chem. Issue 1, p. 3 (1992).
[384] L. V. Prokhorov, *Vestnik Leningr. Univ.*, Series 4: Phys. Chem. Issue 3, p. 3 (1990).
[385] L. P. Prokhorov, S. V. Shabanov, *Int. J. Mod. Phys. A* **7**, 7815 (1992).
[386] L. V. Prokhorov, In *Proceedings of the 10th Seminar "Problems in High Energy Physics"* (IHEP, Protvino, 1988), p. 131.
[387] P. Carruthers, *Introduction to Unitary Symmetry* (Interscience Publ., New York, 1966).
[388] J. Frölich, G. Morchio, F. Strocchi, *Phys. Lett. B.* **97**, 249 (1980).
[389] J. Frölich, G. Morchio, F. Strocchi, *Nucl. Phys. B* **190**, 553 (1981).
[390] M. Grater, *Ann. Phys.* **319**, 217 (2005).
[391] V. A. Matveev, A. N. Tavkhelidze, M. E. Shaposhnikov, *Theor. Math. Phys.* (USSR) **59**, 323.
[392] A. S. Kronfeld, G. Schierholz, U. J. Wiese, *Nucl. Phys. B* **293**, 461 (1987).
[393] M. N. Chernodub, M. I. Polikarpov, In: *Confinement, Duality, and Nonperturbative Aspects of QCD*; ed. P. van Baal,; NATO ASI Series, Series B: Physics Vol 368 (Plenum Press, London, 1998) p. 378.
[394] H. Shiba, T. Suzuki, *Phys. Lett. B* **343**, 315 (1995); ibid. **351**, 519 (1995).
[395] Y. M. Cho, *Phys. Rev. D* **21**, 1080 (1980); ibid. **23**, 2415 (1981).
[396] L. D. Faddeev, A. J. Niemi, *Phys. Rev. Lett.* **82**, 1624 (1999).
[397] S. V. Shabanov, *Phys. Lett. B* **458**, 322 (1999).
[398] L. D. Faddeev, A. J. Niemi, *Phys. Lett. B* **449**, 214 (1999).
[399] S. V. Shabanov, *Phys. Lett. B* **463**, 263 (1999).
[400] L. D. Faddeev, *Quantization of Solitons*, IAS preprint, IAS- 75-QS70 (IAS, Princeton, 1970).
[401] L. D. Faddeev, A. J. Niemi, *Nature* **378**, 58 (1997).
[402] H. Gies, *Phys. Rev. D* **63**, 125023 (2001).
[403] A. Shibata, S. Kato, K.-I. Kondo, T. Murakami, T. Shinohara, S. Ito, *Phys. Lett. B* **653**, 101 (2007).
[404] S. V. Shabanov, *The Proper Field of Charges and Gauge-Invariant Variables in Electrodynamics*, JINR preprint, E2-92-136 (JINR, Dubna, 1992).
[405] L. V. Prokhorov, D. V. Fursaev, S. V. Shabanov, *Theor. Math. Phys.* **97**, 373 (1993).
[406] G. S. Bali, *Phys. Rept.* **343**, 1 (2001).
[407] J. Greensite, *Prog. Part. Nucl. Phys.* **51**, 1 (2003).
[408] E. Seiler, *Gauge Theories as a Problem of Constructive Quantum Field Theory and Statistical Mechanics* (Springer, Berlin, 1982).
[409] J. Glimm, A. M. Jaffe, *Quantum Physics: A Functional Integral Point of View*, 2nd ed. (Springer, New York, 1987).
[410] D. B. Leinweber, *Visualizations of QCD*, web reference: http://www.physics.adelaide.edu.au/dleinweb/VisualQCD/Nobel/.
[411] F. Bissey *et al*, *Phys. Rev. D* **76**, 114512 (2007).

[412] K. Rummukainen *et al.*, *Nucl. Phys. B* **532**, 283 (1998).
[413] M. Laine, K. Rummukainen, *Nucl. Phys. B* (Proc. Suppl.) **73**, 180 (1999).
[414] I. D. Ado, *Izv. Fiz.-Mat. Obsch.* (Kazan'), **7**, 1 (1935).
[415] I. D. Ado, *Transl. Amer. Math. Soc.*, **9**, 308 (1962); *Uspekhi Mat. Nauk.*, **2**, 159 (1947).
[416] K. Huang, *Quarks, Leptons and Gauge Fields* (World Scientific, Singapore, 1982).
[417] L. V. Prokhorov, In: *Proc. Intern. Seminar* "Path Integrals: Theory and Applications", Eds. V.S. Yarunin, M.A. Smondyrev, (JINR, Dubna, 1996), p. 192.
[418] F. A. Berezin, *Introduction to the Algebra and Analysis with Anticommuting Variables* (Publ. Depart., Moscow State Univ., Moscow, 1983) (in Russian).
[419] R. Floreanini, R. Jackiw, *Phys. Rev. D* **37**, 2206 (1988).
[420] E. C. Titchmarsh, *The Theory of Functions* (Oxford University Press, Oxford, 1932).
[421] J. R. Klauder, B. Skagerstam, *Coherent States* (World Scientific, Singapore, 1985).
[422] C. Becchi, A. Rouet, R. Stora, *Phys. Lett. B* **52**, 344 (1974).
[423] C. Becchi, A. Rouet, R. Stora, *Commun. Math. Phys.* **42**, 127 (1975).
[424] I. V. Tyutin, *Gauge Invariance in Feild Theory and Statistical Physics in Operator Formalism,* Lebedev Physics Institute preprint **39** (Moscow, 1975).
[425] M. Henneaux, C. Teitelboim, *Quantization of Gauge System* (Princeton Univ. Press, Princeton, 1992)
[426] F. G. Scholtz, S. V. Shabanov, *Annals Phys.* **263**, 119 (1998).
[427] I. A. Batalin, G. A. Vilkovisky, *Phys. Lett. B* **69**, 309 (1977).
[428] K. Fujikawa, *Nucl. Phys. B* **223**, 218 (1983).
[429] J. Govaerts, *Hamiltonian Quantization and Constrained Dynamics*, Leuven Notes in Mathematical and Theoretical Physics, Series B, Vol. 4 (Leuven Univ. Press, Leuven, 1991).
[430] F. G. Scholtz, G. B. Tupper, *Phys. Rev. D* **48**, 1792 (1993).
[431] F. G. Scholtz, S. V. Shabanov, *Supersymmetric quantization of gauge theories*, Electronic preprint, arXiv:hep-th/9509015, (1995).
[432] F. G. Scholtz, S. V. Shabanov, *SUSY quantization of Yang-Mills theory and possible applications*, In: *The Proceedings of 2nd Sakharov Conference on Physics*, ed. I. M. Dremin, A. M. Semikhatov (World Scientific, Singapore, 1997), p. 558 (Electronic preprint, arXiv:hep-th/9606074, 1996).
[433] A. A. Slavnov, *Lorentz invariant quantization of the Yang-Mills theory free of Gribov ambiguity*, Electronic preprint, arXiv:0902.1847 (2009).

Index

www.ingramcontent.com/pod-product-compliance
Lightning Source LLC
LaVergne TN
LVHW062203200926
848484LV00006B/1309
* 9 7 8 0 5 2 1 8 9 5 1 2 5 *